EARTH LAB

Exploring the Earth Sciences

THIRD EDITION

Claudia Owen

University of Oregon and Lane Community College

Diane Pirie

Florida International University

Grenville Draper

Florida International University

Australia • Brazil • Japan • Korea • Mexico • Singapore • Spain • United Kingdom • United States

Earth Lab: Exploring the Earth Sciences, Third Edition
Claudia Owen, Diane Pirie, Grenville Draper

Publisher: Yolanda Cossio
Acquisitions Editor: Laura Pople
Developmental Editor: Samantha Arvin
Editorial Assistant: Kristina Chiapella
Media Editor: Alexandria Brady
Marketing Manager: Nicole Mollica
Marketing Assistant: Kevin Carroll
Marketing Communications Manager: Belinda Krohmer
Content Project Manager: Michelle Clark
Creative Director: Rob Hugel
Art Director: John Walker
Print Buyer: Becky Cross
Rights Acquisitions Account Manager, Text: Roberta Broyer
Rights Acquisitions Account Manager, Image: Mandy Groszko
Production Service: Lynn Steines, S4Carlisle Publishing Services
Photo Researcher: PrePress PMG
Copy Editor: Sue Dillon
Cover Designer: Jonathan Peck, Dovetail Publishing Services
Cover Image: Tinyevilhog/Getty Images, "High altitude climb in La Paz, Bolivia"
Compositor: S4Carlisle Publishing Services

About the Cover: Sure footing is needed on the loose slate which is the surficial country rock on Chacaltaya mountain in the Bolivian Cordillera Real, 10 miles north of LaPaz as the condor flies. The granite plutons beneath were responsible for tin reserves once mined in this previously glaciated and rugged terrain.

Library of Congress Control Number: 2010920865

ISBN-13: 978-0-538-73700-5
ISBN-10: 0-538-73700-X

Brooks/Cole
20 Davis Drive
Belmont, CA 94002-3098
USA

Cengage Learning is a leading provider of customized learning solutions with office locations around the globe, including Singapore, the United Kingdom, Australia, Mexico, Brazil, and Japan. Locate your local office at **www.cengage.com/global**

Cengage Learning products are represented in Canada by Nelson Education, Ltd.

To learn more about Brooks/Cole, visit **www.cengage.com/brookscole**

Purchase any of our products at your local college store or at our preferred online store **www.CengageBrain.com**

Printed in the United States of America
1 2 3 4 5 6 7 14 13 12 11 10

Contents

SECTION 3 Maps and Time

LAB 6

LAB 7

LAB 8

SECTION 4 Earth's Dynamic Interior

LAB 9

LAB 10

SECTION 5 Earth's Surface

LAB 11

LAB 12

LAB 13

LAB 14

SECTION 

Weather, Climate, and Their Effects

LAB 15

LAB 16

LAB 17

LAB 18

Preface

The Earth is our home. "On it everyone you love, everyone you know, everyone you ever heard of, every human being who ever was, lived out their lives," wrote astronomer and author Carl Sagan. It behooves us, then, to know our home. This book will help you in this exploration.

Earth Lab: Exploring the Earth Sciences provides college students in introductory-level physical geology, Earth science, and environmental geology courses with hands-on experience handling natural materials, gathering data, and experimenting with the behavior of natural processes such as the movements of landslides, rivers, groundwater, and glaciers. We designed the simple, student-tested experiments and exercises to enhance students' comprehension by seeing and doing, and to help them experience ideas and processes that are often hard to conceptualize or verbalize. The manual has an informal, accessible style, avoiding technical language except where the aim is to teach terminology. Graphic illustration is an essential means of learning and communication in the sciences. For this reason, we have made *Earth Lab* rich in visual content, including hand sample photos, maps, block diagrams, and field photos that are integrated into the exercises so that students might learn to see the world through the eyes of an Earth scientist, to reason critically, and to solve problems creatively.

The intent of *Earth Lab* is to guide the student toward a better understanding of the Earth Sciences through active participation. Even though they might not become scientists, such an understanding will help students to be active in discussions where science is part of social change. How we interact with our environment and the technology we've created increasingly shapes our lives. Science courses prepare students to become part of the conversation surrounding major scientific developments and the effects of science and technology. In addition, a scientific understanding of Earth's natural environments may be critical to our well being on this fragile planet.

HANDS-ON ACTIVITIES

In teaching geology lab classes, we find that students' natural inquisitiveness is engaged when we encourage them to handle samples, conduct experiments, and use the various tools of science. Students blossom with the experience of doing science. The more the students engage with the material, both physically and mentally, the more their depth of understanding increases.

Wherever possible, we have included activities working with materials in ways that are fun and informative, facilitating a deeper understanding and appreciation of the subject.

SPECIAL FEATURES

The following features illustrate the pedagogy of the book.

- Diagrams that guide—Examples: Rock and mineral mazes that visually guide students through the steps in categorizing minerals or identifying rocks, focusing on methods rather than solely memorization. We have tested this technique of teaching rock identification side-by-side with other techniques and see that students find these guides easier to understand and quicker to grasp.
- Rich visual content—Examples: hand sample photos, colorful maps, block diagrams, and field photos that are well-integrated with exercises. Students interpret updated aerial, satellite, and photographic images in their study of processes and landforms.
- Progressing complexity—Stepwise learning features from simple, idealized diagrams progressing to complex, real-world cases. Examples: topographic and geologic maps. Students start

by working with simplified map representations, and finish by building their own maps and examining topographic and geologic maps of actual areas.

- Experiments—Example: Testing the relative density of warm and cold, fresh and salty water using colored dyes in a mixing experiment to illustrate deep ocean currents. Although experiments take extra setup and cleanup, we feel that it is worth the effort to do them because of the resulting enthusiasm students show and the insight they gain into the processes that coexist and interact on the Earth.
- Graphs—Reading and creating graphs are crucial tools of science, and we have included interpretation and use of graphed material in many places in the book to encourage independent thought based on data analysis.
- Experiencing processes—Examples: Stream and glacier flow, mass wasting in action. Students create landslides and simulate glacial and fluvial flow and then measure and make comparisons after having seen them in action.
- Activities—Example: Through simple exercises with compasses and models, students can build a basis for interpretation of maps, relationships of rock layers, and simple stratigraphy.

With activities like these, which are the basis of the book, labs become a pleasure for the students and the instructor.

NEW TO THIS EDITION

The third edition maintains the important qualities of the second edition, but has some changes in structure and rearrangement of chapters. While continuing to include the traditional skills of rock and mineral identification, topographic map analysis, and geologic map interpretation, we now include environmental and resource-related activities in many labs and have also introduced innovative pedagogy for many of the exercises.

- A new first chapter introduces plate tectonics, the rock cycle, and maps so that following chapters may incorporate these important, broadly applicable ideas and tools. This provides a framework in which to place the material that follows and starts students thinking about how seemingly disparate concepts relate.
- Minerals reside in one chapter and the identification tables are now integrated with a step-by-step method.
- We combined parts of the Rock Structures chapter from the second edition with geologic maps and moved it to follow the discussion of geologic time. Where appropriate, we also moved some of the material from this former chapter to the various rock chapters to reinforce unifying ideas among minerals, rocks, and their occurrences.
- For those students taking an Earth Science course, we have extended the atmosphere chapter into two separate chapters; the first on climate and the atmosphere and the other on wind and weather. We have updated greenhouse gas data and climate information.
- Rather than combine geologic resources into one final chapter, we have incorporated those concepts within various chapters as students encounter related subjects. This encourages exposure to these important ideas without requiring an instructor to allow time for an extra lab.
- In all of the chapters, we have updated and enhanced many of the diagrams, replaced photographs, enhanced the quality of existing maps and introduced new ones, and improved exercises after working through them with students.

The new edition has fewer, more effective, and often shorter chapters with improved hands-on exercises. By being more concise, choosing the most effective exercises, and eliminating those that are less practical or less productive, we maintain a wide variety of activities to choose from within each chapter.

MESSAGE TO INSTRUCTORS

The content of this book directly supports its exercises and lab activities, and we do not intend it to cover every subject comprehensively. We encourage instructors to coordinate the use of this lab book with an Earth science, physical geology, or environmental geology text that can complete the theoretical explanations necessary for a broader understanding of the material. We have also designed the book for flexibility. Most of the 18 labs have more activities than can be completed in a single lab session, so instructors can pick and choose between activities or, in some cases, expand specific topics to cover more than one lab period. Those instructors inclined to emphasize mineral and rock identification have a rich body of instructional materials to choose from, ranging from high-quality photographs to reference materials and charts. In addition, the chapters—geologic time, structures and geologic maps, oceans, climate, and weather—are particularly extensive, so that instructors can expand any of these subjects to two labs if that is their desired emphasis. The wide selection of activities provides instructors with more than enough options to design labs based on their own particular resources and preferences.

Although we provide an instructor's manual (see below) we have included instructions right here in the book so instructors will know what materials are needed and how to set up a lab activity without needing access to the instructor's manual. However, we do recommend the instructor's manual as a resource for more complete and useful information.

Instructor's Manual

Accompanying this book is an online instructor's manual that provides answers to all questions in the book, sample data for experiments, and example graphs. Many of the exercises and activities require particular materials to be present in the laboratory for use by the students. The instructor's manual provides a list of materials needed to do each lab activity, alternates that may be used, and a source guide for the resources needed. We provide helpful hints for instructors to enhance lab activities and avoid common pitfalls. Also included are suggestions of methods to shorten or expand selected activities to emphasize certain themes.

ACKNOWLEDGMENTS

We, along with scores of other instructors, have used many of the labs in this book at Florida International University (FIU) for over 19 years and at Lane Community College for 5 years. Such experience has allowed us to incorporate improvements in the material for this edition. Students have also tested some of the material at the University of Oregon (UO). At all three campuses combined, thousands of students have tested material from this book. In addition, we have received very helpful feedback from instructors at the numerous schools using the previous editions. All of their feedback has helped us shape and hone the book to ensure that the content and activities are sound, effective, and help students focus on the important concepts.

We thank the numerous lab instructors and countless students at FIU, UO, and LCC for their patience and feedback in producing this manual. We particularly thank Mary Baxter, Deborah Arnold, Lois Geier, and the many teaching assistants at FIU who not only taught many of the labs, but also suggested improvements. Marli Miller has been very generous in her contribution of many fine photographs for this book. Roger Cole, R.V. Dietrich, Fabian Duque, Mike Gross, Andrew Macfarlane, L. J. Mayer, James 'Stew' Monroe, Sue Monroe, Bogdan Onac, Derek Owen, Orene Owen, Bernard Pipkin, Stephen Porter, Leslie Sautter, Matthew Sullivan, Johnathan Turk, and Hugh Willoughby have helped provide images in this book, for which we thank them. We especially thank David Blackwell for his extensive help with critique and suggestions and for the use of many rock samples from his collection. The Geology Departments of FIU and UO and Science Division at LCC provided scanners, maps, samples for the mineral and rock photographs, and map images. We are indebted to the FIU Maps and Image Librarian, Jill Uhrovic, for her help in acquiring many difficult to find and out-of-print maps, and to all those at the many state and federal agencies, such as Robert Kimmel and Madeleine Zirbes, who provided digital files. We would also like to express appreciation to our spouses and family members who have endured our lack of attention during the many months of propelling this project to completion, and the friends who have tolerated hearing "just a few more chapters" too many times.

We thank the folks at Brooks/Cole—Editor Laura Pople, Assistant Editor Samantha Arvin, Editorial Assistant Kristina Chiapella, Media Editor Alexandria Brady, Production Project Manager Michelle Clark, Marketing Communications Manager Belinda Krohmer, Permissions Account Managers Mandy Groszko and Bob Kauser, service vendor S4Carlisle Publishing Services, and contributing photographer Dr. Parvinder Sethi.

We would especially like to thank Gloria Britton of Cuyahoga Community College; David T. King, Jr. of Auburn University; Patrick Kinnicutt of Central Michigan University; Bogdan P. Onac of University of South Florida; K. Panneerselvam of Florida International University; Greg W. Scott of Lamar State College-Orange; Mona-Liza Sirbescu of Central Michigan University; and David Thomas of Washtenaw Community College for their detailed reviews of the second edition of *Earth Lab*. We also thank the previous reviewers of *Earth Lab*: Elizabeth Catlos at Oklahoma State University; John Degenhardt of Texas A&M University; Nathan L. Green of the University of Alabama; Diann Kiesel at University of Wisconsin; Carrie E. Schweitzer of Kent State University; Gloria C. Mansfield of the University of Tennessee at Martin; and E. Kirsten Peters at Washington State University. We thank the following organizations for providing maps, photos, or data: Defense Mapping Agency, U.S. Geological Survey, Grand Canyon Natural History Association, Nevada Bureau of Mines, National Aeronautics and Space Administration, National Oceanic and Atmospheric Administration, and the Agriculture Stabilization and Conservation Services of the U.S. Department of Agriculture, Maine Geological Survey, National Earthquake Information Center, National Snow & Ice Data Center, Digital Globe.

Claudia Owen
Diane Pirie
Grenville Draper

LAB 1

Introduction to Plates and Maps

OBJECTIVES

- **To understand the theory of plate tectonics**
- **To learn the differences among the types of plate boundaries, including the general types of physiographic features associated with each**
- **To learn the three categories of rocks and how the rock cycle can recycle them**
- **To understand the differences among some basic types of maps**
- **To be able to read maps using the information in legends, scales, and keys**
- **To be able to use a compass to find bearings**

The Earth is a unique system in which everything interacts, from the level of atoms to the sphere of human life. In the following chapters, we start by studying the small pieces—from minerals to rocks, from internal structures and surface features to oceans and atmosphere—progressing to a consideration of our planet as a complete system. Along the way, we will learn to use maps as a means of visualizing the Earth's surface, and study geologic time as a way to chronicle Earth's changes. Before we stride into the details, a quick overview of the entire perspective will help place the smaller pieces within the whole framework. This chapter introduces the overarching concept of plate tectonics, the rock cycle, and provides a basic background on maps so that as you examine the pieces of the geological puzzle, you can see their contribution to the whole.

INTRODUCTION TO PLATES AND THE THEORY OF PLATE TECTONICS

Earth's internal and external structures interact to produce the ocean basins and the continents, the deepest abyss and the highest mountains. To understand this interaction, we start with a look at the division of Earth's surface layer into pieces called *plates* (■ Figure 1.1) and their movement, which is the field of **plate tectonics**, with its basic principle, the *theory of plate tectonics*.

A *theory* is a well-tested hypothesis that accounts for a large body of data collected over a long time (see Appendix). Occasionally, in science, a theory explains such a wide variety of data, phenomena, and observations that it supplies a **unifying theory** or guiding principle for its area of science. The *theory of plate tectonics* provides such a unifying theme for geology. It helps to explain and predict a wide variety of geologic processes, including earthquakes, volcanoes, mountain building, and even the locations where we find some rocks and fossils.

1. Scientific theories evolve to become comprehensive explanations of the natural world through testing and refining of facts and observations. Scientific methods must support and test observations and predictions. With this in mind, circle the letter of the statement that would most likely evolve as an integral part of a theory:

a. Star charts predict human behavior.

b. Deep ocean earthquakes generate tsunamis.

c. Dogs are people's best friends.

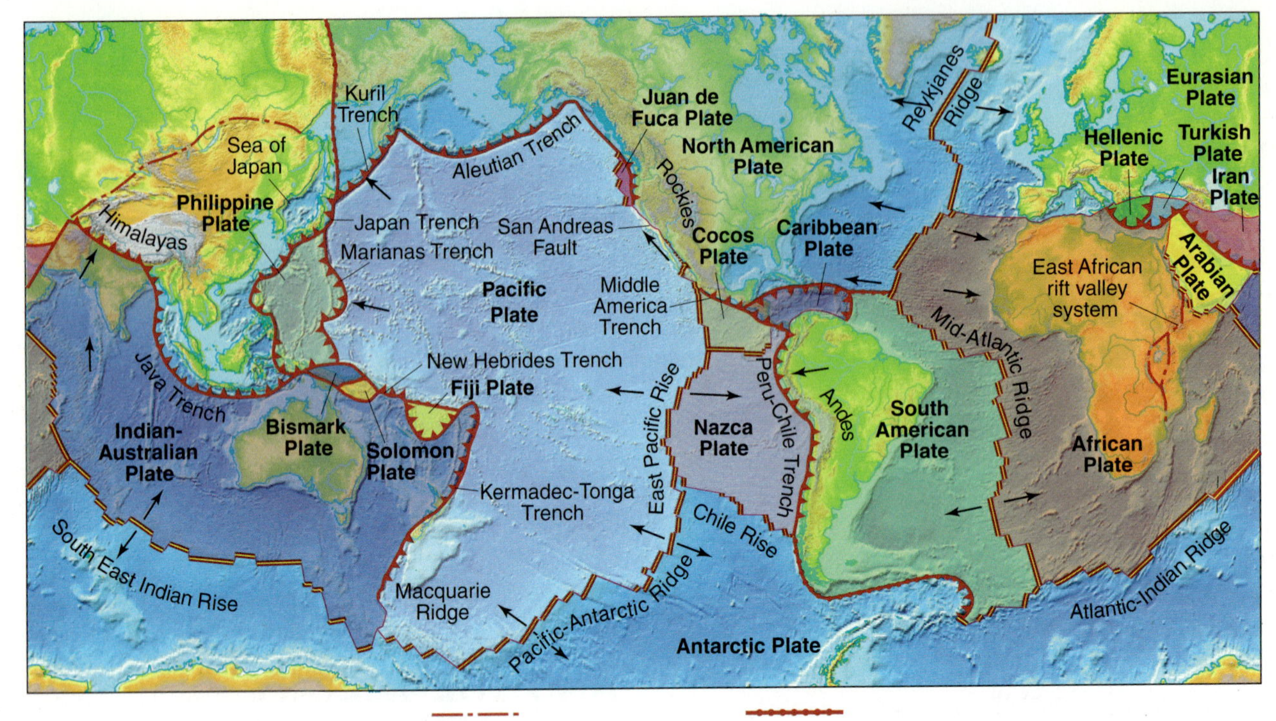

Figure 1.1

Plates and Plate Boundaries

World map of the plates and plate boundaries. Arrows indicate the plate movements. Underneath the plate colors, notice that warm colors (yellow to orange to purple and white) indicate higher elevations and cool colors (green to blue) indicate lower elevations. *Source:* From MONROE/WICANDER/HAZLETT, *Physical Geology*, 6e. 2007 Brooks/Cole, a part of Cengage Learning, Inc. Reproduced by permission. www.cengage.com/permissions

Why does this observation hold promise as a future part of a theory?

____

____

Since plate tectonics involves the upper layers of the Earth, let's study these first. The uppermost, roughly 100-km-thick layer composed of strong, brittle rock is the **lithosphere.** It is distinct from the more plastic *asthenosphere* below (■ Figure 1.2). Both the *crust* (oceanic or continental) and the uppermost part of the *mantle* comprise the lithosphere. Entirely within the Earth's mantle and below the lithosphere is the **asthenosphere**, which is a soft, weak layer that deforms readily allowing the lithosphere to move across the surface.

The **theory of plate tectonics**, simply put, states that Earth's lithosphere is divided into separate **plates**, each of which moves as a unit over the asthenosphere relative to the other plates. The body of evidence that supports this theory is as large and varied as the phenomena that it helps to explain. The lithospheric plates move across the surface at speeds of a few to several centimeters per year—about as fast as your fingernails grow. These plates can include oceanic crust, continental crust, or a combination of both.

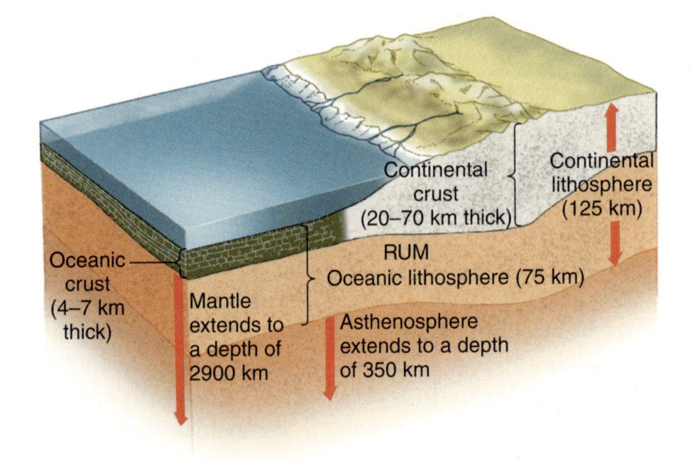

Figure 1.2

Uppermost Layers of the Earth

Oceanic and continental crust, and lithosphere. The chemically distinct layers, crust, and upper mantle overlap with the physical layers, lithosphere, and asthenosphere. The lithosphere includes the crust and rigid upper mantle (RUM).

Earth has two types of crust: the layer we live on, the **continental crust**, and the layer that lies directly beneath the oceans, the **oceanic crust**. Oceanic crust averages 8 km thick, and continental crust averages about 45 km thick. Differences in chemical composition, and therefore density, of oceanic and continental crust produced large elevation differences. The highest point on continental crust is Mount Everest in the Himalayas where the continental crust is doubly thick; the deepest place on Earth's surface, in oceanic crust, is the Mariana Trench near Guam where the **oceanic lithosphere,** made of oceanic crust and uppermost mantle, is sinking downward into the asthenosphere. Old oceanic lithosphere, is actually denser than the asthenosphere, giving it a tendency to sink. The elevation and the depth of Earth's surface are a direct result of plate tectonics, as you will discover as you do the exercises in this lab and in Lab 9, which more fully explores plate tectonics.

Three types of boundaries occur between the plates and are distinct because of the relative movement of the plates on either side of the boundary (■ Figure 1.3). Plate boundaries strongly influence the shape of the land surface and the configuration of the ocean floor, producing what we call **physiographic features,** such as *oceanic trenches, island arcs,* narrow seas, ocean basins, and mountain ranges. To understand these influences, we will study the differences among the types of plate boundaries: divergent, convergent, and transform.

Types of Plate Boundaries

- **Divergent boundaries** occur where plates move away from each other and **magma**, molten rock, forms and solidifies between the two plates that moved away. This type occurs as either continental or oceanic divergent plate boundaries. Divergence breaks up or rifts apart continents and, with further divergence, creates oceanic crust (Figure 1.3). Because the process of divergence makes oceanic crust, most current divergent plate boundaries are in the oceans.
- **Transform boundaries** occur where plates slide past each other and come in two varieties: oceanic (Figure 1.3) and continental. Oceanic **transform faults,** or the fracture zones that form transform plate boundaries, most commonly occur as dislocations in oceanic divergent plate boundaries, but can also occur between convergent and divergent boundaries or as dislocations in convergent plate boundaries (Figure 1.3). Transform movement *conserves* crust, neither creating nor destroying it.
- **Convergent boundaries** exist where plates move toward each other and can be ocean–ocean convergent boundaries (■ Figure 1.4a), ocean–continent convergent boundaries (Figure 1.4b), or *continental collisions* (continent–continent convergent boundaries; Figure 1.4c). Where oceanic lithosphere descends under another plate, a process called **subduction** occurs. The place where the slab of oceanic lithosphere subducts is a **subduction zone**. Subduction creates volcanic arcs, builds new continental crust, and destroys oceanic crust. The lower density of **continental lithosphere** (lithosphere made of continental crusts and uppermost mantle) makes it too buoyant to sink. Continental collision thickens continental crust, creates mountains, and builds continents.

2. Fill in the blanks and answer these questions using the discussion of plate boundaries above and Figures 1.1 through 1.4:

a. At divergent plate boundaries, a large or deep gap never has a chance to form because

____

comes up from below and fills in before a gap develops.

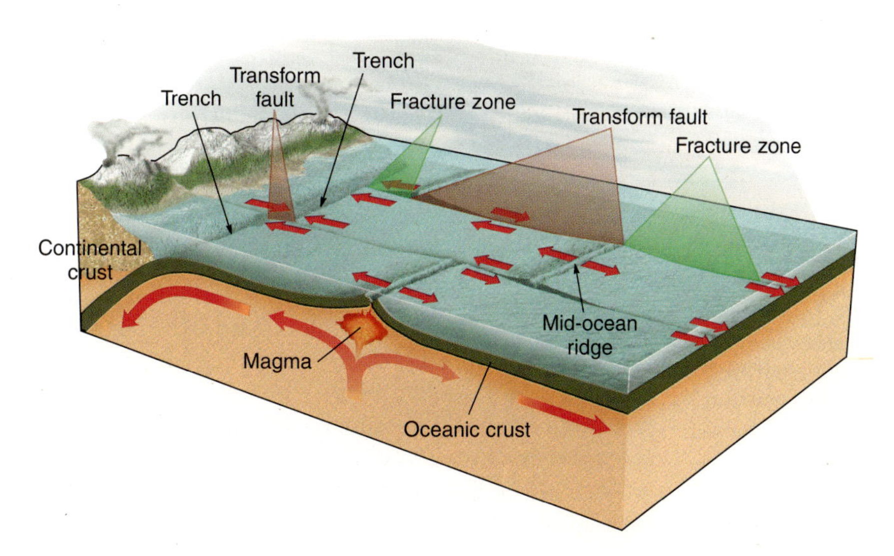

Figure 1.3

Types of Plate Boundaries

Block diagram illustrating different types of plate boundaries. Fracture zones are not plate boundaries, because the plate on both sides of the fracture zone moves in the same direction, as a unit. *Source:* From MONROE/WICANDER/HAZLETT, *Physical Geology,* 6e. 2007 Brooks/Cole, a part of Cengage Learning, Inc. Reproduced by permission. www.cengage.com/permissions

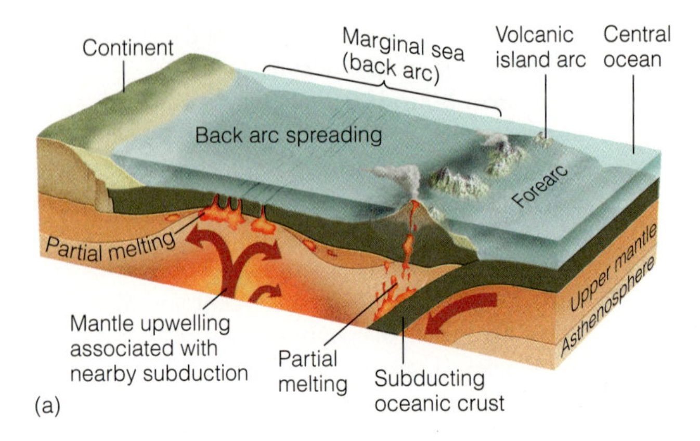

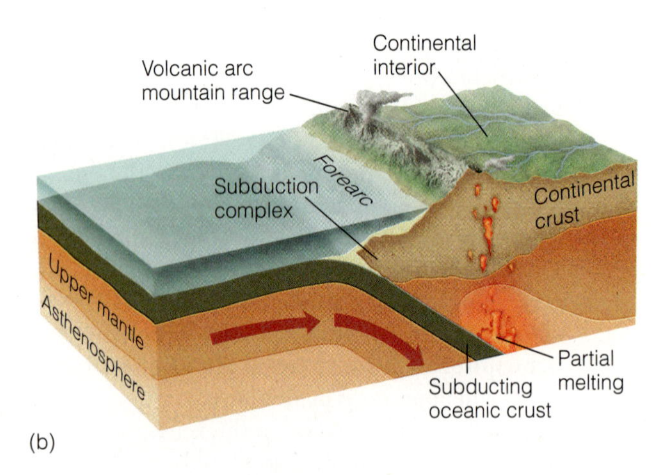

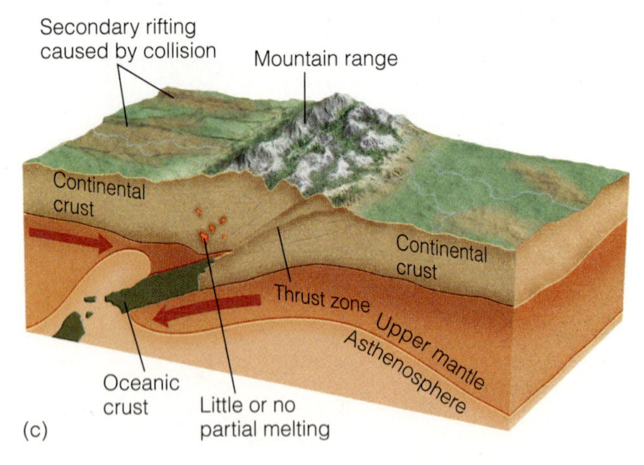

Figure 1.4

Convergent Plate Boundaries

Block diagram showing different types of convergent plate boundaries. (a) Ocean-ocean convergent plate boundaries have subduction, generate magma at depth, and produce volcanoes in a string of islands all of the same approximate age, known as an **island arc**. (b) Ocean-continent convergent plate boundaries also have subduction, generate magma at depth, and produce a string of volcanoes on a continent, known as a **continental volcanic arc**. (c) Continent-continent convergent plate boundaries do not have subduction and generate at best small amounts of magma at depth. The magma does not rise to produce volcanoes, but these plate boundaries still produce mountains—large ones. From MONROE/WICANDER/HAZLETT, *Physical Geology*, 6e. 2007 Brooks/Cole, a part of Cengage Learning, Inc. Reproduced by permission. www.cengage.com/permissions

b. Divergent plate boundaries also create

_______________________________.

c. Most divergent plate boundaries occur in

_______________________________.

d. What type of lithosphere can subduct?

____

e. Look for volcanoes in the figures. Volcanoes form where _______________

crust and _______________ crust (i) converge (ii) diverge or (iii) slip past each other along transform faults (circle one), and where _______________

crust and _______________ crust (i) converge (ii) diverge or (iii) slip past each other along transform faults (circle one).

f. What types of convergent plate boundaries produce mountains? Keep in mind that islands are mountains in the sea.

____

____

____

Plate Boundaries and Physiographic Features

3. Determine which symbol shown in ■ Table 1.1 corresponds to which type of plate boundary in Figure 1.1. Write the plate-boundary types next to the appropriate symbol in Table 1.1.

4. In this exercise, use the plate map in Figure 1.1 and/or a globe or other geographical information source. Find the features listed in ■ Table 1.2. For any that are not on the figure, write them in on Figure 1.1 and fill in Table 1.2. *Hint:* use the arrows showing plate movement direction on Figure 1.1 to tell you the plate boundary types.

5. Based on these observations, what hypotheses can you make about the relationships of plate boundaries and each of the following types of physiographic features? Find an example of each feature in Table 1.2 and its type of plate boundary. Then make a general statement that summarizes the relationship between the feature type and the boundary. Include "because" in your statement.

Table 1.1

Plate-Boundary Symbols

Symbols for the Three Types of Plate Boundaries as shown in Figure 1.1. Follow instructions in Exercise 3.

Symbol	Type of Plate Boundary
(line with triangles)	
(double line)	
(single line)	

Example: *Long narrow seas* such as the Red Sea and the Gulf of California: Long narrow seas occur at continent-continent divergent plate boundaries because oceanic crust fills in between the two pieces of continent as they split.

a. *Oceanic trenches* such as the Peru Chile Trench and the Aleutian Trench (*Hint:* list two types of boundaries):

____

____

____

____

b. *Volcanic island arcs* such as the Aleutians:

____

____

____

____

c. *Continental volcanic arcs* such as the Andes:

____

____

____

____

d. *Mid-ocean ridges* and *rises* such as the Mid-Atlantic Ridge and the East Pacific Rise:

____

____

____

____

6. On another sheet of paper, sketch an ocean-continent convergent plate boundary and label each of the following: both types of crust, lithosphere, asthenosphere, subduction zone, oceanic trench, continental volcanic arc.

Table 1.2

Physiographic Features at Different Plate Boundaries

Follow instructions in Exercise 4. For Crust Type, choose ocean-ocean (o-o), ocean-continent (o-c), or continent-continent (c-c).

Physiographic Feature	Plate Boundary Type	Crust Type	What Plates Are on Either Side of the Boundary?
Peru Chile Trench			
Andes Mountains			
Aleutian Trench			
Aleutian Islands			
Himalayas			
Mid-Atlantic Ridge			
East Pacific Rise			
San Andreas Fault			
Red Sea			

Another subject that is basic to geology is the study of rocks. The next section of this lab will introduce the three categories of rocks and the concept of the **rock cycle**, which explains how types of rock can continually regenerate into other types of rocks, if given enough time. Plate tectonics is one of the major factors driving these changes.

ROCK CYCLE

■ Figure 1.5 illustrates the **rock cycle**, which shows the relationships among the three major groups of rocks. **Igneous rocks** form by the solidification of **magma** (molten rock). **Sedimentary rocks** form when a process called *lithification* consolidates sediment. **Sediment** consists of broken pieces of rock or minerals, the remains of organisms, or material chemically precipitated at Earth's surface. **Metamorphic rocks** form because of transformation in the solid state during changes in temperature and pressure, usually in the deep crust or upper mantle.

Given sufficient time, usually millions of years, and the necessary geologic processes, commonly involving plate tectonics, rocks from each group can change into the other types. Follow these processes in the diagram in Figure 1.5 as we describe them in this paragraph. Both divergence and subduction (and occasionally deep areas of continental collisions) can generate magma that cools and hardens into igneous rocks. Convergence lifts rocks into mountain ranges. High mountains and steep slopes undergo weathering and erosion more readily and these processes produce sediment. As gravity, streams, glaciers, wind, and waves move sediment down toward or within

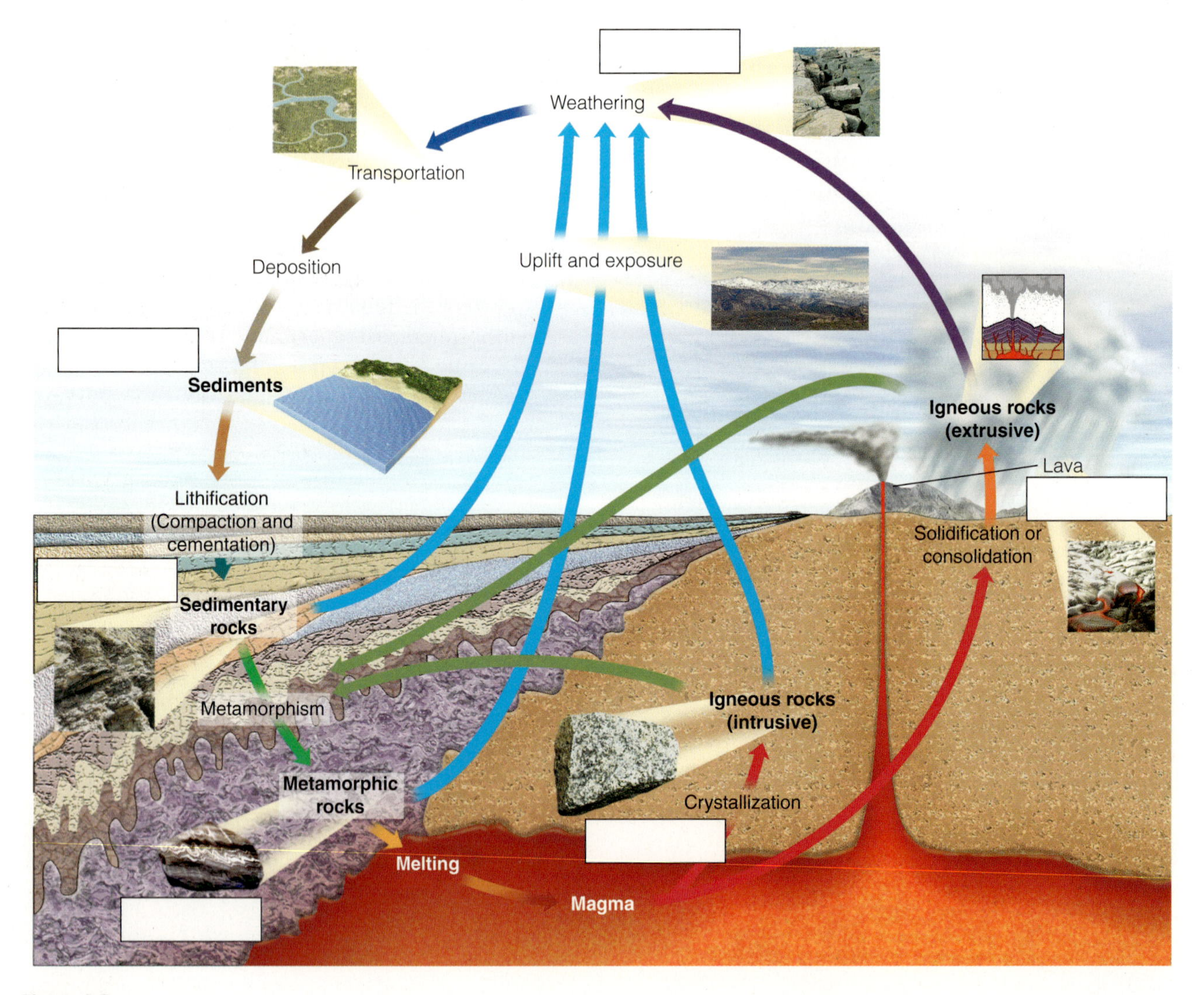

Figure 1.5

Rock Cycle

Rocks can change from one category or rock type to another if given sufficient time and the right processes. Follow instructions in Exercise 7.
Source: From WICANDER/MONROE, *Essentials of Physical Geology*, 5e Brooks/Cole, a part of Cengage Learning Inc. Reproduced by permission. www.cengage.com/permissions

the oceans, or toward lower elevations on continents, sediment accumulates. Where more sediment buries the deposits, sedimentary rocks form. Igneous or sedimentary rocks become metamorphic rocks (1) if further burial pushes sedimentary rocks down far enough to reach the hotter parts of the crust, (2) if convergence forces rocks deep beneath mountain ranges where Earth's heat warms them, or (3) if convergence forces them down into subduction zones where the pressure changes them.

7. Examine the rock cycle and the materials your instructor provides that come from the rock cycle.

a. Record each sample number and describe the sample in the appropriate row in Table 1.3. Use your powers of observation to collect data (see "The Scientific Method" in the Appendix) including notes on relative weight, color, and surface textures. In later labs you will learn a methodology for detailed description of specific rocks, but this is not necessary here.

b. Determine where each sample fits in the rock cycle and place the sample number on the rock cycle diagram.

c. For two of the samples that are far apart on the rock cycle diagram, describe what actions could change one sample into the other. Starting with sample ________

and ending with sample ________

____

____

8. Clearly label one location on each diagram of Figures 1.3 and 1.4 where the rock categories, igneous (label IG), sedimentary (label SED), and metamorphic (label META), would occur (12 labels in all).

In the process of examining plate tectonics and the different rocks on Earth's surface, we need a way to communicate how we visualize the Earth's surface and locate surficial features. This leads us to the third introductory topic in this chapter, which is an introduction to maps.

TYPES OF MAPS

Whether studying in a geography class, planning a vacation, or visiting friends in unfamiliar towns, you have probably used maps many times. You could also use them in many other ways: you may want to consult a flood zone map before building a house, use a topographic map to plan a hike in an unfamiliar area, or even consult a geologic map before investing in a mining company. Maps are very useful in the Earth sciences to display observations and data by location—that is, in a geographic context. Simply put, we need to know *where* rocks or sediments are sampled, *where* wind or ocean velocities are measured, or *where* plate boundaries are located.

Maps are among the most important of the visual and conceptual tools Earth scientists use to study the physical surface of the Earth. Maps can show the location and distribution of almost anything, but Earth scientists use certain kinds of maps frequently:

Planimetric maps are the simplest type of map. They depict the location of major cultural and geographic

Table 1.3

Rock-Cycle Materials

Follow instructions in Exercise 7 to fill in descriptions of rocks and other material from the rock cycle.

Place in the Rock Cycle	Sample Number	Description
Volcanic Rock		
Intrusive Rock		
Weathered Material		
Sediment		
Sedimentary Rock		
Metamorphic Rock		

features such as towns, rivers, roads, and railroads. Most people are familiar with this kind of map. Highway maps, city plans, maps located using an Internet search engine, and the campus map in your college handbook are examples.

Topographic maps are more complicated and contain elevation information in addition to the features that planimetric maps show. They also present data on the shape of the land surface. Features such as valleys, steep slopes, and mountain peaks—which may only be sketchily indicated on a planimetric map—are clearly represented in shape and form on a topographic map using various types of colors or symbols (Figure 1.8, on p. 10). Lab 6 covers topographic maps in more detail. Figure 9.10a (p. 207) gives an example with topographic information using color-coding and shaded relief. Elevations are color-coded where greens represent lowlands, warmer colors, uplands, and whites and lavenders, the highest elevations. Shaded relief supports the color-coding because the shading applied to the map helps define the shapes of mountains and valleys as if light from the upper left illuminates the surface. Some of these coloring techniques have become commonplace, so maps may not include an explanation of what the colors mean.

Bathymetric maps do for the ocean floor or a lake bottom what topographic maps do for the land (Figure 1.1, and Figure 14.3, on p. 319). In coastal areas, bathymetric maps are particularly useful for marine navigation because they show possible submerged nautical hazards. They also show the submarine canyons, trenches, ridges, and plains that make up the ocean floor. Figure 1.1 shows the bathymetry using shaded relief.

Geologic maps show the distribution of rock masses in patterns (inside back cover) or colors or both (■ Figure 1.6). Geologic maps may be **lithologic** as in Figure 1.6a showing the rock types, or **chronologic** showing the age of the rocks as in Figure 1.6b, or both. The rock information commonly overlays a topographic map as in Figure 8.13a on p. 177. Such maps are essential for geologists, for mineral exploration, and for construction projects. Lab 8 explains geologic maps and their applications more thoroughly.

Weather maps show the distribution of aspects of the weather, such as storm systems, precipitation, and temperature, including warm and cold fronts (Figure 16.16, p. 383). Some include variations in atmospheric pressure, wind direction, and speed. You will learn more about weather maps in Lab 16.

Other maps: Many more types of maps exist than we can discuss here. Some examples include maps of ocean currents (Figure 14.17, p. 332), maps showing climate zones (■ Figure 1.7), vegetation maps, land use maps, and maps showing the plate boundaries of the Earth's lithosphere (Figure 1.1), and so on.

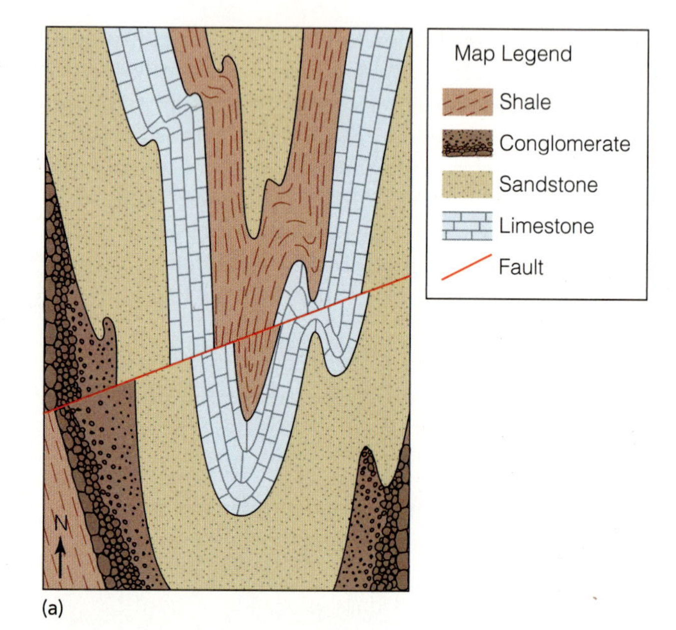

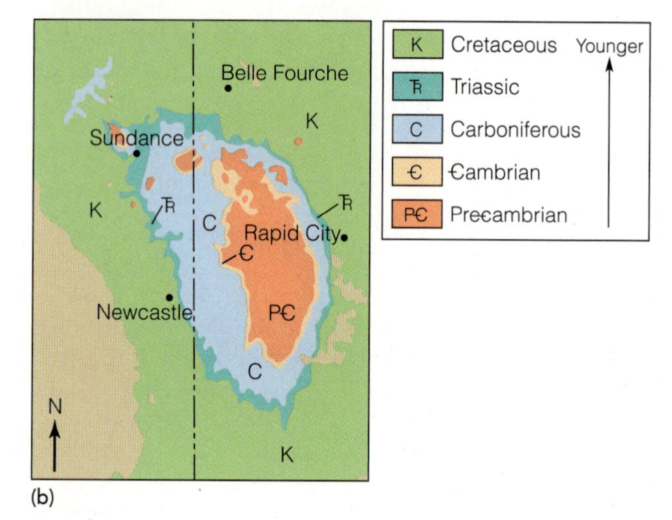

Figure 1.6

Geologic Maps

Two geologic maps using two different methods for labeling rocks. (a) Common geologic symbols in this map represent specific rock types. (b) Colors in this map represent rocks of different ages, formed during different periods of geologic time. From WICANDER/MONROE, *The Essentials of Physical Geology*, 5th ed. 2009 Brooks/Cole, a part of Cengage Learning, Inc. Reproduced by permission. www.cengage.com/permissions

MAP LEGEND

Generally, a **legend** (also known as a **key** or **explanation**), which explains the meaning of different symbols and colors on a map, should accompany the map, whatever its type. One example is the map explanation in Figure 1.7. For another example, the inside back cover of this book illustrates the symbols that may appear on a topographic map such as ■ Figure 1.8. Sometimes maps do not have legends if the symbols are standard and well known. Although the

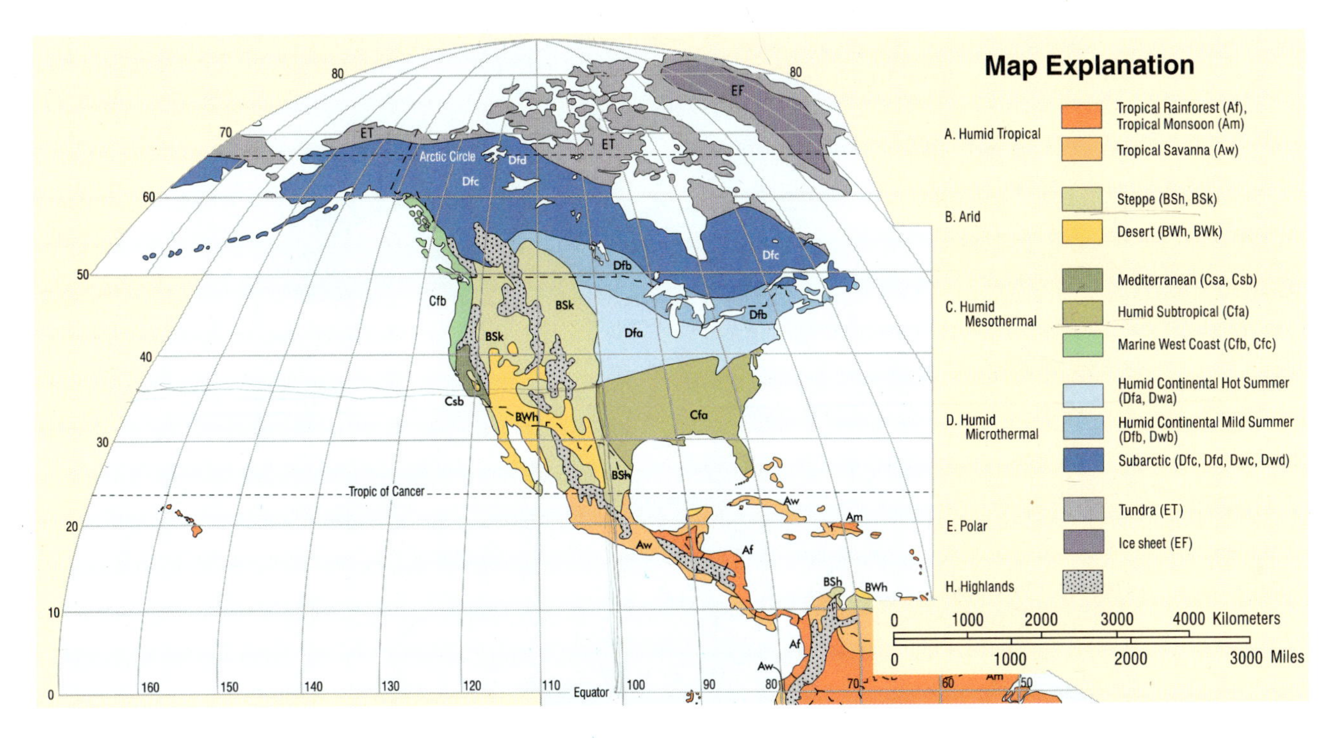

Figure 1.7

Climate Map

Modified Köppen climate classification system for North America. *Source:* From GABLER/PETERSEN/TREPASSO. *Essentials of Physical Geography* (with CengageNOW Printed Access Card). 8e. 2007 Brooks/Cole, a part of Cengage Learning, Inc. Reproduced by permission. www.cengage.com/permissions

information contained in the legend may vary, good map legends generally include the following:

North: All maps should contain some indication of their orientation, so most maps have at least a north arrow. Customarily, in the Northern Hemisphere, if no other indication of north is present, the top of the map is north as in Figure 1.1. The north arrow generally indicates the direction of geographic north, which is the direction toward the northern end of Earth's axis of rotation, called **true north.** You may think that north is just north, but there are other types of north arrows. **Magnetic north** is the direction a compass or a magnetized needle points. The angle between true north and magnetic north is the **magnetic declination;** east declination is to the right or east of true north, and west declination is to the left or west. The magnetic field is not static, but migrates slowly so that the magnetic declination changes from year to year by a fraction of a degree. **Grid north** is the orientation of the "north–south" set of grid[1] lines of a regional coordinate system used on the map, which for various reasons may not be exactly north–south. These three types of north are commonly not coincident. Figure 1.8b has a north arrow showing all three types of north.

On maps that show a large area, north may vary from one part of the map to another. Lines of longitude (explained more later) commonly indicate north as in Figure 1.7; each line of longitude (the north–south lines labeled 50 to 150 in Figure 1.7) shows a different orientation for north.

Scale: Features on a map represent features on the Earth. A scale tells us how much smaller, in numerical terms, the items on the map are compared to the actual size of those features on the Earth. We talk more about scale later in this chapter.

Location: Maps usually contain information indicating where the map area sits on Earth's surface. This information may be in the map key or directly on the map. Showing and labeling lines of latitude and longitude on the map (Figure 1.7) can uniquely establish the location of the map. Tick marks along the map boundaries may replace the lines (Figures 14.3 and 14.5, pp. 319 and 321). A small inset map of a larger, easily recognized area may show the location of the map's borders (Figure 1.8d).

Symbols: Most maps will have additional symbols that the legend needs to define. Examples are the different color-coded lines for different kinds of roads on a highway map; different colors representing ranges of elevation on some kinds of topographic (Figure 1.1) and bathymetric maps; and symbols for swamps, springs, mountain peaks, cities or populated areas, city size, political boundaries, cold fronts, wind direction markers, and ocean currents. The possibilities are limitless. Some publishers use standardized legends and do not key all features on every map; instead, they provide a separate legend page or pamphlet such as the one for U.S. Geological Survey topographic maps reprinted in part on the inside back cover of this book.

[1] A *grid* is a system of lines intersecting at right angles to form rectangles.

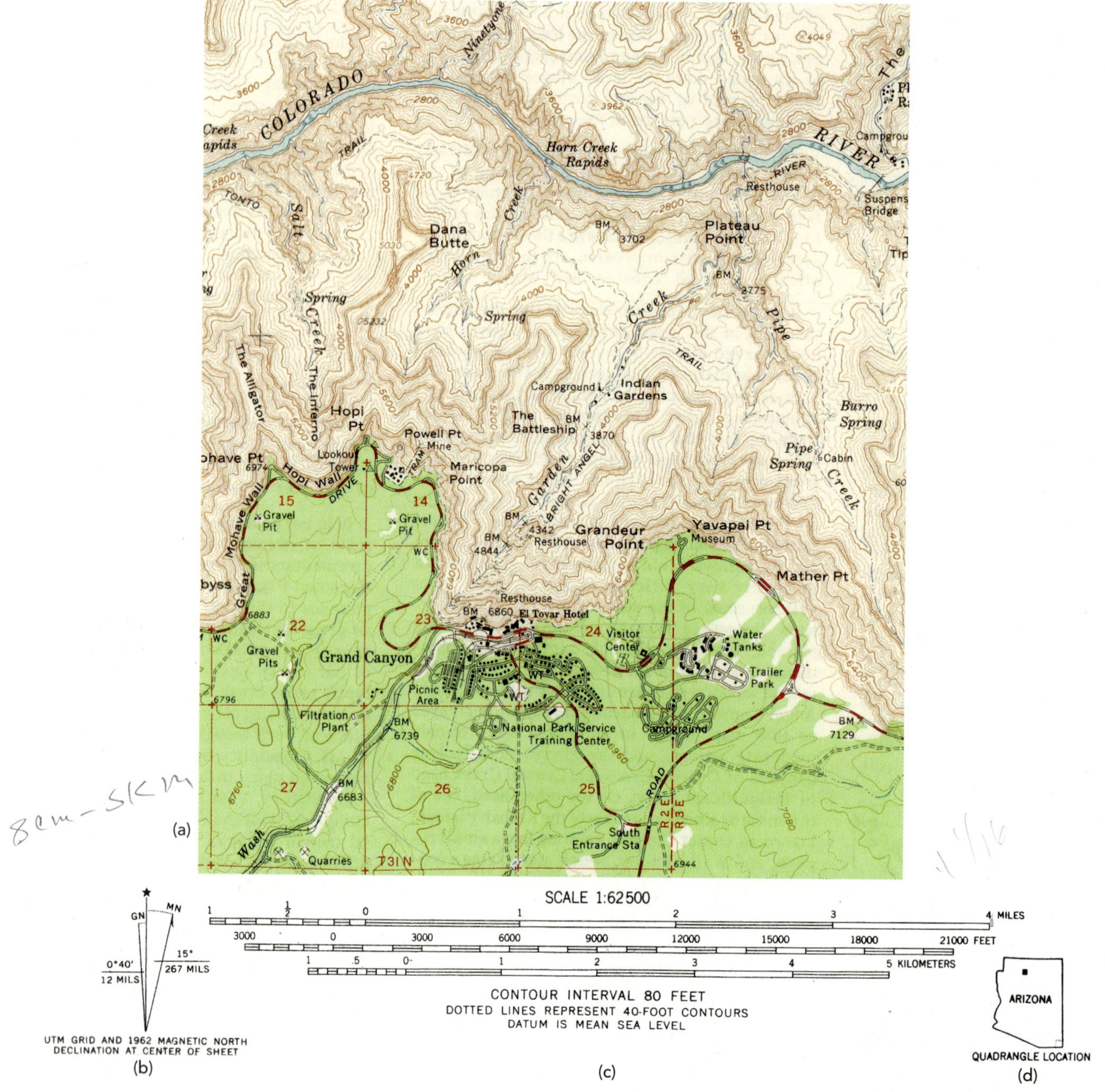

Figure 1.8

Grand Canyon Topographic Map

(a) Map showing elevations of areas near the South Rim of the Grand Canyon, Arizona. (b) North arrows, including grid north (GN), true north (★), magnetic north (MN). Magnetic declination of 15°E was for 1962; declination changes through time. (c) Scale and contour intervals. (d) Quadrangle location map.

9. Examine the various maps that your instructor provides. Determine what type of map each one is. On a separate piece of paper, write down the name of each map, what type it is, which of the items listed above are in the legend, and what additional features the legend includes.

10. Use the map legends in Figure 1.6a and b.

a. What symbol represents sandstone?

¢

b. How is a fault depicted on this map?

red-line

c. What color and symbol represent Precambrian? orange / PE

d. Which is younger, Triassic or Carboniferous? Triassic

e. On maps such as these, with no north arrow, how would you determine what direction is north?

top of map

11. Examine Map 1.7.

a. What is the significance of the colors in the map?

Type of climate

______

b. What letter symbol represents the driest climate? BWh What color represents subarctic climate? purple What letter symbol represents ice? EF
If you live in North America, what climate zone do you live in?

Humid subtropical

12. Compare the maps in Figure 1.8 and Figure 8.13a (on pp. 176–177).

a. What outstanding feature do both maps show? ______
In what state is this?

______

b. List three features that both maps show clearly, but that vary slightly in how they appear from map to map. ______

______ ______

c. For each map, list a feature that the one map displays in most detail or most prominently:

Figure 1.8 ______

Figure 8.13a ______

d. How does each map show or indicate the area's most impressive feature?

Figure 1.8 ______

Figure 8.13a ______

13. Examine the map in Figure 1.8.

a. What is the name of the symbol used to show elevation? (refer to the inside back cover if necessary) ______

b. Locate an open pit mine or quarry. Circle it on the map and draw its symbol here. ______

c. On another sheet, name each of the three types of north shown on the topographic map in Figure 1.8.

______

______

______

SCALE

A **map** is a representation of part of the Earth's surface depicted smaller than the actual surface. We normally have no trouble understanding size differences in most two-dimensional representations, such as a sketch of an object like an apple, whether it is drawn to scale or not. However, items on a map are less easily recognized and more complex in concept, so we need ways to describe the scale specifically to show the relationship between the actual place and its two-dimensional representation. Maps express quantitative scales in the following ways:

Statement (or **verbal scale**): On many simple maps, an ordinary statement expresses the scale. Examples are "One inch equals one mile" or "1 cm = 100 m."

Representative fraction (or **ratio scale**): A more common way of expressing a map scale is to state it as the fraction or ratio of corresponding distances on the map and on the Earth's surface, using the same units of measure for both distances. For example, if two towns are 2 cm apart on the map, and 1 km apart on the Earth's surface, the scale is 2 cm:1 km, or 2 cm:100,000 cm, which in turn reduces to a representative fraction of 1:50,000. Similarly, a map 1 inch to the mile has a representative fraction of 1:63,360 because there are 63,360 inches (5280 ft/mi × 12 in/ft) in 1 mile. The U.S. Geological Survey commonly uses a scale of 1:62,500 as a close approximation of 1″ = 1 mile (Figure 1.8c).

Scale bar (or **graphic scale**): A simple and very effective way of representing the scale of a map is to draw a line representing distances on the Earth's surface. For example, a line, or bar, 2 cm long and labeled "1 km" illustrates the scale of a 1:50,000 map. Figure 1.8c gives examples of three scale bars: one in miles, one in feet, and one in kilometers, but all at the same scale or ratio. The advantage of this visual method of representation is that people instantly grasp the scale without needing calculations. A further advantage is that the scale bar remains an accurate indicator of scale even if you reduce or enlarge the map. You must always recalculate a verbal or ratio scale when enlarging or reducing a map. Next, let's practice expressing scales by different methods.

14. Assume the line just below is the length of a road on a map at the scale in Figure 1.8. Determine the length of the road four ways as described below and answer the questions.

a. Use the scale bar in Figure 1.8c to determine the length of the road in miles.

______________ *Hint:* Read the scale carefully. On a scale bar, it is common to place smaller divisions to the left of 0; this aids measuring distances that are not exactly whole numbers of units, such as 2.7 miles.

b. Use the scale bar to determine the length of the road in kilometers.

____

c. Use the verbal scale of "1 mile is 1.014 inches" and an appropriate ruler to measure the road in miles.

______________ Show any calculation steps needed.

d. Compare your answer in **c** to that in **a.** Is it twice **a**, half **a**, close to **a**, or the same as **a**? (Circle one.)

e. What is the representative fraction (ratio scale) of the map in Figure 1.8?

____

f. Use the ratio scale and an appropriate ruler to measure the road in kilometers.

______________ Show the calculation steps needed.

g. Compare your answer in **f** to that in **b.** Is it twice **b**, half **b**, close to **b**, or the same as **b**? (Circle one.)

Since maps have specific purposes, the map's use will determine its scale. In general, a map is **small scale** or **large scale.** *Large* and *small* here refer to the representative fraction. Thus, a map with a scale of 1:1,000,000 is small scale, whereas one of 1:5,000 is large scale (because the fraction 1/1,000,000 is less than 1/5,000). Objects on a small-scale map appear smaller than the same objects on a large-scale map. A small-scale map covers a larger area than a large-scale map on the same size sheet, because shrinking the scale shrinks the map and creates more room on the page. What scale you use depends on your purpose. If you want to look at ocean currents, you may use a map with a small scale similar to Figure 1.7; if you plan a hike, a moderate-scale map such as Figure 1.8 would be appropriate; if you hired a landscape architect to plan your yard, she would use a large-scale map such as 1 inch = 10 feet (1:120).

h. Use the representative fraction to calculate the relationship between inches on the map and feet on the ground and then write out the statement scale: ______________

______________ Show the calculation.

15. Examine the maps in Figures 1.1 and 1.7. Which is the smaller-scale map? ______
Which scale is smaller between Figures 13.16 and 13.19, on pp. 312 and 316? ______

16. Calculate the representative fraction of the following verbal scales, showing how to do each calculation. ■ Table 1.4 provides some useful conversion factors.

Table 1.4

Conversion Factors

Some useful conversion factors for linear measurements.

Unit	Number of Equivalent Units
1 km =	1000 m = 100,000 cm = 1,000,000 mm
1 m =	100 cm = 1000 mm = 3.281 ft
1 cm =	10 mm = 39.37 in
1 mi =	5280 ft = 63,360 in = 1.609 km
1 ft =	12 in = 30.48 cm

a. 5 cm equals 10 km ______________

b. 1 inch equals 4 mi ______________

c. 4 mm equals 100 m ______________

d. 2 ⅝ inches equals 4 mi ______________

e. Which of the scales above has the smallest scale: a, b, c, or d (circle one)?

f. Which has the largest scale: a, b, c, or d (circle one)?

g. Which is most useful for finding a path for a short hike: a, b, c, or d (circle one)?

h. Which is most useful for a cross-country road trip: a, b, c, or d (circle one)?

17. Determine the ratio scale for the map in Figure 1.7. Show your work.

18. Draw scale bars below, in both kilometers and miles, for a map with a scale of 1:12,000 and for a map with a scale of 1:100,000.

a. 1:12,000; kilometers:

b. 1:12,000; miles:

c. 1:100,000; kilometers:

d. 1:100,000; miles:

19. Which scale in the previous question is for a larger-scale map? ______________

REFERRING TO LOCATIONS ON A MAP

A map marks the locations of features within a particular area. People use a number of different systems on maps to designate a particular location on the ground. These systems include latitude and longitude, UTM grids, and Township and Range Systems. As you will see, different systems are useful in different situations.

Global Location System: Latitude and Longitude

The fundamental system for locating features in relation to the entire Earth's surface is latitude and longitude. The poles of Earth's axis of rotation, otherwise known as the *geographic* or *North* (N) and *South* (S) *Poles,* are the basis of this system. A family of circles passes through the N and S Poles, which divide the Earth into vertical segments, a little like segments of an orange. The half circles between the N and S Poles indicate the east-west position of a location and are **meridians,** or **lines of longitude** (■ Figure 1.9). One of these meridians, the **Prime Meridian** or **Greenwich Meridian,** has the value 0° and (for historical reasons) passes through Greenwich, United Kingdom. We identified the other meridians using their angle from the Prime Meridian, measured in degrees in the plane of the equator, as shown in Figure 1.9 for 55°W longitude. The measurements are toward the east or west, making the highest numbered meridian 180° from the Prime Meridian.

A second set of circles, perpendicular to the meridians, have centers along the Earth's rotational axis. The circles in this set are **parallels,** or **lines of latitude** (Figure 1.9). The equator is the origin (0°) of the latitudinal system, and the angle, north or south, from the equator specifies each parallel (Figure 1.9). The distance on the Earth between lines of latitude is constant, with a spacing of 1° corresponding to about 111 km. Contrast this with lines of longitude where the spacing diminishes to 0 at the poles and is widest at the equator.

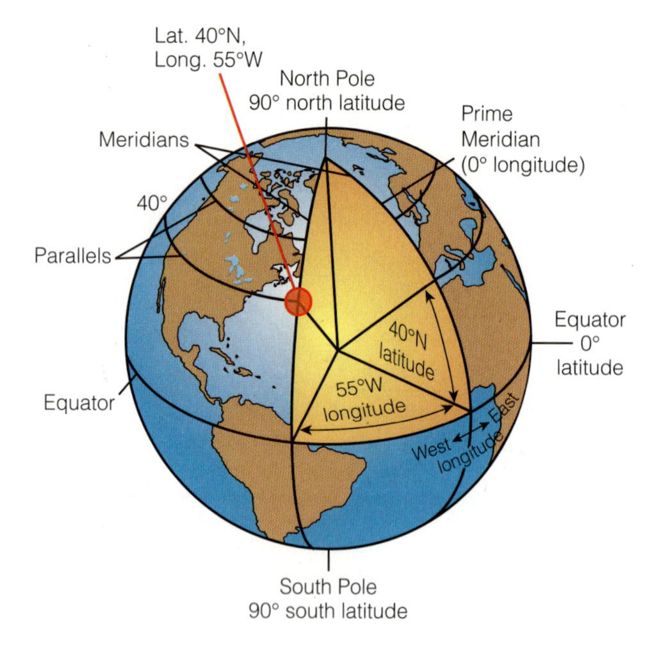

Figure 1.9

Latitude and Longitude

Lines of longitude (meridians) are circles through the North and South Poles. Lines of latitude (parallels) are circles with their centers lining up along the axis of rotation. The angle of a line of longitude is measured east or west from the Prime Meridian. The angle of a line of latitude measured north or south from the equator.

The crisscrossing lines of latitude and longitude form a reference grid on the surface of the Earth that you can use to specify the location of any point. You can specify longitude and latitude angles either as degrees/minutes/seconds[2] or as decimal degrees. Longitude divides the Earth into eastern and western halves or hemispheres, specified either as E or W, or with a sign: east positive (+) and west negative (−). With latitude, N or S indicate the Northern or Southern Hemispheres, respectively, or north is positive (+), south is negative (−).

20. Using a globe, map, atlas, or the Internet, find the latitude and longitude for each location in Table 1.5 and enter them in the table.

[2] A degree has 60 minutes (60′), and a minute has 60 seconds (60″).

Table 1.5

Latitude and Longitude Practice

Latitude and Longitude of Six Cities. Follow instructions in Exercise 20.

Location	Latitude	Longitude
Kansas City		
London		
Beijing		
Nairobi		
Rio de Janeiro		
Honolulu		

To convert from a decimal degree to minutes, multiply the decimal part by 60. Then the whole number part of your answer is minutes. Next, to get seconds, take the decimal part of the answer and multiply by 60 again. For example: For 45.07°, multiply $0.07 \times 60 = 4.2$, which gives you 4 minutes. Then multiply $0.2 \times 60 = 12.0$, which gives you 12 seconds. Thus, 45.07° = 45°4′12″.

To convert from minutes and seconds to decimal degrees, divide minutes by 60 and seconds by 3,600 and add the two decimals. For 119°34′21″, first divide 34/60 = 0.567, then divide 21/3,600 = 0.006, and then add 0.567 + 0.006 = 0.573. So, 119°34′21″ = 119.573.

21. Use Table 1.6:

a. In the first row in Table 1.6, convert the latitude and longitude from decimal degrees to degrees, minutes, and seconds. Then look this place up in on a map, atlas, or globe and name what is located there.

b. In the second row in Table 1.6, convert each latitude and longitude from degrees, minutes, and seconds to decimal degrees.

Table 1.6

Latitude and Longitude Conversions

Converting decimal degrees to degrees, minutes, and seconds and vice versa. Follow instructions in Exercise 21.

Decimal Latitude	Decimal Longitude	Latitude in Degrees/ Minutes/Seconds	Longitude in Degrees/ Minutes/Seconds	What Is at This Location?
40.714°N	74.006°W			
		37° 45′ 22″N	119° 35′ 34″W	

Look up and list the feature that is located there in Table 1.6.

22. Use Figure 1.7:

a. What climate zone is present at latitude 36°N, longitude 102°W?

Humid Subtropical - Stepp

b. What is the latitude and longitude of the northernmost occurrence of humid mesothermal climate classification in North America?

122 W 47 N

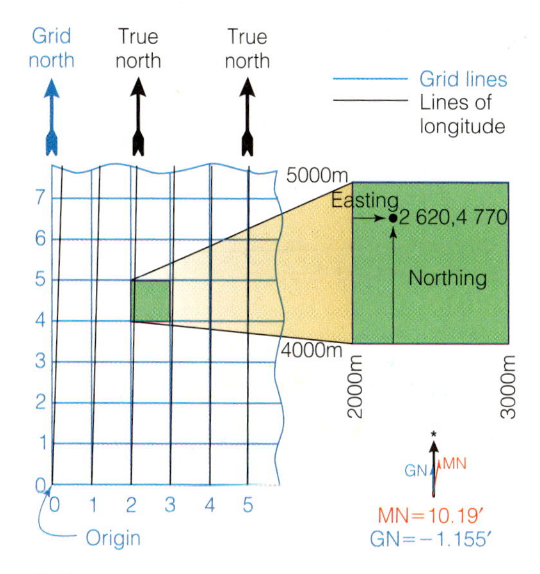

Figure 1.10

UTM System

Example grid system based on 1000-m spacing. See text discussion about how to associate points with numbers on the grid. The north arrow has three parts: * for true north, representing the North Star, Polaris, GN = grid north, and MN = magnetic north. Positive declination angles are clockwise and negative ones are counterclockwise.

Map Grids and Orientation

Specifying the location of points on a map using latitude and longitude is useful for small-scale maps, but is often clumsy and inconvenient on large-scale maps. On maps of a scale of 1:250,000 or larger, such as in the topographic maps we use in this book, a local rectangular grid is often more convenient. The grid is prepared by placing two sets of straight, parallel, equally spaced lines on the map. One set of lines is oriented north–south (or at least approximately so), the other set east–west. The spacing between the lines depends on the territory or country concerned. The most widely used spacing (including that used in the United States) is 1,000 m (1 km).

In the grid system, lines running north and south, with numbers increasing eastward, are **eastings;** the E–W lines, with numbers increasing northward, are **northings** (■ Figure 1.10). The origin of this coordinate system sits outside the region of interest so that all points in the area will have positive northing and easting values. You can easily specify points within the map area using their coordinates, much the same way as you would indicate the position of a point on a graph (Figures 1.10). Specify eastings (*x* coordinates) first and northings (*y* coordinates) second. You can indicate points between grid lines using a decimal-like subdivision of the grid squares, but without writing a decimal point (Figure 1.10).

Grids and UTM One such system seen on topographic maps is the **Universal Transverse Mercator** (or UTM) coordinates (■ Figure 1.11). The UTM grid has zones 6° of longitude wide. Numbering starts at the 180° meridian and progresses eastward. Grid zones are 8° of latitude in the north–south direction, and these are labeled with letters starting with C between 72° and 80°S. For example, Grand Canyon, Arizona is in Zone 12S (S here does not mean south), most of Florida is in Zone 17R, and Maine is in Zone 19T. Within each zone, the origin is in the southwest corner of the zone. UTM coordinates are in meters.

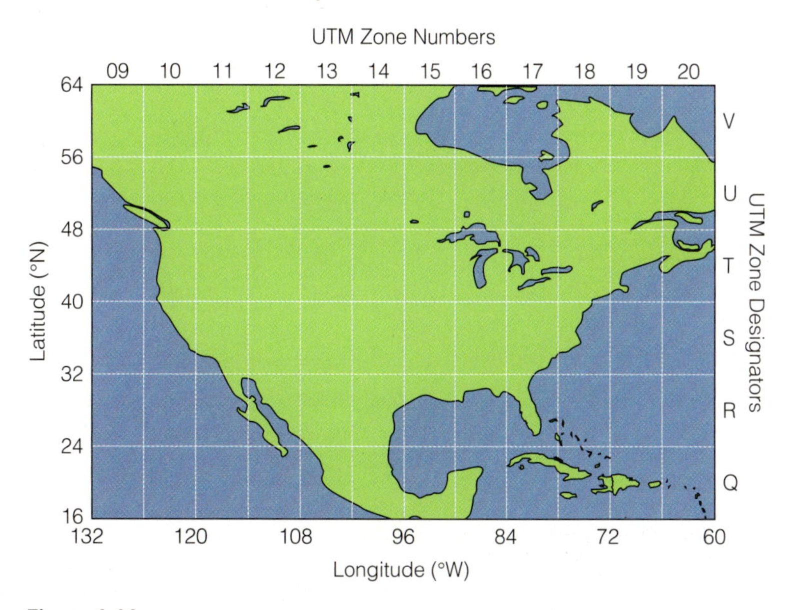

Figure 1.11

UTM Grid for North America

Universal Transverse Mercator coordinates for North America showing the labeled grid over a larger area.

The UTM grid does not converge at the poles as lines of longitude do. This means that "north–south" UTM grid lines, *grid north,* are not actually exactly north–south (Figure 1.10).

23. For each of the following UTM locations, place an x on ■ Figure 1.12 and note what is at that location:

a. 502100E, 5178400N

____

b. 502400E, 5182700N

____

24. For the dots labeled **v** and **w** in Figure 1.12, write out their locations using the UTM system.

v. ___________________________

w. ___________________________

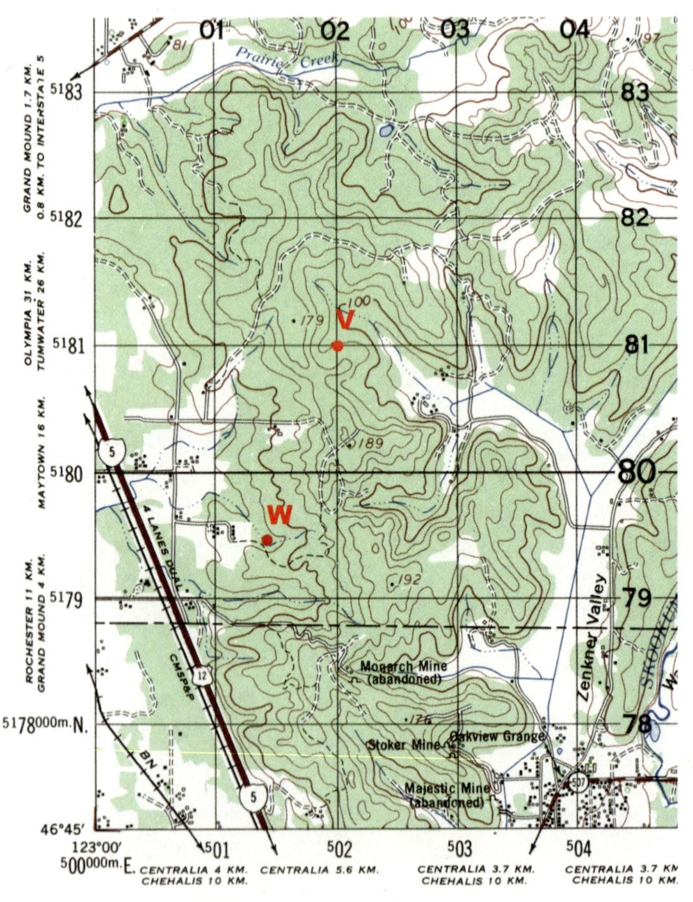

Figure 1.12

UTM on a Topographic Map

A small section of a Tenino, Washington, topographic map for UTM reading practice in Exercises 23 and 24.

Township and Range System In parts of the United States, maps have another type of grid system, known as the **Township and Range** System.[3] An east–west line through the origin or reference point of this system is the **base line,** and a north–south line through the point is the **principle meridian,** not to be confused with the Prime Meridian. In the Township and Range System, the reference point may be in the middle of the area of interest, with locations measured east or west and north or south from that point. The spacing of the grid is a 6-mile square (36 sq. mi) called a **township.** ■ Figure 1.13 shows the township numbering system, where *T* and *R* stand for "township" and "range." For example, T3S, R2E—or Township 3 South, Range 2 East—expands at the right in Figure 1.13. Thirty-six squares called **sections** divide each township, where each section is 1 mile square. The numbering of sections starts in the NE corner and wraps back and forth across

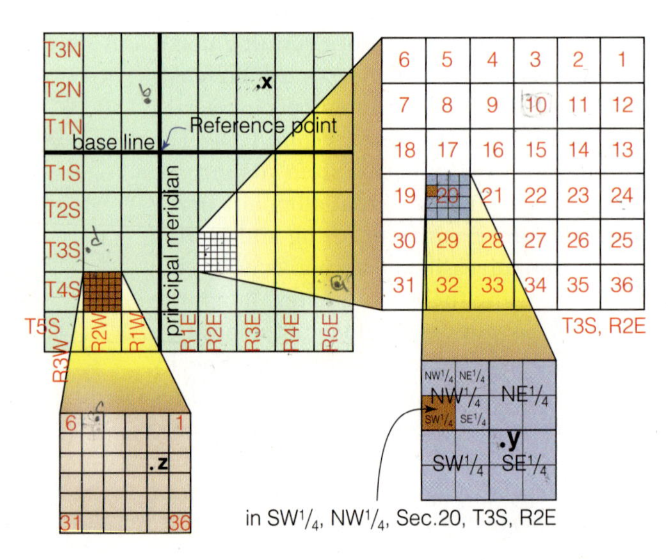

Figure 1.13

Township and Range System

See text above for a description of this system and see instructions for Exercises 25 and 26.

[3] Also called the Land Office Grid System or Public Land Survey.

the township grid, as shown in Figure 1.13. You can further subdivide each section into quarter sections and each quarter section into quarters again. Figure 1.13 shows Section 20 subdivided into 16 quarter–quarter sections. The tiny point on the green small-scale map in T3S, R2E is a quarter–quarter section (¼-mile square area, 1/16 of a square mile) designated as SW¼, NW¼, Sec. 20, T3S, R2E. You would read this "the southwest one quarter of the northwest one quarter of Section 20 in Township 3 South, Range 2 East."

25. For each of the following township and range locations, place a dot on Figure 1.13 and label it with the appropriate letter:

a. Sec. 10, T4S, R5E

b. Sec. 26, T2N, R1W

c. NE¼, Sec. 5, T4S, R2W

d. SW¼, SE¼, Sec. 18, T3S, R2W

26. For the dots labeled x, y, and z in Figure 1.13, write out their locations using the Township and Range System. For x, specify the section and township; for y and z, also specify their quarter–quarter sections.

x. Sec. 10 T2N R3E

y. NW¼, SE¼, Sec. 20 T3S R2E

z. Sec 14 T4S, R2W

27. Use the Township and Range System in the map in Figure 1.8.

a. What is the township and range location of the Grand Canyon Visitor's Center in Figure 1.8? Specify it to the nearest the quarter–quarter section.

NE¼, SE¼, Sec 24, T30N, R1E

b. What is located at NE¼, NE¼, Sec. 26, T31N, R2E.

National Park Service Training Center

Using a Compass Compasses are essential tools for map makers, earth scientists, surveyors, sailors, and hikers. However, many people now rely on a GPS (Global Positioning System), which is more expensive than a compass, needs batteries, and may have electronic failures. A compass is commonly used to find a **bearing**, or a direction measured from north. In the next exercises, you will learn to use a compass to take a bearing. ■ Figure 1.14 shows two types of compass roses with various bearings indicated. For the style of compass rose illustrated in Figure 1.14a, the bearings correspond to north at 000°, northeast at 045°, east at 090°, and so forth, measured in degrees clockwise from north. The style shown in Figure 1.14b has bearings measured either east or west from either north or south; no bearing has a value greater than 90°. For example, the northeast direction is N45°E, and 50° counterclockwise from S is S50°E; due west is N90°W, etc.

To take a bearing, hold your compass level (Figure 1.14c) and point it in the direction you want to measure. Sight along the notches on the compass, as your instructor directs. If your compass is similar to the one shown in the figure, you should rotate the compass rose until the red arrow aligns with the magnetic compass

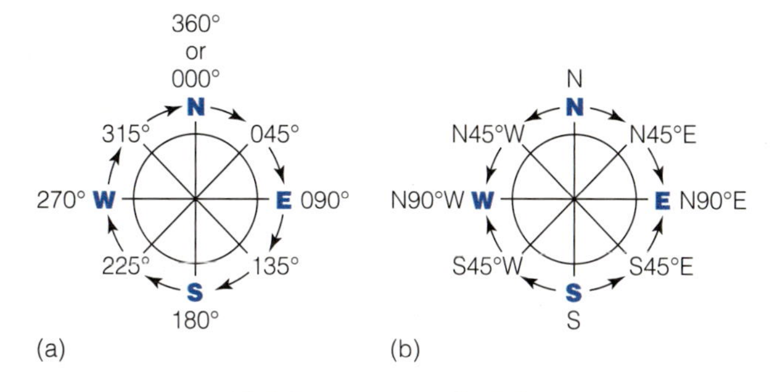

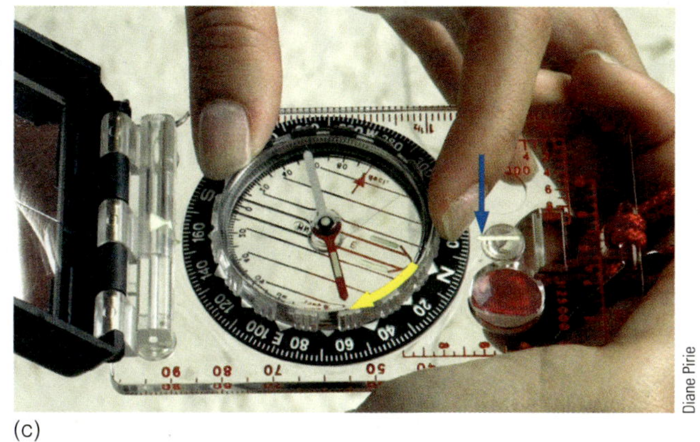

Figure 1.14

Using a Compass

(a) One type of compass rose, showing the directions of the compass and their angular relationship with respect to north. For this style of compass rose, measure a bearing in degrees clockwise from north. Give the answer as a three-digit number. (b) This style of compass rose measures angles in degrees from either north or south in either easterly or westerly directions. Place the number of degrees between letters for the quadrant in which the bearing falls. For example, write a bearing 15° from north in the westerly direction as N15°W. (c) A compass. To take a bearing, rotate the dial (compass rose) until the red-etched arrow aligns with the red end of the magnetic compass needle, as shown by the yellow arrow. Read the bearing at the tick indicated by the blue arrow.

needle, as the yellow arrow shows in Figure 1.14c. Read the bearing off the compass on the side away from you at the tick mark, where the blue arrow in the figure points. Some compasses differ from this configuration, so check with your instructor for procedures for their use.

28. If you have Internet connections in your classroom (or you can do this as homework), find the magnetic declination at your school, using a declination calculator on-line. The NGDC (National Geophysical Data Center) and the Canadian government each have one.

a. You will need your school latitude __________ and longitude __________, which you can look up in an atlas or online. What is the magnetic declination?

____

b. Find the magnetic declination for Los Angeles, California, at 34°03′N, 118°15′W. __________

c. Find the magnetic declination for Washington, D.C., at 38°54′N, 72°2′W.

____

d. Find the magnetic declination for Calgary, Alberta, Canada, at 51°42′N 114°8′W.

____

29. In class, adjust the magnetic declination (if not already set) of the compass you will use for the next exercise as your instructor directs, using the declination you found for your school.

30. Your instructor will designate points A, B, and C. Use a compass to take a bearing for each of the following:

a. Bearing from points A to B. _______

b. Bearing from points B to C. _______

c. Bearing from points C to A. _______

31. You may work in pairs or small groups for this exercise. Use a large space (out-of-doors, a gymnasium, or an open area free of obstacles and suitable for walking across) that your instructor designates, where your compass can function. Your goal is to pace out an equilateral triangle that is 50 paces on a side. Your instructor may give you a different number of paces to fit the space available. Your instructor will also suggest where to start.

- Mark your location with a survey flag if outside, or colored tape if inside, and set your compass to north (360° or N).Turn toward the north and sight along the compass. Pick out a landmark in the distance. The landmark can be a signpost, a tree, a building feature, etc.
- Walk out 50 paces (double steps—counting each time your right foot touches the ground).
- Next, set your compass to 120° or S60°E and pace out another 50 double steps.
- Then direct your compass to 240° or S60°W and walk out another 50 paces. At this point, you should have completed walking a triangle and should end up fairly close to your starting point. See how close your group can come to your starting point and compare results with other groups.

a. What factors determine the accuracy of this exercise?

____

____

b. What would aid in better accuracy (include tools and skills)?

____

____

LAB 2

Minerals

OBJECTIVES

- To learn the definition of a mineral
- To become familiar with the physical properties of minerals
- To determine these properties for unknown minerals
- To develop a logical and systematic approach to mineral identification
- To become familiar with some common minerals and ones used as resources

The importance of minerals in geology and other geosciences should become clear as you read and work through this lab. Some minerals are valuable resources, which we use as raw materials to make products. Some are economically important, some are vital to specific technologies, and some cause pollution when mined. Minerals are also the principal building blocks of rocks and make up the very surface on which we live. In their purest form, minerals can grow into beautiful crystals (■ Figure 2.1). In this lab we identify minerals through their properties and then we look at some of their uses and discuss the impact of mining.

DEFINITION OF A MINERAL

Many people think of a *mineral* as something contained in a multivitamin capsule. Mineral in this sense is really an abbreviation of "mineral salt," something that is derived from a mineral. Geologists define a **mineral** as a naturally occurring, usually inorganic, crystalline solid with a strictly defined chemical composition and characteristic physical properties.

Let's first consider the idea that a mineral is crystalline. In a **crystalline** solid, atoms occur in an orderly arrangement with a distinct structure. In contrast, an **amorphous** solid such as glass has haphazardly arranged atoms. A **crystal** is a single grain of a mineral in which the structural planes of atoms extend in the same directions throughout the grain. The orderly arrangement of atoms controls many of the properties of a mineral, such as the external shape of well-formed crystals (Figure 2.1) and the way a mineral breaks. It even influences the *hardness* and *density* of a mineral. We can use these properties and several others to help distinguish and identify different minerals.

PROPERTIES

In the process of exploring the large variety of minerals that exist in nature, mineralogists recognized that various properties help to distinguish minerals, so we begin with mineral properties.

Crystal Habit

When a crystal grows freely, its external shape reflects its internal order, and it may display **crystal habit**, as seen in Figure 2.1. The nature and symmetry of specific crystal habits help us identify minerals. ■ Figure 2.2 illustrates some common terms used to describe single-crystal habits. When you describe the habit of a crystal, first determine the external shape of the individual crystals from Figure 2.2, especially their relative lengths, widths, and heights. If the crystal is grouped with others, the sample is an **aggregate.** In this case, the next step is to determine the aggregate type; ■ Figure 2.3 illustrates some of these.

Figure 2.1

Crystals

(a) Crystal aggregate of stilbite on mordenite, showing sheaf-like groups of pink stilbite crystals. (b) Chemically zoned tourmaline crystals growing with quartz. (c) Teal colored dioptase (hydrous copper silicate) and white calcite (calcium carbonate) crystals.

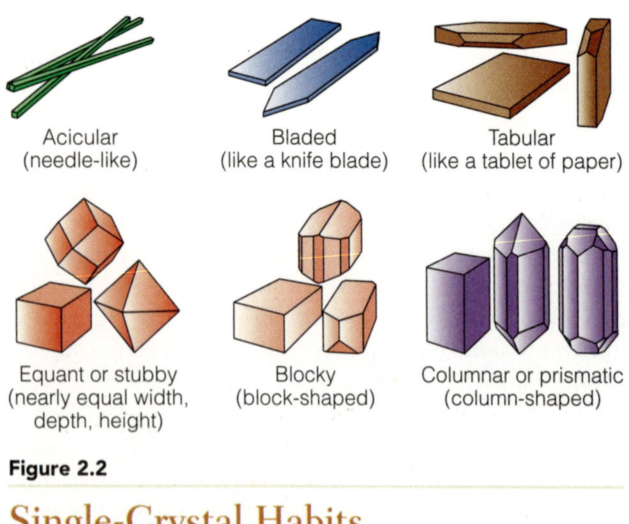

Figure 2.2

Single-Crystal Habits

Determine the crystal habit of an individual grain by observing the relative length, width, and height of the crystal.

1. Look at an aggregate of several crystals in a hand sample your instructor provides.

a. Describe the shape of the *individual* crystals, using the shapes and terms illustrated in Figure 2.2.

____

b. Describe the aggregate type by referring to Figure 2.3.

____

2. Examine additional samples showing a variety of crystal habits—some individual crystals and some aggregates. List each sample number and one or more terms from Figures 2.2 and 2.3. Keep in mind that single crystals will only have habits from

Figure 2.3

Mineral Aggregate Habits

Samples with multiple crystals: micaceous foliate muscovite schist; reticulate jamesonite needles; granular massive green olivine in peridotite and pink calcite in marble; bladed aggregate of blue kyanite crystals; acicular radiating aggregate of the zeolite mineral natrolite; fibrous asbestos; equant aggregate of pyrite crystals; desert rose–gypsum; quartz geode; dogtooth calcite; dendritic pyrolusite; and botryoidal hematite.

Figure 2.2, but for aggregates you should consider both figures.

____

____

____

____

____

____

3. The aggregate habit of the pink mineral, stilbite, in Figure 2.1a is called sheaf-like. What is the single-crystal habit of this mineral (use Figure 2.2)?

________________ What single-crystal habit do the colored tourmaline crystals in Figure 2.1b display?

____

Luster

Luster describes how *light reflects* from a fresh surface. Although you can see some aspects of luster in the photographs in Figures 2.8 and 2.9, on pages 26–29, photographs do not reproduce luster very well. You should look at actual samples to see luster properly. Luster has two broad classifications: **metallic** and **nonmetallic**. Lusters of each type are listed in the exercise below. You will understand the various luster terms better after you do the next exercise in which you associate each luster's appearance with simple descriptions.

4. Examine the set of samples provided to demonstrate luster. In the blank following each luster description, record the number of the sample that has that luster. Remember that you are looking for surface shine or appearance rather than color, transparency, or opaqueness. Metallic lusters include **metallic** (galena and pyrite, shiny metal __________) and **submetallic** (magnetite and graphite, dull metal shine __________). Nonmetallic lusters include **vitreous** (quartz, shines like glass __________), **splendent** (biotite, a bright shine like patent leather __________), **resinous** (sulfur, shines like hardened tree sap, similar to some types of plastic __________), **adamantine** (garnet, gem-like shine __________), **greasy** (common opal __________), **waxy** (shines as wax shines __________), **pearly** (talc crystals, shine like pearls __________), **silky** (satin spar gypsum __________), and **dull/earthy** (kaolin __________).

Color

On a fresh, unaltered surface, color might help you to identify a mineral, but beware: color is a very unreliable property to use in identifying many minerals. Impurities within a mineral may give rise to a variety of colors. Quartz, for instance, has many varieties: amethyst (purple), rock crystal (colorless), rose quartz (pink), smoky quartz (gray), citrine (yellow or orange), and milky quartz (white) (Figure 2.8j–t, p. 27). In some cases a single crystal can display color variation that develops as the mineral grows (Figure 2.1b). In addition, more than one mineral may have the same color. For example, both amethyst and some varieties of fluorite are purple (Figures 2.8n and 2.9g, pp. 27 and 29). Also, quartz, K feldspar, calcite, and gypsum all have pinkish varieties.

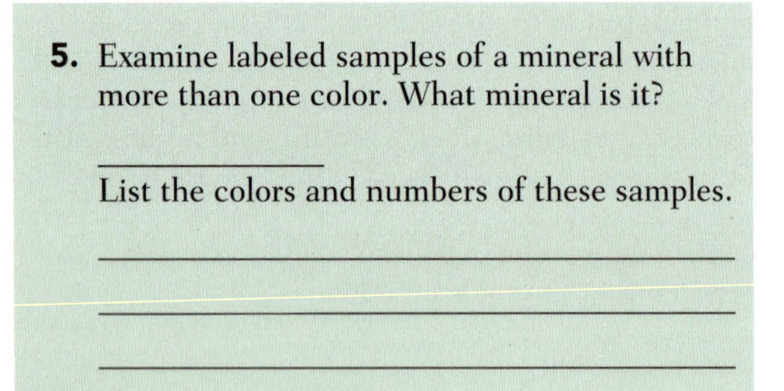

5. Examine labeled samples of a mineral with more than one color. What mineral is it?

____

List the colors and numbers of these samples.

____

____

____

Hardness

Hardness is the resistance of a mineral to scratching (abrasion). In Mohs scale of hardness (H = 1 to 10), minerals with higher hardness will scratch minerals of lower hardness (■ Table 2.1). Determine the hardness of samples by scratching them with an object of known hardness. For example, if a mineral scratches glass (H = 5.5), its hardness is greater than 5.5 (>5.5). If glass scratches it, its hardness is less than 5.5. You can narrow down the range of possible hardnesses by testing against additional objects of known hardness (Table 2.1). *When you think you see a scratch,* check to make sure that you have not simply left powder from the scratching object behind; wipe away any powder and look at the surface closely. Also, the physical nature of a mineral specimen may prevent correct determination of hardness if, for example, a mineral is splintery or granular and falls apart when tested. When you start identifying minerals later in this lab, you will have many opportunities to practice testing their hardness.

Table 2.1

Mohs Scale of Hardness

Mohs Hardness	Mineral	Common Object
1	Talc	
2	Gypsum	Fingernail ~2.2
3	Calcite	Penny ~3
4	Fluorite	
5	Apatite	Nail ~5.2, Knife ~5.1
6	Orthoclase (a feldspar)	Glass 5.5
7	Quartz	File ~6.5, Streak plate ~7
8	Topaz	
9	Corundum	
10	Diamond	

Claudia Owen

Note: Hardness may vary in some minerals (1/2 to 2 points) from crystal face to crystal face as seen in kyanite, which has a hardness of 5 parallel to its length and 7 across the length.

Imagine that you are recording the properties of a sample you think is fluorite. You may have noticed that fluorite has a hardness of 4 on the Mohs scale. However, this does not mean you should record 4 for its hardness. Instead, you should use the evidence obtained by the scratching tests to provide a range of possible hardness. Do not jump to conclusions when testing minerals, but carefully record your observations. What if you thought the mineral was a purple variety of quartz instead of fluorite and decided—knowing the hardness of quartz is 7—that you would just write down 7? This conclusion would mislead you in the mineral identification. In fact, this is an example of changing the data to fit a hypothesis—a definite scientific "no-no." Refer to Appendix A for a discussion of scientific methods.

Streak

Streak is the color of a mineral when powdered (■ Figure 2.4). The color of the powder is less variable than the color of a mineral, so streak is a more reliable property than color. Use a porcelain streak plate to obtain a small amount of powder from a specimen. Look at the example of the streak for hematite shown

Figure 2.4

Streak

Gypsum (white streak), malachite (green streak), hematite (red-brown streak), galena (gray streak), and chalcopyrite (greenish-black streak) each have a different *streak*, which is the color of the powder left on a piece of porcelain called a *streak plate*.

in Figure 2.9f on page 28. Notice that even though one sample of hematite is silvery gray, both samples have a reddish brown streak. Since the hardness of porcelain is about 7 on the Mohs scale, the streak of a mineral with a hardness greater than 6 cannot be easily determined and can be said to have **no streak**. Softer minerals should have streak, but keep in mind that it may be hard to see a white streak on a white streak plate. Make sure that you have powdered the mineral rather than the streak plate.

Broken Surfaces of Minerals

The broken surfaces of a mineral often have characteristics that can help in mineral identification.

Cleavage If a mineral breaks along parallel repetitive planes, the mineral has **cleavage**. A cleavage plane is a plane of weakness in the atomic structure of the mineral (■ Figure 2.5). A mineral may have many or no cleavage planes, and these planes may be perfectly planar, as in muscovite, or slightly irregular, as in pyroxene or hornblende. A mineral with no planes of weakness (such as quartz) will fracture. This does not necessarily mean that the mineral is exceptionally hard or tough, only that the inherent weaknesses in the crystal structure are not planar.

It is easy to detect cleavage if you look for reflections off the mineral's surface. Because cleavage planes are parallel to atomic planes and this arrangement repeats in the mineral, multiple reflections from the same cleavage direction will light up simultaneously, even though the surfaces are at different levels. These different level steps may appear similar to a miniature staircase.

A complete description of cleavage includes the number of **cleavage directions** (planes with different orientations) and an expression of the *angles* between planes, if more than one direction. For instance, galena

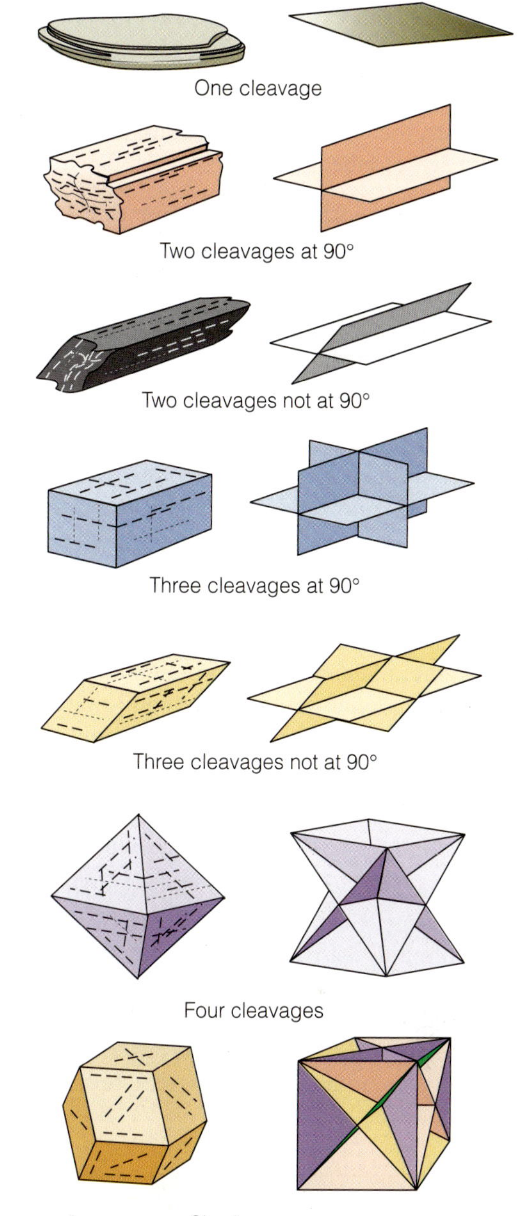

Figure 2.5

Types of Cleavage

Different minerals have different numbers (1, 2, 3, 4, and 6) and angles (90° or not) of their cleavage directions. A mineral with 3 cleavages at 90° has *cubic cleavage*; one with 4 cleavages has octahedral cleavage and with 6 has dodecahedral cleavage. Five cleavage directions are not possible in minerals.

and halite both have three cleavage planes at 90° to each other (Figure 2.9c and h, on pp. 28 and 29, which is equivalent to *cubic cleavage* (Figure 2.5). In some cases, the angle between the cleavages helps to distinguish minerals. Augite (pyroxene group) and hornblende (amphibole group) have similar hardnesses, color, and luster; both have two cleavages, but the angle between the cleavages is near 90° for augite and much more oblique (≠90°; in this case 56° and 124°) for hornblende (Figure 2.8c and d inset on p. 26).

Fracture If the broken surfaces of a mineral are irregular and nonplanar, the mineral is said to have **fracture.** Fracture should be described with one of the following terms:

Conchoidal: A smooth, curved surface that looks like the inside of a clam shell (■ Figure 2.6).

Fibrous: Fracture surface has the appearance of many fine threads lying parallel to each other. A good example is asbestos (fibrous habit in Figure 2.3); another is satin spar gypsum (Figure 2.9b right, on p. 28).

Hackly: A sharp, irregular surface, the same as *jagged.* Both garnet (Figure 2.8b on p. 26) and wollastonite (Table 5.1, p. 98) may produce a hackly fracture.

Uneven/Irregular: General terms that can be applied to the fracture of many different minerals that otherwise defy definition.

Parting: Generally not a common characteristic, **parting** is a roughly planar break in a mineral that is not as well developed as cleavage. The deformation of a mineral can produce it. A clear example of parting can most often be seen in hand specimens of corundum.

6. Examine the samples with unknown fracture or cleavage. In the appropriate place in ■ Table 2.2, list the minerals that show cleavage; then list those minerals that have fracture. Describe the cleavage using one of the choices in Figure 2.5 and describe the fracture using one of the fracture terms above.

Diane Pirie

Figure 2.6

Conchoidal Fracture

The curved, shell-like, broken surfaces on this sample are a good example of conchoidal fracture. Rock crystal is the name for clear colorless samples of quartz such as this.

Table 2.2

Cleavage and Fracture

Samples with Cleavage	Cleavage Describe the Cleavage
Samples with Fracture	**Fracture Describe the Fracture**

Distinguishing Cleavage from Crystal Faces

A **crystal face** is a planar surface of a well-formed crystal that grew as the crystal grew (Figures 2.1, 2.8b left, and 2.8j and n, pp. 26–27). A student may mistake a well-formed crystal of quartz that has flat crystal faces for a sample displaying cleavage. This is incorrect. By now you realize that quartz has conchoidal fracture, not cleavage. The flat surfaces on quartz did not break along planes of weakness, but grew that way when the mineral formed.

How can you tell the difference between cleavage and crystal faces? Since cleavage is an inherent planar weakness in the mineral, you will almost always see multiple examples of a particular cleavage plane exhibited. These may appear as a step-like surface feature (augite, Figure 2.8c, p. 26; and galena, Figure 2.9c, p. 28). Look closely at samples for these steps—you may want to use a hand lens.[1]

7. Examine the two samples provided for this exercise. On a separate piece of paper, sketch each sample and indicate any cleavage steps in your sketch. Label crystal faces with the letters XL, fracture surfaces with the letters FR, and cleavage steps with the letters CL. Write the sample number next to each sketch.

---

[1] When using a hand lens, touch the lens to your eye lashes, then move the sample close until it is in focus. Tilt your head back to let in light.

Density (Specific Gravity)

Density is defined as the mass per unit volume of a substance. Another way to express this property is **specific gravity:** the ratio of the weight of a mineral specimen to the weight of an equal volume of water. In hand specimens, we are usually concerned more with relative density, comparing the density of one mineral specimen to another. You can make general comparisons of hand samples by using the "heft test," in which you compare two specimens of about the same size by picking them up to determine which is heavier.

8. Examine and heft the samples provided. While looking at them, judge their relative density. Looking at the samples allows your eyes to judge the size while your hands judge the weight; together, the two give you an estimate of the density. Now rank the samples in order of their density from most dense to least dense and record the samples in order below:

____

____

____

____

____

____

9. Check your density rankings using an appropriate reference or with densities provided by your instructor. Table 2.3 (on pp. 30–34) includes the density measurements of minerals. List these numbers after your answers for the previous exercise.

Special Properties

Some minerals have special properties, such as effervescence in acid, magnetism, unusual visual properties, or a distinctive smell, taste, or feel. **Special properties** are those that only a few minerals possess.

Effervescence Effervescence, seen in carbonates (primarily calcite; Figure 2.9a, p. 28), is a fizzing or bubbling that takes place when a dilute solution of hydrochloric acid (HCl) is applied. This property is especially useful in distinguishing calcite from other common rock-forming minerals. Dolomite will effervesce only when it is powdered. The chemical reaction that occurs when HCl is applied to calcite ($CaCO_3$) is:

$$\underset{\text{(from HCl)}}{2H^+} + \underset{\text{(from CaCO}_3\text{)}}{CO_3^{--}} \Rightarrow \underset{\text{(water)}}{H_2O} + \underset{\text{(escaping gas in fizzing)}}{CO_2}$$

Magnetism The mineral is attracted to a magnet. Magnetite (Fe_3O_4) is strongly magnetic (Figure 2.9e, p. 28). Magnetism is especially useful in distinguishing magnetite from other common rock-forming minerals. Hematite (Fe_2O_3) and magnetite often occur together in specular hematite samples; what looks like a pure hematite sample may be magnetic as a result.

Feel Some minerals have a diagnostic feel (such as the greasy feel—not greasy luster—of graphite).

Fluorescence Minerals may fluoresce when they are placed under ultraviolet (UV) light. The short-wave radiation of the UV is absorbed by the mineral and radiated back as longer-wave visible radiation. One mineral that often shows fluorescence is calcite; another is fluorite.

Double Refraction Visible in clear calcite crystals, **double refraction** occurs when light entering a crystal is broken into two rays (■ Figure 2.7). You see a double image when looking through the crystal. If you rotate the crystal, one image will stay fixed and the other will move around it.

Smell Some minerals, such as sulfur, possess a distinctive odor.

Taste A few minerals have a characteristic taste. Halite tastes salty. Sylvite tastes bitter. Taste is a valid mineral property and can be quite helpful in identifying some minerals.

When you work on mineral identification next, be sure to look for special properties.

MINERAL IDENTIFICATION

Now that you are familiar with mineral properties, you are ready to start testing and identifying minerals. By approaching mineral *identification* systematically and logically, even a beginner can distinguish a surprising number of minerals. The identification tables (■ Table 2.3, on pp. 30–34) will help to guide you in your search through properties to find the name of a mineral.

Recognizing common rock-forming minerals is necessary to distinguish and interpret rocks correctly—a skill needed for the upcoming three labs. Your instructor

Figure 2.7

Double Refraction

Calcite (Iceland spar) displays double refraction.

may even ask you to identify a few minerals you have never seen before to determine how well you have mastered the techniques of mineral identification.

Strategy for Mineral Identification

Mineral identification is most successful when done systematically, using multiple properties of minerals in combination with mineral identification tables such as Tables 2.3. Minerals generally have no "fingerprint," or single property that sets them apart from others, but you can tell them apart by *combinations* of physical properties. *Use of photographs*, such as those in ■ Figure 2.8 and ■ Figure 2.9, *cannot replace close examination of a mineral's physical properties*. In nature (and perhaps in your tests or quizzes), different specimens of the same mineral can display quite a wide variety of appearances, but many of their specific physical properties will remain the same.

Diane Pirie

(a) **Olivine:** an aggregate of grains showing the typical olive green color.

Diane Pirie

(b) **Almandine garnet**: *Left:* a trapezohedral single crystal, equant habit. *Right:* a broken sample showing hackly fracture and adamantine and waxy luster.

Diane Pirie

(c) **Augite (pyroxene)**: a fragment showing two cleavages at nearly 90°.

Claudia Owen

Diane Pirie

(d) **Hornblende (amphibole):** fragments illustrating vitreous luster and two cleavages at an oblique angle (inset).

Diane Pirie

(e) **Biotite mica** (black) has splendent luster and one perfect cleavage. The white grains in this sample are quartz.

Diane Pirie

Claudia Owen

(f) **Muscovite mica**: Top surface has pearly luster; bottom shows cleavages on edge. *Inset:* thin cleavage sheets illustrate the one perfect cleavage of mica.

Figure 2.8

Silicate Minerals

(g) **Potassium feldspar** (**orthoclase** or **microcline**) varies in color from white to salmon pink and has two cleavages at 90°.

(h) **Plagioclase feldspar** of different chemical compositions: darker ones are usually Ca-rich. *Upper right:* dark gray Ca-rich. *Left:* light gray Na-Ca. *Lower right:* white Na-rich with twinning (stripes of reflected light). Twinning, if visible, distinguishes plagioclase from potassium feldspar.

(i) **Kaolin:** an extremely fine-grained aggregate showing earthy luster and no visible cleavage.

(j) **Quartz:** hexagonal colorless quartz crystals: (*rock crystal*) with columnar habit.

(k–t) Varieties of quartz.

Figure 2.8

Silicate Minerals—*Continued*

Diane Pirie

(a) **Calcite**: a broken fragment showing three cleavages at an oblique angle. The bubbles result from reaction with dilute hydrochloric acid.

Claudia Owen

(b) **Gypsum:** common habits. *Top:* massive aggregate, *alabaster; lower left:* a single crystal cleavage fragment, *selenite*, showing a scratch made by fingernail; *lower right:* fibrous *satin spar* with silky luster.

Diane Pirie

(c) **Galena**—lead ore: a fragment with three cleavages at 90° (cubic cleavage). *Left side:* metallic luster shows where sample is freshly cleaved. *Right side:* weathered surfaces are submetallic.

Diane Pirie

(d) **Pyrite**: three samples showing brassy color and metallic luster. *Upper left:* granular massive aggregate of pyrite. *Right:* aggregate of crystals. *Lower left:* pyrite cube.

Diane Pirie

(e) **Magnetite**—iron ore: granular aggregate, strongly magnetic with submetallic luster.

Diane Pirie

(f) **Hematite**—iron ore. *Left: oolitic hematite*—an aggregate of spheres of fine-grained hematite crystals. The grains and streak are deep reddish brown. *Right: specular hematite* is silvery yet has a reddish brown streak, although darker than the oolitic sample.

Figure 2.9

Nonsilicate Minerals

(g) **Fluorite** showing four cleavages and displaying a variety of colors. Its colors and vitreous luster are similar to quartz, but its cleavage distinguishes it from that harder mineral.

(h) **Halite**: transparent to translucent with three cleavages at 90° (cubic cleavage).

Figure 2.9

Nonsilicate Minerals—*Continued*

A useful strategy is to use mineral identification tables methodically (Table 2.3, Sections A through E on pp. 30 through 34):

1. Test mineral properties as the headings in Table 2.4 indicate.
2. Then follow the arrows, starting on the left of Table 2.3 to find the mineral name.

Materials needed:

- Assorted unknown minerals
- Mineral testing kit
- Squeeze bottle of 10% hydrochloric acid solution. (Use as little acid as possible—one drop is enough—and clean off the sample with a tissue or paper towel when you finish.)

10. Fill in the mineral table (Table 2.4 on pp. 35–36) with the samples you are given by testing and analyzing in the following manner:

a. Begin testing the mineral properties of an unknown sample (one with a number only) and filling in the properties in the Mineral Identification Chart in Table 2.4 on pages 35–36. If you can guess the mineral name based on your observations of Figures 2.8 and 2.9, you may lightly pencil it in, but remember that this is only a hypothesis (see Appendix) and may change as you work on part **b**.

b. Next, use Table 2.3 to discover the mineral name. Then go back to part **a** for the next mineral sample.

c. After filling in the table for all the samples, highlight characteristic distinguishing properties of each mineral in the mineral table (Table 2.4).

CLASSIFICATION OF MINERALS

Recall that the definition of a mineral refers to its strictly defined chemical composition. It is chemical composition that we use to classify minerals. The majority of minerals belong to the mineral classes listed in Table 2.5. Silicates are the most abundant minerals and comprise a large proportion of the rock-forming minerals, which make up the majority of rocks at the Earth's surface.

Minerals of only one element, such as native gold, belong to a class called the *native elements*. The Periodic Table of the Elements in Table 2.6 shows which elements are metals, metalloids, and nonmetals. Only a few of these occur as native elements. All other minerals are chemical compounds, which we classify primarily

Table 2.3

Minerals Identification Tables

Using the Mineral Identification Tables

The Tables 2.3 A–E on the following pages will help you identify unknown mineral specimens. The two main subdivisions are minerals with metallic luster (Table 2.3A) and minerals with nonmetallic luster (Table 2.3B–E). Each section starts with the hardest mineral and ends with the softest.

Section A: Minerals with metallic and submetallic lusters*

Table 2.3A is divided into subdivisions based on streak. Generally, minerals with metallic luster have streak colors that are diagnostic.

Luster	Streak	Properties	Mineral
Metallic to Submetallic Luster →	**Dark Streak**: green-black, dark brown, gray-black →	H = 6–6½; **Pale brass yellow**; "Fool's gold"; S = green-black; D = 5.0; F = irregular; Habit = equant, **cubic crystals** with striations common or pyritohedral crystals (having 12 pentagon-shaped sides)	**Pyrite** (Fig. 2.9d) *Fe sulfide*
		H = 6; Black; S = black to gray-black; D = 5.2; F = irregular; L = submetallic; Habit = equant; **Strongly magnetic**	**Magnetite** (Fig. 2.9e) *Fe oxide,* major ore of iron
		H = 3½–4; **Bright brass yellow** where fresh; tarnishes to iridescent purple "peacock ore"; S = green-black; D = 4.1–4.3; brittle	**Chalcopyrite** (Fig. 2.4) *Cu, Fe sulfide,* major ore of copper
		H = 2½; **Gun metal gray**; CL = 3 at right angles (perfect cubic CL); **S = lead gray; D = 7.6**; L = very **bright metallic** on fresh surfaces, submetallic where tarnished	**Galena** (Fig. 2.9c) *Pb sulfide,* lead ore
		H = 1–2; Iron black/steel gray; Soils fingers; S = dark gray; D = 2.2; L = metallic to submetallic; **greasy feel**	**Graphite** *Native element (C)*
	Medium Streak: → pale brown, red-brown, yellow-brown, yellow	H = 5–6½, may flake; Steel gray; L = bright metallic and steel gray; **S = red-brown to brown**; D = 5.3; F = irregular; Habit = micaceous (called "micaceous" hematite)	**Specular hematite** (Fig. 2.9f) *Fe oxide,* major ore of iron
		H = 5–5½; Dark to brown to black; **S = yellow-brown**; D = 3.3–5.5; F = irregular; L = metallic to submetallic, luster may be obscured by alteration. "Limonite" is often used to name any hydrous iron oxide.	**Limonite** (mineraloid) *Hydrous Fe oxide,* ore of iron
		H = 2½–3; Shades of yellow; S = gold yellow; L = metallic; plates, flakes, or nuggets; D = 19.3 when pure; very malleable and ductile; color yellow becomes paler with increasing silver content	**Gold** *Native element (Au),* ore of gold
		H = 2½–3; **Coppery (orange-red-brown) colored**; Malleable and ductile; S = copper brown; D = 8.9; L = metallic, but surface is often tarnished and may be oxidized to blue; Habit = dendritic	**Copper** *Native element (Cu),* ore of copper
	Light Streak: pale brown to light yellow →	H = 3½–4; Dark brown to yellow; variegated appearance common; CL = 6 good planes, **S = yellow** to brown and lighter than sample; D = 3.9–4.1; L = resinous, adamantine, or submetallic	**Sphalerite** *Zn sulfide,* ore of zinc

Nonmetallic Luster continue to Table 2.3B-E

* The abbreviations are CL = cleavage, P = parting, F = fracture, L = luster, S = streak, D = density (in g/cm^3) or specific gravity (no units), ~ = approximately. Properties that are especially diagnostic are shown in **bold**.

Table 2.3

Minerals Identification Tables—*Continued*

Section B: Nonmetallic minerals with cleavage or parting that are harder than glass*

Luster
- **Metallic Luster** → Go to Table 2.3 Section A
- **Nonmetallic Luster** → **Cleavage**
  - **Without cleavage** → Go to Table 2.3 Sections D and E
  - **With cleavage** → **Hardness**
    - **Softer than glass** → Go to Table 2.3C
    - **Harder than glass** → (green rows below)
    - **About as hard as glass** → (blue rows below)

Hardness	Properties	Mineral
Harder than glass	**H = 10, hardest natural substance**; CL = 4 ≠ 90° perfect, (octahedral CL); Habit = equant, octahedrons; Pale yellow to colorless, many pale shades; D = 3.5; **L = adamantine** to greasy in natural crystals; high brilliance	**Diamond** *Native element (C)*
Harder than glass	**H = 9** on a fresh surface; P = basal; Gray to brown to pink, almost any color; D = 3.9–4.1; L = vitreous to adamantine; Habit = prismatic, hexagonal crystals narrow toward the ends; Gem varieties include ruby (red) and sapphire (blue)	**Corundum** *Al oxide*
Harder than glass	**H = 8; CL = 1**, basal; Colorless, yellow, white, pink, blue, green; D = 3.4–3.6; L = vitreous; semiprecious gem: transparent golden and blue varieties	**Topaz** *Al, F, OH silicate with single tetrahedra*
Harder than glass	H = 7½–8; CL = 1 imperfect, uneven F is more apparent; Blue-green or yellow; D = 2.6–2.8; gem varieties: emerald (deep green), aquamarine, morganite (rose beryl), and golden beryl	**Beryl** *Be, Al ring silicate*
Harder than glass	**H = 7** perpendicular to blades and **H = 5** parallel to blades; CL = 2 at 74°; **Blue** to gray to green; D = 3.6–3.7; L = vitreous to pearly to silky; **Habit = bladed**. Note different hardness in different directions	**Kyanite** (Fig. 2.3) Al silicate
Harder than glass	**H = 6–7**; **CL = 1**; Colorless to brown or pale green; D = 3.2; L = vitreous; **Habit = fibrous to acicular**	**Sillimanite** (Table 5.1) Al silicate
Harder than glass	**H = 6**; **CL = 2** good planes at **~90°**; White, cream, gray, salmon to dark pink; D = 2.5–2.6; L = vitreous. A pink color often distinguishes K-rich feldspars. Amazonite is a rare blue-green variety of an alkali feldspar called microcline	Alkali **Feldspars** (Fig. 2.8g) *K, Na, Al Tectosilicate*
Harder than glass	**H = 6**; **CL = 2** good planes at **~90°**; White, gray, greenish or bluish gray; D = 2.6–2.8; L = vitreous. An iridescent play of colors may be seen in some plagioclases, especially labradorite. Parallel, regular **striations** on a cleavage plane (= **twinning**) are common in the plagioclase series. Albite (Na-rich) and anorthite (Ca-rich) are the end members	Plagioclase **Feldspars** (Fig. 2.8h) *Na, Ca, Al Tectosilicate*
About as hard as glass	H = 5–6; **CL = 2 at ~90°** (not perfectly planar); Dark green to black; S = pale green to gray if any; D = 3.2–3.3; L = vitreous (slightly duller than hornblende); Habit = stubby	**Augite** (pyroxene) (Fig. 2.8c) *Single-chain silicate*
About as hard as glass	H = 5–6; **CL = 2 at approx. 120° and 60°**; Greenish dark gray to **black**; S = pale gray if any; D = 3.0–3.3; **L = vitreous**; Habit = prismatic with diamond-shaped cross-section	**Hornblende** (amphibole) (Fig. 2.8d) *Double-chain silicate*
About as hard as glass	H = 5–6; CL = 2 at ~120° and ~60°; Medium to dark **green**; S = light gray if any D = 3.1–3.3; L = vitreous; **Habit = acicular** with diamond-shaped cross-section	**Actinolite** (Table 5.1) (amphibole) *Double-chain silicate*
About as hard as glass	**H = 5** parallel to blades and **H = 7** perpendicular to blades; CL = 2 directions at 74°; **Blue** to gray to green; D = 3.6–3.7; L = vitreous to pearly to silky; **Habit = bladed**	**Kyanite** (Fig. 2.3) *Al silicate*
About as hard as glass	H = 5–5½; CL = 2 at 84°, splintery to hackly; White, colorless to gray; D = 2.8–2.9; L = vitreous to pearly; Habit = prismatic to fibrous, splintery	**Wollastonite** (Table 5.1) *Ca silicate*

* The abbreviations are CL = cleavage, P = parting, F = fracture, L = luster, S = streak, D = density (in g/cm^3) or specific gravity (no units), ~ = approximately. Properties that are especially diagnostic are shown in **bold**.

(*Continued*)

Table 2.3

Minerals Identification Tables—*Continued*

Section C: Nonmetallic minerals with visible cleavage that are softer than glass*

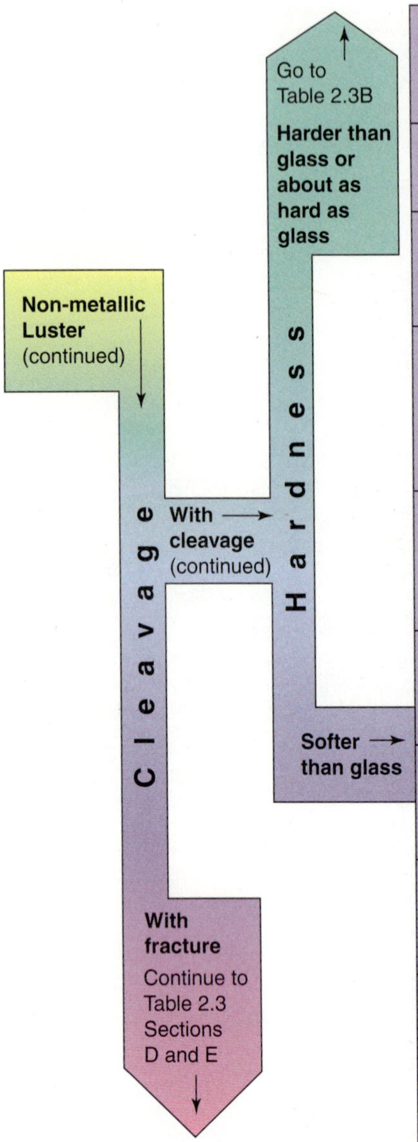

Non-metallic Luster (continued) → **Cleavage** → **With cleavage** (continued) → **Hardness**

- **Harder than glass or about as hard as glass**: Go to Table 2.3B
- **Softer than glass** → see table below
- **With fracture**: Continue to Table 2.3 Sections D and E

Properties	Mineral
H = 4; CL = 4 ≠ 90° perfect (octahedral CL); Highly variable color, violet, pale green, yellow are common; S = white; D = 3.2; L = vitreous to waxy; Transparent to translucent; Habit = equant, cubes common; May fluoresce under black light	**Fluorite** (Fig. 2.9g) *Ca, F halide*
H = 3½–4; CL = up to 6 good planes; Dark brown to yellow; **L = resinous to adamantine to sub-metallic**; Variegated appearance common; **S = Yellow**; D = 3.9–4.1	**Sphalerite** *Zn sulfide,* ore of zinc
H = 3½–4; CL = 3 ≠ 90°, perfect rhombohedral CL; Yellowish white to pink, gray or light brown; Color varies with impurities. D = 2.8; L = vitreous. **Effervesces in dilute HCl when powdered** (Fig. 4.15c inset p. 82).	**Dolomite** *Ca, Mg carbonate*
H = 3; CL = 3 ≠ 90°, perfect rhombohedral CL; Colors vary: colorless to white to yellow to pink. S = white; D = 2.7; L = vitreous; **Strongly effervesces with dilute HCl**; *Iceland spar*: transparent with double refraction (Fig. 2.7) Fine-grained varieties do not show cleavage, but all will effervesce freely with dilute HCl	**Calcite** (Fig. 2.9a) *Ca Carbonate*
H = 2½–3, marked with fingernail by creasing sheets rather than scratching; **CL = 1** perfectly planar; Dark reddish brown to **black**; S = Light Brown, may break apart when testing; D = 2.8–3.2; **L = splendent** to vitreous; Elastic, flexible, and transparent in thin sheets	**Biotite (mica)** (Fig. 2.8e) *K, Al sheet silicates*
H = 2–2½; **CL = 1** perfectly planar; Colorless to white to light greenish brown; S = white, may break apart when testing; D = 2.7–2.9; **L = resinous** to vitreous to **pearly**; Sheets elastic, flexible; Transparent	**Muscovite (mica)** (Fig. 2.8f) *K, Al sheet silicates*
H = 2½; **CL = 3 perfect at 90°** (cubic CL); Colorless to white; S = white; D = 2.1–2.3; L = waxy to vitreous; **Salty taste**; Transparent; **Dissolves in water and on fingers**; Tan to reddish with impurities	**Halite** (Fig. 2.9h) *NaCl halide*
H = 2–2½; **CL = 1** perfect, but flakes are small compared to micas; **Medium to dark green**; S = white to pale green, may break apart when testing; D = 2.6–3.3; L = vitreous to pearly. **Thin sheets are flexible but not elastic**	**Chlorite** *Mg, Fe, OH sheet Silicate*
H = 2; **CL = 3 perfect at 90°** (cubic CL); Colorless to white to pale blue; S = white; D = 2.0; L = waxy to vitreous; **Bitter salty taste**; Transparent; **Dissolves in water and on fingers**; Yellow or reddish with impurities; Much rarer than halite	**Sylvite** *KCl halide*
H = 2, Can scratch with fingernail; CL = 1 perfect (in sheets), and 2 more irregular CL; Colorless to white; S = white; D = 2.3; L = vitreous to pearly; Flexible in thin sheets; can shave with knife (sectile); transparent to translucent	**Gypsum (selenite)** (Fig. 2.9b lower right) *Ca sulfate*

* The abbreviations are CL = cleavage, P = parting, F = fracture, L = luster, S = streak, D = density (in g/cm^3) or specific gravity (no units), ~ = approximately. Properties that are especially diagnostic are shown in **bold**.

Table 2.3

Minerals Identification Tables—*Continued*

Section D: Nonmetallic minerals with apparent fracture (without visible cleavage) that are harder than glass*

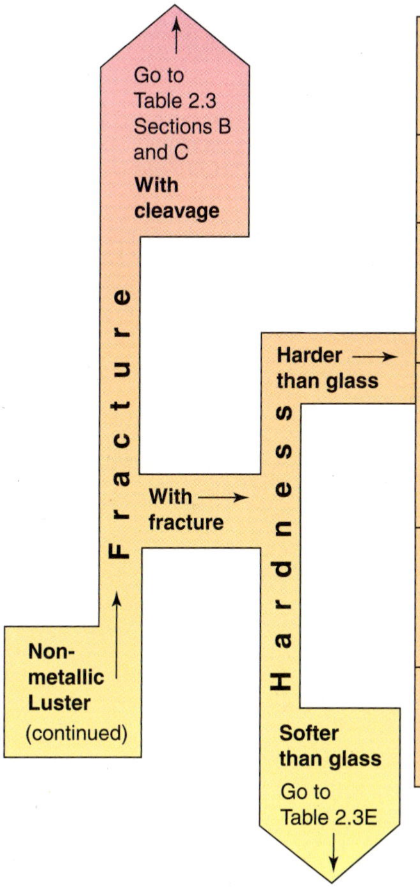

Non-metallic Luster (continued) → Fracture → (Go to Table 2.3 Sections B and C — **With cleavage**) / **With fracture** → Hardness → **Harder than glass** → / **Softer than glass** Go to Table 2.3E

Properties	Mineral
H = 9 — test on a fresh surface; P = basal; Gray to brown to pink; almost any color; D = 3.9–4.1; L = vitreous to adamantine; Habit = prismatic, hexagonal crystals narrow toward the ends. Gem varieties include ruby (red) and sapphire (blue)	**Corundum** *Al oxide*
H = 7–7½; Red-brown to brownish black; L = resinous to vitreous; D = 3.6–3.8; Habit = prismatic, obtuse angled prisms with common **crossing twins**	**Staurolite** (Table 5.1) *Fe Al nesosilicate*
H = 7–7½; F = conchoidal; Black common, but may be green, yellow, red, pink, or blue; D = 3–3.3; L = resinous to vitreous; **Habit = prismatic with triangular cross sections.** Tourmaline may be transparent and of semiprecious gem quality, e.g., rubellite (red or pink)	**Tourmaline** (Fig. 2.1b) *Chemically complex ring silicate*
H = 6½–7½; **F = hackly** to conchoidal; Black, dark brown, red, tan, or green; D = 3.5–4.3; L = adamantine to vitreous to waxy on parting; **Habit = equant** well-formed **dodecahedral (12-sided) crystals.** Colors vary with composition. **Almandine is deep red to brown.** Grossular garnet is often tan, pale yellow, pink, or green	**Garnet** (Fig. 2.8b) *Mg, Fe, Mn, Al nesosilicate*
H = 7; F = conchoidal; Color varies widely, colorless to smoky in common rocks; D = 2.65; L = vitreous; (waxy in microcrystalline varieties). Hexagonal crystals show striations perpendicular to prism (Fig. 2.8j). Quartz may be colorless or have many varieties of color (Fig. 2.8k-t)	**Quartz** (Fig. 2.8j–t) *pure silica tectosilicate*
H = 6½–7; **F = conchoidal** and may appear irregular due to small grain size; Light to dark **olive to yellow green**; D = 3.3–4.4; L = vitreous; Habit = good crystals rare and equant, **granular massive aggregates are usual.**	**Olivine** (Fig. 2.8a) *Mg, Fe nesosilicate*

* The abbreviations are CL = cleavage, P = parting, F = fracture, L = luster, S = streak, D = density (in g/cm^3) or specific gravity (no units), ~ = approximately. Properties that are especially diagnostic are shown in **bold**.

(*Continued*)

Table 2.3

Minerals Identification Tables—*Continued*

Section E: Nonmetallic minerals with apparent fracture (without visible cleavage) that are softer than glass*

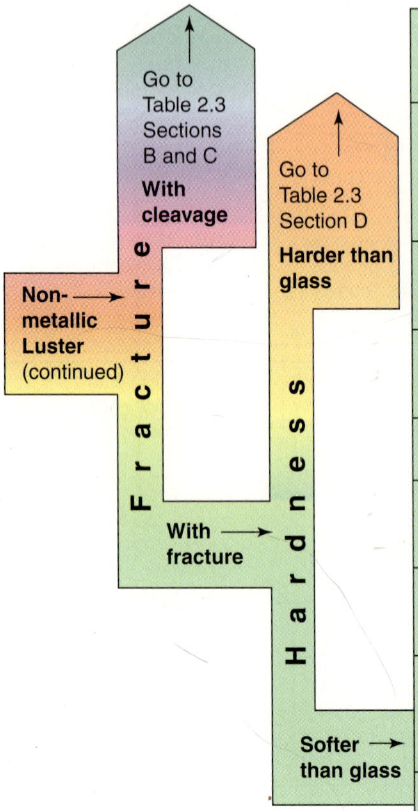

Non-metallic Luster (continued) → Fracture

Fracture: With cleavage → Go to Table 2.3 Sections B and C

Fracture: With fracture → Hardness

Hardness: Harder than glass → Go to Table 2.3 Section D

Hardness: Softer than glass →

Properties	Mineral
H = 5; **F = conchoidal**; poor basal cleavage; Variable color, commonly green to red-brown; S = White; D = 3.1–3.2; L = vitreous; **Habit = hexagonal prisms** or granular aggregates	**Apatite** *Phosphate*
H = 3½–4; F appears uneven or earthy, CL may obscure due to small crystal size (CL = 3 rhombohedral); Yellowish white to pink, gray, or light brown; S = White; D = 2.8; L = vitreous; **Effervesces in dilute HCl when powdered** (Fig 4.15c inset p. 82); Color varies with impurities	**Dolomite** (Fig. 4.15c, p. 82) *Ca, Mg carbonate*
H = 3–5; Light green to dark **green** to nearly black; color often **variegated**; S = White; D = 2.5–2.6; L = **greasy to waxy**; Habit may be fibrous = asbestos (Fig. 2.3)	**Serpentine** (Fig. 5.10b) *Mg OH sheet silicate*
H = 2–5½; Yellow ocher to dark brown; **S = yellow-brown**; D = 3.3–5.5 or less if porous; F = irregular; L = earthy. "Limonite" is often used to name any hydrous iron oxide	**Limonite** (mineraloid) *Hydrous Fe oxide*
H = 2–3; **apple green**; S = pale green; D = 2.2–2.8; L = greasy to waxy to earthy; ore of nickel	**Garnierite** *Ni Mg OH sheet silicate*
H = 2; **F = fibrous**, cleavage is not apparent; Satin spar is the fibrous variety; White to pink; S = White; D = 2.3; **L = silky**; habit = fibrous; can shave with knife (sectile)	**Gypsum (satin spar)** (Fig. 2.9b) *Ca sulfate*
H = 2; F = uneven, cleavage is not apparent due to its small crystal size; White to pink; S = White; D = 2.3; L = vitreous to pearly and **compact and massive**; can shave with knife (sectile)	**Gypsum (alabaster)** (Fig. 2.9b) *Ca sulfate*
H = 1–6, Hardness is variable due to variations in grain size; F = uneven ; **Red-brown**; **S = Red-brown**; D = 4.8–5.3; L = dull to earthy; Small (egg-shaped) ooids in oolitic hematite; massive in red ocher; major ore of iron	**Oolitic hematite** and **red ocher** (Fig. 2.9f) *Fe oxide*
H = 1½–2½; F = uneven to conchoidal; **Bright yellow** (when pure); S = Pale yellow; D = 2.1; L = resinous to vitreous; **Distinctive odor**	**Sulfur** *Native S*
H = 1–5; L = dull to earthy; Highly variable color even in one sample: white, gray, yellow-brown, and red-brown; S = variable but commonly red-brown to yellow brown; D = 2–2.6; Habit = **spherical concretionary grains** (pisolites). Bauxite is a mixture of 3 Al hydroxide minerals so is not a distinct mineral but a composite.	**Bauxite** *Al oxides and hydroxides*, ore of aluminum
H = 1–2; **F = uneven**; microscopically has one perfect cleavage; White when pure, variable when impure; S = white; D = 2.6; **L = dull**, earthy; **Powdery** clay, earthy smell; **No effervescence**	**Kaolinite** (Fig. 2.8i) *Al OH sheet silicate*
H = 1; **F = earthy**; (CL is obscured by fine grain size); White to tan; S = white; D = 2.5; **L = earthy; Strongly effervesces with dilute HCl**; soft, powdery; fine grained and earthy	**Chalk (Calcite)** (Fig. 4.14, p. 81) *Ca Carbonate*
H = 1; CL = 1 perfect (may be microscopic); White to gray-green; S = White, flakes when powdered; D = 2.7–2.8; **L = pearly** in coarse varieties to dull when fine grained; **Greasy feel**. Soapstone is a compact and massive variety. Talc containing tremolite (H = 5-6) appears harder	**Talc** (Table 5.1) *Mg OH sheet silicate*

* The abbreviations are CL = cleavage, P = parting, F = fracture, L = luster, S = streak, D = density (in g/cm^3) or specific gravity (no units), ~ = approximately. Properties that are especially diagnostic are shown in **bold**.

on the basis of their nonmetal and metalloid elements (Table 2.5). Notice also in Table 2.5 that all the mineral classes ending with *-ates* contain oxygen, and those ending with *-ides* do not contain oxygen unless *ox* is part of the name. In the chemical formulas of compounds, the first part usually lists the metals; the second part, the nonmetals and metalloids. You can tell a mineral's class from the second part of the formula. ■ Table 2.7 lists some minerals (with formulas) belonging to mineral classes that are not silicates. Many non-silicate minerals are important resources, which we will discuss in the last part of this chapter.

Table 2.4

Mineral Identification Chart of Unknown Samples

Record the number of the sample.	Describe the specific metallic or nonmetallic luster of the sample. Also list single-crystal or aggregate habits using Figures 2.2 and 2.3, if applicable.	Test with common objects, (Table 2.1) list possible **range**.	Scratch the sample on a streak plate and record the color of the powder.	For samples with cleavage, write **CL:** and choose from Fig. 2.5. For samples without cleavage, write **FR:** and choose from conchoidal, fibrous, hackly, or uneven.	Describe predominant mineral's color(s).	List special properties such as magnetism, bubbling in acid, etc (see p. 25). If density is unusually high or low, write high or low.	Determine the mineral's name by looking up its properties in Table 2.3.
Sample Number	**Luster, Crystal Habit** (if applicable)	**Hardness**	**Streak**	**Cleavage or Fracture**	**Color**	**Special Properties and Unusual Density**	**Mineral Name**
#1	Doll / special	5	No streak	2 cleavege concoil	Black	Black	Augite
#2	Vitrious	3	white streak	3 cleaveg e	clear	HCL	Calcite
#3	sub-metallic/dull	6	No streak	1 cleavage, concoil	Silver	None	Feldspar (orthoclase)
#4	Vitrous	2	No streak	1 cleavage	clear	Thin micacious	Muscovite
#5	Doll / submetallic	6	No streak	conchodal fracture	Red/dark/ brown	Cristolyzed	Garnet
#6	Vitrious	7	white	conchoidal fracture	clear	fools	Quartz
#7	Metallic	6-6½	black streak	conchoidal fractur-	Goldish/ yellow	fools gold	Pyrite
#8	sub metallic / Dull	7	No streak	conchoidal fracture	Greenish / Brown	Green mineral	olivine
#9	Doll / sub metallic	6	Black streak	irregular uneven fracture	Black	magnetic	Magnetite

(Continued)

Table 2.4

Mineral Identification Chart of Unknown Samples—*Continued*

Sample Number	Luster, Crystal Habit (if applicable)	Hardness	Streak	Cleavage or Fracture	Color	Special Properties and Unusual Density	Mineral Name
#10	Metallic	2	Black/ Metallic	uneven-fracture	Grey, silver	shiny	Galena
#11	earthy, dull	1	white	Conchoidal	white	white powder	chalk (calcite)
#12	sub-metallic dull	6	Blue/Navy	2 cleavege	Black	White stars/ agregates	Hornblende
#13	vitrous	2 1/2	white	2 cleavege, uneven fracture	clear	salty	Halite
#14	Vitrious	6	No streak	2 cleavege	white,	Vers like	Feldspar (Plagioclase)
#15	Dull to Vitrious	2	White	uneven cleavege	light salmon	'Round, light.	Gypsum
#16	Dull, to sub metallic	>2	No streak	1 Perfect cleavege	Dark caramel	sheet-like.	Biotite (Mica)

Table 2.5

Composition of Mineral Classes

Nonmetals, metalloids, and metal elements are listed in the Periodic Table in Table 2.6

Mineral Classes	Chemical Makeup
Silicates	Contain silicon (Si) and oxygen (O) at least
Carbonates	CO_3 plus metal(s)
Sulfates	SO_4 plus metal(s)
Sulfides	S plus metal(s)
Oxides	O plus metal(s) without other nonmetals or metalloids (no Si, C, P, S, V, or W)
Hydroxides	OH plus metal(s) without other nonmetals or metalloids
Phosphates	PO_4 plus metal(s)
Halides	F, Cl, Br, or I plus metal(s) without other nonmetals or metalloids
Native elements	Occur in elemental form (one element only)

11. Look at the mineral formulas listed in Table 2.8 and fill in the mineral class name based on the chemical formulas. Refer to Table 2.5.

Of the different mineral classes we have just studied, silicates are by far the most abundant minerals and are therefore the most common in rocks. We next look at the classification of silicate minerals.

SILICATE MINERALS

The most important and most complex mineral class is the silicates. The variety of crystal structures of silicate minerals leads to a wide variety of physical properties seen in this large mineral class. Silicates belong to different families based on their crystal structure, which makes sense because this classification also groups minerals according to their properties.

Table 2.6

Periodic Table of the Elements

Only about a dozen elements are common in minerals and rocks, but many uncommon ones are important sources of natural resources. For example, lead (Pb) is not found in many minerals, but it is present in the mineral galena, the main ore of lead. Silicon (Si) and oxygen (O), in contrast, are important elements in most of the minerals in Earth's crust.

Key: MAIN GROUP METALS; TRANSITION METALS; METALLOIDS; NONMETALS

Uranium / 92 — Atomic number / U — Symbol / 238.0289 — Atomic mass number

Period	1A (1)	2A (2)	3B (3)	4B (4)	5B (5)	6B (6)	7B (7)	8B (8)	8B (9)	8B (10)	1B (11)	2B (12)	3A (13)	4A (14)	5A (15)	6A (16)	7A (17) Halogens	8A (18)
1	Hydrogen 1 H 1.0079																	Helium 2 He 4.0026
2	Lithium 3 Li 6.941	Beryllium 4 Be 9.0122											Boron 5 B 10.811	Carbon 6 C 12.011	Nitrogen 7 N 14.0067	Oxygen 8 O 15.9994	Fluorine 9 F 18.9984	Neon 10 Ne 20.1797
3	Sodium 11 Na 22.9898	Magnesium 12 Mg 24.3050											Aluminum 13 Al 26.9815	Silicon 14 Si 28.0855	Phosphorus 15 P 30.9738	Sulfur 16 S 32.066	Chlorine 17 Cl 35.4527	Argon 18 Ar 39.948
4	Potassium 19 K 39.0983	Calcium 20 Ca 40.078	Scandium 21 Sc 44.9559	Titanium 22 Ti 47.867	Vanadium 23 V 50.9415	Chromium 24 Cr 51.9961	Manganese 25 Mn 54.9380	Iron 26 Fe 55.845	Cobalt 27 Co 58.9332	Nickel 28 Ni 58.6934	Copper 29 Cu 63.546	Zinc 30 Zn 65.39	Gallium 31 Ga 69.723	Germanium 32 Ge 72.61	Arsenic 33 As 74.9216	Selenium 34 Se 78.96	Bromine 35 Br 79.904	Krypton 36 Kr 83.80
5	Rubidium 37 Rb 85.4678	Strontium 38 Sr 87.62	Yttrium 39 Y 88.9059	Zirconium 40 Zr 91.224	Niobium 41 Nb 92.9064	Molybdenum 42 Mo 95.94	Technetium 43 Tc (97.907)	Ruthenium 44 Ru 101.07	Rhodium 45 Rh 102.9055	Palladium 46 Pd 106.42	Silver 47 Ag 107.8682	Cadmium 48 Cd 112.411	Indium 49 In 114.818	Tin 50 Sn 118.710	Antimony 51 Sb 121.760	Tellurium 52 Te 127.60	Iodine 53 I 126.9045	Xenon 54 Xe 131.29
6	Cesium 55 Cs 132.9054	Barium 56 Ba 137.327	Lanthanum 57 La 138.9055	Hafnium 72 Hf 178.49	Tantalum 73 Ta 180.9479	Tungsten 74 W 183.84	Rhenium 75 Re 186.207	Osmium 76 Os 190.2	Iridium 77 Ir 192.22	Platinum 78 Pt 195.08	Gold 79 Au 196.9665	Mercury 80 Hg 200.59	Thallium 81 Tl 204.3833	Lead 82 Pb 207.2	Bismuth 83 Bi 208.9804	Polonium 84 Po (208.98)	Astatine 85 At (209.99)	Radon 86 Rn (222.02)
7	Francium 87 Fr (223.02)	Radium 88 Ra (226.0254)	Actinium 89 Ac (227.0278)	Rutherfordium 104 Rf (261.11)	Dubnium 105 Db (262.11)	Seaborgium 106 Sg (263.12)	Bohrium 107 Bh (262.12)	Hassium 108 Hs (265)	Meitnerium 109 Mt (266)	Darmstadtium 110 Ds (271)	Roentgenium 111 Rg (272)	— 112 Discovered 1996	— 113 Discovered 2004	— 114 Discovered 1999	— 115 Discovered 2004	— 116 Discovered 1999		

Lanthanides	Cerium 58 Ce 140.115	Praseodymium 59 Pr 140.9076	Neodymium 60 Nd 144.24	Promethium 61 Pm (144.91)	Samarium 62 Sm 150.36	Europium 63 Eu 151.965	Gadolinium 64 Gd 157.25	Terbium 65 Tb 158.9253	Dysprosium 66 Dy 162.50	Holmium 67 Ho 164.9303	Erbium 68 Er 167.26	Thulium 69 Tm 168.9342	Ytterbium 70 Yb 173.04	Lutetium 71 Lu 174.967
Actinides	Thorium 90 Th 232.0381	Protactinium 91 Pa 231.0388	Uranium 92 U 238.0289	Neptunium 93 Np (237.0482)	Plutonium 94 Pu (244.664)	Americium 95 Am (243.061)	Curium 96 Cm (247.07)	Berkelium 97 Bk (247.07)	Californium 98 Cf (251.08)	Einsteinium 99 Es (252.08)	Fermium 100 Fm (257.10)	Mendelevium 101 Md (258.10)	Nobelium 102 No (259.10)	Lawrencium 103 Lr (262.11)

Note: Atomic masses are 1993 IUPAC values (up to four decimal places). Numbers in parentheses are atomic masses or mass numbers of the most stable isotope of an element.

Table 2.7

Nonsilicate Mineral Classes, Common Rock-Forming Minerals and Some Resource Minerals

Mineral	Chemical Formula	Mineral Class
1. **Calcite**[1] Varieties: Iceland spar, calcite chalk	$CaCO_3$	Carbonates Basic unit: (CO_3)
2. **Dolomite** Malachite Azurite	$CaMg(CO_3)_2$ $Cu_2CO_3(OH_2)$ $Cu_3(CO_3)_2(OH)_2$	
3. **Gypsum** Varieties: alabaster, satin spar, selenite	$CaSO_4 \cdot 2H_2O$	**Sulfates** Basic unit: (SO_4)
Galena 4. **Pyrite** Sphalerite	PbS FeS_2 ZnS	**Sulfides** S plus a metal(s)
Corundum 5. **Magnetite** 6. **Hematite**	Al_2O_3 Fe_3O_4 Fe_2O_3	**Oxides** O plus a metal(s)
Limonite	$FeO \cdot OH \cdot nH_2O$	**Hydroxides** (OH) plus metal(s)
Apatite	$Ca_5(PO_4)_3(F,Cl,OH)$	**Phosphates** Basic unit: PO_4
Fluorite 7. **Halite**	CaF_2 $NaCl$	**Halides** Halogen element plus a metal
Metals: Silver Ag Gold Au Platinum Pt Copper Cu Iron Fe	Nonmetals: Diamond C Graphite C Sulfur S	**Native elements** Occur in elemental form

[1]Important rock-forming minerals are numbered and shown in bold.
[2]Limonite is not a mineral—it is mainly a field term referring to natural iron hydroxides.

Table 2.8

Class Identification

Mineral Name	Formula	Mineral Class
Sylvite	KCl	Halide.
Chalcopyrite	$CuFeS_2$	Sulfide
Sulfur	S	Native element
Tremolite	$Ca_2(Mg,Fe)_5Si_8O_{22}(OH)_2$	Silicate
Rutile	TiO_2	oxides
Anhydrite	$CaSO_4$	Sulfate

The basic structural unit for the silicates is the silica tetrahedron (SiO_4), as shown in Figure 2.10a. This building block is often illustrated as lines connecting the oxygen atoms, with no atoms shown, as in Figure 2.9b and as in each of the diagrams for the different silicate families in Table 2.9. These lines connect to make a triangular pyramid called a **tetrahedron,** thus the name *silica tetrahedron.* Imagine an oxygen atom at each corner and a silicon atom in the center. The silicate families are based on the arrangement of the silica tetrahedra (plural) in the atomic structure of the mineral. Silica tetrahedra are negatively charged so the minerals incorporate positively charged ions (**cations**) to balance the charge and to hold the structure together between the isolated tetrahedra or chains or sheets. The silicate structures shown in Table 2.9 lack the cations. To a large extent, the internal structure of the silicates influences each mineral's physical properties, as we shall see. First let's try to understand the structural types involved.

Materials needed:

- Colored paperboard printed with tetrahedron models as in Figure 2.11a, scissors, and tape

Or

- Half toothpicks, stale mini-marshmallows, and gumdrops

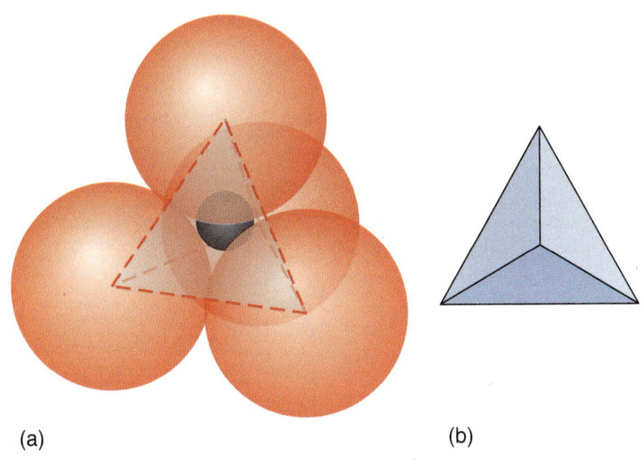

Figure 2.10

Two Ways of Illustrating a Silica Tetrahedron

(a) This figure illustrates the configuration of the atoms (red for oxygen and gray for silicon). Dashed lines from the center of one oxygen atom to another show the location of the tetrahedron, illustrated in (b). (b) Diagrammatic representation of a silica tetrahedron: The corners are at the centers of the oxygen atoms, and the center of the tetrahedron corresponds to the location of the silicon atom.

12. Work together in a group. Each student should build a model of a tetrahedron. Then join your models together to create the structures representing various silicate families. As you make more complicated structures up through phyllosilicates, either build more tetrahedra or join yours with those of people in other groups. The model materials described here do not hold together well enough to make complete tectosilicates by themselves. If your instructor asks you to attempt a tectosilicate model structure, you will need additional support or a different medium for your model. Once you have built your models, answer the following questions:

a. What family of silicates has isolated silica tetrahedra?

____

b. When you make a ring or a single chain of tetrahedra (as in Figure 2.11c), how many corners do the tetrahedra in the middle of the chain share? _____

c. How does this affect the number of oxygen atoms compared to silicon atoms in the structure? Remember that the corners of a tetrahedron represent oxygen atoms and the center represents silicon. The ratio of silicon to oxygen (a) goes up, (b) goes down, (c) stays the same. _____

d. Look at the formulas of the chain and cyclosilicates in Table 2.9 to assess your idea about the silicon to oxygen ratio.

What ratio do nesosilicates have? _______
What ratio do single-chain (check diopside) and cyclosilicates have? _______

e. As you build more complicated structures, how does the silicon to oxygen ratio change?

__________________ Why is this true?

____

____

____

13. Let's see how different silicate structures influence a mineral's properties.
Name a mineral with single chains.

____

What type of cleavage does it have?

____

What is an example of a sheet silicate?

____

What type of cleavage does it have?

____

14. Look through Table 2.3 on pages 30–34.

a. What sub-tables of Table 2.3 (A, B, C, or D) have most of the silicates (not counting sheet silicates)? ______ This is because the strong bonds in these silicates make them hard.

b. What sub-tables of Table 2.3 list the sheet silicates? ______ How do you account for this?

____

____

____

____

c. Notice that some sheet silicates are listed in tables with fracture rather than cleavage. This is because they tend to

Table 2.9

Common Rock-Forming Minerals—Silicate Mineral Class

Mineral	Chemical Formula	Silicate Family Name	Silicate Structure
8. **Olivine**[1]	$(Mg,Fe)_2SiO_4$	*Nesosilicates* (isolated Si tetrahedra)	
Topaz	$Al_2SiO_4(F,OH)_2$		
9. ***Garnet***[2] Varieties: almandine, grossular	$(Mg,Fe,Ca,Mn)_3(Al,Fe,Cr)_2(SiO_4)_3$		
Tourmaline	$(Na,Ca)(Li,Mg,Al)_3(Al,Fe,Mn)_6(BO_3)_3(Si_6O_{18})(OH)_4$	*Cyclosilicates* (ring silicates)	
Beryl	$Be_3Al_2(Si_6O_{18})$		
Pyroxenes		Inosilicates (chain silicates)	
10. **Augite**	$(Ca,Na)(Mg,Fe,Al)(Si,Al)_2O_6$	*Single chain*	
Diopside	$CaMgSi_2O_6$		
Amphiboles:			
11. **Hornblende**	$(Ca,Na)_{2\text{-}3}(Mg,Fe,Al)_5Si_6(Si,Al)_2O_{22}(OH)_2$	*Double chain*	
Actinolite	$Ca_2(Mg,Fe)_5Si_8O_{22}(OH)_2$		
Talc	$Mg_3Si_4O_{10}(OH)_2$	*Phyllosilicates* (sheet silicates)	
12. **Kaolin**	$Al_2Si_2O_5(OH)_4$		
Chlorite	$(Mg,Fe)_3(Si,Al)_4O_{10}(OH)_2 \cdot (Mg,Fe)_3(OH)_6$		
Micas:			
13. **Biotite**	$K(Mg,Fe)_3(AlSi_3O_{10})(OH)_2$		
14. **Muscovite**	$KAl_2(AlSi_3O_{10})(OH)_2$		
15. ***Feldspars:***		*Tectosilicates* (framework silicates) Every tetrahedron in this structure is connected to four others: many are not shown	
Alkali feldspars	$(K,Na)AlSi_3O_8$		
Plagioclase feldspars	$(Na,Ca)Al_{1\text{-}2}Si_{3\text{-}2}O_8$		
16. **Quartz**	SiO_2		
Varieties: amethyst, rock crystal, smoky quartz, milky quartz, rose quartz. Microcrystalline and amorphous varieties: chert, flint, jasper, agate, chalcedony (Amorphous solids are not tectosilicates; in fact, they are not even minerals, and are called *mineraloids* instead. Recall that a mineral must be crystalline.)			

[1]Important rock-forming minerals are numbered and shown in bold.
[2]Mineral group names are shown in italics.

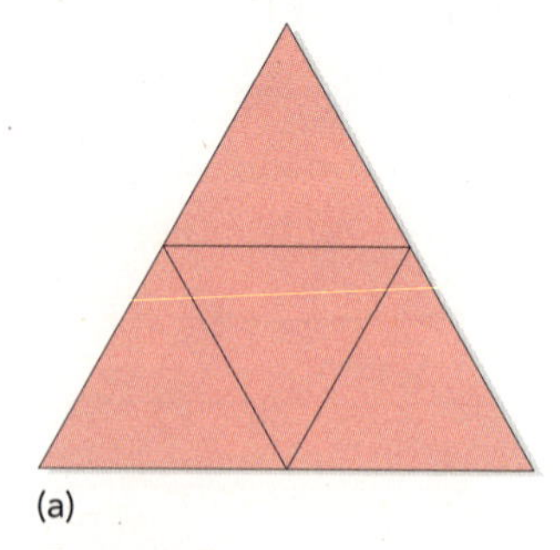

(a)

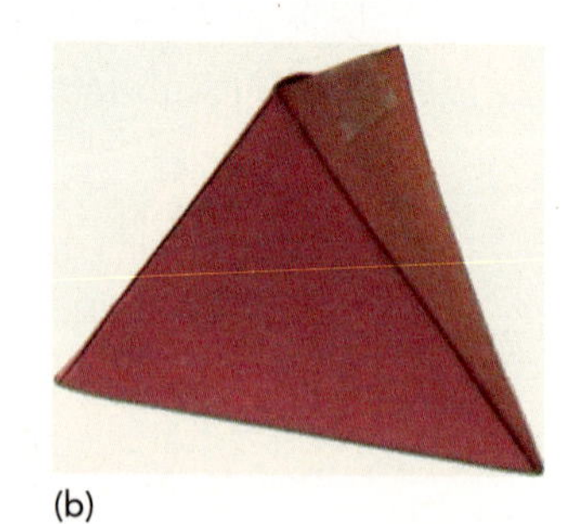

(b)

(c)

Figure 2.11

Silica Tetrahedron Model

(a) To build a silica tetrahedron, cut out a shape like this, then fold along the black lines and tape it into a triangular pyramid. (b) Model of tetrahedron after folding and taping. (c) Chain of silica tetrahedra using stale mini-marshmallows for oxygen and gumdrops for silicon. Toothpicks represent the edges of the tetrahedra, not chemical bonds.

be too fine-grained for us to see the individual cleavage sheets. These minerals tend to be very soft. List two such sheet silicates that are commonly pale in color:

_______________ _______________

Silicate minerals of various structures make up the most common rock-forming minerals, but with the exception of quartz, feldspar, and kaolin, they are not commonly used as resources. On the other hand, nonsilicates may be important resources, especially carbonates and sulfates, which we use for in building materials; and oxides and sulfides, which we use to produce metals.

ECONOMICALLY VALUABLE AND USEFUL MINERALS

Minerals provide us with many of the resources we need to produce the common objects we take for granted in the industrial world. A mineral or rock that is mined at a profit and can be used to make metal is known as an **ore**. Other useful minerals that are valuable are called **mineral resources**. First let's examine the common resources used to make building materials (such as in walls, wire, and sidewalks) or simple tools (such as in eating utensils, dishes, and cans) and learn which minerals provide the material in these and other common items.

15. Fill in ■ Table 2.10 with the mineral properties of some resource mineral samples your instructor provides and then identify the minerals using Table 2.3.

Scavenger Hunt

16. Your instructor will divide the class into teams, giving each team a name or letter. As a team, number individual, colored sticky notes from 1 to 11 and write your team name on each; then wait for the signal to start. At the start, try to find the objects listed in ■ Table 2.11 somewhere in the classroom. When you find an item, place a sticky note with the corresponding number on that item. You may not use items others have already used. At the ending signal, the team with the most correctly labeled objects wins.

17. Next, fill in Table 2.11 with the information about the objects you found in your scavenger hunt. Also add the remaining items you did not find in time (if any). Fill in blanks in the table describing each substance and its components. Then use ■ Table 2.12 to help you determine what resource materials (most of these are minerals) would be used to make these objects.

18. In the appropriate locations in the last column of Table 2.11, enter the sample

Table 2.10

Resource Minerals and Their Properties

Sample Number	Luster, Crystal Habit (if appl.)	Streak and Hardness	Cleavage/ Fracture	Color	Other Properties	Mineral Name

Table 2.11

Common Objects and the Resources They Are Made From

Object #	Object Name	Name of Metal or Primary Substance	Description (What Is the Substance Like?)	Made From/Of	Resource Material(s)	Sample #s
1		Chromium stainless steel	Magnetic, silvery, does not tarnish	Chromium, Iron, and carbon	Chromite Hematite/magnetite Coal	
2		Nickel stainless steel	Non-magnetic, silvery, dense, does not tarnish	______, chromium, iron, and carbon	Garnierite/Pentlandite Chromite Hematite Coal	
3		Aluminum		Aluminum	Bauxite	
4		Brass		Copper Zinc	Chalcopyrite, native copper Sphalerite	
5		Gold	Golden metal that does not tarnish	Gold	Native gold	
6		Copper	Reddish metal	Copper	Chalcopyrite	
7		Concrete		Lime and aggregate	Calcite Gravel	
8		Sheet rock	Wallboard or plasterboard	Plaster	Gypsum	
9		Porcelain		Kaolinite and feldspar	Kaolinite Feldspar	
10		Glass		Silica	Quartz	
11		Graphite		Graphite	Graphite	

numbers of minerals you have studied in previous parts of this lab and from Table 2.10 of the resource minerals you just identified. The common objects require the use of minerals such as these to make them.

Now that we have looked at some useful minerals, we will take a moment to consider the economic side of mineral resources.

Scarcity versus Abundance of Mineral Resources

Mineral resources are **nonrenewable** in the sense that they have a finite supply. When that supply is used up, the only way to continue to use that resource is to recycle. The United States Geological Survey (USGS) and other government agencies commonly report the abundance of different elements or commodities in terms of reserves and projected lifetimes. **Reserves** are the quantity of a resource that has been found and is economically recoverable with existing technology. To calculate *projected lifetimes* we use the current rate of production of the resource, assuming that it will continue at the same rate. **Projected lifetimes** are the reserves of the material divided by the rate of production:

$$\text{Projected lifetime} = \frac{\text{Reserves}}{\text{Production rate}}$$

19. According to the USGS, in 2008 the worldwide reserves of tin were 5.6 million metric tons, with a production rate of 0.333 million metric tons per year. What is the projected lifetime of tin? ________________. Show your calculations.

20. In a group, discuss the results of your calculations and answer the following questions:

a. As these ores become scarcer, how would increasing prices influence their rate of use? ______________________________

Table 2.12

Resource Rocks and Minerals and Common Substances

Resource Material	Used to Provide	Material or Substance Produced
Hematite or magnetite	Iron	Steel (with carbon), iron, alloys, stainless steel (with chromium ± nickel)
Chalcopyrite, native copper, malachite, or azurite	Copper	Metallic copper, wire, coins, brass (with zinc), bronze (with tin)
Sphalerite	Zinc	Metallic zinc, galvanized metal, coins, and brass (with copper), white pigment
Chromite	Chromium	Chrome plating and stainless steel (with iron and carbon)
Garnierite or pentlandite	Nickel	Stainless steel (with iron, carbon, and chromium) and coins
Cassiterite (tin oxide)	Tin	Tin plate, tin coating inside "tin" cans, bronze (with copper), solder (with lead)
Bauxite	Aluminum	Metallic aluminum, cans, wire, airplanes
Native gold	Gold	Metallic gold, coins, jewelry, electronics
Pyrite	Sulfur	Sulfuric acid
Graphite	Graphite	"Lead" for pencils, lubricant
Diamond	Diamond	Diamond for cutting, abrasion, jewelry
Coal	Carbon	Fossil fuel, steel (with iron, ± chromium, ± nickel)
Gravel	Aggregate	Concrete (with cement)
Calcite	Lime	Cement (with alumina, silica)
Clay	Clay and alumina	Ceramic
Quartz	Quartz or silica	Glass, glazes, elemental silicon, porcelain (with kaolinite and feldspar)
Gypsum	Gypsum	Plaster, wallboard (or sheet rock or plasterboard)
Kaolinite	Kaolinite	Porcelain (with quartz and feldspar)
Feldspar	Feldspar	Porcelain (with kaolinite and quartz)

b. As a result of increasing price, some part of the resource that was not economically valuable would become valuable enough to mine. How would this change influence the quantity of reserves? (*Hint:* Review the definition of *reserves* and notice the economic component of the definition.)

____

c. How would an improved mining technique or new technology influence the quantity of the reserves? ________________

d. How would the projected lifetimes change as a result of these three factors?

____

Although projected lifetimes give us an estimate of how long a resource will last, these numbers are generally too small. This is because as the resource becomes scarcer, its price increases and its production rate slows. At the same time, the size of the reserves expands because resources that were once not economic deposits become economic as the price rises. Both of these results cause an increase in the projected lifetimes. Nevertheless, increasing scarcity of the resource and higher prices will make the resource unavailable or undesirable for certain uses.

21. In your same group, discuss the following questions: What is likely to happen if we run out of a resource or if its quantity drops to a very low amount? Think about its use and what could be done about its shortage, economics, price, and mining. Might there be environmental consequences? Select a specific resource or two as you explore the possibilities and consider current events in the media. One example is how petroleum production affects pricing, policies, and products. Describe your conclusions in detail on a separate sheet of paper.

Mining of Minerals and the Environment

Although mining provides us with mineral resources, it can also have a severely negative impact on the environment. This is especially true of surface mining. Not only are the surface vegetation and soil stripped away, but the crushing of rock and its interaction with air and water produce toxic substances that may be released into groundwater, streams, and lakes as a result of mining activities.

Mining sulfide deposits leads to water pollution when sulfide minerals interact with rainwater. Sulfur in the minerals may be oxidized in water, producing sulfuric acid (H_2SO_4). If this happens, **acid mine drainage** results (■ Figure 2.12). If the acid water gets into streams, lakes, and rivers, it may kill fish and other aquatic organisms directly. In lower concentrations, it can destroy their eggs. It also causes toxic metals to leach out of the mine area and surrounding rocks, adding metallic poisons to the water. The water is commonly red to orange-brown due to the presence of very fine iron oxides (Figure 2.12). In addition, some mining processes involve toxic substances directly, such as mercury or cyanide, that aid the mineral extraction. These, too, can leach into waterways with devastating effects.

Andrew Macfarlane

Figure 2.12

Acid Mine Drainage

Water draining from the Mancita mine in Peru contains acid resulting from oxidation of the sulfide minerals galena, sphalerite, pyrite, and chalcopyrite. Veins were mined to obtain lead, zinc, copper, and silver from deposits associated with shallow intrusions of magma. The orange color comes from very fine iron oxide particles suspended in the acidic water. Other metal contaminants in the water may include lead, zinc, copper, antimony, arsenic, and manganese.

22. At the Bohemia mines near Cottage Grove, Oregon, gold was mined extensively over the last 150 years. Associated with the gold were pyrite, the minerals listed below, and other sulfides. What metals (some of which are toxic) could acid mine drainage leach into the water in this area? *Hint:* Use the chemical formulas of these minerals in Table 2.7 and the periodic table in Table 2.6.

Chalcopyrite Cu Copper

Galena Pb lead

Sphalerite Zn Zinc

Cinnabar (HgS) Hg Mercury

Champion Creek, Oregon, is currently a site of mine reclamation where environmental engineers are dealing with mine waste problems, which have led to contamination. High levels of all the metals from the minerals listed above are present in the water and soil of the area. Downstream, the creek flows into a local reservoir used as a municipal water supply.

23. What problems might arise as a consequence of acid mine drainage and toxic metal contamination in local streams?

acidic water can break down minerals and make toxic water

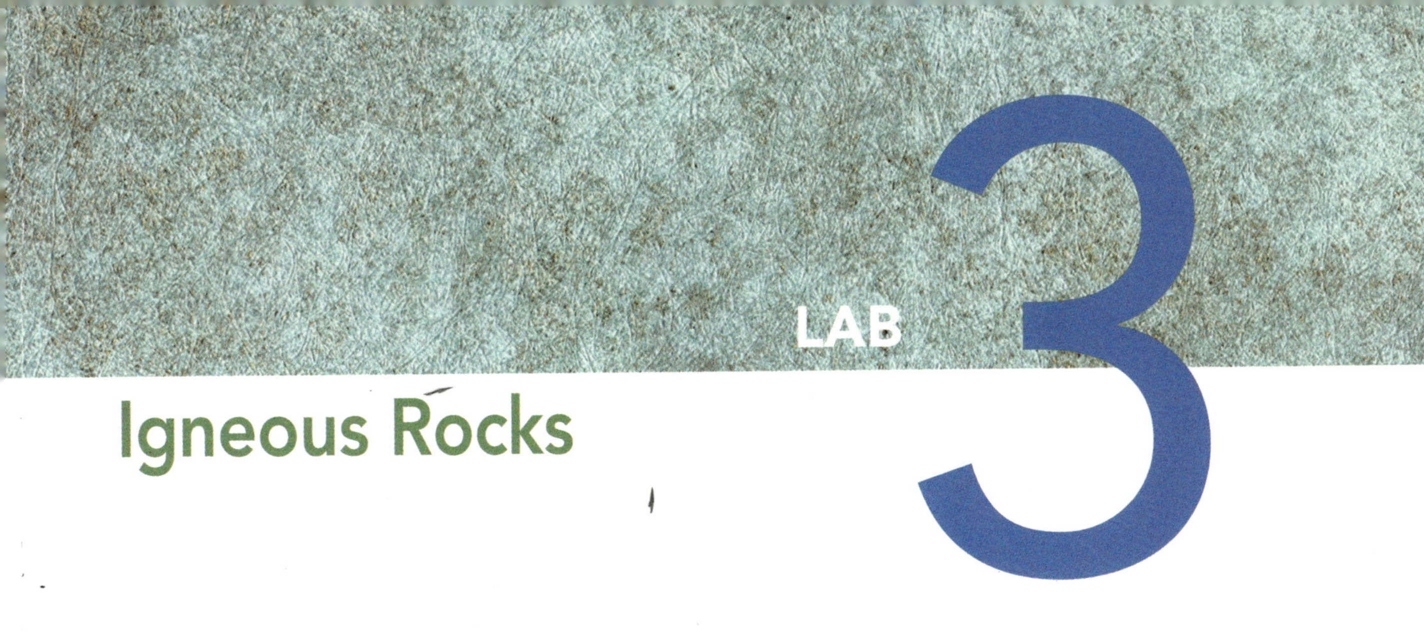

Igneous Rocks

OBJECTIVES

- To recognize minerals in igneous rocks and learn how a rock's color is related to its mineral composition
- To recognize textural features of igneous rocks and understand their origin
- To be able to identify common igneous rocks based on mineral content, color, and texture
- To understand how igneous rocks form
- To recognize various masses of igneous rocks including types of volcanoes and plutons
- To understand the association of hydrothermal veins and their resource minerals with igneous intrusions

As we saw in the discussion of the *rock cycle* in Lab 1, igneous rocks form by solidification of magma into volcanic rocks above or plutonic rocks below Earth's surface. Differences in the way the magma cools and in its chemical composition lead to the wide range of igneous rocks that develop. Minerals present in an igneous rock reflect the chemical composition of the magma that formed the rock. The **texture** of the rock is the arrangement and size of grains, as well as the presence or absence of glass and holes in a rock and reflects the cooling history (■ Figure 3.1). We use both mineral content (or chemical composition) and texture to classify igneous rocks. This method of classification conveys the history of the rocks and the source of their magma.

We find igneous rocks within various volcanic and plutonic bodies or rock masses. Some of these can be gigantic volcanic edifices or extensive crystallized intrusions, but some are much smaller. We discuss the various igneous rock masses at the end of this chapter, but first we need to understand how to classify igneous rocks using mineral and chemical composition and texture.

MINERALOGICAL AND CHEMICAL COMPOSITION OF IGNEOUS ROCKS

The minerals that make up igneous rocks are mostly silicates. They are often subdivided into two groups: light-colored, or *felsic* minerals; and dark, or *mafic* minerals. The *fel* in **felsic** stands for feldspar; and the *si* in felsic indicates the high **silica** (SiO_2) content of felsic minerals, which lack the iron that causes dark coloration. The common felsic minerals in igneous rocks are

plagioclase feldspar	**quartz**
alkali[1] **feldspar**	**muscovite** (some)

Mafic minerals are high in iron and magnesium. The *m* in mafic stands for magnesium, and the *f* stands for iron (chemical symbol Fe, Table 2.6 on p. 37). Iron tends to darken the overall shade of the minerals and rocks containing it. The common mafic minerals are

olivine	**hornblende**
pyroxene	**biotite**

Since these minerals crystallized from magma, their proportions in an igneous rock indicate the composition

[1] Alkali feldspars are potassium (K) or sodium (Na) feldspars. In the periodic table of the elements (Table 2.6, p. 37), alkali comes from the alkali metals in the first column, which includes potassium and sodium.

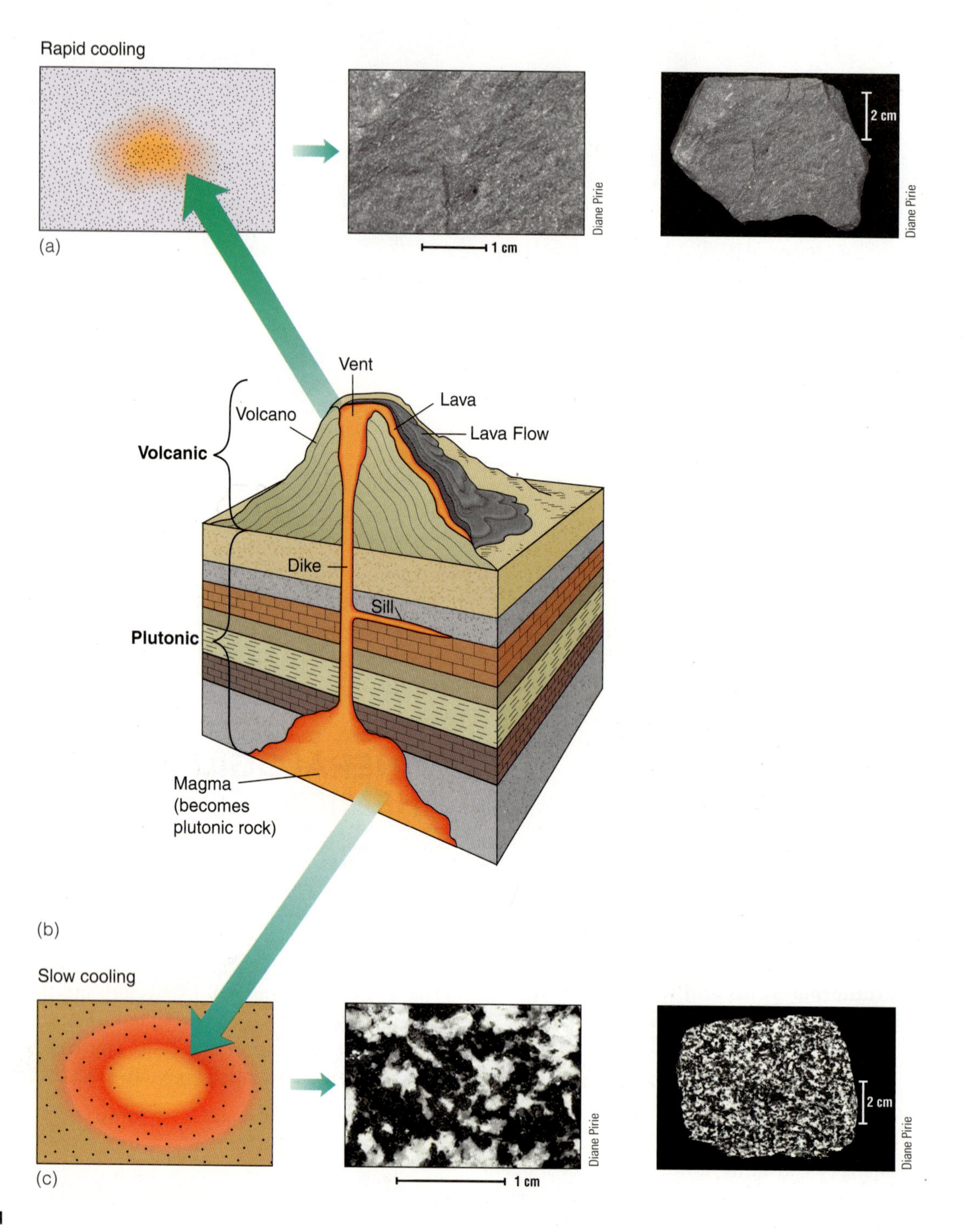

Figure 3.1

Volcanic and Plutonic Features

Lava flows are volcanic or extrusive and cool quickly. The rocks in plutons, dikes, and sills are intrusive or plutonic. As magma of the same composition cools, intermediate in this example, the more rapidly cooled lava forms a fine-grained rock, andesite, and the slowly cooled magma underground forms coarse-grained diorite. Thin plutons such as dikes and sills tend to cool more quickly than deep-seated thicker plutons and are finer grained. *Source:* Based on *Earth Science Today* B. Murphy/D. Nancy p. 29. Copyright © 1999, Brooks/Cole. All rights reserved. *Physical Geology: Exploring the Earth*, 5th ed., and J. S. Monroe/R. Wicander Copyright © 2004, Brooks/Cole. All rights reserved.

of the magma. This magma composition suggests the chemical nature of Earth's internal process and the interactions of lithospheric plates that generate magma. Therefore, students should carefully observe the minerals and their proportions in igneous rocks. If the rock is almost all mafic minerals, it is **ultramafic** (■ Table 3.1); if mafic minerals predominate, it is **mafic**. If roughly equal proportions of mafic and felsic minerals are present,

Table 3.1

Magma Types and Compositions for Various Igneous Rocks

As silica content increases, the rocks contain an increasing proportion of felsic minerals and decreasing proportion of mafic minerals. Texture terms for the rocks are listed in blue in the left column.

	Magma Composition			
Magma Type	**Ultramafic**	**Mafic**	**Intermediate**	**Felsic**
Silica Content (Increasing silica content →)	<45%	45–53%	53–65%	>65%
Mineral Content (if minerals are present)	Olivine Pyroxene (Ca Plag. Feldspar) Mafic Minerals	Ca Plag. Feldspar Pyroxene (Olivine) Mafic Minerals > Felsic Minerals	CaNa Plag. Feldspar Pyroxene Hornblende (Biotite) Mafic Minerals ≈ Felsic Minerals	Na Plag. Feldspar K Feldspar Quartz (Hornblende) (Biotite) (Muscovite) Mafic Minerals < Felsic Minerals
Overall Rock Color (Increasing darkness)	Green to dark green-gray	Dark	Medium	Light (if minerals are present)
Plutonic Rocks (Intrusive Rocks) Coarse or pegmatitic	Peridotite	Gabbro	Diorite	Granite Pegmatite Granite
Volcanic Rocks (Extrusive Rocks) Fine grained	Ultramafic lava flows are rare (komatiites)	Basalt	Andesite	Rhyolite
Vesicular		Scoria		Pumice
Pyroclastic		Breccia, Tuff		
Glassy		(rare as purely glass)		Obsidian

(Texture; Increasing grain size ↑)

the rock is said to be **intermediate**. If felsic minerals predominate, it is **felsic**. For example, the granite in ■ Figure 3.2, which is felsic, is made up of the felsic minerals K-feldspar, plagioclase, quartz, and a small amount of the mafic mineral biotite. With few exceptions, dark igneous rocks are either ultramafic or mafic, medium-colored are intermediate, and light-colored are felsic (Table 3.1). Two notable exceptions to this rule are (1) felsic minerals may be darkish, such as smoky quartz in Figure 3.2; and (2) volcanic glass is dark whether it is mafic or felsic.

Igneous rocks and the magma from which they formed contain somewhere between about 40% to 80% silica (SiO_2). Notice that you can determine the silica content of a rock from its mineral content and sometimes even from its color (Table 3.1). Silica content in magma is closely related to the viscosity of the magma. **Viscosity** describes a fluid's resistance to flow. Felsic magma is much more viscous and mafic magma more fluid. When magma has a combination of high viscosity and high gas content, explosive eruptions may result. Thus, silica content influences more than the color of the rock; it and the magma's gas content also influence the behavior of the magma and the nature of volcanic eruptions. The way the magma crystallizes underground or erupts at the surface influences the texture of the rock as well.

IGNEOUS TEXTURES REVEAL HOW IGNEOUS ROCKS FORM

Texture is the key to understanding an igneous rock's history. The texture of your sample will tell how the magma moved, its crystallization, release of gases,

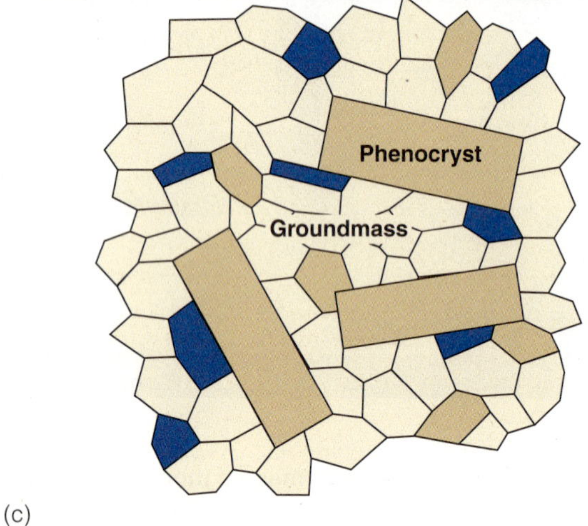

Figure 3.2

The Minerals in Granite and Porphyritic Texture

(a) Porphyritic granite. Granite typically consists of (b) the minerals K-feldspar, plagioclase, quartz, and biotite (and sometimes hornblende, not shown). (c) Porphyritic texture with coarse groundmass.

whether the rock was formed dramatically in a violent volcanic eruption, as a lava flow, or quietly deep underground. At the end of this lab you should be able to trace the journey of your sample back to before its formation. To do this, you must first learn to identify and analyze basic textures in igneous rocks.

Recall that the term **texture** refers to the arrangement and size of mineral grains in a rock. Additional aspects of texture include the presence of glass, the proportion of glass to crystals, the presence and proportion of cavities, and the occurrence of broken rock fragments. Each of these aspects reveals something of the rock's history.

Textures of Plutonic Rocks

Rocks and their constituent minerals melt and crystallize at high temperatures, generally between 800°C and 1,200°C. Crystals formed by the solidification of magma typically have an interlocking texture. Grain size is an aspect of texture that reveals how rocks cool. A plutonic rock will cool very slowly underground, allowing the crystals to become large and well formed (Figure 3.1c). Shallower intrusions, which cool more quickly, will tend to develop finer-grained rocks. The presence of water dissolved in the magma can also have an important influence on the grain size of the rock, as we will see shortly. The following are the principal textures found in plutonic rocks.

Coarse-Grained (Phaneritic) Textures have visible crystals all about the same size (Figures 3.1c and 3.8–3.11). Igneous rocks of this type cooled slowly deep in the Earth, where it is warm. Coarse-grained textures are therefore typical of intrusive or plutonic rocks. Granite, diorite, and gabbro are common plutonic rocks with coarse-grained textures.

Pegmatitic (pronounced pĕg-mə-tĭt´-ĭc) ***Texture*** has grains several centimeters across. We call rocks with this texture *pegmatite* (■ Figure 3.3).

Porphyritic Texture occurs in either plutonic (Figure 3.2a and c) or volcanic rocks (discussed below). It indicates larger crystals—**phenocrysts**—embedded in a more finely crystalline **groundmass**. For plutonic rocks the groundmass is coarse grained. The two grain sizes must be distinctly different, not gradational. Such a texture usually indicates a *change* in the rate of cooling. At first, cooling takes place slowly at depth, forming some large crystals. The magma and large crystals then rise to an environment where faster cooling takes place. In this second stage of cooling, the finer groundmass crystals grow around the larger grains. If the second stage occurs beneath the surface, the magma becomes completely solid without ever erupting, so the rock is plutonic or intrusive, and the groundmass is likely to be coarse grained, surrounding even coarser-grained phenocrysts (Figure 3.2a).

Figure 3.3

Pegmatite

Dikes and close-up of the very coarse-grained plutonic igneous rock, **pegmatite**. Here the fluid-rich magma that formed the pegmatite intruded along cracks in the dark rocks, forming numerous dikes (Figure 3.1). Cascades, Washington. *Inset:* Close-up of a rock similar to those in the dikes. The very large crystals in this pegmatite are K feldspar (white, above), quartz (white, without cleavage, lower right), silvery light pale gray books of muscovite (upper left and middle right) and black tourmaline (middle right).

Porphyry: The noun *porphyry* applies to a rock in which phenocrysts comprise 25% or more of its volume.
Porphyritic: The adjective *porphyritic* describes rocks with less than 25% phenocrysts (Figure 3.2a).

Rocks with porphyritic texture and rocks with other textures may display geometrically shaped crystals (■ Figure 3.4). A crystal that displays well-formed, planar surfaces, or **faces,** is termed **euhedral.** A somewhat imperfectly formed shape is termed **subhedral.** If the crystal lacks crystal faces, it is described as **anhedral.** Porphyritic rocks commonly contain some euhedral phenocrysts (Figure 3.4).

Growth of Crystals from a Melt

To gain a better understanding of the igneous processes that form the different cooling textures just discussed, let's perform some melting and crystallization experiments. Real rocks are difficult to melt and crystallize in a laboratory because of the high temperatures involved. Instead, we will experiment with artificial "igneous rocks" using thymol, an organic substance that has a low melting (or freezing) point. Your instructor may provide prepared samples in Petri dishes for Parts 1, 2, and 3, or you may have the opportunity to make them yourself. In either case, you will need to read how to conduct the experiment in order to interpret the results.

Materials needed:

- Hotplate
- Thymol
- 3 Petri dishes
- Labels
- Spatula
- Ice bath

The hotplate, ice bath, and a place to set aside Petri dishes while they cool should all be located in a well-ventilated area or in a fume hood.

Before starting the experiment, ask your lab instructor what setting to use for the hotplate: It should be just warm enough to melt the thymol. Set the hotplate to the correct temperature and turn it on. Do *not* turn up the temperature to speed the process as this tends to vaporize the thymol excessively.

Part 1: Slow Cooling Use a Petri dish that already contains crystals of thymol. If one is not available, use a clean spatula to place solid thymol in a clean, dry dish—use enough to cover the bottom completely when melted. You may add more thymol if needed once melting has begun. Label the dish "slow cooling" or "Part 1." Put the dish on the hotplate and watch until the thymol has almost completely melted. Do *not* allow the thymol to vaporize. (If vaporization occurs easily or quickly, the hotplate is set too hot.) Put the dish with liquid thymol aside in a ventilated area so that it may cool slowly, undisturbed.

Part 2: Rapid Cooling Repeat the procedure used in Part 1 with another Petri dish, but do not set it aside when the thymol has liquefied. Instead, transfer the dish to the ice bath and observe the crystallization. Do not breathe the fumes and be careful not to get water into the dish. Look back at your dish from Part 1 occasionally to see how the slow cooling is progressing.

Part 3: Slow Followed by Rapid Cooling Repeat the procedure used in Part 1 and observe the formation of the first crystals. After a few large crystals have formed, but some liquid is still left, transfer the dish to the ice bath.

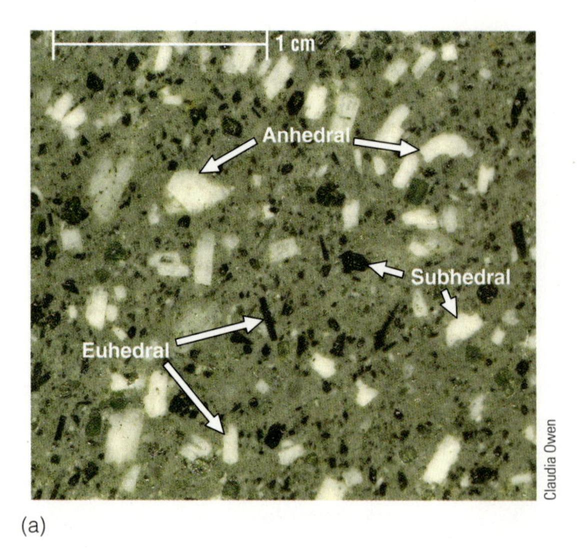

(a)

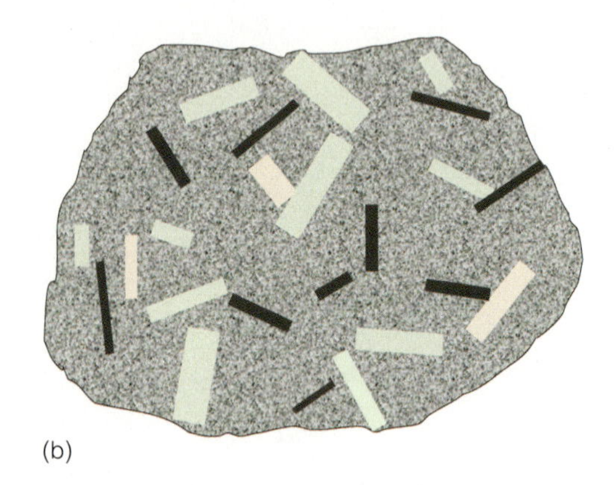
(b)

Figure 3.4

Porphyritic Texture in Volcanic Rocks and Crystal Shape

(a) **Andesite porphyry** with a medium-gray groundmass and anhedral, subhedral and euhedral phenocrysts of plagioclase (white) and hornblende (black). (b) porphyritic texture with fine-grained groundmass.

Table 3.2

Results of Cooling Experiments

Experiment	Size(s)	Texture	Description: shape (triangular, circular, square, etc.) perfection of form (euhedral, subhedral, or anhedral), arrangement of crystals
Part 1 (slow cooling):			
Part 2 (rapid cooling):			
Part 3 (slow followed by rapid cooling):			

1. Examine the three dishes of crystallized thymol and enter your observations in Table 3.2.

a. Determine the relative crystal sizes (small, medium, or large) in the experimental products you just made (or ones provided for you).

b. What texture do you observe for each? Use the texture terms you have learned.

c. Briefly describe the crystals from Parts 1, 2, and 3. Are they triangular, circular, square, or some other shape?

d. Describe the perfection of their form. Are the crystals euhedral, subhedral, or anhedral (Figure 3.4a)?

2. Draw some conclusions from the experiment.

a. How does the rate of cooling affect crystal size and shape?

____

____

____

b. Are your observations of cooling rates and crystal sizes for Part 1 and Part 2 consistent with your observation of Part 3?

______ Explain.

____

____

____

____

c. The experiments are intended to be analogies of natural processes in the Earth. Discuss what these are. For each part of the experiment, explain what would happen naturally in the Earth to make similar events, results, and textures.

   i. Part 1 (slow cooling):

   ____

   ____

   ii. Part 2 (rapid cooling):

   ____

   ____

   iii. Part 3 (slow followed by rapid cooling):

   ____

   ____

   ____

Clean up from your experiments as instructed.

These experiments illustrated some textures of plutonic rocks. When viewed closely or under a microscope some volcanic rocks have similar textures in addition to special textures created in the process of eruption and cooling.

Textures of Volcanic Rocks

A volcanic rock generally cools so quickly at the Earth's surface that its crystals are invisible to the unaided eye (Figures 3.1a) or crystals do not grow at all. When rocks crystallize the grains interlock, although you can only see this in the coarser plutonic-rock textures or under a microscope for volcanic rocks.

Fine-Grained (Aphanitic) Texture occurs in rocks whose crystals are generally too fine-grained to be seen without a hand lens. The small size of the grains means that they have cooled quickly (Figure 3.1a on p. 46, example rocks with aphanitic texture in Figures 3.12 on p. 55, and 3.14 on p. 56). Fine-grained textures are therefore typical of extrusive (volcanic) rocks that form from *lava flows* or *lava domes*. Some shallow intrusive bodies such as thin dikes and sills (Figure 3.1b) may have fine-grained texture because they are thin enough and near enough to the surface to cool rapidly.

Porphyritic Texture is even more common in volcanic rocks than in plutonic rocks. Porphyritic volcanic rocks still have *phenocrysts*, but they are embedded in a more finely crystalline or glassy *groundmass* (Figure 3.4). The sample shown in Figure 3.4a has enough phenocrysts (>25%) to be called a *porphyry*. Porphyritic volcanic rocks undergo a slow first stage of cooling underground which forms the larger crystals. The second stage for volcanic rocks occurs during eruption of a lava flow or a lava dome, when the groundmass cools quickly at the Earth's surface. All igneous rocks start with magma underground; it is whether the magma ever reaches the surface that determines if a rock is plutonic or volcanic.

Glassy (or Hyaline) Texture is the texture of glass, which has vitreous luster (Figure 3.15 on p. 56). Glassy texture results when lava is cooled so quickly that minerals have no opportunity to form, so the solid lacks a crystalline structure (review the definition of a mineral in Lab 2). Glassy texture frequently forms from highly viscous lava (often felsic). A **viscous** liquid is resistant to flow. Atoms are less mobile in viscous magma and may not be able to join together into crystals before they become solid, in which case a glass forms. *Obsidian* and *pumice* are common rocks with glassy texture (Figures 3.15 on p. 56 and 3.16b on p. 57).

Volcanic rocks may display other textures when magma is rich in gas. Gases may form bubbles in the liquid, which will produce holes in the resulting solidified rock, or gases may produce explosive eruptions, resulting in entirely different textures.

Vesicular Texture refers to the presence of small cavities called **vesicles**, which were originally gas bubbles in the liquid magma (■ Figure 3.5). The gas was initially dissolved in the magma, but came out of solution because of

Diane Pirie

Figure 3.5

Vesicular Texture

Vesicular basalt with more rock than holes (vesicles). This sample is also porphyritic, but the olivine phenocrysts are difficult to see.

(a)

(b)

Getty Images

Grenville Draper; inset: Diane Pirie

Figure 3.6

Volcanic Ash

(a) The central eruption of Mount St. Helens on May 18, 1980, produced billowing clouds of volcanic ash that later fell over a large swath of Washington, Idaho, and Montana. (b) Volcanic ash and cinders erupted from Volcán Colima, Mexico. *Inset:* Volcanic ash from Mount St. Helens.

the pressure decrease during eruption or as the magma rose before eruption. This process is similar to the formation of bubbles and foam when you open a container of a warm, carbonated soft-drink, thus reducing the pressure. If vesicular texture is a minor feature of a volcanic rock, similar to the holes in Swiss cheese, the word *vesicular* modifies the rock name, as in *vesicular basalt* (Figure 3.5). Rocks where the vesicular texture is a more prominent feature, especially where the rock has more empty space than rock, have special names such as *scoriaceous basalt*, *scoria* (Figure 3.16a, p. 57), and *pumice* (Figure 3.16b).

In some cases, secondary minerals may fill the vesicles long after the solidification of the original rock. The filled area is an **amygdule** and the rock is said to have an **amygdaloidal texture**.

Pyroclastic Textures **Tephra,** made of volcanic ash and larger rock fragments, is violently ejected from volcanoes during some eruptions (■ Figures 3.6a). This debris and the rocks that form from it have a **pyroclastic texture** (Figure 3.17). Such eruptions are dangerous, so the recognition of pyroclastic texture in existing rocks helps volcanologists assess areas of future volcanic hazards. Rocks with pyroclastic texture do not have the compact, interlocking-grain texture of rocks that crystallized directly from magma, but instead may have a more "powdery" texture or appear to be made of broken pieces stuck together. A pyroclastic rock is said to be welded if the loose pyroclastics fuse together while the components are still hot (Figure 3.18, p. 58). Although pyroclastic material may be mafic to felsic, it is more commonly intermediate to felsic. The combination of high viscosity and high gas content of more felsic magmas tends to make them explosive.

Your new knowledge of the various plutonic and volcanic textures we have just covered will be valuable in the next section when you learn how to identify and classify igneous rocks. Along with chemical and mineralogical composition, rock textures form the basis of the classification system.

CLASSIFICATION AND IDENTIFICATION OF IGNEOUS ROCKS

Scientists design classification systems to help them understand natural phenomena. The classification of igneous rocks does this by emphasizing composition and texture, both of which are integral in understanding the history of the rock's formation. Table 3.1 on page 47 and Figure 3.20 on page 59 can help you classify and identify an unknown igneous rock sample. The texture of the rock is a good place to start the classification.

To identify an igneous rock, use the following strategy in conjunction with Figure 3.20:

- Examine the texture of the rock: Is it coarse-grained or fine-grained, or does it have a mixture of grain sizes (porphyritic)? Is it glassy, vesicular, or pyroclastic? Locate the appropriate path in Figure 3.20.
- For some volcanic rocks, such as obsidian, tuff, and breccia, the texture and grain size are sufficient to identify the rock. For other rocks, use their shade (dark, medium, or light) and Table 3.1 to aid with identification of the specimen's minerals and magma type (ultramafic, mafic, intermediate, or felsic). Then locate the appropriate branch in the maze, which will lead to the rock name. You can also double-check your rock name using the following rock descriptions.

Plutonic Igneous Rocks

Plutonic rocks most commonly have coarse-grained or pegmatitic textures. Their compositions vary (Table 3.1) and may indicate the plate tectonic setting where they solidified (■ Figure 3.7). The following list describes a number of common plutonic rocks.

Granite is the most abundant plutonic igneous rock in continental crust (Figure 3.8). The dramatic cliffs of Yosemite National Park in California (Figure 3.30 on p. 66) and the faces at Mt. Rushmore are granite. Alkali feldspars and quartz are diagnostic for granite (Figure 3.2a, b and ■ Figure 3.8). These feldspars are commonly white, light gray, and/or pink, even salmon red, often with two visible varieties. The quartz usually looks like smoky gray glass (*smoky quartz*). Granite is generally a light-colored rock and is felsic. The dark minerals such as hornblende

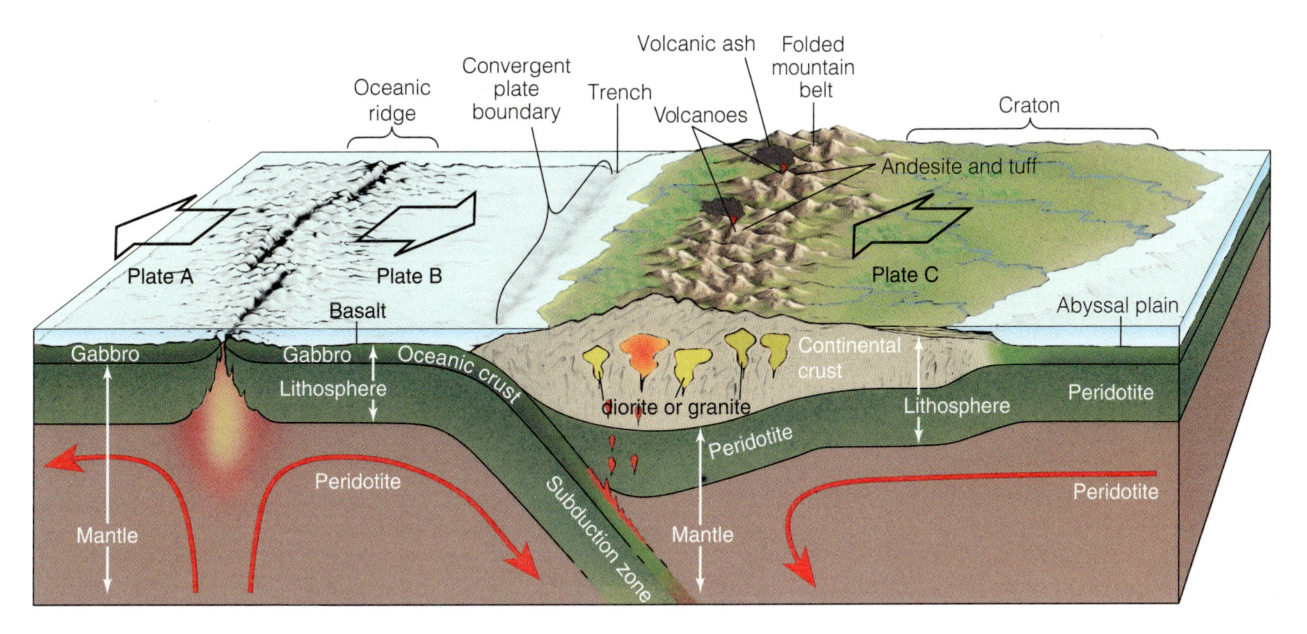

Figure 3.7

Plates and Igneous Rocks

Earth's uppermost rock layers divided into three plates (see also Labs 1 and 9), two of which are converging at a subduction zone. Igneous rocks make up substantial amounts of the crust and upper mantle. Continental crust consists of granite in large part; diorite may occur here as well. Stratovolcanoes above the subduction zone are substantially andesite, tuff, and volcanic ash. Oceanic crust is substantially gabbro, with basalt at the surface of the seafloor; and the upper mantle is mainly peridotite. Adapted from: *Environmental Science*, 8th ed., by Miller and Spoolman.

left and *right* Claudia Owen; *center* Diane Pirie

Figure 3.8

Granite

Three granites showing a range of grain sizes and colors. Two colors of feldspar are visible in these samples: pink to peach K feldspar, microcline (which is very pale in the middle sample), and light gray to white Na-rich plagioclase. The other minerals are gray vitreous quartz (smoky quartz), black splendent biotite and black vitreous hornblende.

and biotite may visually stand out, but they comprise only a small percentage of the total rock volume.

Diorite commonly crystallizes near convergent plate boundaries from intermediate magma produced where oceanic lithosphere sinks beneath other lithosphere (Figure 3.7). It is made of dark green to black pyroxene or hornblende ± biotite and white to gray plagioclase feldspar in similar proportions. This equal amount of felsic and mafic (light and dark) minerals often give this rock a "salt-and-pepper" appearance (■ Figure 3.9) and make diorite medium colored and intermediate in composition. The absence of quartz (except in quartz diorite) helps distinguish diorite from granite. The presence of pyroxene (usually augite) is also an indication although some diorites have hornblende instead of pyroxene.

Gabbro is the most abundant plutonic rock in oceanic crust (Figure 3.7). It is a coarse, dark mafic rock composed primarily of plagioclase feldspar, which is usually translucent and may be gray, and crystals of pyroxene, which are often large and irregularly shaped (■ Figure 3.10). Olivine is a minor constituent. Some feldspar grains in gabbro may show very straight lines visible when light reflects off the cleavage surface. These lines are caused by a kind of *twinning* in plagioclase (Figure 2.8h, p. 27).

Peridotite is primarily made of olivine and pyroxene. Its green color and coarse-grained texture clearly indicate its identity (■ Figure 3.11). Another way to distinguish peridotite from gabbro is the lower proportion of cleavage surfaces that catch the light.

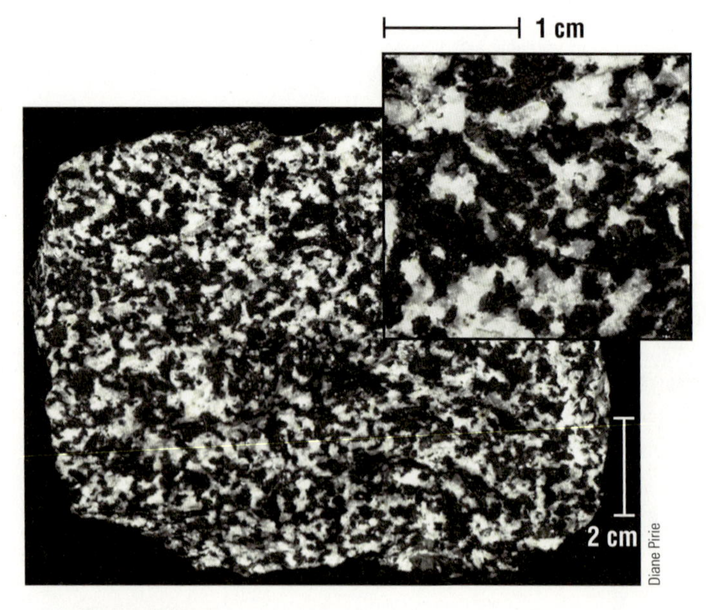

Figure 3.9

Diorite

In this sample, the black grains are hornblende, and the white to pale gray ones are calcium-sodium plagioclase. An occasional grain of black splendent biotite appears in the inset.

Figure 3.10

Gabbro

This sample is made up of plagioclase and the pyroxene, augite. Both plagioclase and pyroxene have cleavage, so most of the rock reflects light off small, flat cleavage surfaces as one turns the sample in the light. The white rectangular grains are plagioclase grains reflecting light from cleavage planes. *Inset:* Euhedral tabular plagioclase grains (rectangular in cross section) are gray and somewhat translucent. The pyroxene grains are more opaque, anhedral, and greenish black.

Figure 3.11

Peridotite

Left: Peridotite containing olivine and pyroxene. Notice the cleavage visible on some of the pyroxene where the light is reflecting off the flat surface. Olivine does not have cleavage, so a large proportion of the rock does not reflect light in this manner. In this sample, as in many peridotites, the olivine has been partially to totally replaced by serpentine, making it dark greenish black instead of green. *Right*: two fragments of peridotite (green *xenoliths*) contained in basalt (fine-grained gray). This sample, also shown in the *inset*, has a type of peridotite (dunite) with a very high proportion of olivine: Olivine is green, and pyroxene is dark green to black. Basaltic magma from the mantle carried these peridotite fragments upward to be erupted with lava.

This is due to the abundance of olivine in the rock, which has fracture, not cleavage. The absence of felsic minerals means that peridotite is ultramafic. Peridotite is the rock that makes up most of the upper mantle (Figure 3.7).

Pegmatite is an extremely coarse-grained rock that has crystals at least several centimeters across (**pegmatitic** texture). It almost always forms as dikes of fluid-rich magma (Figure 3.3 and its inset on p. 49). Occasionally, crystals may be over a foot, and even several feet long. For example, cleavage sheets of large white mica crystals from quarries near Moscow, Russia, were large enough to use for windows in that city in the 17th and 18th centuries—hence the name "muscovite." Granite pegmatite generally consists of potassium feldspar, muscovite, and quartz (Figure 3.3, inset). This type of pegmatite is felsic. Pegmatites may contain unusual minerals and gems such as tourmaline (Figure 2.1b on p. 20), beryl (aquamarine), and topaz.

Volcanic Igneous Rocks

Plutonic and volcanic rocks are chemically equivalent, but the textures of volcanic rocks are more varied than those of plutonic rocks. The following three volcanic rocks are simply finer grained than their plutonic cousins. However, the subsequent ones have special textures.

Rhyolite is a fine-grained, light-colored, felsic rock (■ Figure 3.12). **Porphyritic rhyolite** has a fine-grained groundmass. Either may be light tan, pink, beige, yellowish, or light gray. Porphyritic rhyolite has at least a small percentage of quartz phenocrysts that help to differentiate it from andesite. Crystals of transparent, colorless potassium feldspar (with cleavage) are commonly visible.

Andesite is sometimes fine-grained, but more commonly is porphyritic. It is the most common lava-flow rock in stratovolcanoes, or composite cones (Figure 3.24, p. 63). Gray, dark tan, brown, mauve, or purplish in color, andesite is intermediate in chemical composition (Figure 3.1a, right, p. 46). **Porphyritic andesite** (■ Figure 3.13) has phenocrysts of white plagioclase feldspar and/or black mafic minerals such as pyroxene and hornblende (see also Figure 3.4 on p. 50).

Basalt is a dark-gray to black, fine-grained rock that makes up much of the surface of the ocean floor beneath any sediment (Figure 3.7). It is also the major rock of shield volcanoes (Figure 3.22, p. 62) and *flood basalts*. Basalt is dense, mafic, and has few phenocrysts (■ Figure 3.14). The *phenocrysts* in **porphyritic basalt** may be olivine or plagioclase feldspar. Olivine phenocrysts may be hard to see because of a lack of color contrast of the clear green olivine in the black rock matrix.

Obsidian is an often black, volcanic glass with conchoidal fracture and vitreous luster, which are diagnostic (■ Figure 3.15). In spite of its dark color, obsidian is generally felsic (Table 3.1). Since obsidian is a glass, with no minerals, it is transparent on thin edges, but the minor amount of iron in it strongly influences the color: black where unoxidized and red where oxygen has mixed into the lava. *Flow banding* and red streaks may be present in some samples as in the right-hand sample in Figure 3.15.

Figure 3.12

Rhyolite

This sample is fine-grained pink rhyolite with a scattering of very small phenocrysts.

Figure 3.13

Andesite

Porphyritic andesite with hornblende phenocrysts and a fine-grained, medium-colored groundmass. White shows where light reflects off hornblende cleavage.

Claudia Owen

Figure 3.14

Basalt

This sample is a fine-grained, mafic igneous volcanic rock with a few green olivine phenocrysts that are quite difficult to spot.

Claudia Owen

Figure 3.15

Obsidian

A glassy felsic igneous volcanic rock with conchoidal fracture. Obsidian is commonly black (*left*) but may also be streaked with red flow banding (*right*).

Scoria: The vesicles making up a high proportion of the volume of scoria give it a sponge-like appearance. Scoria may be reddish brown, but is more commonly black (■ Figure 3.16a) and is usually mafic. The rock part of scoria (that is, the material between the holes) is commonly fine grained, but can also be glassy. Notice that vesicular basalt (Figure 3.5) is similar, but has fewer holes and is consequently denser than scoria.

Pumice is a highly vesicular and glassy, felsic to intermediate igneous volcanic rock. Pumice is composed of more vesicles than rock, and is essentially solidified lava "foam" (Figure 3.16b). Because of the high vesicle content, pumice is able to float on water. Generally, pumice ranges from light gray to tan to yellowish. The rock part of pumice is glassy, which you can see with a hand lens in bright light (see Figure 3.16b, inset). Pumice is usually felsic, having the same chemical composition as obsidian, granite, and rhyolite (Table 3.1 on p. 47).

Volcanic ash consists of loose pyroclastic volcanic particles consisting of silt- to sand-sized fragments that explosively erupted from a volcano (Figure 3.6). The inset in Figure 3.6b shows a close-up of loose volcanic ash.

Tuff is a pyroclastic volcanic rock consisting of dust- to pebble-sized tephra and/or pumice fragments that lithified (became stuck together to form rock) after settling (■ Figure 3.17).

Welded tuff is a special kind of tuff that was deposited when the tephra particles were still molten. As a result, it often has visible areas with a glassy and streaked appearance (■ Figure 3.18a). The streaks are commonly pumice fragments that have flattened and lost their gas after the ash was deposited (see the black bands of obsidian in Figure 3.18a). Welded tuff results from high-speed, pyroclastic flows (Figure 3.18b), and therefore is another indicator of volcanic hazard.

Volcanic breccia is a coarse rock with angular volcanic fragments either cemented together (■ Figure 3.19) or held together by lava. As they move, lava flows produce some volcanic breccias by incorporating cooling crusts broken off the top and front of the flow.

Igneous Rock Identification

As you identify igneous rocks, you may find that the rock maze for identification (■ Figure 3.20) will help you get started, but do not depend on it entirely. You will want to start practicing igneous rock identification without it once you have a good idea of how to proceed.

3. Examine the unknown rocks.

   a. For each rock, fill in ■ Table 3.3, "Igneous Rock Identification Form," as indicated. Fill in the columns of *Texture*, *Made of*, and *Magma type*, and then determine the *Rock name*.

   b. Remember that the goal of geologists is not just to identify rocks, but to understand what they can tell about the geologic history of an area. Use the texture to tell you how the rock formed from

(a)

Derek Owen

(b)

Claudia Owen

Figure 3.16

Highly Vesicular Rocks

(a) Very **vesicular** black (*front*) and red (*back*) mafic **scoria**. Because the vesicles constitute a major proportion of the volume of the rock, the sample has quite low density, although not as low as pumice and will probably sink in water. (b) The gray sample of **pumice** on the left has unusually large vesicles and shows the glassy texture very well (*inset*). In many samples, such as the paler pumice on the right, however, the vesicles are much smaller, and the glassy texture can only be seen with a hand lens.

Diane Pirie

Figure 3.17

Tuff

Tuff with numerous light gray pumice fragments embedded in a matrix of beige ash.

the discussion earlier in this chapter and briefly indicate this in the *Origin* column. Determine and state whether the rock is volcanic or plutonic.

Geologists gain more information about the origin of a rock by studying its occurrence in the context of its rock mass, its situation in the Earth.

IGNEOUS ROCK MASSES

Igneous rocks occur in variously shaped bodies that largely depend on whether the rocks are volcanic or plutonic. Some volcanic rocks form broad, flat masses such as lava flows and volcanic ash layers, which are similar in form to sedimentary layers. Other masses of igneous rocks are tall mountains, vast volcanic fields or small volcanic edifices. Masses that originally formed underground may be flat or occupy large volumes of the crust; however, the characteristics and shapes depend on how and where magma extruded or intruded. Figure 3.27 on page 64 illustrates some of these igneous masses.

Extrusive Rock Masses

Molten lava solidifies into lava flows, lava domes, or pyroclastic material depending on how it erupts.

- **Lava flows** are extruded magma that has solidified in tongue shapes and as sheets. Fluid lava will tend to flow farther, forming long tongues or occasionally to spread out in incredibly extensive, flat lava sheets called **flood basalts**. More

Claudia Owen

(a)

Peter Lipman/USGS

(b)

Figure 3.18

Welded Tuff and Its Source: Pyroclastic Flow

(a) **Welded tuff** formed from the lower part of a pyroclastic flow similar to the one shown in (b). Black obsidian bands in (a) formed when the weight of material above flattened still hot pumice as the flow came to rest and cooled. (b) Churning **pyroclastic flow** traveling at more than 100 km (60 mi) per hour down the slope of Mount St. Helens on August 7, 1980. Welded tuff results from such flows.

Diane Pirie

Figure 3.19

Volcanic Breccia

This volcanic breccia has angular *clasts* (broken pieces) of scoria in which the vesicles (holes) have been filled with a dark-colored mineral. The clasts are *cemented* (naturally held together) by aphanitic cement.

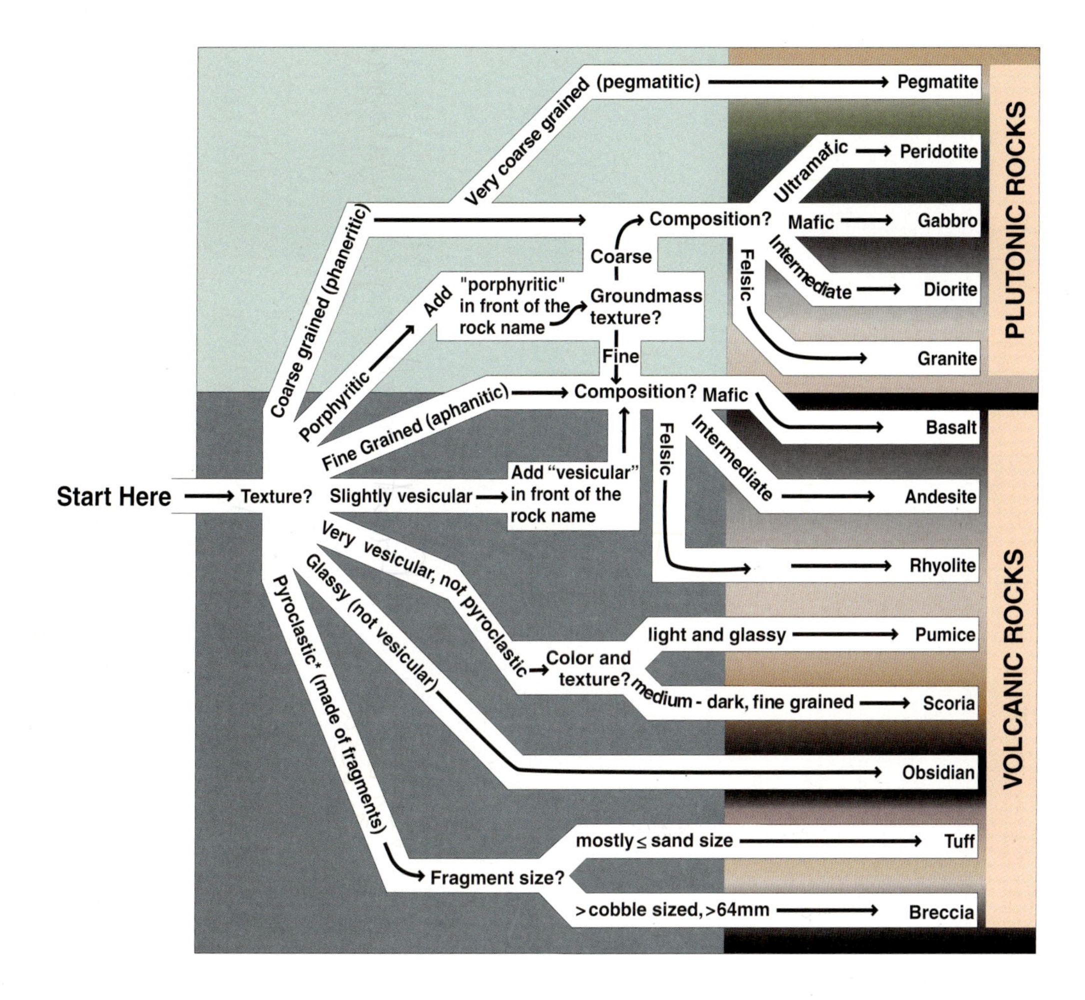

Figure 3.20

Identification of Igneous Rocks

Maze for identification of igneous rocks. At "Start Here," choose a texture for the rock that is appropriate for the whole sample, not just pieces in it.* At each junction in the maze, decide which path to take by answering the question. (Texture and composition terms are discussed earlier in the text.) Once you choose a path, it will lead you to the rock name. After identifying several samples, test yourself to see if you have learned the technique by identifying some rocks without the chart.

* Note that pyroclastic rocks have fragments in them that may have any of the other volcanic textures, but pyroclastic texture takes precedence.

viscous lava will make thick, short tongues covering smaller areas. Basaltic lava tends to produce either fluid, ropy flows known as **pahoehoe** or rough flows called **aa** (■ Figure 3.21, p. 62).

- Sometimes lava is so viscous that it cannot spread out, but forms a dome-shaped body instead. When this happens it creates a **lava dome**.
- When volcanic eruptions are explosive, a spray of magma and particles of rock spew out of the volcano, producing **volcanic ash** and other pyroclastic deposits (Figure 3.6, p. 52) that may lithify into tuff (Figures 3.17 and 3.18, pp. 57 and 58). Volcanic ash is a layered deposit (Figure 3.6b) that may extend over a wide region and become buried within volcanic or sedimentary sequences. Volcanic ash, or tuff, deposits may be interlayered with lava flows and *lahars* depending on the sequence and type of eruption.
- Where large quantities of pyroclastic material become mobilized by water a **lahar**, or volcanic mudflow can form. Lahar deposits commonly consist of a large range of fragment sizes in a matrix of volcanic ash that may have become cemented together.

Where a **volcanic vent** (the opening where volcanic eruptions occur, Figure 3.27, p. 64) erupts frequently or repeatedly, a hill or mountain, a **volcano**, can build up. A volcano may be built entirely of lava flows, entirely of pyroclastic deposits, or a mixture of both.

Table 3.3

Igneous Rock Identification Form

Sample Number	Texture Use appropriate terminology from Figure 3.20	Made of: List coarse minerals (and their colors in parentheses) and color of any fine grained or glassy material	Magma Type (Ultramafic/Mafic/Intermediate/Felsic)	Rock Name	Origin	
					How did the rock form?	Volcanic/Plutonic
set 41	Fine grained	Olivine	ultramafic	Peridotite	slow cooling	Plutonic
set 2 49	Coarse grained	feldspar + Qz	Felsic	Granite	slow cooling	Plutonic
set 2	Fine grained	feldspar + Qz	Felsic	Rhyolite	fast cooling	Volcanic
set 59	Vesicular	Feldspar Qz	Felsic	Scoria	fast cooling	Volcanic
set 2 83	Fine grained	hornblend Feldspar	Intermediate	Andesite	fast cooling	Volcanic
set 2 219	coarse grained	amphibole	Intermediate	Diorite	slow cooling	Plutonic
Set 2 34	Very vesicular	Feldspar/quartz	Felsic	Pumice	fast cooling	volcanic

Table 3.3

Igneous Rock Identification Form —*Continued*

Sample Number	**Texture** Use appropriate terminology from Figure 3.20	**Made of:** List coarse minerals (and their colors in parentheses) and color of any fine grained or glassy material	**Magma Type** (Ultramafic/Mafic/ Intermediate/ Felsic)	Rock Name	Origin	
					How did the rock form?	Volcanic/ Plutonic
set 2 11	Glassy	Quartz/Feldspar	Felsic	Obsidian	fast cooling	Volcanic
set 2 27	fine grained	amphibole feldespar	Intermediate	Porphyritic Adesite andesite	fast cooling	Volcanic
set 36	fine grained	plagioclase Pyroxene	Mafic	Basalt	fast coolings	Volcanic
set 71	coarse grained	Plagioclase Pyroxone	Mafic	Gabbro	slow cooling	Plutonic

- A **shield volcano** is made of basaltic lava flows and very little ash (■ Figure 3.22). Shield volcanoes can be very massive and are gently sloping with an angle of about 10°. The tallest (not highest) and most extensive mountain on Earth, Mauna Loa in Hawaii, is a shield volcano with its base at the bottom of the sea.
- A **cinder cone** is a small volcano made up entirely of pyroclastic material, ash, and cinders, especially scoria (■ Figure 3.23). Many cinder cones have lava flows that flowed out from under the cinders at the base of the cone.
- A **stratovolcano**, also called a **composite volcano**, is made of intermediate to felsic interlayered pyroclastic deposits and lava flows (■ Figures 3.24 and 3.6a). Stratovolcanoes commonly make impressive snow-capped volcanic peaks with steep (about 30°) slopes, but are nevertheless typically much smaller in volume than shield volcanoes.
- A **caldera** is a large round depression (Figure 3.23) formed after a major eruption when rocks above collapse into the emptied magma chamber. **Ring dikes** may form where some remaining magma push upward into the ring of cracks around the caldera rim (Figure 3.27, p. 64).

4. Examine the Geologic Map of Mauna Loa in ■ Figure 3.25.

a. Describe the shape of the rock masses shown in color on the map.

Orene Owen

Figure 3.21

Aa and Pahoehoe Lava

Two basalt lava flows on the big island of Hawaii show the two flow styles: aa on the left and pahoehoe on the right.

Bernard Pipkin

Figure 3.22

Shield Volcano

The shield volcano in the distance is Fernandina in the Galápagos Islands.

Roger Cole

Figure 3.23

Cinder Cone and Caldera

Wizard Island is a cinder cone in the caldera of Crater Lake, Oregon, at Crater Lake National Park. The cone of the small volcano is composed of scoria. Notice the lava flow extending to the right at the base of the cone. Crater Lake fills the caldera that formed when Mt. Mazama collapsed into the magma chamber after a large eruption had emptied the chamber.

b. What type of rock masses are these? extrusive

c. What color represents the youngest rock masses? red

d. Find a single young, continuous rock mass clearly separated from the others. What is the approximate length of this rock mass? 50 mile

e. Do you think this lava was fluid or viscous? viscous

f. Why do you think this?

5. Compare the scales and slopes of Mount St. Helens, Capulin Mountain, and Mauna Loa in the topographic profiles in ■ Figure 3.26. **A topographic profile** is a side view of the terrain.

a. On Figure 3.26c, roughly sketch Mount St. Helens and Capulin Mountain profiles next to and at the same scale as Mauna Loa. Review scales as needed from Lab 1. Only the part of Mauna Loa above sea level is shown in Figure 3.26c!

Jennifer Adleman/Alaska Volcano Observatory/USGS

Figure 3.24

Stratovolcano

Augustine Volcano, Alaska. A steam plume is visible extending from the summit. Taken November 04, 2006

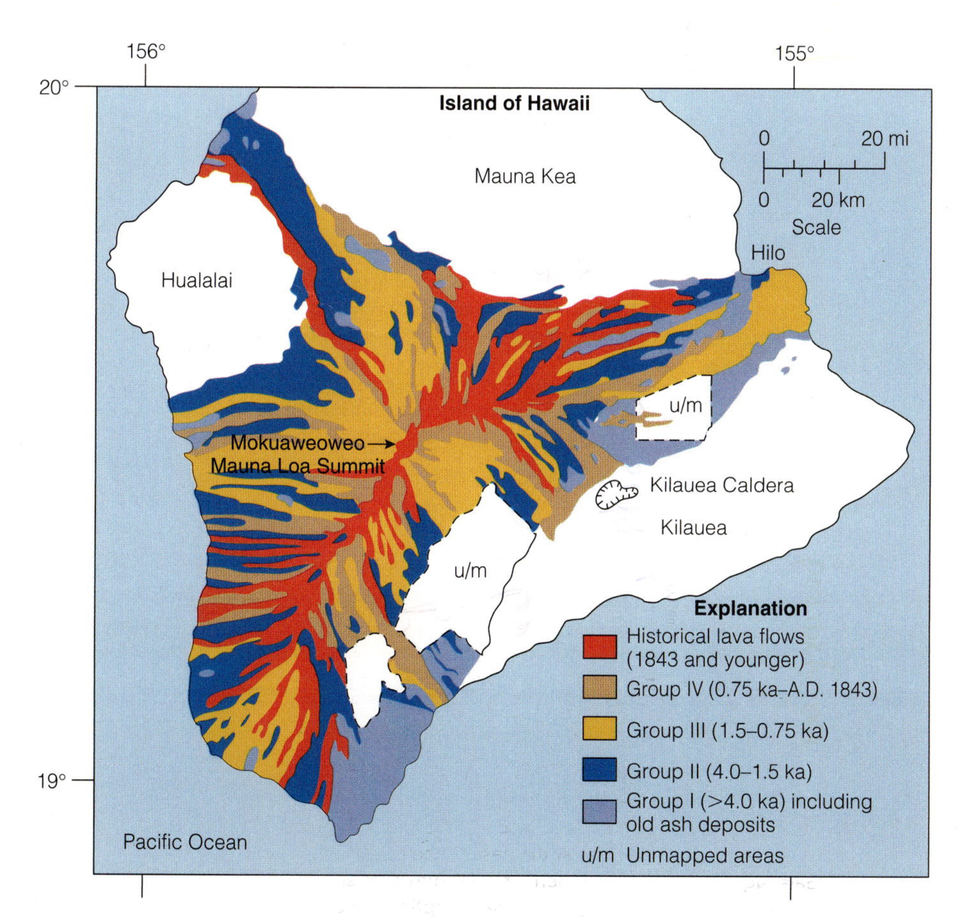

Figure 3.25

Geologic Map of Mauna Loa

Mokuaweoweo is the summit caldera of Mauna Loa. The age unit *ka* means thousands of years. 0.75 ka = 750 years; 1.5 ka = 1,500 years.

b. Measure the approximate slope of the side of each mountain using a protractor and identify what type of volcano each is:

Volcano	Slope Angle	Volcano Type
Mount St. Helens	30	Strato
Capulin Mountain	10°	cylnder
Mauna Loa	50°	shield

c. How do you know which volcano is which type?

Because of its forms layers

Intrusive Rock Masses

Intrusive igneous rock masses, called **plutons** or **intrusions,** have a variety of shapes and sizes and are classified based on these characteristics, as shown in ■ Figure 3.27. Some intrusives, such as dikes and sills, are roughly planar in form; that is to say they are sheet-shaped, tabular, or two dimensional:

- **Dikes** are roughly tabular rock masses that cut across layers or through unlayered rocks (Figure 3.27 and ■ Figure 3.28).
- **Sills** are fairly planar masses that intrude parallel to layers (Figure 3.27 and ■ Figure 3.29). Unlike lava flows, which bake only the rocks below them, a sill will bake the rocks both above and below.
- **Laccoliths** also intrude parallel to layers but they bulge upward to make a dome shape, doming the layers above them (Figure 3.27).

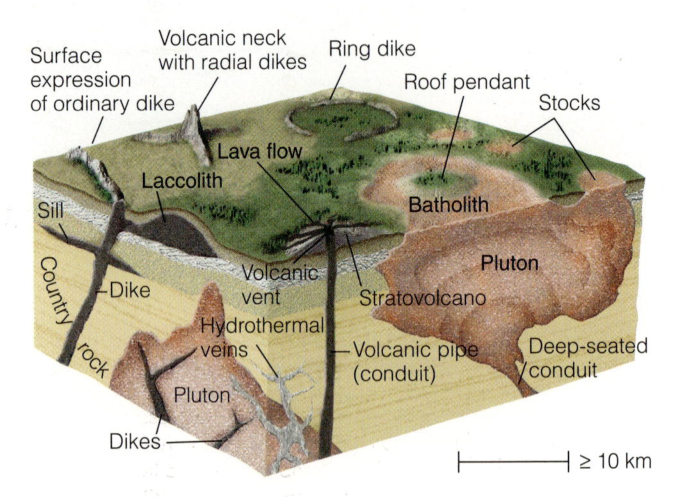

Figure 3.27

Volcanic and Plutonic Igneous Rock Bodies

Block diagram showing various igneous rock masses: volcanic neck, volcanic vent, lava flow, and the associated stratovolcano (or composite volcano) are volcanic features. Stock, batholith, laccolith, dike, sill, and volcanic pipe are plutons. A volcanic neck is an erosional remnant of a volcanic pipe. Sills and laccoliths are **concordant** and intruding parallel to layers, while other plutons are **discordant**, cut across layers. From MONROE/WICANDER/HAZLETT, Physical Geology, 6e. 2007 Brooks/Cole, a part of Cengage Learning, Inc. Reproduced by permission. www.cengage.com/permissions

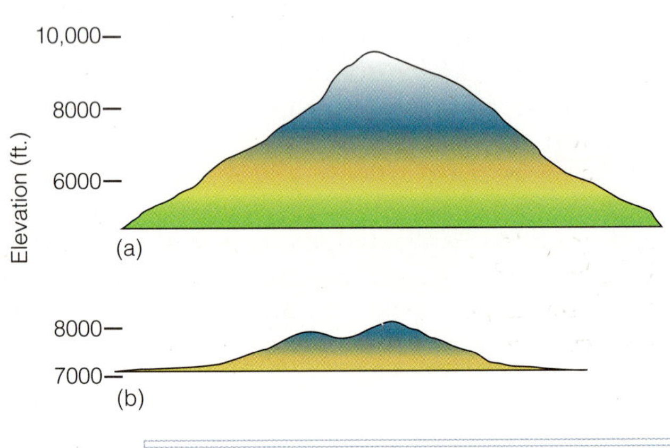

Elevation (ft.)
10,000
0
(c) 0 10 20 30 40

Figure 3.26

Topographic Profiles of Three Volcanoes

Unexaggerated topographic profiles (side views) of (a) Mount St. Helens before its eruption in 1980; (b) Capulin Mountain; (c) Mauna Loa. Unexaggerated means the vertical scale equals the horizontal scale. The three volcanoes are drawn at three different scales, as indicated by the elevation information on the left of each.

Figure 3.28

Dike and Veins

A dike cuts across metamorphic rocks with numerous white veins at Butte Creek, California. The dike shows chilled margins where it came in contact with the country rock. The veins are associated with copper and gold and formed when hydrothermal solutions moved through cracks in the rocks. The outcrop location is a few miles north of the northwest corner of the map in Figure 3.32.

Figure 3.29

Sill

White porphyry sill intruded between darker tilted sedimentary layers in upper Cretaceous rocks, Matanuska River, Cook Inlet region, Alaska, taken in 1910. The sill is not horizontal since the sedimentary layers it intrudes are not horizontal. The dikes cut across the layers.

Large Intrusive Masses are more funnel-shaped or cylindrical in form:

- **Stocks** are intrusions of relatively small size (Figure 3.27), with an outcrop area less than 100 km^2 (or 40 mi^2).
- **Batholiths** are intrusions of large size, with an outcrop area greater than 100 km^2 (or 40 mi^2). Very large batholiths are often composite, consisting of a collection of smaller plutons. One example of such a large batholith occurs in the Sierra Nevada, California, including the granite batholith in Yosemite National Park (■ Figure 3.30).

Where intrusions come in contact with the **country rock** (the rock they intrude), the edges are cooled more quickly than the centers, producing generally finer-grained igneous rock at the margin of the intrusion which is referred to as a **chilled margin** (Figure 3.28). The country rocks are metamorphosed to rock called **hornfels** where they come in contact with an intrusion (see Figure 5.2, p. 97). The intrusion sometimes envelops pieces of the country rock or brings up rock pieces from deeper within or below the Earth's crust. These pieces of foreign rock, embedded in igneous rock, are known as **xenoliths** (Figure 3.11 right-hand sample, p. 54).

In the next exercises we look at rock masses on a field sketch and a map. Although geologic maps may appear quite complicated, a user armed with even a little knowledge of rocks can extract information from them. You may feel more comfortable with the map after reviewing the introduction to maps in Lab 1.

Claudia Owen

Figure 3.30

Batholith at Yosemite

A composite panorama of Sierra Nevada Batholith at Yosemite National Park, California. All the rocks visible are plutonic, part of the batholith, and most are granodiorite, which is intermediate between granite and diorite. This batholith is a combination of many smaller plutons, but is vastly larger (> 10,000 square miles) than the minimum of 40 square miles required for it to be a batholith.

6. ■ Figure 3.31 is a sketch from the notebook of field geologist Dr. Ohio Smith. From the information in the sketch determine what kind of igneous rock mass the basalt represents. Is it a flow, dike, sill, stock, laccolith, or batholith? Flow

Give three reasons for your interpretation.

Because it's only

baked zone

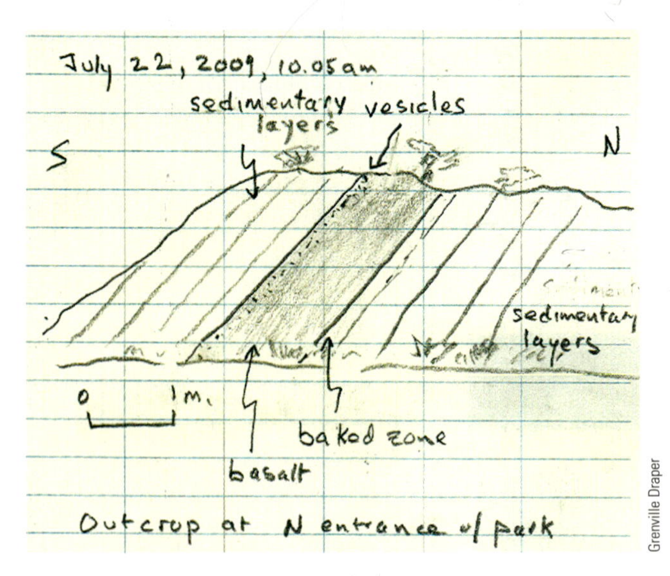

Grenville Draper

Figure 3.31

Geologist's Outcrop Sketch

From the notebook of field geologist Dr. Ohio Smith.

7. Examine the geologic map and explanation of the area near Chico, California, in ■ Figure 3.32.

a. What are the red-orange rocks, **KJqd**, on this map?

quartz diorite

tonalite, trondhjemite

b. Judging by the size and shape of the masses of **KJqd**, what types of plutons are they (refer to Figure 3.27)? Record the area and type for each pluton listed below.

KJqd Pluton	Approximate Area (square km)	Pluton Type
Granite Basin	~30 km²	stock
Concow	30 km³	stock
Bald Rock	336 km³	batholith

c. How many periods of plutonic activity are visible on the part of the map shown? 5 Volcanic activity? 7

d. What is the name of the rock unit **Pv**[b] on the northwestern part of the map?

Pliocene volcanic rocks

Is it intrusive or extrusive?

extrusive

Name the rock that makes it up:

Basalt What type of

igneous rock mass is it most likely to be?

lava flow

e. What material makes up the unit **Ptu**? Note that the term volcaniclastic is a synonym of pyroclastic:

of lahan, volcanic sediment tuff

f. What part of the map has the most extensive deposits of **Ptu**? north / south / east / west (Circle one). Describe what you would expect to see in the rocks there?

mainly of volcanic ash

g. The purple unit, **um**, on the map is now largely a metamorphic rock called serpentinite, but it was once plutonic. What name would be appropriate for the rock before it was metamorphosed? (*Hint:* read the map explanation and refer to Table 3.1.)

perodite

What figure in this chapter shows a picture of a similar rock? fig 3.11

HYDROTHERMAL VEINS

Hydrothermal veins are found in all kinds of rocks, but are particularly common in metamorphic rocks around large igneous intrusions, as seen in the white veins in Figure 3.28, which are close to the plutons on the map in Figure 3.32. **Veins** are open fractures that have been filled by minerals deposited from hot water that flowed through the vein. They differ from dikes in that dikes are fractures filled with solidified magma. Many veins form in the final stages of pluton cooling. Igneous rocks, like all cooling solids, will contract and that contraction often causes fracturing. The hot water that flowed into the fractures is originally dissolved in the magma, or is heated groundwater derived from rain. This hot water contains dissolved minerals, most commonly quartz, feldspar, or calcite, which are then precipitated on the walls of the fracture. This often continues until the fracture is completely sealed. Veins can be tabular, but more often are irregular in shape and thickness. Veins can range in size from a hairline fractures to meter scale bodies.

The fluids that fill hydrothermal veins often contain elements that cannot easily fit into the crystal structures of common rock-forming minerals. These elements can be substances that have economic uses, such as metals and gemstones. Metals such as gold, silver, copper, zinc, lead, tin, and molybdenum are often obtained from hydrothermal vein deposits. Most metals occur combined with other atoms, such as oxygen or sulfur, as oxide and sulfide minerals. Gold and sometimes silver and copper may precipitate as metals. Gems found in hydrothermal veins include varieties of beryl such as emerald and aquamarine. Minerals such as pyrite, calcite, and quartz (called **gangue** minerals) are not economically desirable.

8. Examine the samples from hydrothermal veins provided by your instructor.

a. Use the tables and techniques you learned in Lab 2 to identify these vein minerals, and enter the information in Table 3.4. Some of the vein minerals may not be valuable. For these, write *gangue* in the column labeled "Valuable gem or metal..."

b. Study the sample of a hydrothermal vein and describe the vein's size and overall shape.

Huge, Crystal, Calcit

c. Why are hydrothermal veins important to society?

It's where we get our metal

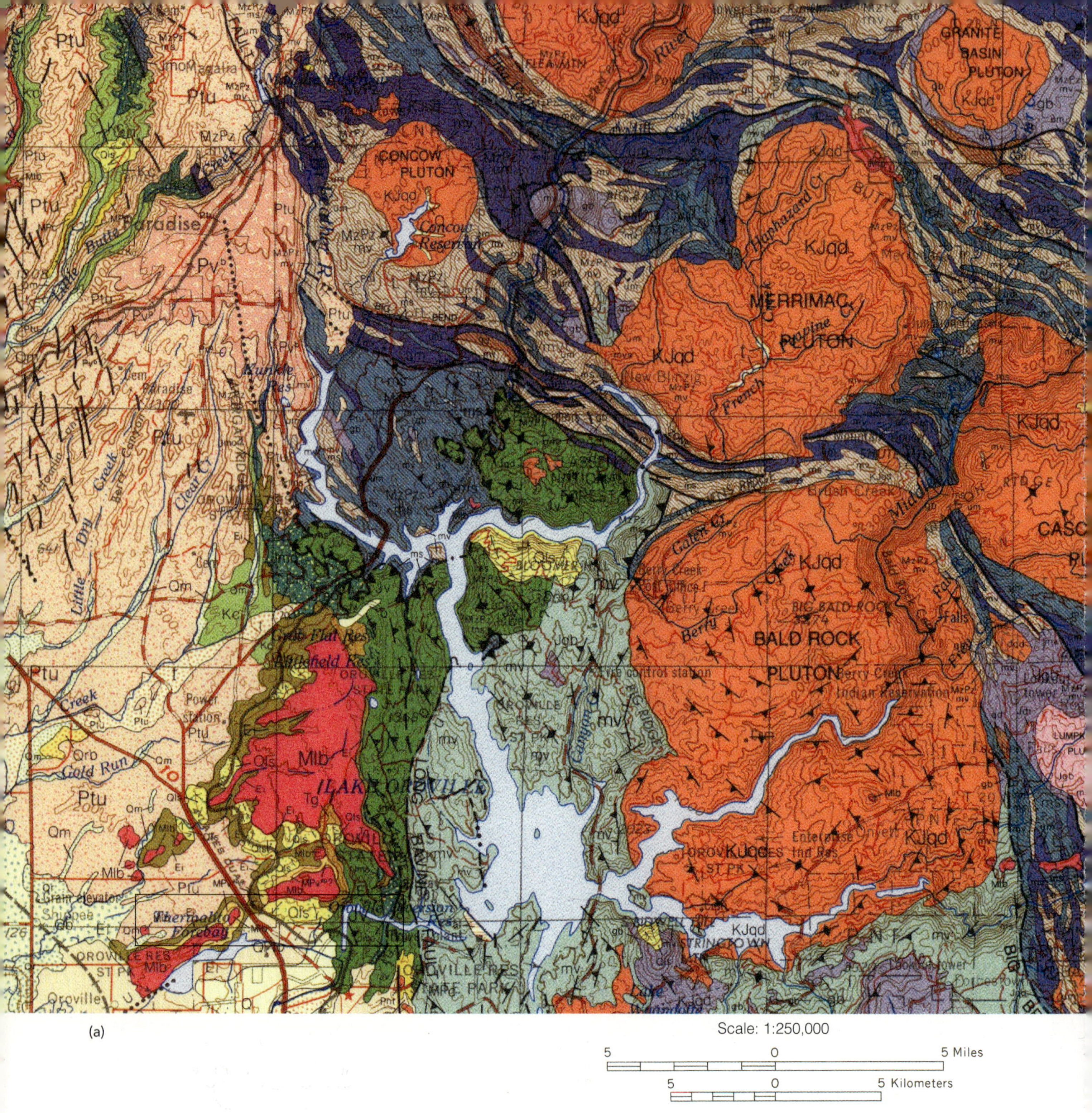

Figure 3.32

Geologic Map of Plutons

(a) Geologic map of an area east of Paradise, near Chico, California and on the facing page (b) map explanation. *Source*: Part of the Chico Sheet, Geologic Map of California, Olaf P. Jenkins Edition, compilation by John L. Burnett and Charles W. Jennings, 1962, USGS

ABBREVIATED EXPLANATION

Approximate stratigraphic relationships only; see Geologic Map Explanation for more accurate age designations and unit descriptions.

CENOZOIC — QUATERNARY — Holocene

- t — Dredge or mine tailings
- Qb — Basin deposits (*Alluvium*)
- Qls — Landslide deposits

Pleistocene

- Qm — Modesto Formation (*Alluvium*)
- Qr — Riverbank Formation (*Alluvium*)

TERTIARY — Pliocene

- Qrb — Red Bluff Formation (*Coarse red gravel, sand, and silt*)
- QPto / Pnt — Tuffs of Oroville (*Volcaniclastic sediments and tuff*) Pnt-Nomlaki Tuff
- Ptu / Pnt — Tuscan Formation (*Lahars, volcaniclastic sediments, and tuff*) Pnt-Nomlaki Tuff

Miocene

- MPc — Miocene-Pliocene channel deposits (*Fluvial conglomerates and sandstone*)
- MPv — Miocene-Pliocene volcanic rocks (*b–basalt; a–andesite; af–andesite flows; ap–andesite pyroclastic rocks; t–dacitic tuff-breccia*)
- Mlb — Lovejoy Basalt

Eocene

- Ei — Ione Formation (*Quartzose sandstone, claystone, and conglomerate; mostly nonmarine*)
- Tg — "Auriferous" Gravels

MESOZOIC — CRETACEOUS

- Kc — Chico Formation (*Sandstone, conglomerate, and siltstone; marine*)

JURASSIC

SMARTVILLE COMPLEX

- Jmo — Monte de Oro Formation (*Sandstone and slate; marine*)
- Jv — Jurassic volcanic rocks (*Pyroclastic rocks and flows*)
- mv — Volcanic rocks

MESOZOIC PLUTONIC ROCKS

- KJqd — Quartz diorite, tonalite, trondhjemite, quartz monzonite and similar rocks
- Jdi — Diorite

PALEOZOIC

PALEOZOIC AND MESOZOIC ROCKS

MzPz

- ms — ms–metasedimentary rocks
- mv — mv–metavolcanic rocks
- qd — qd–metadiorite
- gb — gb–gabbro
- um — um–ultramafic rocks

(b)

MAP SYMBOLS

Anticlinal fold
Showing direction of plunge; dashed where inferred; dotted where concealed by younger rocks.

Synclinal fold
Dashed where inferred; dotted where concealed by younger rocks.

Monoclinal fold

Strike and dip of foliation
General strike and dip of foliation in metamorphic rocks.

Strike and dip of dikes

Vertical foliation

Strike and dip of overturned beds

Strike and dip of beds
General strike and dip of stratified rocks.

Vertical beds

? ?

Contact
Observed or approximately located; queried where gradational or inferred.

U
?—?
D

Fault
Solid where well located; dashed where approximately located or inferred; dotted where concealed by younder rocks or water; queried where continuation or existence is uncertain. U, upthrown side; D, downthrown side (relative or apparent).

? ?

Thrust fault—barbs on the upper plate. Generally dips less than 45°, but locally may have been subsequently steepened. Dashed where approximately located or inferred; dotted where concealed by younger rocks or water; queried where continuation or existence is uncertain.

Figure 3.32

Table 3.4

Hydrothermal Vein Samples

Sample Number	Key Properties of the Mineral			Minerals Found in Sample	Valuable Gem or Metal Obtained from the Mineral
156	Hardnem = 7	concordial		gabna	copper
	Hardness = 6	2 = 90°			Zinc
	Hardness = 9	redish			Laardus
181-5	Hardnem = 7	conchoid		Beryl	emerald
	Hardnem = 6	2 = 90°		Tupaz	Tupaz
	Hardnem = 8	bladed crystal			

LAB 4

Sedimentary Rocks

OBJECTIVES

- To understand how sediments and sedimentary rocks form
- To understand what sedimentary structures and fossils reveal about a sedimentary rock
- To recognize depositional environments and the economic resources of sedimentary rocks
- To recognize the common types of clastic, chemical, and biochemical sedimentary rocks

Sedimentary rocks form by the *lithification* of *sediment*. Since sedimentary rocks form on the Earth's surface, they give geologists clues to the nature of past environments. In addition, economic resources—such as fossil fuels, many construction materials, and soil—come from sedimentary rocks and from the processes that form them. The methods by which these rocks form also help us to classify and organize them into easily recognizable categories for identification.

FORMATION OF SEDIMENTARY ROCKS

The formation of sedimentary rocks, as seen in the rock cycle in Lab 1 (Figure 1.5, p. 6), starts with weathering and erosion of pre-existing rocks, which either produce loose rock and mineral particles or material in solution. Deposition of these products creates the variety of *sediments* that make up sedimentary rocks.

Types of Sediment

The three major types of sediment (■ Figure 4.1a–c) are: **clastic,** or loose material from rock and mineral particles; **chemical,** from precipitation at the Earth's surface; and **biochemical,** or **organic,** from organisms and their remains. These three types of sediment become the three types of sedimentary rocks upon lithification (Figure 4.1d–f).

Clastic, Fragmental, or Detrital[1] Sediment forms when mechanical weathering breaks rocks and minerals down into pieces that are then eroded, transported, and deposited (Figure 4.1a). This sediment consists of loose grains or fragments, called **clasts.** The rock *texture* (arrangement of grains) resulting from lithification of clasts is known as **clastic texture.** In clastic rocks, we use the term **matrix** for finer-grained material that surrounds larger pieces. The conglomerate in Figure 4.1d has clastic texture and abundant matrix material.

The size, shape, and mineral makeup of particles further subdivide clastic sediments. ■ Table 4.1 shows the classification of clasts by grain size. It lists rocks made of these clasts. The term **sorting** describes the size distribution within sediment where **poorly-sorted** sediment has a wide mixture of sizes and **well-sorted** sediment has clasts with similar sizes (■ Figure 4.2). Another aspect of clastic sediment is how well rounded or angular the grains are. Conglomerate (Figure 4.1d) is a rock with **rounded** grains, and breccia (Figure 4.6, on p. 78) has **angular** grains with sharp corners. The sorting

[1] The terms *fragmental* and *detrital* are synonyms for clastic. *Siliciclastic* is slightly more precise than clastic as it excludes fragmental limestones, which clastic and fragmental may not.

(a) (b) (c) (d) (e) (f)

Figure 4.1

Types of Sediment and Sedimentary Rocks

(a) Clastic sediment from a beach made up of well-worn and rounded clasts (fragments) of mainly silicate rocks that resulted from weathering, erosion, transport, and deposition. (b) Chemical sedimentation forms polygons through evaporation of brine and crystallization of salts, near Badwater, Death Valley National Park, California. (c) Biochemical sediment from the seafloor beneath the Gulf Stream, Atlantic Ocean. This sample consists of white planktonic (open sea) foraminifera (one-celled animals), bryozoan stalks, clear pteropods' coiling shells, and a clear three-pointed sponge spicule. These are sand-sized shells. (d) Clastic sedimentary rock, conglomerate, with a clastic texture, pebble clasts, and visible matrix. (e) Chemical sedimentary rock, rock salt, with a crystalline texture. (f) Biochemical sedimentary rock, fossiliferous limestone, with a bioclastic texture.

and rounding indicate how **mature**—far traveled—a sediment is. More angular, coarser, and poorly sorted sediment is more **immature.** Well rounded, finer, and well-sorted sediment is more **mature.**

Common minerals in clastic rocks include:

- **quartz**
- **feldspar**
- **clay** including **kaolin**
- **iron oxides** (giving red, orange, yellow, and tan coloring)
- **muscovite, biotite,** assorted **mafic minerals** are accessory minerals (commonly present in small amounts)

Notice that most of the common minerals in clastic rocks are silicates, which is why geologists may call clastic rocks **siliciclastic.** Mineral content also indicates the maturity of sediment, depending on how resistant a mineral is to physical and chemical breakdown. Quartz is one of the most resistant minerals and feldspar tends to weather to clay in moist climates; therefore, rocks with quartz or clay are *mature* and ones with feldspar are moderately mature in dry climates and immature in wet climates. Sediments with rock fragments and mafic minerals are *immature* and have not traveled far from their source. A proper description of clastic sediment would then include size, sorting, rounding, mineral composition, and maturity.

Chemical Sediment forms when compounds precipitate from water. Evaporation of a desert lake, for example, forms salt deposits, which are chemical sediment (Figure 4.1b). Chemical sedimentary rocks may have

Table 4.1

Clastic Classification

Classification of clastic sediment and the corresponding clastic sedimentary rocks. Photographs show the actual size of each sediment. Scale 1:1

Clast Type	Grain Size	Grain Diameters (Photos show actual size)	Rock Name: Rounded Clasts	Rock Name: Angular Clasts
Boulder	Very coarse (beach ball and basketball size)	>256 mm		
Cobble	Coarse (fist and softball size)	64–256 mm		
Pebble	Moderately coarse (pea and apricot size)	4–64 mm	Conglomerate	Breccia
Granule	Medium coarse (rice size)	2–4 mm		
Sand	Medium (salt-grain size)	1/16–2 mm	Sandstone	
Silt	Fine (slightly gritty)	1/256–1/16 mm	Siltstone	
Clay	Very Fine (smooth, powder size)	<1/256 mm	Shale	

Claudia Owen

(a) (b) (c) (d)

Claudia Owen

Figure 4.2

Sorting

Examples of four rocks made of sediment having different degrees of sorting. (a) *Very poorly sorted* sediment makes up this sample of glacial tillite. (b) Outwash (water distributed) sediment from the same glacier produced the *poorly sorted* sediment in this rock. (c) *Moderately sorted* sediment in a granule to pebble conglomerate. (d) *Very well sorted quartz* sand in a sandstone.

Table 4.2

Biochemical and Chemical Classification

Classification of biochemical and chemical sediment and the corresponding textures and sedimentary rocks.

Sediment Composition	Texture	Rock Name	Sediment Type
Calcite (or aragonite)	Bioclastic, crystalline, oolitic, or microcrystalline	Limestone	Biochemical or chemical
Dolomite	Crystalline	Dolostone	Chemical
Halite	Crystalline	Rock salt	Chemical
Gypsum (or anhydrite)	Crystalline	Rock gypsum (or rock anhydrite)	Chemical
Silica or quartz	Amorphous, cryptocrystalline	Chert	Chemical or Biochemical
Carbonaceous/organic material	Amorphous or bioclastic	Coal	Biochemical

interlocking crystals that fit together so closely that there is no space between grains. This arrangement of crystals is **crystalline texture** and is seen in rock salt (Figure 4.1e). *Ooids* in oolitic limestone (Figure 4.15a, inset, p. 82) are chemical sediment having **oolitic texture,** or texture consisting of chemically precipitated sand-size spheres. Minerals and materials likely in chemical sediment are listed in Table 4.2 along with the rocks they make up.

Biochemical Sediment results from the accumulation of parts of living things. While the organisms live, they extract material from their surroundings and produce shells, skeletons, wood, and living tissue. When the organisms die, surface processes transport their remains, especially the harder parts, and deposit them as sediment (Figure 4.1c). Biochemical sediments may vary widely in appearance depending on what organisms and by-products are preserved. Generally, you can recognize them by their **bioclastic texture** (Figure 4.1f), which consists of an abundance of organisms' remains and few siliciclastic grains. Finer biochemical sediment may make up a matrix between larger pieces. The minerals and materials that constitute biochemical sediment are listed in Table 4.2 along with the rocks they compose.

The three types of sediment may accumulate and settle separately or mix in various combinations. Deposits of these sediments have numerous uses. Some of the largest quantities of resource materials come from clastic sediment such as gravel, sand, and clay. These represent important components in construction material, glass, and ceramics. Chemical sediments provide the salt used in food products and road deicing and the gypsum used in construction (gypsum or wallboard). Some common types of resources made from biochemical sediments are crushed shells used as fertilizer and in cement making, peat (a precursor to coal) used in soil amendments, and coal used as a fuel.

Sedimentary Analysis

1. Use a hand lens or binocular microscope to examine the samples of sediment your instructor provides. For each sample, identify the sediment and write a brief description by following the instructions and headings in Table 4.3. The things you should include in your description depend, in part, on the type of sediment described in the table.

a. After recording the sample number, list the name of the clast size (from Table 4.1) and/or record the size of organisms' remains in column 2 in Table 4.3.

b. Note the mineral composition and the presence of any broken pieces of rock (rock clasts) in column 3 of the table. Use the mineral identification skills you acquired in Lab 2 (such as visual identification and dilute HCl testing). Take a small amount of sediment, place it between two glass slides, and rub them together. If you see any new scratches on the glass, or if larger clasts are pulverized, this will tell you the hardness of the larger grains.

c. Describe the other aspects of the sample and record the type of sediment in the next two columns as indicated in Table 4.3.

d. Use Tables 4.1 and 4.2 to name the sedimentary rock these sediments might

Table 4.3

Types of Sediment

Sample Number	Grain Size and/ or Size of Organic Debris	Minerals, Rock Clasts	Additional Description Including Colors and for . . . *Clastic sediment* include: sorting, rounding, and maturity. *Chemical sediment* include: grain shapes. *Biochemical sediment* include: organism name and/or description.	Type of Sediment	Potential Sedimentary Rock
1	Medium	SiO_2 Sand	white, tan, orange, Moderately sorted	Clastic	Sandstone
2	fine	Silt Quartz, feldspar.	light grey, hints of black. very-well sorted	chemical	siltstone.
27	Med	Granule	White, tan, orange, clay.	clastic	Breccia

become if lithified, and enter this in the last column in Table 4.3.

Details of the sizes of grains in a sediment may help to indicate how the sediment formed. ■ Figure 4.3 shows the size distribution of two sediment samples; one is wind-blown sand and the other is sediment transported by a stream. In the next exercise, you will make a similar graph.

2. Table 4.4 shows the results of sieving a clastic sediment sample. Shaking the sediment through sieves divides the sample into size fractions from coarse to very fine, much in the same way that you might sift flour or drain noodles. The larger mesh sieves catch the larger sediments and the finer sediments fall through to the next finer sieve. ■ Table 4.4 gives the weight of each size fraction. The sizes are in micrometers (μm); 1 mm is 1,000 μm.

a. To complete Table 4.4, for each size fraction, calculate the weight percentage by dividing its individual weight by the total weight of the sediment sample, and multiplying by 100.

b. Use these tabulated results to plot a histogram graph similar to those in Figure 4.3 on the graph paper provided in ■ Figure 4.4 with weight percentage on the vertical axis and the corresponding size on the horizontal axis. Draw columns for the weight percent of each size fraction.

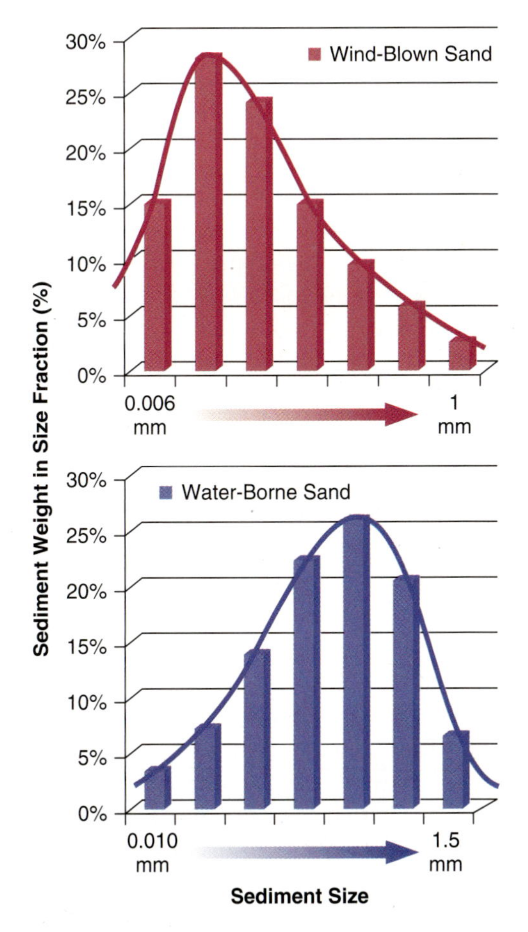

Figure 4.3

Sediment Graphs

Sediment Size Distributions for Wind-Blown and Water-Borne Sand.

Table 4.4

Sediment Size Fractions

Size fractions of clastic sediments and their weight percentages.

Fraction Size Range (microns, µm)	Weight (grams)	Weight Percent
62–88	1.5	
88–149	9.0	
149–250	6.4	
250–355	3.4	
Total	20.3	100%

c. Draw a smooth line through the top center of each column as shown in the graphs in Figure 4.3.

d. Compare your graph with the graphs in Figure 4.3. Which graph is more similar to your graph?

____

Add this information to the title of your graph.

3. The photos in Table 4.1 for granules, sand, and silt are from one sample of river sediments. Compare these photos with the data in Table 4.4. Could the data in Table 4.4 and your graph in Figure 4.4 represent the river granules, sand, and silt in Table 4.1? Yes / No ______ Explain how you can determine this.

____

____

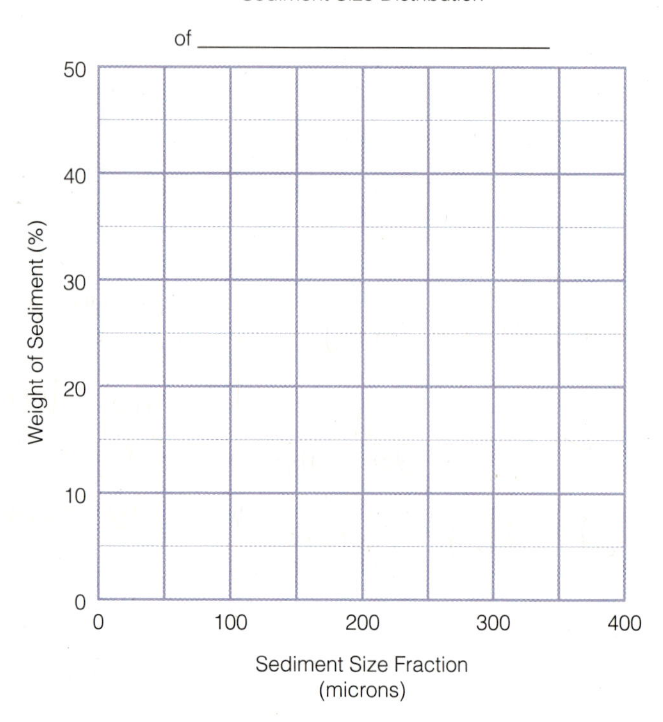

Figure 4.4

Sediment Histogram Graph Paper

Lithification

Once sediment has accumulated, additional sediment usually buries it, causing lithification into a sedimentary rock. **Lithification** of sediment involves bringing the grains closer together by **compaction,** and "gluing" them together by **cementation.** Burial under more sediment provides the pressure for the compaction to take place. Compaction can make a rock out of material such as clay, which tends to stick to itself when squeezed.

Compaction alone is insufficient to cause sand or pebbles to hold together to form a rock. For this to happen, cementation is needed: natural mineral **cement** holds the grains (including matrix) together and partly fills the space between grains. Cementation occurs when water containing dissolved substances moves through *pore spaces* between the sediment grains and precipitates minerals such as quartz, calcite, or iron oxide. These minerals, filling in the pores between the grains, cement sediment particles to each other, thereby lithifying the sediment.

Fossils

Fossils are natural remains or traces of life preserved in a rock at the time the rock lithified. The letters in parentheses in the remainder of this paragraph refer to the fossils pictured in ■ Figure 4.5. Fossils are a principal component of biochemical sedimentary rocks, but may also appear in other sedimentary rocks. A rock containing numerous fossils is **fossiliferous** (Figure 4.11 on p. 79). A fossil can be the actual remains of the organism such as shells (a), bones or leaves (carbonized leaf in d), their impressions called **molds** (a and b), or some part of the organism that fossil *replacement* has preserved (c and e). It can also be a sign of the organism's activity, called a **trace fossil,** such as *footprints* (f) or animal *burrows* (g). **Bioturbation** occurs where organisms have stirred up the sediment, causing obliterated or disturbed *bedding* layers. Burrows are common in bioturbated sediment.

Figure 4.5

Fossils

(a) Internal and external molds of the gastropod (snail) *Turritella sp.* A small amount of the original shell material is still preserved.
(b) Trilobite, *Metacryphaeus venustus,* Paleozoic, early to middle Devonian Period (411 to 392 million years old). The trilobite is a mold that is 2.25 inches long. (c) Ammonite, relative of the squid and octopus, which lived during the Cretaceous Period. Pyrite replaced the original fossil shell. (d) Impressions (molds) and carbonization of fern leaves on black shale, Pennsylvanian Period. (e) Silica replacement. A slice of petrified wood across the trunk of a tree, *Araucaria species,* Triassic, showing tree rings and well-preserved bark. Cross-section is of trunk 6″ across. (f) Dinosaur footprint on arkose, *Archisauripus sillimani,* Triassic Period, front foot about 6″ long. (g) Fossil burrows preserved on the underside of a sedimentary layer are a kind of trace fossil.

CLASTIC SEDIMENTARY ROCKS

Clastic or fragmental sedimentary rocks can vary widely, but all have *clasts* (fragments) lithified together. Many have **bedding**—layers—which we discuss in more detail later. The classification of clastic (or detrital[1]) rocks depends on the clast size, as outlined in Table 4.1. When determining clast size in rock classification, look at the largest clasts that make up the majority of the rock and use the clast types named in Table 4.1. We describe some common clastic sedimentary rocks next, and suggest common environments in which their sediments may have come to rest (Table 4.5, p. 86).

Coarse-Grained Clastic Rocks

Breccia has granule- to boulder-sized clasts (larger than 2 mm, Table 4.1) that are angular (■ Figure 4.6). Most breccias are *massive* (without layers or *bedding*). Breccia

[1] The terms *fragmental* and *detrital* are synonyms for clastic. *Siliciclastic* is slightly more precise than clastic as it excludes fragmental limestones, which clastic and fragmental do not in some uses.

Figure 4.6

Breccia

This sample is a pebble breccia with angular clasts in a finer-grained gray or yellow matrix. Some iron staining colors the matrix yellow ocher in places.

Figure 4.7

Conglomerate

Pebble conglomerate in which pebbles are the largest clasts, surrounded by a matrix of sand. Some of the pebbles in this sample are chert, granite, and porphyritic volcanic rocks.

is likely to form in an environment where sediment has traveled only a very short distance, such as a rockslide or landslide (*mass wasting*). Breccia is **immature,** meaning the sediment has not traveled far from the sediment source rock. Glacial transport (Lab 17) may also produce very poorly sorted deposits with angular grains, a kind of breccia called **tillite** (Figure 4.2a).

Conglomerate also has particles that are greater than 2 mm across, but these particles are rounded (■ Figure 4.7). The large clast size in a conglomerate indicates the clasts probably were transported only a short distance, yet the rounded shape suggests farther travel than that for breccia. Moving water can tumble rocks to make large rounded clasts seen in Figure 4.1a and d. Some conglomerates were once streambed gravels; others formed from wave-washed boulders at the base of a sea cliff (Table 4.5, p. 86). Some tillites may have mostly rounded grains, which makes them conglomerates.

Figure 4.8

Quartz Sandstone

(a) Light tan laminated quartz sandstone with a small amount of yellow iron oxidation especially concentrated along certain layers. (b) Friable (crumbly) ferruginous quartz sandstone with enough iron oxide cement to make it a deep red-orange.

Medium-Grained Clastic Rocks (Sandstones)

Quartz sandstone is a rock made of quartz sand grains cemented together (■ Figure 4.8). The quartz particles appear transparent-to-translucent and sometimes *vitreous*, looking like glass, when viewed with a hand lens. Iron oxide may stain them red or orange (Figure 4.8b). When a rock breaks between grains and easily crumbles, it is **friable.** Sandstone feels like medium- to coarse-grit sandpaper. Because quartz is resistant to chemical and physical weathering, it lasts a long time at the surface of the Earth. Sandstone with a high percentage of quartz is *mature*. Quartz sandstones are common along passive continental margins (without plate boundaries, Figure 14.2, p. 318). Pure quartz sandstone is useful in making glass and producing silicon for computer chips.

Arkose is a sandstone that contains abundant feldspar in addition to quartz (■ Figure 4.9). Feldspar grains are usually weathered and appear opaque, chalky, and white, light pinkish, yellowish, or grayish in color. If unweathered, the feldspar grains may have visible cleavage that flashes in the light. Weathering of granite in a dry climate is a source for quartz, feldspar, and often biotite and muscovite. Arkosic sediment is moderately immature.

Graywacke is a "dirty" or impure sandstone characterized by angular grains of quartz, feldspar, and small fragments of rock set in a matrix of finer particles (■ Figure 4.10). All these features of graywacke indicate it is immature. Graywackes are common along active continental margins (with plate boundaries, Figure 14.2) where transport distances tend to be short.

Fossiliferous sandstone contains abundant fossils, probably deposited in a shallow marine environment, such as the sample in ■ Figure 4.11.

Fine-Grained Clastic Rocks

Shale and **mudstone** are soft, fine-grained clastic rocks made of clay (and silt) compacted together. Claystone is composed exclusively of clay. These rocks are usually gray-to-black, buff, greenish, or reddish. Shale is **fissile,** which means it splits into platy slabs parallel to bedding (■ Figure 4.12). Mudstone is a clay (and silt) rock that

(a)

(b)

(a) Diane Pirie; (b) Claudia Owen

Figure 4.9

Arkose

Many arkoses or arkosic sandstones form from the weathering of granite in a dry environment. Both of these particular samples have some granules, but sand predominates. (a) Red arkose rich in quartz and feldspar. The red color comes from iron-oxide stain and cement. (b) Pink potassium feldspar and light gray to white quartz make up the majority of the clasts in this granule-bearing arkose. Yellow colors come from iron-oxide stain.

Claudia Owen

Figure 4.10

Graywacke

Graywacke sandstone with abundant rock fragments (lithic grains) and high matrix content. Inset shows medium to dark gray volcanic fragments, and lighter quartz and feldspar grains.

Claudia Owen

Figure 4.11

Fossiliferous Sandstone

Sandstone with abundant clam fossils.

Figure 4.12

Shale

Multiple samples of shale showing its laminar bedding and its variety of colors. From left to right: The bottom gray and tan sample has fossil leaves (black). The second sample (gray) has some mud cracks visible. Iron oxide gives the next sample its red color. The top sample at the right shows color mottling in tans and light grays and is a softer variety.

Claudia Owen

does not split into layers, but breaks into chunky pieces. Many shales or mudstones crumble or fall apart when water is added. Shale or mudstone may contain well-preserved fossils and form in quiet environments such as floodplains, lakes, lagoons, estuaries, and the deep sea (Table 4.5, p. 86).

Siltstone is a fine-grained clastic sedimentary rock consisting of silt-sized particles, which need magnification to make them visible (Table 4.1). It has a slightly gritty feel to your teeth (touch also, but fingers are less sensitive), compared with smooth-feeling claystone and very gritty sandstone.

BIOCHEMICAL AND CHEMICAL SEDIMENTARY ROCKS

Whether biological or chemical processes formed a sedimentary rock, we use chemical or mineralogical composition to categorize it (Table 4.2, p. 74). For example, **limestone** is any sedimentary rock made almost entirely of calcium carbonate, usually calcite, regardless of whether it is made of shells, tiny calcite spheres (*ooids*), limestone boulders, or limey ooze. Table 4.2 lists some examples of rocks classified by composition. Limestone is a major constituent in making cement.

The names of the following rocks may reflect both their composition and the biochemical and/or chemical processes that formed them.

Carbonates

Carbonates are sedimentary rocks containing carbonate minerals such as calcite or dolomite. There is a very wide range of rock types within this group due to various conditions of formation. So even after you determine if the rock is a carbonate, you will still find many rock names to choose from.

Biochemical Limestones **Fossiliferous limestone** is limestone that has numerous visible fossils in it, generally with other calcium carbonate sediment, *matrix*, between the fossils (■ Figure 4.13a). More specific names include the following: **Coralline limestone** is composed almost entirely of coral (Figure 4.13b). Coralline limestone indicates a warm, shallow-sea environment. **Bryozoan limestone** consists of calcareous remains of bryozoans, which appear lumpier than coral. Bryozoans live in warm, shallow, quiet lagoons. **Coquina** is composed almost entirely of coarse calcium carbonate shells and shell fragments (Figure 4.13c) that accumulate in a beach environment where wave action breaks the shells.

Chalk is a powdery limestone (■ Figure 4.14) containing calcite from shells of microscopic oceanic plants and animals (Figure 4.14 inset). These remains may mix with very fine-grained inorganic calcite sediment. Chalk is white or lightly colored, and it is soft and porous.

Chemical and Mixed-Origin Carbonates **Micrite limestone** is exceptionally fine-grained and homogeneous and exhibits subconchoidal fracture. It may form as chemical precipitate or as lithified biochemical ooze of microscopic calcium carbonate shells.

Oolite or **oolitic limestone** contains small spherical bodies, or **ooids**, usually ½ mm to 1 mm in diameter, cemented together (■ Figure 4.15a). These calcium carbonate spheres (Figure 4.15a inset) are precipitated in subtidal areas or shallow seas where each grain is washed back and forth and accumulates calcium carbonate layers around a small particle. Although it looks somewhat like sandstone, testing with dilute HCl easily distinguishes oolite from sandstone.

Travertine is limestone deposited out of solution from groundwater. It generally has a crystalline texture with visible crystallized layers (Figure 4.15b). Most travertine occurs as flowstone, a rock that forms from flowing water on the walls or floors of a cave. Depending on the environment of formation, the layers may have formed horizontally or vertically. If the travertine was part of a stalactite[2] or stalagmite[3] (see Figure 13.13b, on p. 307),

[2] *Stalactites* are calcite "icicles" hanging from the roof of a cave.

[3] *Stalagmites* are upright calcite "posts" on the floor of a cave.

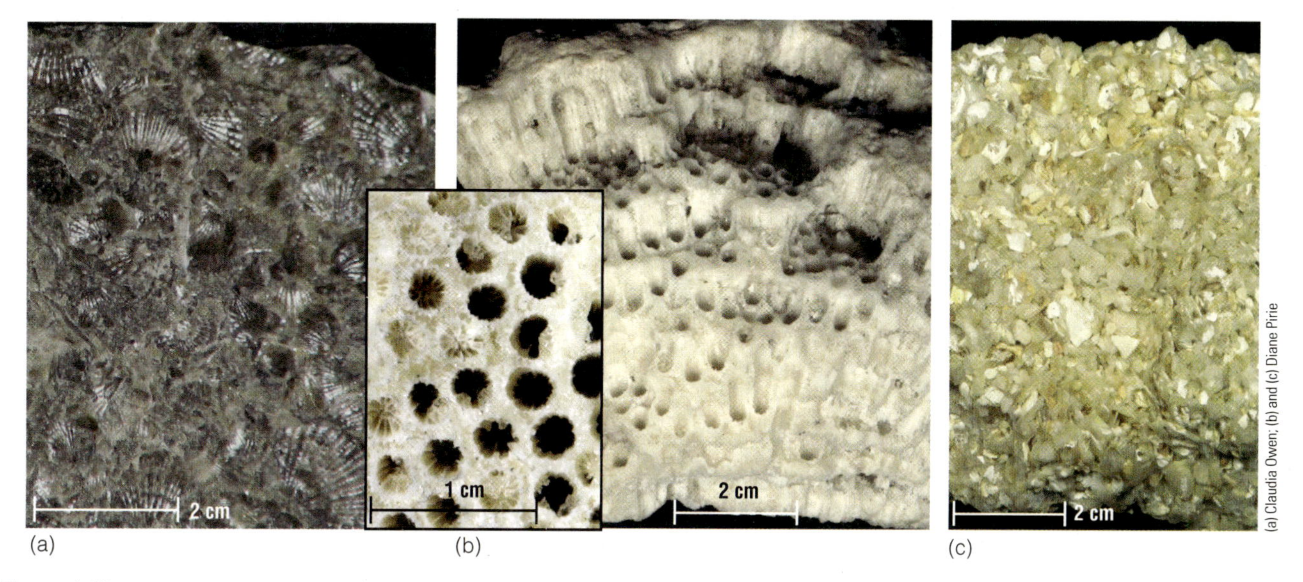

Figure 4.13

Fossiliferous Limestones

(a) Pearly-shell gray limestone with numerous brachiopod fossils with the matrix between them. (b) Coralline limestone. Inset shows detail of the coral structure. Such structures are absent in bryozoan limestones. (c) Coquina made of clam shell fragments cemented together. The broken shells indicate the sediment was deposited and formed in a high-energy beach environment.

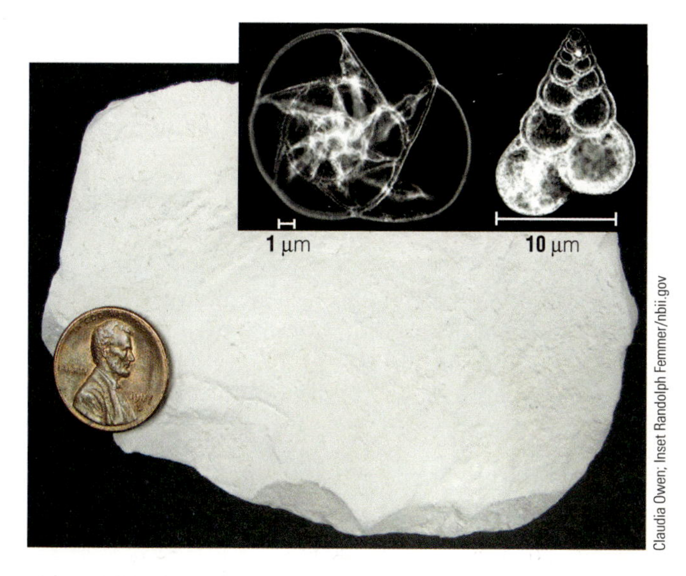

Figure 4.14

Chalk

Hand sample of a biochemical limestone, chalk, which is made up of microscopic calcium carbonate shells loosely cemented to produce a powdery rock. Inset: Calcium carbonate tests (shells) of two species of marine foraminifera.

the layers may form concentric circles in cross-section. *Tufa* is a fine-grained limestone, which forms from surface waters and calcareous springs, but is softer, spongier, or more porous than travertine.

Dolostone (Figure 4.15c) is very similar to limestone except that it reacts much more slowly with dilute hydrochloric acid because of its large amount of dolomite. Powdered dolomite, such as the powder left on a streak plate, reacts more vigorously (Figure 4.15c inset). Dolostone forms by sedimentary chemical alteration of limestone or calcareous sediment. The limestone texture may be obscured or obliterated: fossils appear as faint shadows or as molds. Small amounts of iron may make dolostone tan, but other colors are possible.

Fossil Fuels

Fossil fuels are energy resources preserved in rocks. The fossil in fossil fuel is organic matter that was once part of a living organism. Ancient plants and plant-like single-celled organisms performed **photosynthesis,** combining water and carbon dioxide and using sunlight to form organic molecules (carbohydrates). Large quantities of organic molecules—preserved, compacted, and sealed off by burial—create a potential energy resource. If the buried organic material has a sufficient concentration of stored energy, we can later extract and burn it for energy. **Oil** and **natural gas** are *hydrocarbons* and form from the organic matter of microscopic aquatic organisms. They tend to occupy the pore spaces of **permeable** sedimentary rocks (rocks that allow fluids to pass through them) such as sandstone and limestone, and become trapped beneath impermeable rocks such as shale.

Figure 4.15

Chemical Carbonates

(a) Oolitic limestone, also called oolite. *Inset:* Close-up of the spherical, sand-size ooids cemented together with calcite in oolitic limestone. (b) Travertine precipitated from freshwater in caves. The bedding in this travertine may have formed vertically, incorporating holes as the travertine formed. (c) Dolostone made of the mineral dolomite (calcium-magnesium carbonate). *Inset:* Powdered dolomite effervesces in dilute hydrochloric acid.

Coal Coal forms by accumulation, burial, and lithification of plant matter in a wetland environment (Table 4.5, p. 86). Conditions of formation must include rapid deposition, a wet and reduced-oxygen environment to keep organic sediment from further decomposing, and burial and compression by additional sediment. Increased compression of the organic matter also increases its hardness and carbon content (see ■ Figure 4.16). Coal is the most abundant fossil fuel and is often used in power plants to generate electricity. We next discuss three types of coal and the sediment that can become coal—peat.

Peat is a dark brown, fibrous, decomposed, and compacted residue of swamp plants (Figure 4.16a). Because peat is generally porous and crumbly, you will need to be careful with the samples so future students can view them intact. Peat forms as a step in the formation of coal, but is generally not considered coal itself. It is, however, a fossil fuel and can be burned to provide heat.

Lignite is smooth-surfaced, brown-to-almost black coal, and forms from further alteration of peat (Figure 4.16b). Softer than bituminous coal, lignite commonly has dusty, dirty surfaces that soil fingers. It is also called "dirty coal" because it puts out a lot of pollution when burned.

Bituminous coal, or "soft coal," yields bitumens (hydrocarbons) upon heating. It is dark brown to black, and glossy (Figure 4.16c), but not as shiny as anthracite. It is harder than lignite, but fractures and falls apart more readily than anthracite. It results from deeper burial of lignite, with the corresponding higher pressure and temperature.

Anthracite is "hard coal" with a high percentage of carbon (Figure 4.16d). It is black and has an adamantine-to-submetallic luster. It is often grouped with metamorphic rocks, because in deposits of anthracite, slate (a low-grade metamorphic rock; see Lab 5) is a common neighboring rock. Anthracite forms by low-temperature metamorphism of other coal.

4. Examine the samples of coal and peat provided. Match the samples with the descriptions above, and list the sample numbers and names in order of their temperature and pressure of formation.

\# 18 Name Peat

\# #126 Name Lignite (brown coal)

\# #20 Name Bitumous coal

\# 031-A Name Anthracite (hard coal)

5. In what environment does the organic matter found in coal accumulate?

wetland environment.

What other rocks might you expect to form above or below the coal (what might be deposited in a similar environment (Table 4.5, p. 86)?

above - shale

below - laminar bedding.

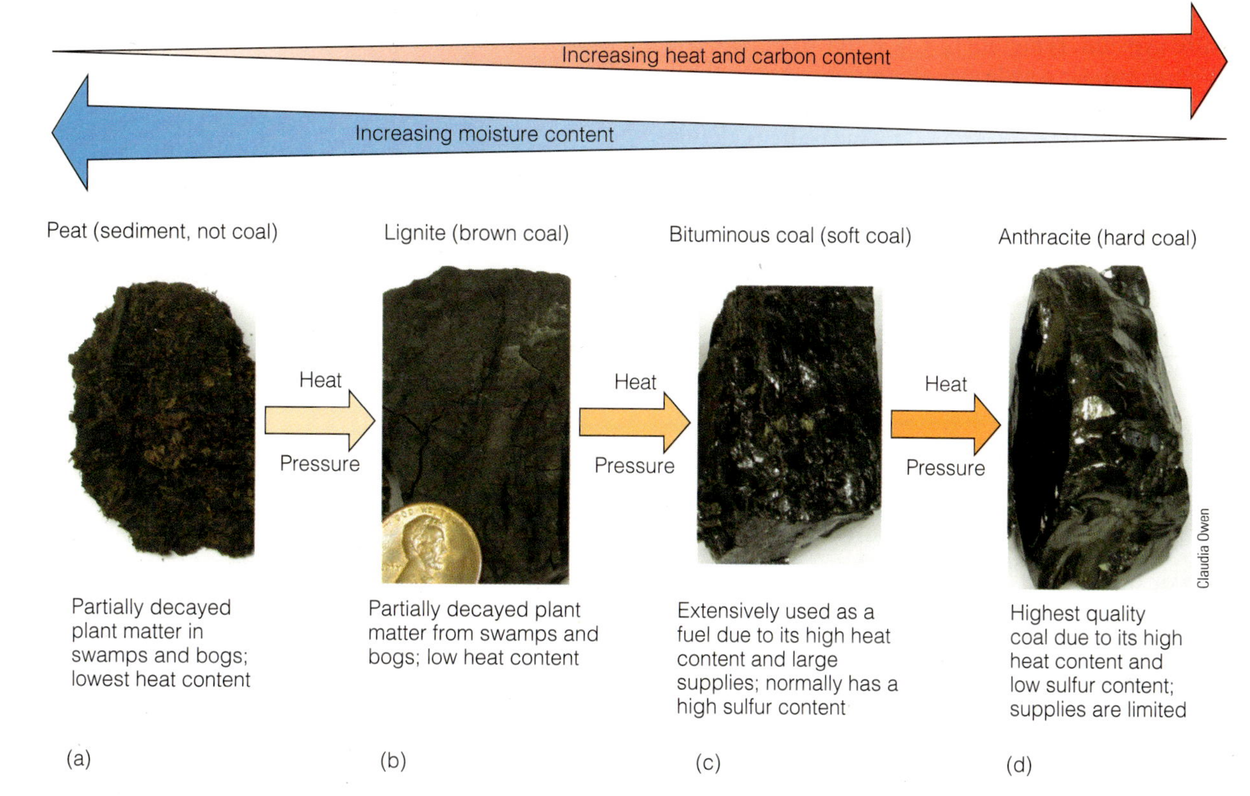

Figure 4.16

Coal Types

Diagram showing the formation of coal from (a) peat to (b) lignite to (c) bituminous coal to (d) anthracite due to increasing heat and pressure. Moisture content decreases and carbon content increases as the temperature and pressure increase. Peat is a type of sediment; lignite and bituminous coal are both sedimentary rocks; and anthracite is a low-grade metamorphic rock.

Evaporites

Evaporites are rocks formed by evaporation and include rock salt, rock gypsum, rock anhydrite, and some crystalline limestones. Evaporite minerals may include halite, gypsum, anhydrite, and sometimes calcite. An interlocking crystalline texture is common for evaporites (Figure 4.1e, p. 72). This texture is very similar to the coarse-grained (*phaneritic*) texture of igneous rocks, but with completely different minerals.

Rock salt is made of halite, which precipitates from water when evaporation concentrates salt (sodium chloride). This may occur in desert lakes (Figure 4.1b) or in an arid marine environment where seawater in a confined basin evaporates. Rock salt can vary in color, but generally is translucent and light colored (■ Figure 4.17). Use the properties of the mineral halite to help you identify rock salt. You know that salt is important in your diet, but did you know large salt beds and salt domes also trap oil and gas reserves?

Rock gypsum is a common evaporite in which gypsum makes up a high percentage of the rock

Figure 4.17

Rock Salt

Three colors of rock salt, all with crystalline texture. A close up of the gray sample is shown in Figure 4.1e, on p. 72.

(■ Figure 4.18). Use your mineral identification skills to recognize gypsum; remember, it is softer than a fingernail. Rock gypsum is a hydrous calcium sulfate with three varieties: alabaster, selenite, and satin spar (Figure 2.9b, p. 28). Anhydrite, the *anhydrous* (lacking water) calcium sulfate, is commonly massive in evaporite beds.

Evaporite Experiment Place a watch glass on a hotplate and half fill it with seawater or saltwater. Allow the water to evaporate without boiling. While you are waiting for it to evaporate, start the next exercise set, then come back and answer these questions.

6. After the water has evaporated, examine the residue on the watch glass under a microscope or with a hand lens.

a. Name and describe the texture and general appearance.

___

___

___

b. Where did the solid material in this sediment come from?

___

c. What natural sedimentary processes do you think this experiment is supposed to simulate?

___

What might be a natural energy source for these processes?

___

Figure 4.18

Rock Gypsum

Rock gypsum showing crystalline texture in the inset.

Chert

Chert is a hard, silica-rich, and smooth rock that has a variety of colors and conchoidal fracture (■ Figure 4.19). It is composed of amorphous silica or super-fine-grained—*cryptocrystalline*—quartz. Chert may form chemically as silica precipitate or silica replacement of limestone. Banded chert (with bedding, Figure 4.19b) is commonly biochemical, formed from deep-sea **silica ooze**, which is sediment made of microscopic silica shells of one-celled organisms such as diatoms or radiolaria (Figure 4.19a inset). Some colored cherts have special names: **flint** is black or dark gray; **jasper** is red; **carnelian** is yellow-to-orange and translucent.

Banded iron-formation is a rock made of layers of chert, often jasper, and hematite or magnetite (■ Figure 4.20). It is an extremely important resource for iron, and is economically important because it is the main ingredient in steel, which is essential in heavy industry. Almost all banded iron-formations are 2 billion years old or older. They probably formed as a result of oxygen increasing due to early photosynthesis in a marine environment, which caused iron oxide to form. Single-celled photosynthesizing cyanobacteria produced

Figure 4.19

Chert

(a) Massive gray chert with inset of a diatom. (b) Bedded chert: this sample is pure chert interbedded with siliceous mudstone.

Figure 4.20

Banded Iron-Formation

Red layers of chert interbedded with silvery specular hematite (dark gray) in a rock formed about 2 billion years ago.

large amounts of oxygen, allowing the reaction between iron and oxygen to precipitate lots of iron out of the seawater. Iron oxide does not dissolve in seawater, so when formed, possibly seasonally, it precipitated, and it sank, creating extensive layers.

7. List five resources of sedimentary origin that you either used or came in contact with today:

____

____

____

Now that we have considered how sedimentary rocks form and explored their various types, let's study how to determine where sediment deposition took place.

SEDIMENTARY ENVIRONMENTS

Sedimentary rocks contain a wealth of information about Earth's surface environments where the sediment accumulates, known as **sedimentary environments** or **depositional environments** (■ Table 4.5). Since some sedimentary rocks date back millions and even billions of years, they can become a kind of time machine, giving us a glimpse of the past long before humans existed.

Table 4.5 summarizes some depositional environments, in three broad categories: **marine** environments involve deposition in the ocean; **coastal,** also called **transitional,** occur where sediment may be under sea water or exposed to air at various times; and **terrestrial** environments, also called *continental* or *nonmarine,* occur on land or in lakes, streams, or swamps. When you identify rocks later in the lab, you can use this table, along with your observations of each rock's characteristics, to help determine the origin and history of the rock. Features that are useful are the texture, including the grain size of the rock, which we have already discussed, and sedimentary structures and fossils, which we discuss in the following sections.

Structures of Sedimentary Rocks

Larger features of sedimentary rocks that can be seen from more than a few feet away are **sedimentary structures.** They can suggest the depositional environment as indicated in the lower half of Table 4.5. **Bedding,** which forms by the deposition of sedimentary particles one layer after another, is the most common type (■ Figure 4.21). Igneous rocks may have flow banding or layers of lava or ash; metamorphic rocks (Lab 5) may have layers of minerals, or *foliation;* but sedimentary bedding is much more common and is usually easy to distinguish from igneous or metamorphic layering.

Bedding may be visible as a result of changes in grain size, constituents, orientation, or color and may be as thin as a few millimeters (**laminar** bedding, Figure 4.21b) or as thick as the height of a two-story building. Changes in bedding units may result from environmental and climate changes or the arrangement or orientation of grains. Sedimentary rocks without visible bedding in hand sample are **massive** Figures 4.6 and 4.10, pp. 78 and 79). As you examine and describe sedimentary rocks at the end of this lab, be sure to describe any bedding you see in the samples. Note what differs in the beds, such as color, grain size, clast type, and cement composition. Other structures common in sedimentary rocks may give us additional facts about the environment of deposition.

Cross-bedding occurs when layers of sediment form at low angles (up to about 35°) to horizontal at deposition (■ Figure 4.22a and c). Migrating sediment in underwater bars or wind-blown dunes becomes cross-bedded as sloping beds accumulate on the lee—protected side—of a dune or bar. The action of wind or water currents produces cross-beds with the layers tapering downward in the direction of the current. This feature of cross-bedding preserves ancient current directions. Erosion typically removes the uppermost layers. Cross-bedding is also an **up indicator;** that

Table 4.5

Sedimentary Environments

The rocks, structures, and fossils likely to be found in different depositional environments.

Depositional Environments

Terrestrial | Transitional | Marine

Mountains — Downhill and downstream — Coast — Offshore

		Land-slide	Alluvial Fan	Playa	Dunes	River Channel	Floodplain	Swamp	Delta	Lagoon	Tidal flat	Beach or bar	Reef	Continental shelf	Continental rise	Abyssal plain
Rock Types	Brevccia	yes	yes			headwaters							lime-stone			
	Conglomerate		yes			upstream						yes				
	Sandstones: Quartz (clean) sandstone				yes							yes				
	Sandstones: Arkose		yes			yes										
	Sandstones: Graywacke (dirty sandstone)					yes			green and red					near shore	turbidites	
	Siltstone						yes	yes	yes	may be sandy	yes					
	Shale			yes			brown or black	brown or black	green or red	brown or black	yes			brown or black	brown or black	red, brown or black
	Limestone									yes	lime mud	coquina	yes	warm clear shallow water		
	Coal						isolated fragments	yes	landward part							
	Rock salt & gypsum			yes						restricted						
	Chert													nodules in limestone		bedded
Structures	Ripple marks				asymmetrical (current ripples)	asymmetrical (current ripples)			asymmetrical (current ripples)	symmetrical ripples near shore	yes	symmetrical (oscillation ripples)		asymmetrical ripples near shore	asymmetrical (turbidity current)	
	Cross bedding		rare		yes	yes			yes			yes				
	Graded bedding					yes			yes						yes	
	Mud cracks			yes			rare				yes					
	Laminar bedding			yes			yes	yes		yes	yes			yes	yes	yes
	Bioturbation				yes			yes	yes	yes	yes	yes		yes	rare	rare
	Marine fossils									brackish water fossils	yes	coquina or broken shells	reef fossils	yes		
	Freshwater fossil shells					broken and rare		rare	rare and fragmented							
	Plant fossils						fragments	yes	landward part							
	Oolites									strong agitation			back reef	strong agitation		

(a)

Parvinder Sethi

(b)

Claudia Owen

Figure 4.21

Bedding

(a) Sequence of thinly-bedded coastal, intertidal layers of mudstone and siltstone, from the Cretaceous Period, exposed in south-central Utah. (b) A hand sample of bedded sandstone with laminations.

is, since the top is different from the bottom, cross-bedding can tell you which way was up when the rock originally formed.

8. Look closely at Figure 4.22a, a photo of lithified, wind-blown dunes from the Arapaho Formation in Zion National Park.

a. With a colored pencil or highlighter, outline the top and bottom of the center bed of cross-bedding (the one where the tree tops end) with a thick line.

b. Then, more lightly, trace the individual cross-beds within it.

c. Draw an arrow indicating the direction, left or right, of the wind that deposited this bed.

d. Why is the top of this bed flatter than the bottom?

____

e. Did the wind blow consistently in the same direction for all of the beds in the photograph? ______

Graded bedding occurs in a *single* layer or bed of sediment or sedimentary rock in which the largest grains are concentrated at the bottom, gradually decreasing in size upward to the smallest at the top of the bed. Graded bedding is an up indicator (Figure 4.22c and d). Graded bedding indicates an ebbing (diminishing) current because a whole layer of graded bedding is deposited in a single episode. It might occur where a river enters a lake or floodplain or as a result of a submarine landslide, which generates a sediment-laden current (**turbidity current**) that sweeps down the continental slope to slowly settle out on a submarine fan. Graded bedding suggests a transition from a high-energy to a low-energy environment.

Ripple marks are wavy sedimentary structures that may be either symmetrical or asymmetrical. Oscillating currents, such as coastal waves, form **symmetrical ripples.** Symmetrical ripples have cusp-shaped crests and rounded troughs, making them good up indicators (Figure 4.22c). One-directional currents form **asymmetrical ripples** (Figure 4.22b) and may also produce mini cross-bedding.

Mud cracks (Figure 4.22c and e) indicate a drying environment, wet to dry, that can occur in tidal flats, floodplains, and desert lakes (Table 4.5). They are up indicators because the cracks open upward and mud layers may curl up along the edges. Rain may make small indentations—**raindrop prints**—as it falls on soft sediment. When preserved in rocks these are also up indicators. Raindrop prints and mud cracks can be in the same rock.

Make Your Own Sedimentary Structure

Materials needed:

- Jar, 250 ml (8 oz.) or larger, with lid
- Water
- Sediment of various sizes

Fill the jar roughly halfway with sediment and most of the way with water, or use a jar already filled with sediment and water. Cover the jar with the lid and shake it until the sediment is well-mixed with the water. Then set the jar down.

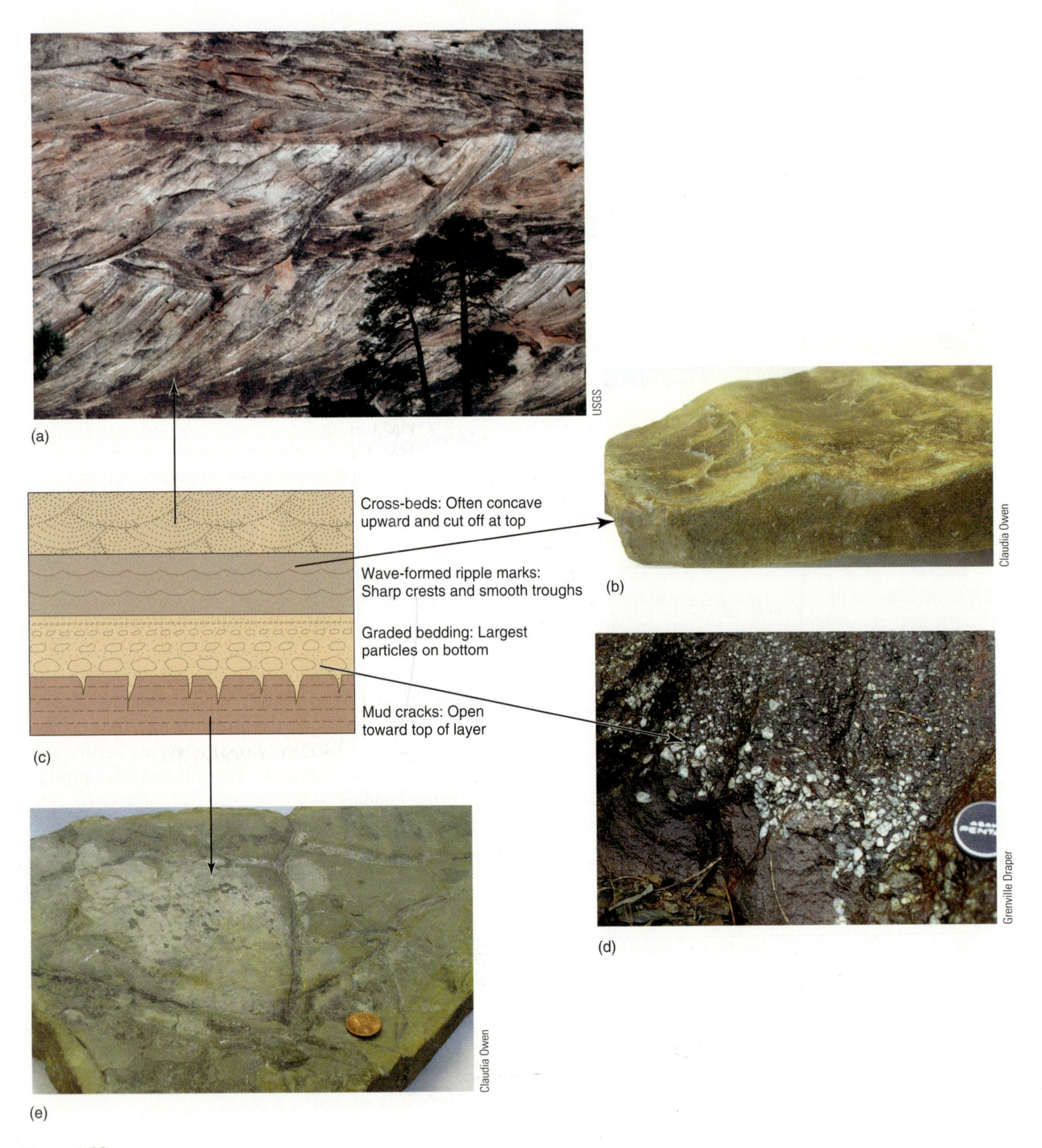

Figure 4.22

Sedimentary Structures

(a) Cross-bedding in the Arapaho Formation in Zion National Park, Utah. (b) Ripples in sandstone. (c) Sketch of the side view of some sedimentary structures that are also up indicators. The cross-bedding, ripple marks, graded bedding, and mud cracks shown each have some aspect that distinguishes the top from the bottom. (d) Graded bedding of milky quartz clasts in a darker sand matrix in a conglomerate bed of the Wilhite Formation, Tuckaleechee Cove, Tennessee. (e) Mud cracks in shale. After the mud cracked, sediment filled the cracks. In all of these photos, the top is up.

9. Observe what happens in the jar.

   a. Sketch the bed deposited in the jar, paying particular attention to the grain-size variation of the sediments. Include a 1" scale bar on your sketch of the sediment.

   b. What sedimentary structure did you make? ______________________

   c. What does this experiment tell you about the speed of large and small grains as they fall through water?

      Large- fast, Small slow.

   d. What natural environment could produce the mixing of sediment you achieved by shaking the jar?

      riverbeds.

   e. What natural environment would allow mixed material to settle all at once?

      Lake, lagoon, [illegible]

Fossil as Clues in Depositional Environments

A fossil can tell you the depositional environment of the sediment with which it accumulated, sometimes the climate at that time, and the history of the rock in which it occurs. It may also indicate the age of the rock. For example, the trilobite in Figure 4.5b on p. 77 tells us the rock is Devonian (early to middle Devonian, 411 to 392 million years old) because the organism did not exist before or after that time period and it is now extinct. Fossils also give us information about the evolution of organisms and animal behavior, two topics that are beyond the scope of this book. Trilobites, such as the one in Figure 4.5b, can now be found in sedimentary rocks in mountainous areas. Because trilobites were similar to some tropical crabs existing today, we know they swam in a warm, shallow sea. That they are now in rocks high in the mountains tells us that the sediment was buried, turned into rock, and then forced upward by plate tectonics—a history of millions of years. Table 4.5 proposes some ideas about depositional environments that you can obtain from fossils. When working with a fossil, use what you know about modern organisms to make an educated guess about the type of environment the fossil suggests.

10. Examine the samples of sedimentary structures and/or fossils provided by your instructor.

    a. Fill out ■ Table 4.6 with the sample number and name of each structure or fossil and sketch it. If it is a fossil, what organism is it, or is it similar to an organism living now?

    b. If the structure is an up indicator, show which way was up on your sketch.

    c. If the structure reveals the direction of the current, draw an arrow that direction on your sketch.

    d. Include a 1" scale bar on your sketch. Note the rock type.

    e. From the fossils or structures in your sketches in Table 4.6, what additional information can you add about the depositional environment in which the organism(s) lived or the structure formed? Consider various environments based on habitats of similar living organisms and climates conducive to these organisms, or how structures form (bottom part of Table 4.5).

IDENTIFICATION AND DESCRIPTION OF SEDIMENTARY ROCKS

Sedimentary rocks, like igneous rocks, can be identified using a combination of texture and composition. Proper description of sediments and sedimentary rocks also includes general observation and description of sedimentary structures and fossils. All of these characteristics help to interpret a rock's environment and history.

In the next exercise you will need to determine the name of each sedimentary rock; the maze in ■ Figure 4.23 can help you. Begin where it says "Start Here." Notice that right away you need to make decisions based on

Table 4.6

Fossils and Sedimentary Structures

Space to sketch and describe fossils and sedimentary structures of samples provided by your instructor.

Sample	2 Sample Number ________ 3		Sample Number ________ 43 3	
Fossil or Sedimentary Structure Name	Brachiopod	Parallel bedng	Carbonaceous shale	Ferruginous Sandstone
Description/ Sketch	* Hardend shell	*Layers form of sediments *Load casts *Cross-bedding	*white organic material	*Sediments of iron
Rock Type	Biochemical	Chemical [illegible]	Shale	Chem[illegible]
Sedimentary Environment and Other Information	Environment-ocean			

the composition of the rock. Farther into the maze you will have to make decisions involving the rock's texture and do additional composition testing. At each fork in the maze, observe a property of the rock or make the test requested to decide which way to go. You may want to try using the maze with a known sample or two first, progress to unknowns, and then test yourself on samples without the maze to see how much you have learned of the identification process.

11. Examine each unknown rock specimen with and without a hand lens. Fill in Table 4.7, "Sedimentary Rock Identification Form," as the column headings indicate. We have listed some additional suggestions next:

Texture: Choose one of the following textures: (1) **clastic** (made of broken pieces of silicate rocks or minerals); (2) **bioclastic** (made of remains of organisms); (3) **crystalline** (interlocking crystals); or just (4) **fine-grained** (the grains are too small to see with the naked eye); (5) **cryptocrystalline** or **amorphous** (very smooth). For clastic textured rocks, also describe the sorting and rounding of the grains.

Grain size: Use the terms for grain types in Table 4.1 on p. 73.

Fossil size: If fossils are present, measure and record the long dimension of an average fossil in centimeters or millimeters.

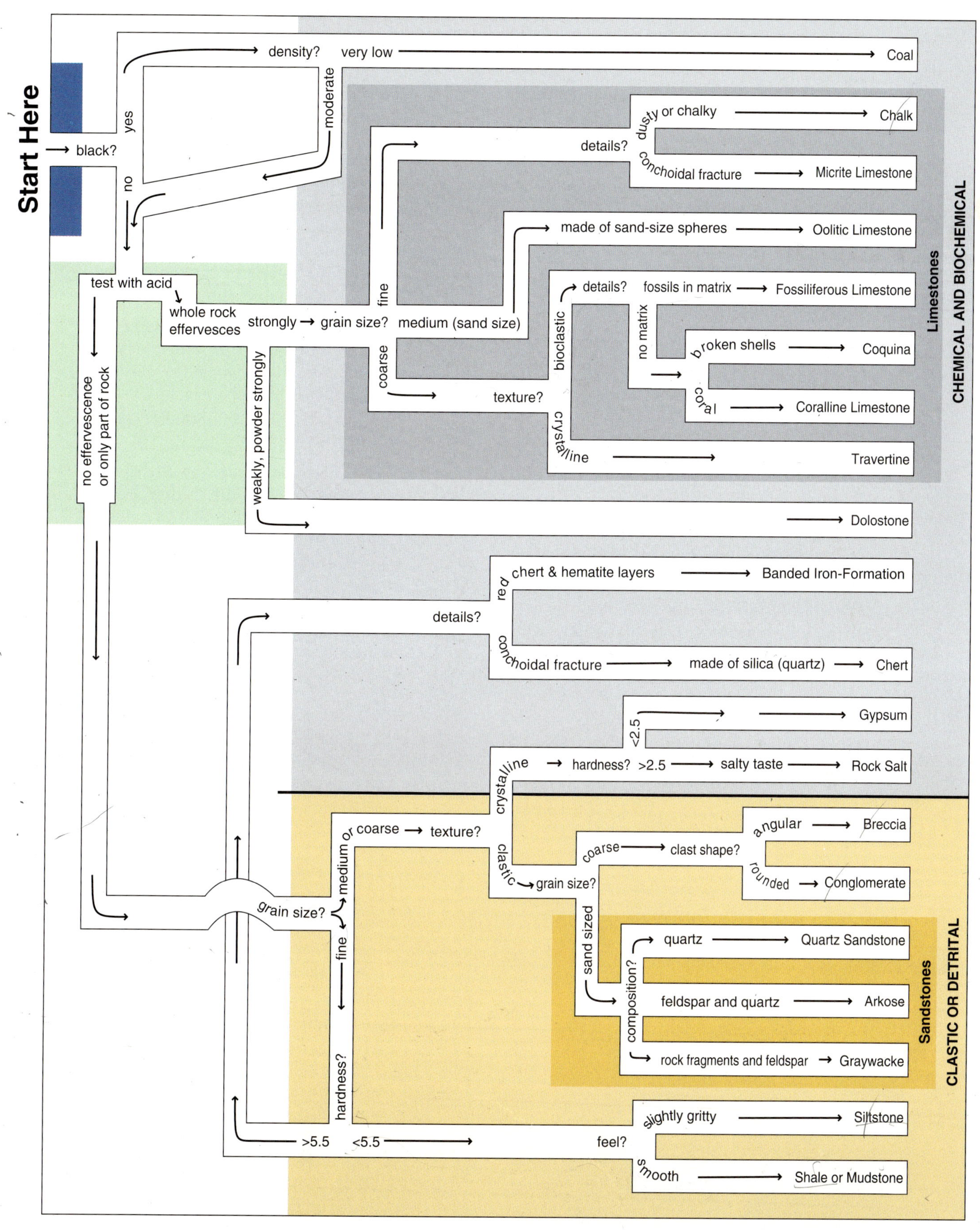

Figure 4.23

Maze for Identification of Sedimentary Rocks

Use the instructions in the text.

Structures and fossils: List any sedimentary structures, features, or fossils. Include bedding (parallel, cross, graded), fossils, mud cracks, ripple marks, raindrop prints, and so forth. Describe any *fossils* present, identifying what animal or plant is preserved, if possible, or describing its characteristics.

Composition: Indicate minerals or rock clasts in the rock or its chemical composition. For example, limestone is predominantly calcite, a chert conglomerate contains chert clasts, and coal is made of organic material. Composition is especially important for the classification of nonclastic sedimentary rocks. Determine composition by testing for the distinguishing properties of the minerals commonly found in sedimentary rocks listed on page 72 and in Table 4.2 (p. 74). For coarse clastic rocks, you should attempt to identify the rock type of the clasts.

Color: Carefully describe the color, giving some specifics. The color of any sedimentary rock may simply help to distinguish the rock from another similar rock, or it may have a greater significance. For example, reddish to yellowish colors indicate the presence of oxidized iron and may suggest a non-marine origin for the rock. Black often results from abundant organic matter, and gray and greenish color hints at a lack of oxygen in some lake or marine environments.

Rock name: Identify the rock using the maze in Figure 4.23, and check your identification against the descriptions earlier in this chapter and, if available, against samples of known sedimentary rocks.

Origin: Briefly indicate how the rock may have formed as the column heading suggests. Keep in mind that virtually all sedimentary rocks form by weathering, erosion, transport, deposition, burial, and lithification, so give details of these steps.

Table 4.7

Sedimentary Rock Identification Form

Use the instructions within the text in Exercise 11.

Sample Number	Texture (see instructions) Grain and Fossil Size (use terms from Table 4.1)	Structure and Fossils (describe bedding, describe fossils, etc.)	Composition and Color (minerals, rock clasts, or chemical makeup)	Rock Name	Origin	
					How Did the Rock Form? (describe sediment source, depositional environment, lithification)	Clastic/ Biogenic/ Chemical
176	Bioclastic	Broken shell	coarse calcium carbonate	Coquina	high energy coastal environment	Biogenic
77	clastic coarse grain	No Fossils	rock fragments rounded rock	Conglomerate	terrestrial coastal	clastic
95	fine grain clastic	No particular structure	silt-size particles	Siltstone	terrestrial or marine	clastic
92	oolitic texture sand	spherical bodies	calcium carbonate	oolitic limestone	coastal	Biogenic Chemical
74	clastic coarse grain	no bedding	Rock clasts	Breccia	terrestrial environment	clastic
24	crypto-crystalline		silica a form of Qz	chert	marine	Biogenic Chemical
290	fine grain or cryptocrystalline	Fossils	calcium carbonate	marine limestone	marine	Biogenic or Chemical

(Continued)

Table 4.7

Sedimentary Rock Identification Form—*Continued*

Sample Number	Texture (see instructions) Grain and Fossil Size (use terms from Table 4.1)	Structure and Fossils (describe bedding, describe fossils, etc.)	Composition and Color (minerals, rock clasts, or chemical makeup)	Rock Name	Origin	
					How Did the Rock Form? (describe sediment source, depositional environment, lithification)	Clastic/ Biogenic/ Chemical
73	Bioclastic fine to medium	Fossil	Calcium carbonate	Bryozoan Limestone	Marine	Biogenic
296	Bioclastic	fossil	calcium carbonate	Coralline Limestone	Marine	Biogenic
63	clastic medium grained		Quartz	Sandstone	terrestrial, coastal	clastic
218	fine grained	w	Calcium carbonate	Chalk	marine	Biogenic
197	clastic fine-grain	No bedding	clay, blue	Shale	terrestrial (lakes) or marine (open ocean)	clastic

LAB

Metamorphic Rocks

OBJECTIVES

- To understand how metamorphic rocks form
- To identify metamorphic minerals and understand how composition affects the rocks
- To recognize metamorphic textures and understand their origin
- To recognize and be able to identify major metamorphic rock types
- To understand the concepts of metamorphic grade and zones

We do not actually see **metamorphism,** the creation of metamorphic rocks, because it takes place entirely underground, generally at high pressure. This makes metamorphic rocks the most mysterious of the three classes of rocks. The heat from Earth's interior that drives plate tectonics is also the cause of metamorphism. In addition to the change the heat produces directly, plate movement can cause stress and deformation and can move rocks deeper within the lithosphere where the pressure is high. Each of these processes is capable of transforming solid rock. The information extracted from metamorphic rocks is especially useful to **structural geologists**—geoscientists who study tectonics and rock deformation. **Deformation** is the change in shape of rocks after they have formed.

METAMORPHIC PROCESSES AND TYPES OF METAMORPHISM

The motion of the lithospheric plates may drag rocks from Earth's crust into a subduction zone, deform them beneath a volcanic arc, or involve them in a continental collision. Such *tectonic activity*[1] subjects the rocks to substantial increases in temperature and pressure. At the new temperature and pressure, the original minerals in the rocks can become unstable, break down, and recrystallize into a new set of minerals, without melting. This process, **metamorphism,** forms metamorphic rocks and changes both the minerals and the texture of the rock. The new minerals in a metamorphic rock may only be stable in a narrow range of temperatures and pressures, so we can use them as indicators of these conditions and the depth of metamorphism. The composition of the original rock—the **parent rock** or **protolith**—also influences the mineral constituents of the final metamorphic rock.

The temperature of metamorphism can range from about 200°C to about 900°C. The upper limit of metamorphism is melting, which produces igneous rocks, and is usually lower than 900°C. However, the

[1] Changes in the broad architecture of the outer part of the Earth, especially movement and deformation of plates.

melting temperature depends on pressure and chemical composition. Temperatures in the Earth increase about 25°C for every kilometer of depth, and this increase in temperature with depth is one of the major agents of metamorphism.

For a rock at depth, the weight of the overlying mass of rock causes **confining pressure,** which squeezes the rock in all directions. The general effect of confining pressure is to reduce the volume, causing the rock to have higher density (■ Figure 5.1a and b). Confining pressure directly relates to depth of metamorphism and is similar to the pressure scuba divers experience as they descend into deep water. A common unit for atmospheric pressure is the bar, which is approximately the same as an atmosphere. Standard pressure is one atmosphere (1 atm) or 1.01325 bars at Earth's surface at sea level; but varies slightly depending on the weather conditions. Greater pressures may be measured in kilobars (kb); 1 kb is 1,013.25 times atmospheric (atm) pressure. Pressure increases about 3 kb for every 10 km depth. The rocks we see at Earth's surface rarely reached depths greater than 40 km (12 kb).

> **1.** Use the caption of Figure 5.1a. Calculate the pressure in kilobars and list it in the blanks above the right two cups. Indicate how many times this is than standard pressure at sea level.

As the temperature and pressure of metamorphism increase, the **metamorphic grade** of the rock also increases. You can think of metamorphic grade as an approximate measure of the amount or degree of metamorphism. The three metamorphic grades relate closely to the temperature of metamorphism: **low grade** is mainly low temperature and low pressure, **medium grade** is moderate temperatures and pressures, and **high grade** is for rocks that have undergone metamorphism at high temperatures and pressures. Pressure is of less importance than temperature, so a rock metamorphosed at high temperature and moderate pressure, for example, would still be high grade.

Movement of lithospheric plates can produce a kind of pressure called **differential pressure**—also called **differential stress** or sometimes **directed pressure** (see Figure 5.1c and d)—for which, unlike confining pressure, the squeezing of the rock is not equal in every direction. Differential pressure flattens or stretches rocks, resulting in the formation of a parallel arrangement of minerals called **foliation**. Sometimes differential pressure may cause flattening or stretching of pebbles or fossils in a rock.

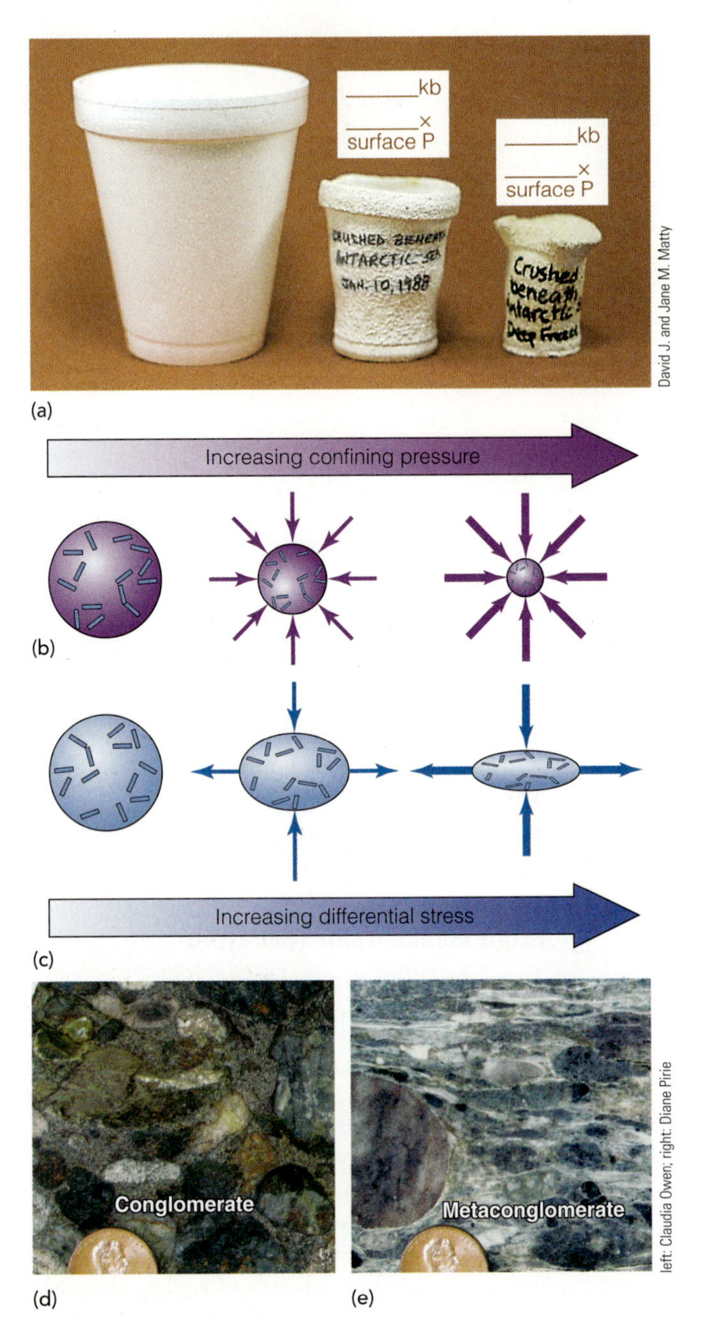

Figure 5.1

Types of Pressure

The difference between confining and differential pressure. (a) Confining pressure (called hydrostatic pressure in the ocean) compressed the middle and right Styrofoam cups when they were lowered to 750 m and 1,500 m depth in the ocean, respectively. The pressure, equal in all directions, increases at a rate of ~100.5 bars/km in the ocean, causing the cups to shrink and become denser, but retain their basic shape. The confining pressure within the crust affects rocks in a similar way at a higher rate. (b) Increasing confining pressure, from left to right, causes a reduction in volume and consequent increase in density. (c) Increasing differential stress from left to right causes increased flattening of a rock. (d) Unstressed conglomerate, a sedimentary rock. (e) Stretched pebble conglomerate that has experienced differential stress, causing the pebbles to become flattened.

Regional Metamorphism

Mountain building and plate tectonic processes acting over large regions deep within Earth's crust (Labs 1 and 9) produce **regional metamorphism.** Two important types of regional metamorphism are *subduction zone metamorphism* and *orogenic metamorphism*. Because of the deformation involved in these types of metamorphism, regional metamorphism may also be referred to as **regional dynamothermal metamorphism** ("dyanamo-" refers to the deformation and "-thermal" refers to the heat). This type of metamorphism commonly produces the rocks: slate, schist, and gneiss.

Subduction zone metamorphism occurs in subduction zones at convergent plate boundaries. The *subducting* (descending) plate is cold in relation to its surroundings. As a result, minerals that form at high pressures but low temperatures characterize this type of regional metamorphism.

Orogenic metamorphism results from mountain building during plate collisions. Differential pressure, combined with a wide range of temperatures and confining pressure at moderate to great depths, produces this type of metamorphism. Temperature increases with depth, as does the confining pressure.

Contact Metamorphism

Heat from an intrusion produces **contact metamorphism**, a variety of *thermal metamorphism* (■ Figure 5.2). The metamorphic rock produced around the pluton is the **contact aureole.** The aureole may be a few centimeters to several kilometers thick, depending on the size, composition, and depth of the intrusion. The greater the size and the higher the temperature of the intrusion and its surroundings, the thicker the contact aureole will be. Most contact metamorphic rocks are hornfelses, quartzites, marbles, or skarns, depending on the composition of the rock. These rocks are not foliated because contact metamorphism is due to heat, not differential stress. The type of metamorphism—regional or contact—will influence the texture of the rock.

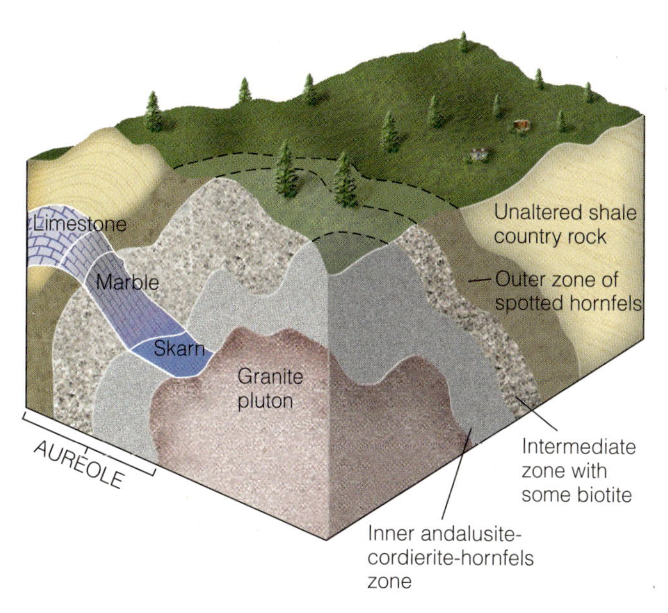

Figure 5.2

Contact Metamorphism

Increased baking toward the intrusion (granite pluton) resulting from **contact metamorphism** of shale *country rock* produces hornfelses with distinct mineral assemblages in different zones **(contact aureole).** A limestone layer has also experienced contact metamorphism, becoming marble. Near the intrusion, skarn occurs, where silicate-rich fluids from the magma react with the limestone. **Skarn** is a contact metamorphic rock rich in calcium-silicate minerals. From WICANDER/MONROE, *Essentials of Physical Geology*, 5e 2009 Brooks/Cole, a part of Cengage Learning Inc. Reproduced by permission. www.cengage.com/permissions

CLASSIFYING AND NAMING METAMORPHIC ROCKS

As for the other rock types, the classification of metamorphic rocks involves both texture and composition, but the varied features and differing origins of metamorphic rocks make them more complicated to classify than igneous and sedimentary rocks. The deformational history of the rock affects its texture, the parent-rock affects its chemical composition, and the temperature and pressure of metamorphism and the parent-rock affect its mineral makeup. As a result, we classify and name metamorphic rocks according to three different aspects of the rock: texture, mineral content, and sometimes the rock's protolith (parent-rock). It is common to list the major distinctive minerals in a metamorphic rock at the beginning of the rock name. The convention for listing minerals in a metamorphic rock name is to list the most common ones last, with hyphens between the mineral names.

COMMON MINERALS OF METAMORPHIC ROCKS

The **parent-rock**, or **protolith** (the rock from which a metamorphic rock forms), significantly influences both the mineral makeup and the chemical composition of the final metamorphic rock (■ Table 5.1), but the minerals in the rock also depend on the temperature and confining pressure of metamorphism. In the table, *carbonate minerals and calcsilicates* are from parent-rocks

Table 5.1

Metamorphic Minerals

Associations among parent rock, chemical composition, and metamorphic minerals. Photographs show common porphyroblasts (large crystals) and metamorphic minerals not shown in Lab 2. For other minerals refer to Lab 2 for photographs. Crystal habits in metamorphic rocks can influence whether a visible foliation develops or not in a rock undergoing dynamothermal metamorphism. Equant and blocky minerals tend not to produce a foliation, but platy, prismatic, or acicular minerals can. Minerals in bold are common rock-forming minerals that are important constituents in their parent rocks.

Rock Composition (parent rocks)	Metamorphic Mineral (bold = abundant)	Photographs of Porphyroblasts and New Metamorphic Minerals	Habit
Carbonate and calcsilicates (limestone & dolostone)	**Calcite**	See Figure 2.9a, p. 28	Stubby
	Dolomite	See Figure 4.15c, p. 82	Stubby
	Diopside (Ca pyroxene)	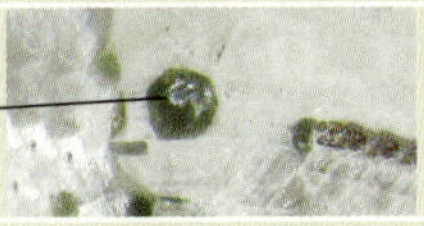	Stubby
	Actinolite (green amphibole)	See mafic rocks (right)	Prismatic or acicular
	Garnet (grossular, Ca-rich)		Equant
	Wollastonite (calcium silicate)		Primatic or acicular
High Aluminum, Metapelities (shale, mudstone)	Chlorite	See mafic rocks (right)	Platy or tabular
	Muscovite mica	See Figure 2.8f, p. 26	Platy or tabular
	Biotite mica	See Figure 2.8e, p. 26	Platy or tabular
	Garnet (almandine-pyrope, Fe-, Mg-rich)		Equant
	Staurolite		Primatic
	Kyanite		Bladed
	Sillimanite		Prismatic or acicular
Silica-rich (quartz sandstone & chert)	**Quartz**	See Figure 2.8j-t, p. 27	Stubby (in metamorphic rocks)

Table 5.1

Metamorphic Minerals—*Continued*

Rock Composition (parent rocks)	Metamorphic Mineral (bold = abundant)	Photographs of Porphyroblasts and New Metamorphic Minerals	Habit
Felsic (granite, rhyolite, tuff, and obsidian)	**Quartz**	See Figure 2.8j-t, p. 27	Stubby
	K feldspar		Stubby
	Plagioclase feldspar	See Figure 2.8h, p. 27	Stubby
	Muscovite	See Figure 2.8f, p. 26	Platy or tabular
	Hornblende	See Figure 2.8d, p. 26	Prismatic
	Biotite	See Figure 2.8e, p. 26	Platy or tabular
Mafic (basalt and gabbro parent rocks)	Chlorite		Platy or tabular
	Epidote		Stubby
	Actinolite (green amphibole)		Prismatic or acicular
	Hornblende (black amphibole)	See Figure 2.8d, p. 26	Prismatic
	Plagioclase feldspar	See Figure 2.8h, p. 27	Stubby
	Blue amphibole (glaucophane-crossite, blue-black sodium-bearing amphibole)		Prismatic
	Green pyroxene (omphacite, sodium- and aluminum-bearing)		Blocky
Ultramafic (peridotite)	Serpentine		Platy or fibrous
	Talc		Platy or tabular

Claudia Owen; Diane Pirie

such as limestone and dolostone, and in some cases, their interaction with granitic intrusions. *High aluminum minerals* are from parent rocks such as shale. Likewise, during metamorphism, the metamorphic varieties of *felsic minerals* form from felsic rocks, *mafic* and associated minerals from mafic rocks, and *ultramafic minerals* from ultramafic rocks. Table 5.1 lists minerals in each of these groups. It should also be noted that the precise mineralogy of metamorphic rocks may reveal temperature and pressure, which we will discuss later.

A characteristic of many metamorphic minerals is that they have crystal habits that aid in the formation of foliation, such as platy (tabular), bladed, prismatic, or acicular habits (see Lab 2). Table 5.1 also lists the habits of some common metamorphic minerals.

TEXTURES OF METAMORPHIC ROCKS

The **texture** of a metamorphic rock (■ Figure 5.3)—the shape, size, orientation, and arrangement of grains—is closely related to the rock's history, just as we have seen for igneous and sedimentary rocks. As for all rocks, the classification of metamorphic rocks depends in part on their texture.

The grain size of metamorphic rocks varies from fine- to coarse-grained. **Fine-grained** rocks have grains too small to see unaided. **Medium-grained** rocks have visible grains that are smaller than 2 mm. In **coarse-grained** rocks, the grains are larger than 2 mm. These grain-size divisions are the same as for sedimentary rocks, for which medium grains are sand-sized. The following sections describe some textures found in metamorphic rocks.

Foliation Textures

Differential pressure during dynamothermal metamorphism causes the arrangement of platy minerals, such as micas, to orient parallel to planes in the rock (Figure 5.3a and c). This parallelism of mineral grains, or **preferred orientation,** is **foliation**. A number of common metamorphic rocks have foliation, as seen in Figures 5.5–5.8 (pp. 103–105). Table 5.1 lists some platy metamorphic minerals. **Lineation** is another type of preferred orientation. A **lineated** rock has prismatic or acicular minerals (Table 5.1), such as hornblende, arranged parallel to a line within the rock and may be combined with foliation (see Figure 5.8a). Some rocks with foliation, as we will discuss later, are *slate*, *schist*, and *gneiss*.

Experimental Formation of Metamorphic Textures

2. Take a half-fist-size blob of Play-Doh® and push the edges of a number of pennies (or buttons, or washers) into it randomly, so about half the penny is sticking out (see ■ Figure 5.4). The pennies represent platy materials. Now squeeze the Play-Doh® against the tabletop with your hand. This is a simulation of deformation during metamorphism of a rock.

a. Examine Figure 5.1 to help you determine what type of pressure you applied to the Play-Doh®.

maximum stress

(a) Foliation

(b) Granoblastic Texture

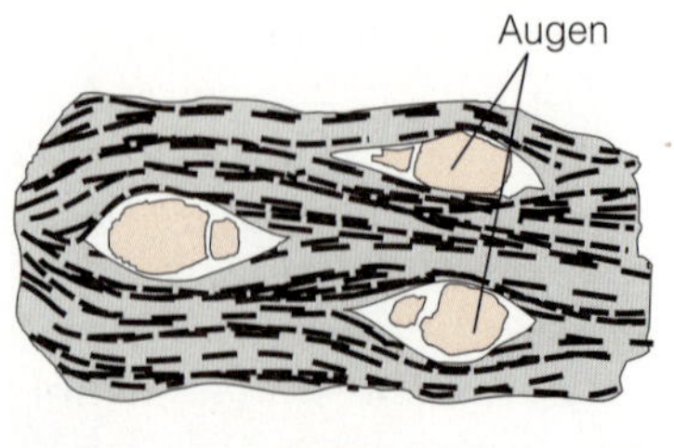

(c) Porphyroblastic Texture with Foliation

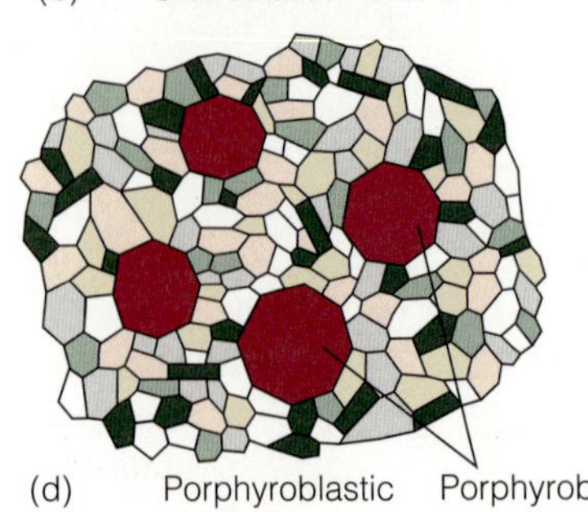

(d) Porphyroblastic Texture

Figure 5.3

Metamorphic Textures

(a) Foliation—interlocking mineral grains showing parallelism to planes in the rock. (b) Granoblastic texture—interlocking grains arranged at random in the rock. (c) Porphyroblastic texture—large crystals surrounded by smaller crystals in a foliated rock where the porphyroblasts are elongated and eye-shaped *(augen)*. (d) Porphyroblastic texture in an otherwise granoblastic, or nonfoliated, rock.

Figure 5.4

Play-Doh® Deformation Experiment

Play-Doh® and pennies, buttons, or washers in the configuration suitable for Exercises 2 and 3 in Experimental Formation of Metamorphic Textures.

b. Sketch and describe the orientation of the pennies/buttons/washers in the Play-Doh®.

c. What is the name of the texture you created? foliated

3. Compare your pennies/buttons/washers in Play-Doh® with the marble and schist provided by your instructor. Which of these two rocks exhibits some similarity to your simulated Play-Doh® rock? ____________ Why?

____

____

____

4. Collect some cards and pencils for the next exercise.

a. Spread the cards across the tabletop randomly. What type of texture is this?

____

b. Scatter some pencils randomly across a large tray. What type of texture is this?

____

c. Make a bundle of the pencils. What type of texture is this? ____________

d. Roll out the bundle of pencils onto a tray. What two textures do the pencils represent? ____________

____

Porphyroblastic Texture

This type of texture (Figure 5.3c, d) occurs in metamorphic rocks that contain large crystals (**porphyroblasts**) surrounded by smaller ones, and is similar in appearance to *porphyritic texture* in igneous rocks. Porphyroblastic texture can occur in either foliated rocks (Figure 5.3c), granoblastic rocks (Figure 5.3d), or rocks that are otherwise just fine-grained (Figure 5.10c). Porphyroblastic texture generally occurs because some metamorphic minerals grow much faster than others do.

Foliated Rocks

Metamorphic petrologists, the geoscientists who study metamorphic rocks, classify some metamorphic rocks based on the details of their foliation texture (■ Table 5.2).

Slate is so *fine-grained* that minerals are not visible without magnification. This low-grade metamorphic rock breaks along fairly smooth planes of foliation called **slaty cleavage** (■ Figure 5.5a). It may have a wide range of possible colors.

Phyllite is *fine- to almost medium-grained*. The cleavage of phyllite has a sheen and may have a **crenulated** (slightly rippled) surface (Figure 5.5b). Micas, chlorite, and possibly graphite lying parallel to the foliation are responsible for the sheen and the cleavage of this low-grade rock.

Schists are *medium- to medium-coarse-grained* and have a texture called *schistosity*. **Schistosity** occurs where individual grains are visible, are abundantly platy, tabular, or acicular, and are arranged parallel to each other, forming a foliation. Another defining characteristic of schistosity is the property of breaking parallel to the foliation that makes the rock flattish. The foliation surface often contains mica and has a sparkly look, like glitter (■ Figure 5.6). Schists vary from moderately low to medium grade depending on their minerals. **Mica schist** has both biotite and muscovite. It is common to see schists with *porphyroblasts* of garnet, staurolite, or other minerals. **Garnet–mica schist** is schist with garnet commonly as porphyroblasts, and biotite and muscovite micas bending around the garnet (Figure 5.6b). **Staurolite-mica schist** (or *staurolite-garnet-mica schist*) has micas (and garnet) with staurolite porphyroblasts (see staurolite in Table 5.1). The parent-rocks of these schists are shales or mudstones (Table 5.1).

Table 5.2

Foliated Metamorphic Rocks

We define most of these rocks based on their texture. Amphibolite, an exception, is commonly foliated, but its name reflects its minerals rather than its texture. Mineral names appear in front of the rock names as modifiers. Textures in **bold** are diagnostic.

Texture	Grain Size	Color or Luster	Other Identifying Characteristics	Rock Name and Grade	Common Parent Rock
Good **slaty cleavage**	Very fine-grained (not visible)	Dull luster, gray, black, red, tan, or green	Generally flat	**Slate** low grade	Shale (high aluminum)
Cleavage (may be wavy)	Fine-grained (not, to barely visible)	Silky sheen on flat surfaces, gray to green	Chlorite & muscovite barely discernible with hand lens	**Phyllite** low grade	Shale (high aluminum)
Schistosity (may be wavy or folded) ± **porphyroblastic**	Medium- (to coarse-) grained (visible grains, usually <~4 mm)	Flat surfaces sparkly. Colors variable including greenish, silvery white, gray, black	Minerals visible: biotite, muscovite, *may* see quartz, garnet, staurolite, kyanite	**Schist** **Mica schist** **Garnet-mica schist** Staurolite-mica schist medium grade	Shale (high aluminum)
Schistosity or gneissic banding. (Some are **porphyroblastic**.) Hornblende parallel to foliation or lineation	Medium- (to coarse-) grained (visible grains, usually <~4 mm)	Black and white or gray	Hornblende, and plagioclase, may see quartz, garnet, biotite, epidote	**Amphibolite** medium to high grade	Basalt or Gabbro (mafic)
Schistosity. Chlorite, muscovite, actinolite parallel to foliated	Medium-grained (visible grains)	Green	Actinolite, epidote, chlorite, may see muscovite, quartz, plagioclase	**Greenschist** low grade	Basalt or Gabbro (mafic)
Foliated or lineated	Medium-grained (visible grains)	Blue or blue gray ± green or white	Blue amphibole, may see white mica, lawsonite, epidote, garnet	**Blueschist** high pressure, low temperature	Basalt or Gabbro (mafic)
Schistosity	Medium-grained	White and pearly	Talc, hardness = 1	**Talc Schist,** Medium grade	Peridotite (ultramafic)
Gneissic banding. Layered, banded or streaked; breaks across foliation	Coarse- (to medium-) grained (usually > 1 mm)	Bands or separation of black and white or pink	Light colored bands of feldspar & quartz, black bands of biotite or hornblende, ± garnet, ± feldspar augen	**Gneiss** or **Augen gneiss** high grade	Granite or shale or basalt

Greenschist and blueschist have mafic protoliths (parent rocks) and talc schist has an ultramafic protolith (Table 5.2). **Greenschist** contains chlorite, muscovite, and actinolite (Figure 5.6c). Moderately low temperature and pressure metamorphosed this mafic rock. **Blueschist** (Figure 5.6d) contains blue amphibole (Table 5.1) and was metamorphosed at high pressure and low temperature in a subduction zone. **Talc schist** is composed of foliated talc, which has a hardness of 1, with crystals big enough to see without magnification. A fingernail can easily scratch it (Figure 5.6e).

Gneiss is *medium- to coarse-grained*, generally coarser than schist. Its foliation, called **gneissic banding,** consists of alternating segregations, streaks, or bands of granular minerals and platy or elongate minerals, often making light and dark streaks (■ Figure 5.7). Gneissic banding is the defining characteristic of gneiss, with many of the segregations typically wider than about 2 mm or 3 mm. Gneiss does not generally tend to break along its foliation as schist does, so it is generally blockier and not at all flaky as schist can be. A very common type of gneiss consists of feldspar and quartz,

(a) (b)

Figure 5.5

Metamorphic Rocks with Slaty Cleavage

(a) **Slate:** These samples are fine-grained and show their foliation at the edges and by their flatness. Slate has a large number of possible colors. Some of the common colors—red-brown, gray, or gray-green—are shown here. (b) **Phyllite:** Graphite imparts the same gray color to these particular phyllites, but the left sample appears lighter because light reflecting off its foliation shows the sheen. Wavy foliation **(crenulations)** is visible in the inset sample.

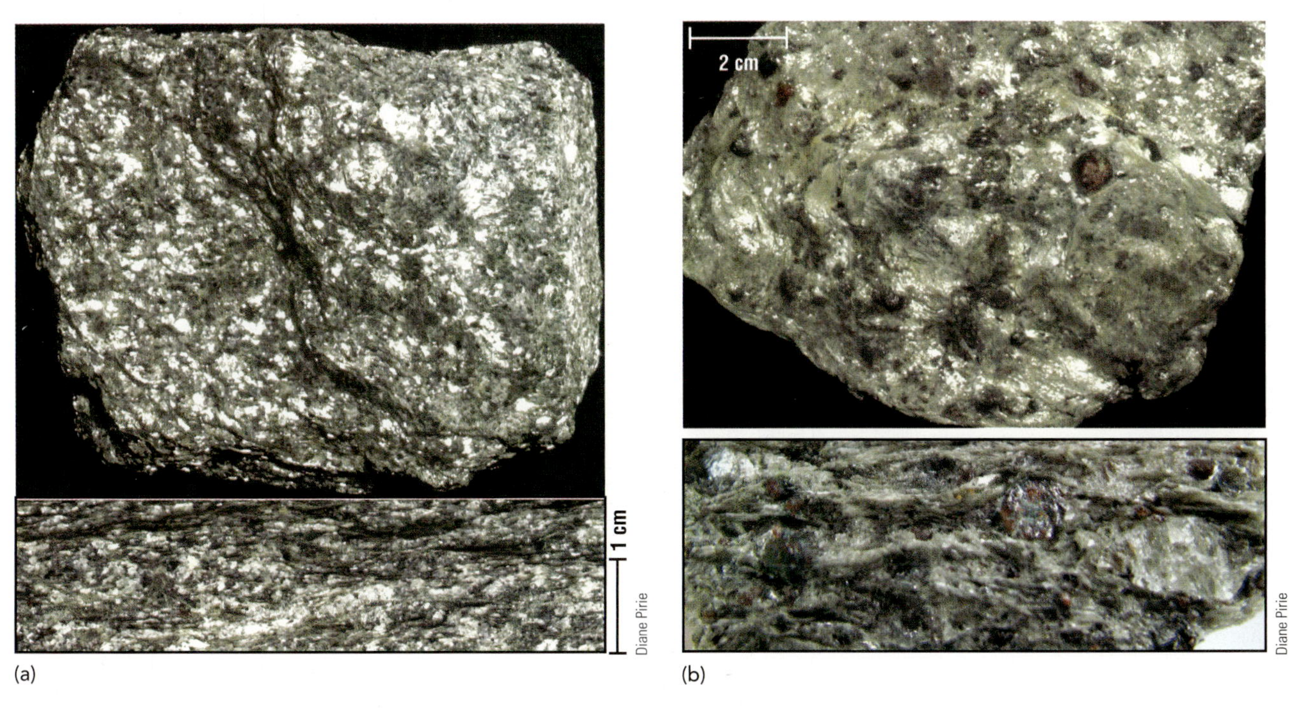

(a) (b)

Figure 5.6

Schists

(Continued)

(a) On the sparkly schistosity surface of this **mica schist,** two types of mica are visible: biotite (black) and muscovite (silvery/pearly). Both micas reflect light as bright white spots. *Inset:* A view along the edge of the schistosity where micas, quartz, and feldspar are visible. The reflections are much reduced in this view along the mica cleavage edges. (b) **Garnet-mica schist,** a foliated metamorphic rock with equant porphyroblasts of red-black garnet and two types of mica. Muscovite is much more abundant than biotite in this sample. *Inset:* This view, parallel to the schistosity, shows the garnets more clearly.

(c)

(d)

(e)

Figure 5.6

Schists—*Continued*

(c) **Greenschist** is a low-grade schistose metamorphic rock with chlorite and muscovite visible in this sample, shown here with some crenulations along its foliation. Actinolite is also present but not visible at this scale. (d) **Blueschist** is a high-pressure, low-temperature foliated metamorphic rock, shown here with blue amphibole, white mica, and garnet clearly visible; the mica and blue amphibole form the foliation. Some blueschists have an almost gneissic banding instead of schistosity as seen in the inset. (e) **Talc schist** is a medium grade schistose metamorphic rock made completely of talc. You can easily scratch it with a fingernail (*inset*).

(a)

(b)

(c)

Figure 5.7

Gneiss

(a) Dramatically foliated and folded **gneiss** with some eyes, but not quite enough to call it an augen gneiss. This rock shows the type of intense deformation that can occur during metamorphism. (b) Lineated **gneiss** clearly showing typical gneissic banding with streaks of light and dark minerals. The black minerals are biotite and hornblende, and the white and light gray minerals are plagioclase feldspar and quartz. An unusual feature of this gneiss is that it is lineated and not foliated. (c) **Augen gneiss,** with pink potassium feldspar *augen* embedded in a matrix of dark biotite and hornblende, and lighter quartz, plagioclase, and more alkali feldspar. The porphyroblasts are very apparent in this view although the typical gneissic banding is not.

with streaks of black minerals such as hornblende and biotite. If the felsic minerals predominate, this rock is probably **granite gneiss** formed from deformation of granite. **Augen gneiss** is a type of gneiss that has large feldspar porphyroblasts with tapered concentrations of light-colored minerals around them, which makes the rock look as though it has eyes (*augen* means "eyes" in German; see Figure 5.7c).

Amphibolite is classified based on its mineral content but is commonly foliated or lineated. Its protolith is mafic and it consists of hornblende and plagioclase (■ Figure 5.8a), and may contain additional minerals such as garnet (Figure 5.8b) or epidote.

Nonfoliated Rocks

Nonfoliated metamorphic rocks have such a distinctive mineral makeup that we use the mineral composition to classify them (■ Table 5.3). In some cases these rocks may be foliated.

Granoblastic Texture Not all metamorphic rocks have minerals with preferred orientation. If the mineral grains are visible and are randomly oriented in the rock, the texture is called **granoblastic,** or *nonfoliated* (see Figure 5.3b). Granoblastic texture is visible in the marble and quartzite in Figure 5.9a, b, and d. It occurs in rocks that have formed under conditions where differential pressure was weak to absent. Granoblastic texture may also result if the rock is composed of only equant or blocky minerals such as calcite, quartz, and feldspar that do not easily form a foliation (Table 5.1). Metamorphic rocks that are commonly granoblastic include *marble, quartzite, eclogite,* and *skarn.*

Marble is a medium- to coarse-grained rock made of calcite, which reacts strongly to dilute hydrochloric acid. **Dolomite marble** consists of dolomite, which reacts when powdered. In both types of marble, the grains are large enough to see, with obvious cleavage. Marble is generally granoblastic (■ Figure 5.9a, b), but may have a foliation (Figure 5.9c). The sample in Figure 5.9b has compositional layering inherited from impurities in its protolith. Marble forms by metamorphism of limestone (or dolostone for dolomite marble), but more prominent crystalline grains and a lack of sedimentary textures and fossils easily distinguish it from these rocks.

Quartzite is a fine- to coarse-grained, usually granoblastic rock made of crystalline quartz (Figure 5.9d). Most quartzite is medium-grained and may occasionally be foliated. With a hand lens, you should be able to see the vitreous luster and conchoidal fracture of quartz. The parent-rock of quartzite is a rock rich in quartz or silica, such as quartz sandstone or chert.

Eclogite is a medium- to coarse-grained rock containing red garnets and a distinctive sodium- and aluminum-rich green pyroxene (Table 5.1). Eclogite in Figure 5.9e has giant garnet porphyroblasts. White mica may also be visible. Eclogite is mafic rock, gabbro or basalt, metamorphosed at high pressure, and may be associated with blueschist and serpentinite in areas of subduction zone metamorphism. Some eclogites are foliated.

Skarn is a granoblastic rock composed of calcium-rich silicate minerals and forms where fluids from igneous, usually granite, intrusions react with limestone

Figure 5.8

Other Foliated Metamorphic Rocks

(a) This **amphibolite** is both schistose and lineated and contains hornblende and plagioclase feldspar. On the foliation surface the lineation is visible as hornblende crystals arranged approximately parallel to each other, up and down. (The hornblende is black, with grains reflecting the light looking bright white.) (b) **Garnet amphibolite** containing hornblende, plagioclase, garnet, biotite. This could also be called a garnet-amphibolite gneiss (c) **Stretched-pebble conglomerate:** the former pebbles are flattened and elongated due to deformation, and the foliation wraps partly around them. The pebbles may once have been chert but are now quartzite.

Table 5.3

Nonfoliated Metamorphic Rocks

We define these rocks based on their mineral content. Some rocks in this category may be foliated.

Hardness	Minerals	Color	Texture and Grain Size	Other Characteristics	Rock Name and Grade	Common Parent Rock
H>5½	Quartz (> 90%, commonly 100%)	Light shades, various colors possible—tan, pink, green, purple	**Granoblastic** Medium-grained (fine to coarse)	Generally has one homogeneous color throughout	**Quartzite** low to high grade (use grain size to estimate grade)	Quartz—sandstone, or chert
	Hornblende and plagioclase, ± biotite	Black and white	Granoblastic or Lineated and/or **foliated** Medium to coarse-grained	May contain garnet or epidote	**Amphibolite** medium grade	Basalt, gabbro (mafic)
	Green pyroxene (with high Na and Al) and garnet (red)	Bright to dark green with red—brown spots	**Granoblastic** Medium to coarse-grained (visible grains)	High density for a rock ($D=3.3\ g/cm^3$)	**Eclogite** high pressure	Basalt, gabbro (mafic)
	May have spots (porphyroblasts)	Dark to medium gray to tan—variable	Fine-grained or porphyroblastic	Very tough	**Hornfels** or spotted hornfels low to high grade	Shale, basalt…
	Wollastonite, garnet, green pyroxene, calcite	Variable: white, red-brown, green, white	**Granoblastic** Medium to coarse-grained	Mineral variable, mostly calcium-rich silicates	**Wollastonite Skarn** medium to high grade	Limestone intruded by granite
2½<H<5½	Calcite (> 90%)	Light shades; white, white with gray streaks, or pink are common	**Granoblastic** Medium to coarse-grained	(Reacts with HCl) rarely fossils may be preserved	**Marble** low to medium high grade (use grain size to estimate grade)	Limestone
	Dolomite (> 90%)	Light shades; white, white with gray streaks, or pink are common	**Granoblastic** Medium to coarse-grained	(Reacts with HCl if powdered)	**Dolomite Marble** low to medium high grade (use grain size to estimate grade)	Dolostone
	Serpentine	Light to dark green to black, commonly streaked or mottled	Slickensides look similar to foliation Fine-grained	Waxy or greasy luster; some varieties fibrous (asbestos)	**Serpentinite** low grade	Peridotite (ultramafic)
H<2½	Talc	Blue-green, gray, variable	Fine-grained	Waxy	**Soapstone** medium grade	Peridotite (ultramafic)

or dolomite, forming a contact metamorphic rock (Figure 5.2). Skarns (Figure 5.9f) vary widely in their mineral composition, but tend to contain calcite, calcium garnet, calcium amphibole, or pyroxene, and sometimes wollastonite (Table 5.1). Some skarns contain valuable minerals, such as scheelite, a tungsten mineral.

Nonfoliated Fine-Grained Texture Other nonfoliated rocks exist without visible minerals and are simply **fine-grained** (Figure 5.10a). These rocks are likely to form where heat and confining pressure cause metamorphism without the presence of differential pressure, and where the temperature or the time span of metamorphism was insufficient to grow larger crystals. Contact metamorphic environments commonly produce such conditions. Hornfels and some quartzites are fine-grained and nonfoliated.

Greenstone is a fine-grained nonfoliated green rock with a basalt protolith (■ Figure 5.10a).

Serpentinite is a rock made up of serpentine minerals, which are soft (H = 3–5), have greasy to waxy luster, and get their name from their green to nearly black and

(a) Parvinder Sethi (b) Diane Pirie (c) Claudia Owen

(d) Diane Pirie

(e) Grenville Draper

(f) Claudia Owen

Figure 5.9

Granoblastic Rocks

The mineral content of these metamorphic rocks defines them. (a) Typical coarse-grained **marble** entirely composed of calcite (cleavage visible), showing an interlocking *granoblastic* texture. (b) Pink **marble** with compositional variation of actinolite (green) and black biotite. Calcite here is pink, white, and gray. The actinolite and biotite show a slight preferred orientation/ vague foliation. (c) Highly deformed, well-foliated, medium-grained gray **marble.** (d) Three colors of **quartzite,** with granoblastic texture composed primarily of quartz. Quartzite may also be gray, purple, and other colors. (e) **Eclogite:** high-pressure metamorphic rock with large (>2″) red garnet porphyroblasts in a green pyroxene (omphacite) matrix. (f) Diopside-garnet-wollastonite **skarn.** The white mineral with silky luster is wollastonite. Some of the white is also calcite. The red-brown is garnet and the green is diopside. Wollastonite formed by a decarbonation chemical reaction as temperature increased: calcite + quartz reacted to form wollastonite + carbon dioxide: $CaCO_3 + SiO_2 \rightarrow CaSiO_3 + CO_2$.

(a)

(b)

(c)

Figure 5.10

Other Nonfoliated Rocks

(a) Fine-grained, low-grade **greenstone.** (b) This **serpentinite** sample has *slickensides* on the front surface, showing a region of highly foliated serpentine that formed along a fault zone or a zone of high shear stress. The scaly, or mottled, coloring common to serpentinite is visible along the upper-left side of this metaperidotite sample. (c) Nonfoliated **spotted hornfels.** The spots are cordierite porphyroblasts that are paler blotches surrounded by fine-grained biotite hornfels. This is a contact metamorphosed shale.

often variegated "serpent-like" coloring. Serpentinite is fine-grained with properties like its serpentine minerals (Table 5.1). It can be nonfoliated and very dark green, but can have smooth, slick surfaces called *slickensides* with a lighter or mottled green color (Figure 5.10b). Serpentinite forms from either low-grade or low temperature, high-pressure metamorphism of ultramafic igneous rocks (peridotite).

Soapstone is a very soft, commonly fine-grained rock made of talc. It differs from talc schist only in its texture and grain size. The color of soapstone can vary even within one sample. Soapstone results from the metamorphism of peridotite, at a slightly higher temperature than serpentinite.

Hornfels is a fine-grained or porphyroblastic rock produced by contact metamorphism. It is also very tough because its minerals tend to be intricately interlocking, a feature only visible under a microscope. Hornfels are difficult to recognize in a lab because they are rather nondescript—fine-grained and nonfoliated. Their color may vary but they are commonly black and may be mistaken for basalt. Basalt is generally denser and has more magnetite than hornfels. In the field, however, the proximity to an intrusive igneous rock is a way to confirm the identification of hornfels. **Spotted hornfels** is a hornfels with porphyroblasts (Figure 5.10c). You will see larger grains, which may look more like lumps than distinct grains, in a fine-grained and usually black matrix. Commonly the porphyroblasts are not distinct because they are full of inclusions of other minerals.

"Meta-" Rocks

Sometimes we use the prefix *meta-* in front of the protolith name to name a metamorphic rock. You would use this naming system to emphasize the parent-rock, history, and chemical composition rather than other aspects of the metamorphic rock such as its foliation or its current mineralogy. This naming convention can be a useful way to group metamorphic rocks.

Metabasalt is metamorphosed basalt (for example, Figures 5.6c and d, 5.8a and b, and 5.10a). Mafic minerals that form in metabasalts include chlorite, biotite, garnet, epidote, three types of amphibole (actinolite, hornblende, and blue amphibole), and pyroxene (Table 5.1).

Stretched-pebble conglomerate (or **metaconglomerate**) is foliated metamorphic rock in which the pebbles or clasts contained within a former sedimentary conglomerate are still visible, but stretched out into a lineation or foliation (Figures 5.1e, and 5.8c).

Metapelite (or **metashale**) is metamorphosed shale or mudstone (for example, Figures 5.5a and b and 5.6a and b). The clay minerals common in shale give this group their high aluminum content, which allows crystallization of aluminum-rich minerals (Table 5.1). Thus, these rocks are usually rich in micas and are commonly well foliated (that is, can form slates, phyllites or schists) if formed under differential pressure.

Metalimestone and **metasandstone**: Metamorphism of limestone over a large range of temperatures produces a rock containing 95% to 100% calcite (*marble*, occasionally called metalimestone, Figure 5.9a–c). If quartz sandstone is metamorphosed, it becomes a rock containing 95% to 100% quartz (called *quartzite*, infrequently called metasandstone, Figure 5.9d).

TEMPERATURE AND PRESSURE OF METAMORPHISM

One of the most important concepts to understand about metamorphic rocks is that *different minerals form at different temperatures and pressures.* As temperature and pressure change during metamorphism, new minerals grow within the metamorphic rock, replacing preexisting minerals, while remaining in the solid state. This means that the minerals can be clues about the pressure-temperature history of the rock.

Minerals Reveal Metamorphic Temperature

A group of minerals that grow or coexist together at the same temperature and pressure is a **mineral assemblage.** These minerals are characteristic of specific ranges in temperature. Mineral assemblages operate like medical thermometers used to measure your body temperature in that they record the maximum temperature experienced. We can therefore use mineral assemblages as our thermometers to measure the temperature of metamorphism. Metamorphic petrologists call them *geothermometers.*

High-aluminum rocks such as shales are the most sensitive, when metamorphosed, in showing progressive mineral changes with increasing temperature during regional metamorphism. They may display **Barrovian zones** (■ Figure 5.11), which are mineral assemblages in metashales characteristic of specific ranges in temperature (see Figure 5.11a and b). A single distinctive assemblage represents a **metamorphic zone.** Simply put, a metamorphic zone refers to the group of minerals that would form at a particular temperature range in a rock of a specific composition. The zone names match an especially useful **index mineral,** which is stable at that particular temperature, and usually easy to see *in the field* (in rocks in their natural setting). Each **Barrovian zone** starts with the first appearance, as one is progressing from lower-grade rocks toward higher-grade rocks, of a distinctive mineral such as chlorite, biotite, or garnet. However, this index mineral generally persists into higher temperature zones (Figure 5.11c). The boundaries between the different zones are **isograds** (Figure 5.11a). If the region is uplifted and eroded (Figure 5.11a at the *right*), these assemblages can be viewed and mapped at Earth's surface, as was done in northwestern Michigan (see ■ Figure 5.12). The rocks are now at surface temperature, but the minerals remain showing us the maximum metamorphic temperatures experienced.

Some minerals or mineral assemblages are stable over an extensive range of temperatures and pressures. Quartz remains stable if metamorphosed at 200°C or 800°C. Calcite, in marble, is stable over almost as wide a range of temperatures. Pressure can also vary widely for the stability of these minerals.

5. Read vertically down from a zone in Figure 5.11a to the minerals present in Figure 5.11c to determine which minerals are typically present in a particular zone. This would be the mineral assemblage for that zone.

a. What mineral assemblage is present in chlorite zone?

chlorite, muscovite, calcite

b. What mineral assemblage is present for staurolite zone?

chlorite, muscovite, garnet

c. What mineral assemblage is present for kyanite zone?

muscovite, gneiss

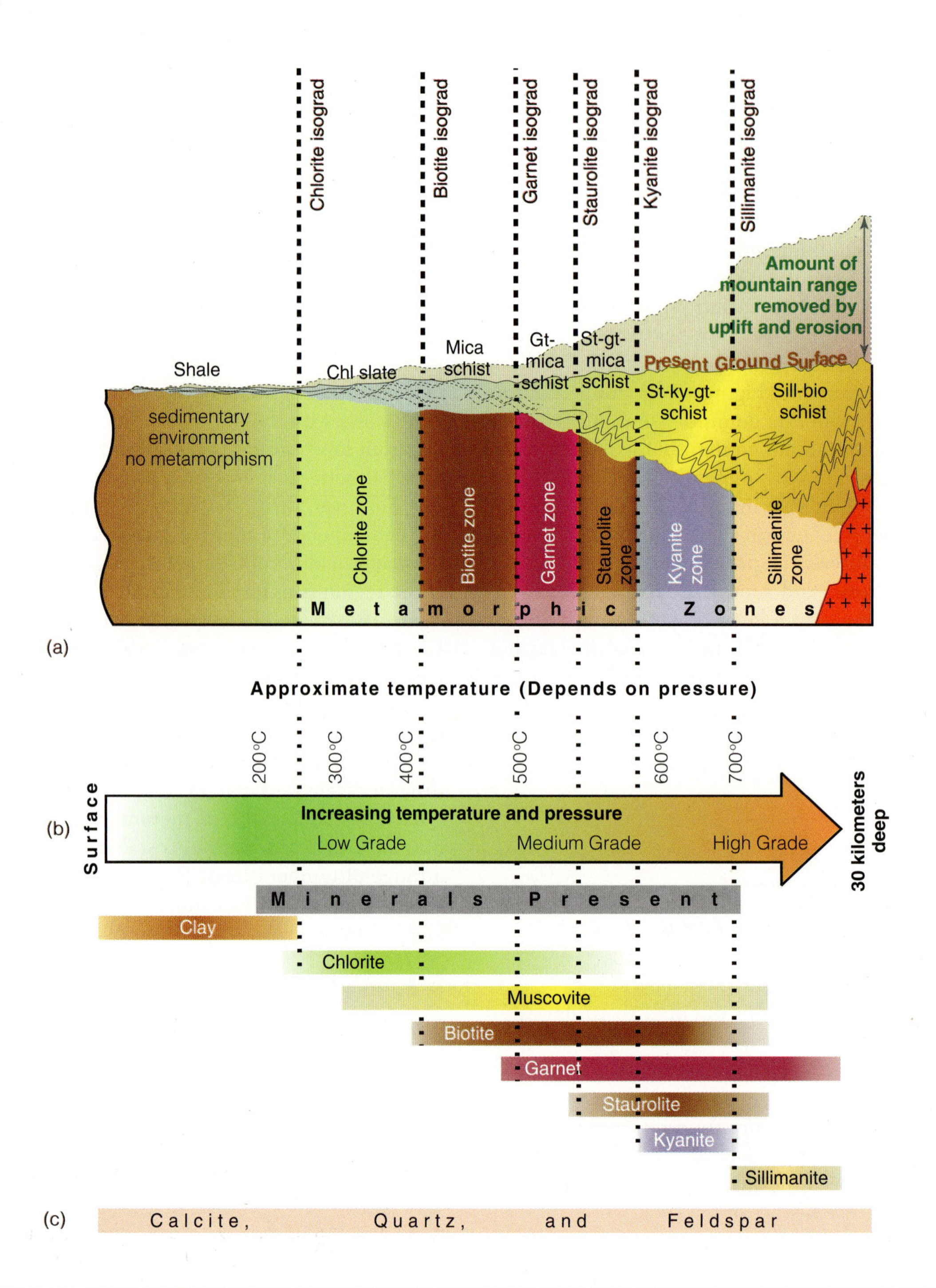

Figure 5.11

Metamorphic Zones

Diagrams illustrating the variation in metamorphic zones with intensity of metamorphism. (a) Geologic cross section of a deeply eroded mountain range at a former convergent plate margin. This cross section is of an area with more rapid uplift on the right than the left side; erosion on the right exposed rocks formed at greater depths. Progressive dynamothermal metamorphism took place in these rocks, producing different mineral assemblages (represented by the metamorphic zones) with increasing temperature and pressure into the core of the mountain range. Compressive stress in the roots of the mountain range also produced foliation and deformed the rocks. As a result, the shale protoliths recrystallized and deformed into metamorphic rocks. Abbreviations used here are chl = chlorite, gt = garnet, bio = biotite, ky = kyanite, st = staurolite, and sill = sillimanite. (b) The progression of temperatures experienced within the mountain range. (c) At these different temperatures different minerals are present. The color streaks represent the range of temperatures at which each mineral is likely to exist. Minerals that characterize various zones occur in a progressive sequence from low to high temperature, from left to right.

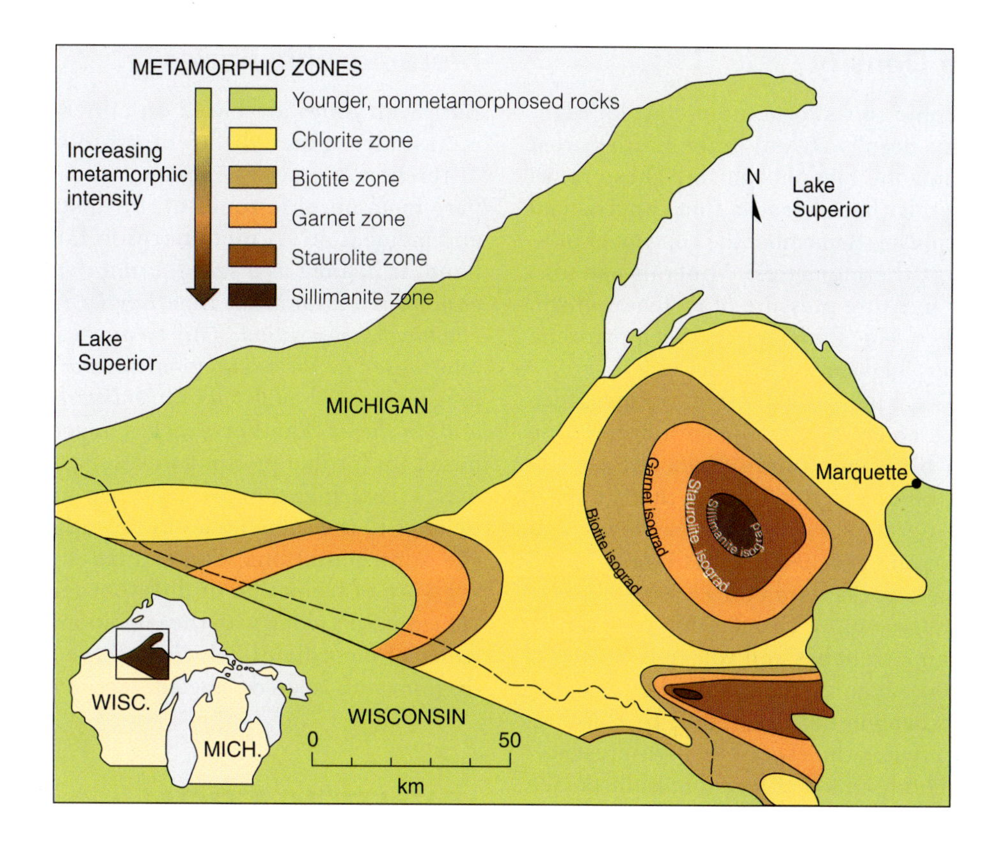

Figure 5.12

Map of Metamorphic Zones

Metamorphic zones in metashales in the Upper Peninsula of Michigan. Metamorphism here took place about 1.5 billion years ago. In this area, low-grade, low-temperature metamorphism occurred at the outside of the "bull's-eye" and high-grade, high-temperature metamorphism occurred in the center. Metamorphosed shales in this region best delineate the *isograds*, which are shown as black lines separating the colored *metamorphic zones*. The **isograds** connect the first appearance of metamorphic index minerals. Moving along an isograd leads you along equal metamorphic temperatures, whereas crossing isograds moves you through changes in metamorphic temperatures. From H. L. James, G.S.A. Bulletin, Vol. 66, Plate 1, p. 1454. Reprinted by permission of Geological Society of America.

6. Do you think either calcite or quartz would make good index minerals? no
Would they make good indicators of pressure, or geobarometers? no
Why or why not?
Because they do not change therefore not good in indicating pressure

Grain Size and Metamorphic Temperature (Grade)

As temperature increases, grain size of minerals in the rocks generally also increases. This is because atoms move faster at higher temperatures and thus can travel farther to attach themselves to a mineral grain. Because more material is available for growth due to the increase of speed and migration of atoms at higher temperatures, mineral grains have the ability to grow larger. At low metamorphic temperatures *(low grade)*, atoms move slowly and the rocks are generally fine-grained (e.g., slate). As grade increases, the grain size increases to produce phyllite, then schist with medium-size grains, or marble with medium-size grains *(medium grade)*. Finally, at the highest temperatures of metamorphism *(high grade)*, atoms within the solid rock become sufficiently mobile to allow for segregation of different minerals into separate bands, such as those seen in gneiss, or to allow for growth of large crystals in a coarse-grained marble.

Other variables also influence grain size, such as, deformation, the rock's water content, distinct growth rates of minerals, and time elapsed during the metamorphism. Thus, there may be considerable variation in grain size of different rocks that experienced the same temperature. Nevertheless, grain size can give you a quick rough estimate of the grade of metamorphism.

Pressure and Density

Certain metamorphic rocks form under special high-pressure conditions deep (>20 km) in the Earth's crust during subduction zone metamorphism. These rocks have denser minerals than do rocks that experienced orogenic metamorphism. Remember the concept of mineral assemblages as thermometers? Minerals can also be barometers, to measure pressure of metamorphism. Blue amphibole, Fe-, Mg- garnet, and a green pyroxene called omphacite are high-pressure minerals.

Blueschist is a high-pressure/low-temperature metamorphic rock characteristically containing blue amphibole. The blue amphibole gives the rock a bluish-gray appearance (Figure 5.6d). Other unusual high-pressure minerals such as *jadeite,* a pyroxene that is one type of the gem jade, occasionally occur in blueschists. These rocks form at high pressure near and within subduction zones[2] at a depth of 20–40 km and between oceanic trenches and volcanic arcs. The subduction of cold oceanic crust cools the normally high temperatures deep in the Earth; thus the subduction zone setting creates the especially high-pressure/low-temperature conditions that form blueschists (see blueschist facies in Figure 5.13).

Eclogite is a rock of basaltic composition that has recrystallized into high-pressure denser minerals: a green pyroxene with high sodium and aluminum content called omphacite and red garnet (Figure 5.9e). Although the pressure of metamorphism of this red and green rock is consistently high, the temperatures of metamorphism of eclogite may vary considerably, as shown in Figure 5.13.

7. Your instructor will explain how to compare density by measurement or by heft test (as in Exercise 8, Lab 2). Measure (in g/ml) or compare (higher, lower) the density of a piece of eclogite ______ and a piece of basalt ______. Explain why these rocks have the relative densities they do.

____

____

____

____

[2] A *subduction zone* is an area at some convergent plate boundaries (Labs 1 and 9) where oceanic crust moves downward into the mantle. Oceanic trenches occur where the crust first starts to descend, and volcanic arcs occur where the subducting plate generates magma that rises to form the volcanoes at the surface.

Metamorphic Facies

Barrovian zones are based on mineral assemblages in metashales. Since different protoliths may produce different mineral assemblages, it is difficult to compare metamorphic zones of, for instance, metashales and metabasalts. A **metamorphic facies,** on the other hand, is *defined as a set of mineral assemblages found commonly associated together in rocks of differing chemical composition. This means that they were metamorphosed at the same conditions of temperature and pressure even though they have different groups of minerals in them.* Whether a rock is a metashale or a metabasalt (or for that matter a metalimestone), if it formed over a particular range of pressure and temperature, it belongs to the same facies as any other rock formed at the same conditions. The estimated temperature and pressure of formation of different facies are shown in ■ Figure 5.13 along with the conditions for the Barrovian zones of metashales, which we have marked in on the diagram with isograds labeled "biotite in," and so forth.

IDENTIFICATION AND DESCRIPTION OF METAMORPHIC ROCKS

In your identification of metamorphic rocks you should use the foliated rocks (p. 102) and nonfoliated rocks (p. 106) classification systems in Tables 5.2 and 5.3, and in Figure 5.14.

8. Examine the unknown rocks and fill in the information in ■ Table 5.4, Metamorphic Rock Identification Forms. Use the maze (■ Figure 5.14) and Tables 5.2 (p. 102) and 5.3 (p. 106) to determine the rock name. Use the column headings and the following guidelines as you fill in the table:

a. *Texture:* Choose one of the following textures: (1) ***slaty cleavage*** (fine-grained and breaks along parallel flat planes), (2) ***slaty cleavage*** with a **sheen** (not sparkles) and possibly crenulations (wavy planes), (3) ***schistosity***, (4) ***gneissic banding***, (5) ***granoblastic***, or just (6) ***nonfoliated, fine-grained.*** Also decide if the rock is ***porphyroblastic*** as well.

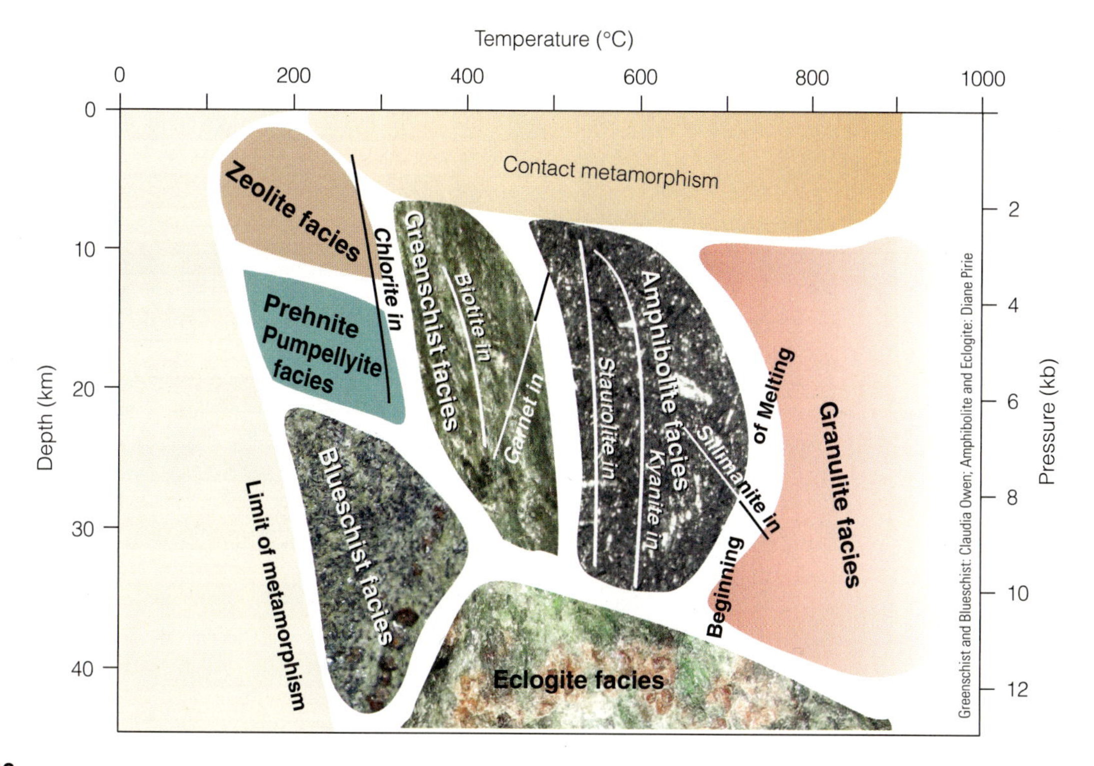

Figure 5.13

Metamorphic Facies

Approximate pressure-temperature fields of metamorphic facies. Greenschist, amphibolite, blueschist, and eclogite are the mafic metamorphic rocks we have discussed in this chapter that give the names to some of the metamorphic facies. These make up the background patterns on the diagram within their respective stability fields. The isograds, such as *garnet in*, mark the fields of the Barrovian zones.

b. *Grain size:* In the same column, also record the grain size: ***fine-grained***, ***medium-grained,*** or ***coarse-grained***.

c. *Composition and color:* Write down the names of all medium- to coarse-grained minerals in the rock and their colors. For example, marble is predominantly calcite (which might be various colors). Also include any minerals you can determine from their properties even if they are fine-grained, such as talc, which can be identified by its softness. Carefully describe the color of fine-grained rocks, giving some specifics, and describe any colors you haven't already mentioned.

d. *Rock name:* Identify the rock using the maze in Figure 5.14, and check your identification against the descriptions earlier in this chapter, and if available, against samples of known metamorphic rocks.

e. *Origin:* Determine the ***parent-rock*** (or protolith) and the ***conditions of metamorphism.*** This second item involves giving the metamorphic grade, the metamorphic zone, the metamorphic facies, or the temperature and pressure of metamorphism to the best of your ability. Also indicate the type of pressure: *confining pressure* or *differential stress*. Did the rock experience contact metamorphism or did it undergo regional dynamothermal metamorphism? Did the rock undergo deformation displayed as foliation, stretched pebbles, stretched fossils, folding, or crenulations?

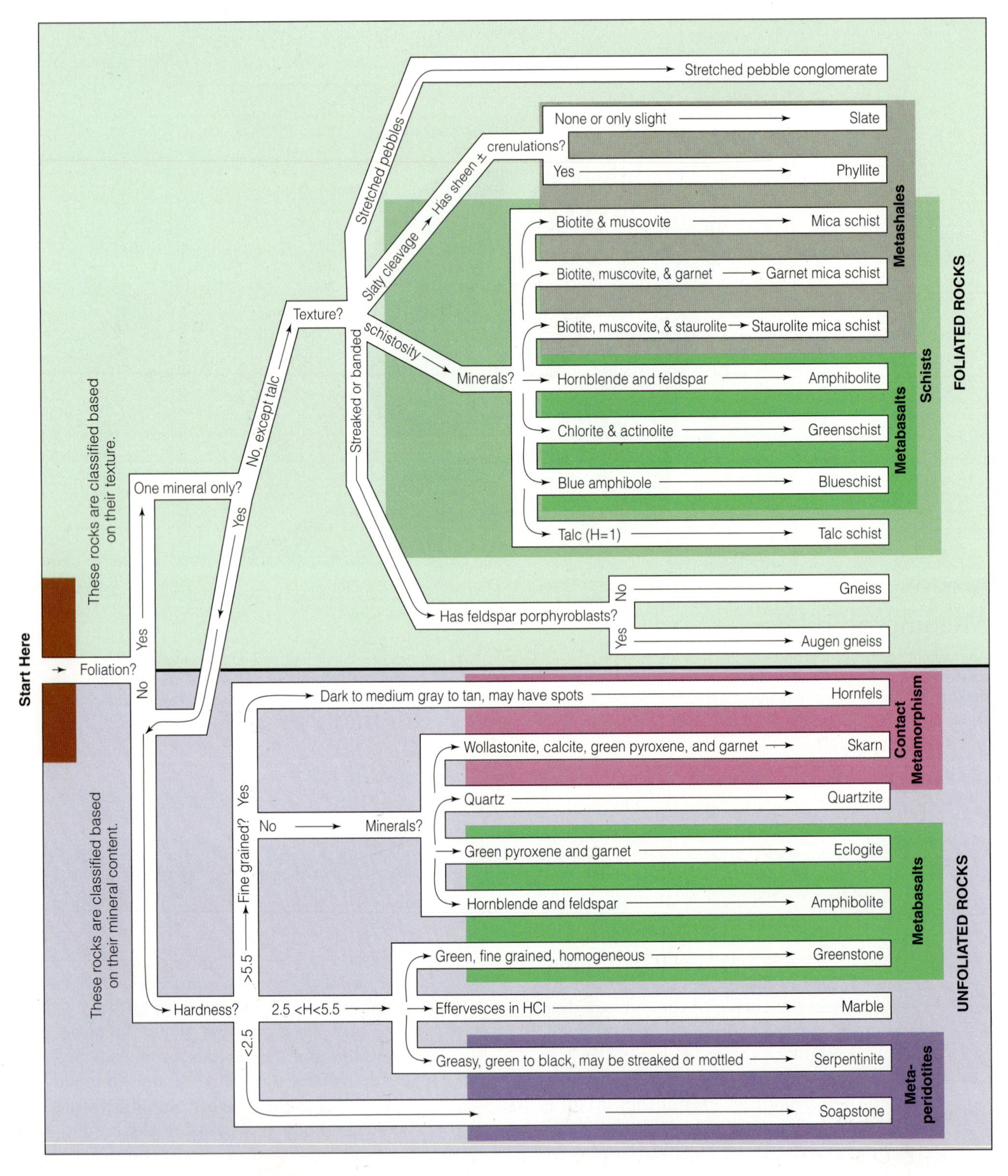

Figure 5.14

Metamorphic Rock Maze

Maze for identification of metamorphic rocks. At the middle left where it says "Start Here," begin following the maze by choosing a path at each question based on the properties of the rock you are identifying.

Table 5.4

Metamorphic Rock Identification Forms

See Exercise 8 in the text for detailed instructions for filling out this table.

Sample Number	Texture (see instructions) and Grain Size	Composition and Color (minerals and their colors, porphyroblasts, overall rock color)	Rock Name	Origin	
				Parent Rock	Conditions of Metamorphism (confining or differential pressure and metamorphic grade, zone, facies, or best guess for pressure and temperature)
17	Foliated	red-black	Gneiss	Shale	med / direct press
54	Non foliated	black-green	Serpentinite	Peridotite	low / confined
82	foliated	grey	Gneiss	Granite	high tem & pressure
286	Foliated	grey, black	Garnet mica shist	Shale	Blueschist facies
61/181	foliated	Black/gray	slate	Shale	low grade
193	un foliated	green pyroxene garnet	Ecologite	Basalt	med to high
79/47	Foliated	Shining red	Mica Schist	Shale	med grade / direct press
75	granular	Black/gray	soapstone Soapst	Peridotite	med grade.

(Continued)

Table 5.4

Metamorphic Rock Identification Forms—*Continued*

Sample Number	Texture (see instructions) and Grain Size	Composition and Color (minerals and their colors, porphyroblasts, overall rock color)	Rock Name	Origin	
				Parent Rock	Conditions of Metamorphism (confining or differential pressure and metamorphic grade, zone, facies, or best guess for pressure and temperature)
184/28	unfoliated	red + green. massive	Quartzite	sandstone	med grad / confined
95	unfoliated	dark	Amphybite	Basalt	med /coarse
03	Foliated	Black/ grey. chlorite	Rhyolite	shall	low grade.
273	unfoliated	coarse crystalline	Marble	Limestone	high tem med / confined

METAMORPHIC ZONES MAPPING EXERCISE

Earlier in this chapter we introduced metamorphic zones so you would be able to estimate the conditions of metamorphism of the rocks you identified in Table 5.4. When geologists find metamorphosed shales in the field, they may map them and the Barrovian zones using this type of information. Let's practice in this next exercise.

9. Figure 5.15 shows a region with mostly metamorphic and a few sedimentary rocks. Ohio Smith has mapped the area, but has used his knowledge of protoliths and metamorphic minerals to show only the parent-rocks. Ohio also collected some samples and indicated the sample numbers on the map and in Table 5.5. This exercise is enhanced by using rock samples, but can be done without them if you start with part **b**.

a. If your instructor provides samples matching those in Ohio Smith's collection. Examine each rock, determine what minerals are present, and match the sample with one of rock names listed in Table 5.5 (see also Figure 5.11). Enter the sample number in Table 5.5.

b. Determine the metamorphic zone of each metashale in Table 5.5. One rock has been done for you. For any rocks that are not metashales, just place a dash or an X in the metamorphic zone column.

c. On the map in Figure 5.15, next to each sample number, place letters indicating the metamorphic zone. For example, you could place the letters **st** for samples in the staurolite zone and **si** for those in the sillimanite zone. One rock has been done for you as an example.

d. Using pencil, draw the metamorphic isograds on the map showing the borders between the metamorphic zones. See Figure 5.12 for an example.

e. Label the isograds, and color and label the metamorphic zones in between them. Use a complete name for each, such as "garnet zone."

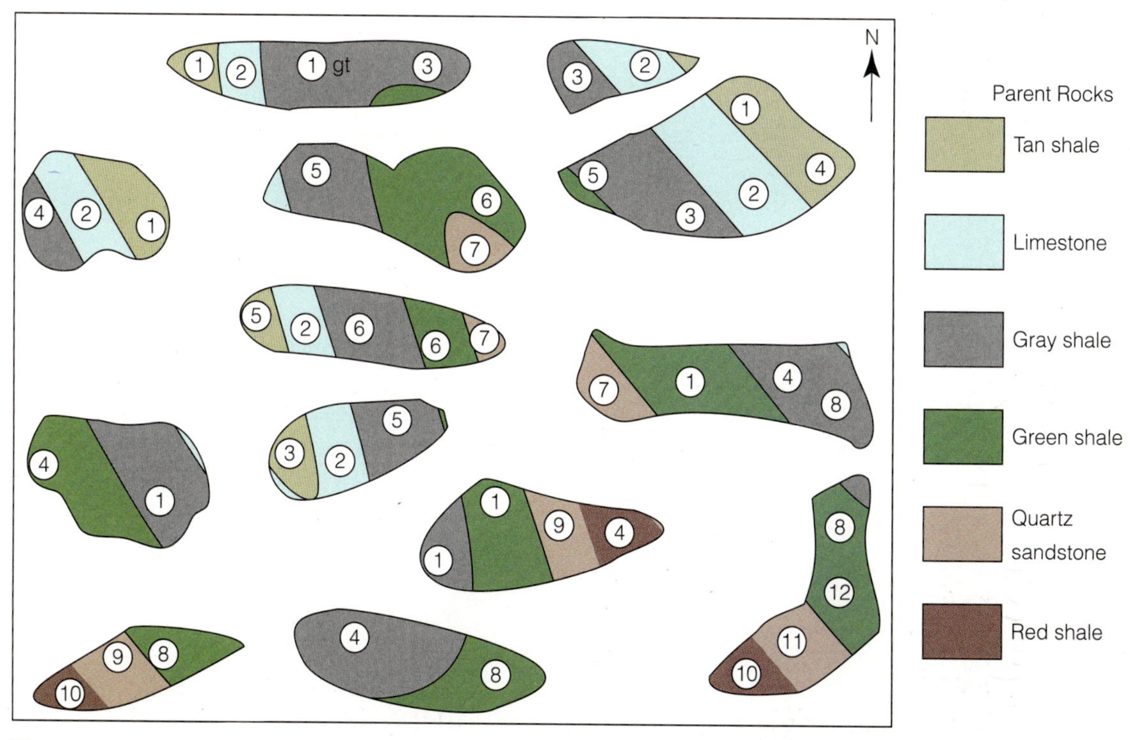

Figure 5.15

Metamorphic Zones and Sample Locality Map

Follow the instructions in the text at Exercise 9.

Table 5.5

Rocks on the Metamorphic Zones Map

This table lists the rocks found on the Metamorphic Zones and Sample Locality Map in Figure 5.15. See Exercise 9.

Number on Map	Sample Number	Rock Name	Parent Rock	Metamorphic Zone
1		Garnet-mica schist	Shale	Garnet zone
2		Coarse-grained marble		
3		Staurolite-mica schist		
4		Mica[1] schist		
5		Kyanite schist		
6		Sillimanite gneiss		
7		Coarse-grained quartzite		
8		Chlorite slate		
9		Medium-grained quartzite		
10		Red shale		
11		Quartz sandstone		
12		Green shale		

[1]Mica schist is also called two-mica schist or biotite-muscovite schist.

f. Which zone was metamorphosed at the highest temperature?

____

Which at the lowest?

____

g. Among the metashale samples, if you used them in part a, is there a relationship between temperature and grain size?

_____ If so describe it. _______________

____

____

____

h. Cooling rate is not relevant in a metamorphic situation as it is for grain size of igneous rocks. Instead, what other factor would be more pertinent to metamorphic grain size than cooling rate? It also has to do with time.

____

____

i. Based on the grain size of the porphyroblasts in the metashale samples for Table 5.5, which two minerals seem to grow the fastest? _______________

____

LAB 6

Topographic Maps

OBJECTIVES

- To be able to define topographic contours and understand their use
- To be able to read and construct simple topographic maps and profiles
- To learn to use aerial photographs to view landforms in three dimensions

Maps are key tools for geoscientists. Topographic maps illustrate the geometry of the Earth's surface and geologic maps show the locations and features of different rock formations. Because Earth's history is so long—4.6 billion years—diverse rocks of vastly different ages and origins occur at Earth's surface. When mapping rocks, geologists include geologic time information, which helps toward understanding the geologic history of a region. In this lab and the two that follow, we will cover topographic maps, geologic time, and then geologic maps.

TOPOGRAPHY AND CONTOURS

The term **topography** refers to the shape of the physical features of the land surface. A topographic contour map, familiar to many because of its widespread use in land development, recreation, and other land use, is a representation of the landscape of a given area. Although drawn on a flat surface, it uses contour lines to depict three-dimensional landforms such as mountains, valleys, depressions, and cliffs.

Each **contour line** (often shortened to **contour**) connects points on a map that have the same elevation on the Earth's surface (see ① on the map in ■ Figure 6.1). A single contour line does not divide or branch, nor does it cross others.[1] Contours must be continuous, stopping only at the edge of a map or where elevation labels briefly interrupt them. The elevations are measured above a level called a *datum plane,* usually

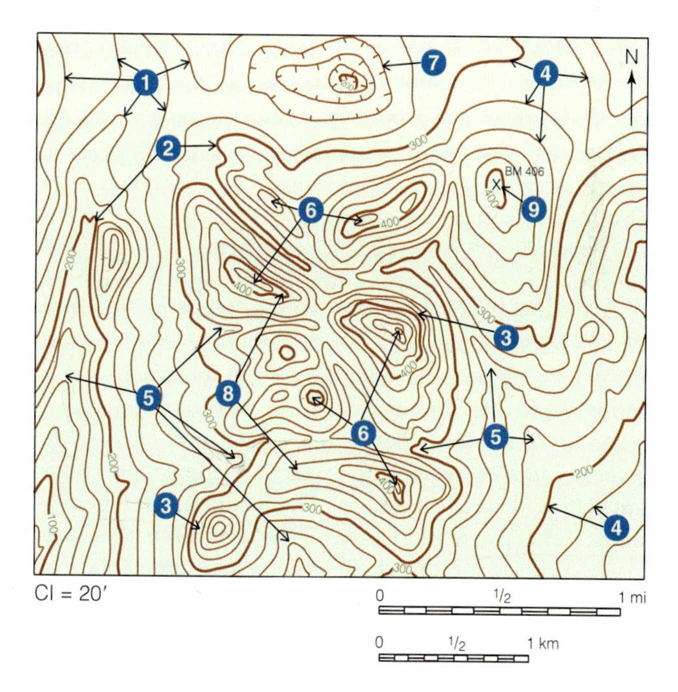

Figure 6.1

Hypothetical Topographic Map

A topographic contour map of a hypothetical area. 1 = contour lines; 2 = index contours; 3 = steep slopes; 4 = gentle slopes; 5 = valleys; 6 = hills; 7 = depression; 8 = ridges; 9 = bench mark.

[1] Contours may appear to join and divide on a vertical cliff, but these must be contours of different elevations. Contours that cross occur only where an overhanging cliff exists (the lower contours will be dashed in this case).

mean (average) sea level. An island's 0-ft contour line would be the shoreline. Assuming that the island is a single high hill, a footpath located exactly 20 ft above sea level and continuing entirely around the island at this same elevation would coincide with the 20-ft contour on a map. In this case, the **contour interval,** or difference in elevation between two adjacent contour lines, would also be 20 ft. The next higher contour would connect all points that were 40 ft above sea level. Contour elevations on a map are always a multiple of the contour interval. An **index contour** (see ② in Figure 6.1) is a heavier fourth or fifth contour line, usually labeled with a number. Index contours help to make the map easier to read.

On a single map, more closely spaced contours indicate steeper terrain ③. Conversely, more widely spaced contours portray gentler slopes ④. However, a map with generally steeper terrain is likely to have a larger contour interval to prevent the contours from being too close together. Check the contour intervals before jumping to conclusions when comparing the slope steepness on two different maps.

If you can imagine a series of horizontal planes at equally spaced intervals above sea level, then contour lines are the traces where these planes intersect the land surface. Contours run parallel to valley walls and hillsides. They curve upstream where they cross a stream **valley** typically making a V shape on the map that points *upstream* ⑤. A series of more or less concentric contours (one inside the other) that form continuous loops represent **hills** ⑥. A hollow topographic **depression,** or basin, appears as one or more lines with hachure marks ⑦ that encircle the basin; the **hachure marks** point in toward the center of the depression. A **ridge** ⑧ is a long, narrow area of higher elevation often found as an extension of a hill or mountain, or as a connection between mountains. It has contours that double back on themselves, with higher elevations toward the center and lower outside ⑧. As you move across contours in one direction on a map, the elevations must continue up or continue down unless you have crossed the same contour twice. In various places on topographic maps are X's that show the locations of benchmarks ⑨. A **benchmark** is a permanent marker on the ground that has precisely surveyed location and elevation measurements. The X, labeled BM, commonly lists the elevation next to it ⑨. Notice the chart of Topographic Map Symbols inside the back cover of the book that illustrates benchmarks and some other features shown on topographic maps.

The difference in elevation between the highest and lowest point in an area is the **relief.** One way to report the relief of an area is by subtracting the two contours that are numerically most different. An area with high relief will have either closely spaced contours or a large contour interval or both. An area with low relief will have a combination of widely spaced contours and/or a small contour interval. The southeast corner of Figure 6.1 has a lower relief than the center. Where the relief is high the terrain is rough and foot travel across contours is difficult. For example, in places near the center of Figure 6.1 you would need to climb up 60 ft for every 0.1 mile of horizontal travel. The southeast corner still has a noticeable slope, but it is much more modest, with only a 15-ft climb over the same distance. Areas of low relief are almost flat and have little change in elevation across the area. Where an area is completely flat, no contours will be present on the map.

Rules of Contours

1. Answer the following questions about contour lines and topographic maps:

a. When walking along a contour line, do you go up, down, or stay level?

level

b. What level (reference plane or datum) is usually zero on a topographic map?

sea level

c. Because of their definition and their geometric relationship to the Earth, contour lines obey certain rules. Some of these rules are obvious from the definition. Some have been discussed in the previous paragraphs. Fill in the blank spaces below to make your own complete set of rules:

Rule 1: Every point on a contour line is the same elevation.

Rule 2: A single contour line never crosses or branches.

Rule 3: Two contours can only appear to across at an overhanging cliff.

Rule 4: Widely spaced contour lines indicate indicate slopes.

Rule 5: Closely spaced contour lines indicate steep slopes.

Rule 6: The V a contour line makes as it crosses a river or stream points up stream.

Rule 7: Hills / ridges / valleys / depressions (circle the correct word) appear as continuous closed loops without hachures (short lines).

Rule 8: Hills / ridges / valleys / depressions (circle the correct word) appear as closed contours with hachures that point toward the center.

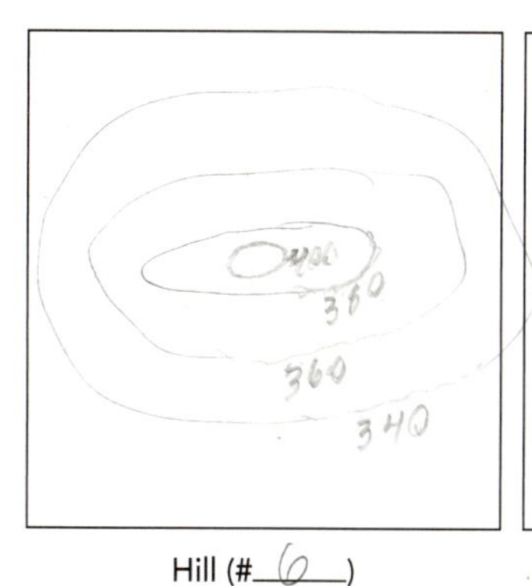

Hill (#_____)

Basin (#_____)

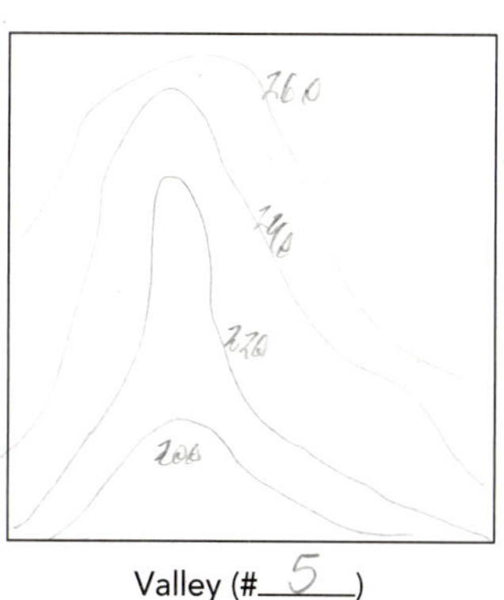

Valley (#_____)

Ridge (#_____)

Figure 6.2

Topographic Features

See the instructions in Exercise 2 to fill in these features.

Rule 9: Contours are continuous and only stop at the __________________ of the map.

Rule 10: As you move across contours in one direction on a map, the elevations must continue up or continue down unless you have crossed ____________________.

d. The change in elevation between any two different, adjacent contour lines is the ____________________, sometimes abbreviated CI.

e. What does the term *relief* mean?

____

____

____

2. On Figure 6.1, draw a route for an easy hike from ❼ to ❸. Relief of the area traversed should be less than 70 ft, but still reflect the shortest possible route. Determine the relief traversed __________________ and using a string and ruler, measure the length of the hike __________________.

3. In each of the boxes in ■ Figure 6.2, sketch a hypothetical contour map of the feature indicated. Use a 20-ft contour interval and label every contour. Label each of your maps with the corresponding feature numbers from Figure 6.1. Your maps should be similar to those in Figure 6.1, but not identical.

Raised and Shaded Relief Maps

Topographic maps drafted onto a plastic sheet and molded into a scale model of the shape of the landscape make **raised relief** maps (■ Figure 6.3). Some beginning students find it difficult to imagine what an area looks like by reading the contours on a two-dimensional map surface. A raised relief map with contours drawn on it creates a sort of mental bridge that aids in this visualization.

4. Use the raised relief map your instructor provides to answer the following questions.

a. Locate a hill or mountain peak on the map and write its name here:

____

Claudia Owen

Figure 6.3

Raised Relief Map

Close-up of a raised relief map of Glacier National Park.

b. Describe the shape of the contours for this hill/peak and their geometric relationship with each other and to the shape of the hill/peak.

____

____

____

c. Find a named feature on a gentle slope or nearly flat area and write its name here:

____

d. How could you recognize that this was a gentle slope (from **c**) if you were using just a flat topographic map?

____

e. Locate a valley. What are two ways you could use contours to tell this is a valley, if it were a regular flat map instead of raised relief?

____

____

Elevations of Contours on a Topographic Map

In the next exercise, you will determine the elevations of different unnumbered contours. Here are some helpful tips to remember: Concentric closed loops without hachures on a contour map signify a hill; therefore, the elevations will increase toward the center in steps equal to the contour interval. When you see a series of nearly parallel contours, with no contours doubling back, you are looking at a continuous surface, which slopes in the same direction. Number the contours here consecutively, all going up or all going down. Follow a contour for which you know the elevation around on the map to areas where you do not know the elevations. Do not assume that all index contours (bold contours) have the same elevation; they often do not.

5. Study ■ Figure 6.4.

a. First, label the elevations of the contours up Antelope Hill from 1500′ in Figure 6.4 using a contour interval of 100′.

b. Is it likely that the top of Antelope Hill corresponds exactly with a

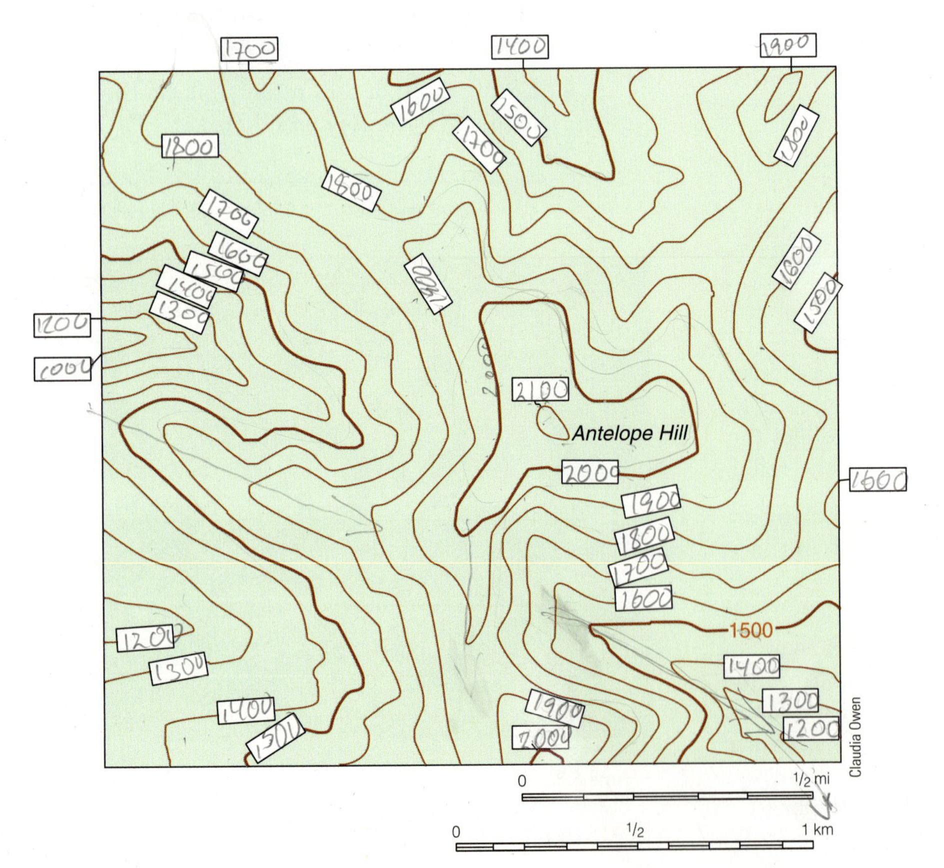

Figure 6.4

Map of Antelope Hill

Obtain the contour interval from the instructions in Exercise 5a: CI = ______

contour? Yes What is the range of possible elevations for this point?

2100

c. Next, label the elevations of other contours on the map. In some cases, you will want to follow a contour of known elevation to help you determine more contour elevations.

d. Draw in the locations of streams on the map in Figure 6.4. To do this, locate valleys and be sure to distinguish them from ridges. Draw arrows on each stream showing the direction of flow. Recall that contours V upstream across stream valleys, which is opposite to the flow direction of the stream.

e. Locate the steepest slope on the map and label this spot with the letter S. Locate the gentlest slope on the map and label it with the letter G.

CONSTRUCTING TOPOGRAPHIC MAPS AND PROFILES

Making a Topographic Map of a Model

Your instructor will supply a plastic terrain model for the following exercises or may ask you to make a topographic feature out of Plasticine®. This may be a model of a volcano or some other topographic feature(s). The purpose of these exercises is to demonstrate the concept of contour lines and therefore improve your understanding of topographic maps. It is important to remember the rules of contours as you draw your map.

Procedure for Map Construction

Materials needed:

- Plastic or Plasticine model of a landscape feature such as a hill, valley, volcano . . .
- Water
- Clear plastic box in which the model will fit that can hold water and comes with a clear plastic lid
- Overhead projector plastic sheet
- Erasable marker
- Clear tape
- Tracing paper

6. Construct a map.

a. Your instructor may ask you to mark a series of equally spaced elevations on the outerside of a clear plastic box, or such a box may already be marked for you. Follow any marking instructions carefully.

b. With the model inside the plastic box, pour water in the box up to the first elevation marked on the side, which represents a 50-ft elevation on the ground.

c. Place the clear top on the box and tape a clean plastic sheet over it. Looking from above, trace the water line onto the plastic sheet. *Label* the contour line with the elevation by leaving a short gap in the contour and inserting "50" there.

d. Repeat this procedure, labeling each of the contours in increments of 50, until the water reaches an elevation just below the top. Verify that the model is not floating due to air bubbles trapped inside it. Pour water into any depressions, such as a volcano's crater, to the same level as the water outside, and draw two contours—one outside and one inside the depression. If so, label both with the correct elevation. What symbol will you use on the contours of a depression?

____

e. If you are doing this activity with lab partners, each partner should trace the contour map onto a sheet of paper for the next exercise. Include the contour interval at the bottom of the map. Do not forget to label the contours on your copy.

Model Map Questions

7. Answer these questions about the model and map you made.

a. Briefly describe the general shape of the model. Use general terms such as *valley, volcano, crater, hill,* and *depression.*

____

____

b. Describe the general shape of the contour lines around any hills in the model. What is their geometrical relationship to one another?

____

____

c. Describe the general shape of the contours in any valleys in the model.

____

____

d. Make a general statement about the closeness of the contours and the steepness of slope.

____

____

Topographic Profiles

A **topographic profile** from a topographic map shows how landforms would appear if you could see them in side view or silhouette (Figure 6.7d). It adds a third dimension to a map, giving you an accurate picture of the topography along the line from which the profile is drawn, making it easier to imagine the area as it actually appears.

Constructing a Topographic Profile In this exercise, you will draw a topographic profile from the map that you constructed of the model, or if your instructor prefers, from the map in Figure 6.14. Follow the procedure described next and watch carefully as your instructor demonstrates the method.

8. Construct a profile.

a. Draw a straight line 6 inches long through a feature your instructor suggests on the map of the model you just made. Clearly mark the ends of the line and label one A and the other A′ (A prime). This is your reference line or **baseline** for your topographic profile. If your instructor prefers, you can use the baseline from A to A′ already drawn on Figure 6.14.

b. First, make a **profile gauge:** take a piece of scratch paper that is longer than the baseline (6″) and lay one edge along the line, as shown in ■ Figure 6.5.

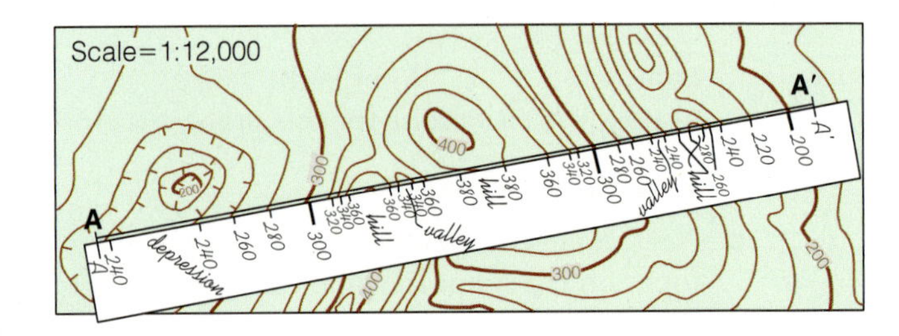

Figure 6.5

First Step in Making a Topographic Profile

This shows the map and **baseline** (A–A′) along which the profile will be constructed, and the **profile gauge**—the strip of paper used to mark off the contour positions.

c. Mark the placement of the ends and label them A and A′ on the scratch paper profile gauge.

d. Where contours intersect the baseline at the edge of the profile gauge paper, make a small tick mark on the paper perpendicular to its edge.

e. Label each tick mark with the elevation of the contour. For your model, these elevations will be 50 ft, 100 ft, 150 ft, and so forth. For the map in Figure 6.14, they will be 650 ft, 600 ft, 550 ft, and so forth.

f. It is usually helpful to label each high point (hill or ridge) and each low point (depression or valley) along the profile gauge, as shown in Figure 6.5. You should be able to recognize when you have reached a high or low point because the same contour elevation crosses the baseline twice in succession and that makes the same elevation repeat along the gauge.

g. Use the lines in ■ Figure 6.6 to construct your topographic profile. Draw a straight vertical line at the left edge of the horizontal lines and label this A (as in Figure 6.7a).

h. Lay the marked edge of the profile gauge horizontally along a line of your profile; match the tick mark labeled A on the gauge with the line labeled A on your profile. Now mark the location of A′ at the other end of the profile, draw a vertical line at this location, and label it A′. You should now have two vertical lines

Figure 6.6

Topographic Profile of Your Model Map

Draw a topographic profile of your model map, or of A–A′ in Figure 6.14 on the lines here as explained in the instructions in Exercise 8.

6 inches apart on your profile in Figure 6.6. These two lines are the edges of your profile.

i. Starting on the left of the lowest line, move upward and label every horizontal line with the elevations 0 ft, 50 ft, 100 ft, 150 ft, and so forth. If your elevations only go up to 500′ or less, you can label every other line; or if 300′ or less, every third line. You are now ready to draw your topographic profile.

j. Find the highest elevations on your profile gauge. Lay your profile gauge horizontally along the line for that elevation in Figure 6.6, matching your A and A′ ends (see Figure 6.7a). Mark a tick on your profile for each tick on your profile gauge of the same elevation. In ■ Figure 6.7, step 2 is shown for a 380-ft contour, step 3 at a 360-ft contour, then skipping down to step 8 at a 260-ft contour (Figure 6.7a, b, and c, respectively). (*Note:* Your contours have different numbers.) Next, move the profile gauge down to the next line and mark ticks on Figure 6.6 for all ticks on the profile gauge at the next highest elevation. Continue this process until you have a mark on your profile for every tick mark on your profile gauge.

k. Draw a continuous, smooth curve through all the ticks in Figure 6.6, as shown in Figure 6.7d. Connect the curve at hilltops *above* the horizontal line and at valley bottoms *below* the line. Your curve should pass through the *intersection* of the horizontal lines with each tick mark.

l. If you drew a profile of your model, you can check your profile against the model. How does your profile compare to the model? How is it similar and how is it different?

____

____

m. Calculate the relief along the profile.

____

Vertical Exaggeration in Topographic Profiles A topographic profile has two scales. If necessary, see the discussion of map scales in Lab 1, page 11. One is horizontal and matches the scale of the map. The other is the vertical scale, which depends on the spacing of profile lines and their relationship to the actual elevation values on the Earth's surface. For example, if you have a series of horizontal lines on your profile 1/10 inch apart and each one represents 50 ft of elevation, the vertical scale of the profile is 1/10″ = 50′, or 1:6,000. Depending on the two scales you used, your profile may be vertically exaggerated, which simply means that the profile is stretched vertically because the vertical scale is larger than the horizontal scale. Vertical exaggeration is useful if you want to see subtle features better, but it also changes the character of the features, making slopes steeper and relief appear greater than it is on the actual landscape.

To determine the amount of vertical exaggeration in a topographic profile, you need to compare horizontal and vertical scales. First, express both scales as fractional scales or in the same units. You can then set up the **vertical exaggeration** as:

$$\text{Vertical exaggeration} = \frac{\text{Equivalent horizontal units on the ground}}{\text{Equivalent vertical units on the ground}}$$

For example, if the horizontal scale is $1'' = 2{,}000'$ and the vertical scale is $\frac{1}{2}'' = 250'$, then convert the vertical scale to $1'' = 500'$, then

$$\text{Vertical exaggeration} = \frac{2{,}000'}{500'} = 4\times$$

Notice in this example that we converted both scales to the same units first. In another example, a profile has a horizontal scale of 1:24,000 and a vertical scale of 1:3,000. The vertical exaggeration can be calculated directly as $24{,}000/3{,}000 = 8\times$.

Profiles that are drawn with the vertical scale identical to the horizontal scale are *normal profiles* and have a vertical exaggeration of 1× or no exaggeration. If the relief is low in an area or if topographic features need to be accentuated, you can increase the vertical scale in relation to the horizontal scale; this makes an **exaggerated profile,** for which the values are > 1×.

9. In this exercise, you calculate the vertical exaggeration of the profile in Figure 6.7d.

a. First, precisely measure the vertical distance between the 400 and the 200 lines.

____

b. Next, calculate the vertical fractional scale for the profile reviewing the explantion for determining the scale on page 11 in Lab 1. Show each step of the calculations:

Vertical fractional scale: _______________

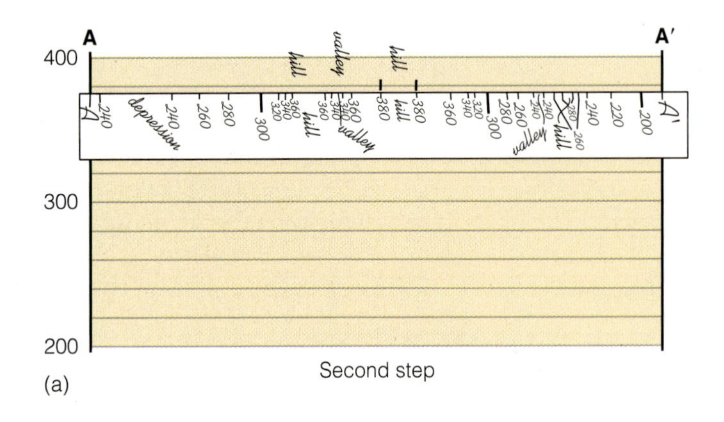

(a) Second step

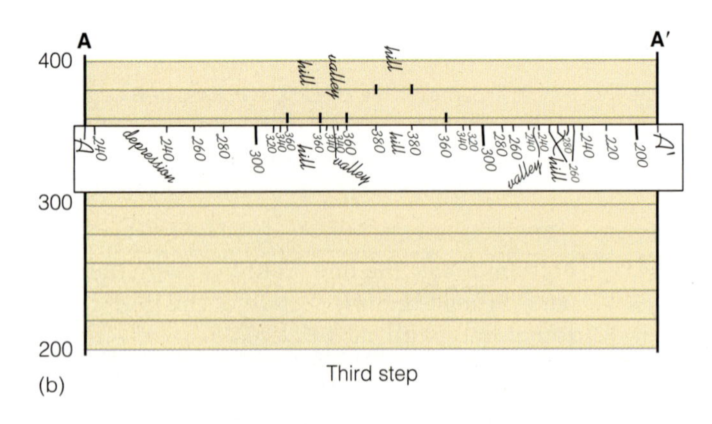

(b) Third step

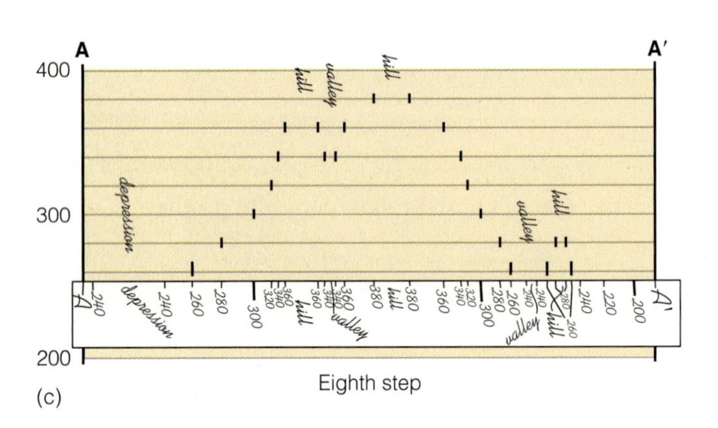

(c) Eighth step

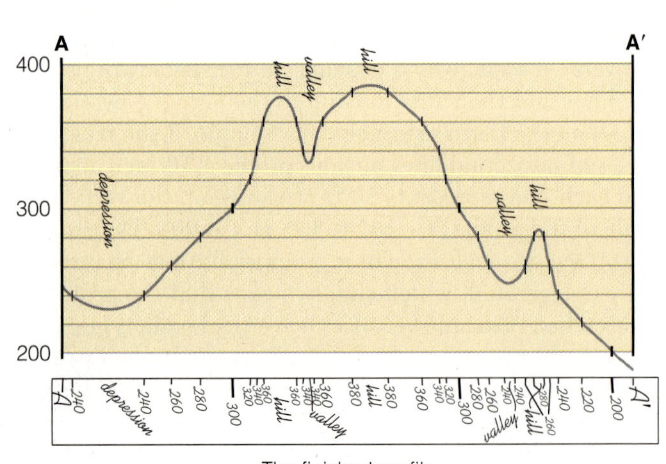

(d) The finished profile

Figure 6.7

Topographic Profile Construction

More steps in making a topographic profile, continued from Figure 6.5 and Exercise 8j and k. (a) Step 2. (b) Step 3 . . . (c) Step 8. (d) the finished profile.

c. The scale of the map is 1:12,000. Next, calculate the vertical exaggeration as described in the previous paragraphs. Show each step of the calculations:

Vertical exaggeration: ____

10. Draw another profile from A to A′ in the map in Figure 6.5 on the lines in ■ Figure 6.8. How does this profile compare with the one in Figure 6.7d?

____

11. Calculate the vertical exaggeration.

a. First, calculate the fractional scale for the vertical direction on your new profile in Figure 6.8. Show each step of the calculations:

Vertical fractional scale: ________

b. Now, calculate the vertical exaggeration. Remember that the scale of the map is 1:12,000. Show each step of the calculations:

Vertical exaggeration: ____

12. In general, if the number for the vertical exaggeration is greater, how will the profile be different?

____

Figure 6.8

A Topographic Profile

Draw a topographic profile as instructed in Exercise 10.

Constructing a Topographic Map from Elevation Information

When constructing a topographic map, **cartographers** (mapmakers) draw contours using elevations of points from surveying techniques, GPS data, and aerial photography. The following exercises should help you construct contour maps from a map with elevations given at various points. Do the exercises in order, as the difficulty increases. *Carefully follow the instructions and start each exercise by writing down the contour interval in the figure caption for that map.*

13. On ■ Figure 6.9 connect the x's of equal elevation to make a contour map with a contour interval of 10 ft as described here. Contours make a V shape when they cross a stream. The V points upstream. Using a pencil with eraser, connect the x's for the 110-, 100-, and 90-ft contours in order, first, keeping in mind that a stream can only cross a single contour elevation once. Notice how these contours V upstream. Next draw in the 80- and 70-ft contours with a V similarly

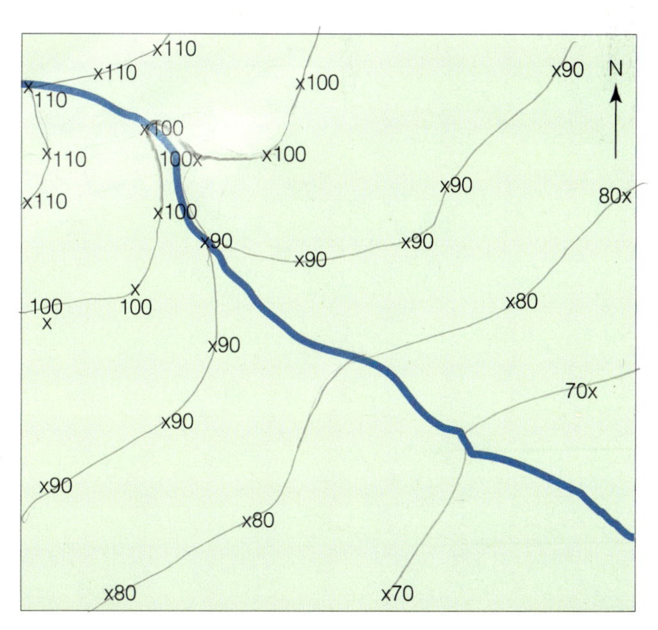

Figure 6.9

Map of a Stream Valley

Follow the instructions in Exercise 13 to make this into a topographic map. Obtain the contour interval from the instructions: CI = 10

shaped to the ones at higher elevation. Label the elevation of each contour in a small gap in the contour line.

14. Cartographers use all elevation data available, not just elevations that exactly match the contour interval, so let's try to do the same. On ■ Figure 6.10, draw in the contours at a 10-ft contour interval and label the elevation of each contour. Here are some suggestions to help you do this:

- Remember that each contour must be a multiple of 10. Notice that this time you will ***not*** be connecting the X's. Each line must pass between the various X's because no elevation is exactly at the contour interval.
- Some students find it helpful to draw in their own X's at the elevation of the contour (every multiple of 10′ elevation in this example) by estimating where these values would occur between elevations given on the map and then connecting those X's. Thus, for example, if a map had an elevation marked on it at 87′ and another at 91′, you would mark an X ¾ of the way between the two elevations to estimate the location of the 90′ contour. After a while, you can do this by eye without actually drawing in your own X's.
- To draw your lines correctly, all of the numbers **higher** than your contour elevation need to be on one side of the line and all numbers lower on the other side.
- Draw your contour lines so they come close to X's with similar elevations and stay farther away from X's with more different elevations.

15. ■ Figure 6.11 has a stream valley near the coast and a contour interval of 20 ft. When you draw the contours on this map, don't forget to make the contours V upstream. The coastline gives you additional information. The line at the seashore is equivalent to a 0-ft contour. Notice that closely spaced contours near and parallel to the shoreline indicate the presence of steep sea cliffs. Label the elevation of each contour.

16. Examine the sketch of the coastal area in ■ Figure 6.12a and the elevations in feet given on the corresponding map in Figure 6.12b. Draw contour lines on the map with an interval of 50 ft. Since the seashore is 0 ft and the cliff is 101 ft high, you must

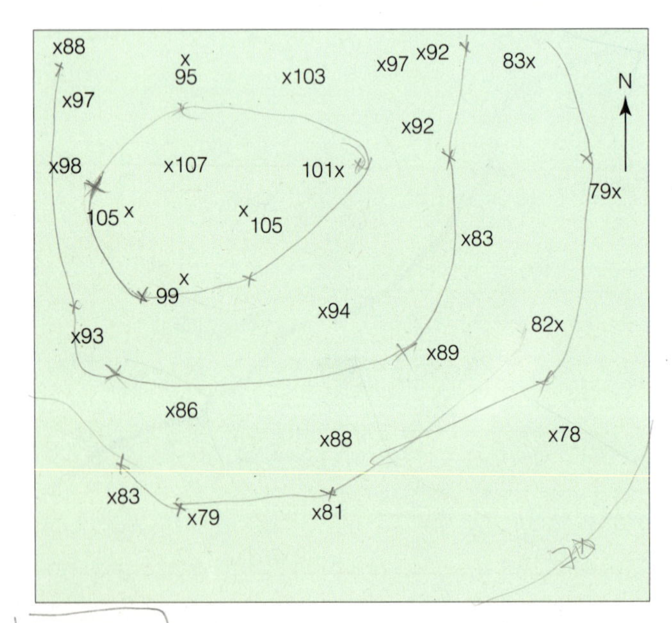

Figure 6.10

Map of a Hill

Use the instructions in Exercise 14 to make this into a topographic map. CI = ______

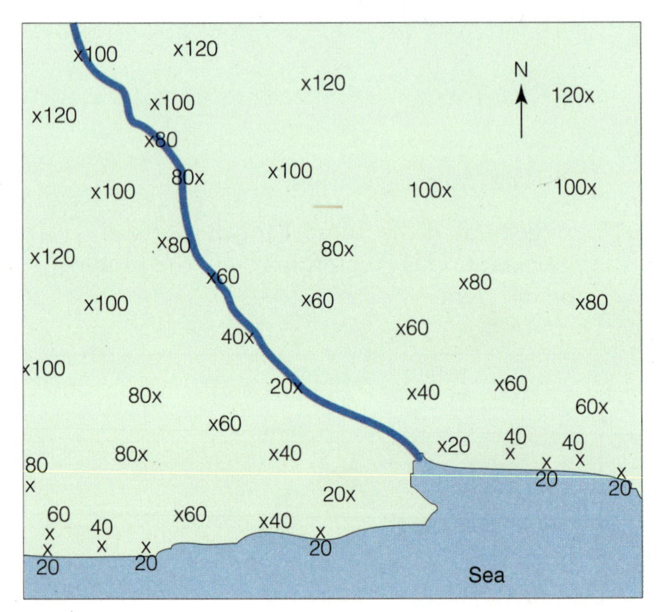

Figure 6.11

Map of a Stream and Sea Cliff

Use the instructions in Exercise 15 to make this into a topographic map. CI = ______

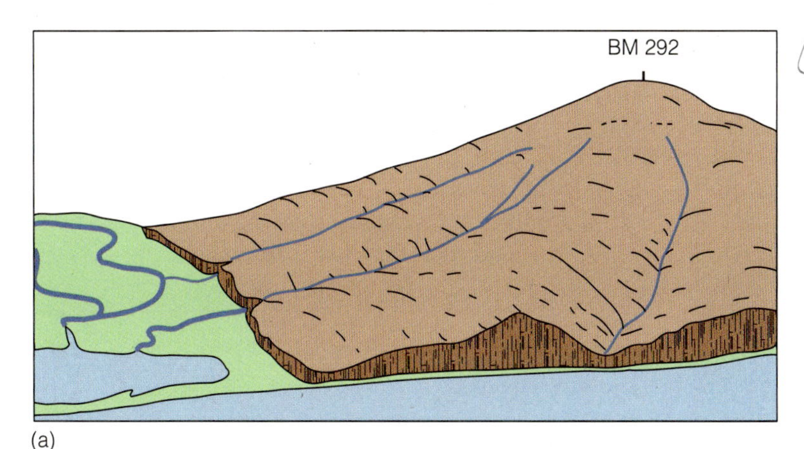

(a)

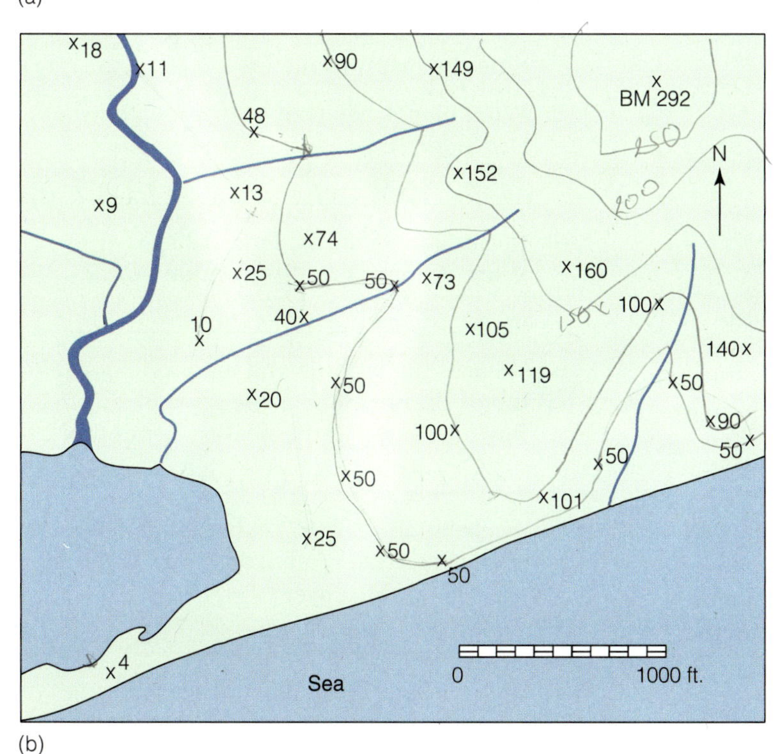

(b)

Figure 6.12

A Coastal Area and Its Map

(a) Perspective view. (b) Topographic map. Follow instructions in Exercise 16 to draw the contours on the map. CI = ______

draw your 50- and 100-ft contours parallel to the shore along the cliff between the sea and the 101-ft elevation point. Label the elevation of each contour.

USING TOPOGRAPHIC MAPS

Scale of U.S. Geological Survey Maps

The U.S. Geological Survey is the government agency responsible for creating official maps of the United States. The most popular varieties of maps are their topographic quadrangle series. The number of degrees or minutes of latitude and longitude that they cover (■ Table 6.1) identify each series. ■ Figure 6.13 shows relative size and scale of USGS topographic maps.

Table 6.1

USGS Topographic Quadrangles

Area of coverage and scale of US Geological Survey topographic maps.

Quadrangle Name	Quadrangle (lat. × long.)	Scale
2-degree sheet	1° × 2°	1:250,000
1-degree sheet	30′ × 1°	1:100,000
30-minute quad	30′ × 30′	1:125,000
15-minute quad	15′ × 15′	1:62,500
7½-minute quad	7½′ × 7½′	1:24,000
Alaska		
15-minute quad	15′ × 15′	1:63,360

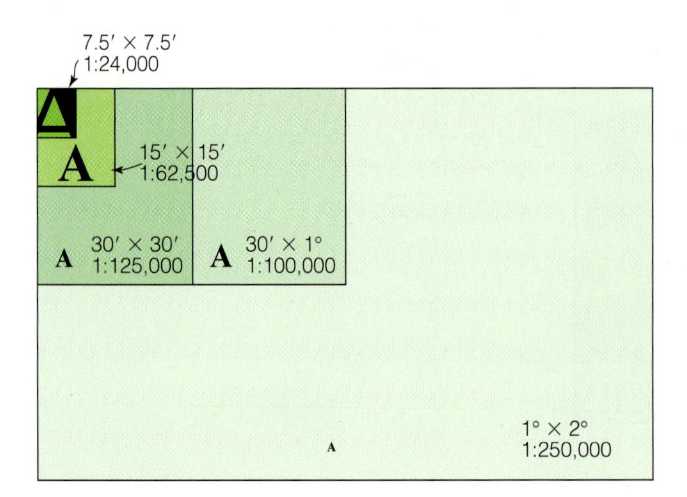

Figure 6.13

USGS Topographic Map Comparison

Comparison of the scales and relative areas covered by standard U.S. Geological Survey topographic maps. The letter A on each map shows the size that a 1,000-ft-tall letter A lying on the ground would occupy on the map. The numbers 7.5′, 15′, 30′ refers to 7½ minutes, 15 minutes, and 30 minutes, respectively, of latitude and longitude covered by each map. The notations 1° and 2° refer to 1 or 2 degrees of latitude and longitude covered by those maps.

In the next section, you will utilize the skills you have acquired in learning to read and interpret topographic maps.Before you answer the following questions, also review latitude and longitude, UTM grids, and the Township and Range System, on pages 13–16.

Exercise Using a USGS Topographic Map

The area around Morro Bay, California in ■ Figure 6.14 has seven to nine distinctive hills, depending on how large you consider a hill, called the 7 Sisters (also called the 9 Morros), which are volcanic plugs eroded from volcanoes. The area also has a spit and a delta, some rolling hills, nearly flat valleys, in addition to some rugged hills and valleys. Apply your knowledge of contours and grid systems to answer questions about this USGS topographic map.

17. Use the map of Morro Bay and vicinity on the Cayucos, California map, in Figure 6.14 to practice some location, distance, and grid-related exercises.

a. What type of USGS topographic map is this map (refer to Table 6.1)?

15 minute quad ______

b. Where in California is Morro Bay located?

SW ______

What was the magnetic declination at the time the map was made? 16 1/2 ______

c. What map is adjacent to this map on the east side?

San Luis Obispo ______

d. Locate lat. 35°20′N, long. 120°50′W. What is present at this location on the map? Morro Bay ______

e. What is the latitude and longitude of the top of Black Hill near the town of Morro Bay, measured to the nearest ½ minute?

35°22′N 120°50′W ______ What is its elevation? 665 ______ What is the name of the elevation symbol you used (refer to the back inside cover)?

Checked spot elevation ______

f. What feature is located at the following UTM coordinates?

i. $^{6}94^{100m.}$E., $^{39}16^{200m.}$N. ______

ii. $^{7}00^{150m.}$E., $^{39}14^{260m.}$N. ______

g. What are the UTM coordinates for the following features?

i. Hollister Peak: ______

ii. The big + sign in southeastern Morro Bay: ______

Figure 6.14

Topographic Map of Morro Bay Area, California (*facing page*)

Morro Bay and the surrounding region from the USGS Cayucos 15-minute quadrangle. The edges as well as the middle of a topographic map are rich with information. The center shows topographic information and the edges show grid system information, scale, location information, north arrows, and adjacent maps.

Morro Rock
(STATE PARK)
Pillar Rock
Beacon
Breakwater
SAN BERNARDO
(CANE)
Morro Bay
Water Tank
Black Hill
MORRO BAY
Fairbank Pt
White Pt
STATE PARK
Chorro
Cerro Cabrillo
Banning Sch
Quintana Cem
SAN LUISITO
San Luisito
Creek
Park Ridge
Hollister Peak
Rifle Range
MORRO BAY
MORRO BAY STATE PARK
Baywood Park
Gravel Pit
Windmill
Eto Lake
Warden Lake
Cuesta-By-The-Sea
CAÑADA DE LOS OSOS
LOS OSOS Y VALLEY
Sunnyside Sch
Reservoir
Water Tank
Los Osos Sch
Los Osos Creek
PRIVATE
Hazard Canyon
PECHO Y ISLAY
IRISH HILLS
Clark Valley
Prospect
Springs
JEEP TRAIL
Islay
Creek
Valencia Peak
SAN LUIS OBISPO
SAN LUIS OBISPO (JUNC. U.S. 101) 7 MI.
SANTA MARIA 39 MI.
(SAN LUIS OBISPO 1854 I
(PORT SAN LUIS) 1854 III
T. 30 S.
T. 31 S.
R. 10 E.
R. 11 E.
35°15'
120°45'
3904000m.N.
703000m.E.
INTERIOR—GEOLOGICAL SURVEY, WASHINGTON, D. C.—1971
SCALE 1:62500
4 MILES
5 KILOMETERS
GN
MN
1°14' 22 MILS
16½° 293 MILS
CONTOUR INTERVAL 50 FEET
DATUM IS MEAN SEA LEVEL
DEPTH CURVES IN FEET—DATUM IS MEAN LOWER LOW WATER
SHORELINE SHOWN REPRESENTS THE APPROXIMATE LINE OF MEAN HIGH WATER
THE AVERAGE RANGE OF TIDE IS APPROXIMATELY 4 FEET
CAYUCOS, CALIF.
N3515—W12045/15
1951
CALIF.
QUADRANGLE LOCATION

h. What section is just north of Sec. 27, T. 29 S., R. 11 E.? ____

i. If the township and range system extended into Clark Valley, what section would be just north of Sec. 3, T. 31 S., R. 11 E.? Give the complete designation for this section:

____

j. Use the graphic scale to determine how far it is from the breakwater to the southern border of Morro Bay State Park at the coast, as the crow flies. ________

18. Use the map of Morro Bay to practice reading contours and features of topographic maps.

a. How do the contours at Black Hill show that it is a hill?

____

b. Describe the topography of the area between Warden Lake and Eto Lake. (*Hint:* Is it steep and rugged, gently rolling, nearly flat, or what? Look at the contour arrangement and spacing.)

____

c. Describe the topography on both sides of the Islay Creek valley in Sec. 5 on the southern edge of the map. Contrast this area with the area you described in the last question.

____

d. Locate the largest marsh or swamp on the map. How are people using the land there? What geographic name does this land use have?

____

The marsh has a triangular shape like the capital Greek letter delta (Δ). What geologic feature is this? Include the geographic name of the creek that is creating this marsh.

____

e. Compare the slopes of the sides of the Irish Hills with the slopes of their tops.

____

____

f. Los Osos Valley is a drainage valley with an **intermittent stream.** It flows only when water is available, as indicated by the blue dot-dash line in the valley. What direction—east, west, north, or south (circle one or two)—does the stream flow when water is flowing?

g. What would it be like on the ground in areas colored green on the map?

____

h. What does the green dot pattern on the map mean?

____

i. Note the placement of the roads. What topography must an engineer consider when choosing the path of a road?

____

j. Note the scale, the contour interval, and the relief for the map; use the appropriate units.

Scale = ____________

CI = ________

Relief = __________

k. What is the elevation of the highest point within Sec. 4, T. 31 S., R. 11 E.?

________ What is the elevation of the lowest point in the same section?

________ Calculate the relief of the section. ________

l. You decide to go on a hike along a straight line from the **I** in Islay Creek to the **r** in Springs near Irish Hills. What would your hike be like along the way?

____

____

____

____

What is the highest elevation you would reach along the way? ______ How would this hike compare to taking the unimproved road to the cluster of buildings below the springs and then hiking up to the springs?

____

____

____

AERIAL PHOTOS GIVE A VIEW OF THE THIRD DIMENSION

Aerial photos, photographed in pairs and viewed simultaneously, give a three-dimensional view. These pairs are a major tool in the making of modern topographic maps and are also useful for viewing topography directly. From an airplane, a mounted camera pointing straight down takes the photos. As the plane flies over an area, the photographer shoots the first picture (Perspective 1 in ■ Figure 6.15); a short distance farther, she snaps a second image (Perspective 2). The two photos must overlap, as in Figure 6.15. When two images of the same area have different points of view (as if the right eye saw one and left eye saw the other), viewing them simultaneously with the aid of a **stereoscope** gives a single image in three dimensions (called *stereo*). Your eyes do this naturally, allowing you to see in 3D. One eye sees things from one perspective; the other eye, from a slightly different perspective. Your brain processes the information in such a way that you interpret it as 3D. You may notice that the view with a stereoscope looks like a miniaturized landscape.

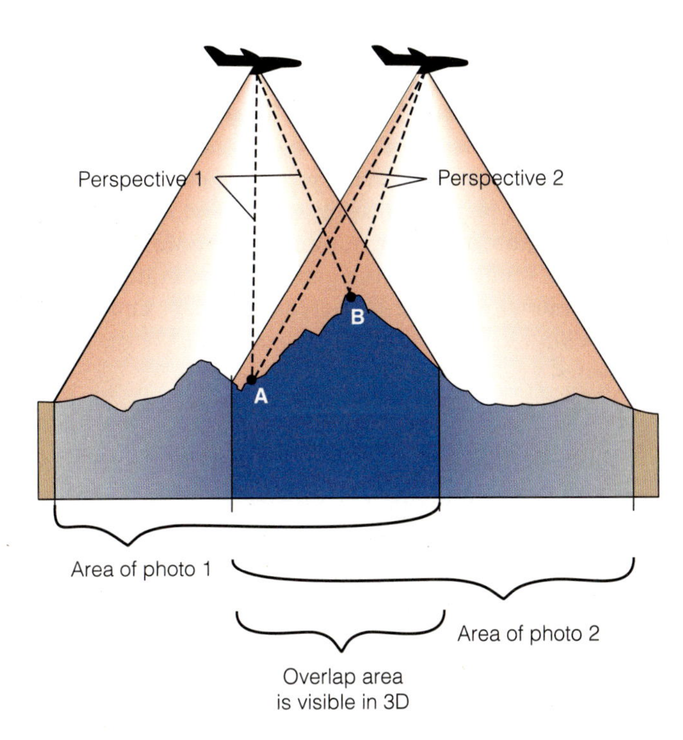

Figure 6.15

Making a Stereo Pair

A photographer in the airplane takes a pair of photographs, one from Perspective 1 and the other from Perspective 2. The angle between a low point A and a high point B is very different in the two perspectives. As a result, in photo 1 points A and B are farther apart than in photo 2. When you view the resulting photo pair with separate eyes, a three-dimensional image appears.

Using a Stereoscope with Pairs of Images on the Same Page

It is quite simple to use a stereoscope with a pair of images that are part of a single page, as are those in this lab book. Adjust the stereoscope to match your eye spacing. Place one lens of the stereoscope over one photo and the other lens over the other photo so that two similar objects are directly below the lenses (■ Figure 6.16). Now look through the stereoscope with your eyes relaxed (as if you are going to sleep). You should be able to see a three-dimensional image appear in the center of three images. Ignore the images on the left and right.

19. In ■ Figure 6.17 you can view a topographic map in stereo (3D) using a stereoscope. This is the same as the first topographic map in Figure 6.1, but with the addition of 3D perspective.

a. How many hills are partially cut off by the east edge of the map? _____
How many valleys slope down to the east on that side? _____

b. How many depressions does the map show without hachure marks? Don't count valleys that slope down off the map edge. _____

c. Label four steep areas each with an S, and three gently sloping areas each with a G.

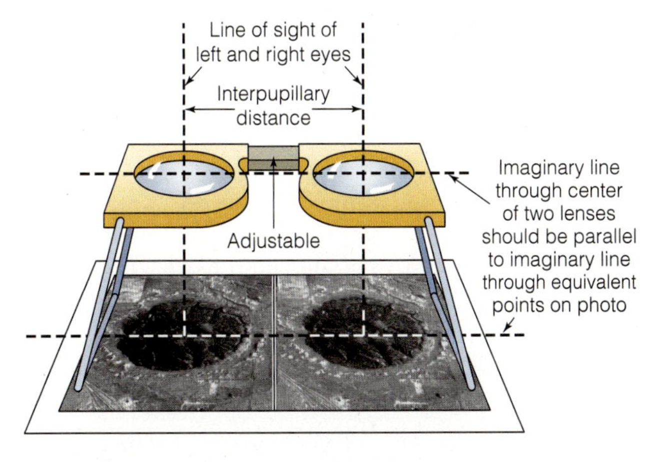

Figure 6.16

Using a Stereoscope

Proper positioning for a stereoscope over a stereo pair.

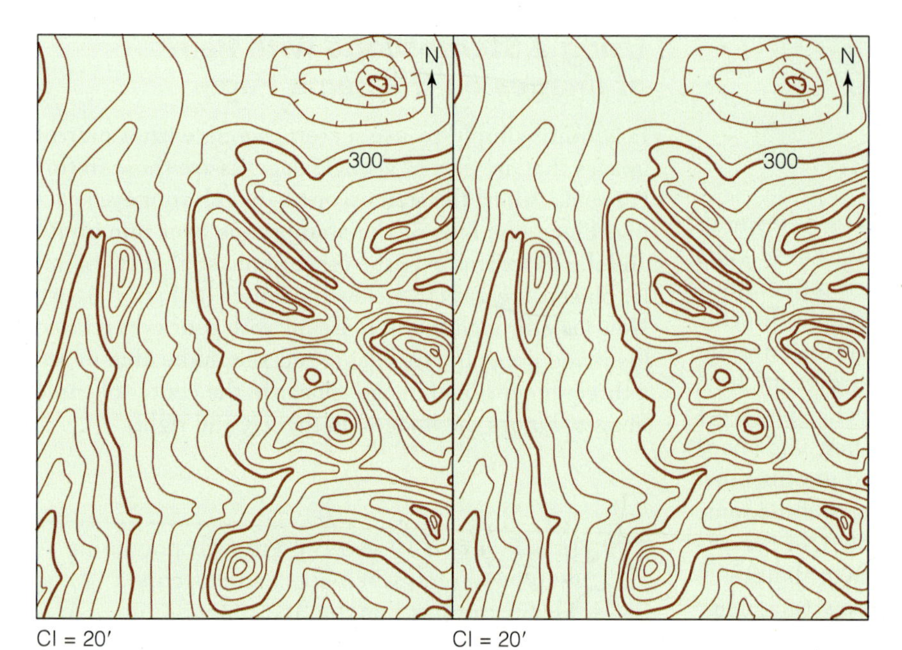

Figure 6.17

Topographic Map Viewable in Three Dimensions

These two images are part of the same map as Figure 6.1, but we moved the line positions depending on their elevations so when you view them with a stereoscope they show the differences in elevation in 3D.

d. The index contour to the south of the north arrow is 300′. In 3D, the north arrow has a particular elevation. Use the nearby contours to determine its elevation. ______

20. Next, view the pair of aerial photos in ■ Figure 6.18.

a. Are the hills or mountains in these photos steep and rugged or gently rounded?

______

b. Does the area have steep gullies/canyons or gently rounded valleys?

______

c. Glacial (ice) erosion formed the topography in this area. Is the glacier still there?

______

d. Read the figure caption. What type of topographic feature do you think an *arête* is? Just use the topography to guide your answer.

______

e. What do you think a *tarn* is?

______

f. Mark a high point on the ridge in the center of one of the pair of photos with an HR and a low spot with an LR.

g. Label a low lake LL and a high lake HL.

Meteor Crater, Arizona Meteor Crater (at approximately 35°N and 111°W) is a large crater formed by a meteorite impact (■ Figure 6.19). This extraordinary feature near Winslow, Arizona, formed when a large mass from outer space, about 25 m in diameter, collided with the Earth's surface in the geologic past. The impact produced the large depression shown and elevated the crater rim above the surrounding plain. Many impact sites have small hills in the center of the crater; these are additional evidence of meteorite impact and result from a kind of "splashing-back" action of the material excavated by impact from the crater.

21. Study the aerial stereo pair of Meteor Crater, Arizona, in Figure 6.19 using a stereoscope.

a. In ■ Figure 6.20, construct a rough sketch of a topographic profile across the crater from west to east (left to right). This is just to visualize the crater and not an exact construction. To make this simple, assume the rim of the crater is 600 ft and the floor is 0 ft.

b. Is the floor of the crater steep or gently sloping?

______

How about the sides of the crater?

______

Figure 6.18

Arêtes and Tarns Near Mount Bonneville, Wyoming

This stereo pair of aerial photographs has arêtes partially encircling tarns. Snow and ice are white; lakes are black.

Figure 6.19

Meteor Crater, Arizona

A pair of aerial photographs showing the location of a *meteorite* impact. A **meteor** is a shooting star—a visible streak of light resulting from a space particle that burns up in the atmosphere, or hits the edge of the atmosphere and passes back into space. Typically these particles are the size of sand grains, although larger ones are possible. A **meteorite** hits the Earth's surface without having been completely vaporized.

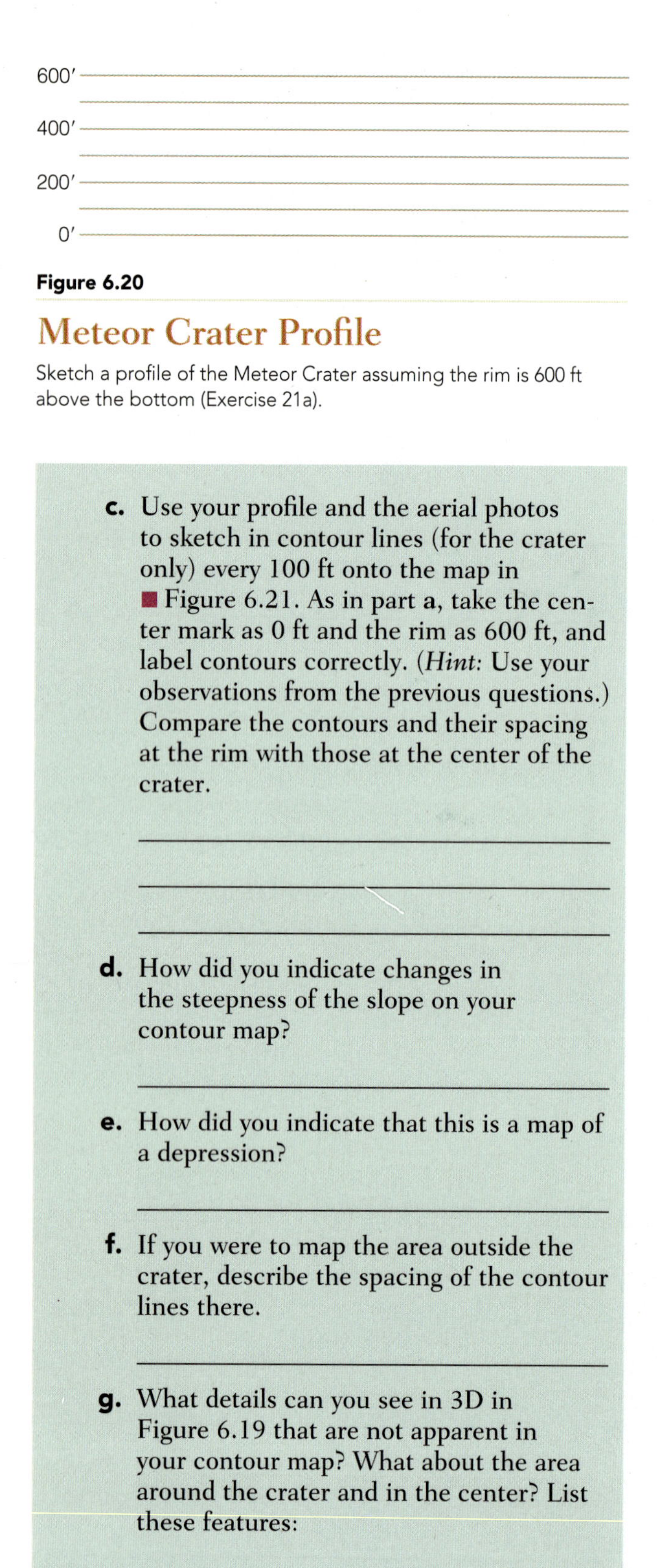

Figure 6.20

Meteor Crater Profile

Sketch a profile of the Meteor Crater assuming the rim is 600 ft above the bottom (Exercise 21a).

c. Use your profile and the aerial photos to sketch in contour lines (for the crater only) every 100 ft onto the map in ■ Figure 6.21. As in part **a**, take the center mark as 0 ft and the rim as 600 ft, and label contours correctly. (*Hint:* Use your observations from the previous questions.) Compare the contours and their spacing at the rim with those at the center of the crater.

____

____

____

d. How did you indicate changes in the steepness of the slope on your contour map?

____

e. How did you indicate that this is a map of a depression?

____

f. If you were to map the area outside the crater, describe the spacing of the contour lines there.

____

g. What details can you see in 3D in Figure 6.19 that are not apparent in your contour map? What about the area around the crater and in the center? List these features:

____

____

____

h. Circle any of the details listed in the previous answer that are evidence of meteorite impact.

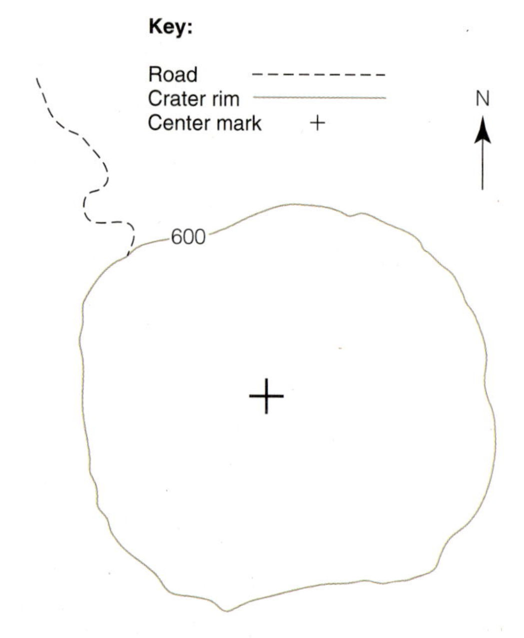

Figure 6.21

Meteor Crater Sketch Map

Sketch a topographic map of Meteor Crater, Arizona (Exercise 21c). The rim contour has been drawn for you, although a special symbol might need to be added. Obtain the contour interval from the instructions: CI = ______

Using the Stereoscope with Pairs of Separate Images

Aerial photos are commonly available at university libraries and commercially, but the pairs are separate photos that you must position properly for viewing. Follow your instructor's methods for using the equipment and supplies that are available to you.

22. Your instructor will provide a pair of aerial photographs. Describe what you see in the stereo pair. What features can you see in 3D that are not evident when looking at a single photo without a stereoscope? Answer any questions provided by your instructor.

____

____

____

____

____

____

____

LAB 7

Geologic Time and Geologic History

OBJECTIVES

- To understand how to use methods of relative dating
- To understand how unconformities develop
- To reconstruct geologic history from cross sections
- To determine the isotopic age of rocks using information about the half-life, and parent and daughter isotopes in the rock
- To use indirect dating techniques to determine the possible numerical age ranges of sedimentary rocks
- To chart the timing of some of the major events in Earth's history

Figure 7.1

Grand Canyon

The Grand Canyon displays rocks of a wide range of ages from the ¼-billion-year-old strata at the top to metamorphic rocks in its depths that are up to eight times that old. Several unconformities occur in the Grand Canyon, one of which is the nonconformity visible here.

One of the goals of the field of **historical geology** is to understand how the Earth has changed over eons of time. Reconstructing Earth's history requires establishing both a sequence of events and utilizing various methods of measuring geologic time. In the 19th century, geologists established a detailed sequence of Earth's history using observations of the relationships between rocks to determine their *relative* age. When you determine the **relative age** of a rock, instead of establishing its numerical age in years before the present, you are comparing its age to other rocks using words such as *older* or *younger*. For example, the rocks at the bottom of the Grand Canyon are much older than the rocks at the top (■ Figure 7.1).

In this lab, we will follow the procedures geologists use to establish the relative ages of rocks. The tools we use can be very powerful: they can reveal the past growth and subsequent erosion of entire mountain ranges, the development of past landscapes, and the evolution of life.

The discovery of radioactivity in 1898 by Henri Becquerel revolutionized the study of geologic time in the 20th century. The natural decay of radioactive isotopes gave geologists a way of measuring the age of certain geologic materials. With this new tool, called **isotopic dating** or **radiometric dating,** geologists could start assigning numerical values to some rock materials to complement the previously determined relative ages. For example, we now know that some of the rocks in the bottom of the Grand Canyon are between 1.706 and 1.697 billion years old with individual dates as precise as ±0.001 billion years. The modern geologic time scale grew out of this combination of numerical and relative ages.

RELATIVE AGE

In Lab 4, we saw that sedimentary rocks have bedding, a kind of planar feature in the rock. A layer of sedimentary rock that is visually separable from other layers is a bed or **stratum**. The plural for stratum is **strata** (Figure 7.1). Because of the way they form, beds or strata generally follow or obey certain laws called the *laws of stratigraphy*.

Stratigraphic Laws

Stratigraphy is the study of strata, or sedimentary layers. The **principle of original horizontality** states that sedimentary layers are deposited horizontally or nearly so. Even after hundreds of millions of years have passed, many of the rocks in the Grand Canyon still have nearly their original horizontal orientation (Figure 7.1) whereas others are inclined or tilted. For inclined beds (excluding cross-beds), this principle implies that after deposition, regional deformation tilted the layers (Lab 8). In some mountainous areas, deformation is so great that tilting is greater than 90°. In this case, we say the rocks are **overturned** (which is a rare occurrence; ■ Figure 7.2).

Sedimentary rock layers are deposited in sequence one on top of the other, so that the oldest rocks are at the bottom of the sequence and the youngest rocks are on top (■ Figure 7.3). This is the **principle of stratigraphic superposition.** Certain sedimentary features,

Figure 7.2

Overturned Rocks

The rocks in the upper center of this photo are overturned (tilted more than 90°) and form part of a syncline in folded rocks.

Figure 7.3

Stratigraphic Superposition

Relative age of sedimentary and volcanic rocks can be determined from their sequence. The oldest occur at the bottom and the youngest at the top in layered strata that have not been overturned. The numbers indicate the order of formation of the layers here. The sill is intrusive and does not obey the law of stratigraphic superposition—it formed at some time after Layer 1. From MONROE/WICANDER, *The Changing Earth*, 5e. 2009 Brooks/Cole, a part of Cengage Learning, Inc. Reproduced by permission. www.cengage.com/permissions

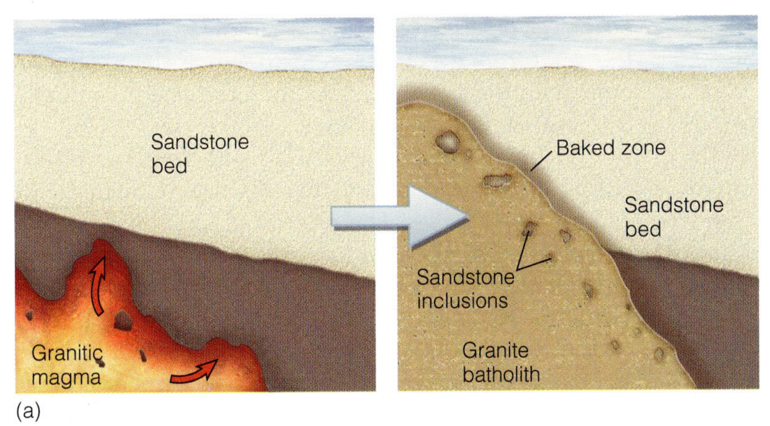

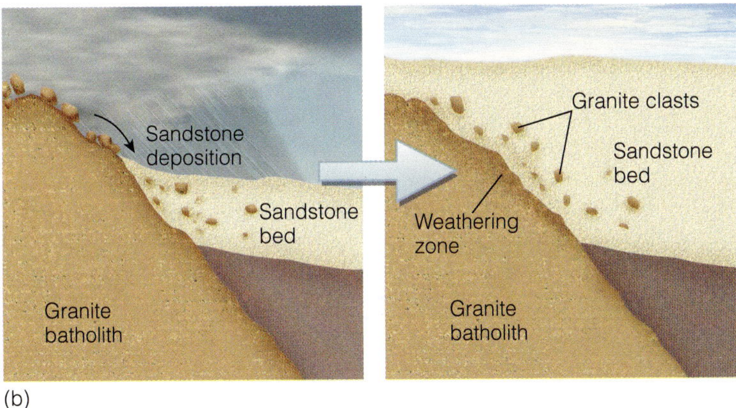

Figure 7.4

Inclusions

Inclusions are older than the rocks that contain them. Sequences (a) and (b) show how inclusions may become incorporated in a rock in two steps. (a) Inclusions of sandstone in granite are older than the granite. (b) Inclusions of granite in sandstone are older than the sandstone. From MONROE/WICANDER, *The Changing Earth*, 5e. 2009 Brooks/Cole, a part of Cengage Learning, Inc. Reproduced by permission. www.cengage.com/permissions

such as graded bedding, cross-bedding, ripple marks, and mud cracks (Figure 4.22, p. 89), may indicate whether a sequence has been overturned after deposition and lithification. Sediments, sedimentary rocks, and volcanic rocks, including lava flows (Figure 7.3) and volcanic ash, obey the law of stratigraphic superposition.

When rocks form, they sometimes incorporate preexisting rocks into them. This leads us to the **principle of inclusions**: the rock containing inclusions is younger than the inclusions it contains (■ Figure 7.4). All rock types obey the law of inclusions.

Cross-Cutting Relationships One geologic feature may cut across another, indicating that the first feature is younger than the second. This is the principle of **cross-cutting relationships.** For example, a dike cutting across a bed of sandstone indicates that the dike is younger than the sandstone. In ■ Figure 7.5, dike D cuts across rocks A, B, and C. Dikes, stocks, and batholiths, discussed in Lab 3 (Figure 3.27, on p. 64), are cross-cutting igneous intrusions younger than the rocks they intrude. Sills and laccoliths are intrusions that parallel the bedding and are more difficult to recognize, but they, too, are younger than the rocks they intrude (Figure 7.3). Faults and erosion can also cut across rocks. Fault F in Figure 7.5 cuts across sedimentary rocks A, B, C, and E, and dike D. Surfaces of erosion **1** and **2** (*unconformities*) cut through the dike and the fault in Figure 7.5. Similarly, folds are younger than the folded layers, and a period of metamorphism is younger than the rocks it metamorphoses.

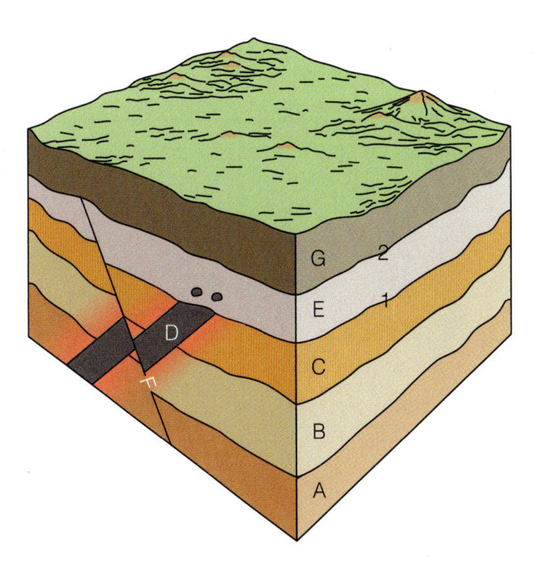

Figure 7.5

Cross-Cutting Relationships

Dike D cuts across and is younger than the beds it cuts and older than those it does not. Erosion cut the top of the dike so this event is younger than the dike. This block diagram has two unconformities, 1 and 2.

Fossil Succession Fossil evidence is another useful tool in determining relative age of rocks. Recall from Lab 4 that a **fossil** is any evidence of preexisting life preserved in a rock. A clamshell in a rock can be a fossil; so can a footprint of a dinosaur. Fossils help us date rocks because of the **principle of fossil succession** (also known as the *law of faunal succession*). This principle states that organisms evolve in a definite order, that species evolve and become extinct, never to re-evolve. Thus, the evolution of a species and its extinction become time markers separating time into three units: one before the organism existed, one during the existence of the organism, and one after the organism went extinct. The presence of a particular fossil species indicates that the rock containing the fossil was formed at a time between the evolution and extinction of that species.[1]

Index, or **guide fossils** are organisms that existed for a short period of time, then went extinct. This short lifetime helps to narrow the possible age of rocks containing them, more than fossils that existed for a long time (■ Figure 7.6). Thus, the fossils present in a rock are characteristic of the time when the rock formed and the *depositional* environment in which the preserved organisms lived. This aids us in determining details of the geologic history of an area.

Rocks with matching sets of fossils from two different localities are probably of similar age (■ Figure 7.7) and are **correlative.** The process of matching them is **correlation.** In addition to fossils, sometimes a unique or distinctive layer called a **key bed,** which forms from a widespread rapid event, is a time marker and helps with correlation. An example is a widespread volcanic ash bed resulting from an explosive volcanic eruption. Correlated areas may be close in distance or regional.

Cenozoic: Quaternary, Tertiary
Mesozoic: Cretaceous, Jurassic, Triassic
Paleozoic: Permian, Pennsylvanian, Mississippian, Devonian, Silurian, Ordovician, Cambrian
Lingula
Atrypa
Paradoxides

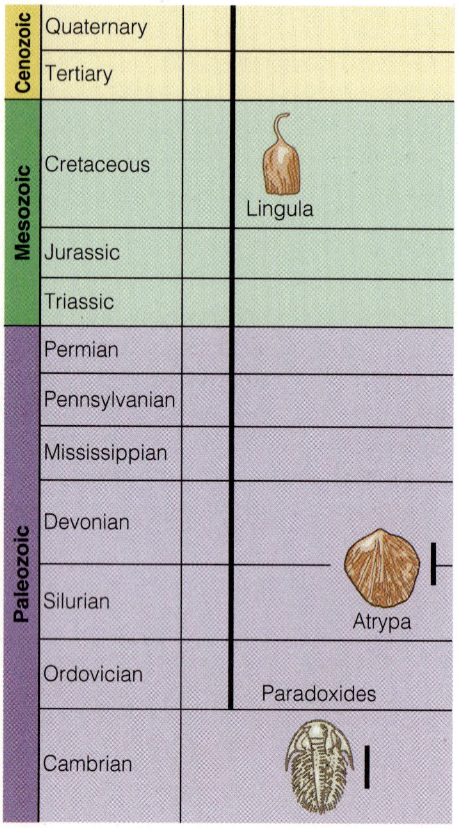

Figure 7.6

Index Fossils

The time ranges for three fossils: The long time range for *Lingula*, including today, makes it unsuitable for dating rocks in which it is found. On the other hand, *Atrypa* and *Paradoxides* both have short geologic ranges, are good indicators of geologic age, so they qualify as *index* or *guide fossils.*

Unconformities

Other features found in rocks that reveal much about the geologic history of an area are *unconformities.* When an area is deformed, uplift and erosion often occur. Later subsidence (sinking of the land) or submergence and *transgression* due to sea level rise and fall may take place, followed by deposition of sediment on the eroded surface. **Transgression** is the progressive deposition of sediment farther and farther inland on a continent

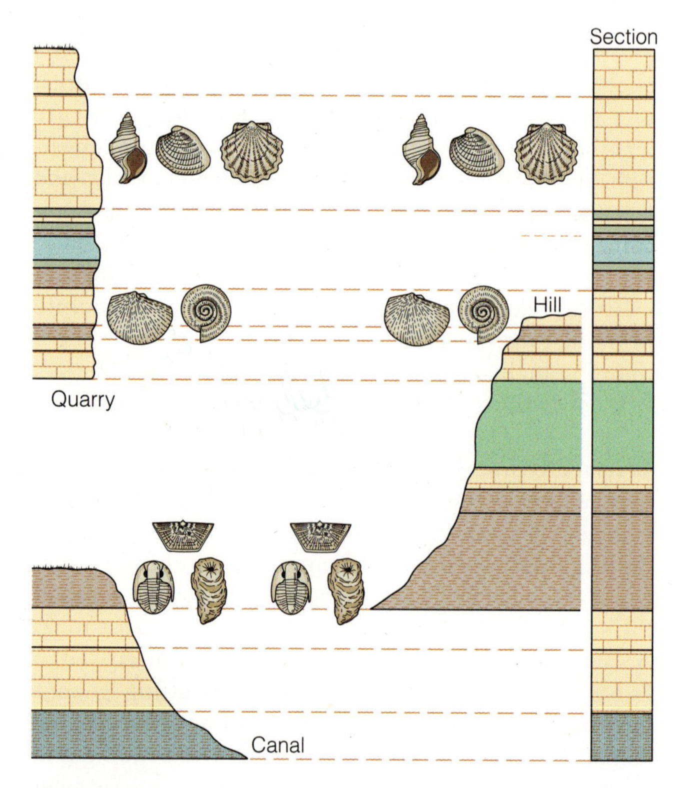

Figure 7.7

Correlation

Matching strata of the same age (correlation) using fossil succession. Rocks exposed in a quarry, on a hill, and in the walls of a canal have been correlated based on equivalent sets of fossils. Once the correlations have been made, a complete section for the region (at the right) shows the relative ages of all of the beds. Adapted from Fenton & Fenton, *Giants of Geology.*

[1] In the exercises that follow, we assume that the absence of a fossil in a column means the organism did not exist on Earth at that time, even though a variety of situations can actually cause its absence. For example, the organism might migrate to or from an area because of a favorable or unfavorable environment. Another possibility is the lack of preservation of remains of organisms.

as sea level rises relative to the land. These sediments are deposited on a surface of much older rocks. Thus, an **unconformity** is a boundary between rocks of distinctly different ages where rocks were either deposited and then eroded, or simply not deposited (■ Figure 7.8).

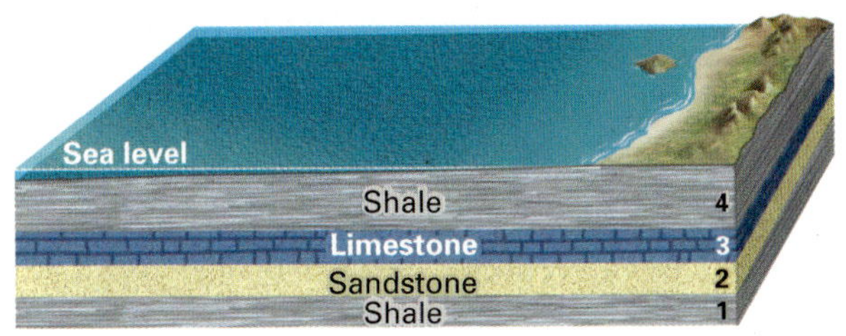

(a) Sediment is deposited below sea level

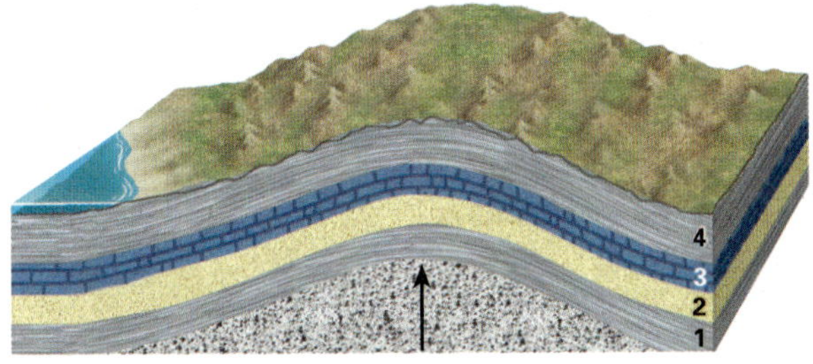

(b) Rocks are uplifted and tilted

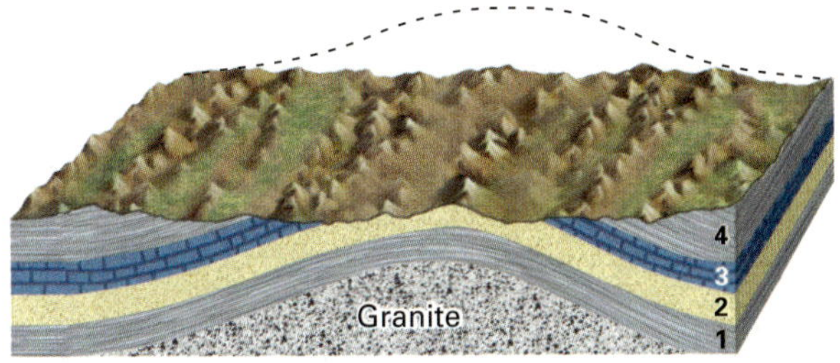

(c) Erosion exposes folded rocks

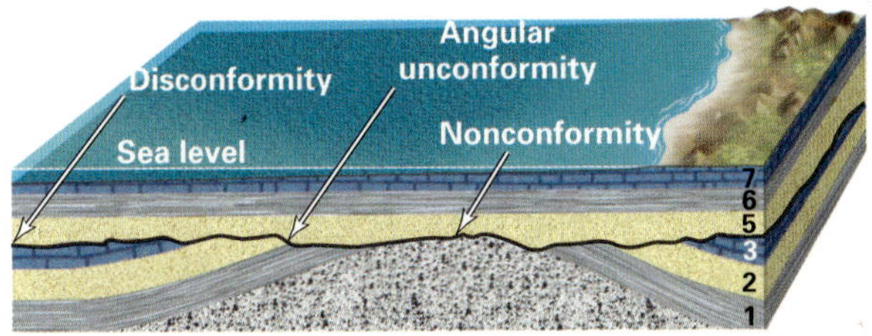

(d) Rocks sink below sea level and new rocks are deposited

Figure 7.8

Development of Different Types of Unconformities

The steps in the formation of three different types of unconformities are shown in (a) through (d). In steps (b) and (c), mountain building prepared the way for angular unconformities and a nonconformity. (d) The nature of an unconformity may change with position. At the left in (d), this is a disconformity, between parallel strata. It becomes an angular unconformity where the rocks were tilted and folded. At the center, it is a nonconformity, where uplift brought plutonic rock (granite) high enough to be eroded before sediments were laid on top. From THOMPSON/TURK. *Earth Science and the Environment* (with ThomsonNOW(T) Printed Access Card) 4e. 2007 Brooks/Cole, a part of Cengage Learning, Inc. Reproduced by permission. www.cengage.com/permissions

The unconformity itself represents the passage of a long span of time and, in some cases, during both the growth (Figure 7.8b) and wearing away (Figure 7.8c) of whole mountain ranges.

For all unconformities, the rocks younger than the unconformity are sedimentary and/or volcanic rocks and are usually close to parallel to the unconformity surface. The older rocks may also be the same rock types (sedimentary and/or volcanic rocks) and parallel to the unconformity surface, making it a **disconformity** (Figure 7.8d). They may be tilted or folded sedimentary/volcanic rocks and not parallel to the unconformity surface, making it an **angular unconformity,** as shown in Figure 7.8d and ■ Figure 7.9. Alternatively, the rocks below the unconformity may be plutonic igneous or metamorphic rocks, making it a **nonconformity** (Figure 7.8d and Figure 7.1 near the bottom of the canyon). An unconformity may change from one type to another, as demonstrated in Figure 7.8.

Unconformities can be obvious or subtle, and they can represent anywhere from modest to vast amounts of geologic time and history. Unconformities where the layers above and below are parallel (that is, *disconformities*) are often quite difficult to recognize, yet they may represent much passage of time. For all unconformities, a period of erosion or nondeposition interrupts an episode of rock formation, followed by a transition to deposition such as subsidence, submergence, and transgression. Since erosion is most effective on steep mountainous slopes, many unconformities denote a period of mountain formation (**orogeny**). Most people think of mountains as permanent, stationary, and unchanging features, which they are when compared to a human lifespan, but

Parvinder Sethi

Figure 7.9

Angular Unconformity

Horizontal layers of gravel and sand deposited unconformably over tilted layers of siltstone and limestone make an angular unconformity, near Zabriskie Point, Death Valley National Park.

over tens of millions of years, mountains form and erode away. Similarly, an unconformity is likely to represent the passage of a great deal of time, a mountain building and a mountain eroding event, that becomes a gap in the rock record.

Exercises in Relative Dating

1. Use stratigraphic superposition to list the order of formation of the sedimentary and volcanic rocks in ■ Figure 7.10, from oldest to youngest. ____________ Does the principle of inclusions give the same results for **C** and **D** as stratigraphic superposition? ______

2. Use relative dating principles to determine the relative age in the following sentences about Figure 7.5 on page 139. Enter *older* or *younger* to complete each sentence:

a. The fault is ____________ than the rocks on either side

b. Erosional surface **2** in Figure 7.5 is ____________ than **E** and **F** and ____________ than **G**.

c. After erosion has occurred, the up side of a fault will expose ____________ rocks at the surface than the down side.

3. What type of unconformity are both the erosional surfaces 1 and 2 in Figure 7.5?

____

4. To help answer the following questions, review graded bedding and cross-bedding from Lab 4 (Figure 4.22, p. 89), if necessary.

a. Locate and label the graded bedding and the cross-bedding in the outcrop of rock illustrated in ■ Figure 7.11.

b. Use the principle of original horizontality and the up indicators you labeled in problem a to help you draw an up arrow on Figure 7.11 pointing in the direction that was up when the sediments were deposited.

c. Have the rocks here been overturned?

______

d. How could the tilting or overturning you see in Figure 7.11 have happened?

____

____

e. On the same figure (7.11), number the rocks in order from oldest (1) to youngest (7), including all layers, even ones that barely show in the cross section.

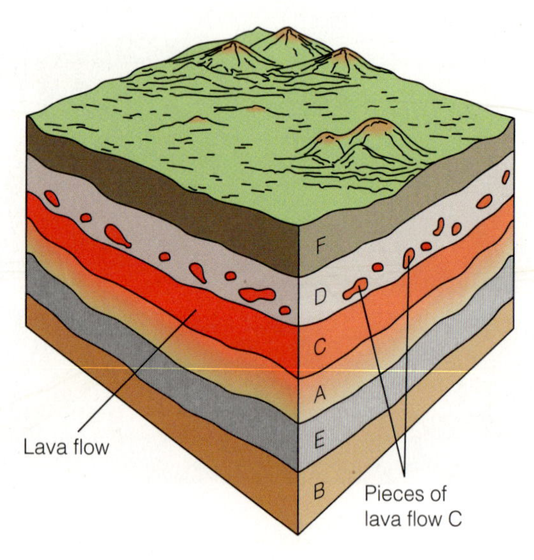

Figure 7.10

Stratigraphic Layers

See instructions in Exercise 1.

Figure 7.11

Using Up Indicators in Relative Dating

Strata containing graded and cross-bedding layers. See instructions in Exercise 4.

The cross-bedded layer is number 5.

The graded bedding is layer number 4.

5. List the four sedimentary rocks and the three dikes in ■ Figure 7.12a in order from oldest to youngest. D, C, B, A F, G, E

6. Label the order of formation of the country rock and the two white dikes from oldest (1) to youngest (3) directly on the photo in Figure 7.12b.

7. List the dike, the country rock, and the veins in the photo in Figure 3.28, on p. 65 in order from oldest to youngest.

____

____

8. List the sill and the country rock in the photo in Figure 3.29, on page 65 in order from oldest to youngest.

____

9. Draw the trace of the angular unconformity on the photo in Figure 7.9 and label it.

10. Using only the cross section in Figure 8.13b, on page 178 of the Grand Canyon, list the rocks between **Pk** and **Mr** in order from oldest to youngest.

____

Check your answers by reading the "Explanation" in Figure 8.13a, on page 177.

11. Examine the same cross section in Figure 8.13b and refer to the explanation in Figure 8.13a as needed. What does the principle of original horizontality tell you has happened to the layers from the Bass Limestone to the Dox Formation in the cross section?

____

12. To answer the following questions examine **p-Ꞓr** in cross sections A-A′ in Figure 8.13b, on page 178.

a. Based on its geometry in the cross section and its age compared with neighboring rocks, what type of rock mass is **p-Ꞓr**? ______________

b. What formation has it intruded?

____

c. How can you tell it is not a lava flow?

____

____

Grand Canyon The Grand Canyon (Figures 7.1 and 8.13) cuts through and displays one of the longest sequences of geologic history anywhere on Earth. As we have seen, the uppermost strata of the Grand Canyon are still nearly horizontal, but deeper into the canyon, tilted and metamorphic rocks also exist.

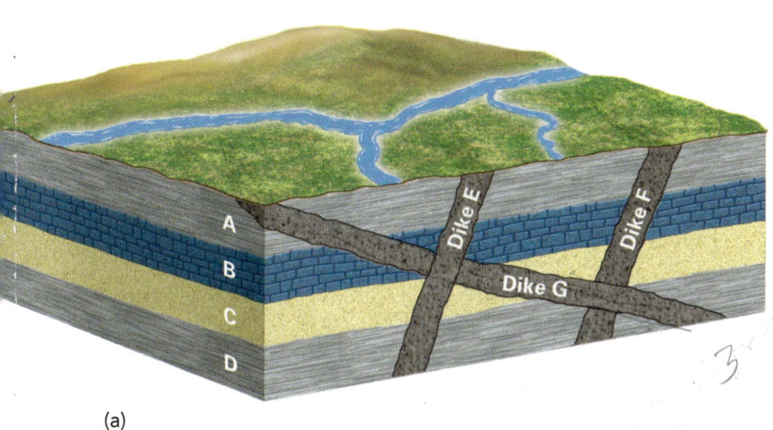

(a)

Grenville Draper

(b)

Figure 7.12

Cross-Cutting Relationships

(a) Block diagram showing a number of dikes that intruded a series of sedimentary layers at different times. (b) Different episodes of intrusion of light-colored rhyolite dikes cutting through a granodiorite pluton of the central mountains of the Hispaniola (Dominican Republic). See instructions for Exercises 5 and 6. From THOMPSON/TURK. *Earth Science and the Environment* (with ThomsonNOW(T) Printed Access Card) 4e. 2007 Brooks/Cole, a part of Cengage Learning, Inc. Reproduced by permission. www.cengage.com/permissions

d. What would you expect the magma to have done to the rocks both above and below the intrusion that would confirm that this is not a lava flow?

____

____

13. Look at the map "Explanation" for Figure 8.13a, on page 177.

a. What do the following symbols stand for?

-Ꞓt ______________________________

p-Ꞓbs ______________________________

b. Examine the contact between these two rocks in cross section A-A′ in Figure 8.13b, on page 178. What is this boundary?

____

c. In the same cross section, what type of boundary occurs between **-Ꞓt** and the sequence **p-Ꞓd, p-Ꞓs,** to **p-Ꞓh**?

____

14. Use the geologic time scale in Figure 7.24 on page 158, Figure 8.13b and Explanation in 8.13a to help with this next question.

a. What geologic periods are missing between the Tapeats Sandstone and the Kaibab Formation?

____

____

b. Between what two formations does this time gap occur?

____

____

c. Examine the cross section in Figure 8.13b to determine what type of unconformity this is. ______________________________

Fossils and Relative Dating

The evidence gleaned from the study of rocks and fossils together helps us to understand the order in which rocks formed and organisms evolved. Within one sequence of rocks, we can use stratigraphic superposition to determine the sequence of fossils found in those rocks. In multiple sequences of rocks, it may be possible to establish both a time line of fossil evolution and a sequence of rock formation. This can help us understand the area's geology, past life forms, and the paleoenvironment.

In this next set of exercises, you will establish the relative age sequence of rocks in the three stratigraphic columns in Figure 7.13.

15. Examine the columnar sections in ■ Figure 7.13 and determine the relative age of the dikes and sedimentary strata using the principles of stratigraphic superposition and cross-cutting relationships.

a. List the sequence of all the rocks, letters, and numbers, from oldest to youngest for each column.

Column 1

____

Column 2

____

Column 3

____

b. Next, fossils can show us the age relationships between the different columns. Which pairs of rock layers in Figure 7.13 from different columns have exactly matching sets of fossils? (_____ _____), (_____ _____), (_____ _____), (_____ _____)
What does this tell you about these rock layers?

____

Notice that even though Figure 7.13 has no *key beds* (time-marking beds), the layers can be correlated by matching the fossils.[2] Observe that *Calamites* and *Pseudoparalegoceras* are missing from Column 2 in Figure 7.13. Absent fossils within a sequence show an unconformity, such as between M and H. In ■ Figure 7.14, the rocks sit along a time line, which shows gaps between the rocks where the unconformities occur.

Techniques of fossil succession and stratigraphy clarify the history of these sections. Figure 7.14 shows correlation for the first two columns of Figure 7.13 using fossils.

[2] This presumes that the fossils were not transported to the location at a later time; evidence for transport may be found as damage to the fossil.

Figure 7.13

Stratigraphic Columns and Their Fossils

Columnar sections of three hypothetical regions with fossils that occur in the layers. In Column 3, "covered" indicates the rocks there are not visible because soil or sediment buries them. See instructions in Exercises 15–18 and 28.

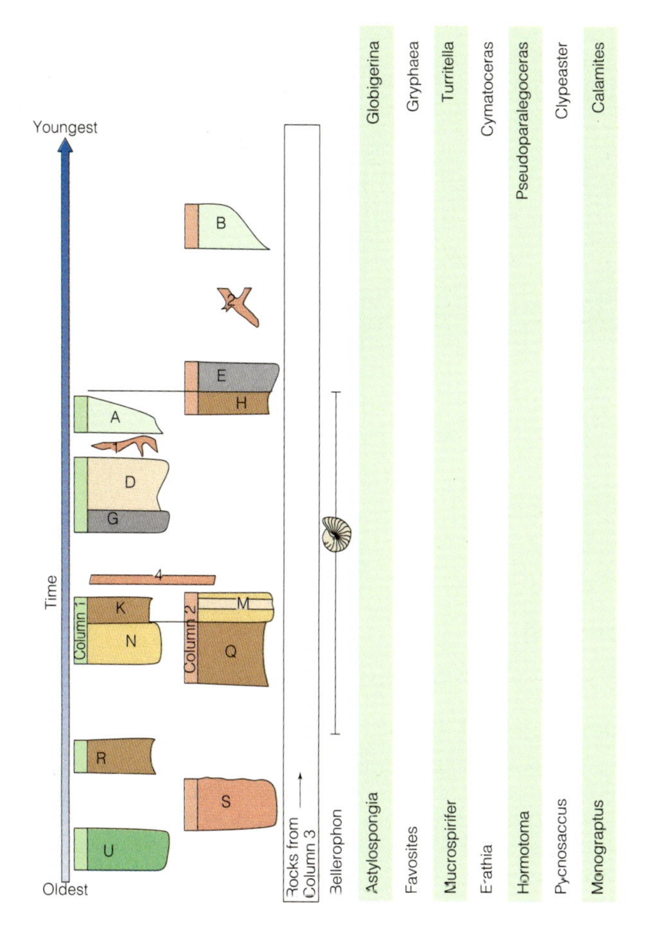

Figure 7.14

Time Line for Stratigraphic Columns

Two of the stratigraphic columns from Figure 7.13 are arranged on a time line with each rock positioned at the time it formed. Time proceeds vertically from oldest at the bottom to youngest at the top. Each dike is also shown at its time of formation. Unconformities occur where there is space (a time gap) between the rock layers. See instructions from Exercises 16–18 and 28.

Materials needed:

- Scissors
- Glue, paste, or tape
- A sheet of paper

16. Work with a lab partner so you will have one whole, uncut Figure 7.13 and one you can cut into pieces.

a. Cut the layers into thin strips along the horizontal lines. Intersperse strips from Columns 1, 2, and 3 by obeying the following rules (but without looking at Figure 7.14):

Rule 1: Keep all rocks in each column in their original stratigraphic order. Notice that within one column the letters are in alphabetical order.

Rule 2: Place rocks with matching sets of fossils on top of each other, in a pile.

Rule 3: Group all rocks with the same individual fossils consecutively in a continuous sequence, vertically, without gaps for any fossil.

b. Since Rocks U and S and the covered area have no fossils, you will not be able to tell exactly where in the sequence these fit. Once all the strips are arranged, glue, paste, or tape them down on the separate sheet of paper.

c. Write in the rock labels C, F, J, and T from Column 3 in Figure 7.14 at the correct position within the time sequence.

d. Use cross-cutting relationships to determine where dike 3 should fit in the time line, and label it on Figure 7.14 as well.

The particular columns in Figure 7.13 represent hundreds of millions of years of time during which many species of organisms evolved and went extinct. For example, *Bellerophon* lived while layers from A to N, H to Q, and J to T were deposited. It probably had not evolved before then, and went extinct after that time. Figure 7.14 shows the range of time when *Bellerophon* lived relative to the stratigraphic columns.

17. Make a time line in Figure 7.14 of each of the other fossil organisms, based on the information you have collected with Figure 7.13. Show when each organism lived by first drawing a short horizontal line at its first appearance then at its extinction. Then join these with horizontal lines with a vertical line from its evolution to its extinction or to the top of the time line if it hasn't gone extinct. If your instructor requests it, draw a cartoon sketch of each fossil on its time line.

18. Examine an uncut version of Figure 7.13:

a. Find two nonconformities on the diagram. Label them "non" in Figure 7.14.

b. Find four disconformities in Figure 7.13 without using any fossil data. Label them "dis" in Figure 7.14.

c. What cross-cutting features give evidence for these disconformities? ______________

d. With additional evidence from fossils from Figure 7.13, or the fossil time line you made in Figure 7.14, what tells you there is an unconformity between **H** and **M** in Column 2?

____

e. Label this unconformity on Figure 7.14 with its correct type as either "dis", "ang" or "non."

Notice that you could not identify some of the unconformities if it were not for the information provided by fossils. The next exercise involves the use of real fossils in correlation.

Materials needed:

- Sets of fossils corresponding to the letters in the stratigraphic columns in Figure 7.15.

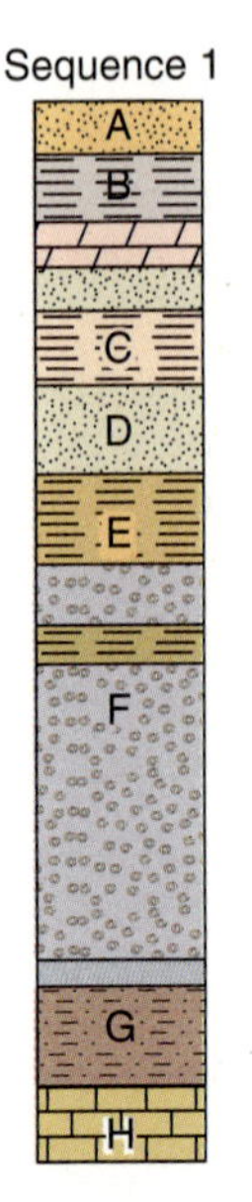

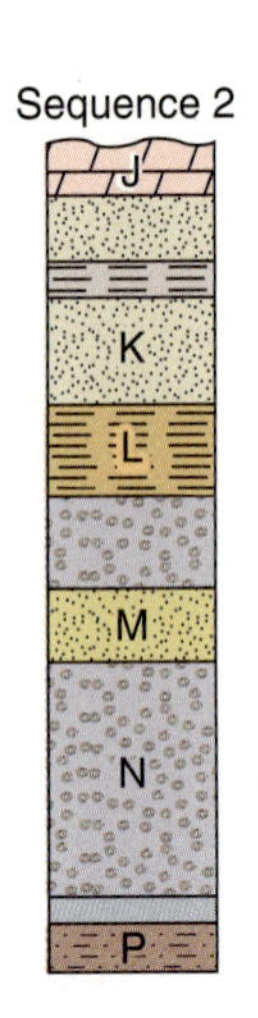

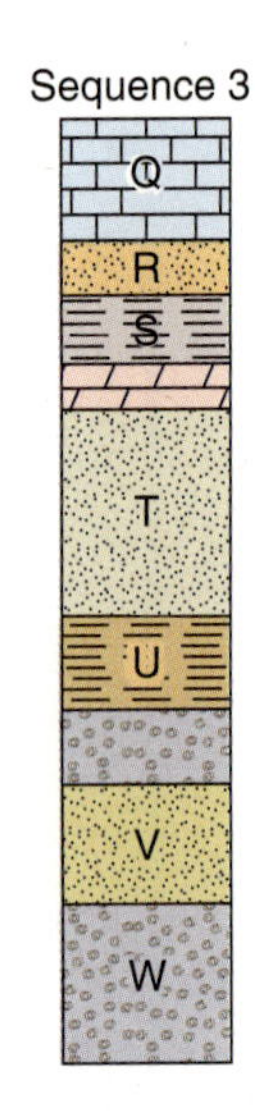

Figure 7.15

Stratigraphic Sequences to Go with Fossils

Columns for three hypothetical sequences of rocks to correspond with fossil sets your instructor provides. See instructions in Exercise 19. From MONROE/WICANDER, *The Changing Earth*, 5e. 2009 Brooks/Cole, a part of Cengage Learning, Inc. Reproduced by permission. www.cengage.com/permissions

19. Your instructor will provide you with sets of fossils from the three hypothetical sequences of rocks shown in ■ Figure 7.15. The letter labeling each set corresponds to a letter in one of the stratigraphic sequences in the figure. Some of the fossils match from one rock sequence to another and some do not.

a. On Figure 7.15 next to the place in each column where they hypothetically originated, quickly sketch small, distinctive cartoons of all the fossils from the sets.

b. Use the fossils to correlate from one stratigraphic sequence to the next by drawing lines from one sequence to another where the fossils match. Some of the lines may need to be diagonal. The rock types may not match even where the fossils do, but use the fossils for your correlations.

c. From the stratigraphic sequences and their correlation, obtain the order of evolution of all the fossils in all the sets. Place the fossils in chronological order of their evolution. List a fossil sample number or cartoon sketch for each fossil in order from oldest to youngest.

____

____

____

d. Write the fossil sample number or sketch the cartoon for the missing fossils in each stratigraphic sequence.

Sequence 1? ______________

Sequence 2? ______________

Sequence 3? ______________

On a separate sheet of paper discuss what these missing fossils tell you.

GEOLOGIC HISTORY

The Earth is constantly changing. Earth movements, volcanic eruptions, intrusions, rivers, glaciers, waves, and other geologic processes, are continually and extensively altering the surface and interior of our planet. **Geologic history** is a study of past geologic processes and events that shaped the Earth. When we see tilted layers, the superposition of strata, cross-cutting relationships, faults, unconformities, and igneous intrusions on geologic maps and cross sections (see also Lab 8), we can combine this information to determine the sequence of past geological events and to interpret geologic history.

The following step-by-step explanation of a geologic cross section demonstrates how to determine the geologic history of an area. Carefully examine the cross section in Figure 7.16f. The principles of stratigraphic superposition, cross-cutting relationships, and original horizontality can tell us what geologic events occurred here. ■ Figure 7.16 shows a series of cross sections (a–f) that illustrate the sequence of events that resulted in the final cross section (f). ■ Table 7.1 summarizes the geologic events that formed the rocks here.

Notice how we can deduce each event, and almost the complete history, by examining the cross section in Figure 7.16f. Especially important clues in the cross section are (1) cross-cutting relationships—where the dike, the unconformity, and the land surface (an erosional surface) cut across other features indicating that the latter must be older—and (2) stratigraphic superposition—the oldest sedimentary or volcanic layers are on the bottom. (3) Inclusions in layer 8 came from rocks that must have formed earlier. This "detective"-type reasoning has enabled geologists to deduce geologic events in small map areas to great expanses on the scale of continents.

20. ■ Figure 7.17 has a series of cross sections with simpler geology. Examine each cross section in Figure 7.17a–d using the following guidelines to write your stratigraphic notes alongside.

i) Deduce the rock names from the rock symbols on the inside back cover of the book.

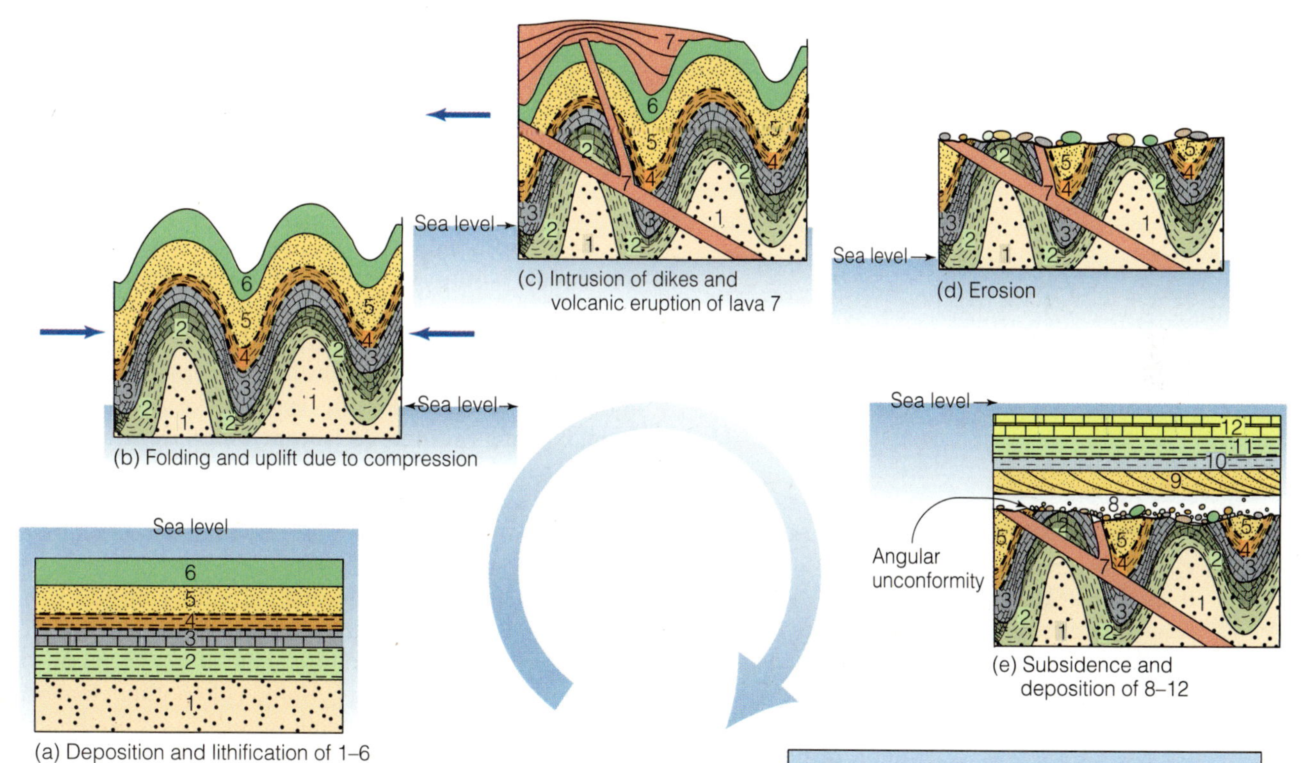

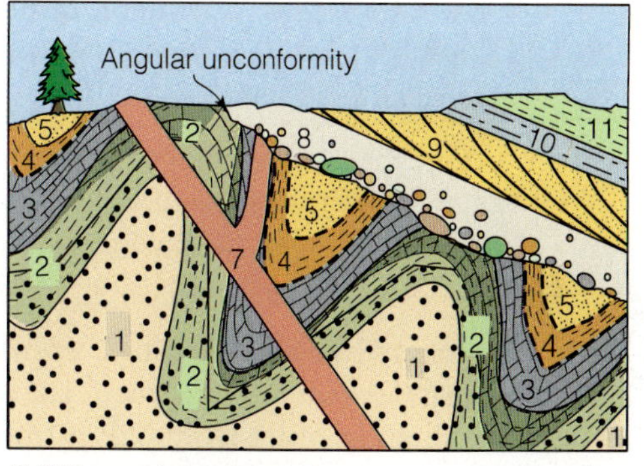

Figure 7.16

Deciphering Geologic History from a Cross Section

The sequence of events that formed the cross section in (f) is illustrated step by step in (a–f) and listed in Table 7.1.

Table 7.1

Geologic History of Figure 7.16f

Sequence of events that formed the rocks in the cross section in Figure 7.16f.

Event Sequence	Complete Description of Geologic Events
Last (youngest)	Erosion removed 12 and parts of other layers
13th	Tilting caused by plate movements
12th	Deposition of shale 11 and possibly limestone 12
11th	Deposition of siltstone 10
10th	Deposition of cross bedded sandstone 9
9th	Deposition of conglomerate 8 as a result of subsidence and/or transgression
8th	Erosion removed 6 and parts of other layers, wearing away mountains to a nearly flat surface A and producing angular unconformity A
7th	Intrusion of 7. Magma intruded and solidified.
6th	Folding and uplift due to compression, mountain building, convergence
5th	Deposition of yellow sandstone 5 and possibly 6
4th	Deposition of red shale 4
3rd	Deposition of limestone 3
2nd	Deposition of green shale 2
1st (oldest)	Deposition of buff sandstone 1

ii) Use the laws of stratigraphy and describe the sequence of events that formed each cross section using words such as deposition of . . . (rock name and letter), intrusion of . . . (rock name and letter), eruption of . . . (rock name and letter), metamorphism of . . . (rock name and letter), tilting, folding, faulting, erosion producing . . . (unconformity type and number), erosion producing surface . . . (number) etc.

iii) Write in the oldest event on the bottom of the list for a cross section in Figure 7.17, then progress upward to younger events. Do this for each cross section, a–d.

21. Deduce the sequence of geologic events in the cross section in ■ Figure 7.18, and describe them in order from oldest to youngest in ■ Table 7.2. Beware: There is a trick in the sequence; carefully examine the graded bedding and the cross-bedding in the cross section. Add some additional details about each event when you write your complete "*description of geologic events*" in Table 7.2. Use Table 7.1 as an example.

ABSOLUTE DATING TECHNIQUES

The term **absolute dating** is commonly used for methods of determining numerical ages of rocks. These ages are usually given in millions of years (Myr). Notice that *absolute* here does not denote *precise* or *accurate*. This means that an absolute date may have error given with a plus-or-minus (±) sign, which indicates the range of possible values. Whereas relative dating relies on stratigraphic superposition and cross-cutting relationships, establishing absolute ages uses *isotopic dating* and *indirect dating* to build the history of an area.

Isotopic Dating

In the Earth, particular *isotopes* of some elements are unstable and break down in a process called **radioactive decay. Isotopes** are different forms of a single element; their atoms have identical numbers of protons but different numbers of neutrons. The decay or breakdown of unstable isotopes occurs at a predictable rate, allowing us to determine the age of rocks. A radioactive atom, called a **radiogenic isotope** or a **parent isotope,** spontaneously converts to another element, known as a **daughter isotope,** by gaining or losing particles from its nucleus. **Isotopic dating** uses this natural radioactive decay to measure the age of rocks.

Radioactive isotopes have a naturally constant rate of decay. **Half-life** is the length of time it takes for half of the radioactive parent isotope of one element to decay into the new daughter isotope of another element (■ Table 7.3). Each radiogenic isotope has a fixed half-life. The sum of parent and daughter atoms remains constant because each parent atom that breaks down becomes a daughter atom. Laboratory measurements establish the rate of decay of radiogenic isotopes and thus allow determination of each half-life.

Processes such as crystallization from magma, or recrystallization during metamorphism, chemically separate parent and daughter atoms from each other (■ Figures 7.19 and ■ 7.20). This is because as different elements, the parent and daughter atoms are different sizes and do not fit into the same crystal structures. During crystallization the parent atoms are fixed in

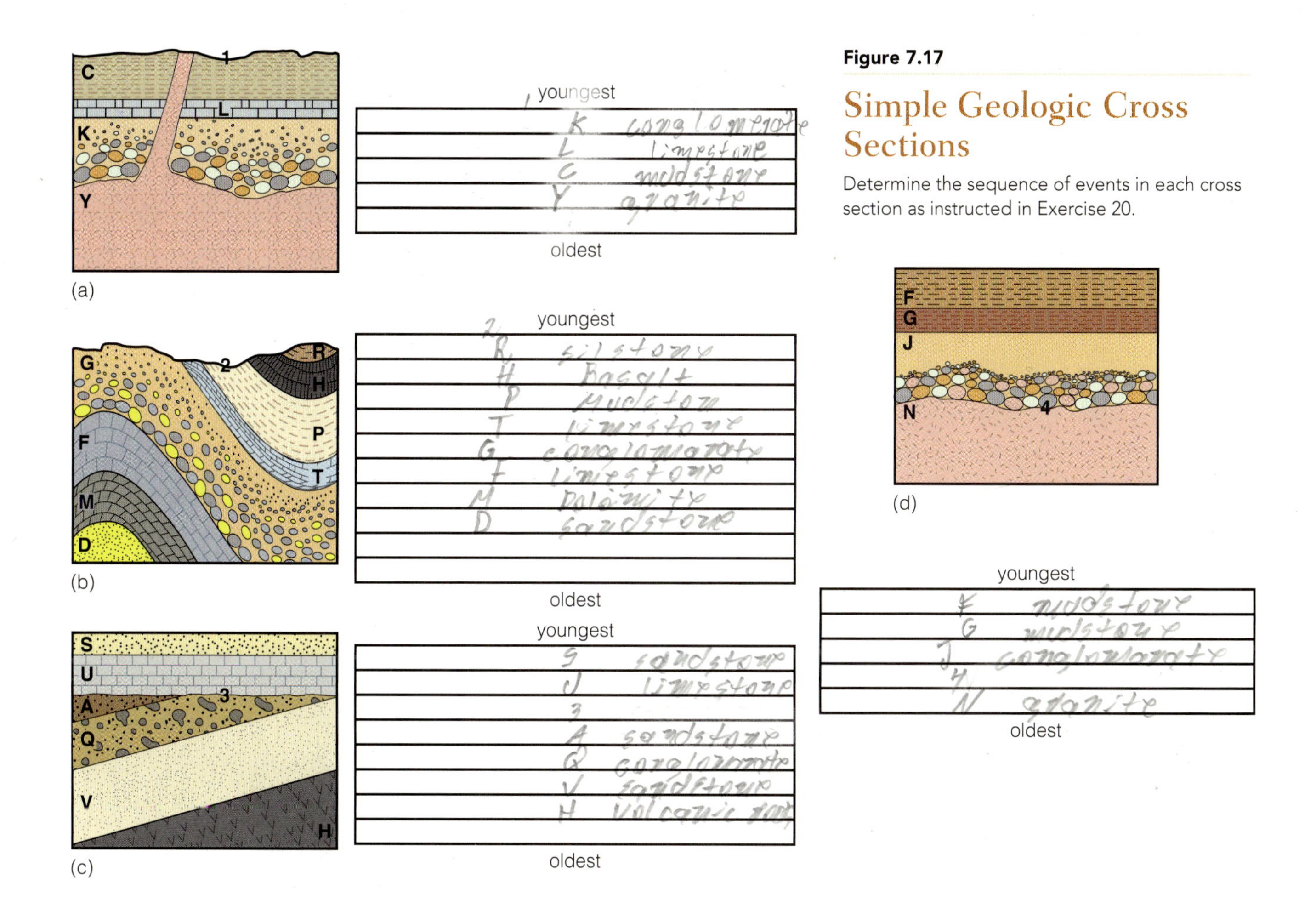

Figure 7.17

Simple Geologic Cross Sections

Determine the sequence of events in each cross section as instructed in Exercise 20.

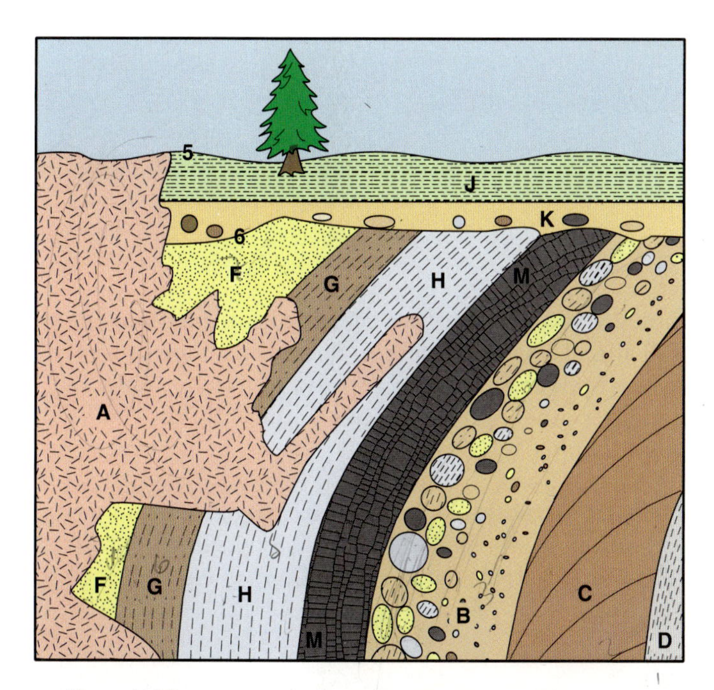

Figure 7.18

Cross Section for Determining Geologic History

See instructions in Exercises 21 and 27.

Table 7.2

Geologic History of Figure 7.18

Sequence of events that formed the rocks in the cross section in Figure 7.18. See instructions in Exercise 21.

Event Sequence	Description of Geologic Events
Last (youngest)	Intrusion of granite
12th	Conglomerate
11th	Depo of siltstone (J)
10th	mudstone
9th	Conglomerate (K)
8th	Erosion produce unconformity
7th	D Mudstone
6th	C Depos of cross bedded sandstone
5th	B Deposition of conglomerate
4th	M Bedded limestone
3rd	H Deposition of mudstone
2nd	G Deposition of siltstone
1st (oldest)	F Massive sandstone coarse-grained

Figure 7.19

Radioactive Decay in an Igneous Rock

(a) Parent and daughter isotopes in the magma. (b) Parent atoms forming crystals excluding daughter atoms. (c) After one half-life has passed in the igneous rock, half the parent atoms have decayed to form daughter atoms in the crystals.

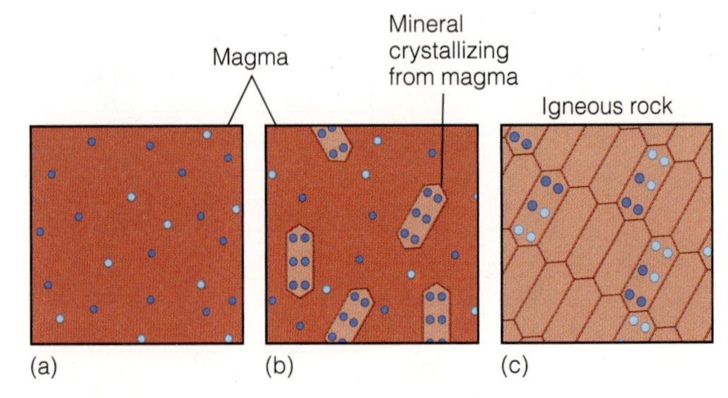

Table 7.3

Radioactive Isotopes

Some radioactive isotopes and their half-lives used for dating rocks. **Age Range** gives the practical range of rock ages that can be dated by this method.

Name of Dating System	Radioactive Isotope (Parent) Used for Dating Rocks	Symbol of Parent Isotope	Daughter Isotope	Symbol of Daughter Isotope	Half-Life	Age Range	Rock Type
Carbon-14 dating	Carbon-14	^{14}C	Nitrogen-14	^{14}N	5730 yr ±30	100 and 70,000 yr	Sedimentary
Potassium-argon dating	Potassium-40	^{40}K	Argon-40	^{40}Ar	1.3 Byr	50,000 yr to 4.6 Byr	Igneous/ metamorphic
Uranium-lead dating	Uranium-238	^{238}U	Lead-206	^{206}Pb	4.56 Byr	10 M to 4.6 Byr	Igneous/ metamorphic
Uranium-235 dating	Uranium-235	^{235}U	Lead-207	^{207}Pb	704 Myr	10 M to 4.6 Byr	Igneous/ metamorphic
Rubidium-strontium dating	Rubidium-87	^{87}R	Strontium-87	^{87}Sr	48.8 Byr	10 M to 4.6 Byr	Igneous/ metamorphic

place by the structure of the mineral, and the radiometric clock starts ticking. At this starting time, no daughter isotopes are present.[3] As time passes, the parent atoms decay and daughter atoms begin to appear in the mineral at a predictable rate, based on the half-life. Every time a parent decays, a daughter forms from it and is trapped where the parent was in the crystal.[4] Thus, the ratio of daughter atoms to parent atoms is a measure of the time elapsed since crystallization. To date a rock, we need two pieces of information: the half-life of an appropriate radioactive isotope present in the rock and the current ratio of parent atoms to daughter atoms present. Both of these things can be measured.

Half-Life Activity—A Radiometric Game The following activity is intended to help you understand the half-life used in isotopic dating.

[3] In practice, daughter atoms can be inherited from previous minerals, so this is not always perfect. If this happens, the measured age of the rock is older than the actual age.

[4] Again, in practice, daughter atoms may sometimes leak out, especially if they are gases (for example, ^{40}Ar). If this happens, the measured age of the rock is younger than the actual age.

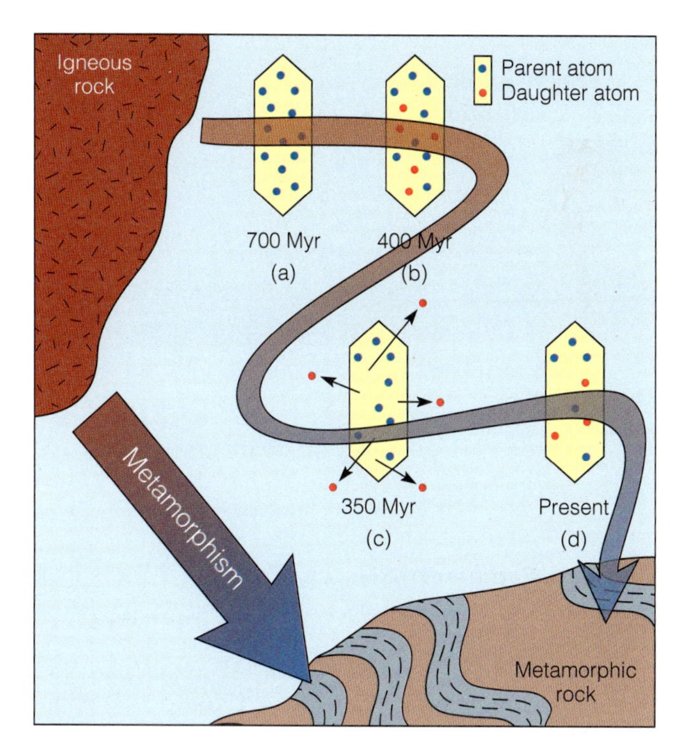

Figure 7.20

Radioactive Decay in a Metamorphic Rock

(a) Crystallization from magma produces a crystal with only parent atoms (blue). (b) Passage of time converts some parents to daughter atoms (red). (c) Metamorphism expels the chemically incompatible daughter atoms from the crystal. (d) Passage of more time converts more parents to daughters. Dating the crystal would give the time of metamorphism. If this rock was chemically isolated with the daughters remaining in the rock during metamorphism (although not in the crystal with the parents), and if the original igneous rock had no daughter atoms, dating the whole rock would give the time of igneous crystallization.

Materials needed:

- 50 dice or 100 pennies (or sugar cubes with a mark on one side)—count them to confirm the number
- A cup or container for shaking the dice (or cubes or pennies)
- A box to roll or toss them into

22. Start the game as follows.

a. Shake up 50 dice or 100 pennies or sugar cubes (objects) in the cup or container and pour them into the box. Remove all dice that turn up a 6, pennies that come up tails, or cubes that show the mark on top. In Table 7.4, record the number of objects removed and the number remaining. The number removed represents the number that decayed into daughters; the number remaining represents the number of parents remaining.

b. Roll the remaining objects into the box and again remove all "daughter" objects. Record the number removed this time, the total number removed so far, and the number remaining.

c. Repeat step **b** until you have reached the fifteenth toss or you are down to three or fewer remaining objects, whichever comes first.

d. In Figure 7.21, number the left side of the graph paper from 0 to 50 or 0 to 100 in steps of 10 depending on how many objects you started with. Then graph the number of objects remaining (parents remaining) with one color or symbol and the total number of daughters removed (second to the last column in Table 7.4) with another color or symbol.

e. Complete the legend by drawing the color or symbol for each. Draw two smooth curves through the points. Points should fall on or close to your curve.

f. Estimate the half-life for your "isotope" in terms of numbers of tosses to the nearest tenth of a toss by finding where half the objects were remaining (read at the line on the graph corresponding to half the objects). ______

g. Share your value with others in your class and calculate the class average half-life obtained: ______ and the range of values obtained: ______

What might account for range of values obtained?

____

____

Table 7.4

Simulation of Half-Life

Results from "Radiometric Game." See instructions in Exercise 22.

Toss Number	Objects Removed This Toss	Total Objects Removed	Objects Remaining
0	0	0	50 or 100 (circle which)
1			
2			
3			
4			
5			
6			
7			
8			
9			
10			
11			
12			
13			
14			
15			

h. Would radioactive isotopes, which exist as vast numbers of atoms, be subject to this same error or variation? ______ Why or why not?

____

____

i. If we assume that each toss took you 3 minutes, how long is a half-life for your "isotope"? ____________ Check to make sure you put away all 50 or 100 objects.

Example Radiometric Dating Problem 1 Let's look at an example rock and try to determine its age using radiometric dating.

A granite contains crystals with 150 ppma (parts per million by atomic proportions) of ^{235}U and 1,050 ppma of ^{207}Pb. How old is the rock?

Calculations:

Add up the parents and daughters. This equals the original number of parents present at the time of the chemical reactions that formed the minerals in the rock:

$$\begin{array}{r} 150 \\ +1{,}050 \\ \hline 1{,}200 \end{array}$$

Figure 7.21

"Radiometric Game" Graph

See instructions in Exercise 22.

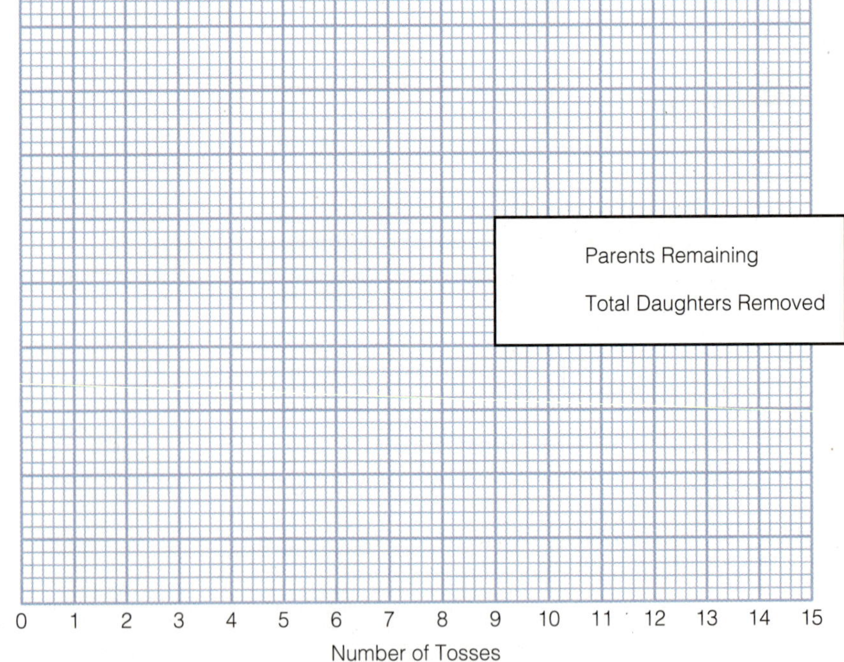

Therefore, when the rock formed, 1,200 ppma radioactive parent atoms were present. We know from the definition of the half-life and from Table 7.3 that the half-life for the decay of ^{235}U to ^{207}Pb is 704 Myr, which means that the uranium parents will decay by half in 704 Myr. Keep on dividing the parents in half, until you get the number found today:

Time (in 1/2 lives)	Parents (ppma)	Calculations and Notes
0	1,200	$\frac{1,200}{2^h} = 150$ where h is the number of half-lives $1,200/2^3 = 150$ ← stop here—this is how many parents are present today—it took three half-lives of time to get this many parent atoms
1	600	
2	300	
3	150	

Count how many times you divided by 2. We divided the parents in half three times to get to the present level of uranium. Thus, three half-lives have passed since the rock formed.

$$3 \text{ half-lives} \times 704 \text{ Myr/(half-life)} =$$

Answer:

$$2,112 \text{ Myr} = 2.11 \text{ billion years (Byr)}$$

23. Determine the age of shale with leaf fossils containing 21 ppma of ^{14}C and 651 ppma of ^{14}N. Look up the half-life in Table 7.3. Write out the complete step-by-step procedure used to determine your answer:

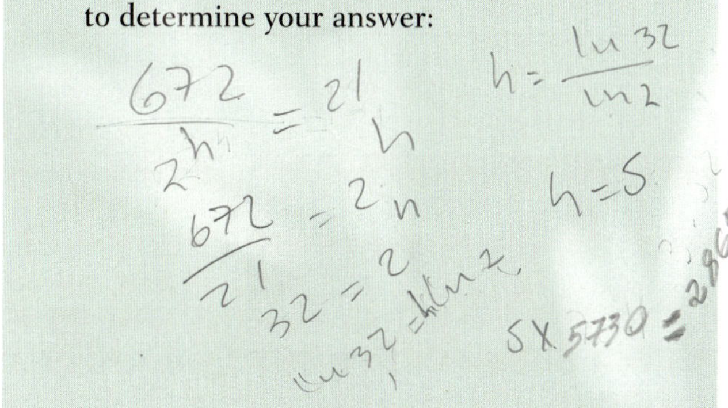

When determining the age of a rock by isotopic dating, if the number of half-lives is a whole number, you need only divide by 2 the right number of times to determine the answer. However, to get an accurate age with partial half-lives, you need to use a mathematical expression or a graph. The following equation will give the age in half-lives since Time 0. Time zero is the time of the chemical reaction that formed the substance being measured:

$$h = \text{age in half-lives} = 3.322 \times \log [(p + d)/p],$$

where p is the number of parent atoms and d is the number of daughter atoms. Parent and daughter atom figures need to be in the same units. Try this equation with the numbers used earlier to see whether you get the same result.

Clastic Sedimentary Rocks and Isotopic Dating

The following problem involves a clastic sedimentary rock, which will be treated differently due to the nature of its formation.

Example Radiometric Dating Problem 2 In this next example, the number of half-lives does not come out as a whole number, so use the equation.

A sandstone contains grains with 2.03 ppma ^{238}U and 0.17 ppma ^{206}Pb. How old is the rock?

Use the equation to determine the age of the grains.

$$p = 2.03 \text{ ppma}; \; d = 0.17 \text{ ppma}$$
$$3.322 \times \log [(2.03 + 0.17)/2.03] = 0.116$$

This means that 0.116 half-lives are needed to bring this number of parents down to the present level at 2.03 ppma. Look up the half-life for ^{238}U in Table 7.3.

$$0.116 \text{ half-lives} \times 4.56 \text{ Byr/(half-life)} = 529 \text{ Myr}$$

Age of the grains:

$$529 \text{ Myr or } 0.529 \text{ Byr}$$

Notice we have not finished the problem yet because this is the age of the grains, not the age of the rock. From the calculations, the minerals in this rock are 529 Myr old—the age of the clastic grains that came from a former igneous or metamorphic rock that became weathered and eroded to make the sediment in this clastic sedimentary rock. This rock must be younger than 529 Myr, and that is all you can say. The dating technique does not tell you how much younger—just younger.

Carbon-14 dating can give ages of sedimentary rocks and fossils because both of these can have chemical reactions involving carbon at about the time they form. Carbon-14 has a short half-life compared to the age of many rocks and cannot be used if the sample is over 70,000 yr (Table 7.3). For substances older than 70,000 yr, isotopic dating generally applies to igneous and metamorphic rocks. We can, however, combine isotopic and relative dating. Using what we call *indirect dating*, we can determine ranges of ages for materials that do not have isotopic age measurements, especially sedimentary rocks and fossils.

24. Determine the age of a conglomerate that has andesite clasts containing crystals with a $^{87}Rb/^{86}Sr$ ratio of 63.90 and a $^{87}Sr/^{86}Sr$ ratio of 0.73. (In this problem, the units are the same for the parent and daughter atoms, so you can use the values as in the example even though they are ratios.) Refer to Table 7.3 for additional information you may need. Use the equation to determine the age of the andesite in half-lives: 0.01639, and in millions of years: 0.8 Byr. How old is the conglomerate?

0.8 Byr

Indirect Dating

We saw earlier that isotopic dating does not generally work on old sedimentary rocks. On the other hand, indirect **dating**—combined isotopic and relative dating—produces numerical dates for rocks that cannot be directly dated by radiometric methods (■ Figure 7.22). Isotopic dating can give the age of volcanic rocks in layered sequences with sedimentary rocks, plutonic rocks that cut across sedimentary rocks, and even igneous or metamorphic rocks below a nonconformity (Figure 7.22), or where the sedimentary rock has been metamorphosed. Then the relative dating techniques of stratigraphic superposition and cross-cutting relationships can give a range of possible ages for the sedimentary rocks.

Example Problem 3—Indirect Dating Use the cross section in Figure 7.22 and the isotopic ages of the igneous rocks to determine the age of the layers labeled A.

Answer:

The Layers A are above and therefore younger than the Lava flow at 600 Myr. They are cut by the dike, which is 150 Myr. Thus, the ages of Layers A are between 150 and 600 Myr.

Sometimes indirect dating gives a wide range of possible ages, as in this case. But over time, as more radiometric dates are determined, it may be possible to narrow the range.

25. Use the cross section in ■ Figure 7.23 and the isotopic ages of the igneous rocks labeled in the cross section to determine the age of the Dakota Sandstone. 146 My. Explain your reasoning.

Because the cretaceous period started 146 Mys

26. How old are the Morrison Formation 202-146 and the Wasatch Formation 66 - 1.8 in Figure 7.23? Explain your reasoning.

Fig 7.24

Figure 7.22

Indirect Dating

Isotopic dating of metamorphic and igneous rocks allows estimation of a range of ages for sedimentary rocks using indirect dating. See Example Problem 3.

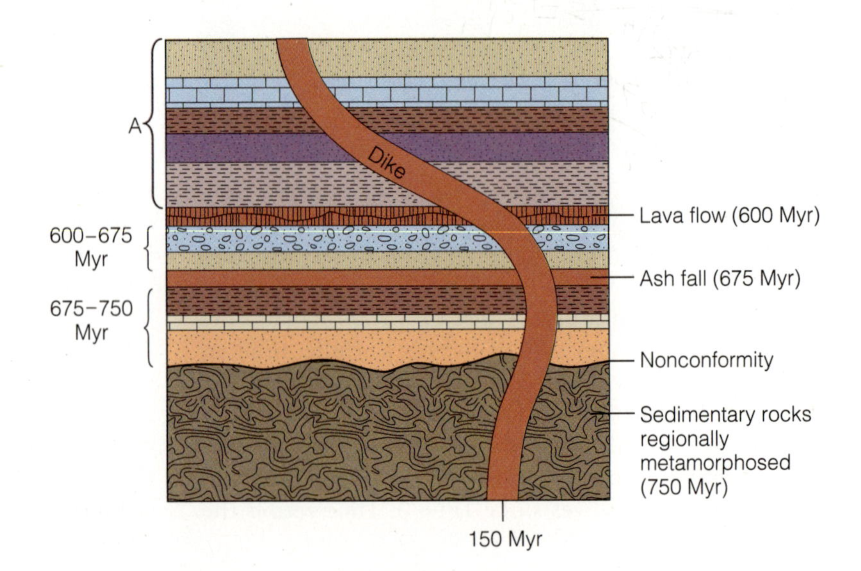

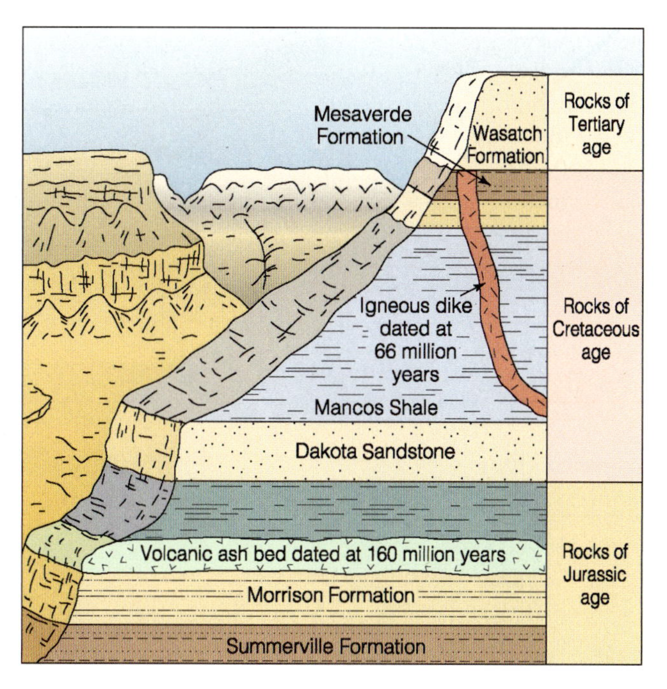

Figure 7.23

Indirect Dating Example

Cross section and perspective drawing of formation in Colorado with isotopic ages of associated igneous rocks. See instructions in Exercises 25 and 26.

a. Enter this information on the left side of the time line in Figure 7.14 to help you determine the possible range of ages for the sedimentary rock layers and fossils there.

b. What is the possible age range using indirect dating of *Elrathia* based on this new information?

____

c. What is the indirect age of: *Mucrospirifer*?

____

Calamites?

________________________ layers A, E, and H? ____________________

d. If you found a rock containing the fossil *Gryphaea,* how old would the rock be?

____

27. If Layer **M** in Figure 7.18 is a basalt lava flow dated at 240 Myr and Intrusion A is a granite dated at 184 Myr, how old is:

Conglomerate B? 240 - 184

Shale H? more than 240

Conglomerate K? 240 - 184

28. The ages of the dikes, other igneous rocks, and metamorphic rocks from Figures 7.13 and 7.14 have been determined using isotopic dating:

Rock	Isotopic Age (years)
1	240 million
2	140 million
3	70 million
4	370 million
S	560 million
U	640 million

By now, you probably have some respect for the talents needed to collect and process the information to complete the geologic history for an area. You may have experienced some frustration with these indirect ages because you know that *Elrathia,* for example, is older than *Mucrospirifer,* but the indirect ages do not show the difference. Inquisitiveness of the investigators and their desire to complete the "puzzle" of the area can often overcome the frustrations over apparent conflicts and missing data. It is only through careful work with more indirect dating that the age ranges can be narrowed. Congratulations—you have now completed a geologic time scale for the rocks and fossils in Figures 7.13 and 7.14!

GEOLOGIC TIME SCALE

The geologic time scale (■ Figure 7.24) was assembled over many years using data from many regions of the Earth in a similar manner to the way you worked with the stratigraphic columns in Figures 7.13, and 7.14. At first, the time scale delineated age that was purely relative without any specific times or dates. It was only later, when isotopic dating techniques became available, that the geologic time scale acquired its numerical ages, the numbers along its edge (Figure 7.24). For example, rocks of the Devonian period were known to be younger than Cambrian rocks and older than Permian rocks, but how much younger and older, and even an approximate age

Figure 7.24

Geologic Time Scale

Some important geologic and biological milestones are listed at the right. See instructions in Exercise 29. From MONROE/WICANDER, *The Changing Earth*, 5e. 2009 Reproduced by permission. www.cengage.com/permissions

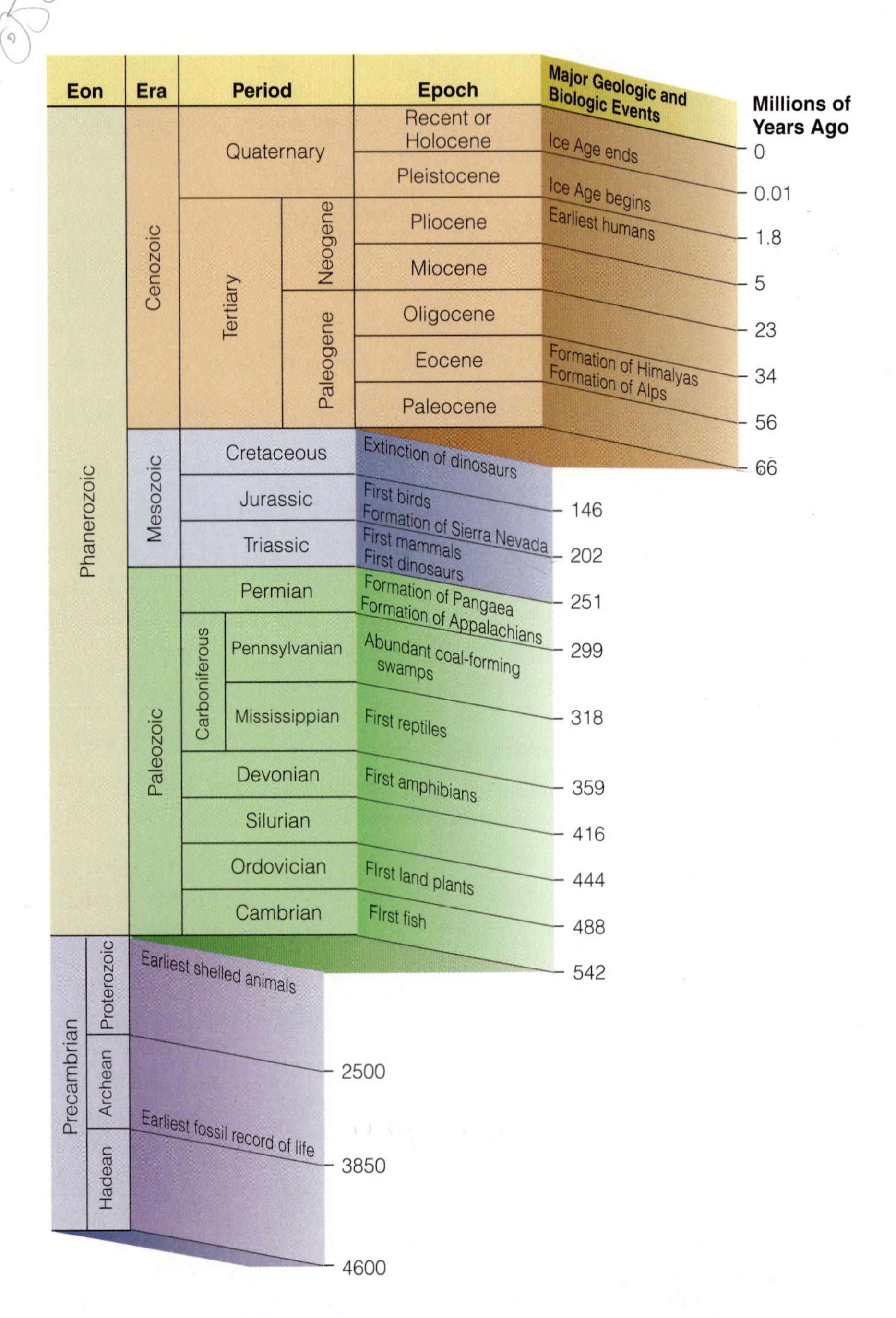

for Devonian was unknown. Indirect dating techniques indicated the numbers on the geologic time scale in many places all over the world. As **geochronologists** (people who study geologic time and measure isotopic dates) find and date new rocks and fossils and improve isotopic techniques, they continue to revise the numbers on the time scale.

The Earth is approximately 4,600 million, or 4.6 billion, years old. Life on Earth began sometime around 3.6 to 3.8 Byr ago. However, *Homo sapiens* such as us have been around for only about 200,000 yr—a tiny fraction of the age of the Earth or the existence of life on Earth. If we compare the time that life has existed on Earth to one day (24 h), modern humans have only been around for about 4 seconds. Let's see whether we can gain a better understanding of the scale of the vastness of geologic time by relating it to life on Earth.

Geologic Time and Life Forms Exercise

This exercise is easier to complete if pairs or small groups work together.

Materials needed:

- Paper tape (such as adding machine tape) or a 46 m section of sidewalk
- Metric measuring tape or meter rule
- Scissors, if paper tape is used
- Colored markers or pencils or colored chalk
- Calculator

29. Use the geologic time scale in Figure 7.24 and either the paper tape and colored markers or the sidewalk and chalk.

a. Using a scale of 1 cm = 1 Myr, measure and cut a length of paper tape to represent the length of time, past to present, since the appearance of the first amphibians. How long ago was this?

_____________ In what period did this happen? _____________________
If marking a sidewalk with chalk instead of using tape, first measure and mark off the length representing Earth's entire history, and then measure and mark the length for the amphibians.

b. On the tape, make a time line using the same scale with the eras, periods, and epochs that occurred during the existence of amphibians. Label each, and label the first amphibians. Or, on the sidewalk, label all the eras, periods, and epochs.

c. Label the numerical ages of the boundaries of each time unit.

d. Mark in a colored line that represents the appearance and extinction of both the dinosaurs and the humans. In what Era did the dinosaurs live?

_______________ If the tape you are using is wide enough, mark the time of each organism in Figure 7.24 on the tape measured in **a.** If space isn't available on the tape, then cut additional lengths as needed for additional organisms. Conserve resources by fitting as many organisms as reasonably possible on one strip of tape. Label these events.

e. Make a legend and a scale showing what each color and mark represents. Give your chart a title that describes it clearly and concisely. If using a sidewalk, write the title and scale directly on the sidewalk.

f. State in percentages what portion of the geologic time scale each species represents:

Amphibians _______%

Dinosaurs _______%

Humans _______%

g. Write one general conclusion concerning the relationship between geologic time and these life forms based on your results in this activity.

____

____

____

____

30. Raven Ridge is a series of ridge-backs cutting through the Colorado-Utah border (■ Figure 7.25). The rock beds, primarily marine in origin, are steeply inclined so that

Figure 7.25

Cretaceous-Tertiary (K–T) Boundary

Astronaut photograph of Ravens Ridge, Colorado showing the location of strata containing the last Cretaceous dinosaurs and the first Tertiary fossils at the K–T Boundary as a dashed line.

at the top of the ridge, from above, you are seeing the bed's edges with younger rocks to the south and older rocks to the north.

a. Find and label the K–T boundary on the geologic time scale in Figure 7.24.

b. Draw the following labels on Figure 7.25: the age of the boundary in Myr, the names of the Era you would expect to see both north and south of the boundary, and word "dinosaur" on the side where dinosaurs would occur in the fossil record.

c. Discuss the following question in a group and summarize your conclusions here: If you were geochronologists trying to narrow down the numerical age of fossils defining the Cretaceous–Tertiary boundary (K–T boundary) such as at Raven Ridge, how would you go about it?

____

____

____

____

LAB 8

Geologic Maps and Structures

OBJECTIVES

- **To learn how to measure the orientation (strike and dip) of planar geologic features such as bedding and foliation**
- **To learn the types of rock structures such as folds, faults, and their symbols on maps and cross sections**
- **To be able to read geologic maps and cross sections and interpret their geologic history**
- **To be able to construct a geologic map and a simple geologic cross section**

INTRODUCTION TO GEOLOGIC MAPS AND CROSS SECTIONS

Geologic maps show the underlying geology of an area: the rocks, their geometry, orientation, and deformation. This lab also explores rock **structures,** which result from deformation of rocks. **Cross sections** of these mapped areas show the rocks in side view, like a slice through a layer cake. As we study geologic maps and cross sections, we will discover how to decipher the geologic history of an area to understand what geologic processes occurred there in the past. Your introduction to skills in this chapter will show how the steps in making a geologic map bring the geologist closer to puzzling out an area's geologic story.

Measuring the Orientation of Planes

Some geologic structures occur as planes or layers, illustrated on geologic maps with special symbols, as shown in Table 8.1, called strike and dip symbols. Sedimentary rocks have layering or bedding that may sometimes be seen in small scale in hand samples, or large scale in rock beds of enormous size; the boundaries between layers are **bedding planes** (■ Figure 8.1c). Planar features include lava flows, and volcanic ash beds in igneous rocks, and foliation in metamorphic rocks.

Deformation often moves bedding planes or volcanic layers from their original horizontal position so that we find them tilted at various angles. Geologists can express the orientation of planar features on a map. Three quantities specify the orientation of a plane—the *strike,* the *dip direction,* and the *dip angle* are also called the plane's **attitude,** or its **strike** and **dip** (Figure 8.1c).

- The **strike** is the direction of any horizontal line on a plane. If the plane is a layer of rock, this measurement is a description of the bed's orientation and can be plotted on a map. Geologists measure strike in one of two ways: They measure it clockwise with respect to north, as indicated on the compass rose in Figure 8.1a; or they measure it either clockwise or counterclockwise from north or south, as shown on the compass rose in Figure 8.1b. Your instructor will suggest which method you should use. Measuring a strike is similar to measuring a bearing, as practiced in Lab 1 (see Figure 1.14c on p. 17), as we write both in degrees from north. However, unlike measuring a bearing, any single strike direction has two possible measurements, from either end

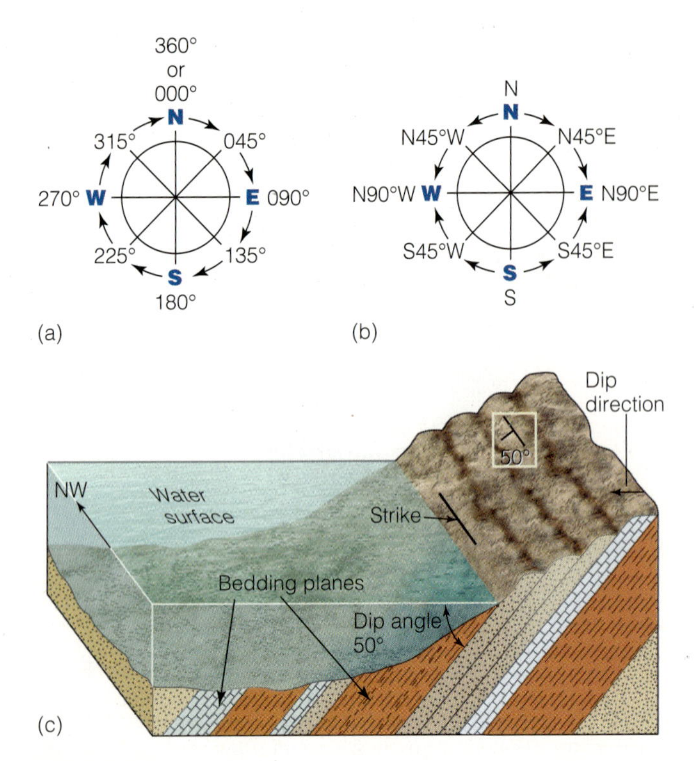

Figure 8.1

Strike and Dip of Sedimentary Beds

(a) and (b) show angles corresponding to the different points of the compass for two styles of compass rose. (a) 360°-style compass rose. (b) Quadrant-style compass rose. (c) Strike and dip of tilted bedding is measured from bedding planes where beds are distinct from one another by characteristics such as particle size, mineralogy, etc. The strike is parallel to the waterline: 135°, or N45°W, depending on which compass rose you use. The dip direction is perpendicular to the strike in the direction the beds tilt downward, in this case SW. The dip angle is measured from horizontal; in this case, it is 50°.

of the strike line, so measurements with exactly 180° difference have the same strike.

- The **dip direction** is the direction of the downward slope of the plane (the direction water would run), always exactly perpendicular to the strike (Figure 8.1), but expressed in general terms (such as NE, NW, and so on).
- The **dip angle** is the angle of tilt, or slope, of the plane measured with respect to the horizontal (Figure 8.1).

Table 8.1 illustrates strike-and-dip symbols (and some of the other typical types of symbols used on geologic maps). First the strike measurement is written, then dip angle, then dip direction. The four ways to write the attitude of the bedding in Figure 8.1 are:

A. 315°, 50°SW;
B. 135°, 50°SW;
C. N45°W, 50°SW;
D. or S45°E, 50°SW.

A specialized compass called a **compass clinometer** is used to measure strike and dip. When looking at an inclined bed, decide which direction is the steepest slope of the bedding plane. This is the *dip direction.* To find the *strike,* hold the long edge of the compass clinometer horizontal (level) against the bed (perpendicular to the dip direction). Your instructor will show you how to read the compass (the method varies with the type of compass). Write down the strike angle. While the compass is still in this position, note the *dip direction* (perpendicular and down the slope). Use the clinometer part of the compass or a separate clinometer to measure the *dip angle.* Again, the procedure varies with the compass or clinometer so follow the directions given by your instructor.

Beds or strata are the most common type of plane, but other geologic features can also be planes: dikes, sills, foliation (slaty cleavage, schistosity, gneissic banding in metamorphic rocks), and faults are all planar features.

1. Your instructor has set up models of dipping planes to simulate tilted bedding planes in sedimentary rocks or igneous or metamorphic planar features as one might see in the field. Your instructor will have fastened these in place with tape or modeling clay or with some other method, and care should be taken to not move them.

a. For these models, use a compass clinometer or a compass and separate clinometer to determine the strike and dip of each model, and enter the measurements in Table 8.2. Be careful not to move the models as this will change the results. As you measure each, also do step b.

b. Without moving them, examine the rocks making up these dipping planes or propped up on models to represent natural geologic features in the field. Determine the type of planar structure for each and also enter this information in Table 8.2.

What Is Shown on a Geologic Map

A geologic map endeavors to illustrate the underlying geology of an area by showing the placement of rock *formations* and their *structures,* and *contacts.*

- A **formation** is a continuous or once continuous layer or mass of rock—sedimentary, igneous, or metamorphic—that a mapper can easily recognize while doing geologic fieldwork. Defined on

Table 8.1

Geologic Map Symbols

Features and symbols likely to occur on geologic maps and their descriptions.

Items on Map	Colors, Codes, and Symbols	Symbols
Strike and dip of bedding Number indicates the dip angle	Longer line showing bedding strike with perpendicular short tick in center showing dip (down direction)	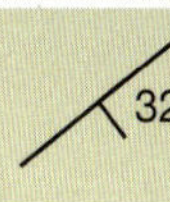 32°
Horizontal bedding	Large plus sign or circle with plus sign	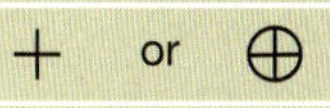 + or ⊕
Strike and dip of vertical bedding	90° dip. Dip line symmetrically crosses the strike line	
Strike and dip of overturned bedding	Dip direction indicated with a J. The down-dip direction is the straight part of the J.	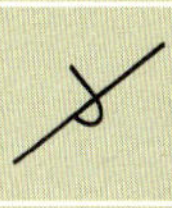
Strike and dip of foliation	Longer line showing foliation strike with triangle at center showing dip	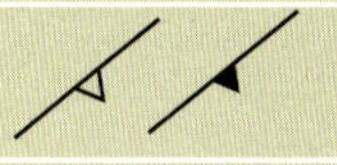
Contacts between rock bodies	Fine black lines, dashed where uncertain, dotted where covered	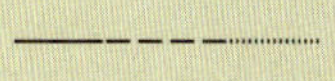
Faults, undifferentiated	Thick black lines, dashed where uncertain, dotted where covered	
Normal faults[1]	Tick on down-dip side, which is also the down-dropped side U on the up-faulted side and D on the down-dropped side	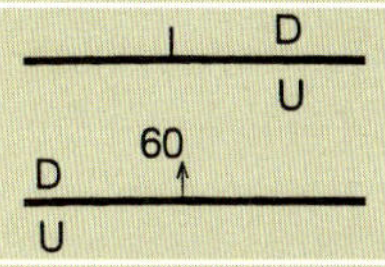 D U 60 D U
Reverse faults[1]	Tick on down-dip side, which is also the up-faulted side U on the up-faulted side and D on the down-dropped side	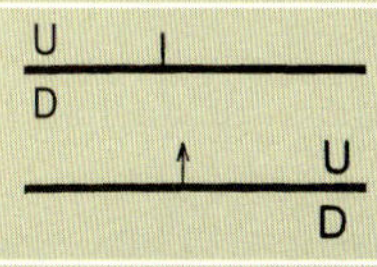 U D U D
Thrust faults[1]	Teeth on the upthrust (hanging wall) side	
Strike-slip faults[1]	Half arrows indicate the sense of motion	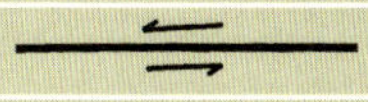
Non-plunging anticline fold axis[1]	Long line along the axis, crossing symbol shows direction of the dip away from the axis	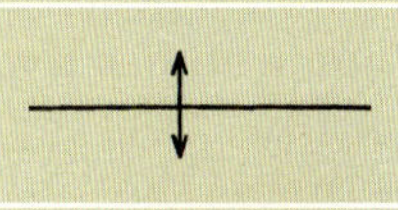
Non-plunging syncline fold axis[1]	Long line along the axis, crossing symbol shows direction of the dip toward the axis	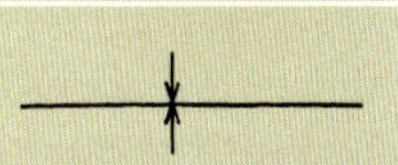
Monocline	Arrows show direction of inclination[1]	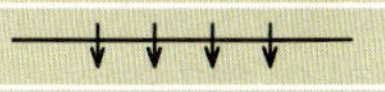
Plunging anticline and syncline[1]	Arrows on the end of the axis show the direction of the plunge (down tilt of the axis)	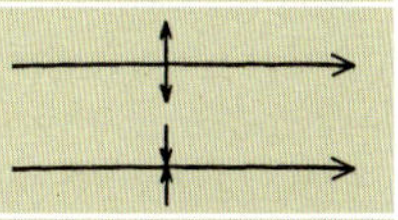
Overturned folds[1]	U shape with arrows pointing in the dip direction; toward the axis for synclines and away from the axis for anticlines	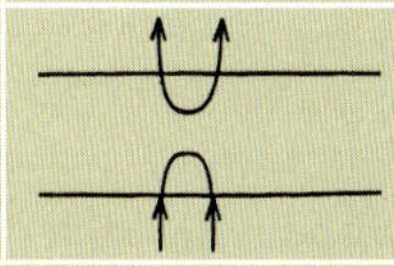

[1] Fault and fold-axis symbols may also be dashed where uncertain and dotted where covered.

Table 8.2

Measurements of Strike and Dip

Strike and dip and planar features in rocks. See instructions in Exercise 1.

Model/ Sample Number	Strike	Dip Direction and Angle	Type of Planar Structure in the Rock

the basis of rock characteristics and stratigraphy, this basic unit is shown on geologic maps as a single entity. Geologists use the term more commonly with sedimentary rocks, but they can use it for igneous and metamorphic rocks as well.

- **Contacts** are the boundaries between rock units or formations, where one formation gives way to the next, depicted as a thin, solid black line on most maps (or dashed line where uncertainly located). Contacts always separate two different rock formations, which usually appear on a map as different colors on either side of the line.
- **Structures** are the physical arrangements of rock masses. They include intrusive bodies, unconformities, attitude of rock layers, and deformational features such as faults and folds. We cover these structures later in this lab.

On a geologic map, you will generally find an explanation of the symbols used for the colors, patterns, and deformational structures on that particular map. The key or legend of the map is an explanation of rock formations in *chronological order*—with the youngest rocks at the top and the oldest rocks at the bottom (Figure 8.2)—and sometimes separated by rock type into igneous, sedimentary, and metamorphic groupings (Figure 8.18, p. 188). Colors usually show the different rock formations. On black and white maps, the cartographer uses different patterns and letter codes to indicate the different rock units. The inside back cover of this book shows some common examples of patterns for different rock types. In the letter system the U.S. Geological Survey employs, the first uppercase letter in for a sedimentary formation indicates the geologic period in which the formation formed; subsequent letters in lowercase are an abbreviation of the formation's name. For example, in the symbol **Tjs** in Figure 8.2, the **T** stands for Tertiary, the **js** stands for Jonston Shale.

Geologists construct geologic maps from natural outcroppings of rocks (Figure 8.2a). **Outcrops** are areas of rocks exposed at the surface that are not covered (Figure 8.2a) with foliage, sediment, or man-made structures. Geologists infer the position and orientation of rock layers (*formations*) from the outcrops that are available, and decipher the unseen structures beneath. They draw the map as if the rocks were entirely exposed at the surface (Figure 8.2b). Some geologic maps (such as Figure 8.15 on p. 180) distinguish the areas of outcrop (solid colors) from covered areas (stippled patterns) where the geologist inferred the existence of formations under the surface.

Cross sections, also known as *structure sections*, are included with many geologic maps. They show the arrangement and sequence of the rocks from a side view, as in the front face and side of the block diagram in Figure 8.2a. A cross section is a topographic profile, but it also shows the underground arrangement of the rocks. A line and letters such as A–A′ or B–B′ (Figures 8.2 and 8.13, p. 176) generally show where a cross section slices through an accompanying geologic map. The information on a cross section comes from the map, in part from strike and dip measurements.

2. Examine the simple geologic map in (■ Figure 8.2b and its key in Figure 8.2c.

a. What type of rock does the symbol **Jrl** represent? Limestone

b. What does the **J** in **Jrl** stand for?
Jurassic

c. How about the **rl** in **Jrl**?
Rand limestone

d. What do the colors represent?
Various rock formation

e. Label a contact in Figures 8.2a and b and name the formations on either side of it.
a) Cretaceous
b) Tertiary

3. For each strike and dip symbol on the map in Figure 8.2b, measure the strike with a protractor and write out the numeric strike and dip.

a. In **Kms**: N 25 E, 38° NW

b. In **Jrl**: N 20 W, 31° NE

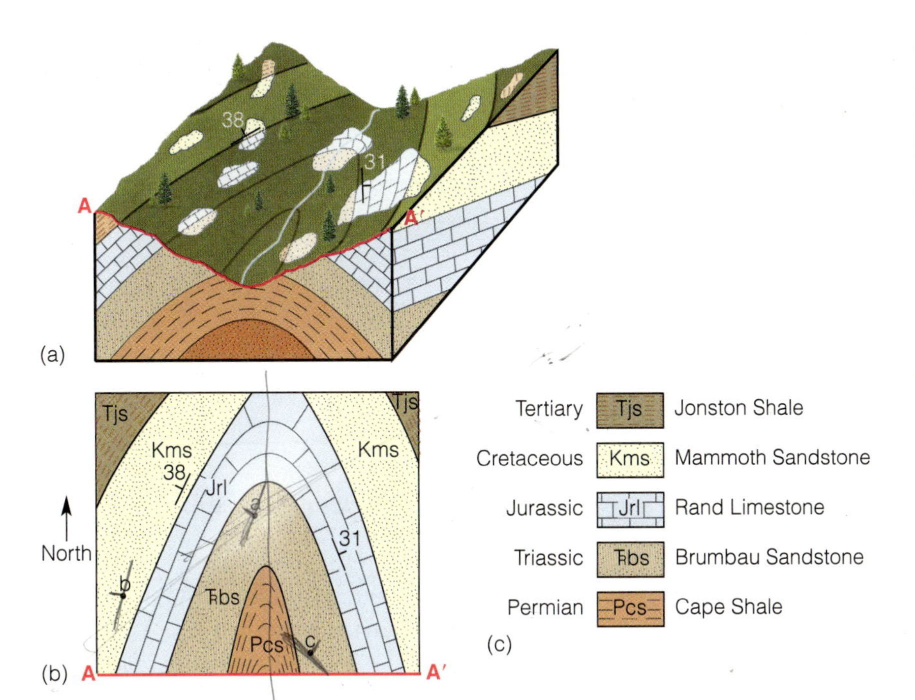

Figure 8.2

Map and Cross Section from Outcrops

(a) A region with scattered outcrops of limestone, sandstone, and shale. Contact lines and cross sections show how the geologist visualizes the underlying geology. The front A to A′ is a cross section, as is the side of the block. (b) Map of the same region inferred from the outcrops. (c) Map explanation (or key) with names and periods of the formations.

4. Put the appropriate strike and dip symbol at each lettered location on the map in Figure 8.2b. The measurements below, first use the 360° compass rose (Figure 8.1a) and then use the quadrant style (Figure 8.1b). Use the style your instructor prefers.

a. 062°, 25°NW S62°W, 25°NW

b. 020°, 43°NW N20°E, 43°NW

c. 345°, 29°NE N15°W, 29°NE

STRUCTURES AND DEFORMATION

Sometimes various processes in the Earth—especially major Earth movements (see plate tectonics, in Labs 1 and 9) and metamorphism—change the shape of rocks. The general term for these changes, is **deformation.** These processes lead to the creation of different geologic *structures.* Different styles and forces of deformation, including folding and faulting, produce different structures.

We have examined strike and dip, but what do they signify? From the principle of *original horizontality* from Lab 7, we would expect sedimentary bedding, lava flows, and volcanic ash layers to be horizontal, yet observations show that they are not always so (Figure 8.6 on p. 168). Deformation can tilt and change the position and orientation of bedding and other structures. One reason we bother to measure and map strike and dip is so that we can better observe the results of deformation, especially over areas larger than an outcrop.

Forces deform rocks in two different ways. **Brittle** deformation breaks rocks. This *style* of deformation occurs when rocks are cold and nearer the surface. **Ductile** deformation occurs when rocks bend, flex, and/or flow, and generally, when the rocks are deep and possibly warm. Rapid movement is also more likely to produce brittle deformation, and conversely, ductile behavior is more likely when the deformation is slow.

Three different *forces* result in deformation, as shown in (■ Figure 8.3. **Compression** squeezes things together, with forces pushing toward each other (Figure 8.3a). **Tension** pulls them apart, with forces moving away from each other (Figure 8.3b). **Shear** is a scissor-like motion causing one rock mass to slide past another (Figure 8.3c).

Folded Layers: Anticlines and Synclines

Folds form whenever deformation compresses rocks parallel to their layering, and the beds are buried deeply enough that they behave in a *ductile* manner (that is, they bend rather than break). The beds may form an upward arch called an **anticline** (■ Figure 8.4a) or may bow downward in the middle and form a **syncline** (Figure 8.4b). Note that anticlines and synclines are bends in rock layers beneath the land surface and are not necessarily directly related to topography. In the block diagram in Figure 8.2, for example, the anticline is a valley. In Figure 8.6, on page 168, the anticline is essentially flat in

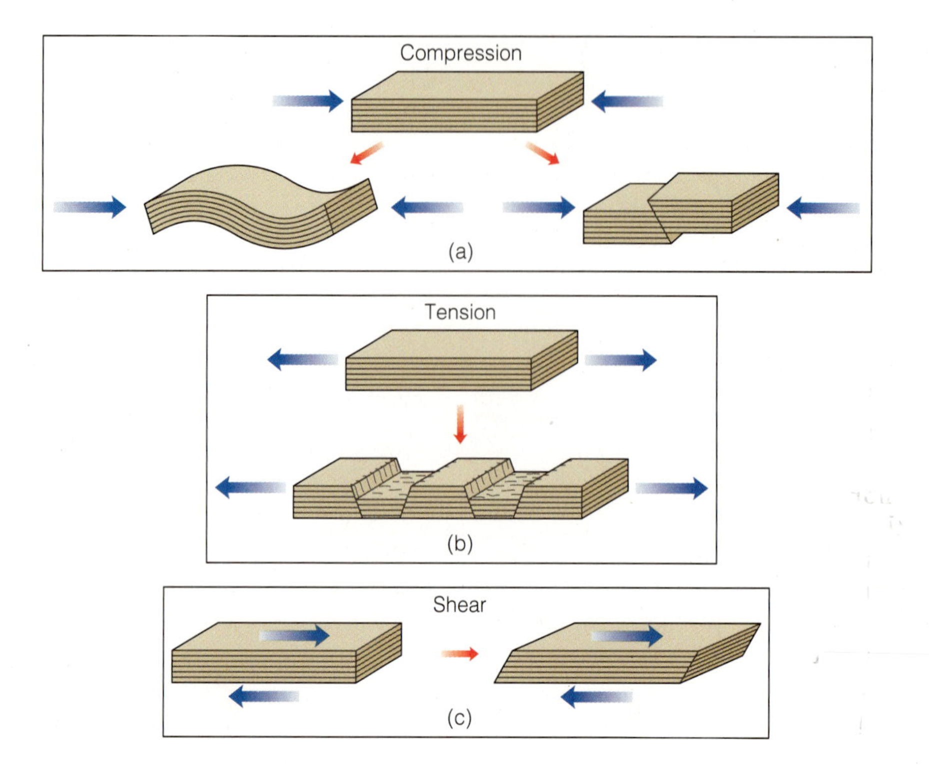

Figure 8.3

Forces That Cause Deformation (blue arrows) and the Resulting Structures

(a) Compression causes either folding or faulting. (b) Tension causes extension such as the faulting shown. (c) Shear causes shearing, which is movement of rock bodies along closely spaced planes.

the foreground and a hill in the distance. Multiple folds of anticlines and synclines alternate unless they are broken by faults. In the foreground in Figure 8.6, notice a syncline is present to the left of the anticline.

The line around which bending takes place (similar to the crease in a folded sheet of paper) is the fold **axis.** The red symbols shown in Figure 8.4 represent fold *axes* (plural of axis) on maps. Fold axes lie in a plane that cuts through the bent part of the fold, called the **axial plane.** The central bend of the fold is the **hinge zone. Limbs** flank the hinge zone on both sides. Where erosion has stripped off some of the uppermost rocks, an anticline will have older rocks exposed in its hinge zone or core and younger rocks in its limbs, whereas a syncline will have younger rocks in the hinge zone.

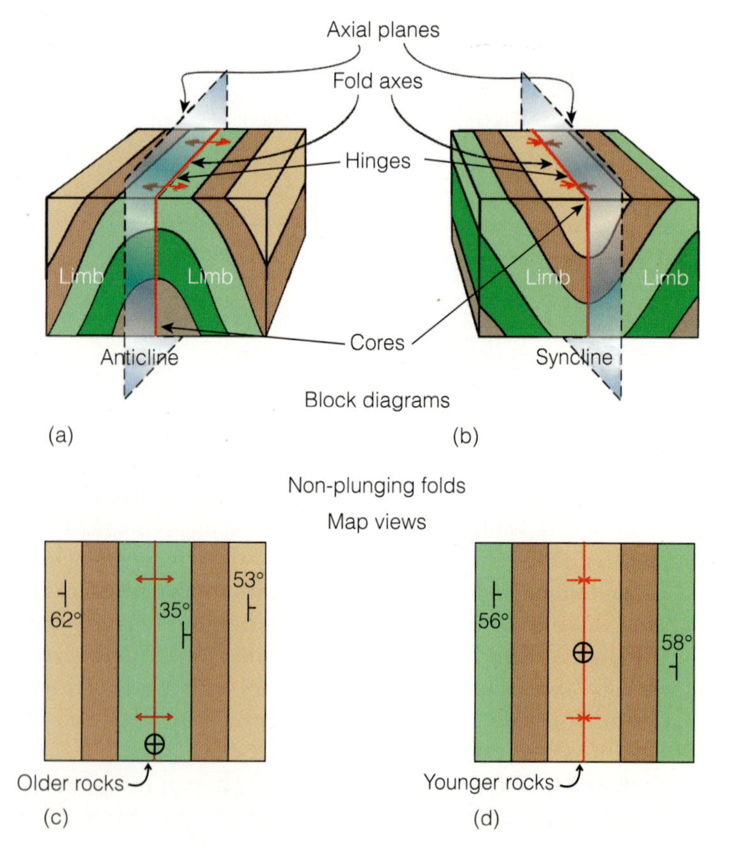

Figure 8.4

Non-Plunging Folds

Block diagrams of (a) an anticline and (b) a syncline; (c) and (d) Maps with fold axis symbol in red and strike and dip symbols in black. (c) Map of an anticline showing older rocks in the core. (d) Map of a syncline showing younger rocks in the core. In both non-plunging folds, the rocks at the core are horizontal because they are at the hinge axis of the fold.

Folds are often too large for us to see them in their entirety at one locality. We, therefore, deduce their presence from observation of strikes and dips of individual beds. When we plot strikes and dips on a map, the presence of folds becomes immediately apparent. Beds *dipping away* from an axis define an **anticline;** beds *dipping toward* an axis define a **syncline.** A **monocline** is an area of tilted rocks in a region with otherwise gently dipping or flat lying strata.

Non-plunging folds are anticlines and synclines that have horizontal fold axes (Figure 8.4). In flat terrain, non-plunging folds show parallel stripes in map view. The map patterns such as the repeating parallel rock layers observed in Figure 8.4c and d will help you to recognize non-plunging folds. Hills and valleys will change this pattern somewhat. Irrespective of topography, areas of non-plunging folds have beds that strike parallel to each other and parallel to the fold axes.

Most folds have axes that are not horizontal because the compression causing the beds to fold and the beds themselves are irregular. Such folds are **plunging folds** and their axes plunge into the Earth at an angle (see ■ Figure 8.5c). Where plunging anticlines and synclines occur, erosion will expose the formations as a series of zigzag patterns (Figure 8.5b). This type of pattern is also visible in ■ Figure 8.6, which shows the Sheep Mountain anticline in Wyoming (center) and a smaller syncline on the left, both plunging toward the viewer. On geologic maps, such a map pattern indicates that plunging folds are present. Beds in plunging folds still dip away from the axis of an anticline and toward the axis of a syncline, but the strikes are not parallel (Figure 8.5a) as they were for non-plunging folds. The dip of beds near the fold axes will point in the general direction of the plunge.

Plunging folds tend to occur in sets of alternating anticlines and synclines plunging in the same direction. Look more closely at Figure 8.5. Arrowheads on the ends of the fold axes portray the plunge direction. Even if the fold axis symbols are not on the map, you can easily locate the axes by looking for areas where formations seem to form a mirror-image repetition of each other. Also, you can see that for anticlines, the bend or "nose" in the outcrop pattern of rock layers points in the direction of the plunge of the fold axis, and the opposite is the case for synclines—the "nose" of a syncline points away from the plunge.

When drawing fold axes on a map of plunging folds, as for non-plunging folds, you should draw the axis where a plane would produce a mirror image of rock layers on either side. For plunging folds, the axial plane (mirror) also connects the "kinks," bends, or "noses" of the folds. The axial planes in Figure 8.5 correspond to the position of the mirror planes for the folds shown in the block diagram. Of course, in nature, multiple folds rarely have such perfectly aligned symmetry.

Overturned folds occur where one limb has been overturned (tilted more than 90°). Table 8.1 on page 163 shows the fold axis symbols for overturned folds.

Because rocks of varying hardness may compose various formations, they erode at different rates; this is **differential erosion.** Harder more resistant rocks tend to stand out as hills or ridges and softer rocks become valleys. Zigzag ridges in the landscape reveal the underlying zigzag pattern of plunging folds. This is particularly common

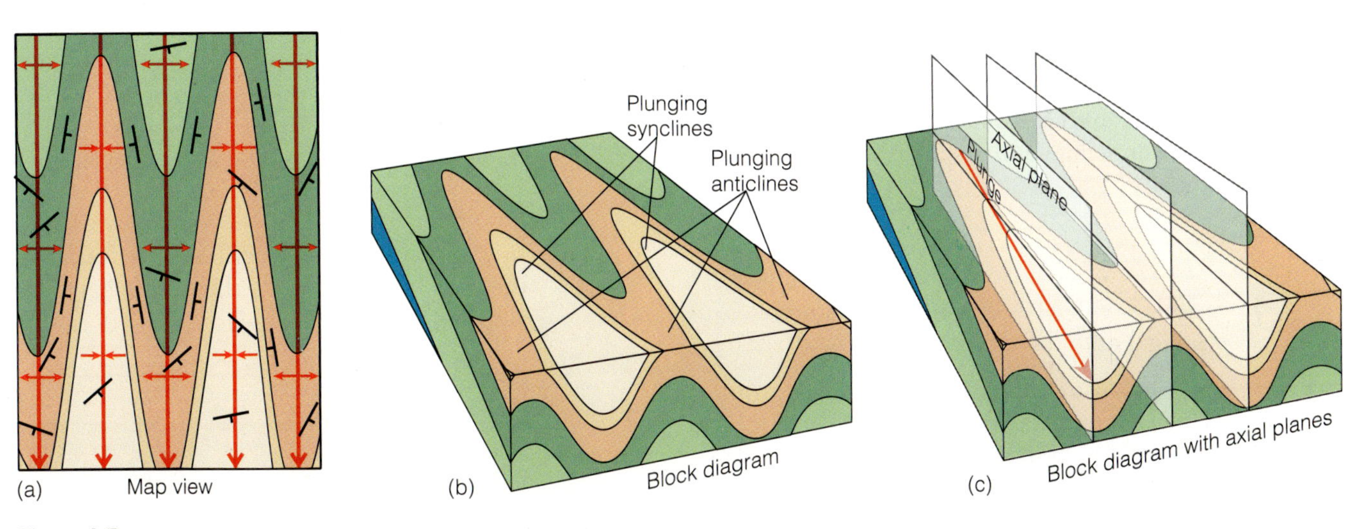

Figure 8.5

Plunging Anticlines and Synclines

(a) Geologic map. (b) Block diagram. (c) Block diagram showing axial planes. The red arrow on (c) is the fold axis and shows the plunge of the folds. Observe the arrowheads on the map in (a) at the end of the fold axes on the downward (plunging) end.

Figure 8.6

Sheep Mountain Anticline in Wyoming

This end of the anticline is plunging down toward the viewer. A syncline is visible in the foreground to the left of the anticline, which is also plunging toward the viewer. The rock units here are mostly marine sediments.

in the Appalachian region of Tennessee, Virginia, and Pennsylvania. In some locations, an anticline may be a hill or mountain where resistant rock makes up the core. In other places along its axis, the same anticline may be a valley, where its core has easily erodable layers.

A **dome** is an anticlinal fold with an elliptical or circular pattern in which the rocks dip away from the center in all directions. A **basin** is synclinal fold with layers in a bowl shape. Here the dips are toward the center. Younger rocks will occur around an older core for the dome but younger rocks will be in the core of a basin. The next exercises should help you visual strike and dip, and anticlines and synclines from symbols and map patterns.

5. Examine a block or clay model of a non-plunging anticline and syncline.

a. Try to visualize how the rock layers would penetrate through the block. Locate the axis of the fold. If you place the model on the table, is the axis horizontal? ______

b. Look at the top of the model (map view) and describe the pattern (zigzags, wiggly, parallel stripes, circular, oval, or whatever) that the rock layers make on this surface.

____

____

6. Examine a block or clay model of a *plunging* anticline and syncline.

a. Try to visualize how the rock layers would penetrate through the block. Locate the axis of each fold. If you place the model on the table, is the axis horizontal?

______

b. In your own words, describe the pattern the rock layers make on the top (map view) of the model.

____

____

____

Strike and Dip of Folds in a Sand Tray Practice visualizing strike and dip, and anticlines and synclines from information on a map using a sandbox model.

Materials needed:

- Three colors or types of small planar items to represent bedding, such as small rectangles cut out of colored index cards, or streak plates, glass plates, small pieces of cardboard, large coins
- A tray (or stream table) with sand of sufficient depth to hold up the small planar items
- Three colored pencils approximately matching the colors on the map in Figure 8.7.

7. Work together in a group to make a model of the map in Figure 8.7.

a. Carefully study the map. Draw, in the sand, the outline of the map and the contacts between the shale, siltstone, and sandstone layers.

b. Decide which color or type of card, plate, or planar item will correspond to each rock type in Figure 8.7. If you are using colored index card pieces, match the color of the formation on the map, if possible.

c. Next, examine the strike and dip symbols on the map. Starting with one strike and dip symbol, locate the appropriate position in your sand model, and position a planar marker well into the sand so that it has the same strike and dip as the symbol represents. Remember that the *dip* is *down* into the ground. Take turns placing markers for the majority of strike and dip symbols on the map. Help each other out if necessary, and ask for help from your instructor if you get stuck. Have your instructor check your model.

d. Determine where all anticlines and synclines are on the map. Draw in the axes for these folds in the sand and on the map using the appropriate symbols in Table 8.1. Draw them freehand, not with a ruler, as fold axes are frequently somewhat curved.

e. Carefully study your model and the map in Figure 8.7 to determine the types of folds. Are the folds plunging (with tilted fold axes) or non-plunging (with horizontal axes)? *Hint:* If the fold axes pass through horizontal beds, they are non-plunging, but if they pass through dipping beds they are plunging. ______________________

f. Notice that for non-plunging folds, all the beds strike in the same direction as in Figure 8.4c and d, but in plunging folds the strikes vary, as in Figure 8.5a. Is the strike of the beds parallel to or at an angle to the fold axes? ________________

g. Fill in the three boxes to make a *key* for the map as follows:

- Decide which rock is youngest (at the top), which is in the middle, and which is oldest (on the bottom). You will have to visualize how the rocks continue down into the ground.
- Write the letter symbol in each box so the rocks are in chronological order from oldest at the bottom to youngest at the top.
- Also in the box, sketch in the pattern used for each rock.
- Use colored pencils to shade each box the approximate color on the map for that rock.

h. What was the direction (in general terms, N–S, NE–SW, or whatever) of compression that affected the area? ________________

8. Examine the map in Figure 8.2b. What structure(s) do you detect?

a. Give its/their full name:

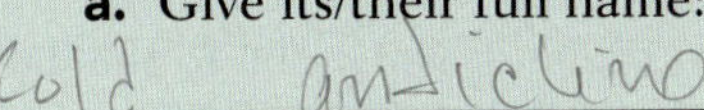

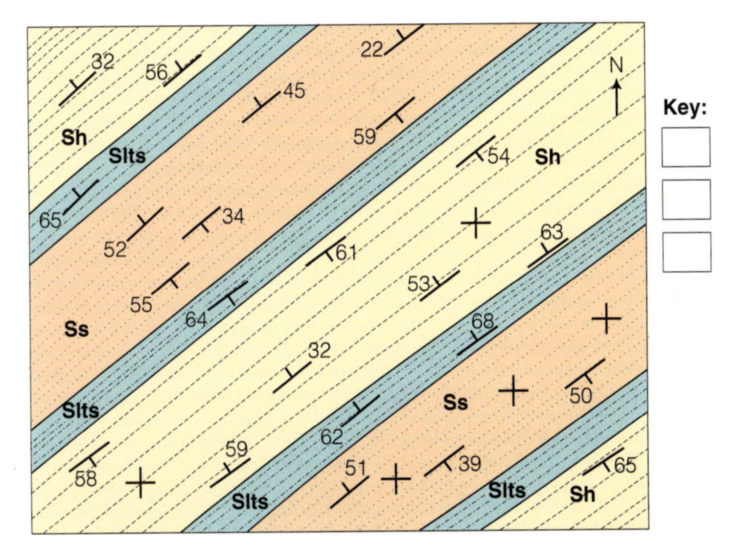

Figure 8.7

Geologic Map of a Sand Tray Model

The region with sandstone (**Ss**), siltstone (**Slts**) and shale (**Sh**) formations shows the strike and dip of bedding.

(a)

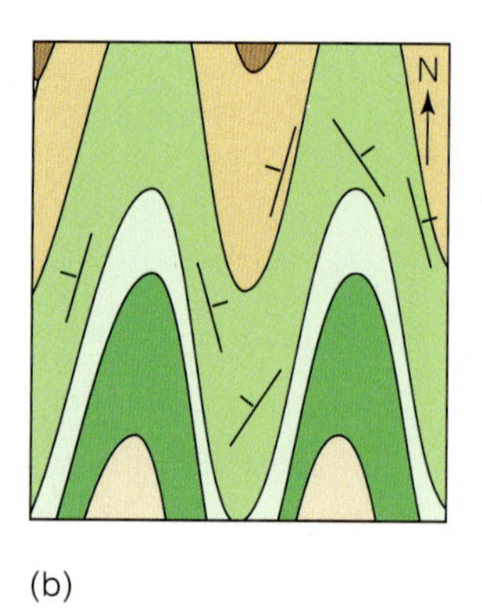

(b)

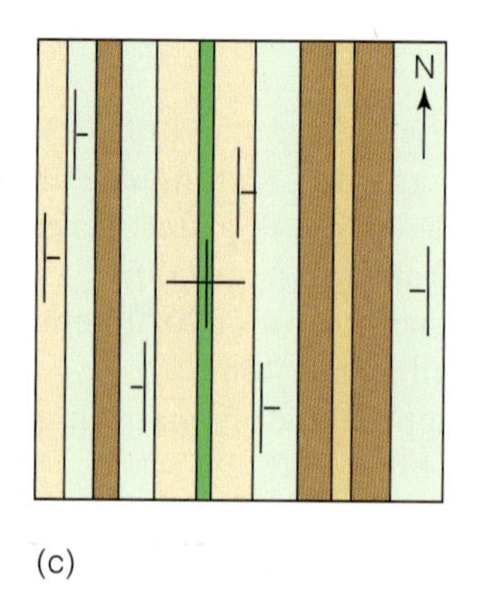

(c)

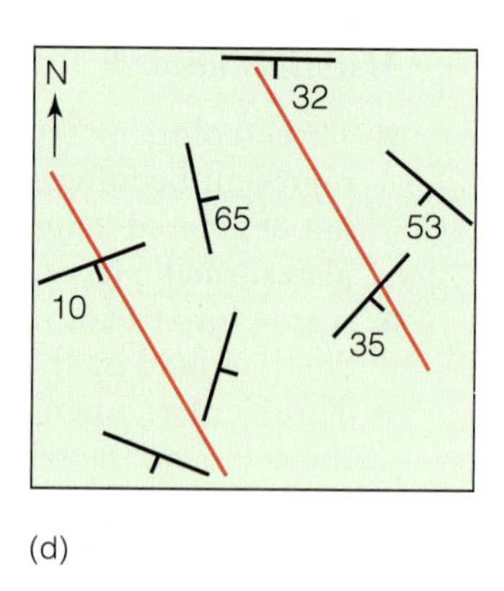

(d)

Figure 8.8

Simplified Geologic Maps of Folds

(a), (b), and (c) show folds, formations (colors), and strike and dip measurements for Exercise 9. (d) Strike and dip symbols may reveal the location of folds. The fold axes need appropriate arrows added.

b. Finish the map by drawing in the appropriate fold axis symbol(s) on Figure 8.2b.

9. ■ Figure 8.8 shows a series of maps with folds.

a. For each map, determine if the folds are plunging or not. If they are, write the compass direction of the plunge in the first row of blanks in ■ Table 8.3. If they are non-plunging folds, write **non-p** in the same row.

c. For each map, write the compression compass direction in Table 8.3.

d. In Figure 8.8d, the fold axis traces are drawn on the figure for you (without the arrows). Finish drawing the complete fold-axis symbols on this map.

Table 8.3

Practice with Folds

Figure 8.8 shows maps of these folds. See instructions in Exercise 9.

	Map			
Fold	**8.8a**	**8.8b**	**8.8c**	**8.8d**
Plunge direction or non plunging				
Compression direction				

Faults

Brittle deformation forms **faults,** which are fractures or zones of fracture for which opposite sides moved relative to each other with the two sides remaining in contact (■ Figure 8.9). The displacement can be from a few centimeters to hundreds of kilometers. Geologic maps usually show faults as thick black lines, and the abrupt displacement of formation contacts commonly make faults easy to recognize. Faults are planar features, so they, too, have strikes and dips. If movement of the rocks on either side of the fault is parallel to the dip of the fault plane, the fault is a **dip-slip fault.** The faults in Figures 8.9a, b, and c are dip-slip faults. **Strike-slip faults** do not have a component of up or down motion, but have slipped horizontally—that is, parallel to the strike (Figures 8.9d, e). An **oblique-slip fault** has movement between the strike and dip of the fault plane (Figure 8.8f).

Dip-slip faults consist of normal, reverse faults, and thrust faults. Geologists often define dip-slip faults in terms of a hanging wall and a footwall. The hanging wall is the block of rock physically above the fault plane, whereas the footwall is below the fault plane. You can also think of the footwall as a ramp and the hanging wall as a block sitting on the ramp. For **normal faults,** the block slid down the ramp, for **reverse** and **thrust faults** it slid up the ramp, but in thrust faults the ramp is gently sloping and in reverse faults, it is steep. Figures 8.9a, b, and c show how to distinguish these faults.

Strike–slip faults also have different types. Figures 8.9d and e show the two possible directions of slip. Imagine that you are standing on one side of the fault and you are looking across to the other side. If the block on the *other* side has moved to your right, then the slip is **right–lateral** (Figure 8.9d); if the block has moved to the left, then it is **left–lateral** (Figure 8.9e).

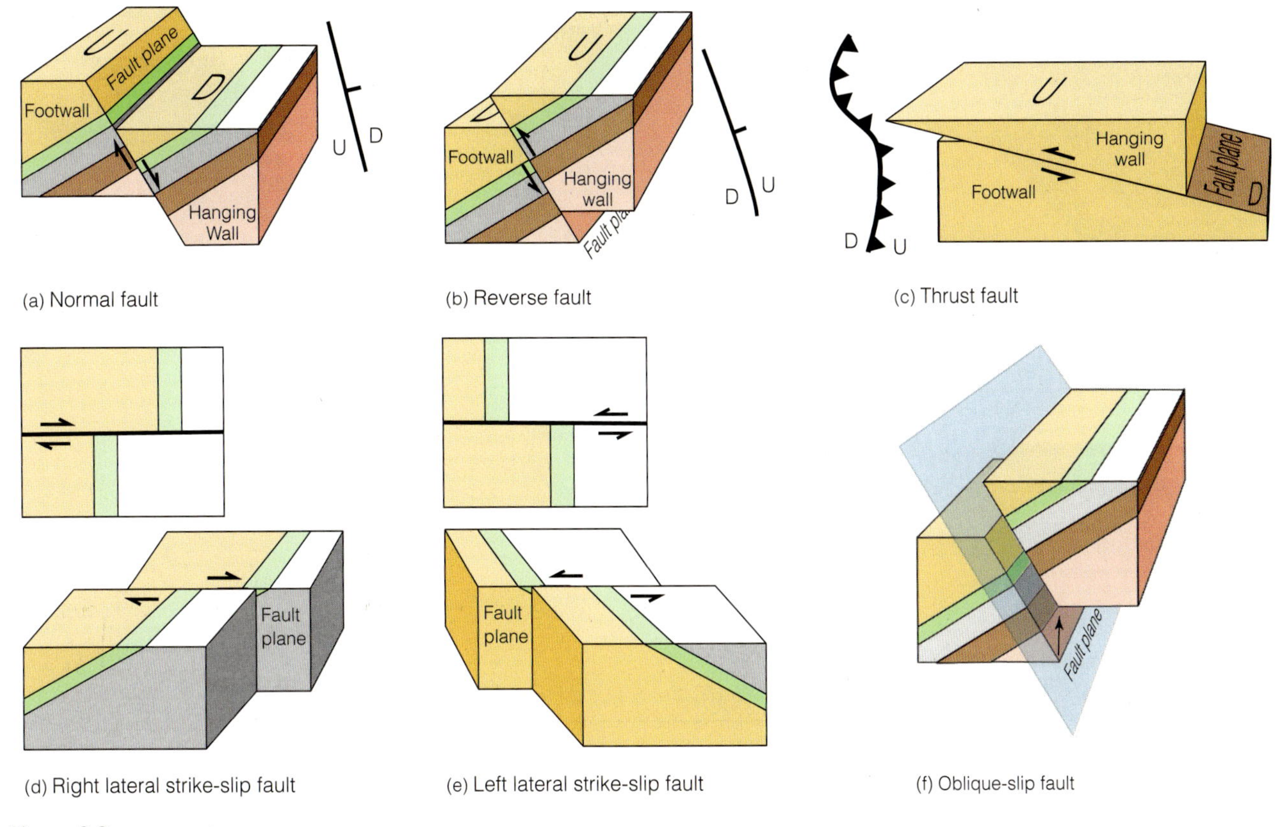

(a) Normal fault (b) Reverse fault (c) Thrust fault

(d) Right lateral strike-slip fault (e) Left lateral strike-slip fault (f) Oblique-slip fault

Figure 8.9

Types of Faults

Block diagrams of faults and their map symbols. U indicates up and D indicates down. (a) **Normal fault.** On geologic maps, a small tick or arrow that shows the down-dip side of the fault trace will occur on the same side as the D. (b) **Reverse fault.** For reverse fault symbols, the small tick in the down-dip direction will occur on the same side as the U. (c) **Thrust fault.** In the map symbol, the up-thrown side of the fault (hanging wall) is decorated with triangular, teeth-like symbols pointing toward the upper plate. (d) **Right lateral strike-slip fault.** (e) **Left lateral strike-slip fault.** (f) **Oblique-slip fault.**

Determining the type of fault is useful for understanding a region's deformational history. This is because normal faults result from extension. Reverse and thrust faults result from shortening commonly seen in areas of compression. Strike-slip faults result from horizontal shear.

10. Examine and label each fault in ■ Figure 8.10.

11. Using blocks provided, arrange the blocks into each type of fault. You may need books to help prop up the blocks so they will sit in the correct position for your fault.

a. For each type of fault, sketch on a separate piece of paper both a map view and a cross section of the fault.

b. On each sketch label the type of fault, label which group it belongs to (dip-slip or strike-slip, if any), and place arrows on each block next to the fault showing the direction of movement (if visible from that view).

c. For which group of faults are the cross sections of little use in seeing the offset of the blocks?

____

d. For which group of faults does the map view fail to show the fault offset?

____

CONTACTS ON GEOLOGIC MAPS

In layered rock bodies, such as sedimentary or volcanic formations, the two-dimensional curving or planar surface of the contact between any two formations

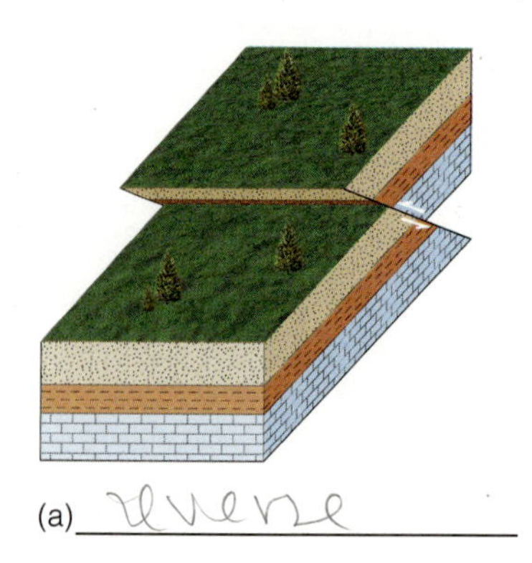
(a) reverse

(b) strike slip

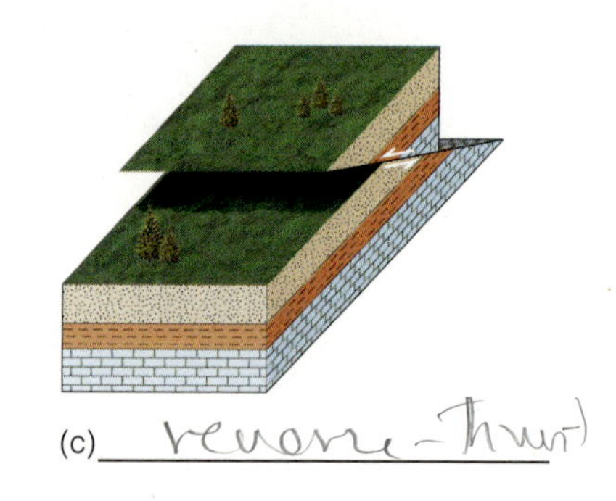
(c) reverse-thrust

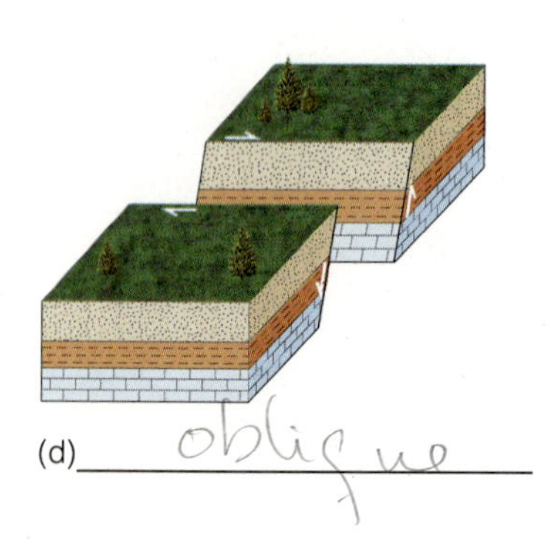
(d) oblique

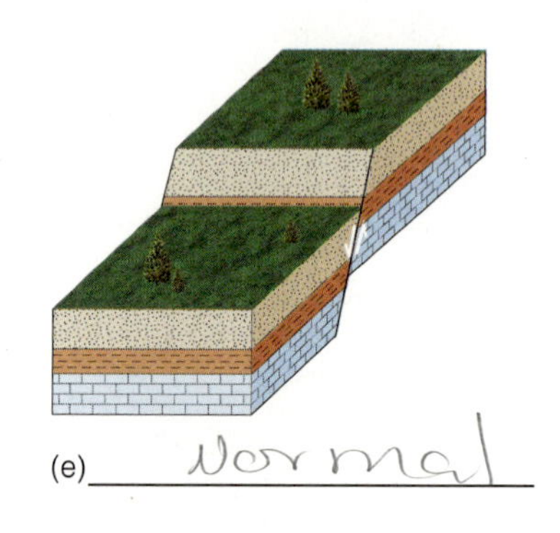
(e) normal

Figure 8.10

Block Diagrams of Some Faults

See instructions in Exercise 10.

intersects the Earth's surface to form a line or curve called the **trace** of the contact.

If a formation is horizontal, the contact traces follow the topographic contours (■ Figure 8.11a). Both contours and contact traces have a winding, or *sinuous,* pattern that follow each other, if the land surface is hilly or irregular (Figure 8.12).

For beds or other planar features that are inclined, or dipping, a contact changes elevations and its trace crosses contours. In Figures 8.11b and c, the map pattern of the beds form a V or arrowhead that points toward the down-dip side. The **rule of V's** states that the V formed by the surface trace of a planar feature *in a valley* points in the

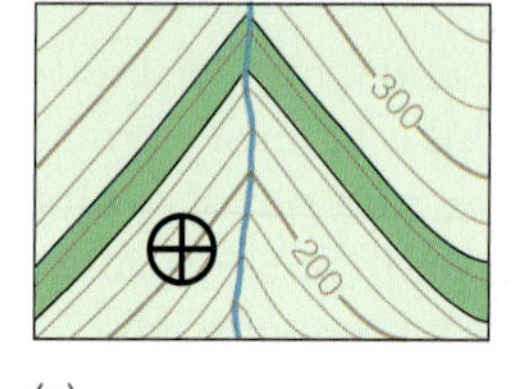

(a)

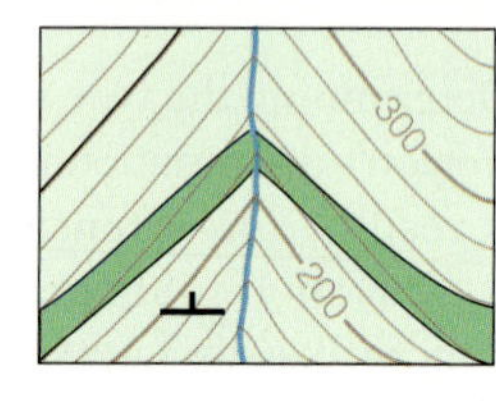

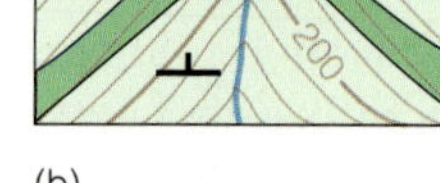
(b)

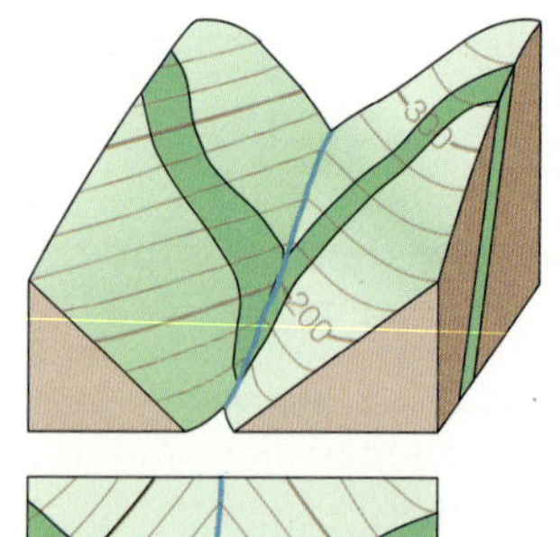

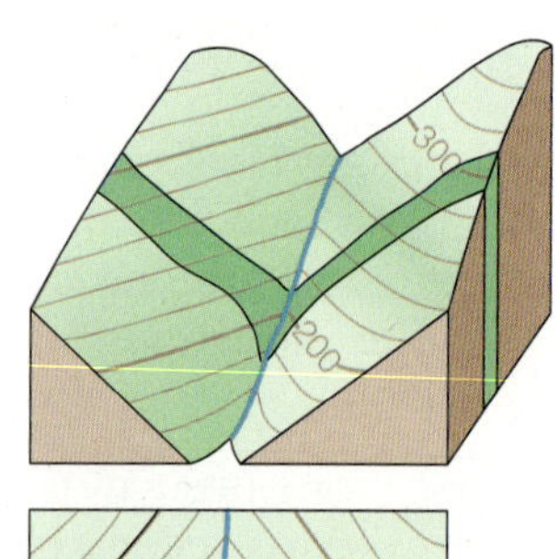

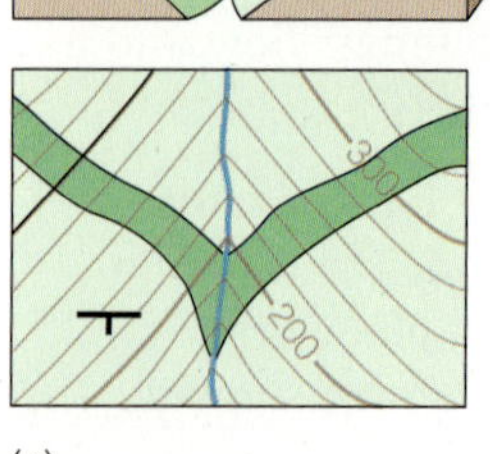

(c)

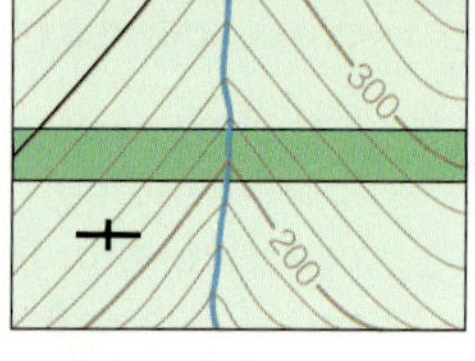

(d)

Figure 8.11

Rule of V's

General rule: The V points in the direction of the dip. All of these block diagrams and map views illustrate areas of identical surface topography but with differing arrangements of rock formations. Superimposed on each are the traces of the contacts of features such as a bed, dike, or sill. (a) Horizontal features: The contacts parallel the contours. Observe the symbol for horizontal beds on the map. (b) Feature dipping upstream: Notice the strike and dip symbol on the map. The V made by the contacts points upstream. (c) Feature dipping moderately to steeply downstream: Observe the strike and dip symbol on the map. The V made by the contacts points downstream.[1] (d) Vertical feature: Look at the vertical strike and dip symbol on the map. Traces on maps of contacts of vertical beds, dikes, or other features do not bend as they cross the stream valley.

direction of dip of the feature[1] (Figure 8.11). Thus, if you determine where valleys are, you can determine the dip direction of the rock beds. This is true whether the feature is a sedimentary bed, a volcanic layer, a sill, a dike, metamorphic foliation, or a fault. Even if the map does not have topographic contours, you can still recognize valleys because rivers and streams occupy them.

In map view, if the beds are vertical planes (Figure 8.11d), the contact appears as a straight line *independent of the topography.* The steeper the dip, the straighter (or less sinuous) the map pattern crossing a valley or ridge is. In general, more sinuous map patterns indicate low dips; straighter map patterns indicate steeper dips.

The rule of V's can help determine the presence of folds. If the V's point away from a region, for instance, it means the dips are away and an anticline is present. V's pointing toward each other indicate a syncline.

Faults are planar features, in the same way that contacts between formations are. Consequently, they obey the rule of V's in exactly the same way. Additionally, steeply dipping faults have straighter map patterns than more shallowly dipping faults do. Different types of faults tend to have different dips, which may aid identification but do not define the fault types. *Normal* and *reverse faults* (Figure 8.9a and b) are usually steeply dipping and so generally have fairly straight map patterns. *Thrusts* dip shallowly (Figure 8.9c), resulting in a more sinuous map pattern. Thrusts often occur in association with major folds whose axes run approximately parallel to the faults (example: Figure 8.15, p. 181, look for an overturned anticline symbol east of the South Fork Thrust Zone). Strike-slip faults are usually very steeply dipping and so tend to have the straightest map patterns of any fault.

Younger rocks on the down side of a fault end up next to older rocks on the up side (unless the rocks have been overturned). Studying the relative ages of the rocks on either side of the faults in Figures 8.9a and b will assist you in confirming this. Therefore, if you can determine the direction of dip and the age of rocks on either side of a fault, you can tell what type of fault it is.

Unconformity Contacts on Geologic Maps

Recall from Lab 7 that an unconformity represents a gap in time and that there are three types of unconformities. The essential characteristic of an *angular unconformity* is that it cuts across the contacts of all of the beds underneath it. This characteristic is easiest to see in cross section (Figure 8.13b, p. 178), but it is also visible on a geologic map (■ Figure 8.12), where the rocks above the unconformity cover the beds below. Often the unconformity surface dips more gently than the underlying beds. As a result, the map pattern of the unconformity may be more sinuous than the contacts underneath it.

Igneous Contacts on Geologic Maps

The map pattern of the contacts of an intrusive igneous rock depends on whether the magma is **concordant** (intrudes *along* existing bedding or foliation) or **discordant** (intrudes *across* existing bedding or foliation).

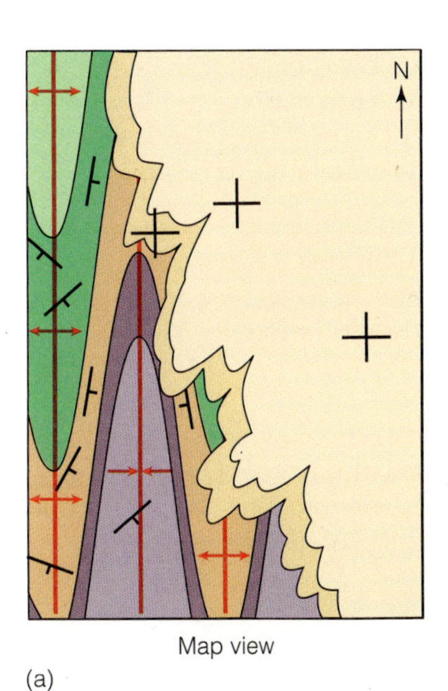

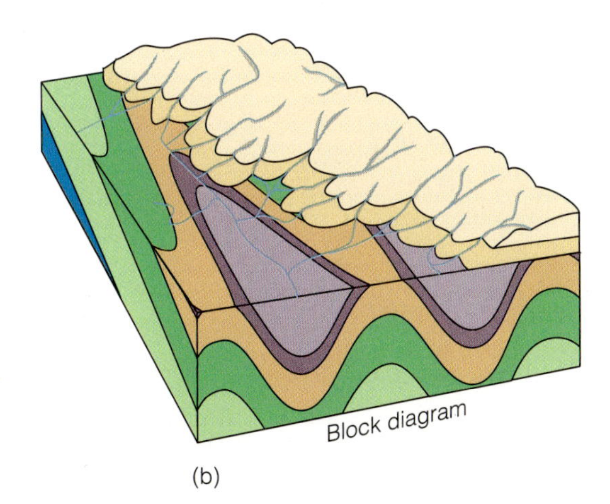

Figure 8.12

Map (a) and Block Diagram (b) of an Unconformity

The plane of the unconformity occurs at the bottom of the yellowish layers. The trace of the unconformity cuts across the contacts of the formations beneath the unconformity and makes a fairly irregular line because it is shallowly dipping or horizontal.

[1] The contact traces of beds crossing a sharp *ridge* or cliff will V in the direction *opposite* to the dip. Exceptions to the rule occur with horizontal beds, as already seen, and when beds dip more gently and in the same direction as the surface slope.

In concordant intrusions, of which sills are the most common, the intrusive contacts are parallel to the contacts of the beds intruded (Figure 3.27, p. 64). Locate the Rama Formation in cross sections **A-A′** in Figure 8.13b. Notice how the top and bottom contacts of the sill are parallel to the bedding contacts. Except at the end of an intrusion where it pinches out, a sill appears to be just another formation within the sequence. Unlike beds, however, sills are not laterally continuous. They often end in a wedge-shaped form, and may turn into dikes, cutting across the layers. In addition, sills are younger than the rocks they intrude. Lava flows, too, are parallel to sedimentary layers unless they lie unconformably on tilted layers.

In discordant intrusions such as dikes, stocks, or batholiths (Figure 3.27, p. 64), the intrusive contact cuts across the existing beds and formation boundaries.

The map views of discordant intrusive contacts are similar in appearance to those of sedimentary unconformities. This similarity of appearance may initially cause confusion in differentiating the two rock bodies on a geologic map. Fortunately, they are easy to identify if a map legend is available. If the rock unit appearing to cut through other contacts is an igneous intrusive rock, such as granite, diorite, or gabbro, then the contact must be intrusive. Sometimes finer-grained rocks such as basalt or andesite may form small intrusions. If, on the other hand, the unit is sedimentary, then the contact must be an angular unconformity or a nonconformity. In either case, cross-cutting rocks are younger than the formations being cut off.

Geologic Map Exercises

12. Examine the Geologic Map of the Bright Angel Quadrangle in ■ Figure 8.13a.

a. Study the contact between the Bright Angel Formation, **Єba** and the Muav Formation, **Єm.** What is the relationship between the contact and the contour lines?

____

b. Are the Bright Angel and Muav Formations approximately horizontal or vertical, or are they dipping at a significant angle?

______________________ If dipping,

what is the dip direction? ______

c. Is the contact between the Vishnu Schist, **p-Єv** and the Brahma Schist, **p-Єbs** gently or steeply dipping?

____

Which diagram in Figure 8.11 applies most closely to this contact? _____

d. Look for foliation symbols in these schists. Is their foliation parallel or at a noticeable angle to the contact?

____

13. Examine the cross section A-A′ in Figure 8.13b.

a. What is the youngest rock cut by the fault just SSW beneath Cheops Pyramid (under the word 'Cheops')? ______________

b. What is the oldest rock that cuts this fault? ______________________

c. What is the possible range of time for movement on this fault?

____

d. What type of fault is this?

____

e. Locate the Phantom Fault in Figure 8.13a. Use the age of the rocks to decide which side went down the most.

____

f. Examine the Phantom Fault and Monocline in the cross section in Figure 8.13b. Notice that the sense of movement of the monocline is opposite to the movement of Hakatai Shale, **p-Єh,** relative to Brahma Schist, **p-Єbs.** This is because the fault moved in opposite directions at different times. What type of fault was it when **p-Єh** moved down relative to **p-Єbs?**

____

What type of fault was it when Muav Formation, **Єm,** moved down relative to Redwall Limestone, **Mr?**

____

g. Sketch a cross section and a map view of the Phantom Monocline. On the map view, draw in the monocline symbol. Refer to Table 8.1 on page 163 if needed.

14. Examine the Bright Angel Quadrangle map in Figure 8.13a and review unconformities on pages 140–141 in Lab 7 if needed.

a. Locate two unconformities in this area and name the rock unit just above the unconformity (*basal unit*) and the type of unconformity for each:

Basal unit: ______________________

Type: ______________________

Basal unit: ______________________

Type: ______________________

b. Describe the ways in which the map (not the cross section) reveals these unconformities.

____

____

____

____

c. Look at the contact between the Redwall Limestone, **Mr**, and the Muav Formation, **Ꞓm**. Is there any indication on the map that signals that this is a disconformity?

______ If yes, how can you tell; if not, why not?

____

____

15. Examine the map of Deer Creek in Figure 8.14.

a. Using only the rule of V's, draw in the appropriate fold axis symbols on the map.

b. In each of the following blanks, write either *older* or *younger*: The rocks in the core of the anticline are ____________ than the rocks in the core of the syncline. We can make a new rule using the rule of V's that the V points toward ____________ rocks.

c. Do the strike and dip symbols show a dip that is consistent with the rule of V's? ______ If not, change the strike and dip symbols on the map to make them consistent with the rule of V's.

d. Add a strike and dip symbol in the pale green Addison Formation at the north end of the map.

16. Next, look at the Geologic Map of Glenn Creek, Montana, in Figure 8.15.

a. Using the rule of V's, determine the dip direction of the contacts.

____

b. What is the dip direction of the Glenn Fault? ______________________

c. What type of fault is the Glenn Fault?

____

d. What is the dip direction of the South Fork Thrust Zone? ____________ What type of fault is this? ____________
Sketch the symbol used for this fault.

____

e. Compare the Glenn Fault with the South Fork Thrust Zone. Which of the two is more steeply dipping?

____

f. For each fault, give its deformational force: tension, compression, or horizontal shear?

Glenn Fault: ______________________

South Fork: ______________________

Thrust Zone: ______________________
Would both forces be likely to occur at the same time? ______

g. On the map, examine pink **p-Ꞓi**. What rocks occur directly above and below its main body?

____

Within what rock does **p-Ꞓi** occur in the northern area next to South Fork Thrust Zone?

____

How does its age compare with all the rocks it is next to? ____________ Is it a sedimentary layer? ______ Explain **p-Ꞓi** and its occurrence in these rocks.

____

____

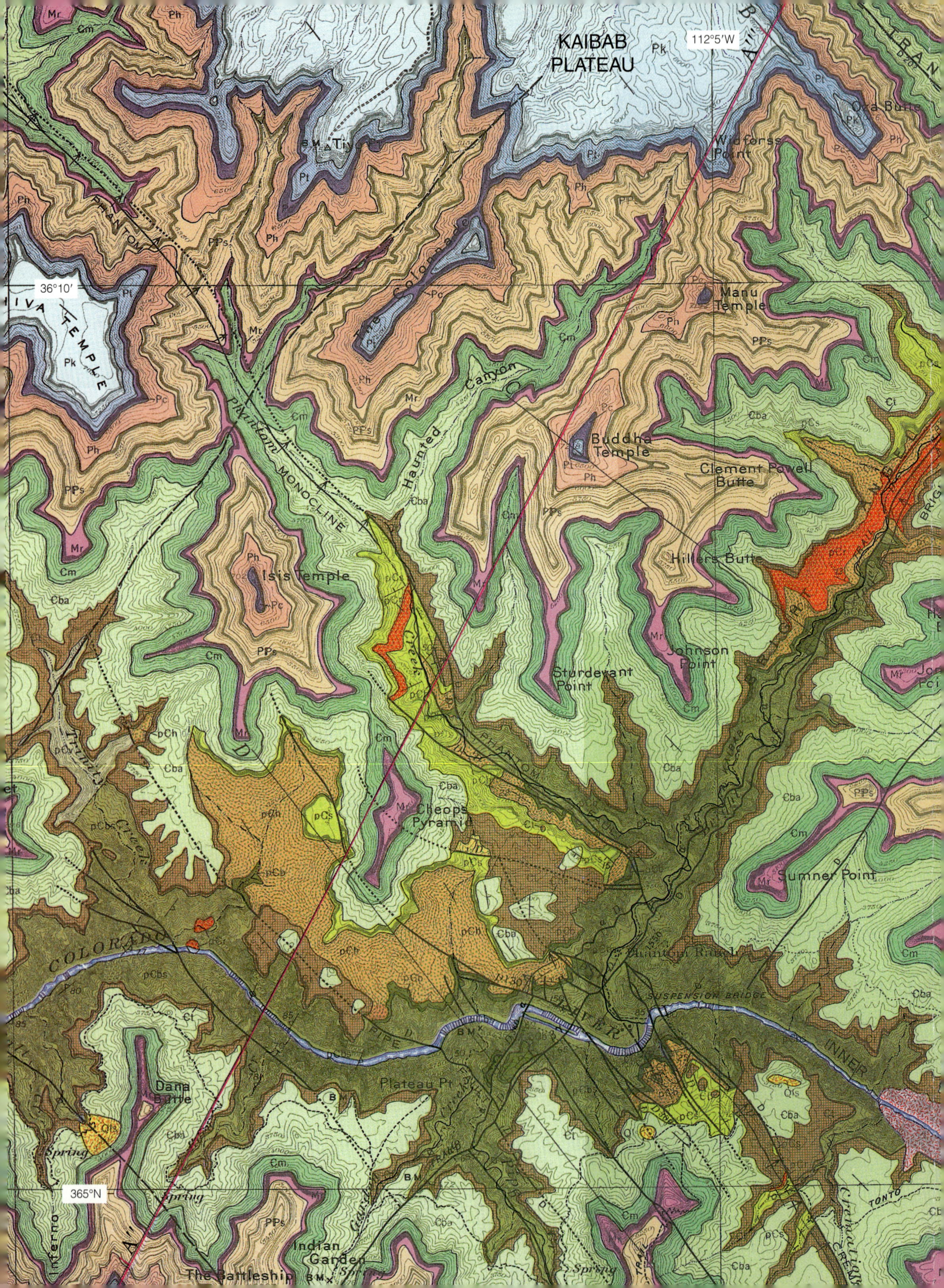

KAIBAB PLATEAU
112°5'W
36°10'
365°N
Manu Temple
Buddha Temple
Clement Powell Butte
Hillers Butte
Isis Temple
Johnson Point
Sturdevant Point
Cheops Pyramid
Sumner Point
Dana Butte
Plateau Pt
Indian Garden
The Battleship
Haunted Canyon
Phantom MONOCLINE
SUSPENSION BRIDGE
COLORADO
INNER
TONTO
Spring

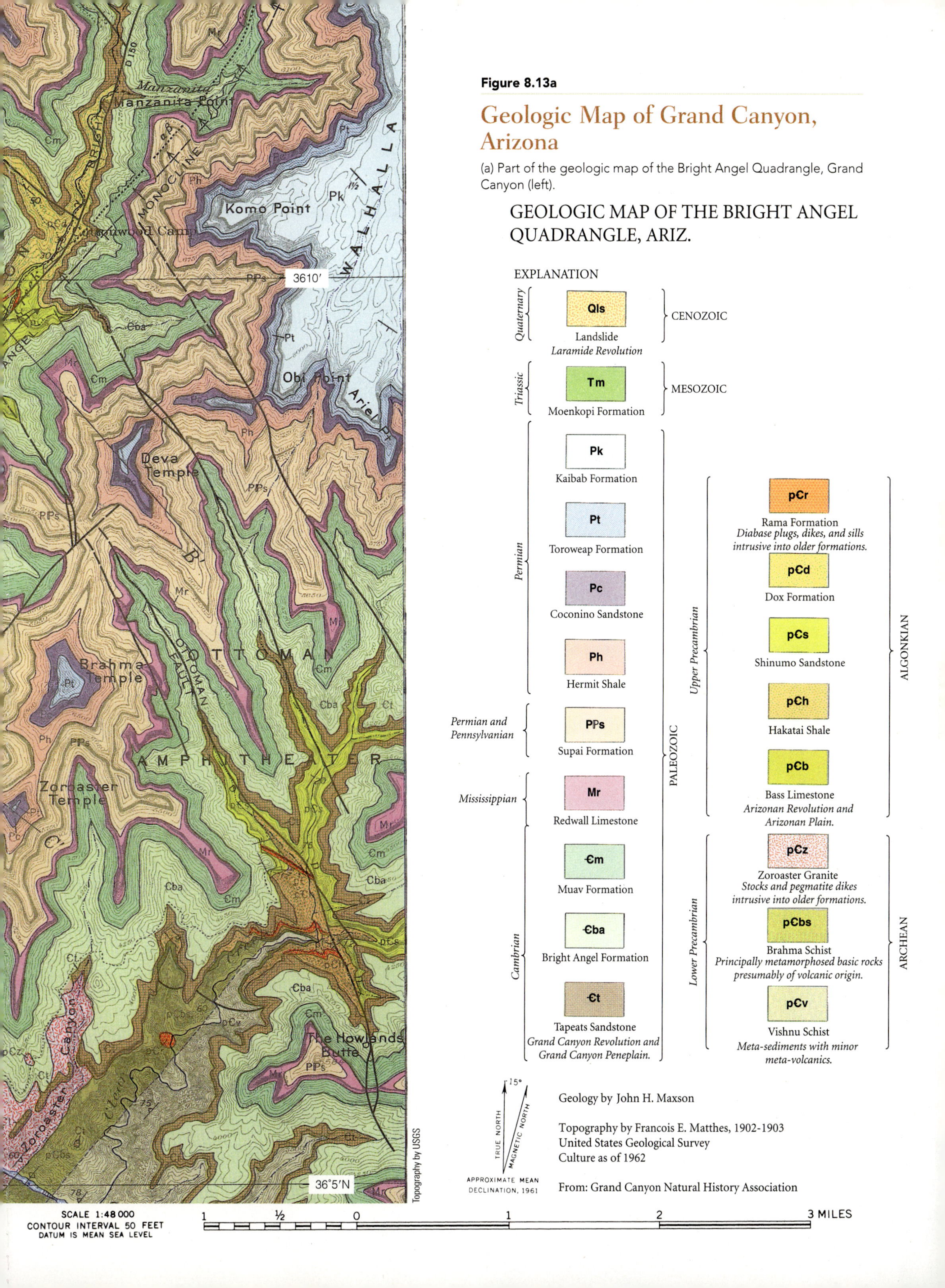

Figure 8.13a

Geologic Map of Grand Canyon, Arizona

(a) Part of the geologic map of the Bright Angel Quadrangle, Grand Canyon (left).

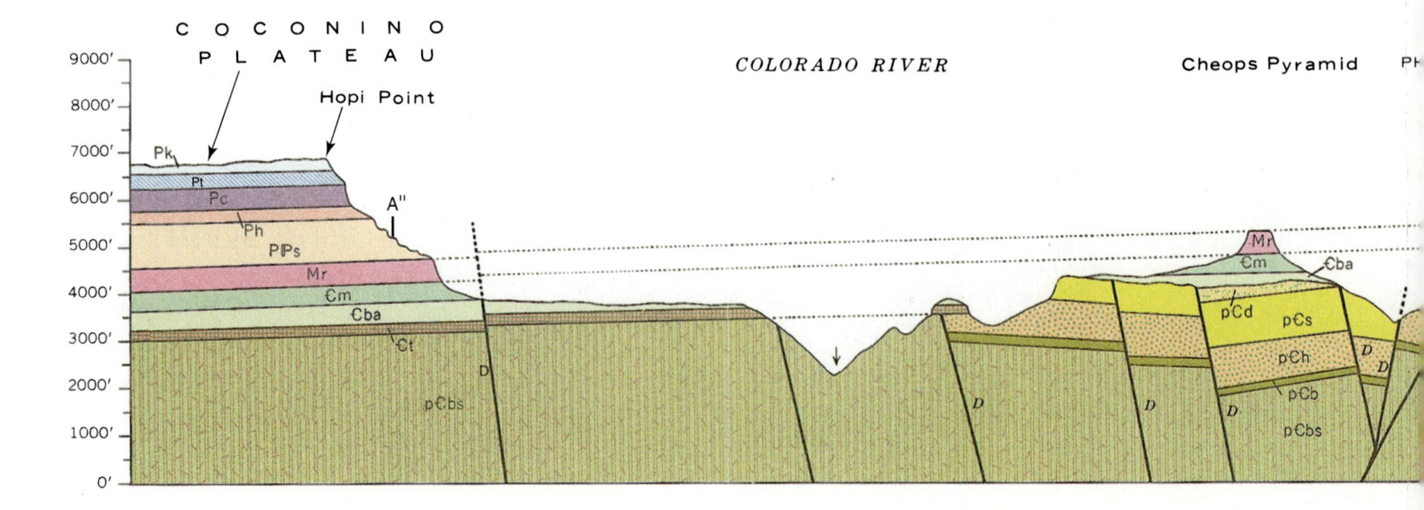

Figure 8.13b

Cross Section Across the Grand Canyon

(b) Cross section for the line A-A''-A'''-A' on the map of part of the Grand Canyon.

17. Compare the northwestern and the central parts of the Geologic Map of the Southern States in ■ Figure 8.16. Keep it in mind that this is a very small-scale map that covers a much larger area than the maps we have been studying so far.

a. In the northwestern part of the map, are the rocks steeply or gently dipping?

____

b. Are the rocks in the central part of the map steeply or gently dipping?

____

What leads you to these conclusions?

____

____

____

c. In the northwest, is the structure centered on pink Middle Ordovician rocks, **Om,** in the vicinity of Nashville and Shelbyville, Tennessee an anticline, a syncline, a dome, a basin, a normal fault, or a thrust fault? ______________

d. Examine the center of the Geologic Map of the Southern States in Figure 8.16, which has SW to NE trending outcrop patterns; this is part of the Valley and Ridge Province. What geologic structures does the overall pattern or arrangement of rocks suggest?

____

e. What is the long, thin geologic structure that occurs at the corner of Alabama and Georgia, at the Tennessee border?

____

Circle all of the following words that apply to it: anticline, syncline, dome,

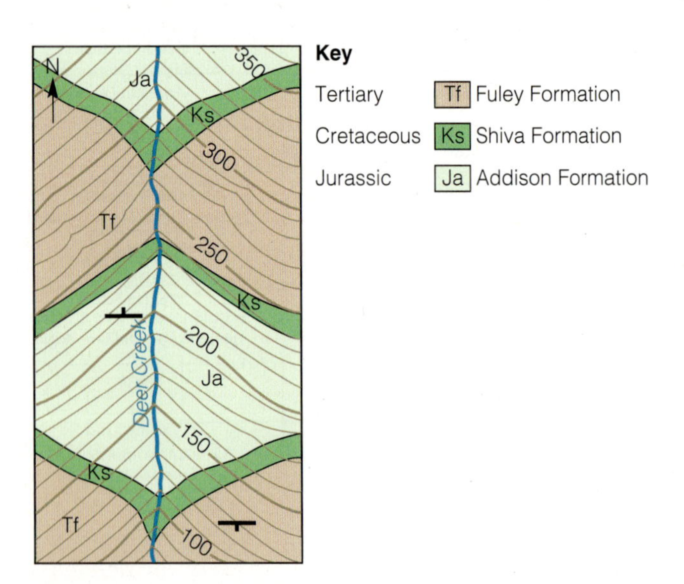

Figure 8.14

Geologic Map of Deer Creek

This region of folds is for Exercise 15.

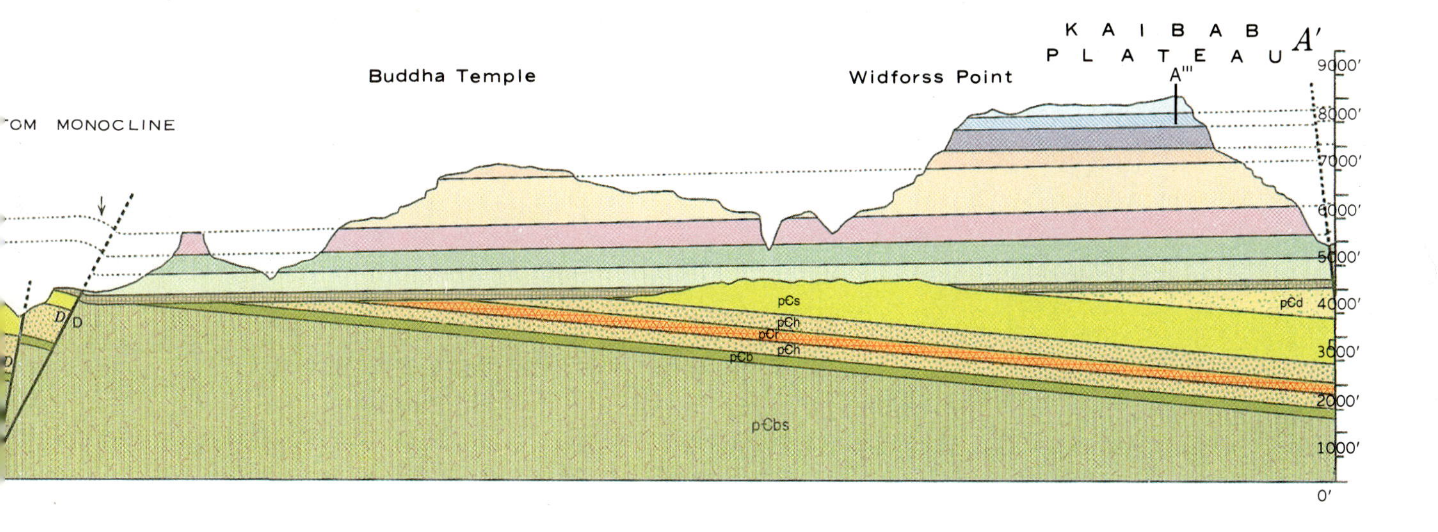

basin, plunging, non-plunging, normal fault, thrust fault, reverse fault, strike-slip fault, compressed, extended, sheared. Give at least two examples of evidence for this structure.

____

____

f. Locate some metamorphic rocks on the map. Name two types: ____________

____________ Are the metamorphic rocks interspersed with the sedimentary rocks? ______ Why or why not?

____

____

The plutonic and metamorphic rocks here are part of the Blue Ridge and Piedmont Provinces.

g. What age are the different granite bodies in the area?

____________________ Are they associated with the sedimentary or the metamorphic rocks? ____________

18. Let's look at unconformities on the Geologic Map of the Southern States in Figure 8.16.

a. Name the formation that is most clearly lying unconformably on other rocks with a very different trend.

____

b. What is the direction of dip of the beds above the unconformity (use the rule of V's)? ____________

c. Compare the strikes and/or dips of the rocks below the unconformity with those of the rocks above the unconformity.

____

____

This unconformity changes type. What two types of unconformity are represented? ______________

d. Name the group that occurs just above a less obvious unconformity.

____

e. What are two pieces of evidence for this second unconformity?

____

____

____

____

Folds on Aerial Photos

Geologists often use aerial photos to assist them in making geologic maps because rock types and structures are often visible due to differential erosion or changes in vegetation. In some places, aerial photos make mapping folds quite easy.

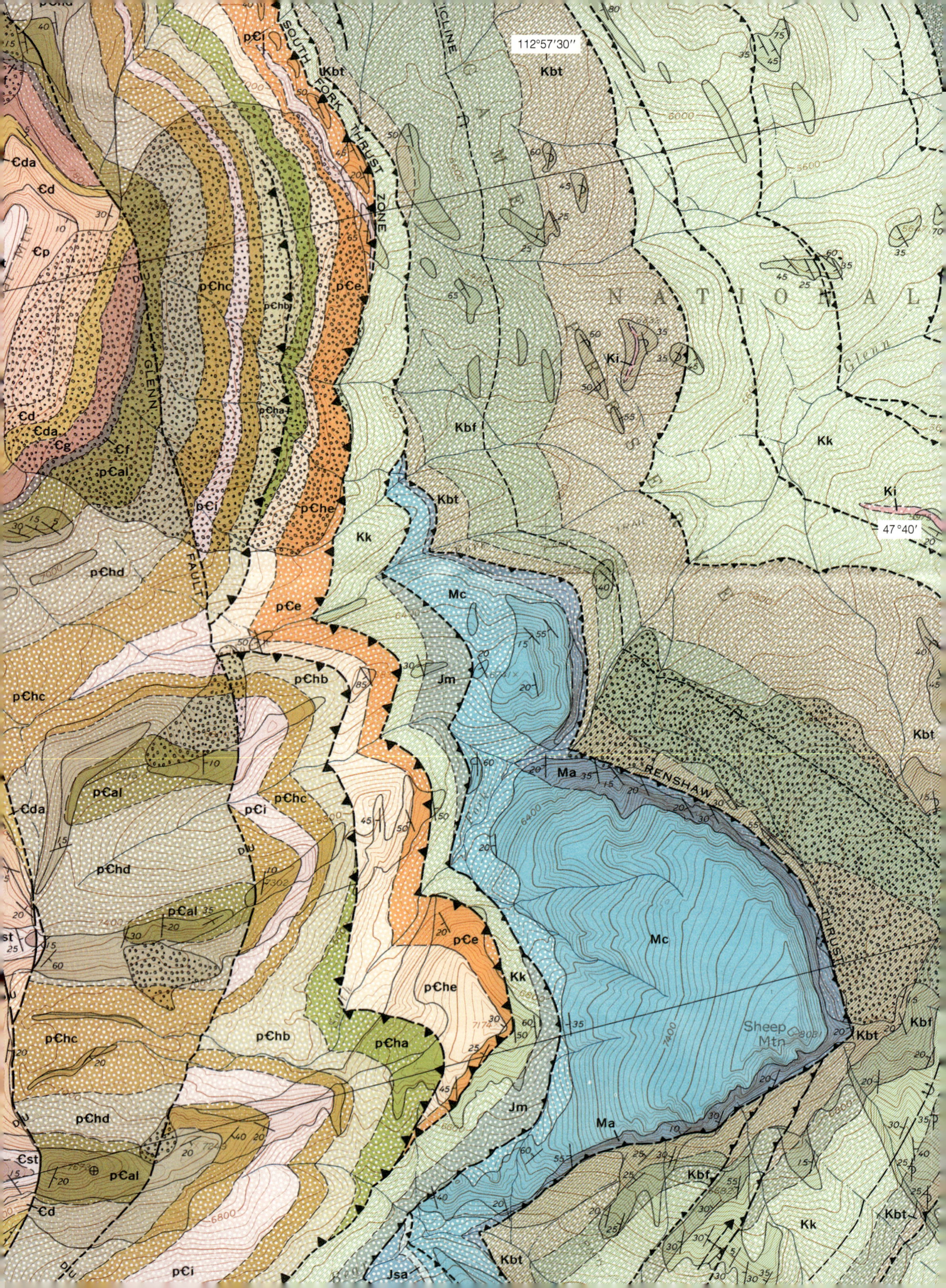

112°57′30″
47°40′
SOUTH FORK THRUST ZONE
GLENN FAULT
RENSHAW THRUST
NATIONAL
Sheep Mtn
Kbt
Kbf
Kk
Ki
Mc
Ma
Jm
Jsa
p€i
p€e
p€he
p€hc
p€hb
p€ha
p€hd
p€al
€da
€d
€p
€g
€f
€st

From: U.S.G.S.–G.Q. 499 Geology by Melville R. Mudge
Base by U.S. Geological Survey

GEOLOGIC MAP OF GLEN CREEK, MONTANA

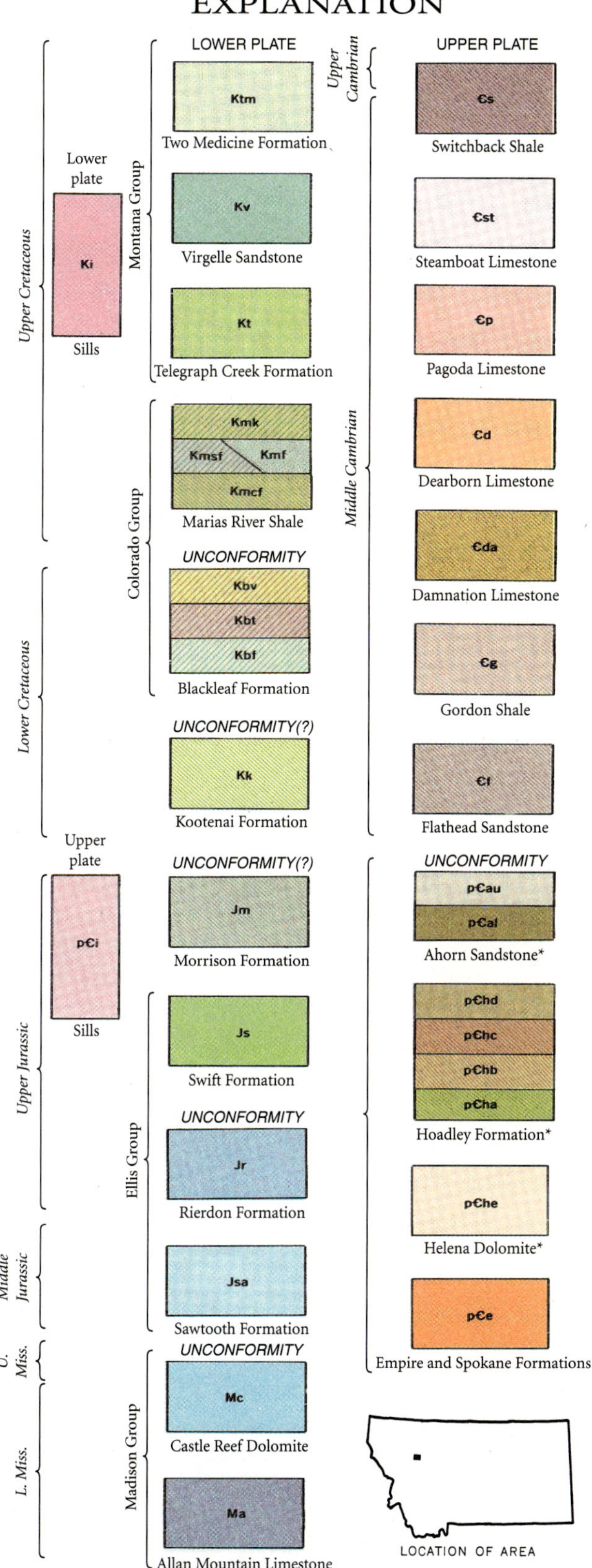

1 ½ 0 1 MILE

SCALE 1:24,000 CONTOUR INTERVAL 40 FEET

Figure 8.15

Geologic Map Near Glenn Creek, Montana

Part of the geologic map of Glenn Creek. Solid colors indicate outcrops. Stippled patterns indicate places where rocks are covered with soil or talus.

19. The Appalachian Mountains, including the part of the Valley and Ridge Province in Pennsylvania shown in ■ Figure 8.17, formed by the collision of Africa into North America.

a. Use a stereoscope to examine the stereo air photos (Figure 8.17a) of this area of Pennsylvania. What makes the pattern of hills and valleys in this area?

____

____

b. All mountains in the air photo pair are formed by erosion-resistant sandstone, and the valleys are of the less resistant shale or limestone.[2] Locate the sandstone in the air photo stereo pair in Figure 8.17a, and then label the areas of sandstone, **ss**, on Figure 8.17b. Shade or color the sandstone layers.

c. Sketch in fold axes on Figure 8.17b using the appropriate symbols.

d. Are the folds plunging or non-plunging?

____

e. Show the direction of plunge, if any, on Figure 8.17b, by adding arrows to the end of your fold axes.

f. Show the direction of shortening (compression) that formed the folds on the figure.

---

[2] Limestone is easily erodable in wet climates but not in dry climates.

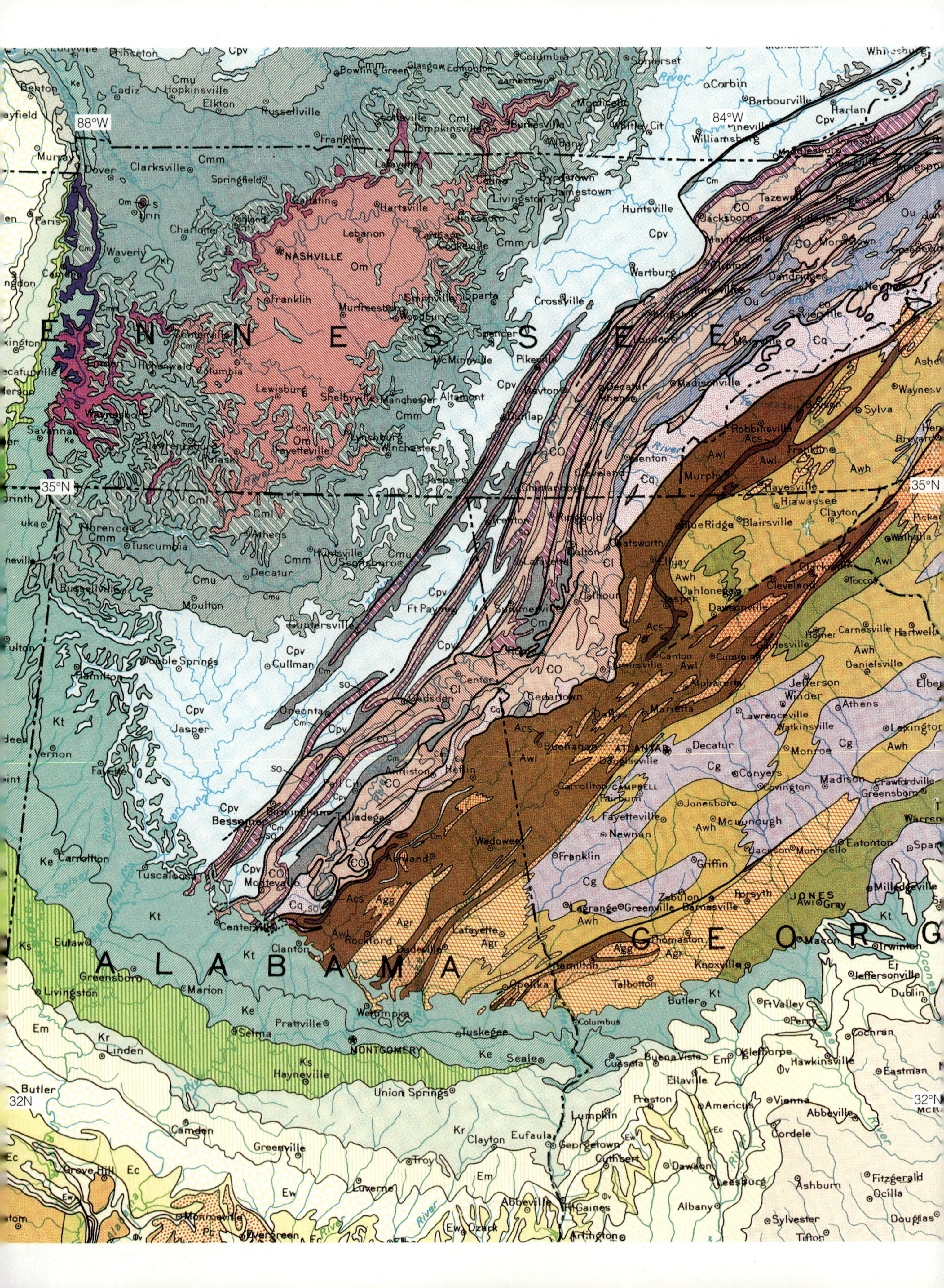

T E N N E S S E E
A L A B A M A
G E O R G
88°W
84°W
35°N
32N
NASHVILLE
ATLANTA
MONTGOMERY
Clarksville
Springfield
Hopkinsville
Russellville
Bowling Green
Glasgow
Columbia
Somerset
Corbin
Barbourville
Harlan
Williamsburg
Lafayette
Livingston
Jamestown
Huntsville
Lebanon
Hartsville
Gallatin
Franklin
Murfreesboro
Smithville
Sparta
Crossville
McMinnville
Shelbyville
Lewisburg
Manchester
Tullahoma
Winchester
Fayetteville
Pulaski
Lynchburg
Waverly
Centerville
Hohenwald
Waynesboro
Savannah
Lawrenceburg
Knoxville
Tazewell
Morristown
Dandridge
Newport
Maryville
Madisonville
Loudon
Kingston
Wartburg
Dayton
Pikeville
Dunlap
Chattanooga
Cleveland
Benton
Athens
Decatur
Robbinsville
Franklin
Murphy
Hayesville
Hiawassee
Clayton
Blairsville
Blue Ridge
Ellijay
Dahlonega
Dawsonville
Chatsworth
Dalton
Ringgold
Lafayette
Summerville
Calhoun
Jasper
Cumming
Canton
Cartersville
Cedartown
Dallas
Marietta
Alpharetta
Decatur
Conyers
Covington
Lawrenceville
Winder
Jefferson
Athens
Lexington
Monroe
Madison
Greensboro
Carrollton
Fairburn
Fayetteville
Newnan
Jonesboro
McDonough
Griffin
Jackson
Monticello
Eatonton
Milledgeville
Franklin
Lagrange
Greenville
Barnesville
Zebulon
Forsyth
Thomaston
Hamilton
Talbotton
Knoxville
Macon
Butler
Columbus
Florence
Tuscumbia
Huntsville
Scottsboro
Decatur
Athens
Moulton
Russellville
Guntersville
Cullman
Double Springs
Hamilton
Jasper
Fayette
Vernon
Oneonta
Gadsden
Centre
Anniston
Heflin
Talladega
Birmingham
Bessemer
Pell City
Ashland
Wedowee
Lafayette
Dadeville
Rockford
Opelika
Montevallo
Columbiana
Clanton
Centerville
Tuscaloosa
Carrollton
Eutaw
Greensboro
Marion
Livingston
Linden
Selma
Prattville
Wetumpka
Tuskegee
Seale
Hayneville
Union Springs
Camden
Greenville
Troy
Luverne
Clayton
Eufaula
Butler
Grove Hill
Monroeville
Evergreen
Ozark
Abbeville
Georgetown
Cuthbert
Dawson
Lumpkin
Cusseta
Buena Vista
Ellaville
Preston
Americus
Oglethorpe
Perry
Fort Valley
Hawkinsville
Cochran
Eastman
Vienna
Cordele
Abbeville
Fitzgerald
Ocilla
Ashburn
Leesburg
Albany
Sylvester
Douglas
Fort Gaines
Arlington
Dublin
Jeffersonville
Twiggs
Gray
Sylva
Waynesville
Hartwell
Carnesville
Danielsville
Homer
Toccoa
Clarkesville
Cleveland
Walhalla
Cpv
Cmm
Cmu
Cml
Om
Ou
Cq
CO
SO
Cm
Awl
Acs
Awh
Agr
Agg
Cg
Kt
Ke
Ks
Kr
Em
Ec
Ew

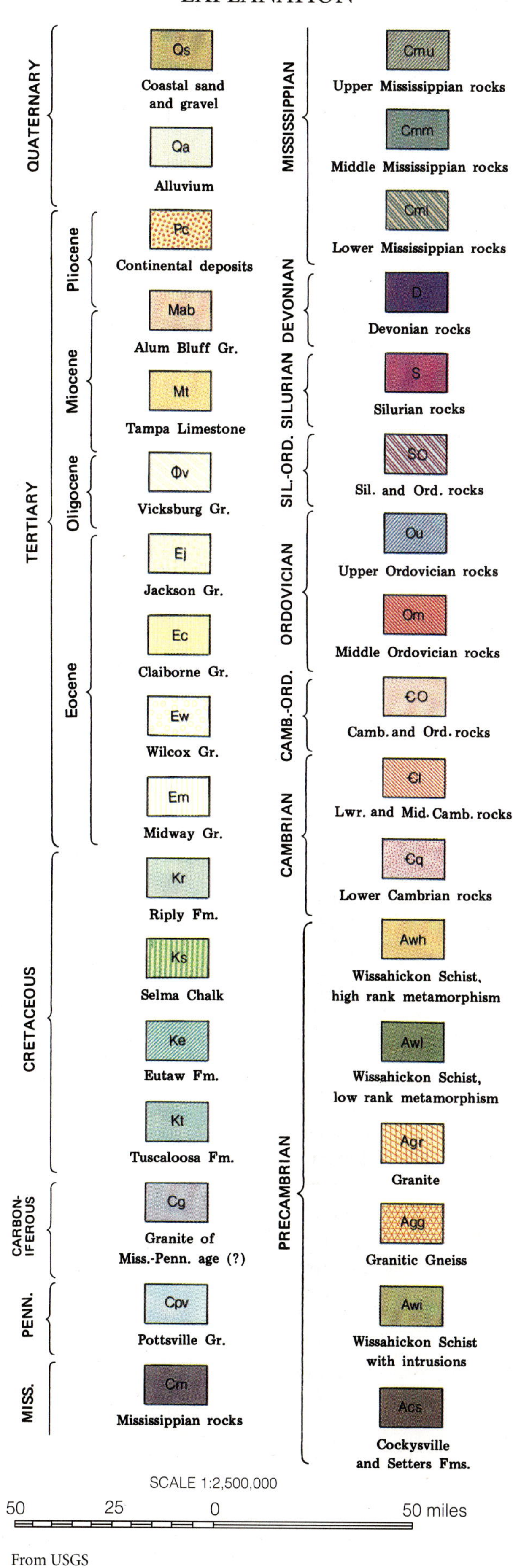

From USGS

Figure 8.16

Part of the Geologic Map of the Southern States

A **group** is a collection of formations that occur together, abbreviated as Gr. here.

INTERPRETING GEOLOGIC HISTORY FROM GEOLOGIC MAPS AND CROSS SECTIONS

Let's put together what you have learned about maps and cross sections in this lab with what you already know about relative geologic time from Lab 7. Now that you have a better understanding of real geologic maps and cross sections, you should be ready to interpret geologic history from them.

20. Examine the Geologic Map of Devils Fence Quadrangle, Montana, in ■ Figure 8.18 and the structure sections B″-B′, C–C′, D-D′. Unless otherwise directed, use all three sections and the map to answer the following questions.

a. Notice that the map shows fold axis symbols in red. However, the major fold in the eastern part of the map has no symbol. Using the age of the rocks, determine whether the large fold in the eastern half of the map is a syncline or anticline.

anticline

b. What two other types of evidence support your answer?

Older rock in the center

Dip point away

c. Draw its fold axis on the map using the correct symbol. What is the direction of plunge, if any?

____

d. What type of forces caused this and the other folds in the area?

________________ What were the compass directions of these forces? __________

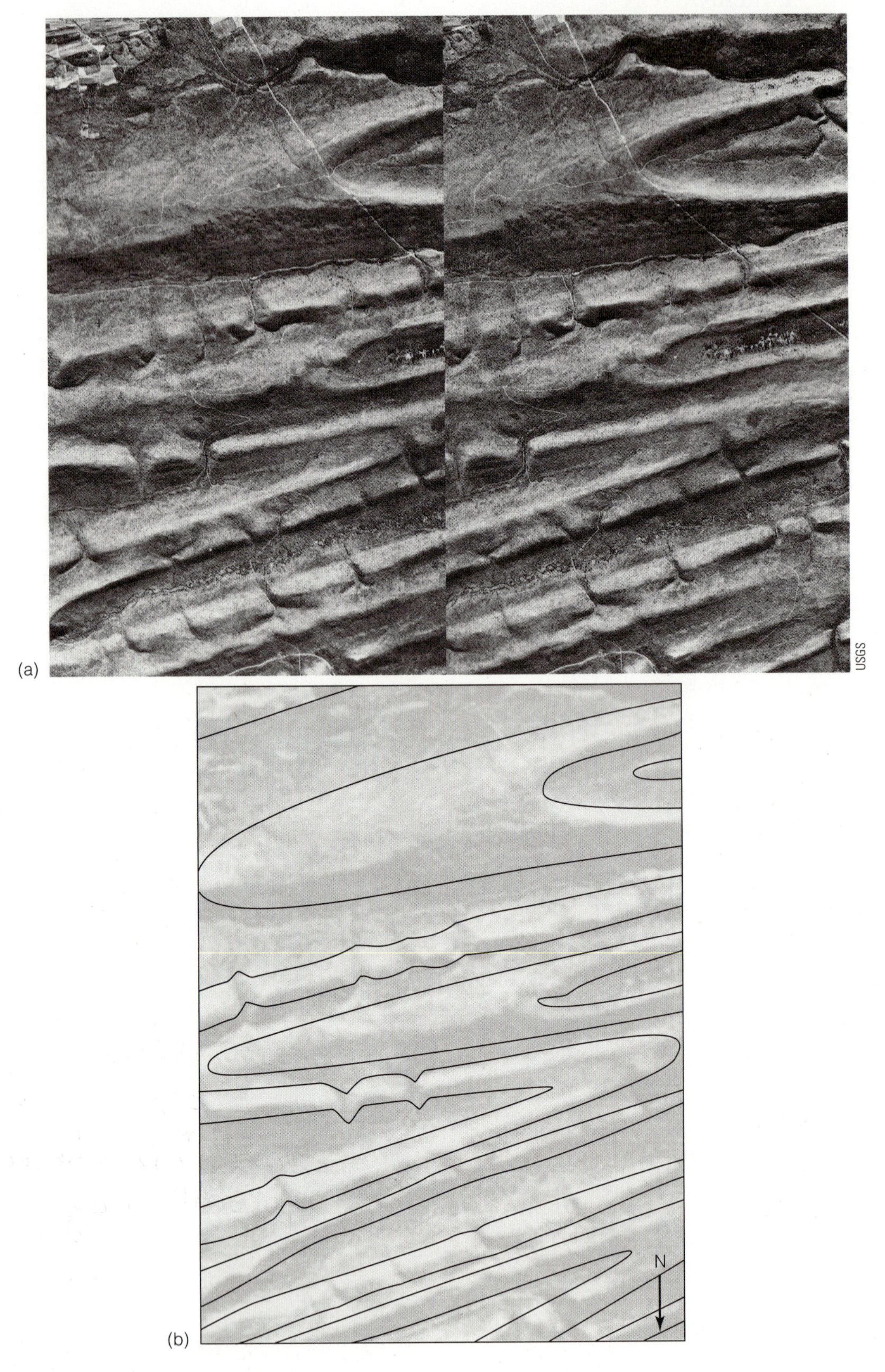

Figure 8.17

Aerial Photos and Geologic Map of Part of the Valley and Ridge Province, Pennsylvania

(a) Air photo stereo pair for Exercise 19. North is down on this pair. (b) Preliminary geologic map of the area made by tracing approximate contact locations on one of the air photos in (a). These photos and map are shown with north pointing down so that features in the photos are lit from above.

e. Find **ad** (orange color) in the southeastern part of the map and use the rule of V's to determine its dip direction in

Section 23. ____________ Look up **ad** in the explanation on the map. What is its rock name and class?

____

Judging from its shape on the map, its dip compared with surrounding rocks, and other factors, which of the following best describes **ad**? *Circle one:*

i. Sediments unconformably overlying other rocks

ii. Discordant intrusive

iii. Concordant intrusive

f. Did the folding you observed in **a, b** and **c** occur before or after the intrusion of **gd**?

______________ Did the folding occur before or after the process forming

ad? ______________

g. Where does **gd** fit in the sequence

of events? ______________

____

Where does **ad** fit in the sequence of events?

____

____

h. What kind of fault is the fault near **Ꞓw** on structure section B″-B′?

____

In the space below, sketch and complete the fault symbol from the map by adding a tic mark or arrows as appropriate to match the type of fault it is.

What rocks are cut by the fault near **Ꞓw** on section B-B″-B′?

______________ Is faulting more likely to have occurred during deposition or after deposition stopped?

____

i. Locate the Horse Gulch Fault on the western edge of the map. What type of movement occurred on this fault?

____

What is your evidence?

____

____

____

j. Draw correct map symbols on the Horse Gulch Fault. Refer to Table 8.1 if needed.

k. Which rocks appear to be more resistant to weathering?

____

____

____

Is this also reflected in topographic

contour patterns? ________ Explain what evidence supports this.

____

l. Examine the areas of **Ttg, Tfg,** and **Qal** on the map. Are these materials intrusive or are they sediments lying unconformably on top of other rocks?

____

m. Use the geologic time scale in Figure 7.24 on page 158 to estimate how much time past between the formation of **Ttg** and the older rocks in the area.

____

n. In Table 8.4, list the sequence of events that occurred in the Devils Fence Quadrangle from oldest at the bottom to youngest at the top. Use your answers to questions **a–m** to help determine the sequence and the details. You need not list each separate formation as a separate event. Instead, list formations that were deposited in a sequence as one event and list the letter symbols for the formation in sequence in the last column of the table as part of the description of the event. Next to each event, explain the event in more detail.

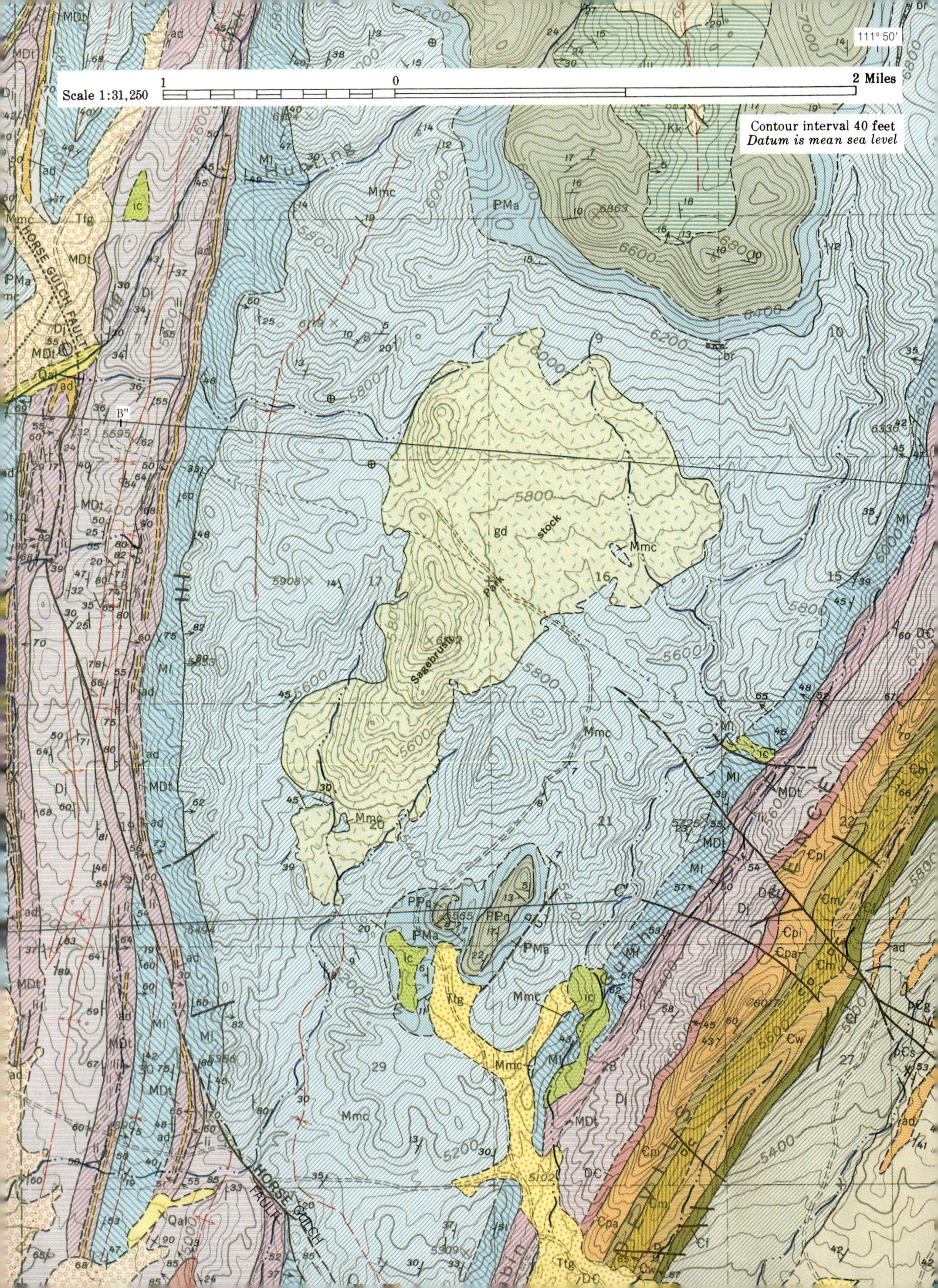

Scale 1:31,250
1
0
2 Miles
111° 50'
Contour interval 40 feet
Datum is mean sea level
HORSE GULCH FAULT
Park
stock
Sagebrush
gd
Mmc
PMa
Ml
MDt
Dj
Tfg
Qal
ic
Kk

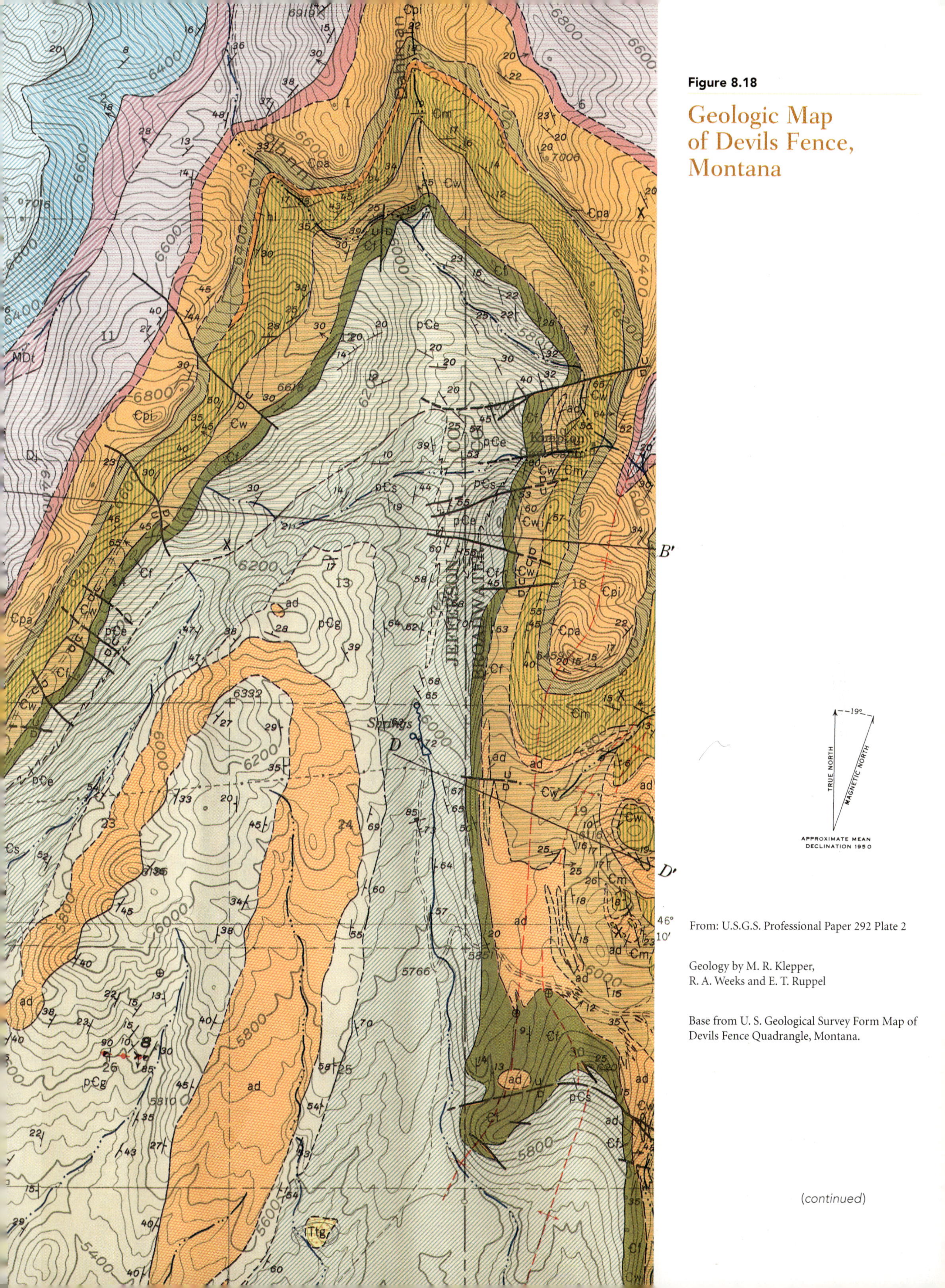

Figure 8.18

Geologic Map of Devils Fence, Montana

From: U.S.G.S. Professional Paper 292 Plate 2

Geology by M. R. Klepper, R. A. Weeks and E. T. Ruppel

Base from U. S. Geological Survey Form Map of Devils Fence Quadrangle, Montana.

(continued)

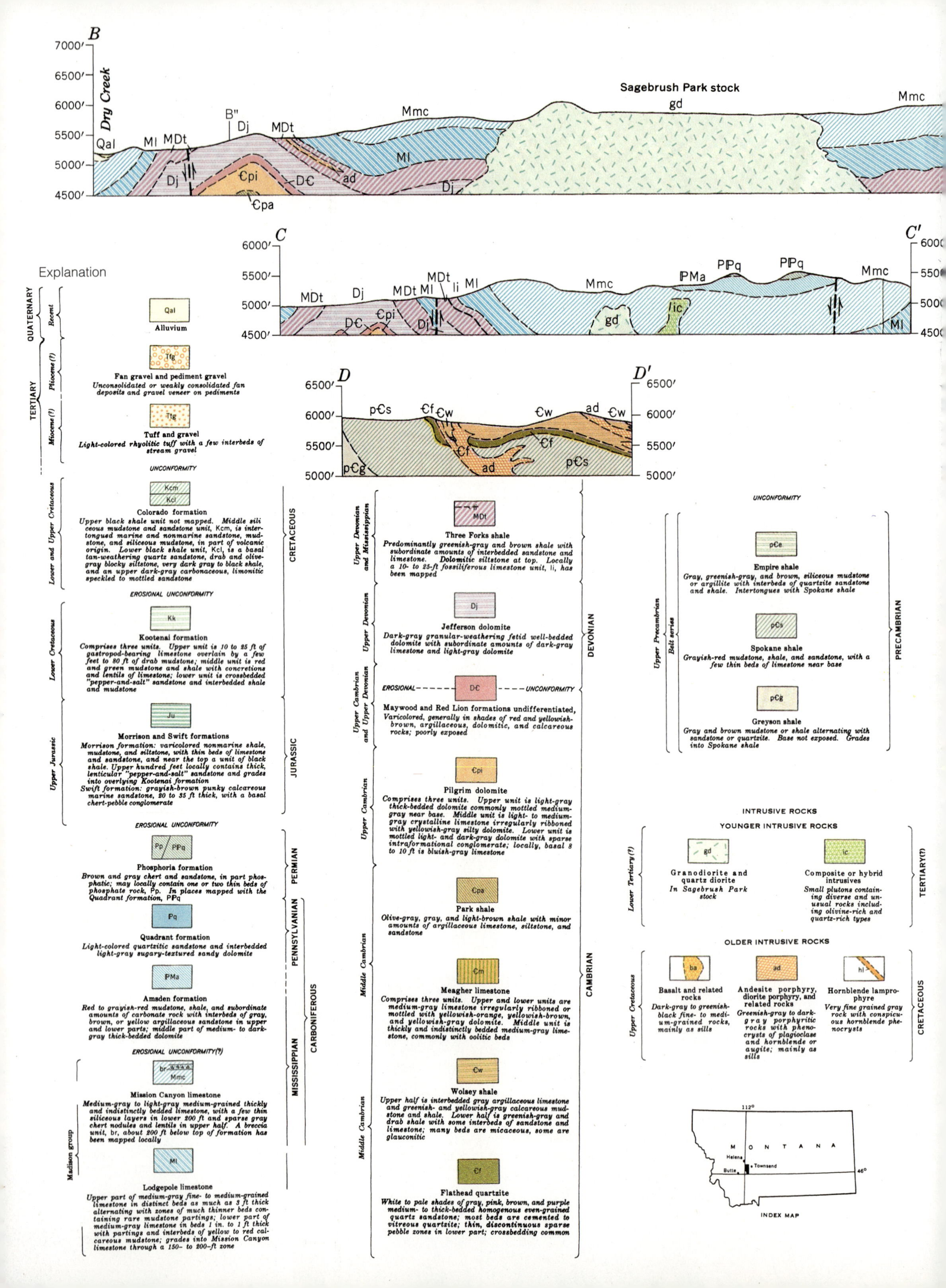

B
Dry Creek
7000′
6500′
6000′
5500′
5000′
4500′
Sagebrush Park stock
gd
Mmc
Qal
Ml
MDt
B″
Dj
MDt
Mmc
Ml
Dj
Єpi
DЄ
ad
Dj
Єpa
C
C′
6000′
5500′
5000′
4500′
MDt
Dj
MDt
Ml
MDt
li
Ml
Mmc
ℙMa
PℙPq
PℙPq
Mmc
DЄ
Єpi
Dj
gd
ic
Ml
D
D′
6500′
6000′
5500′
5000′
pЄs
Єf
Єw
Єw
ad
Єw
Єf
Єf
ad
pЄs
pЄg
Explanation
QUATERNARY
Recent
Qal
Alluvium
TERTIARY
Pliocene (?)
Tfg
Fan gravel and pediment gravel
Unconsolidated or weakly consolidated fan deposits and gravel veneer on pediments
Miocene (?)
Ttg
Tuff and gravel
Light-colored rhyolitic tuff with a few interbeds of stream gravel
UNCONFORMITY
Lower and Upper Cretaceous
Kcm
Kcl
Colorado formation
Upper black shale unit not mapped. Middle siliceous mudstone and sandstone unit, Kcm, is intertongued marine and nonmarine sandstone, mudstone, and siliceous mudstone, in part of volcanic origin. Lower black shale unit, Kcl, is a basal tan-weathering quartz sandstone, drab and olive-gray blocky siltstone, very dark gray to black shale, and an upper dark-gray carbonaceous, limonitic speckled to mottled sandstone
CRETACEOUS
EROSIONAL UNCONFORMITY
Lower Cretaceous
Kk
Kootenai formation
Comprises three units. Upper unit is 10 to 25 ft of gastropod-bearing limestone overlain by a few feet to 80 ft of drab mudstone; middle unit is red and green mudstone and shale with concretions and lentils of limestone; lower unit is crossbedded "pepper-and-salt" sandstone and interbedded shale and mudstone
Upper Jurassic
Ju
Morrison and Swift formations
Morrison formation: varicolored nonmarine shale, mudstone, and siltstone, with thin beds of limestone and sandstone, and near the top a unit of black shale. Upper hundred feet locally contains thick, lenticular "pepper-and-salt" sandstone and grades into overlying Kootenai formation
Swift formation: grayish-brown punky calcareous marine sandstone, 20 to 35 ft thick, with a basal chert-pebble conglomerate
JURASSIC
EROSIONAL UNCONFORMITY
Pp
PℙPq
Phosphoria formation
Brown and gray chert and sandstone, in part phosphatic; may locally contain one or two thin beds of phosphate rock, Pp. In places mapped with the Quadrant formation, PℙPq
PERMIAN
ℙq
Quadrant formation
Light-colored quartzitic sandstone and interbedded light-gray sugary-textured sandy dolomite
PENNSYLVANIAN
ℙMa
Amsden formation
Red to grayish-red mudstone, shale, and subordinate amounts of carbonate rock with interbeds of gray, brown, or yellow argillaceous sandstone in upper and lower parts; middle part of medium- to dark-gray thick-bedded dolomite
CARBONIFEROUS
EROSIONAL UNCONFORMITY(?)
br
Mmc
Mission Canyon limestone
Medium-gray to light-gray medium-grained thickly and indistinctly bedded limestone, with a few thin siliceous layers in lower 200 ft and sparse gray chert nodules and lentils in upper half. A breccia unit, br, about 200 ft below top of formation has been mapped locally
MISSISSIPPIAN
Madison group
Ml
Lodgepole limestone
Upper part of medium-gray fine- to medium-grained limestone in distinct beds as much as 3 ft thick alternating with zones of much thinner beds containing rare mudstone partings; lower part of medium-gray limestone in beds 1 in. to 1 ft thick with partings and interbeds of yellow to red calcareous mudstone; grades into Mission Canyon limestone through a 150- to 200-ft zone
Upper Devonian and Mississippian
MDt
Three Forks shale
Predominantly greenish-gray and brown shale with subordinate amounts of interbedded sandstone and limestone. Dolomitic siltstone at top. Locally a 10- to 25-ft fossiliferous limestone unit, li, has been mapped
Upper Devonian
Dj
Jefferson dolomite
Dark-gray granular-weathering fetid well-bedded dolomite with subordinate amounts of dark-gray limestone and light-gray dolomite
DEVONIAN
Upper Cambrian and Upper Devonian
EROSIONAL
DЄ
UNCONFORMITY
Maywood and Red Lion formations undifferentiated,
Varicolored, generally in shades of red and yellowish-brown, argillaceous, dolomitic, and calcareous rocks; poorly exposed
Upper Cambrian
Єpi
Pilgrim dolomite
Comprises three units. Upper unit is light-gray thick-bedded dolomite commonly mottled medium-gray near base. Middle unit is light- to medium-gray crystalline limestone irregularly ribboned with yellowish-gray silty dolomite. Lower unit is mottled light- and dark-gray dolomite with sparse intraformational conglomerate; locally, basal 8 to 10 ft is bluish-gray limestone
Middle Cambrian
Єpa
Park shale
Olive-gray, gray, and light-brown shale with minor amounts of argillaceous limestone, siltstone, and sandstone
Єm
Meagher limestone
Comprises three units. Upper and lower units are medium-gray limestone irregularly ribboned or mottled with yellowish-orange, yellowish-brown, and yellowish-gray dolomite. Middle unit is thickly and indistinctly bedded medium-gray limestone, commonly with oolitic beds
CAMBRIAN
Middle Cambrian
Єw
Wolsey shale
Upper half is interbedded gray argillaceous limestone and greenish- and yellowish-gray calcareous mudstone and shale. Lower half is greenish-gray and drab shale with some interbeds of sandstone and limestone; many beds are micaceous, some are glauconitic
Єf
Flathead quartzite
White to pale shades of gray, pink, brown, and purple medium- to thick-bedded homogenous even-grained quartz sandstone; most beds are cemented to vitreous quartzite; thin, discontinuous sparse pebble zones in lower part; crossbedding common
UNCONFORMITY
Upper Precambrian
Belt series
pЄe
Empire shale
Gray, greenish-gray, and brown, siliceous mudstone or argillite with interbeds of quartzite sandstone and shale. Intertongues with Spokane shale
pЄs
Spokane shale
Grayish-red mudstone, shale, and sandstone, with a few thin beds of limestone near base
pЄg
Greyson shale
Gray and brown mudstone or shale alternating with sandstone or quartzite. Base not exposed. Grades into Spokane shale
PRECAMBRIAN
INTRUSIVE ROCKS
YOUNGER INTRUSIVE ROCKS
Lower Tertiary (?)
gd
Granodiorite and quartz diorite
In Sagebrush Park stock
ic
Composite or hybrid intrusives
Small plutons containing diverse and unusual rocks including olivine-rich and quartz-rich types
TERTIARY(?)
OLDER INTRUSIVE ROCKS
Upper Cretaceous
ba
Basalt and related rocks
Dark-gray to greenish-black fine- to medium-grained rocks, mainly as sills
ad
Andesite porphyry, diorite porphyry, and related rocks
Greenish-gray to dark-gray porphyritic rocks with phenocrysts of plagioclase and hornblende or augite; mainly as sills
hl
Hornblende lamprophyre
Very fine grained gray rock with conspicuous hornblende phenocrysts
CRETACEOUS
112°
M O N T A N A
Helena
Townsend
Butte
46°
INDEX MAP

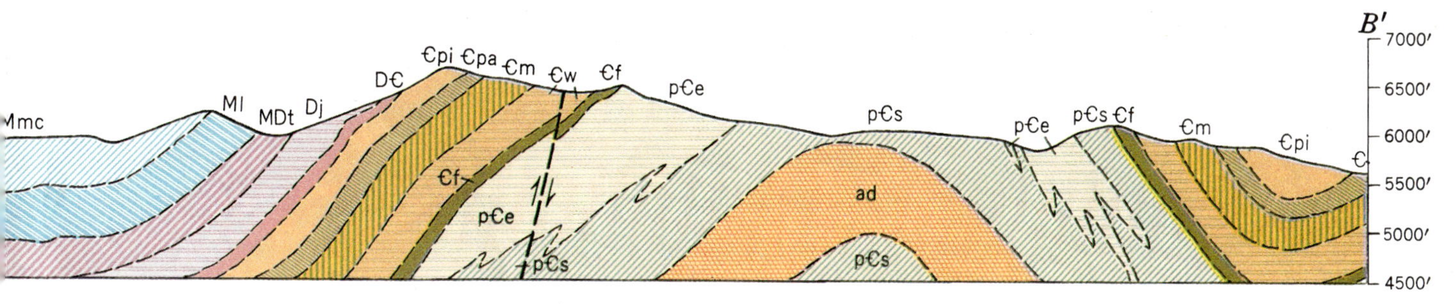

Figure 8.18

Geologic Map of Devils Fence, Montana—(*continued*)

Geologic map and structure sections of part of Devils Fence Quadrangle, Montana. Map scale: 1:31,2500 (1″ is about 1/2 mile). Cross section scale: 1:24,000.

Table 8.4

Sequence of Events for Devils Fence Quadrangle See Exercise 20.

Event Sequence	Geologic Events	Description
Last (youngest)		
7th		
6th		
5th		
4th		
3rd		
2nd		
1st (oldest)		

DRAWING A CROSS SECTION

By now, you know what a cross section is, and you have worked with a number of already constructed ones. Next you can explore how to construct a cross section and learn to draw a simple one. ■ Figure 8.19 shows and describes the method, which you should read through completely before starting. Ask your instructor to demonstrate the technique if necessary. You may want to try drawing a small part of the cross section in Figure 8.19 as practice to see whether you produce the same results.

21. Use the topographic profile in ■ Figure 8.20b to draw a cross section along the line from A to A′ in the map in Figure 8.20a. What structures do you see in your cross section when you are done?

____

____

____

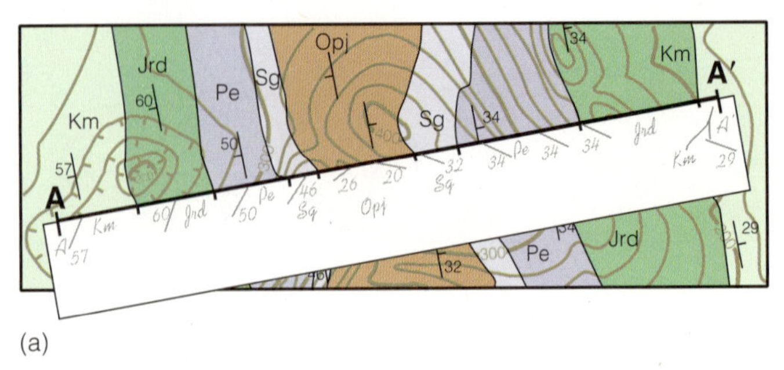

(a)

Line up a thin, straight strip of paper along the cross section line, and mark the contacts and dip angles.

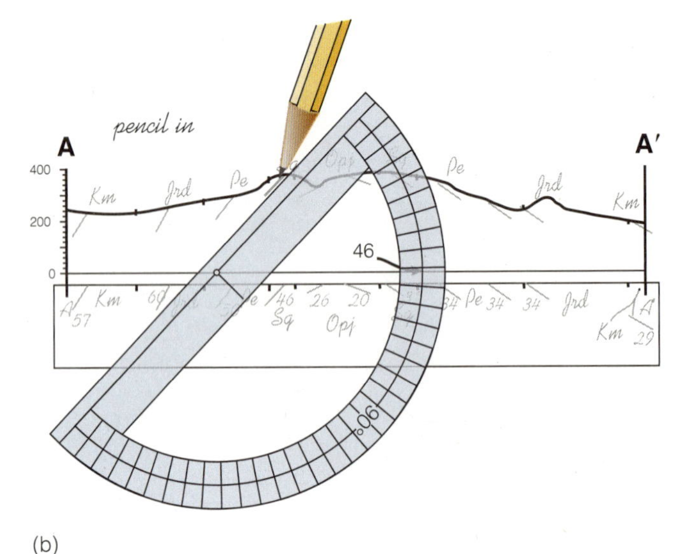

(b)

Use a pencil to transfer the information from the strip of paper to a topographic profile for the same line. Place the information at the land surface. Use a protractor to draw the dip angles accurately.

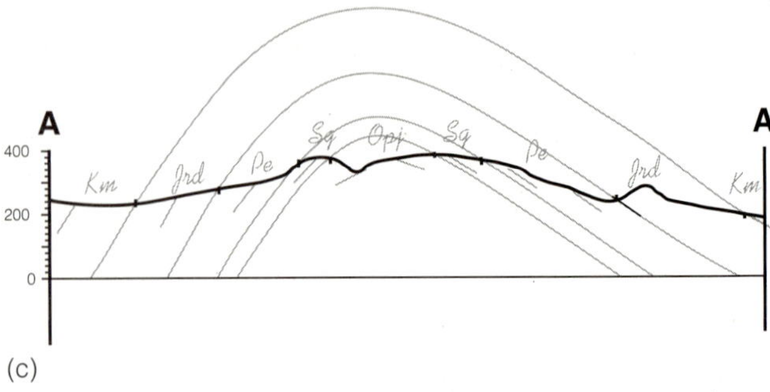

(c)

Continue using the pencil while you sketch in the inferred location of the contacts in the cross section. Maintain constant bed thickness if possible, except where you know the beds change thickness.

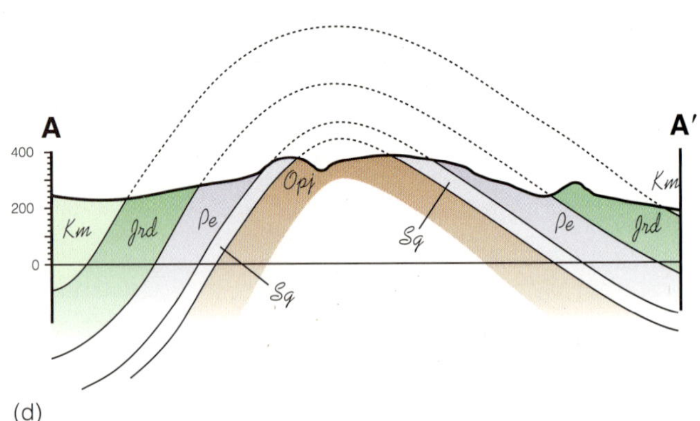

(d)

When you are satisfied with the results, ink in the correct lines and formation names, and erase extraneous marks. Dash in contact lines inferred to have been present above the current land surface before erosion. Color the units below ground level, leaving areas uncolored where information is unavailable or sketchy.

Figure 8.19

Method of Drawing a Cross Section

Method of drawing a cross section on a previously constructed topographic profile. CI = 20 ft. Topographic profile construction is covered in Lab 6.

MAKING GEOLOGIC MAPS FROM FIELD INVESTIGATIONS

Geologists construct geologic maps by making observations at natural rock outcrops and by plotting this information on topographic maps. Field observations at an outcrop may include noting rock type, identifying the formation it belongs to, measuring the dip and strike of beds or foliation, and noting the location of samples collected. If deformational features or structures are present, the mapmaker also measures the strike and dip. For example, if a fault is sufficiently well exposed, the geologist can directly measure the strike and dip of a fault zone in the field, and determine the type of fault. Outcrops of a fold, too, may be sufficient to allow the geologist to determine whether the fold is a syncline or an anticline, non-plunging or plunging. Sometimes, however, it is not until after the data are compiled that a pattern emerges.

The geologist begins to compile all of this information on a topographic map while in the field using the standard symbols in Table 8.1. Aerial photographs of the area can help with interpreting observations and suggesting where the contacts of all the rock bodies occur.

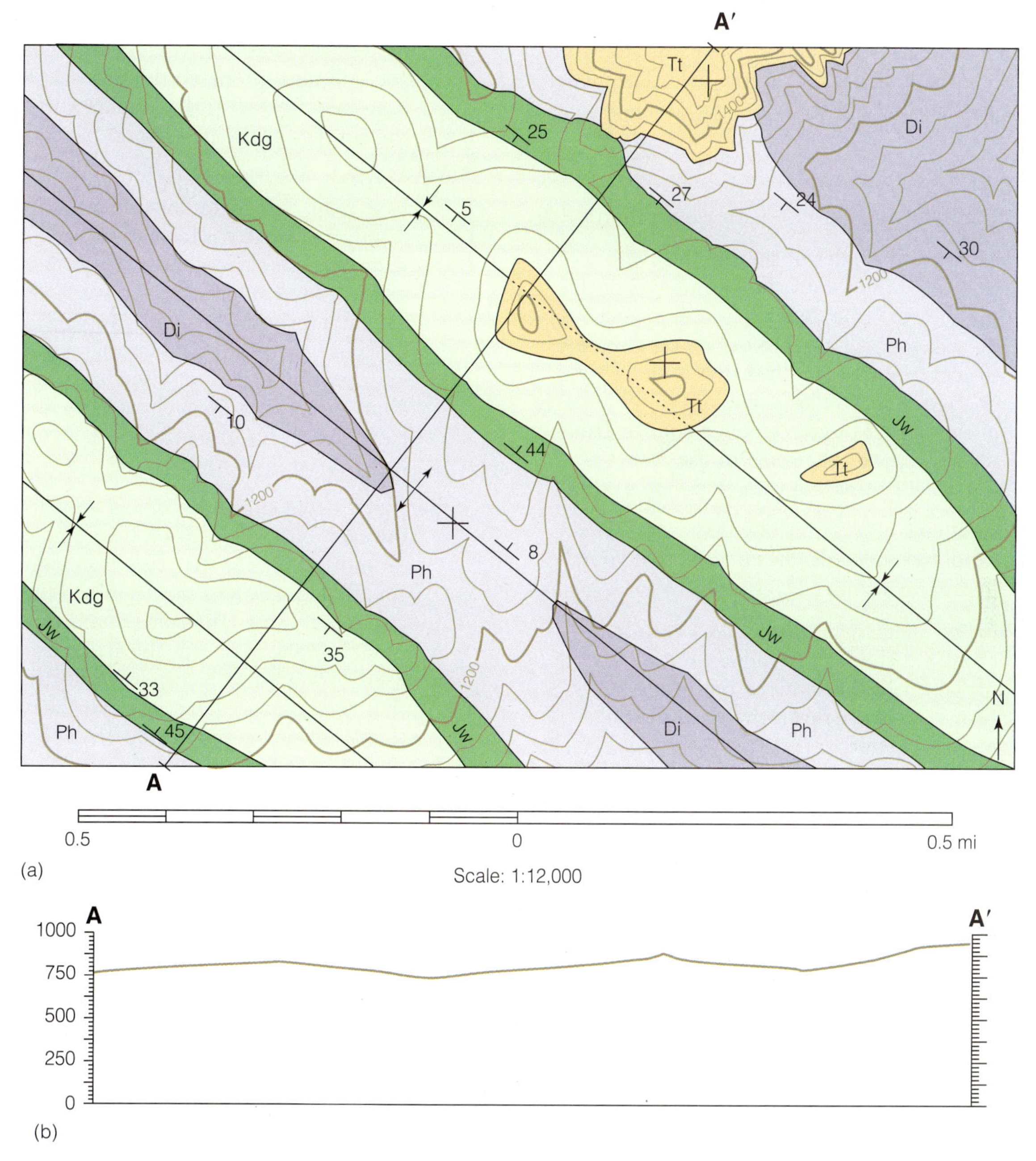

Figure 8.20

A Geologic Map and Topographic Profile for Constructing a Cross Section

(a) A geologic map of a hypothetical area. CI = 40 ft. (b) Topographic profile base for drawing a geologic cross section along the line from A to A′ for the map above.

Often in desert areas, a geologist can easily follow a contact across the landscape, but where soil or vegetation covers the contact, the mapper has to suggest where the contact might be.

MAPMAKING SIMULATION

Your instructor has prepared a simulation of the process of geologic mapping in a sandbox model.

Materials needed:

- Sandbox model with cards, flags, and rocks, as described below
- Stiff acrylic sheet with an optional ruled grid
- Washable overhead marking pens in various colors
- Graph paper with a suitably sized grid

Your instructor has sculpted the sand to form a landscape and inserted small cards, flags, and possibly some actual rocks into the sand. These simulate outcrops, and each one has a number to correspond with observations written in a geologist's notebook.[3] The cards show the orientation of the layers of rock in the field. The model also has a north arrow and scale.

22. Do the following exercise with lab partner(s).

a. Measure the sides of the sandbox with a ruler, and then draw the boundaries of the sandbox to scale onto the graph paper. To do this you will need to decide on a scale to represent the sandbox that is small enough to fit and large enough to fill most of the page. For convenience your instructor may specify the scale to use. What is the scale comparing your grid map to the sandbox? ______________________ The sandbox model also has a scale that your instructor will need to supply: What is it? ______________________ Your map represents both the sandbox and the part of the Earth the sandbox represents. On your map, write down the scale of your map compared to the Earth's surface. Show the calculations here.

b. If the "terrain" in the sandbox is flat, skip this step. Otherwise, place the acrylic sheet over the sandbox. Decide on a reasonable contour interval. Then, with a washable overhead-marking pen, carefully draw topographic contour lines on the acrylic. Take turns doing this so each person has an opportunity to try it. Transfer these contour lines to scale on your grid map of the sandbox. When completed, you will have prepared the topographic base map for the geologic map. Clean off the acrylic for the next group of students.

c. Mark the position of each outcrop on your base map, and write down its identifying number. Transfer information such as rock types, measurements, and other field observation from the geologist's notebook for each location. Also, add to your map any observations regarding rock sample 'outcrops' in the sand. Use colors or appropriate rock symbol patterns for each rock type (refer to the inside back cover for rock symbols needed).

d. Place strike-and-dip symbols on your map for each outcrop location where bedding or another planar structure is present. Small colored cards may indicate the strike and dip by their orientation in the model. Make sure that the strike is correctly oriented with respect to north, and that the tick on each strike symbol is on the downward-dipping side of the bed.

When these steps are completed, you have an outcrop map. Next, you will make an interpretation of the geologic data.

e. Try to evaluate where you should draw the contact traces and decide what kind of contacts they are. Use the rule of V's to help you determine where uncertain contacts should cross valleys. Color the areas on the map between the contacts and/or use appropriate rock symbols from the inside back cover.

f. On the same sheet with your key, interpret the geologic history of the area giving a list of events and a description of each event in the order they occurred.

[3] The "geologist's notebook" here may be in the form of a handout, notes written on the board or an overhead, or an actual notebook. Consult your lab instructor for the form these take.

LAB 9

Plate Tectonics

OBJECTIVES

- To understand divergence, spreading, spreading rate, and the age of rocks on the seafloor
- To understand transform faults and how they move
- To understand how convergence is connected with earthquakes in Wadati-Benioff zones
- To understand how data from Earth's magnetic field support the theory of plate tectonics

PLATE BOUNDARIES

We started this book with a brief introduction to plate tectonics. Because of the unifying nature of the subject, we felt it was important to get a taste of it at the beginning. Answer the following questions as a quick review.

1. Review the three types of plate boundaries from Lab 1 and ■ Figure 9.1 and define them here.

a. Divergent:

____

____

b. Convergent:

____

____

c. Transform:

____

____

2. For each of the following features, list the types of plate boundaries, using the letters D, C, and T, where you would find them. This is review from Lab 1 and Lab 8.

a. Oceanic trenches ______

b. Mid-ocean ridges ______

c. Subduction zones ______

d. Island arcs ______

e. Continental volcanic arcs ______

f. Strike-slip faults ______

g. Normal faults ______

h. Thrust faults ______

i. Anticlines and synclines ______

Now we can develop the concept of plate tectonics in more depth.

Divergent Plate Boundaries

Remember that divergent plate boundaries are also called **spreading centers** (■ Figure 9.2). The spreading creates oceanic crust, which is denser than continental crust. When a break in the continental crust occurs due to divergence, the continent separates into two smaller pieces and new oceanic crust forms between (Figure 9.2). The initial break is a **continental rift.**

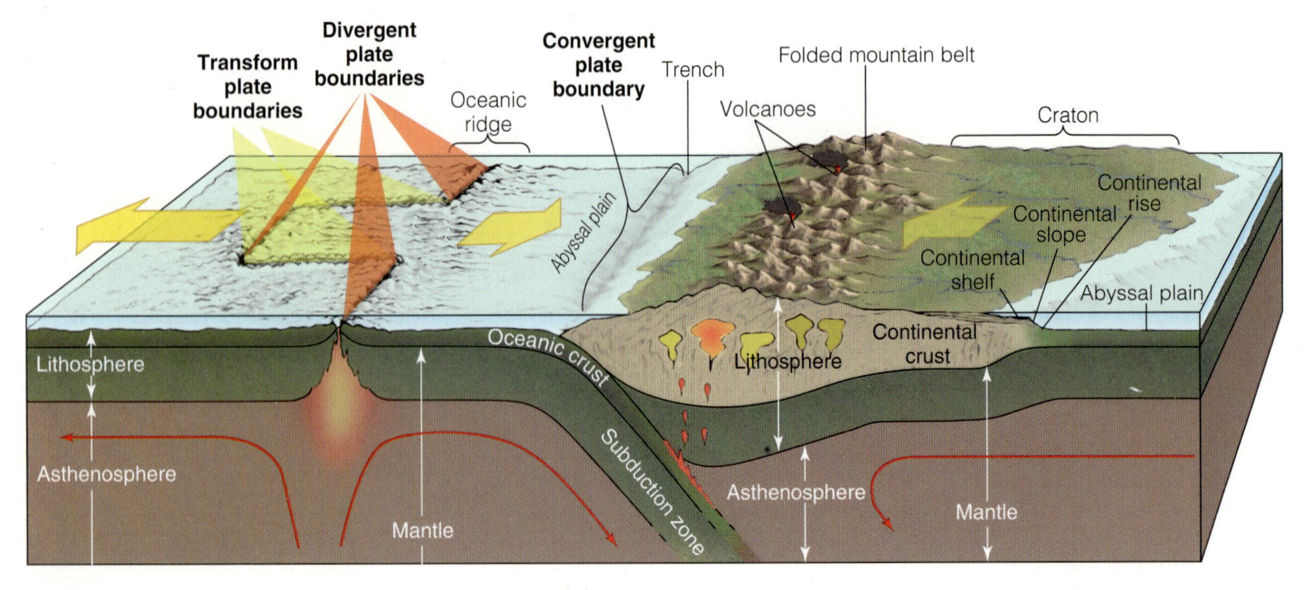

Figure 9.1

Plate Boundaries

Block diagram showing different types of plate boundaries and the features that occur at them.

As a continent diverges, the resulting oceanic crust eventually becomes wide enough to form an ocean basin (Figure 9.2d). If convergence does not subduct rocks within the ocean basin, the divergent plate boundary remains in the middle of the ocean and gives rise to a mid-ocean ridge at that location. Neither the divergent boundary nor the ridge is straight and continuous, both have repeated discontinuities where transform faults occur.

Magnetic Stripes on the Seafloor During the 1960s, oceanographers discovered puzzling, stripe-like patterns of magnetism on the seafloor (Figure 9.4a). The pattern was so distinct that it demanded explanation and led to the modern theory of plate tectonics. The Earth's outer core generates the magnetic field. The field tends to be aligned nearly parallel to Earth's rotational pole, but it also tends to switch direction periodically so that magnetic north becomes magnetic south and vice versa. As lava or other rocks form, they can capture a remnant of the magnetic field called **paleomagnetism** ("paleo" meaning ancient) that can be detected with careful measurement using a **magnetometer.** On land, as one lava flow covers another, a series of rocks will pile up vertically with alternating magnetism (■ Figure 9.3).

On the seafloor, however, as new ocean floor forms at the spreading centers, intrusions and lava flows insert new crust in strips between the diverging plates. The magnetic minerals in these igneous rocks become aligned parallel to Earth's present magnetic field as they solidify and cool. As time passes, the Earth's magnetic field switches direction and the newly erupted basalt at the spreading centers record this reversal. The recurring reversals produce bands of oppositely magnetized rocks that move outward away from the mid-ocean ridge as spreading continues. The age of the oceanic crust is, as a consequence, progressively older away from the ridge centerline (■ Figures 9.4 and 9.5) and the stripes of alternating magnetism reflect this progression of age. In fact, the magnetism helps geophysicists establish the age of the seafloor.

With information about the age of the seafloor, we can also, then, calculate how fast plates move. We calculate speed as distance divided by time, like miles per hour. The age of the seafloor gives us the time, so all we need is the distance to calculate the speed. Let's say two rocks started out at a spreading center and then moved apart. If you calculate speed based on their age and distance apart from each other, you will get the **full-spreading rate** for the two plates containing the rocks. On the other hand, if you calculate speed using the distance of a rock from the spreading center where it formed, you will get the **half-spreading rate.** The half-spreading rate is an approximate measure of the average speed of both plates as they move away from each other.

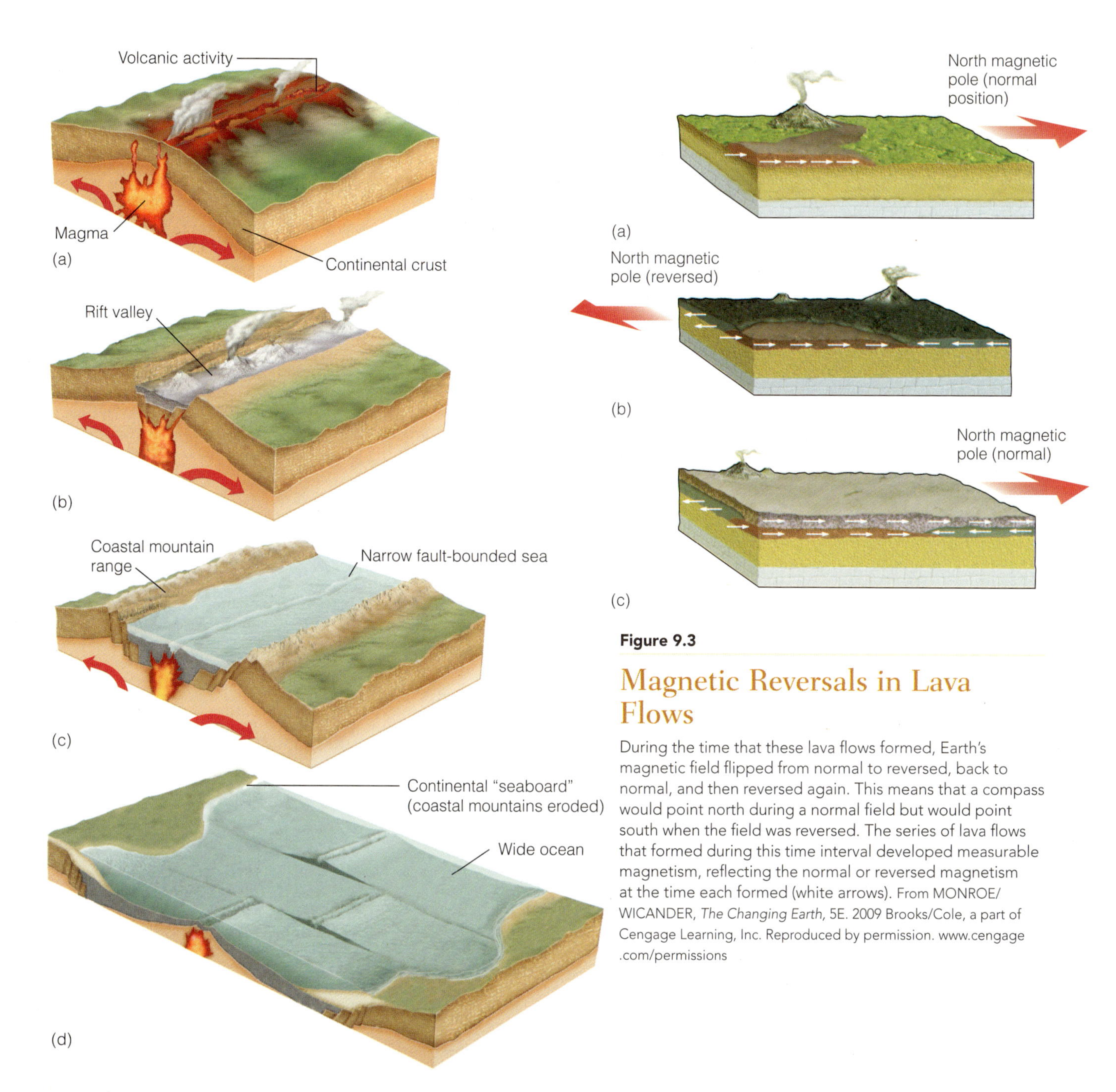

Figure 9.3

Magnetic Reversals in Lava Flows

During the time that these lava flows formed, Earth's magnetic field flipped from normal to reversed, back to normal, and then reversed again. This means that a compass would point north during a normal field but would point south when the field was reversed. The series of lava flows that formed during this time interval developed measurable magnetism, reflecting the normal or reversed magnetism at the time each formed (white arrows). From MONROE/WICANDER, *The Changing Earth*, 5E. 2009 Brooks/Cole, a part of Cengage Learning, Inc. Reproduced by permission. www.cengage.com/permissions

Figure 9.2

Continental Divergence

(a) Divergence stretches and warps the continent, (b) rifts the continent, forming a rift valley, (c) forms oceanic crust between the pieces of continent as a narrow sea or ocean. (d) finally develops a wide ocean after hundreds of millions of years of divergence. The Atlantic Ocean basin and the Mid-Atlantic Ridge formed in this way. From MONROE/WICANDER, *The Changing Earth*, 5E. 2009 Brooks/Cole, a part of Cengage Learning, Inc. Reproduced by permission. www.cengage.com/permissions

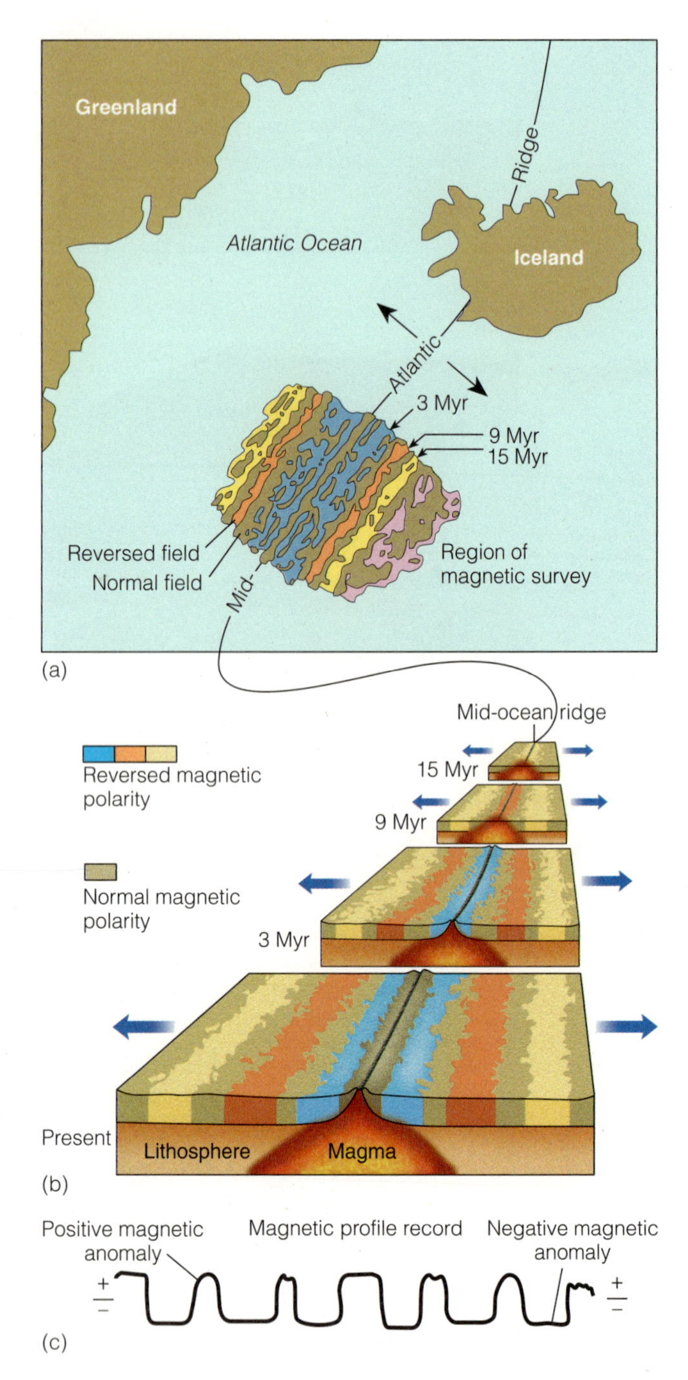

Figure 9.4

Development of Magnetism on the Seafloor

As seafloor spreading occurs, the magnetic field is recorded in the developing seafloor. Ships towing **magnetometers** (instruments that measure the magnetic field) can detect the additions and subtractions from the Earth's overall magnetic field that the seafloor basalts impose on the local magnetic field. (a) Measured magnetic stripes on the Mid-Atlantic Ridge just south of Iceland. (b) Series of block diagrams showing the development of new magnetic stripes as the seafloor spreads. (c) The record of magnetism detected by a ship passing over the mid-ocean ridge. Notice that the pattern is symmetrical, having a mirror image. (a) From GABLER/PETERSEN/TREPASSO/SACK. *Physical Geography,* 9E. 2009 Brooks/Cole, a part of Cengage Learning, Inc. Reproduced by permission. www.cengage.com/permissions (b) From MONROE/WICANDER, *The Changing Earth,* 5E. 2009 Brooks/Cole, a part of Cengage Learning, Inc. Reproduced by permission. www.cengage.com/permissions

3. What are the colored stripes in Figure 9.4?

____

4. Refer to Figure 9.5 to answer the following questions. You may also need to look at Figure 1.1 on page 2.

a. What do the colors in Figure 9.5 represent?

____

b. Locate the East Pacific Rise in Figure 9.5. What type of plate boundary occurs at the rise? ______________ How old are the rocks at the rise? ______________

c. Locate the Mariana Trench in the same figure. What type of plate boundary occurs here?

____

How old are the rocks of the Pacific Plate that are just about to subduct at the Mariana Trench?

____

d. If each vertical line on Figure 9.5 is 30° of longitude, how many degrees of longitude are the rocks at the trench from the spreading center that formed them?

____

5. Explain why the Atlantic Ocean has colored stripes in Figure 9.5.

____

What is the symmetry of the stripe pattern?

____

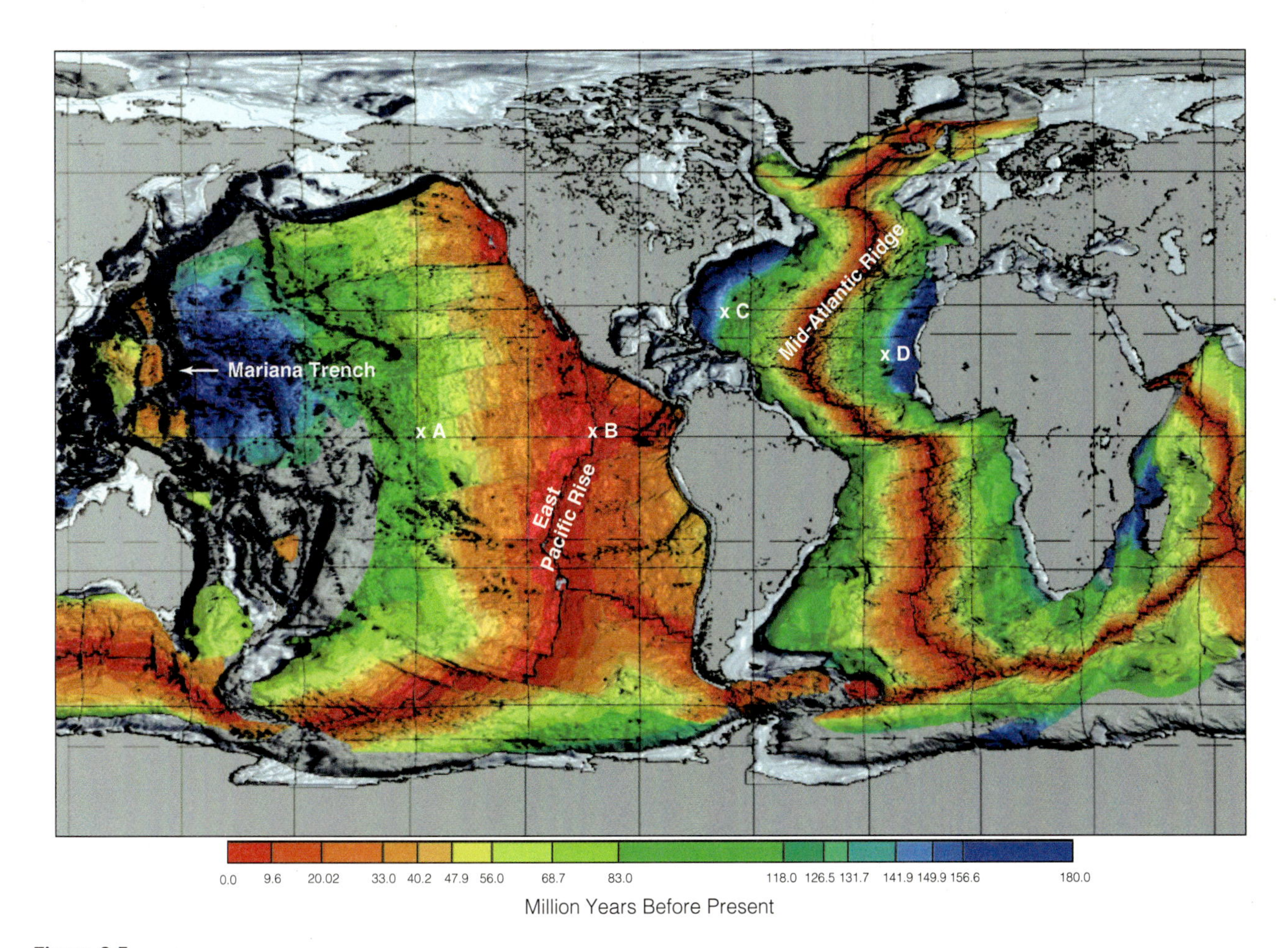

Figure 9.5

Age of the Ocean Floor

Colors represent different-aged rocks on the seafloor, determined from paleomagnetism, with red the youngest and blue the oldest.

6. Use the age of the rocks on the map in Figure 9.5 to help calculate plate speeds.

a. How old are the rocks at point A?

_________ point B? _________

b. Points A and B are 5,200 km apart. How fast is the half-spreading rate for the Pacific Plate between point A and point B? Calculate this in cm/yr. Conversions: 1 km/Myr = 1 mm/yr and 1 cm = 10 mm. Show the calculation method.

Half-spreading rate for Pacific Plate =

____

c. How old are the rocks at point C?

_________ point D? _________

d. Calculate how fast points C and D are moving away from each other. Calculate this in cm/yr. Show how to do the calculations below:

Rate of movement in cm/yr of point C away from point D =

____

Is this the half-spreading rate or the full-spreading rate?

____

e. What is the speed of the North American plate relative to the spreading center?

____

f. The fastest one-half spreading rates are about 12 cm/yr. Were your spreading rates reasonable (i.e., are they less than the fastest rate)? ______

g. Explain why the colored bands in Figure 9.5 are so much wider in the Pacific Ocean basin than they are in the Atlantic Ocean basin.

____

____

Divergent Plate Model In the following group of exercises, you will make a model of a continent that rifts at a divergent plate boundary and that two transform faults cut.

Materials needed:

- Scissors
- Fasteners such as paperclips, or a stapler

7. Work in groups of three or more using Figure 9.6a, b, and c, and referring to Figure 9.7.

a. Before you start cutting, locate and highlight the "stop cutting" symbols on Plates A and B (Figures 9.6a and 9.6b).

b. One person should cut out Plate A and another Plate B from Figures 9.6a and 9.6b.

c. A third person should separate the page containing the base of the model (Figure 9.6c) from the manual and make the folds and cuts indicated on that figure.

d. Fasten the triangles on Plate A and Plate B together, *face-to-face (not end-to-end)*, so that the triangles touch, with the end of each tab flush with the other. Likewise, match and join pinwheels to pinwheels and stars to stars. When assembled, the pages *should not lie flat* but should have the configuration shown in Figure 9.7.

e. Insert the fastened triangle end of the plates from Figures 9.6a and 9.6b into the slot marked with a triangle in the model base (Figure 9.6c). Insert the pinwheel and star ends into their respective slots.

f. Gently pull the three tabs through the slots so they are as far as they will go. Do not crease or flatten, as that prevents smooth movement of the "plates." You have a single continent made up of two plates. You are now ready to watch divergence in action. The continent began to rift 150 million years ago.

8. One person should hold the model base (Figure 9.6c) horizontally at the two ends where it says "hold here." A second person should pull Plate A and Plate B horizontally away from each other, keeping them at about the same level as the model base. The numbers on the plates are the age of the rocks.

a. Starting 150 Myr ago, pull the plates apart until you reach the configuration for 130 Myr ago. The age of the rocks is marked in millions of years on the oceanic crust. On a separate sheet of paper, sketch a map of the two plates.

b. On your sketch, indicate where the plates were oceanic and where they were continental. The parts that were oceanic were probably filled with water.

c. Review the symbols for different plate boundaries from Lab 1. On your sketch, add divergent margin symbols at the spreading centers and transform fault symbols at the transform faults.

d. Place arrows on your sketch showing the plate motion directions and the relative motion along the transform faults (use fault symbols you learned in Lab 8 from Table 8.1, on p. 163).

e. Remember that one continent can be made up of more than one plate, and that one plate can have both continental and oceanic parts. How many plates were there? ________

f. In your model, how many continents were present 130 Myr ago? ________

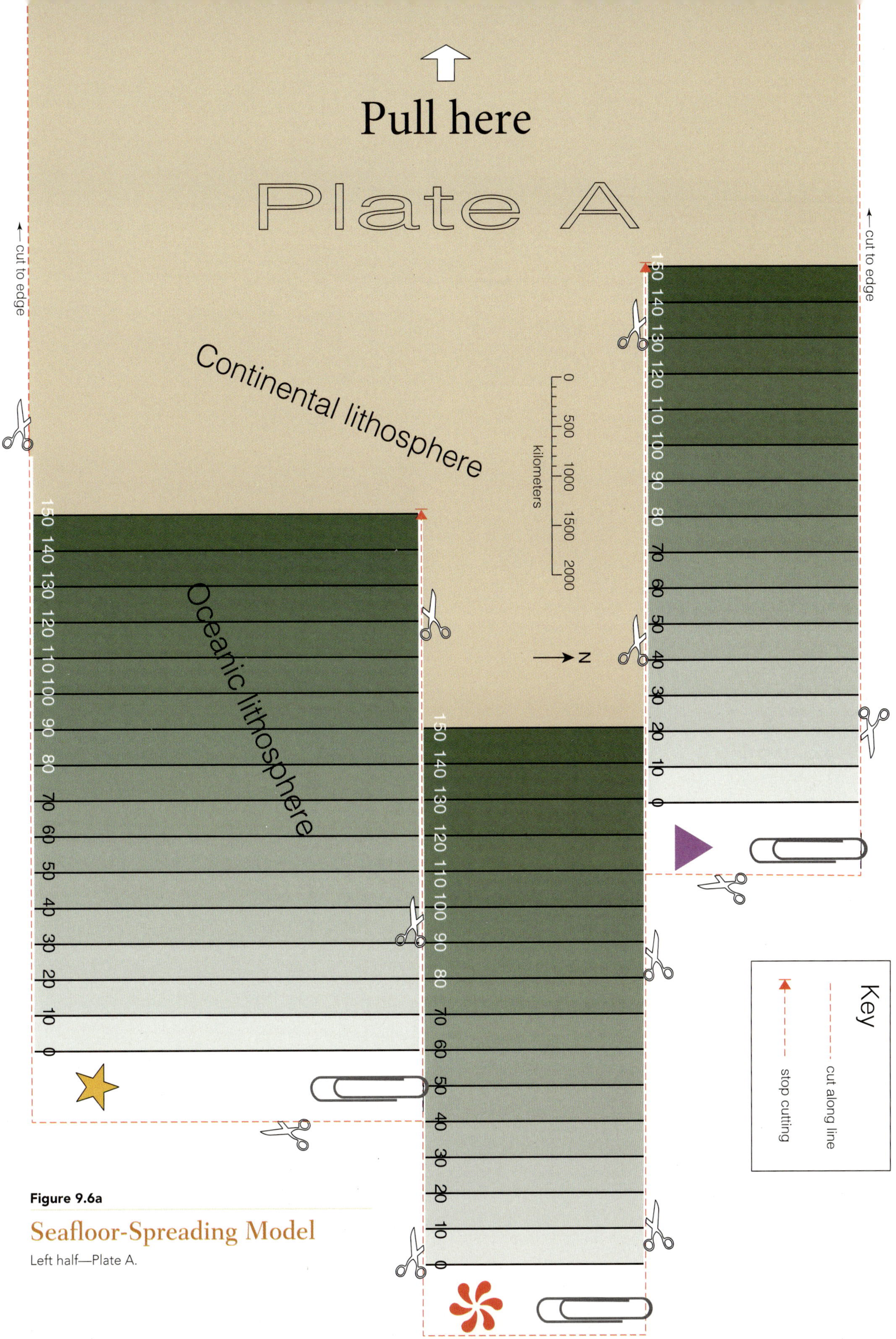
Pull here
Plate A
Continental lithosphere
Oceanic lithosphere
cut to edge
cut to edge
0 500 1000 1500 2000
kilometers
N
150 140 130 120 110 100 90 80 70 60 50 40 30 20 10 0
150 140 130 120 110 100 90 80 70 60 50 40 30 20 10 0
150 140 130 120 110 100 90 80 70 60 50 40 30 20 10 0
Key
cut along line
stop cutting

Figure 9.6a

Seafloor-Spreading Model

Left half—Plate A.

Figure 9.6b

Seafloor-Spreading Model

Right half—Plate B.

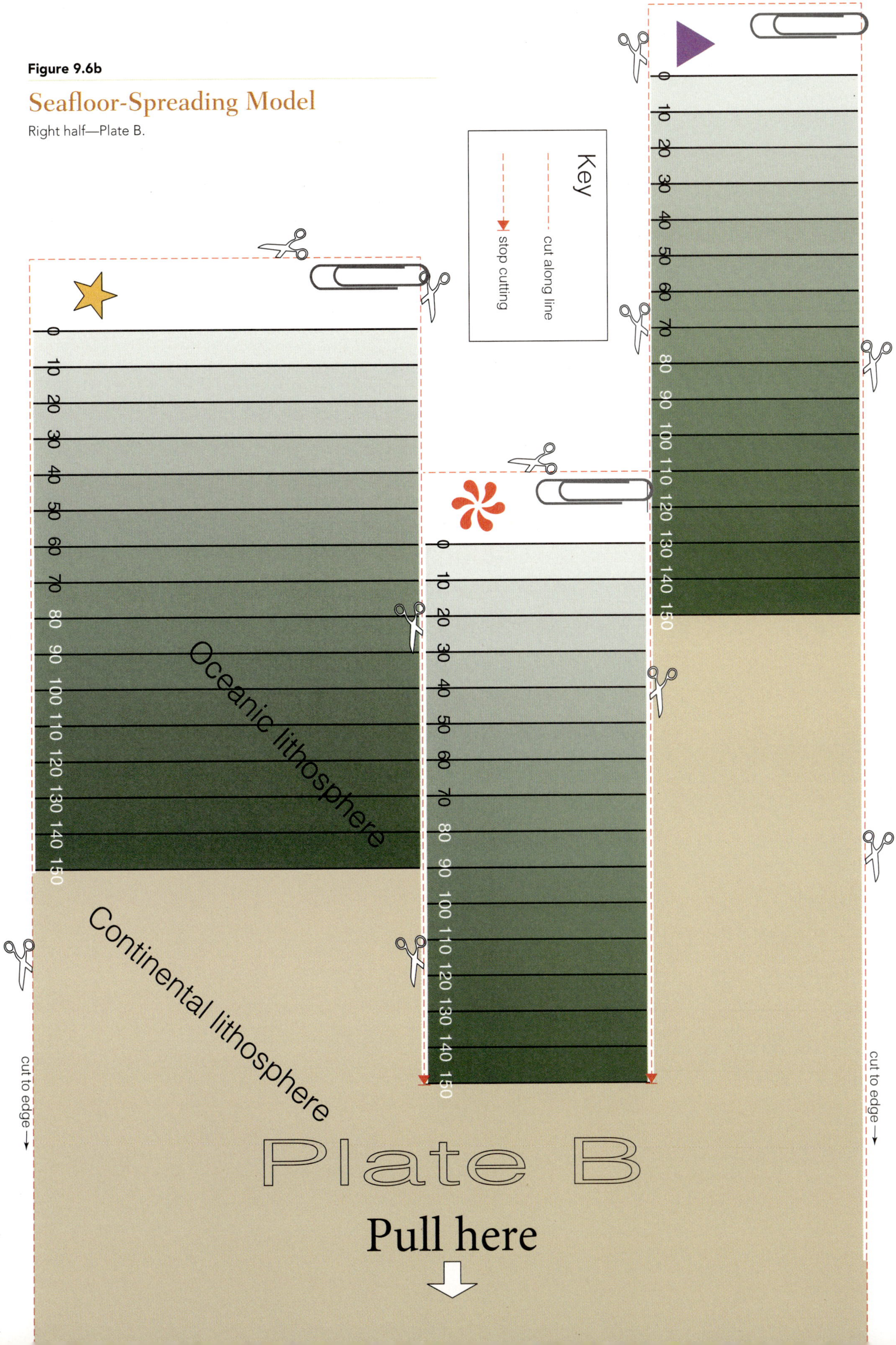

Key

cut along line

stop cutting

0 10 20 30 40 50 60 70 80 90 100 110 120 130 140 150

0 10 20 30 40 50 60 70 80 90 100 110 120 130 140 150

0 10 20 30 40 50 60 70 80 90 100 110 120 130 140 150

Oceanic lithosphere

Continental lithosphere

cut to edge →

cut to edge →

Plate B

Pull here

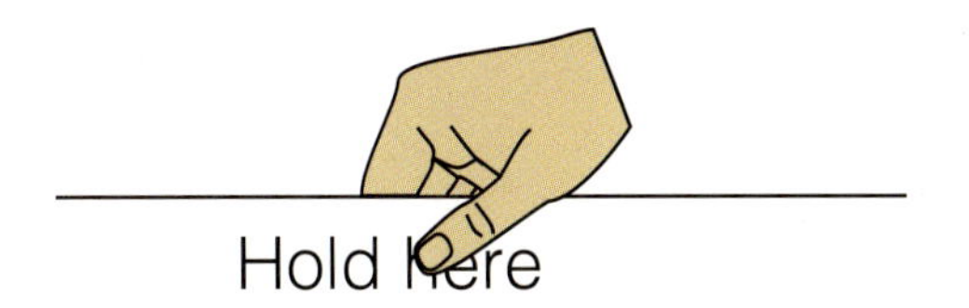

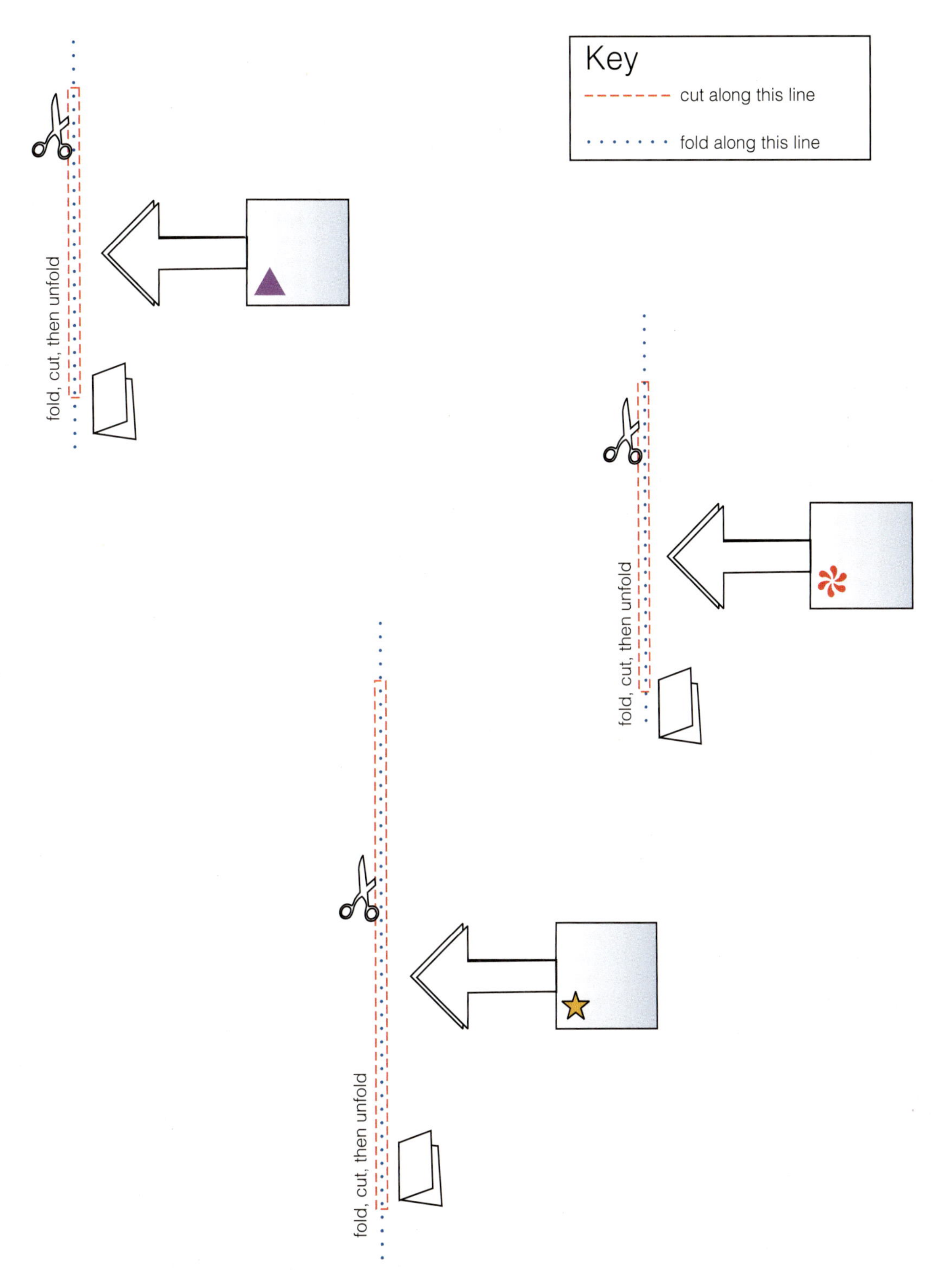

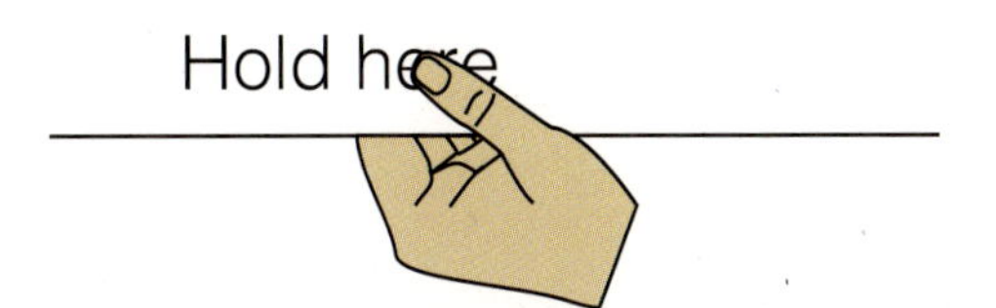

Figure 9.6c

Seafloor-Spreading Model

Base.

Figure 9.7

Photograph of the Constructed Seafloor-Spreading Model

See instructions in Exercise 7 for how to assemble the model.

9. As you proceed with the next exercises using the seafloor-spreading model, match each time step at the left below with the configuration of the landmasses and oceans listed at the right below:

a. 150 Mya _____	1. Two continents with a narrow land bridge
b. 130 Mya _____	2. One continent with narrow sea inlets and a landlocked basin
c. 100 Mya _____	3. Two continents separated by an ocean
d. 0 Mya _____	4. Two continents separated by a narrow ocean
	5. One continent with no seas

10. Continue spreading your model apart until you reach 100 Myr ago. Sketch the new configuration, including oceanic crust and continental crust, spreading centers, transform faults, and movement arrows on your sketch.

11. Spread your model until you reach the present (0 Myr). Sketch the new configuration, including the same information as on previous sketches.

Land organisms can migrate from one continent to another as long as there is some sort of connection ("land bridge") between the continents.

12. On your model, as the continent diverged, when did species of organisms roaming the continent become isolated into two separate populations?

____

13. Study the relationship between the age of rocks and their distance from the spreading center on your model. What general pattern do you see?

____

____

Transform Faults

Transform faults in mid-ocean ridge systems may be a bit puzzling. ■ Figure 9.8 shows two differing interpretations of the same location, one from before geologists understood seafloor spreading and one from after. Before geologists began to understand seafloor spreading in the 1960s, they assumed transform faults displaced the mid-ocean ridges. They believed that the mid-ocean ridges were once continuous and that the transform faults had offset them. For example, Block X in Figure 9.8a appeared to have moved to the right relative to Block Y, making the fault a *right-lateral strike-slip fault* (review faults, in Lab 8, if necessary). However, after studying actual earthquake movements, they realized this predicted motion was wrong. Figure 9.8b illustrates the correct fault movement along an oceanic ridge. Plate C in Figure 9.8b happens to have moved left relative to Plate D, making the transform a *left-lateral strike-slip fault* instead. That means that seafloor spreading must be occurring at each segment of the mid-ocean ridge to account for the actual motion along transform faults.

14. Study ■ Figure 9.9, which depicts a portion of a mid-ocean ridge spreading center in map view. At one time, geologists may have hypothesized that area N on the map moved left relative to area S. It may help you to study your model of Plate A and Plate B and Figure 9.6a and b.

a. Do you agree with their assessment of the fault? ______

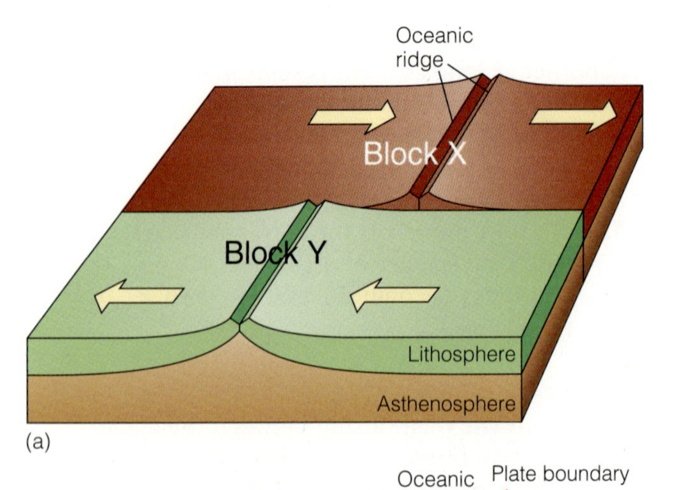

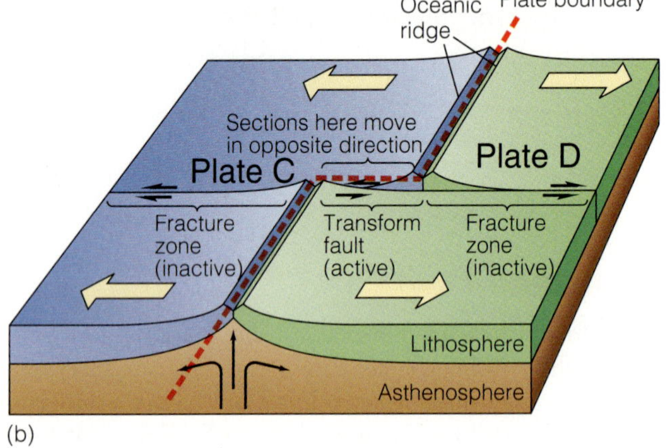

Figure 9.8

Transform Fault Interpretation

Two interpretations of a pair of mid-ocean-ridge segments separated by a transform fault. (a) Early interpretations had Block X moving right relative to Block Y. Earthquake data from along the fault contradict this early interpretation. (b) The motion along a transform fault, according to the theory of seafloor spreading, is consistent with earthquake data that show that the fault in the diagram is a left-lateral strike-slip fault. Plate C (blue) moves left relative to Plate D (green).

b. What motion do you think occurred along the fault?

____

____

Draw in appropriate arrows on the fault in Figure 9.9.

c. What incorrect assumption could lead to the wrong conclusion?

____

____

____

____

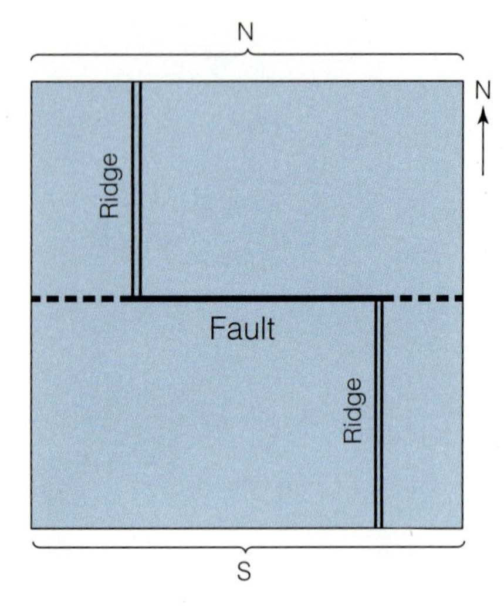

Figure 9.9

Transform Fault Map

Map of a transform fault between two segments of mid-ocean ridge. Follow instructions in Exercise 14.

d. Study the active transform fault and the inactive fracture zones in Figure 9.8b. Use them as a model to highlight and label the inactive fracture zones on Figure 9.9. What does inactive mean?

____

____

15. What type of fault is the northern transform fault on your seafloor-spreading model from Figure 9.6?

____

The southern one?

____

Not all transform faults occur along mid-ocean ridges. Two additional kinds of transform faults are visible in ■ Figure 9.10a. Some of these connect to **triple junctions,** which are places where three plates and three plate boundaries come together.

Unlike other plate boundaries, transform faults do not have specific types of physiographic features that mark their locations. In many cases, however, the fault may appear on the surface as a long valley or will have distinctly different topography on either side.

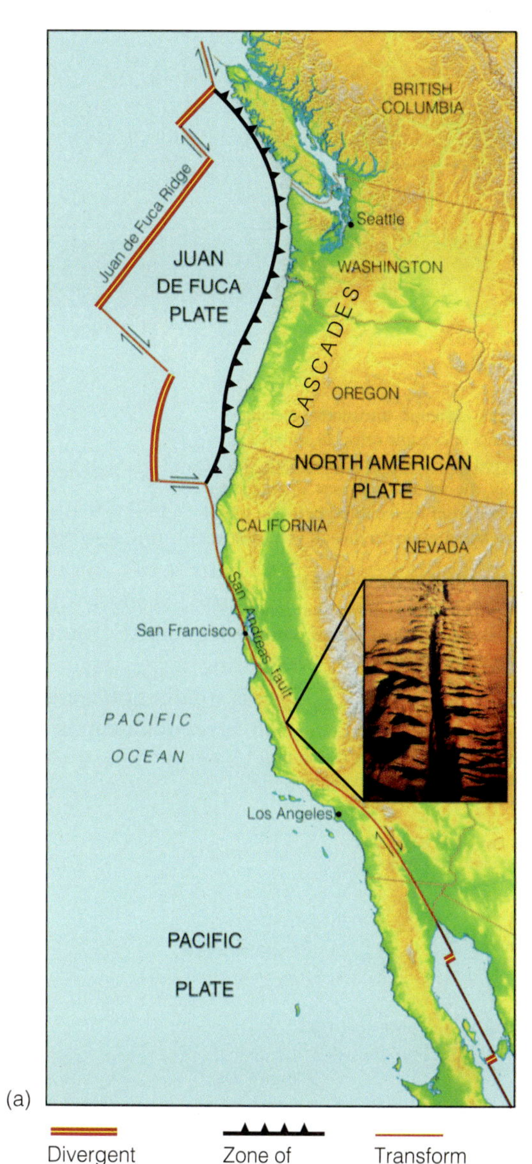

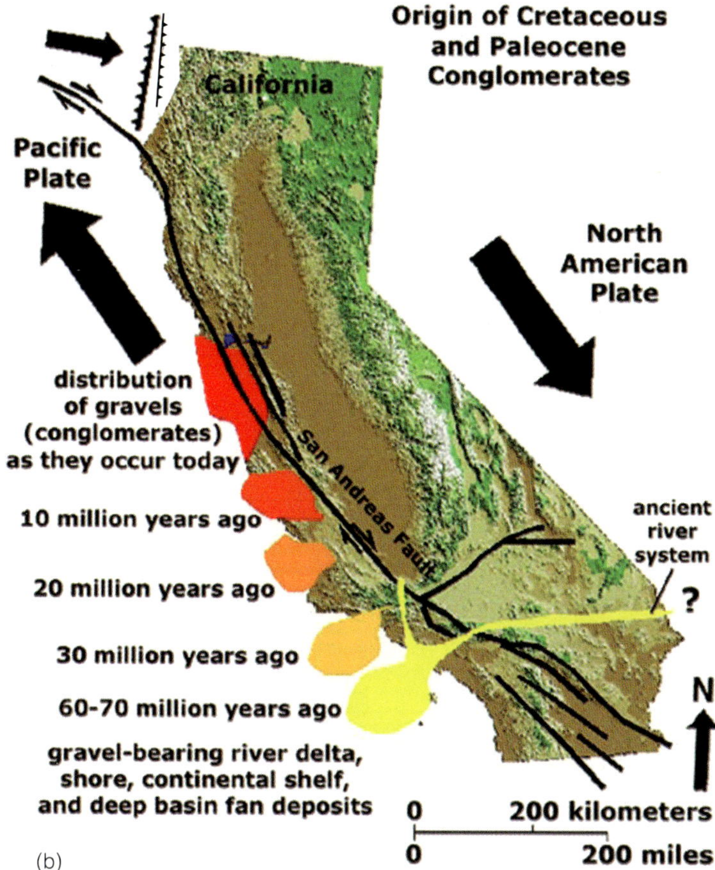

Figure 9.10

Plate Boundaries along the U.S. West Coast

The San Andreas Fault is a continental transform fault. (a) Notice six additional transform faults of different types. Subduction of the Juan de Fuca Plate produces the Cascade Mountain Range. (b) Cretaceous and Paleocene conglomerates found today along the central/northern California coast (red) had a source far to the south (yellow, ancient river system). Movement along the San Andreas Transform Fault system has carried them to their present, more northerly position. (b) From: USGS Bulletin 2195, Rocks and Geology in the San Francisco Bay Region by Philip Stoffer

16. Do the following with Figure 9.10.

a. Use the letters D for divergent, S for subduction, and T for transform. Label each transform fault for the types of boundaries each connects on Figure 9.10a. If one connects two divergent plate boundaries, for example, label it DD, but one might also connect three plate boundaries such as DST. If only one end is shown, add a question mark for the unseen end, such as ?D or ?ST.

b. Examine the photograph inset along the segment of the San Andreas Fault in Figure 9.10a. What landscape features show the position of the fault?

____

____

c. In Figure 9.10b, gravels were transported from the North American Plate at the location labeled "ancient river system" and deposited on the Pacific Plate. The area of yellow shows where the gravels were 60–70 million years ago, the various colors of orange show where they were 30, 20, and 10 million years ago and the red show where they are now. Use the kilometer scale to determine about how fast the Pacific Plate was moving relative to the North American Plate for each time period listed below. Draw a line perpendicular to the fault line and through center of each color patch as the point from which to measure. Give your answers in centimeters per year.

- 0–10 Mya ________________
- 0–20 Mya ________________
- 0–30 Mya ________________
- 0–70 Mya ________________

d. Has the plate speed changed? _______ If so, did it speed up or slow down or did its speed vary in some other way? How did it change?

____

e. Based on what you know about conglomerates and dating techniques, what evidence do you think was collected to produce the data for this map?

____

____

Convergence, Subduction, and Mountain Building

Subduction occurs when a slab of lithosphere descends downward into the asthenosphere. At ocean–continent convergent plate boundaries, the oceanic plate subducts beneath the continental plate (■ Figure 9.11). At ocean-ocean convergent margins, the denser (usually older) oceanic plate subducts beneath the less dense (younger) oceanic plate. In either case, fracturing of the descending plate causes earthquakes along this subduction zone.

As the cold, subducting slab descends deeper into the Earth, rocks on either side warm it. These warm

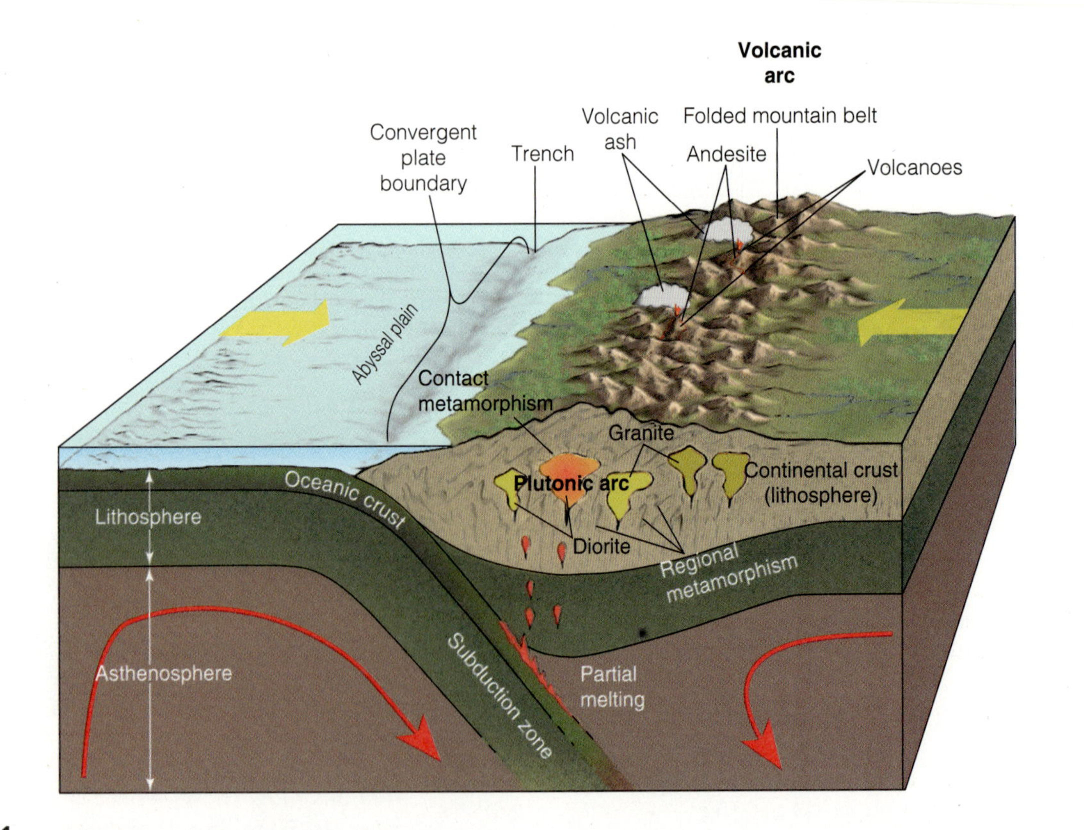

Figure 9.11

Ocean–Continent Convergent Plate Boundary

Volcanic and plutonic arcs with igneous, metamorphic, and deformational features at an ocean–continent convergent plate boundary.

rocks may have temperatures as high as 1,500° C at a depth of about 100 km. Eventually, the cold slab heats up enough to release water that aids melting of the overlying mantle and the resulting magma rises. The magma builds two types of igneous features. One is an arc of intrusions in the crust called a **plutonic arc,** such as granite and diorite plutons, which may join up into a great *batholith* at depth. Second, the magma erupts as lava (especially andesite) or volcanic ash (Figure 9.11). The resulting string of volcanoes at the surface is a **volcanic arc.**

Regional metamorphism generally occurs in the area below the volcanic arc. The differential pressure, and the heat from the magma, help warm the rocks, forming low-, moderate-, and high-grade foliated rocks such as slate, phyllite, schist, and gneiss. The convergence, plutonic activity, and volcanism create a broad band or strip of mountains with the volcanic peaks of this continental-volcanic-arc system on top.

Convergence also leads to the production of sediments as older parts of the arc lose their cover of volcanic rocks, exposing plutonic and metamorphic rocks beneath. At the surface, erosion strips off the volcanoes, shedding volcanic sediments toward the oceanic trench at the plate boundary. The plutonic and metamorphic rocks beneath the volcanoes in turn may shed sediment toward the sea. These sediments are deposited on the continental shelf and in the oceanic trench and may eventually become rocks such as graywacke, volcanic lithic sandstone, arkose, and shale.

Convergence of the North American and Farallon Plates formed the Rocky Mountains of Montana (Figure 9.12) prior to 40 million years ago. The Farallon Plate, an oceanic plate, subducted beneath North America until the Pacific and North American plates came in contact (Figure 9.12b). This left only small fragments of the subducting plate, now the Juan de Fuca and Cocos Plates (Figure 9.12c). The Sierra Nevada Batholith (Figure 9.12d) in eastern California also formed from Farallon Plate subduction. The Cascade Range formed more recently from the subduction of the Juan de Fuca Plate (Figure 9.10a). Let's look at rocks, physiographic features, and structures that formed in these places as a consequence.

17. Carefully examine the Geologic Map of Devils Fence Quadrangle, Montana, and its cross sections and explanation in Figure 8.18, on pages 186–189, to help you answer the following questions:

a. What deformation in the area suggests that convergence occurred here in the past?

____

b. List the volcanic and/or plutonic rocks present that might have originated as magma from above a subduction zone.

____

The *Cascade Mountain Range* spans the distance between northern California and southern British Columbia (Figure 9.12d.)

18. Examine the maps of western North America in ■ Figure 9.12 showing the last 40 million years of evolution of the plate boundaries. Also use the more detailed current plate map of the West Coast in Figure 9.10a.

a. What is an important feature of the Cascade Range shown as black dots on the map in Figure 9.12c and d?

____

b. What plate tectonic action is creating the Cascade Range?

____

____

c. Draw arrows on Figure 9.10a showing the relevant plate motions.

d. Compare the present-day plate configuration to what was happening 40 million years ago. What plate was subducting beneath North America back then?

____

e. Use the present positions of volcanoes in Figure 9.12c and their relationship to the plate boundary to estimate where volcanoes were 40 million years ago. Draw them as a series of black dots on Figure 9.12a.

f. What features/rocks would have been forming back then underneath those volcanoes? See hints in Figure 9.11.

____

____

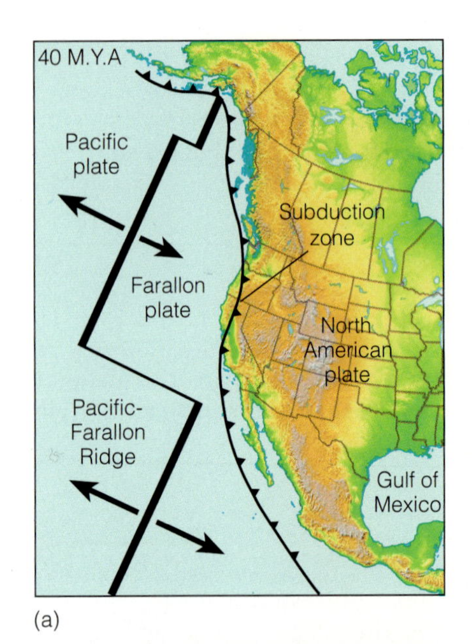

(a)

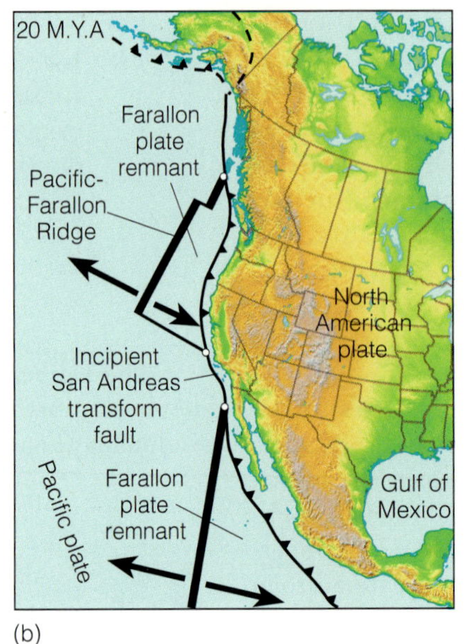

(b)

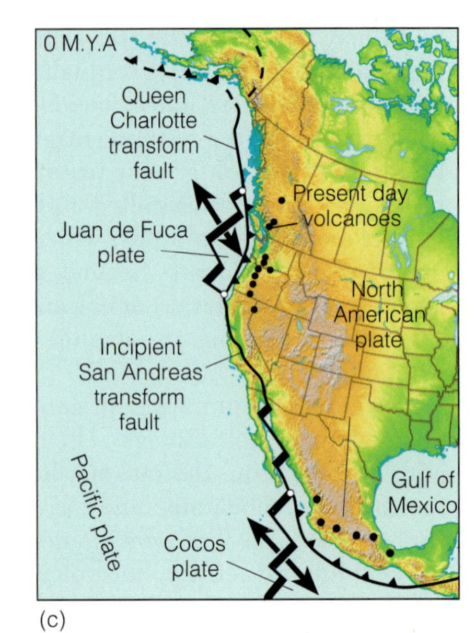

(c)

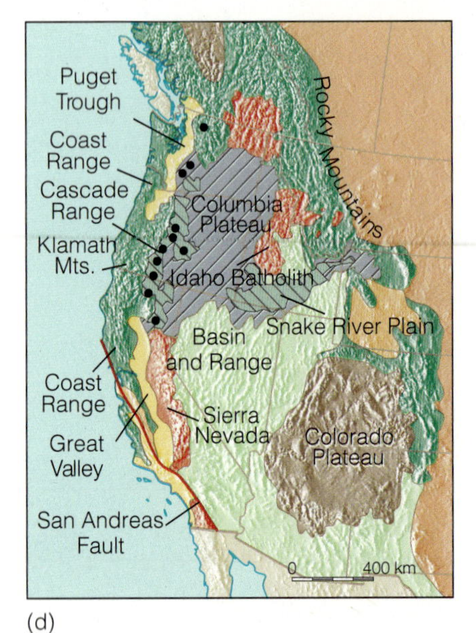

(d)

Figure 9.12

Plate Geometries along the West Coast of North America

Convergence of the Farallon Plate and development of strike-slip movement along the West Coast. As North America drifted west, it overrode and subducted the Farallon Plate, whose remnants are the present Juan de Fuca and Cocos Plates. (a) 40 Myr ago. (b) 20 Myr ago. (c) Today. (d) Locations of mountain ranges and other physiographic features west of the Great Plains. Black dots on (c) and (d) are present-day volcanoes resulting from ocean-continent convergence. From MONROE/WICANDER, *The Changing Earth*, 5E.

19. Carefully examine the Geologic Map of Plutons near Chico, California, in Figure 3.32, on page 68, to help you answer the following questions:

a. This area forms the transition between the plutons of the Sierras and the volcanoes of the Cascades. What rocks indicate the presence of volcanoes in the area as recently as the Pliocene?

____

____

b. What rock categories (volcanic, clastic sedimentary, metamorphic, etc.) in the area suggest erosion has stripped away much of the volcanics, exposing deeper-seated rocks?

____

____

c. What was the magma source for the igneous rocks? Think in terms of plate tectonics.

____

Subduction and Earthquakes Earthquakes (**seismicity**) occur at all plate boundaries (■ Figure 9.13), but deep earthquakes can only occur where cold rocks reach deep levels in the Earth. Subduction takes rocks, which are about 4°C, from the seafloor downward, causing a chilling effect on Earth's interior within the subduction zone and surrounding rocks. This allows brittle deformation (faulting) to take place deep. As we will study more in the following chapter, earthquakes occur along faults and result from brittle deformation.

20. Examine Figure 9.13, which shows where earthquakes have occurred in the past.

a. What do red dots signify?

____

b. Yellow dots? ______________________

c. Dark green dots?____________________

d. Notice that most earthquakes occur at plate boundaries. Use the arrows on Figure 9.13 to determine plate boundary types. List *all* of the types of boundaries including crust type (using abbreviations **o. div., c. div., o-o conv., o-c conv., c-c conv., o. trans.,** and **c. trans.**) that have each depth category of earthquakes:

Shallow (to 50 km): ____________________

Intermediate (50–300 km): ______________

Deep (>300 km): ______________________

e. Only cold rocks experience earthquakes because hot rocks will simply undergo ductile deformation in response to stress, without generating earthquakes. With this in mind, explain the results you obtained for part **d.**

____

____

____

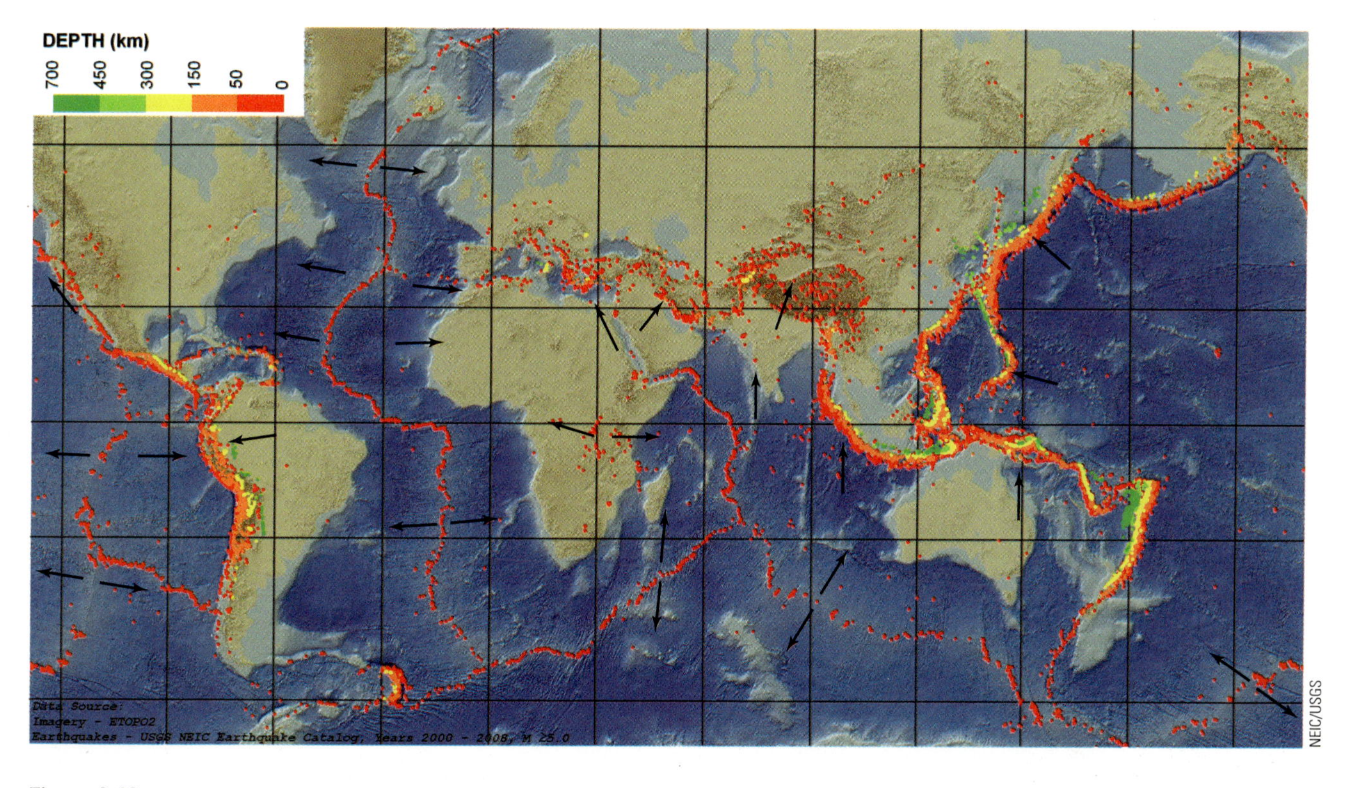

Figure 9.13

Earthquake Map

Map of world seismicity that shows the distribution of shallow, intermediate, and deep earthquakes worldwide. Arrows show the direction of plate movement.

Angle of Subduction The angle of subduction dictates the relative position of an *oceanic trench* and a *volcanic island arc* or a *continental volcanic arc*. This is because magma generation associated with subduction typically occurs at about the same depth for all subduction zones. The **arc-trench gap** is the horizontal distance between the volcanoes and the trench. If the subduction zone dips steeply, the rocks reach a deep enough level for magma to form close to the trench, so the arc-trench gap will be small, such as 50 km. On the other hand, if the subduction zone dips gently, the depth of melt generation is not reached until the subducting slab is quite a distance from the trench, making a large arc-trench gap. When seismologists detect earthquakes within a subducting slab, their **foci** (plural of **focus**—earthquake location at depth) can show us the slab location. This inclined planar distribution of intermediate and deep earthquakes within the subduction zone delineates the **Wadati-Benioff zone** (Figure 9.14a).

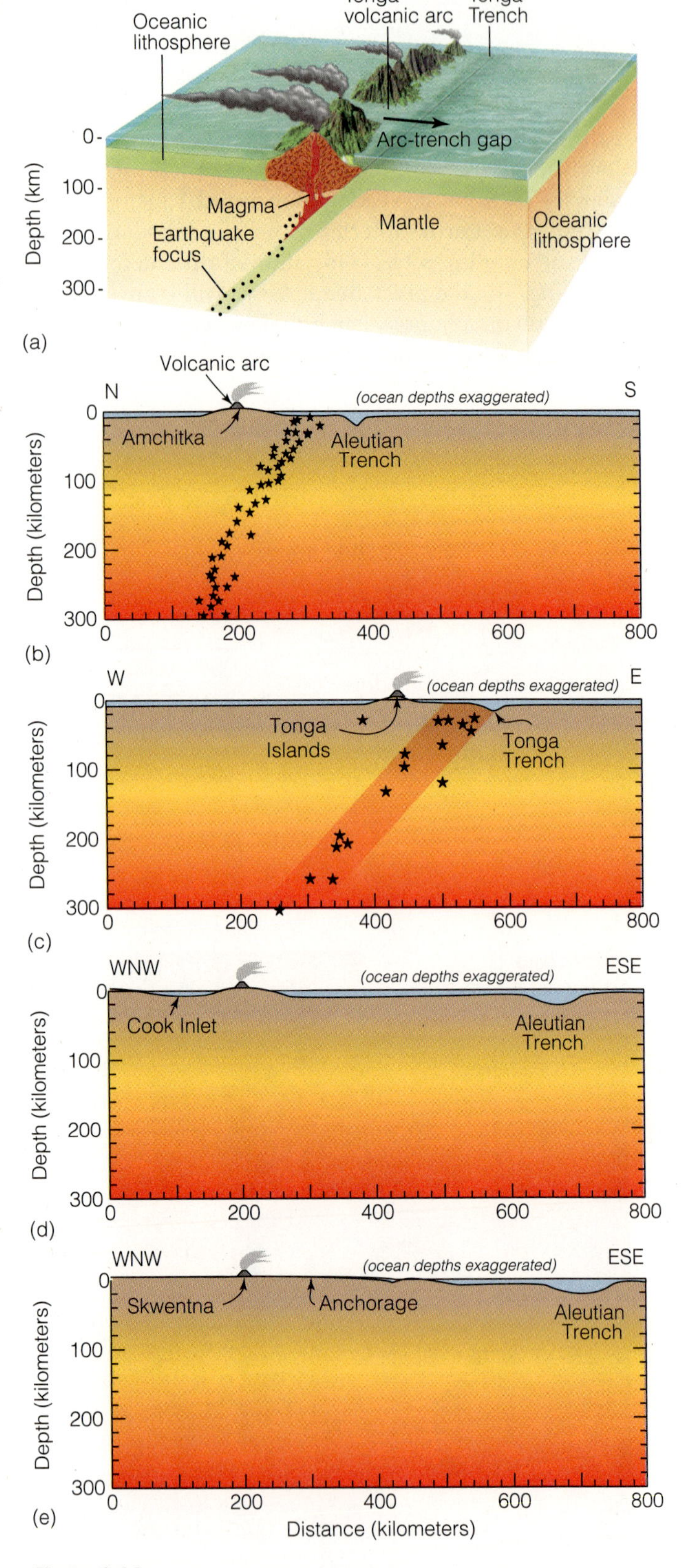

Figure 9.14

Wadati-Benioff Zones and Arc-Trench Gaps

Profiles across segments of the Aleutian and Tonga arcs. Stars show earthquakes in the Wadati-Benioff zones for two locations. (a) From MONROE/WICANDER, *The Changing Earth*, 5E. 2009 Brooks/Cole, a part of Cengage Learning, Inc. Reproduced by permission. www.cengage.com/permissions

21. Examine ■ Figure 9.14, which shows profiles across segments of the Aleutian and Tonga Island arcs. The stars on the diagrams represent the locations of earthquakes, the Wadati-Benioff zone, and are a guide to the location of the subduction zone.

a. Draw the subduction zone in Figure 9.14b using the earthquake foci (Wadati-Benioff zone) at Amchitka as shown for Tonga in Figure 9.14c.

b. Compare the dips of the Wadati-Benioff zones at Amchitka and the Tonga Island arc. Which location has a steeper subduction zone?

____

c. Measure the arc-trench gap across the top of each profile from the volcano to the trench for each profile in Figure 9.14b–e. Amchitka ________ km; Tonga ________ km; Cook Inlet ________ km; Skwentna ________ km.

22. Use the locations of the volcanic islands, the Wadati-Benioff zones, and the fact that most magma rises vertically to help you determine the minimum depth at which melt-generating processes start at the top of the subduction zone in Figures 9.14b and c.

a. At Amchitka? _________ km

b. At Tonga Islands? _________ km

c. Average these two: _________ km

23. Using the same average minimum depth of melting, the positions of the trenches, and the position of the volcanic arcs (at the small volcano), sketch in where you expect the subduction zone to be for Cook Inlet and for Skwentna in Figure 9.14d and e.

24. Based on the previous exercises, circle which one of the following statements is most accurate.

a. All subduction zones dip at the same angle.

b. Subduction zones dip 62° on average.

c. Subduction zones have different dips in different places, and the distance from the trench to the arc is related to the dip.

d. Subduction zones have different dips and the depth to the subduction zone at the volcanic arc is dependent on that dip.

At a convergent plate boundary, two plates move toward each other. Each plate is a slab of lithosphere, which consists of crust at the top and a bit of upper mantle below. As the plates come together, either they collide and crumple, or one plate goes down under the other (subducts). The thing that determines which action will occur depends on the type of crust that makes up part of the lithosphere of the two plates. Oceanic crust at depth is denser than the asthenosphere so it is able to subduct. Continental crust, on the other hand, is less dense so it does not subduct at a convergent plate boundary. The difference in density of oceanic and continental crust is evident in the elevations of their surfaces and the locations of the oceans. The more buoyant continental crust rides higher on the asthenosphere, so it produces land, whereas the denser oceanic crust sinks lower, making ocean basins. Continental crust is also thicker and thus extends downward into the mantle farther than oceanic crust. The thicker the crust is, as in mountain ranges, the deeper it sinks. This creates mountain roots where the continental crust has thickened during a continental collision (■ Figures 9.15 and 9.16).

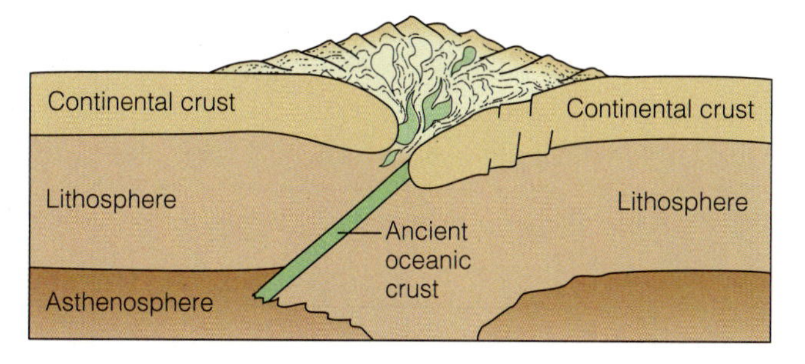

Figure 9.15

Continental Collision

In a continental collision, neither continental plate subducts; instead, both plates crumple, undergoing folding and faulting. The old oceanic part of one of the plates that originally separated the two continents may continue to subduct.

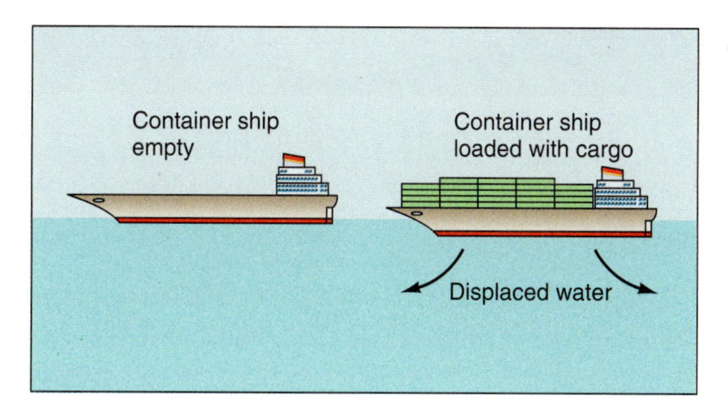

Figure 9.16

Effect of Buoyancy

A filled container ship sits lower in the water than an empty one because it displaces a volume of water of equivalent weight—the more weight, the more water displaced. However, because the containers make it taller, the top of the containers may sit higher than the deck of the empty vessel. This is analogous to oceanic crust and continental crust if we consider the empty ship to be the oceanic crust, and the full ship to be the continental crust. Even if the containers have a low density, their combined weight causes them to displace more water. The water in this analogy corresponds to the asthenosphere, which is soft enough to allow denser or lighter materials to adjust their levels—a process called **isostasy.**

25. Examine the ancient plate boundary shown in ■ Figure 9.17. The actual plate boundary is not labeled so you will need to infer its location by studying the maps.

a. During the Early Carboniferous Period, what type of plate boundary was present between Laurasia and Gondwana? Be specific about both movement type and types of crust involved.

____

____

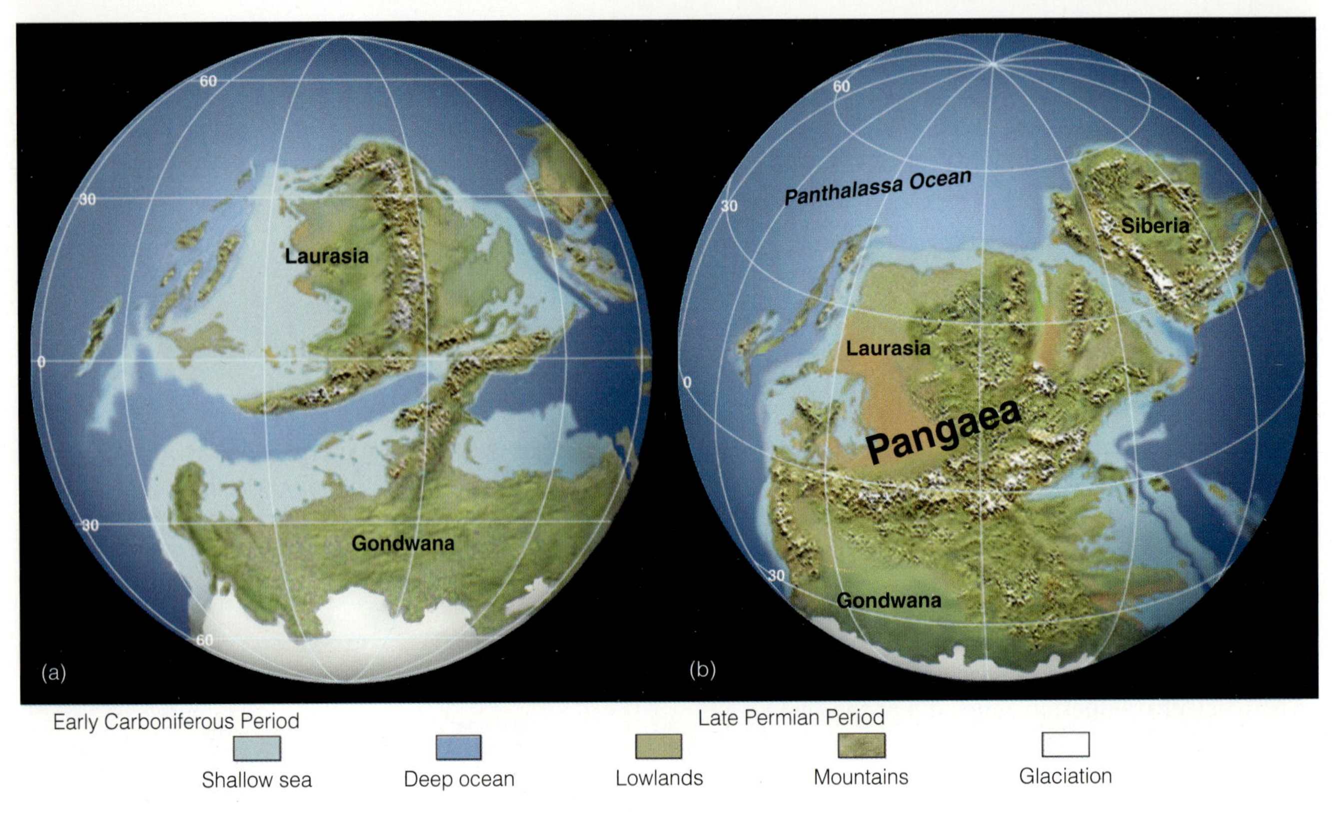

Figure 9.17

Pangaea's Formation and Paleogeography

Paleogeographic reconstructions for (a) the Early Carboniferous Period showing the configurations of Laurasia (which included parts of present-day North America, Europe, and Asia) and Gondwana (which included parts of South America, Africa, India, and other southern continents) and (b) the Late Permian Period showing the configuration of Pangaea, which included most of the continental land masses.

b. What type of crust would have been moving down under the continents?

____

c. By the time of the Late Permian Period, how had the plate boundary changed?

____

____

d. Why did such a large mountain range form and what does that have to do with the density of crust involved?

____

____

____

____

____

e. Where in North America today is it possible to see (■ Figure 9.18) rocks that experienced substantial deformation as a result of the formation of Pangaea? You may want to examine Figures 9.17, above, and 9.20, on page 218. Name the present-day physiographic feature.

____

So you now see that subduction is closely related to the density of the rocks on either side of the convergent plate boundary. The denser plate is the one that subducts, provided it has oceanic crust. Continental crust, with its low density, does not subduct readily. Subducting continental crust would be a bit like trying to sink a cork in a bucket of water—it is just too light to go down.

WANDERING CONTINENTS

From the time of the first maps of South America and Africa, people have noticed the geometric fit of these two continents as if they were giant pieces of a jigsaw puzzle (Figure 9.20). Sir Francis Bacon, the English scientist and philosopher, observed this as early as

Figure 9.18

Physiographic Provinces of Much of North America

Shaded relief map of North America showing various mountain ranges and physiographic provinces. From MONROE/WICANDER, *The Changing Earth*, 5E. 2009 Brooks/Cole, a part of Cengage Learning, Inc. Reproduced by permission. www.cengage.com/permissions

1620 and speculated that the two continents had once been joined. In 1915, Alfred Wegener took the concept of moving continents much further, developing the hypothesis of *continental drift*. Wegener provided data to indicate that the fit is more than geometric; it is also geologic, with matching rocks, evidence of glaciation, and fossils. Since that time, geologists and geophysicists have gathered abundant data to indicate that the continents and the seafloor are and have been moving. With these data, they have established possible positions of the continents in the past.

As we saw in the previous question, at one time, almost all of continental material on Earth was together in one, large supercontinent known as **Pangaea** (Figure 9.17b). At other times, the northern continents of North America, Europe, and Asia made up a supercontinent called **Laurasia,** and the southern continents, including India, made up **Gondwana** or **Gondwanaland** (Figure 9.17a).

Paleomagnetism

An important part of the evidence for the positions of continents comes from **paleomagnetism,** ancient magnetism preserved in rocks. Recall that Earth's magnetic field has switched from a configuration similar to that of the present (**normal**), to a reversed field, and back again, repeatedly. As the seafloor spread, magnetic stripes formed in the seafloor, providing one type of evidence of how the continents moved. However, another type of paleomagnetic information is available to plot the positions of continents. Before we discuss this type of information, let's examine Earth's magnetic field in more detail.

Materials needed:

- Drafting compass
- cm ruler
- Paper
- Bar magnet
- Magnetic compass
- Protractor
- Rigid board (cardboard or poster board)
- Iron filings
- Thumb tacks

26. On the sheet of paper, draw a circle with a radius of 6 cm or larger. Position the magnet in the center of the circle and trace around it so that it can be repositioned if it is inadvertently moved. Note that the Earth has a dipole magnetic field similar to the magnetic field of a bar magnet.

a. Place the magnetic compass on the circle and note the direction of the north end of the compass. Draw this as an arrow on the circle, matching the needle's angle with the arrow's angle.

b. Move the compass to a new position on the circle and draw the new angle of the north arrow. Repeat until you have 20 to 30 arrows on the circle.

c. Move the compass outside the circle and draw additional arrows to get a complete picture of the magnetic field of your magnet.

d. Find where the arrows point directly toward the center of the circle and mark this place the north magnetic pole.

e. Similarly find where the arrows point directly away from the center and mark this place the south magnetic pole.

f. As an extra challenge, measure and label the angle of dip of the arrows with respect to the circle at 12 equally spaced positions around the circle. Your instructor may need to teach you how to measure dips relative to the circle that represents the Earth. When you are done, you should have a reasonable representation of Earth's magnetic field.

g. In the Northern Hemisphere of your "Earth" for this normal magnetic field example, are the dips up into the air (away from the circle) or down into the ground (toward the circle)? ________

h. Place the magnetic compass back on the circle and turn the magnet 180° clockwise or counterclockwise horizontally. What happened to the compass needle? How does the new direction compare to the old direction?

____

____

What does this turning of the magnet represent in terms of Earth's magnetic field?

____

i. Place your drawing on a rigid board and tape it down. To mark the correct placement of the bar magnet on the other side of the board, push a tack through each end of where your magnet was. Tape the magnet on the back of the board in line with the tack holes. Sprinkle iron filings over the front of the drawing and, with it as flat as possible on a firm surface, tap the board lightly until the filings arrange themselves to the magnetic field. What is the relationship between the filings and your drawn arrows?

____

The iron filing pattern is another representation of a magnetic field. When finished observing this magnetic field, carefully remove the magnet from the back without spilling the filings and then tip the filings onto another sheet of paper so you can gather them and return them to their original container.

j. What magnetic mineral might be present in rocks allowing them to become magnetized parallel to Earth's magnetic field? ________

It turns out that rocks can lock in the direction of the Earth's magnetic field when they form, not just whether it is normal or reversed, but also the angle of dip of the field. The angle of dip gives an estimate of the latitude but not the longitude of the rock's location. ■ Figure 9.19 illustrates this concept. Notice that although the continent in Figure 9.19b moved east and west, the rocks do not record that information. They only record its northward or southward movements. In the following exercises, we will use the ability of paleomagnetism to detect northerly and southerly movements of continents, supporting the idea that continents have moved in the past (**continental drift**).

27. The data in ■ Table 9.1 give the magnetic dip found in variously aged rocks at three locations, all on different continents. In order to keep things simple, we will only use normally magnetized rocks. As a rough estimate, assume the dip of the magnetism corresponds directly to the latitude at which the rock formed. If the magnetic dip is up, the rock formed in the Southern Hemisphere, and if the dip is down, the rock formed in the Northern Hemisphere. Vertical magnetic dips are at the poles and horizontal ones are at the equator.

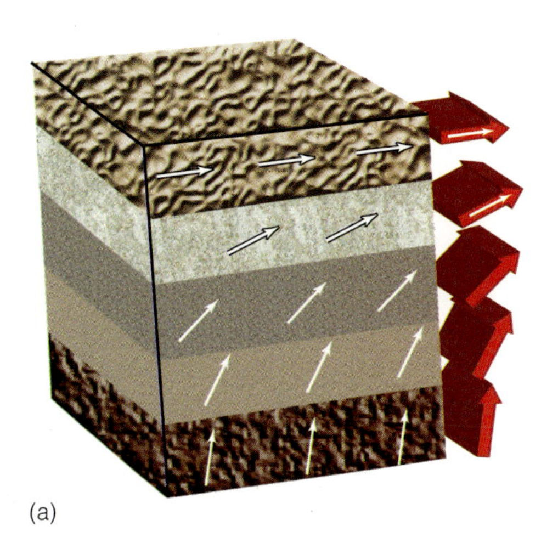

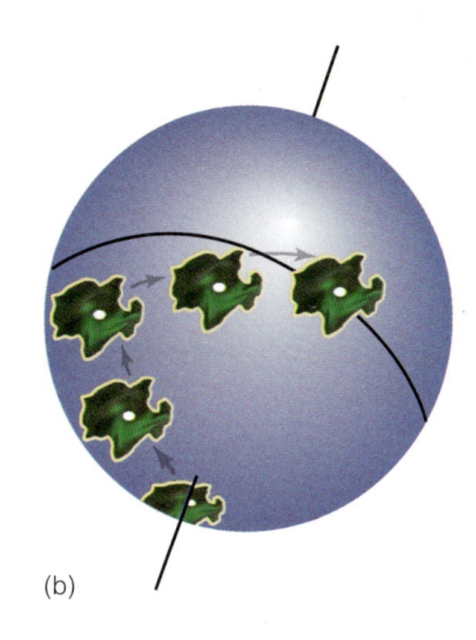

Figure 9.19

Magnetism and Wandering Continent

(a) As a continent moves from near the South Pole to the equator, rocks in one place on the continent develop paleomagnetism that preserves the magnetic field of the place the rocks were when they formed. Each rock in this sequence formed when Earth's magnetic field was normal. (b) A possible path of the continent that contained the rocks in (a). The white dot represents the location of the rock sequence in (a) the first continent position corresponds to the bottom layer in (a). The second one, to the second layer and so forth.

a. For each location and time, draw the line of latitude for that location on the appropriate map in ■ Figure 9.20 for the correct time and label the latitude with the letter of the location, A, B, or C.

b. For each location at each time in Table 9.1, only certain continents existed at the location's latitude. Write those continents down in Table 9.1 in the Possible Continents column.

c. After you have done this for each time and each location, you will find that for one or two locations only one continent contained all the appropriate latitudes at the correct times. You can already narrow down the continent for these locations. Write the continent's abbreviation in the last column in Table 9.1. For the other location(s), find one continent where the location within the continent has not shifted from one time to the next. List that continent in the table.

d. Once you have determined the continent, plot each location's possible approximate position on the map for the Permian (225 Ma, Figure 9.20a) with labels A, B, or C.

e. Could you narrow down each rock to a single continent at a single location?

Why or why not?

____

____

____

Table 9.1

Magnetic Dip for Three Locations

Locations A, B, and C are each on a different continent. Each location has three layers of rock. Paleomagnetism in the rocks indicates the dip of the magnetic field at the time they formed, which matches the latitude of the location. Rocks with an up dip formed in the Southern Hemisphere and those with a down dip, in the Northern Hemisphere. Follow instructions in Exercise 27.

Location	Age Myr	Magnetic Dip Angle in °	Possible Continents (Abbreviated)	Actual Continent
A	225	30 up		
	135	20 up		
	0	2 up		
B	225	50 up		
	135	25 up		
	0	20 down		
C	225	20 down		
	135	50 down		
	0	45 down		

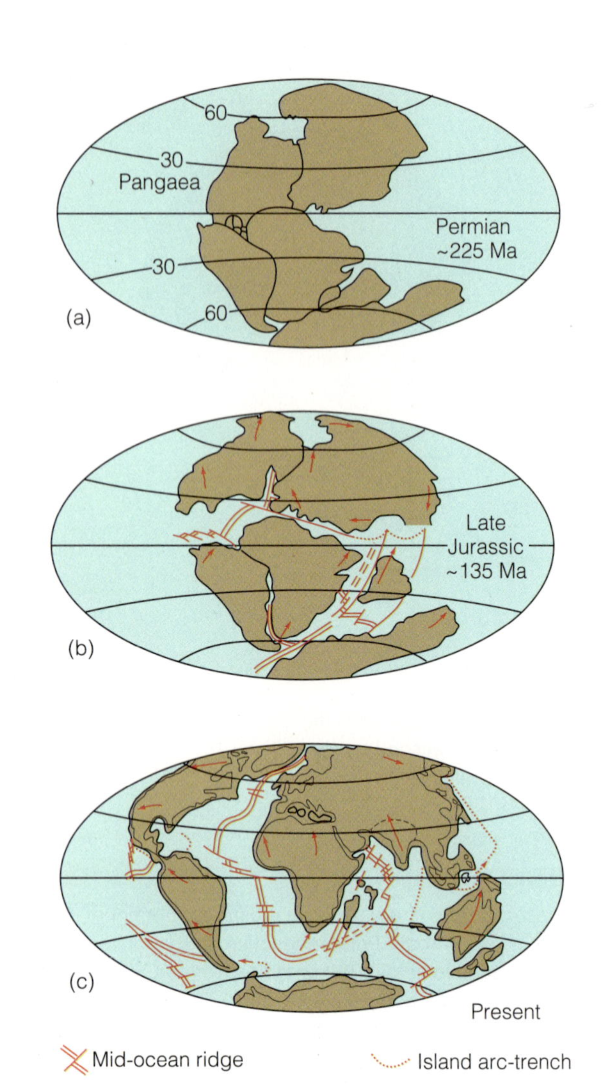

Figure 9.20

Reconstruction of the Breakup of Pangaea

Shown in three steps, a–c. Red arrows show plate motion directions. Ma = million years ago. Follow instructions in Exercise 27.

Continental Drift

Additional evidence for plate movements comes from rocks found on different continents and their depositional environments. Because some depositional environments are dependent on climate, the rocks can indicate something about the continent's position. Figure 9.21 shows the Late Permian configuration of continents and evidence from depositional environments.

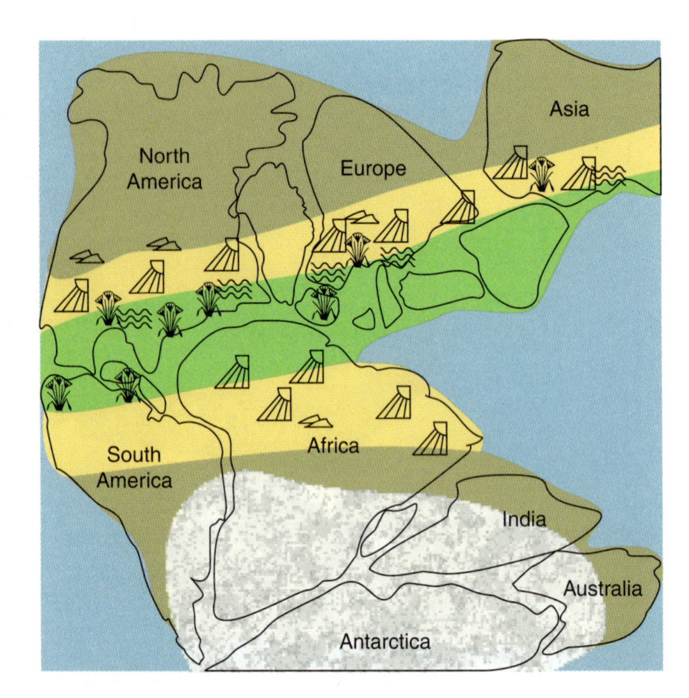

Environment	Rock Name	Sample #
Swamp		
Cross-bedded desert dunes		
Evaporite		
Warm shallow sea		
Glaciated		

Figure 9.21

Late Permian Paleogeography

Map showing climate and other conditions of the Late Permian Period determined from depositional environments of sedimentary rocks. Follow instructions in Exercise 28.

28. Your instructor will provide rocks matching some or all of the environments in ■ Figure 9.21.

a. On the key in the figure, write the rock name and sample numbers corresponding to each environment.

b. Draw in a likely position for the equator on the figure. Draw in a north arrow, and a likely location of the South Pole.

c. What do we call this Permian supercontinent? ____________________

Earthquakes and Seismology

OBJECTIVES

- **To understand earthquake hazards including liquefaction and tsunamis**
- **To understand the origins of earthquakes and seismic P and S waves**
- **To understand elastic rebound and seismic gaps**
- **To learn how to locate an earthquake using three seismograms**

Major and great *earthquakes* are some of the most devastating natural hazards[1] (■ Figure 10.1 and ■ Figure 10.2). They vibrate and shake the ground, sometimes so strongly that people, buildings, bridges, and overpasses cannot stand. Even if you have experienced an earthquake, do you know what causes some places to shake more than others do? You may not realize that a single earthquake has more than one type of vibration. How do *tsunamis* behave? How do seismologists detect earthquakes from remote locations? We will explore these topics in this lab, but first we introduce some terms.

- An **earthquake** is ground vibration resulting from a sudden release of energy in the Earth's crust by natural geologic phenomena such as faulting or volcanism.
- The destructive force of an earthquake is a series of wave movements called **seismic waves** that pass through and vibrate the ground.
- **Seismology** is the study of earthquakes and their waves.
- Ground-moving seismic waves also may generate a series of great ocean waves called **tsunamis.**
- A **seismograph** is an instrument (■ Figure 10.3) that marks the Earth's motion on a record called a **seismogram** (■ Figure 10.4).
- **Seismologists**—scientists who study earthquakes—use seismographs to detect earthquakes and seismograms to measure them.
- The **focus** of an earthquake is the place beneath the surface where faulting generates the earthquake; it is where the earthquake starts (■ Figure 10.5).
- The **epicenter** is the point on the land surface directly above the focus (Figure 10.5). A map can only show the epicenter, not the focus of an earthquake, so people in the media use the term *epicenter* more often, and it is more familiar to most people.

[1] A major earthquake is one with a magnitude greater than 7–8 on the moment magnitude scale and a great earthquake is 8–10.

EARTHQUAKE HAZARDS

The following hazards are a sampling, not a total list, that demonstrate some concepts we will investigate in this lab. Specific details, definitions, and explanations follow in later pages.

A magnitude 7.0 earthquake shook the Haitian capital of Port-au-Prince on January 12, 2010 (Figure 10.1). Movement along a transform fault separating the Gonave Microplate and the Caribbean Plate (Figure 10.1c) caused the earthquake. The quake caused casualties estimated at over 220,000 dead (at the time of writing) and massive damage due to its close proximity (15 miles) to this densely populated city.

(a) (b)

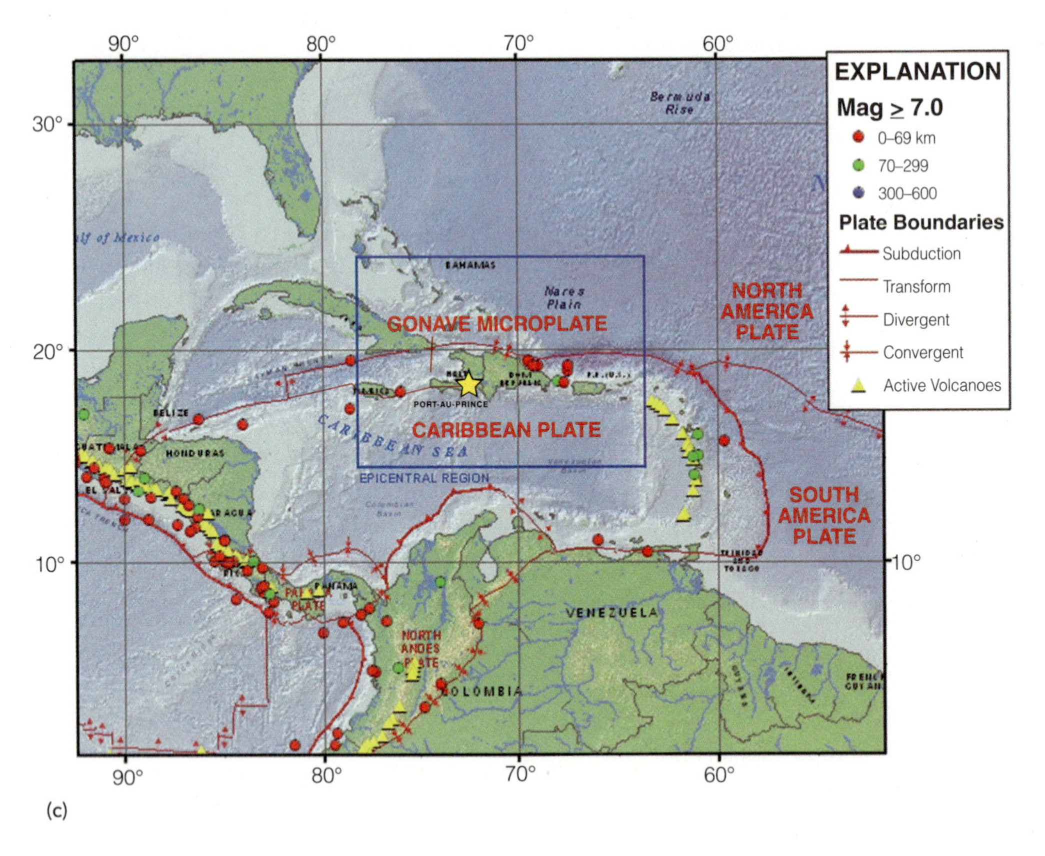

(c)

Figure 10.1

Earthquake in Haiti

(a) The magnitude 7 earthquake near the Haitian capital of Port-au-Prince severely damaged the Presidential palace. (b) Damaged buildings in Port-au-Prince. (c) Faults, plates, and microplates in the Caribbean. ☆ = epicenter of the January 12, 2010 earthquake.

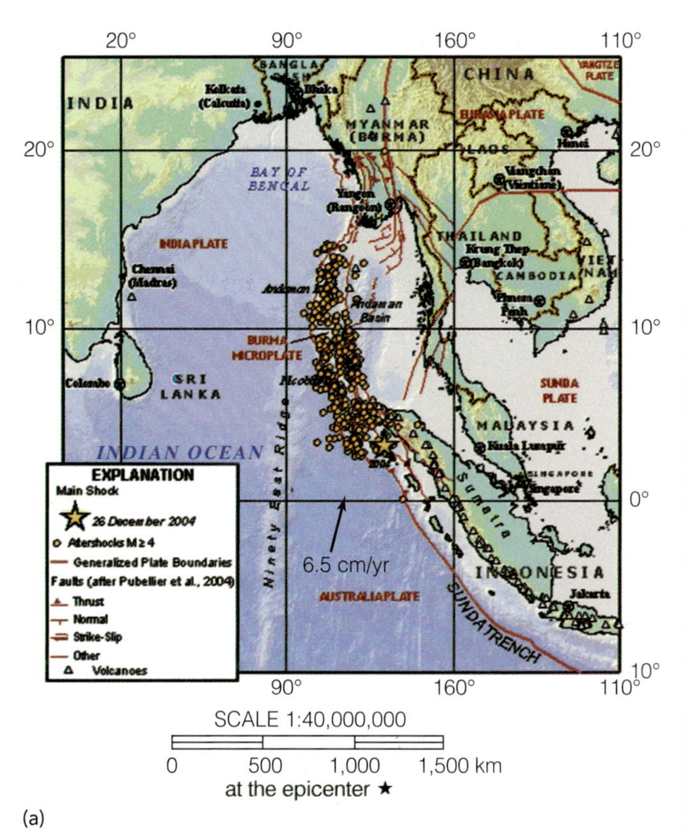

(a)

January 10, 2003

NASA

(b)

(c)

Figure 10.2

Earthquake Near Sumatra, Indonesia

(a) The 9.0 M_w earthquake near the island of Sumatra on December 26, 2004, generated a massive tsunami in the Indian Ocean that killed 230,000 people. The map shows the location of the earthquake (*) and its aftershocks of magnitude 4 and greater. (b) Satellite image of Lhoknga in Aceh Province, Sumatra, Indonesia before the Tsunami of December 2004 taken on January 10, 2003. (c) Similar satellite image of the same area on December 29, 2004, after the tsunami of December 26, 2004, destroyed Lhoknga. The white circular mosque in the city center was the only building not destroyed. (a) From MONROE/WICANDER, *The Changing Earth,* 5E. 2009 Brooks/Cole, a part of Cengage Learning, Inc. Reproduced by permission. www.cengage.com/permissions

On December 26, 2004, a magnitude 9 M_w earthquake struck near the island of Sumatra in Indonesia (Figure 10.2). This earthquake occurred on a *thrust fault* that is part of a subducting, obliquely-convergent plate boundary between the Indo-Australian Plate and a **microplate** (small plate) called the Burma Plate, west of the Sunda Plate. The movement caused an undersea rupture with a maximum length of 1,300 km (800 mi) parallel to the Sunda Trench and a width of over 100 km (62 mi). Most of the movement was concentrated in the southernmost 400 km (250 mi) of the rupture. The thrust fault broke the Earth's surface beneath the sea and the displacement generated an enormous tsunami. Waves reached 34.3 m (113′) high in one location, and affected the entire Indian Ocean basin, with casualties as far away as coastal Somalia, Africa. The earthquake and its resultant tsunami claimed at least 230,000 lives and left millions homeless and without a livelihood. *Aftershocks* extended for over 1,300 km (800 mi) north of the main shock (Figure 10.2a). The locations of the aftershocks helped seismologists determine the region of slip of the fault during the earthquake.

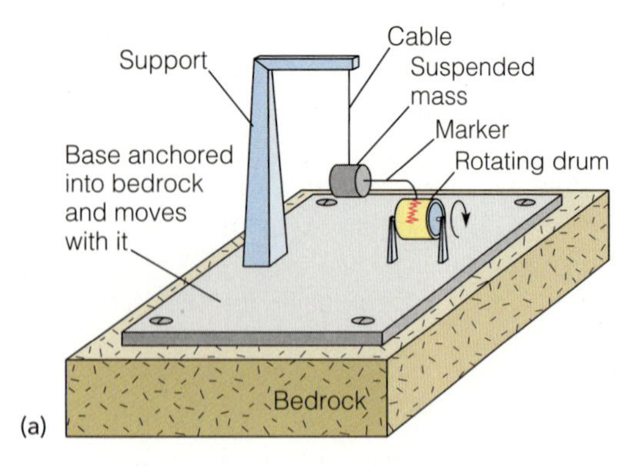

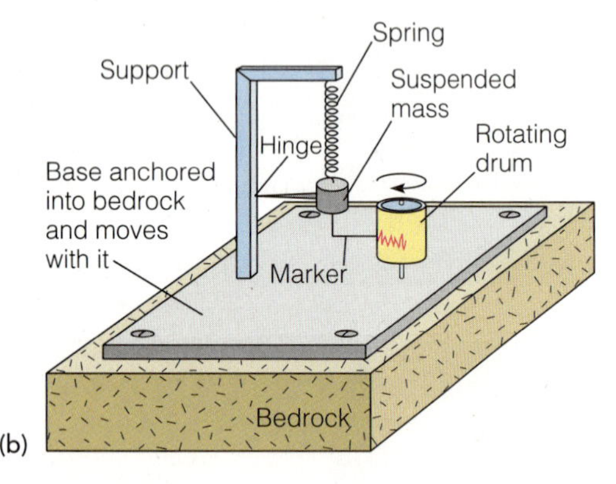

Figure 10.3

Seismographs

The seismographs pictured here work on the principle of inertia. The suspended mass remains stationary, while the other parts of the instrument and the Earth around it move during an earthquake. (a) A horizontal-motion seismograph records the horizontal vibrations that occur during an earthquake. (b) A vertical-motion seismograph records vertical vibrations.

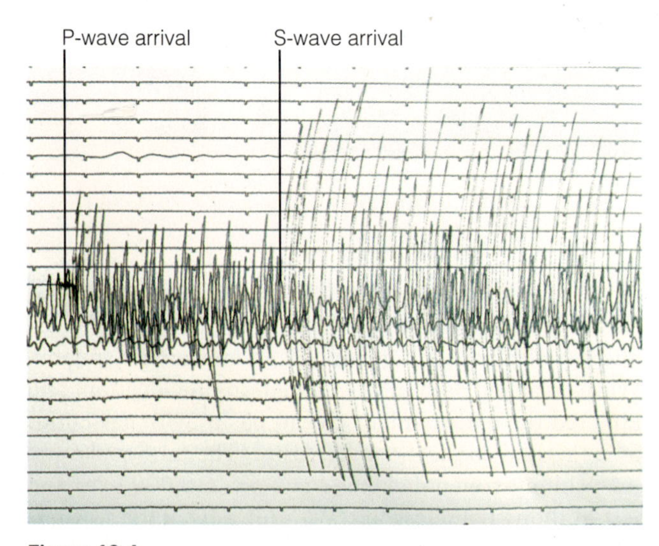

Figure 10.4

Seismogram

From the 1989 Loma Prieta earthquake in California.

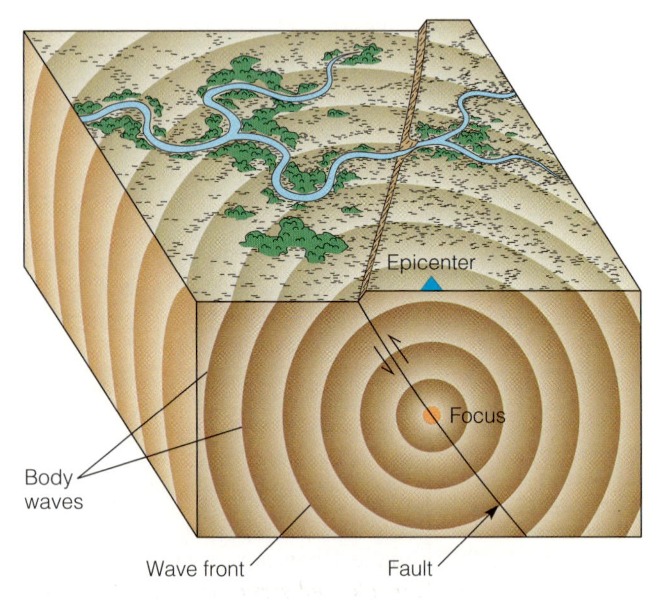

Figure 10.5

Focus, Epicenter, and Body Waves

When an earthquake occurs, the movement starts at the **focus,** generally because of rupture along a fault, and radiates outward in all directions as seismic wave fronts **(body waves).** Directly above the focus, on the ground surface, is the **epicenter** of the earthquake.

One of the largest earthquakes to occur historically within North America, magnitude 9.2 M_w,[2] was similar to the Sumatra event and occurred at Anchorage, Alaska, in 1964. Almost 3 km of coastal bluffs in Anchorage gave way due to a combination of *liquefaction* and landslides. The resulting tsunami hit coastal areas, drowning people as far south as Crescent City, California, where the third wave washed 500 meters inland. In a matter of hours, a tsunami can move across the ocean from where it started (Figure 10.10). Many of the approximately 130 deaths resulted from the large tsunami generated by the sudden upward shift of the seafloor. For large earthquakes in highly populated areas, as in the Sumatra earthquake, the death toll and destruction of property are likely to be much higher.

The famous 1906 earthquake in San Francisco, with an estimated Richter magnitude of 8.2, and moment magnitude 7.9 M_w, released substantially less energy than the Anchorage earthquake. Near San Francisco, a northern 430-km-long portion of the San Andreas Fault (Figure 9.10, p. 207) ruptured and moved as much as 6 m (20 ft). In San Francisco, the death toll from the earthquake was estimated at up to 2,500. Liquefaction caused intensification of the earthquake waves,

[2] USGS reports this number; some papers initially reported a magnitude of 8.4, and later upgraded to 8.5 to 8.6.

damaging structures, especially those built on former marshland. A fire that raged out of control for three days afterward caused 10 times more damage than the actual earthquake (■ Figure 10.6). Earthquakes in the San Francisco Bay Area, including the 6.9 M_w 1989 Loma Prieta quake, continue to endanger residents and challenge government services.

These examples illustrate the major hazards associated with most large earthquakes. The following explains these hazards in more detail:

- **Ground shaking** is a direct hazard of an earthquake. Collapsing buildings (Figure 10.1) and highway structures, and disruption of utilities, access to food and water are the immediate results of this hazard.
- **Aftershocks** are smaller earthquakes that occur after an earthquake. They can cause further damage to already weakened structures. The size and frequency of aftershocks diminish with time after the main shock (■ Figure 10.7). The aftershocks from the December 2004 Sumatra earthquake occurred over a large area (Figure 10.2), reaching moment magnitudes up to 7.1, a sizable earthquake itself.
- **Fire** is commonly associated with earthquakes (Figure 10.6), because during the earthquake electrical short circuits or broken furnaces and stoves can ignite flammable materials such as wood, paper, and gas. Earthquakes often disrupt utilities and transportation, making firefighting more difficult.
- In areas with steep slopes, earthquakes may trigger **landslides** (see Lab 11), on land or under the sea, where rock or debris moves rapidly down the slope in response to gravity.
- **Liquefaction** occurs when an earthquake vigorously shakes water-saturated sediment, which then behaves like quicksand. Resulting compaction of some of the sediment causes displacement of pore water (■ Figure 10.8), which flows to the surface, carrying sand with it. Water and sediment erupt in a kind of miniature volcano called a **sand blow.** Buildings and structures on land where liquefaction occurs are much more likely to sustain damage or to collapse than in areas of bedrock (■ Figure 10.9).
- Although a **tsunami** has nothing to do with tides, its popular name is a *tidal wave*. It is a series of large ocean waves that a sudden disturbance of the seafloor generates (Figures 10.2, 10.13). Tsunamis travel through the ocean in all directions from the *epicenter* (■ Figure 10.10). In the open ocean, even large tsunami waves are not very high (Figure 10.17, on p. 232) and are not easily noticed; but near shore they may build up to heights of tens of meters and can be very hazardous, as in the case of the tsunami that the Sumatra earthquake generated (Figure 10.2). A tsunami is actually multiple waves in a *wave train,* seen along a shore as a repeated retreat

Figure 10.6

San Francisco

From William A. Coulter's panoramic painting of the fire and maritime rescue after the great 1906 earthquake in San Francisco. Some artistic license was taken in this work in the positioning of cityscape features. The evacuation of 30,000 people was the largest of its kind in the United States.

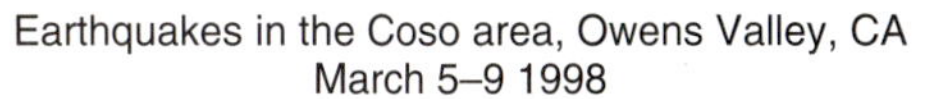

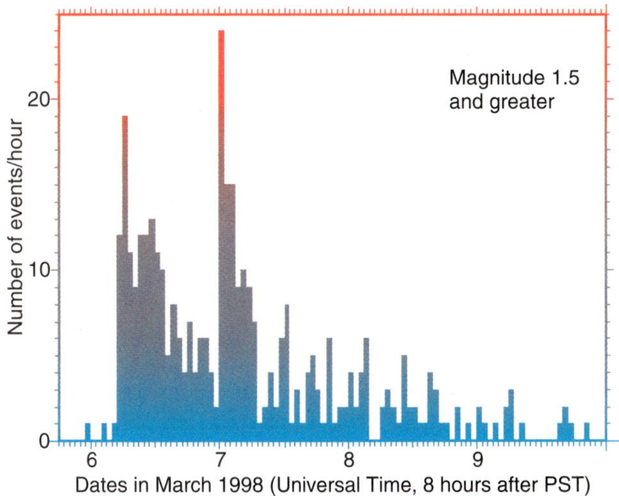

Figure 10.7

Aftershocks

Aftershock series after two earthquakes in the Coso area, Owens Valley, California. A magnitude 5.2 earthquake occurred on the night of March 5, 1998, and a 5.0 on the evening of March 6. Aftershocks from these quakes show the typical decline in frequency after the larger quake. PST = Pacific Standard Time.

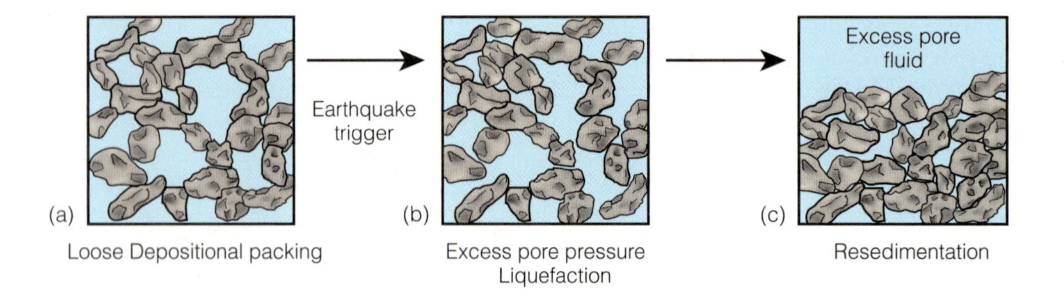

Figure 10.8

Liquefaction

An earthquake causes liquefaction: (a) starting with loose, recently deposited sediment or artificial fill; (b) the earthquake causes shearing of the loose grain–grain contacts and the pore–water pressure increases. As a consequence, material and structures above shift about until hydraulic fracturing allows water and sediment to escape to the surface; and (c) *Pore* spaces between grains of sand decrease because of compaction.

Figure 10.9

Ground Motion on Bedrock versus Saturated Sediment

Ground motion increases, both in strength and duration, on water-saturated and unconsolidated sediments. Seismic wave amplitudes progressively increase in less-consolidated and more-saturated sediments compared to bedrock. Buildings and other construction on weaker sediment are more prone to structural damage.

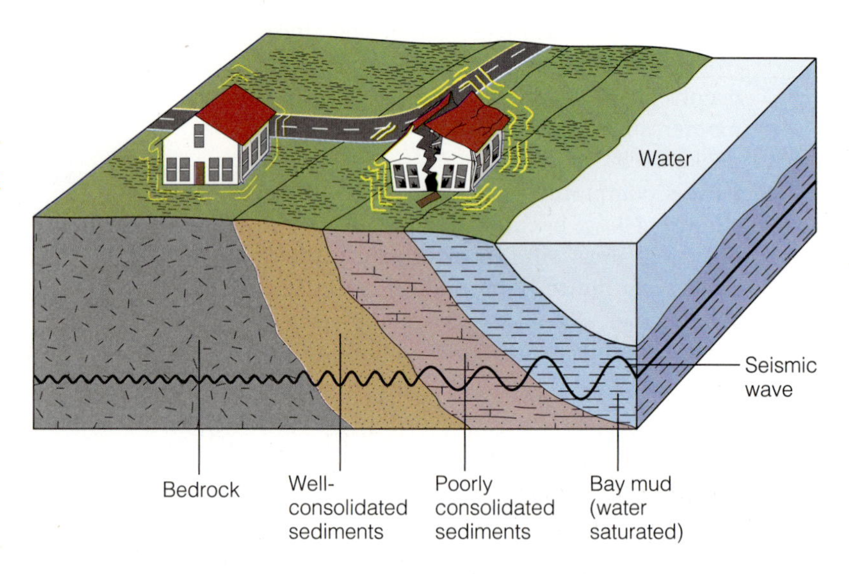

and advance of the ocean. A **wave train** is any succession of equally spaced waves arising from the same source, having the same characteristics, and propagating along the same path. Surges of the sea during a tsunami may occur from 5 minutes to hours apart. The first motion observed on shore may be either advance or retreat of the water. If the sea retreats first, sometimes the empty seabed attracts people on the beach who do not realize that a dangerous wave of high water will follow. Waves that are hours apart are likely to have reflected off a distant shoreline and returned.

Earthquake Hazard Experiment

Read the entire experiment before you start.

Materials needed:

- Two identical pans with 3- to 4-in deep, dry sand
- Something to keep the pans from sliding around, such as a nonskid surface, clamp, or tape
- An earthquake simulation system, such as a cart you can vibrate, or some other method of creating vibrations
- Bricks to simulate buildings
- Graduated beaker

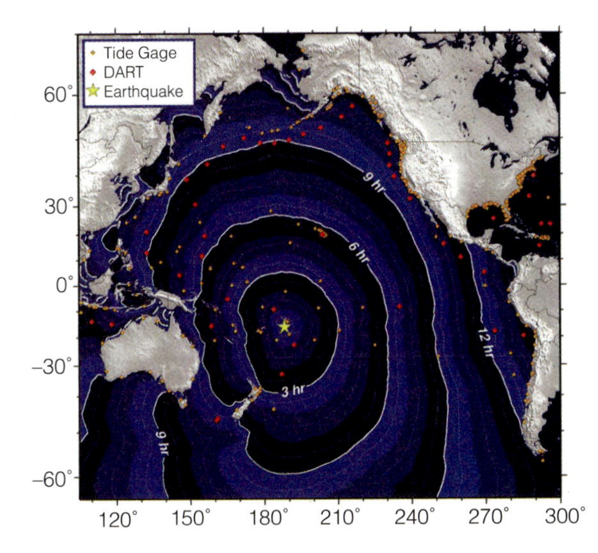

Figure 10.10

Tsunami Travel Times from Samoan Earthquake

Tsunami travel times map calculated for the Pacific Ocean, providing tsunami warnings for an earthquake of magnitude 8.3 that occurred in the area near the Samoa Islands on September 29, 2009. This tsunami killed hundreds of people in the surrounding islands. The highest recorded wave was 14 meters (46') high on Samoa. Orange diamonds are tide gages and red diamonds are DART (Deep-ocean Assessment and Reporting of Tsunamis) buoys.

- Water
- Paper, pen, and tape
- Either a clock, two stopwatches, or two watches

Your lab instructor will show you how to create vibrations to simulate reproducible earthquakes. Your instructor will also indicate how much water to add to Pan 2 after each set of three simulated earthquakes. Each addition of water should be about one-third the amount you would need to saturate the sand in your pan completely, so this volume depends on the pan size and sand volume. **Saturation** occurs when the sand is evenly wet and so wet that water will run off the surface if you tilt the surface.

1. Work in groups of at least four. Prepare the experiment:

a. Label the pans of sand as Pan 1 and Pan 2.

b. Write down how much water to add to Pan 2 at the beginning and after every set of three "earthquakes." ________ ml. Add this amount to Pan 2 now.

c. Compact the surface of the sand and smooth it flat. *Do this between each experiment.* This is a big source of variability, so try to keep the sand as consistently packed as possible.

d. Put identical "buildings" (bricks standing on end) on the sand in each pan. Set up the geometry of the pans so each one receives the same amount of motion from the same earthquake. (You may want to practice making consistent, reasonably sized earthquakes. If your bricks toppled in less than a couple seconds, make your earthquake smaller; if they seem to take a very long time to topple, make your earthquakes larger.)

e. Tape paper to the vibrating surface of the simulator. Choose one person to act as the seismograph (Figure 10.3) and record the earthquake motion. Practice this by holding the point of a pen down on the paper steadily and gently while the "earthquake" vibrates under it. Like a seismogram, the resulting scribbles should record the amount of ground motion each earthquake produces.

f. Assign one person to each pan to measure the "building" toppling time. One person should be sure to observe and note the behavior of the sand and water during the "earthquakes"; if your group is large enough, assign this task to another individual.

2. You will conduct three or four sets of three earthquake simulations for a total of 9 or 12 episodes. You will time and average the sets. Pan 1 will remain dry; only Pan 2 will receive more water after each set of three. Start the experiment.

a. Vibrate both pans at the same time with the same "earthquake." Time how long it takes for each building to topple from the start of the shaking. Record the times for each building to fall for Earthquake 1 in ■ Table 10.1.

b. Determine the earthquake's ground motion by measuring the length of the scribbles made on the paper. Record this in Table 10.1.

c. Remove the buildings, smooth the sand, and replace the buildings in their upright position.

Table 10.1

Simulated Earthquake Data

Relationship between the amount of water in sediment and the time it takes for a "building" to topple during a simulated earthquake. See instructions in Exercises 1–3.

"Earthquake" #	Earthquake's Ground Motion (mm)	Pan 1		Pan 2		
		Time (sec)	Average Time for 3 Quakes (sec)	Water Added So Far (ml)	Time (sec)	Average Time for 3 Quakes (sec)
1						
2						
3						
4						
5						
6						
7						
8						
9						
10						
11						
12						

d. Repeat steps **a–c** two more times with the same size earthquakes, recording the answers for Earthquakes 2 and 3. If earthquakes have ground motion different by more than 0.5 cm, throw out the data and repeat the experiment.

e. Remove the buildings. Use the graduated beaker to add the amount of water to Pan 2 that you recorded in **1b**, spreading it evenly over the surface. Record the amount of water added so far. Keep the sand in Pan 1 dry. Smooth and pack the sand and replace the buildings.

f. Generate, measure, and record three earthquakes with this amount of water in Table 10.1. Remember to generate similar earthquakes throughout the series of experiments. Continue adding water to Pan 2 and making earthquake measurements for both pans in sets of three until you have saturated the sand in Pan 2 and made three measurements for the saturated sand. Depending on the exact amount of water needed to saturate the sand, you may need to generate a total of 9 or 12 earthquakes.

3. Summarize your results as follows:

a. Average each batch of three toppling times and record them in Table 10.1.

b. Describe what happened to or in the sand during the vibrations.

____

____

c. On a separate sheet of paper, list and describe hazards or processes that were occurring during the experiments.

d. Make a graph using the water content of the sand along the horizontal (x) axis in ■ Figure 10.11 and the time it takes for a building to topple along the vertical (y) axis. Label all of the axes of your graph and write a title that provides additional information for the graph.

e. How do the behaviors of dry, damp, and saturated sand differ?

____

____

____

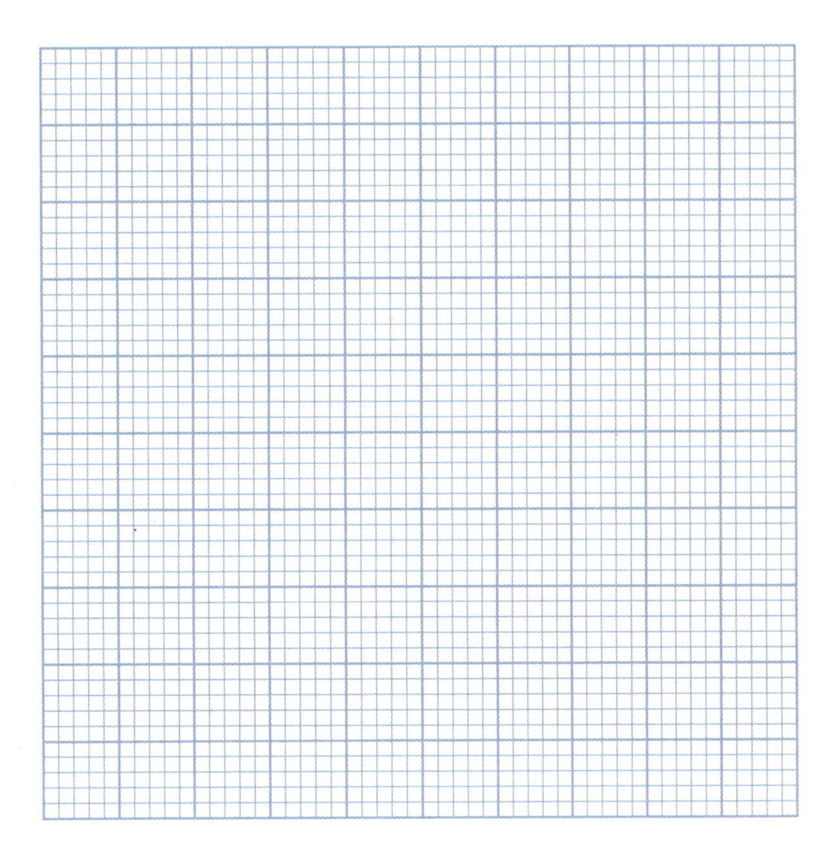

Figure 10.11

Graph Paper

For the earthquake hazard experiment in Exercises 1–3.

f. What was the purpose of having one pan in which you did not add water or change anything during the experiment?

____

____

How did this help you establish clear-cut results? Did this pan teach you anything extra about the experiment? If so, what?

____

____

____

____

g. What are the sources of variability in your experiment? What could you do to improve the results? Add these answers to the sheet of paper you used in part **c**.

Intensity

Ground motion during an earthquake varies from place to place. This is not just the effect of being closer or farther from the epicenter, although this effect is important, but also the influence of liquefaction and other details of the local geology. Seismologists measure earthquake ground-motion intensity using a scale called the **modified Mercalli scale.** For example, four values of the scale given by the USGS are:

> III. Felt quite noticeably by persons indoors, especially on upper floors of buildings. Many people do not recognize it as an earthquake. Standing motor cars may rock slightly. Vibrations similar to the passing of a truck. Duration estimated.
>
> VI. Felt by all, many frightened. Some heavy furniture moved; a few instances of fallen plaster. Damage slight.
>
> IX. Damage considerable in specially designed structures; well-designed frame structures thrown out of plumb. Damage great in substantial buildings, with partial collapse. Buildings shifted off foundations.

XII. Damage total. Lines of sight and level are distorted. Objects thrown into the air.

The complete scale is easily available on the Internet.

Studies using the modified Mercalli scale in connection with geologic mapping have provided insight into the effects of the subsurface and liquefaction on earthquake movement. ■ Figure 10.12 shows the results of one such study. On the geologic map in Figure 10.12a, thicker and thinner alluvium are mapped as well as bay mud and bedrock. **Alluvium** is unconsolidated sediment comparatively young and recently deposited by running water.

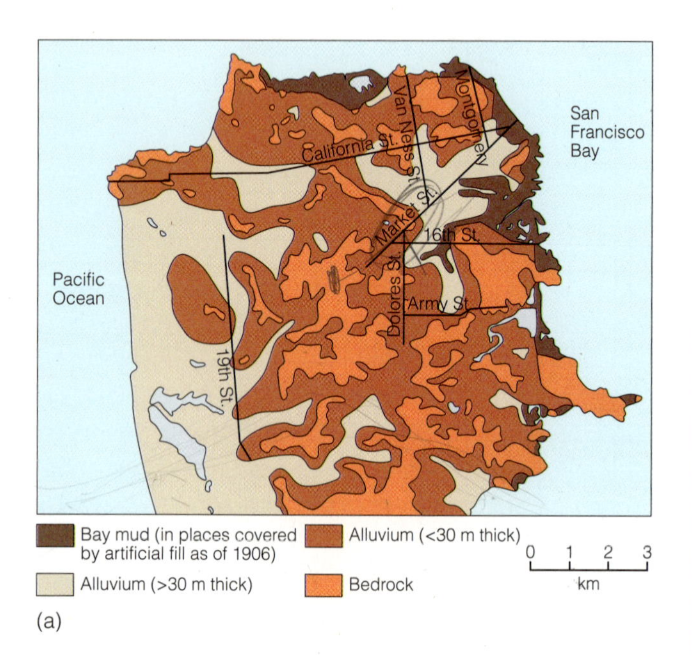

(a)

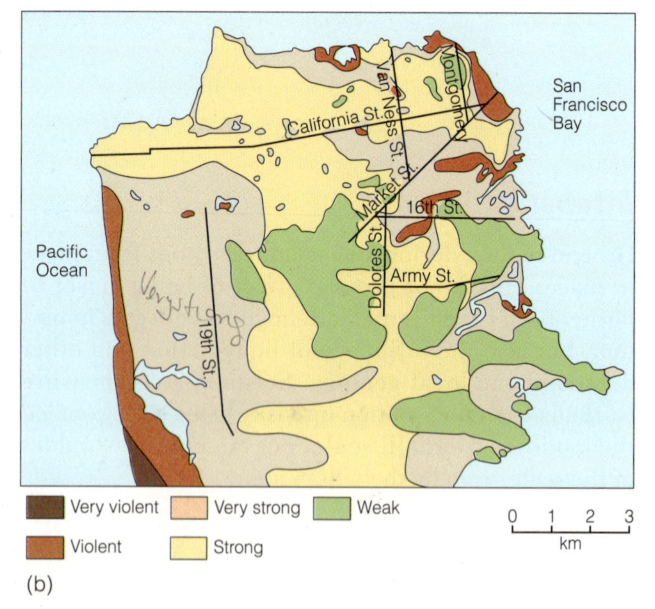

(b)

Figure 10.12

Geologic and Ground Motion Hazard Map of San Francisco

(a) Geology of San Francisco, California. (b) Map of ground motion hazard compiled from data from the 1906 San Francisco earthquake. From MONROE/WICANDER, *The Changing Earth*, 5E. 2009 Brooks/Cole, a part of Cengage Learning, Inc. Reproduced by permission. www.cengage.com/permissions.

4. Use Figure 10.12 to help you answer the following questions:

a. What type of material underlies the location 1 km southeast of the southeast end of Market Street?

Bedrock

What type of ground shaking did this location experience during the 1906 earthquake?

Very Strong

b. What type of material underlies the northeastern end of Market Street?

Bay mud

What type of ground shaking did this location experience during the earthquake?

Violent

c. Choose five or more places for *each* type of surface material in Figure 10.12a, compare the type of ground motion for each material in Figure 10.12b, and then summarize the motion types you found for each material:

- Bay mud Montgomery. Strong
- Alluvium >30 m 19th St Very strong.
- Alluvium <30 m California St Strong
- Bedrock Dolores St Strong

d. Explain these results.

Depends on the type of rock

e. The San Andreas Fault lies west of San Francisco and heads northwest offshore just south of the area shown in the maps. In the southeast corner of the maps of

San Francisco, what type of material occurs underground?

____

What type of ground shaking did this location experience during the earthquake?

____

Why do you think the type of ground shaking here doesn't follow the pattern established in part **c**?

____

____

f. Some of the worst quake damage in San Francisco during the 1906 and the 1989 earthquakes occurred in the Marina District, some distance west of Van Ness St. Mark where you think this district is on the map in Figure 10.12a with the label MD. Then find an area in San Francisco you would consider residing in, with the most stable ground, and mark it on the same map with the label SG.

Tsunamis

Even though famous paintings of tsunamis show curling breakers, tsunami geometry is generally less like surfing waves that curl and form tubes, than they are like the small, shallow waves that lap up at your feet on the beach. Wave geometry is important for many reasons and helps us to understand their movement and power. **Wavelength** is the distance between two equivalent parts of a wave such as two wave crests (tops) or two wave troughs (bottoms). Waves that have moderate to short wavelengths can move through the ocean without touching the bottom and are called **deep-water waves.** The motion in a wave is active to a depth of ½ of its wavelength. However, tsunamis have such extremely long wavelengths that their motion touches the bottom everywhere. That puts them, ironically, into a category called **shallow-water waves,** whose wavelength is more than twice the depth of the water. In other words, the oceans are like a shallow bathtub to a tsunami due to its great size. Shallow-water waves have a back and forth motion rather than the rolling motion that makes curling breakers (■ Figure 10.13).

Shallow-water waves, including tsunamis, travel at speeds that are proportional to the square root of the water depth. The equation to express this relationship is:

$$V^2 = (9.8\text{m/s}^2)(depth)$$

where *V* is the velocity of the tsunami in meters per second. ■ Figure 10.14 shows a graph of this relationship.

Waves break when their wave height reaches 1/7 of their wavelength. **Wave height** is the vertical separation between the trough and the crest of a wave. Tsunamis have extremely long wavelengths compared to their wave heights, even when the waves are extremely high, so they swell upward and downward and do not break. However, waves on top of tsunamis may break if caused by other phenomena, such as wind and obstructions. In addition, the geometry of the seafloor may create a situation where tsunami waves can break near shore, but this will not be the usual situation.

5. Assume the depth of the Indian Ocean is 3,890 m.

a. Use the graph in Figure 10.14 or the equation above to calculate the speed in m/s of a tsunami where the ocean is this deep. 195.3

b. How fast is this in km/h? 702.9

c. How fast is this in mi/h? (1 km = 0.621 mi) 436.5

d. At that speed, how long would it take the tsunami to travel from the main epicenter at the star in Figure 10.2a to Sri Lanka? 1.79 hrs Did the model waves in ■ Figure 10.15 arrive at the time you just predicted? Yes

6. Examine the satellite data for the tsunami in ■ Figure 10.16.

a. The caption mentions that the colors in Figure 10.16a relate to wave heights calculated in a model and the blue in b shows a model-generated graph. What is a model in this context? (Refer to Appendix, pp. 447–449)

A simplification or representation of some aspect of nature.

____

b. Compare the actual satellite measurements in Figure 10.16b with the model estimates. How good was the model at predicting the wave heights and wavelengths?

really good

____

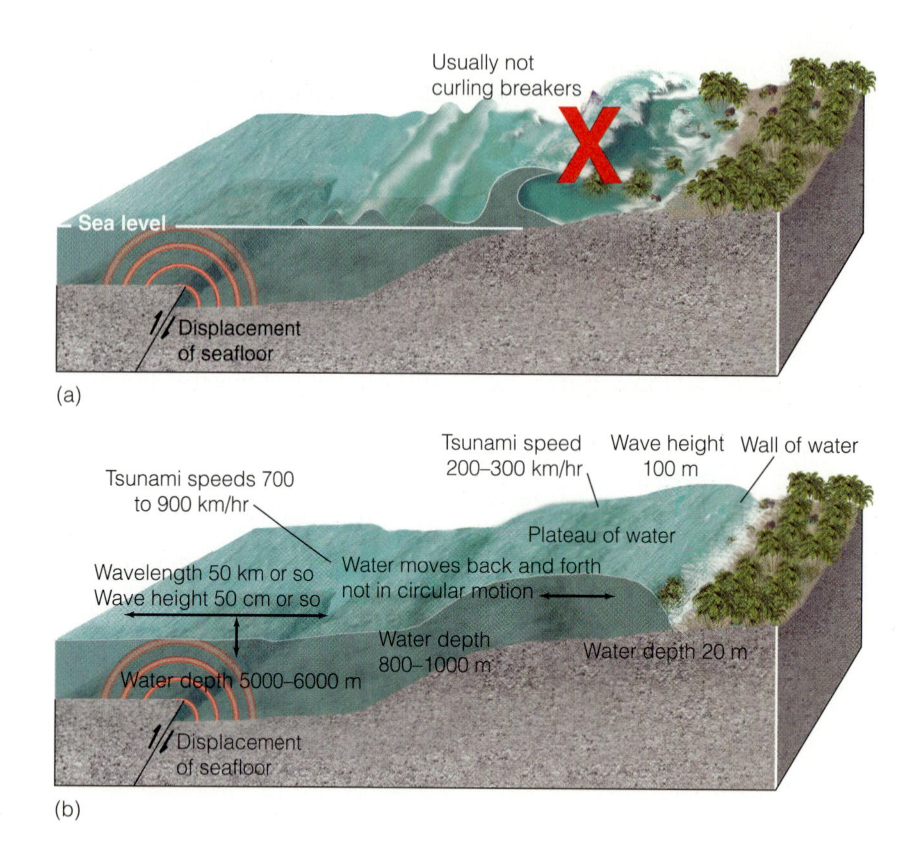

Figure 10.13

Behavior of Tsunamis

(a) Curling breakers as one might typically see at a beach are not likely for tsunamis. (b) The geometry of tsunamis makes them more likely to rise up and inundate an area. Tsunamis in deep water have low height and long wavelength and high speed. As the waves move into shallow water, the wavelength decreases, wave height increases, and the wave inundates the land. **Runup** is the area above sea level that the wave inundates.

c. What was the highest measurement of the tsunami as made from the satellite in the open ocean? 75 cm

d. How would a wave of this height influence a boat or ship in the open ocean?

Nothing

e. Count the number of wave crests in the model from latitude –6° to 4°. 2
Convert the latitude degrees to kilometers using 111 km/°. 1110 Then calculate the wavelength. 555 Km

f. Next calculate the ratio between the wave height and the wavelength of the tsunami in the open ocean. 1 : 740 000

1.35×10^4

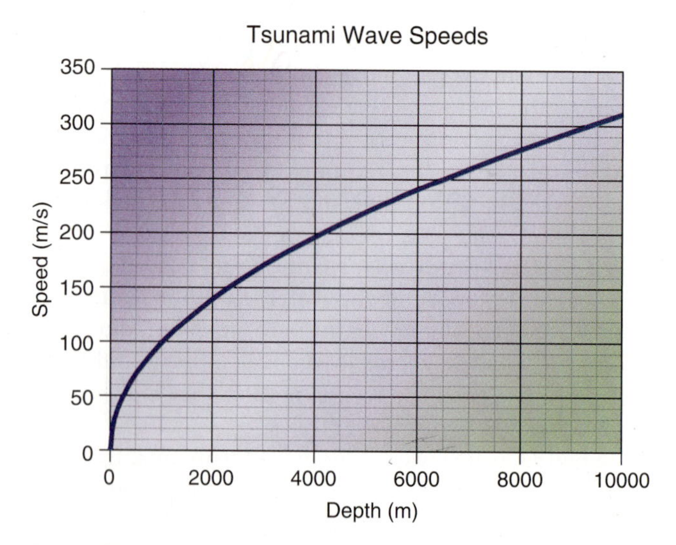

Figure 10.14

Tsunami Wave Speeds

Tsunamis travel at speeds dependent on the water depth.

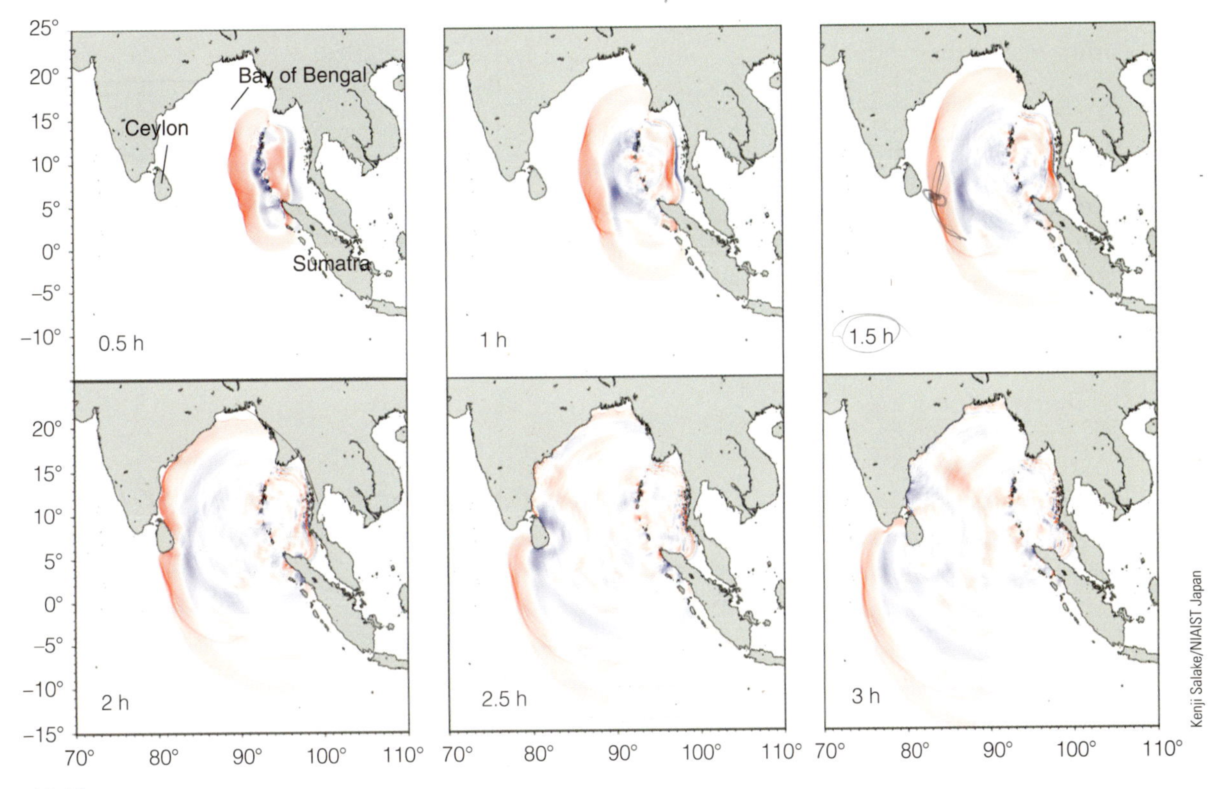

Kenji Salake/NIAIST Japan

Figure 10.15

Computer-Simulated Tsunami in the Indian Ocean

The waves were modeled after a December 26, 2004, tsunami generated near Sumatra. The maps show the wave locations at half-hour increments. Notice that at 2.5 h the waves are reflecting (bouncing) off India and Ceylon so that additional waves hit the coasts around the Bay of Bengal about an hour or two after the first waves arrived.

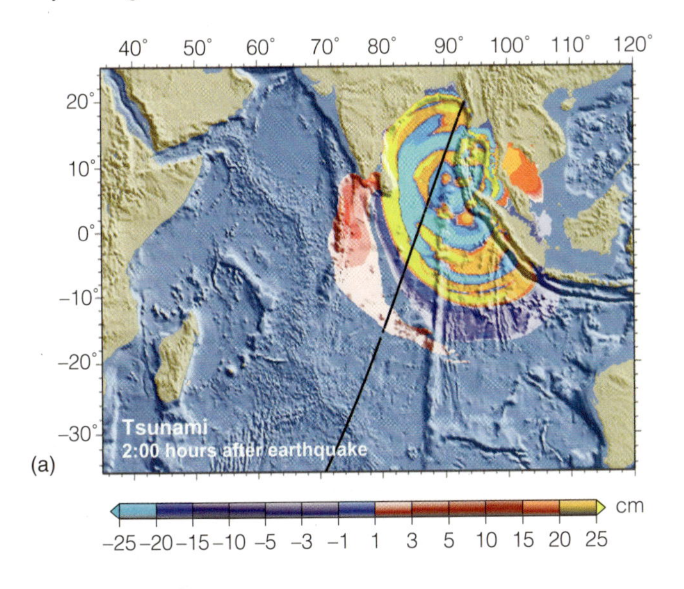

Figure 10.16

Measured and Modeled Amplitude of a Tsunami in the Indian Ocean

Tsunami in the Indian Ocean 2 hours after the initial earthquake hit the region southeast of Sumatra on December 26, 2004.
(a) Map of computer-simulated tsunami in the Indian Ocean. The different colors show the computer-modeled height (*amplitude*) of the sea surface in centimeters relative to mean sea level.
(b) Measured (black) and modeled (blue) amplitude of the tsunami. The measurements are from a radar altimeter on board the Jason satellite along the black line track in (a) compared to the sea surface height from the same track 20–30 days before the earthquake.

Is this greater than 1/7th? Yes

Did the wave break? Yes

g. Assuming the wavelength near shore is about 20 km, calculate the ratio between the wave height and wavelength near shore. Use a wave height of 34.3 m (the highest of the Sumatra tsunami waves). Show your calculations.

20 km : 34.3 m -> 583 : 1

Did the wave break here? Yes

Although these calculations clearly show that tsunami wave geometry is far from a breaking-wave geometry, details of the shape of the sea bottom near the coast may change the shape of the wave. Generally, tsunami waves will not break. One of the deadly features of tsunamis is that the water is more like a plateau than a curling wave and that means the water level stays high for a substantial amount of time.

You may have noticed a big discrepancy in the wave height (amplitude) between the open ocean waves measured by satellite (less than a meter) and the actual maximum measured wave height for the Sumatra tsunami (34.3 m). This is partly due to something called **shoaling amplification.** This occurs as the tsunami wave approaches shallower water. The front of the wave slows down, as you would predict from the velocity equation, but the back of the wave is still moving faster. The water, then, piles up into a taller wave (■ Figure 10.17).

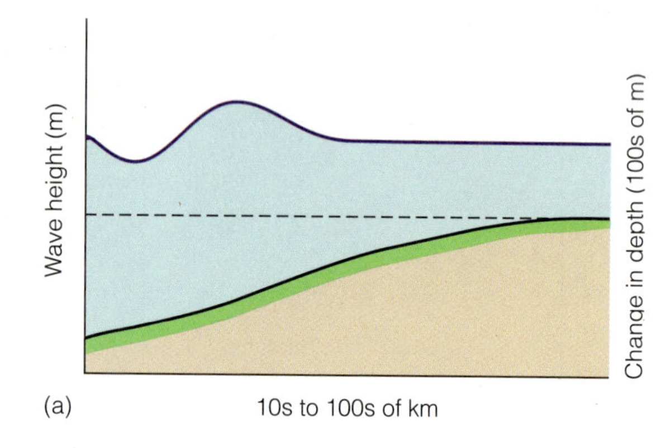

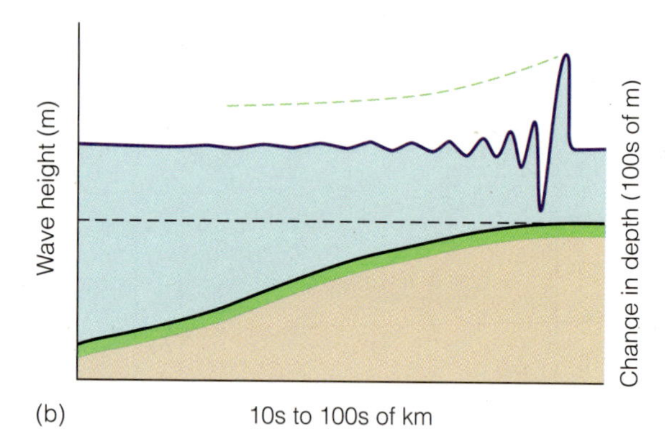

Figure 10.17

Shoaling Amplification of a Tsunami

As tsunamis approach shallower water (called shoaling) the wave in front slows down allowing the back of the wave to catch up somewhat. This increases the amplitude (height) of the wave and shortens the wavelength. This is how a wave that is less than 1 m height in the open ocean can become a wave 10s of m high when it comes on shore.

7. How many times greater is the highest recorded level of the tsunami on land compared to the amplitude of the tsunami you determined from the Jason satellite?

45.7 Explain how this is possible?

Front of wave moves slower, the back moves faster

How would this change the level of danger from the tsunami in the open ocean compared to at the coast?

it most dangerous at the coast

8. Examine the waves in Figure 10.15, especially in the Bay of Bengal from 2 to 3 hours after the earthquake.

a. Once the first set of tsunami waves pass by, has all danger from the tsunami subsided? NO

b. What is happening to the waves that explains your answer?

they are going back and forth.

THE ORIGIN OF EARTHQUAKES

Each of the earthquakes discussed earlier occurred near a plate boundary. The movement of plates past each other is what ultimately causes the vast majority of earthquakes. The forces on the rocks at the boundaries make them bend and flex; this continues until the forces exceed either the strength of the rocks or the friction that normally prevents sliding on fractures. At this point, the rocks break and slip along a fault. The snapping-back action as the bent rocks return to their original shape generates the earthquake (■ Figure 10.18). This snapping back is known as **elastic rebound.** Waves of vibrational energy originate from the point where the break starts; these waves are the earthquake—the shaking of the ground—and are called **seismic waves.**

Seismic Gaps

Plate-boundary faults occur as plates move; however, other acive faults can generate earthquakes because faults tend to stick for a while and then slip. The time of slipping is the time of the earthquake and when elastic rebound occurs. The time while the fault is stuck is known as a **seismic gap** because no earthquakes occur during this time. It is also the time when elastic strain energy is building up. The more elastic energy has accumulated, the larger the earthquake is likely to be. Seismic gaps are used to predict future large earthquakes—an idea called **seismic gap theory.** Seismologists generally cannot predict an earthquake in the short term, but seismic gap theory does allow prediction of earthquakes on the order of tens of years. This is very helpful for engineering and city planning.

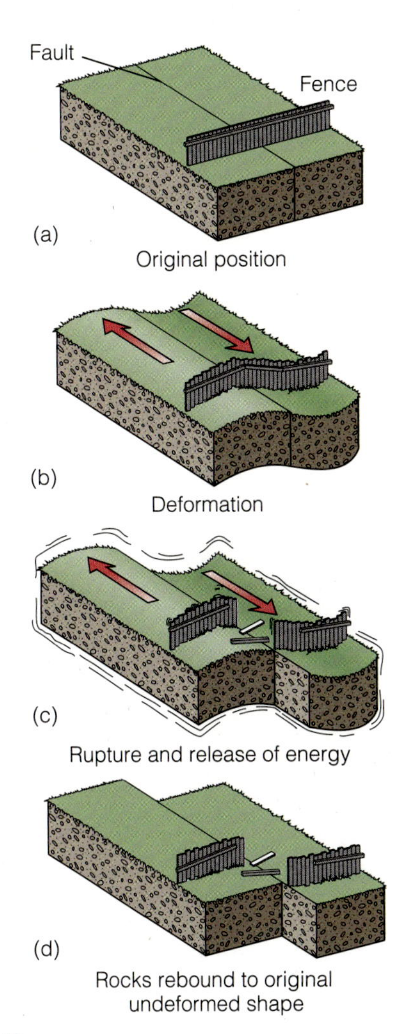

Figure 10.18

Elastic Rebound

Elastic rebound theory states that earthquakes occur when accumulated stress is suddenly released and rocks snap back elastically to their original shape. (a) Before deformation. (b) Moving blocks deform the rocks near a fault so the rocks bend out of shape. (c) When the rocks finally break, the earthquake occurs because the rocks snap back to their original, unbent shape. This diagram is a snapshot of the rocks in the process of rebounding. (d) After the earthquake, the rocks have returned to their undeformed shape, except for the area of rupture of the fault.

Stick-Slip Experiment

Materials needed:

- Wooden block with a rubber band or spring attached through an eyelet or with a staple on the shortest face of the block. The rubber band or spring should be quite stretchy.
- Board with sand paper attached. Your instructor will have chosen a combination of strength and stretchiness of the band or spring and roughness of the sandpaper so that the band will noticeably stretch but not break before the block slides on the sandpaper.
- A weight, such as a rock
- Ruler
- Long strip of paper
- Tape

9. Work with at least one other person. Place the wooden block on the board. Tape the long strip of paper down alongside the block on the board or next to it as shown in ■ Figure 10.19.

Figure 10.19

Slip-Stick Experimental Setup

A spring or rubber band pulls a wooden block across a sandpaper surface. Ruler and paper are available for marking the location of the block and measuring the stretch on the rubber band or spring.

a. Mark the position of the block and the length/position of the unstretched band/ spring on the paper.

b. Take the elastic band/spring, slowly stretch it out, and observe the stress accumulating in the band. Have a partner hold a ruler near the stretched elastic band to estimate the maximum length the band stretches.

c. At a certain point, the block will suddenly begin to slide and the tension in the band/ spring decreases. This illustrates the principle of stick-slip on a fault where friction prevents motion on a fault until the stress parallel to the fault plane reaches a critical value, and then the block moves. Mark the new position of the block.

d. Measure and record how far the block moved and record the maximum stretch of the band/spring in Table 10.2.

e. Do the experiment at least five more times, recording the block movement, and band/spring stretch for each. If you run out of room on the board, move the block back to its starting position to continue the experiments.

f. Average the block movements and band/ spring stretch amounts for the 6 experiments you just completed, and enter the result in Table 10.2.

g. Put the weight on top of the wooden block. Repeat the steps above.

h. Perform the experiment again with smoother sandpaper or by pulling faster and briefly describe this set up in the last column heading in Table 10.2.

i. How do the results compare among the three configurations? Make quantitative comparisons.

____

____

____

j. What part of the experiment corresponds to each of the following?

plate motion ______________________

earthquake ______________________

fault ______________________

elastic rebound ______________________

____

seismic gap ______________________

k. Based on the data you collected for configuration A, when would you predict the next "earthquake" ______________________

______________________? What would you measure to aid the prediction?

____

Table 10.2

Stick-Slip Experiment Data

See instructions in Exercise 9.

Trial	A. Block, Slow Stretching (cm)		B. Block and Weight, Slow Stretching (cm)		C. Describe the Configuration: ______________ (cm)	
	Distance Block Moved	Amount Band or Spring Stretched	Distance Block Moved	Amount Band or Spring Stretched	Distance Block Moved	Amount Band or Spring Stretched
1						
2						
3						
4						
5						
6						
Average						

I. Discuss the results with your classmates and instructor. How else is this experiment analogous to natural situations and earthquake hazards or risk? Write a summary of your discussion point by point on a separate piece of paper.

Movement on the San Andreas Fault One way to help establish the risk of earthquakes in an area is to analyze past movement along a fault. Let's do this for the San Andreas fault.

10. Examine the photograph in ■ Figure 10.20 to answer the following questions.

a. Locate, draw in, and label the San Andreas Fault on the photograph.

b. Locate, draw in with a different color, and label two different offsets of the largest stream by the fault.

c. Measure the amount of offset for each channel in meters using the scale on the photograph. Notice that the scale is only accurate at the position of the fault.

Left channel: __________

Right channel: __________

d. Geologists have determined the age of the sediments in the two offset parts of the stream as 10,000 to 3,680 yrs for older alluvium in the left channel and 3,680 yrs to present in the right channel. Use the older ages for each part of the channel and the distance the channel has shifted to calculate the speed of movement of the fault in cm/yr for each channel.

Left channel: ______________

Right channel: ______________

e. Explain why it makes sense to use the older ages for a rate of movement rather than the younger ages.

____

____

f. Review faults in Lab 8 if necessary; what type of fault is this?

____

g. What type of plate boundary is this?

____

h. If the average amount of movement on this segment of the San Andreas Fault is 9.5 m, what is the average length of time between movements? Use the average of your two speeds calculated in part **d** in this calculation. ______________

Figure 10.20

Stream Offset by the San Andreas Fault

The scale is only appropriate for the middle of the photograph along the San Andreas Fault. See instructions in Exercise 10.

Seismic Waves

An earthquake releases energy in the form of seismic waves, which spread out in all directions from the focus (Figure 10.5), similar to ripples from a stone dropped in a pond. There are important differences, however. In the case of the stone, the waves you see in the water move outward in a circle away from where the stone fell; in contrast, the waves from an earthquake's focus move outward in three dimensions in a spherical pattern. These are called **body waves.**

Body waves, which move through the Earth and are generated at the focus, are of two types: **P waves** and **S waves.** *P* stands for primary, or push-pull, waves and S stands for secondary, or shear, waves (■ Figures 10.21a and b). P waves are called primary because they are the first to arrive. They are faster than all other seismic waves. The P wave is a compressional wave with the vibration direction parallel to the direction it travels. This means that the wave causes material to compress and expand alternately as it moves (Figure 10.21a). The S wave vibrates perpendicular to its travel direction like the waves in a vibrating guitar string (Figure 10.21b). Bodies of water and the liquid outer core have no S waves because S waves do not travel through liquids or gases. The way these waves transmit energy can be demonstrated with a Slinky®.

11. To produce a P wave, you and your lab partner will begin by spreading a Slinky across a tabletop. While one of you holds one end of the Slinky firmly, the other holds the other end and gives it a sharp push-pull movement, parallel to its length. In the gap in ■ Figure 10.22a, sketch how the Slinky looked as the P wave moved through it.

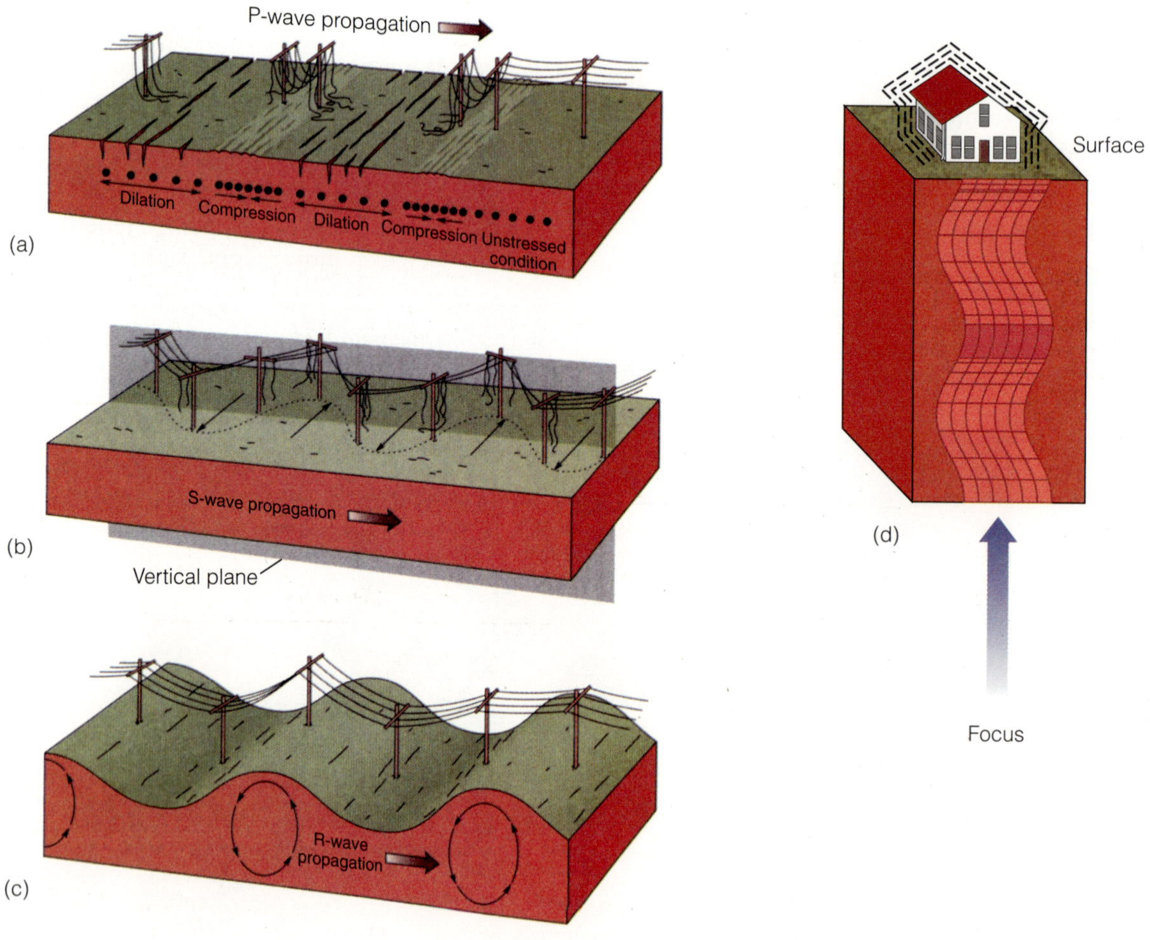

Figure 10.21

Seismic Waves

Elastic deformation of material and damage caused as different seismic waves travel through it. The earthquake epicenter and focus is somewhere to the left. (a) Primary waves (P waves) compress and expand the material. Particles vibrate parallel to the direction that the wave is traveling. (b) Secondary waves (S waves) shear the material perpendicular to the wave travel direction. (c) Rayleigh waves are surface waves that produce a rolling motion like ocean waves, only retrograde—that is, backward from the rolling of ocean waves. These waves diminish with depth. (d) P and S waves traveling from a focus below.

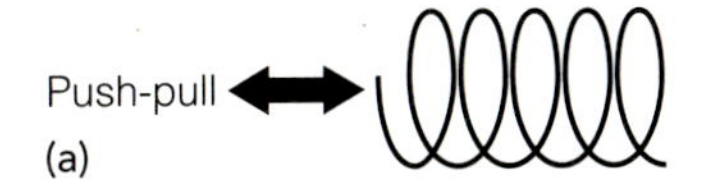

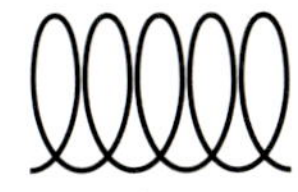

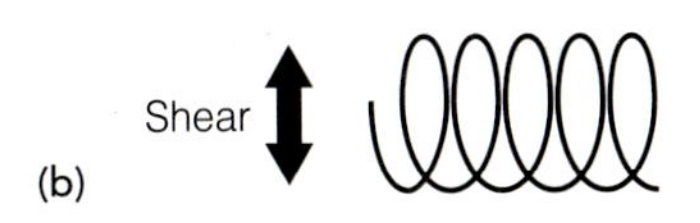

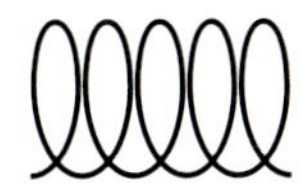

Figure 10.22

Slinky Experiment

(a) P wave. See instructions in Exercise 11. (b) S wave. See instructions in Exercise 12.

12. Produce an S wave by sharply "waving" one end of the stretched Slinky perpendicular to its length. In the gap in Figure 10.22b, sketch how the Slinky looked as the S wave moved through it.

Another kind of seismic wave is a surface wave (Figure 10.21c). **Surface waves** travel along Earth's exterior. P and S waves generated them where they reach the surface. The waves you see in a pond are an example of surface waves. Surface waves are even slower than S waves and are last to arrive. During earthquakes, people have reported seeing waves in parking lots that looked like the ocean or feeling as though they are on a boat; these are surface waves. These waves are especially destructive because they cause different parts of a building or structure to move in different directions at the same time.

When all of the seismic waves are recorded, the resulting seismogram looks something like ■ Figure 10.23. Time progresses from left to right with a sequence of waves: P, S, then surface waves. The diagram shows the first arrival of each wave.

MAGNITUDE

Magnitude is one way that seismologists measure and compare the size of different earthquakes. The **Richter magnitude** was originally based on a measure of the maximum amount of ground motion, the maximum **amplitude** (half the height of the squiggles in Figure 10.23) of the seismic waves, at a 100-km distance from the epicenter. The amplitude is essentially a measure of the amount of shaking an earthquake produces. The largest earthquakes have thousands of times more amplitude (shaking) than small earthquakes, so the magnitude scale is compressed using logarithms. This means that the amplitude increases by a factor of 10 for each single-digit increase in the magnitude. The original scale developed by seismologist Charles Richter was dependent on the local area, the instrument used, and the size of the earthquake, so it has been modified considerably since its invention.

Moment magnitude (M_W) is now more universally used because it is more useful for comparisons of very large earthquakes. It depends on the strength of the rocks, the area of the fault along which movement occurred, and the amount of movement. Whichever method of determining the magnitude is used, the energy released during an earthquake relates to the magnitude.

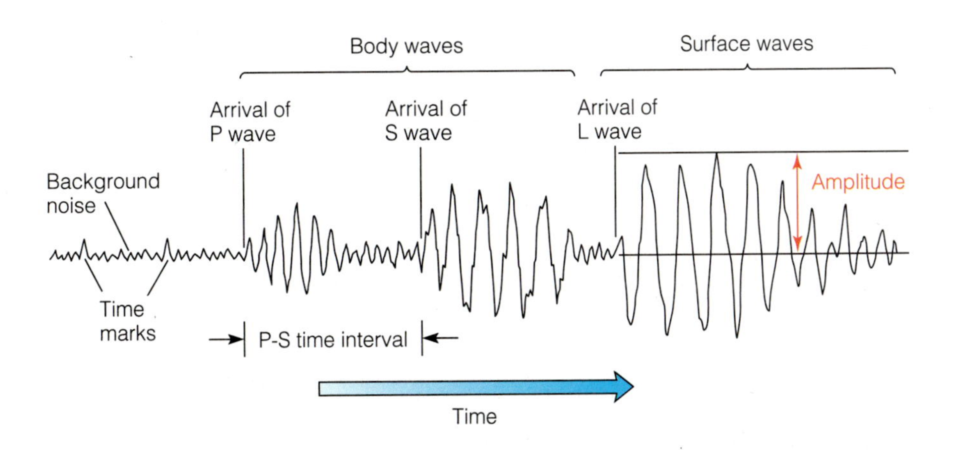

Figure 10.23

Parts of a Seismogram

A diagram illustrating the parts of a seismogram showing P-wave arrival time, the S-wave arrival, and the arrival of the first surface wave (L). The P-S time interval is used to determine the distance to an earthquake. The amplitude is helpful in determining the magnitude of the earthquake.

Imagine you just visited (or live in) earthquake country and experienced an earthquake. Maybe it was a magnitude 4. You say, "That wasn't bad. What's the big deal?" It is important to understand how much more powerful a 5 would be, or a 6 or 7. When comparing two earthquakes, such as Earthquake B and Earthquake A, with magnitudes M_B and M_A, seismologists can estimate the energy released by Earthquake B (E_B) compared to the energy released by Earthquake A (E_A) using following approximation:[3]

$$E_B \approx [30^{(M_B - M_A)}] \times E_A \qquad (1)$$

This means that the ratio of the two earthquakes' released energy is about

$$\frac{E_B}{E_A} \approx 30^{(M_B - M_A)} \qquad (2)$$

An earthquake's energy is important because it is directly proportional to the destructiveness of the earthquake. In fact, the small *w* in Mw stands for work done during the earthquake. This is the sense of *work* physicists use where work is a type of energy. Let's compare a magnitude 4 earthquake with an earthquake that people at rest indoors generally feel—a magnitude 3. The energy released by a 4 is approximately $30^{(4-3)} = 30^1 = 30$ times greater than the energy released by a magnitude 3 quake. A magnitude 7 releases about 810,000 times the energy of a magnitude 3! Another way to look at this is that it would take about 810,000 magnitude 3 earthquakes to release the same amount of energy as a magnitude 7.

13. Calculate how many times as much energy is released the from the first earthquake below than from the second. These you may be able to calculate by hand.[4]

a. A magnitude 8 earthquake compared to a magnitude 6 ____________

b. A magnitude 8 earthquake compared to a magnitude 3 ____________

c. A magnitude 7 earthquake compared to a magnitude 5 ____________

[3] The number 30 comes from rounding the square root of 1,000.

[4] If your calculator does not handle very large numbers, use the following equivalent equation, or work the problem by hand.

$$\frac{E_B}{E_A} \approx 30^{(M_B - M_A)} \times 10^{(M_B - M_A)} \qquad (3)$$

Calculate the last expression by hand, remembering that 10 to any power *x* is 1 followed by *x* zeros; for example, $10^4 = 10{,}000$.

14. Use a calculator and the y^x button to calculate the following:

a. How many times more energy was released by the Anchorage, Alaska, earthquake of 1964 (moment magnitude 9.2 Mw) compared with the Sumatra 2004 quake (moment magnitude 9.0 Mw)? (Remember to subtract first.)

____

b. Using the same equation, compare the energy released by the Sumatra earthquake with the San Francisco earthquake in 1906 (7.9 Mw).

____

15. Consider what it would be like for a great earthquake to hit your area.

a. On a separate sheet of paper make a rough-sketch map of your area showing the locations of major highways; public transportation systems; residential areas; water, gas, and sewer plants; power generation plants; and dams. Also, include any geographic features such as lakes, oceans, hills, mountains, or slopes.

b. Considering all these systems and their interactions with the daily lives of those in your community, write an essay on the back of your sketch sheet describing what might happen if a 8.0 earthquake hit your area today. Note potential problems with any facilities shown in your sketch and what emergency response would result.

c. Also address in your essay what your local government planners should do to help diminish the problems you discussed in your answer to part **b.**

LOCATING AN EARTHQUAKE

A seismograph detects seismic waves and writes the results on a seismogram. To determine the distance from the seismograph to the earthquake focus, we can use the fact that P and S waves travel at different speeds. If a seismograph were located at an epicenter of a shallow earthquake, all of the waves would arrive at nearly the same time because the earthquake was so close. In contrast, a seismograph at some distance from the epicenter, would receive the S waves later than the

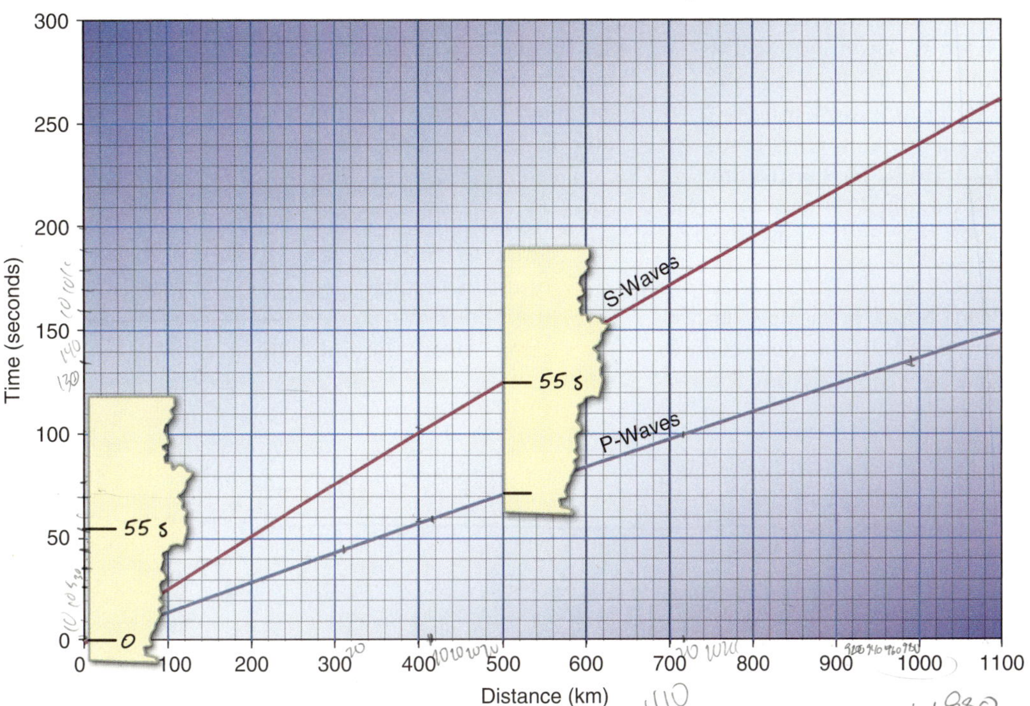

Figure 10.24

Travel-Time Curves for P and S Waves

If the velocities of these waves were constant, the travel time curves would be straight lines. However, notice that the curves have gentler slopes at greater distances—the waves speed up with distance. This is because waves that travel farther move through deeper parts of the Earth where the waves can travel faster in the denser more elastic material there. Use the S-P time lag measured on a piece of scratch paper to find the distance to the earthquake for a seismogram.

P waves, because the faster P waves would pull ahead of the slower S waves.

The amount of time the S waves lag behind the P waves is not exactly proportional to the distance because the wave speeds are not constant. However, with the graph in ■ Figure 10.24, we can use the separation between the P and S curves to get the distance from the earthquake epicenter (or more precisely from the focus).

Distance to the Earthquake

Seismograms such as the one in Figure 10.23 contain a record of the arrival times for different seismic waves. A seismologist measures the time between the P- and S-wave arrivals (P-S time interval) from the seismogram and then calculates the approximate distance to the focus using P and S wave travel time curves in Figure 10.24. Since most earthquakes occur within shallow levels of the crust, distances to the focus are approximately distances to the epicenter as well. With readings from at least three seismograms for the same earthquake from three different seismograph locations, the seismologist can determine the location of the epicenter of the earthquake by drawing circles on a map around each seismograph location (■ Figure 10.25).

Let's locate an earthquake, starting with the time lag between the P and S waves.

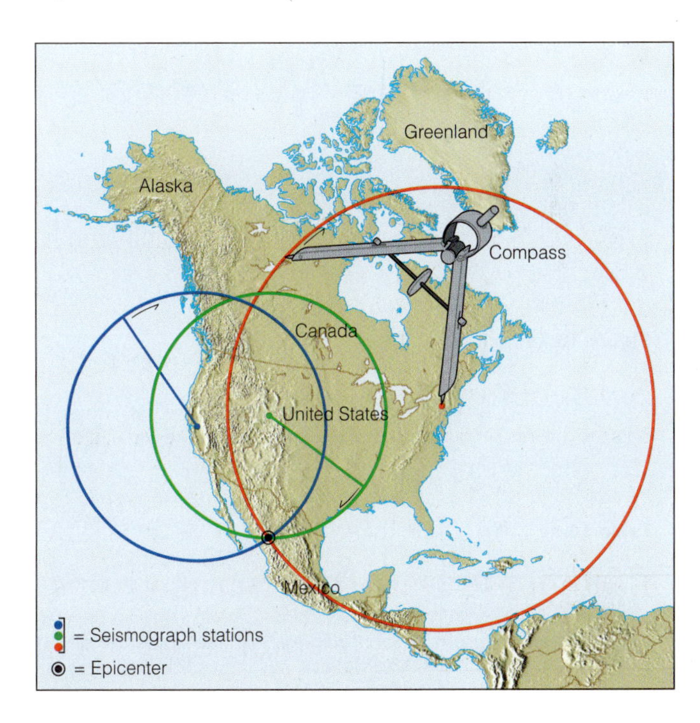

Figure 10.25

Triangulation to Locate an Earthquake Epicenter

Using a compass to pinpoint the location of an earthquake. Once the distance to the epicenter is determined for three seismograph stations, one can use the distance as the radius of a circle centered at each station. The point of intersection of all three circles is the epicenter.

16. Examine ■ Figure 10.26 and find the P- and S-wave arrivals for each seismogram, as illustrated for Seismograph 1. Use the scale on the seismograms to determine how much time passed between the arrivals. This is the **S-P time lag.** Fill in ■ Table 10.3 with the time lags for each seismograph.

With the difference in time for each seismograph we can determine the distance to the focus for each seismograph. For example, if the time between the P and S wave arrivals for Figure 10.23 is 55 seconds, measure this on a piece of scratch paper and then find that spacing between the two curves in Figure 10.24. Be sure to keep your paper vertical as you move it up the P-wave curve until you reach a spot where the S wave arrived 55 seconds later. For this example the distance would be 500 km.

17. Measure the S-P time lag for each seismograph on the time scale in Figure 10.24, then find this vertical spacing on the graph between the blue P-wave curve and the red S-wave curve. Enter the numbers in Table 10.3.

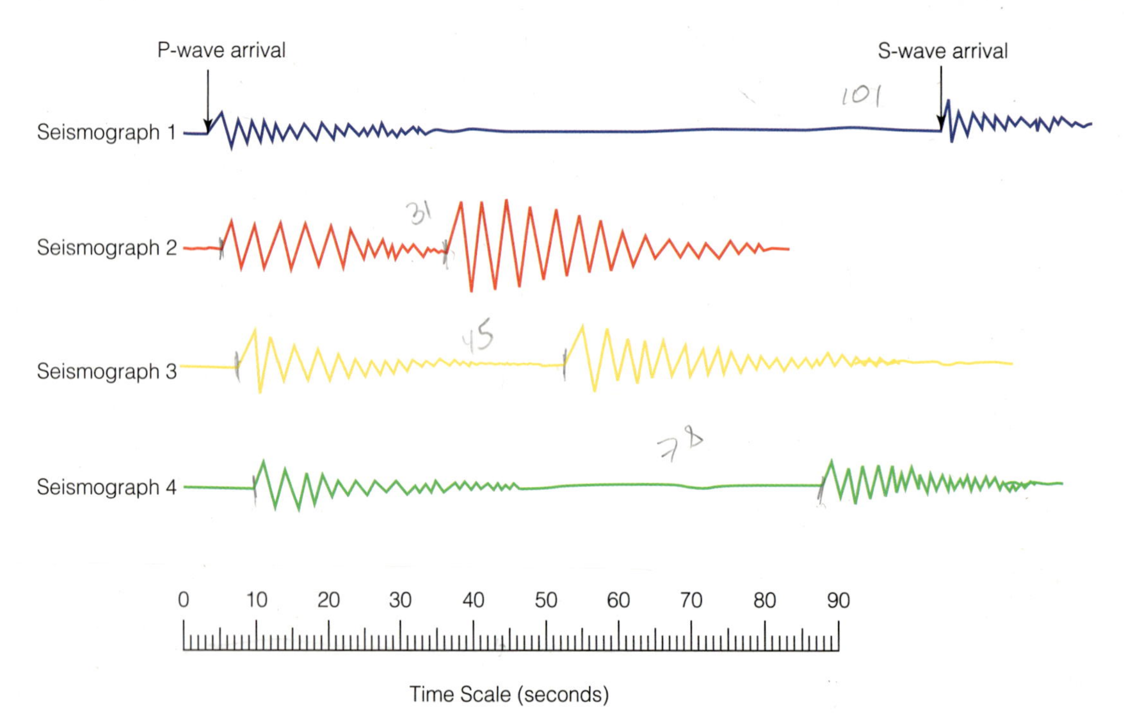

Figure 10.26

Seismograms

Short segments of four simplified seismograms of the same earthquake. Only the P- and S-wave parts of the seismograms show.

Table 10.3

Earthquake Data from Seismograms

Measure seismograms to locate an earthquake epicenter and determine the time the earthquake occurred. See instructions in Exercises 16–20.

Seismograph	$(S - P)_D$ Time Log (s)	*D* Distance (km)	P-Wave Arrival Time as Hours, Minutes, and Seconds (h:min:s)	P-Wave Travel Times (s)	Time of Earthquake (h:min:s)
1			3:56 a.m. Central Standard Time		
2			3:10:45 a.m. Central Standard Time		
3			3:11:04 a.m. Central Standard Time		
4			4:12:30 a.m. Eastern Standard Time		

Locating the Earthquake from Three Seismographs

Once you know the distance to an earthquake's focus from three seismographs, you can approximately locate the earthquake's epicenter on a map. Set a drafting compass so that the distance between the pencil and the point of the compass equals the map distance between the earthquake's epicenter and one of the seismographs.[5] Place the point of the compass on the map at the seismograph location as shown in Figure 10.25. Draw a circle around the seismograph—the earthquake occurred someplace on this circle. Next, set the compass spacing equal to the distance for the second seismograph and draw a circle around that seismograph. The earthquake's epicenter should fall on this circle, too, at one of the two points where this circle intersects the previous one. Repeat for the third seismograph to pinpoint which of the two intersections is the correct one.

[5] We will approximate distances to the epicenter as being equal to the distances to the focus. If the focus is very deep and the seismographs are all close to the epicenter, the distances will all be too large. On the other hand, for many earthquakes the focus is quite shallow compared to the distance from the seismograph to the epicenter, which makes the method demonstrated here a good approximation.

18. Follow the instructions above using the distances you determined in Table 10.3 and the map in ■ Figure 10.27.

a. Draw the circles on the map. Before you draw each circle, double check which distance goes with which seismograph. All three circles should intersect at one point, which is the epicenter as illustrated in Figure 10.25. Go ahead and draw a circle for the fourth seismogram. This one should confirm your earthquake location.

Figure 10.27

Seismograph Locations

Map of the eastern half of United States showing the locations of four seismographs:
1 = Minneapolis, Minnesota;
2 = St. Louis, Missouri;
3 = Bloomington, Indiana;
4 = Bowling Green, Ohio.
Use this map to locate the earthquake following instructions in Exercise 18.

b. Clearly label the epicenter on the map.

c. In what state is the epicenter?

Missouri

In practice, the circles do not always intersect at a single point because seismic waves travel at different speeds through different rocks. In such a case, the region where they nearly intersect is likely to contain the epicenter.

Notice in ■ Figure 10.28 that the earthquake you just plotted occurred in the New Madrid seismic zone, which is where two great earthquakes occurred in 1811 and 1812.

Time of the Earthquake

Another question people often ask is, "When did the earthquake happen?" We have a pretty good idea from the time of arrival of the P waves given in Table 10.3 that the earthquake occurred shortly before 3:11 a.m. Central Standard Time, but the P waves had already traveled for some time before they got to the seismographs. In fact, we have four different times recorded because each seismograph was a different distance from the earthquake's focus. When did the P waves leave the focus? We have determined the distance to the earthquake's focus, and we have a graph showing the travel time for the P wave. Just use the graph to determine how long the P wave traveled, and then subtract this from its arrival time.

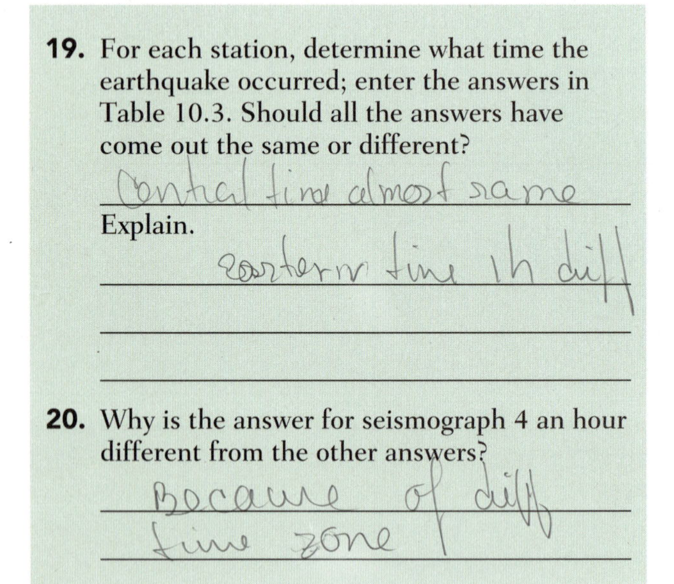

19. For each station, determine what time the earthquake occurred; enter the answers in Table 10.3. Should all the answers have come out the same or different?

Central time almost same

Explain.

eastern time in diff

20. Why is the answer for seismograph 4 an hour different from the other answers?

Because of diff time zone

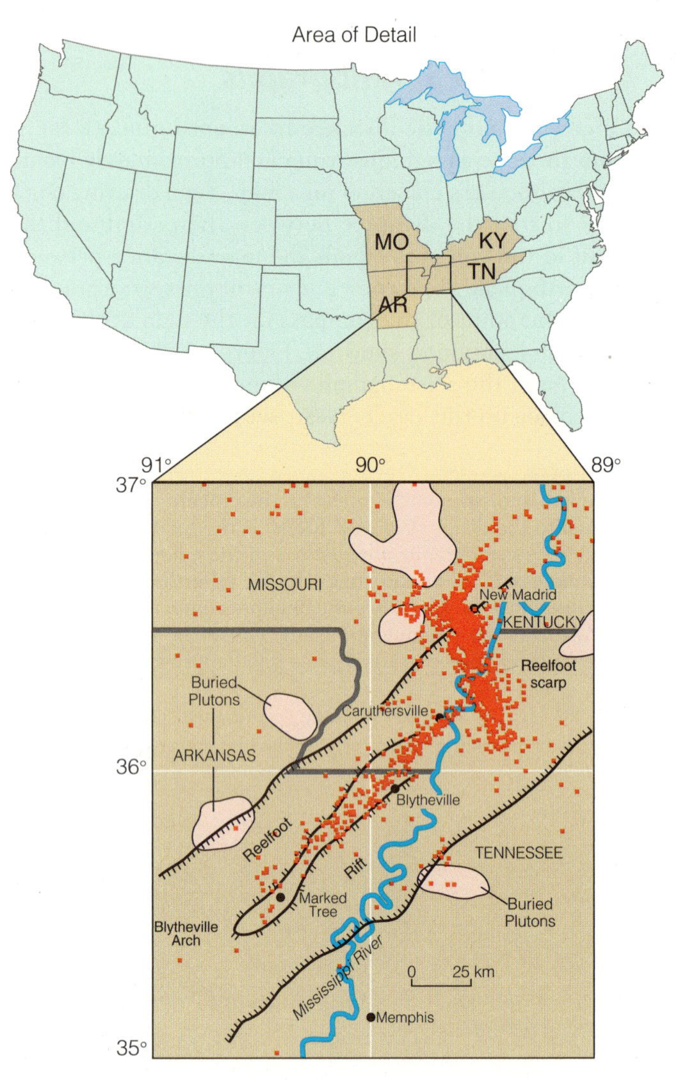

Figure 10.28

New Madrid Seismic Zone

Map of parts of Missouri, Kentucky, Arkansas, and Tennessee showing the New Madrid seismic zone. This is an old rift beneath thousands of feet of sedimentary rocks and is a zone of weakness in the North American Plate, where numerous earthquakes occur (red squares). Earthquakes of magnitude greater than 8 occurred here in the past.

LAB 11

Landslides and Mass Movements

OBJECTIVES

- **To understand the hazards connected with mass wasting**
- **To understand the influence of slope, material, water, and pore pressure on mass movements**
- **To distinguish the types of movement involved in falls, slides, and flows**
- **To recognize the different types of mass wasting and understand their causes and effects**

Loose material at Earth's surface is likely to move downhill on sloping surfaces. The steeper the slope, the faster the material is likely to travel. **Mass wasting** occurs when sediment, soil, and rock move down slope due to the force of gravity. Another term for mass wasting is **mass movement.** Mass wasting can be just a few feet a year or as rapid as an avalanche. When the movement is too fast it may cause a serious threat to humans, but even slow mass wasting can be damaging to property.

MASS WASTING IS HAZARDOUS

Mass wasting becomes a problem, and sometimes a deadly one, in inhabited areas of hilly or mountainous terrain.

- Steep slopes and heavy rain often give rise to *mudflows* (commonly called mudslides in news reports). In December 2003, in the central Philippines, heavy rains triggered multiple mud and debris flows on hillsides destabilized by illegal logging activities, killing as many as 200 people. Devastation due to mass movement is common also in the northern Philippines, where landslides triggered by prevalent wet storms buried more than 160 people and washed away entire villages in October 2009.
- After intense seasonal rains in 1972, a slope steepened during road and building construction resulted in a landslide in Hong Kong. This, and one other landslide, killed 138 people in Hong Kong that year (■ Figure 11.1).

Hong Kong Government Information Services

Figure 11.1

Landslide in Hong Kong

Po Shan Road landslide, June 1972. Road and building site construction steepened the hillside, and combined with saturating rains, triggered this landslide. The hill slope above the road was 37°, an angle too steep for the material making up the hillside. Approximately 40,000 m^3 of volcanic material and soil moved about 270 m down slope. The slide killed 67 people, injured 20, destroyed 2 buildings, and severely damaged a third.

- In 1938, an earthquake triggered numerous landslides in Kansu Province, China, that killed as many as 200,000 people. Mass wasting is widespread in China. The September 2008 collapse of a mine waste reservoir in northern Shanxi province caused a mudflow, that killed more than 250 people. A landslide containing millions of cubic meters of rock flooded a valley in June 2009, burying workers in an ore plant in Wulong county.
- The largest landslide in recorded history occurred on May 18, 1980, when the north side of Mt. St. Helens slid away, moving a volume of 2.8 km^3 of material and releasing a volcanic eruption. The combination of mass wasting, a landslide, and a debris avalanche killed only 5 to 10 people because authorities had evacuated the area in anticipation of an eruption. Figure 11.13 on page 253 shows mudflows associated with this same eruption.
- An avalanche of rock and snow triggered by an earthquake in 1970 swept down the mountain side of Huascarán, Peru, at an average speed of greater than 100 miles per hour, burying two towns and killing about 23,000 people (■ Figure 11.2). Parts of this landslide achieved much higher speeds, reaching at least 250 miles per hour.

Figure 11.2

Debris Avalanche

In 1970, this *debris avalanche* from an oversteepened slope on the volcano Nevado Huascarán, Peru, flowed about 11 miles at more than 100 miles per hour. An offshore earthquake triggered movement of a large mass of glacial ice and rock about a mile long, which became a debris avalanche, picking up about 80 million yd^3 of water, mud, and rock debris.

Many less-dramatic mass-wasting events can still be extremely destructive. Even slow soil creep causes substantial property damage. An understanding of the types of materials and conditions leading to these events can help us make informed decisions concerning construction, emergency management, disaster mitigation, and future land-use development.

FACTORS INFLUENCING MASS WASTING

A number of factors influence mass wasting, but the most obvious involve slope steepening, either by undercutting the base of a slope or by adding material to the top of a slope. Other factors include the material constituting the slope and water saturating the slope. A number of different events or processes can trigger mass wasting. Some of these depend on changing the slope or water content, and others relate to disturbances of the slope such as shaking of the ground during an earthquake. Whatever the trigger, the slope will not fail unless its angle exceeds its stability.

The **angle of repose** is the steepest angle at which loose material will stand without mass-wasting movement (■ Figure 11.3). The steep side of a sand dune is commonly at the angle of repose. When the force driving material down a slope equals the force of friction holding it back, the slope has attained the angle of repose. The force of rolling friction is commonly less than the force of sliding friction for approximately round objects such as sand grains.

Angle of Repose Experiments

Materials needed:

- 200 or more marbles or M&M's®
- Containers
- Four types of sediment as listed in Table 11.1
- Clinometer or protractor

1. Determine the angle of repose: Separately pile marbles or candy and the first three sediments listed in Table 11.1 as steeply as possible in suitable individual containers. For each material, use the clinometer or protractor to measure the angle of repose three times; record your findings in ■ Table 11.1. For the damp sand, stir the sand thoroughly and pour it out of the container into a pile in another container; do not tap or pack the sand. Average the results for each type of material. Sweep up any spilled sand.

Marli Miller

(a)

Parvinder Sethi

(b)

Figure 11.3

Angle of Repose

Angle of repose of (a) sawdust and (b) sand near Stovepipe Wells, Death Valley, California. In (b), the downwind sides of the sand dunes, in bright sunlight, known as the **slip faces,** have achieved the angle of repose and are steeper than other parts of the dunes. On close examination of the slip face, especially of the dune in the foreground, you can see mass wasting that resulted from *grain flow.*

Table 11.1

Angle of Repose

Measurements of angle of repose in several materials. See instructions in Exercise 1.

Material	Angle of Repose (measure 3 times)			Average for the Material
Marble or candy				
Rounded pebbles				
Crushed rock or angular pebbles				
Dry sand				
Damp sand, not packed				

a. Describe how you achieved the angle of repose. Did you need different techniques for different materials? If so, describe the differences.

____

____

____

____

____

b. Do you have enough data to tell how the size of sediment influences the angle of repose? ________ If so, describe the relationship.

____

____

c. How does roundness or angularity influence the angle of repose?

____

____

d. How does water content influence angle of repose?

____

____

e. Measure and label the angle of repose for the sawdust in Figure 11.3a. Of the sediments that you measured in part **a**, which does it closely compare to? ________ What are the implications for this and other types of industries such as rock mining?

____

____

Angles and Forces

The steeper the slope, the more likely it is that mass wasting will occur. This is because the maximum force of gravity acts in a vertical direction, and so the greater the slope, the greater is the component (or part) of gravity acting parallel to the slope. When the slope is gentler, the force of gravity tends to hold material in place. Figure 11.4 is a series of *vector* diagrams showing the forces that result along a slope due to gravity. A **vector** has magnitude (a numerical value) and direction. The figure shows three vectors for each slope: The vector directly downward is the force of gravity. The one parallel to the slope is that component of the force of gravity that acts along the slope and tends to cause movement or mass wasting. We can call this the **driving force** (D). The vector perpendicular to the slope is the **normal force** (N) and is the component of gravity that acts to hold material in place. The stronger the normal force, the greater the friction is that holds the material in position and prevents movement. The force of friction is the product of the **coefficient of friction** (f) times the normal force. The coefficient of friction depends on the material.

$$\text{Force of friction} = \text{coefficient of friction} \times \text{normal force}$$

$$F = f \times N$$

When the force acting parallel to the slope is greater than the force of friction, mass wasting will occur.

$$\text{Driving force} > \text{force of friction}$$

$$D > F \quad \textit{results in mass wasting}$$

2. Decide which of the diagrams in Figure 11.4 show stable configurations and which show unstable arrangements. In order to decide, you will need to determine the force of friction. For each diagram, the coefficient of friction is in Table 11.2.

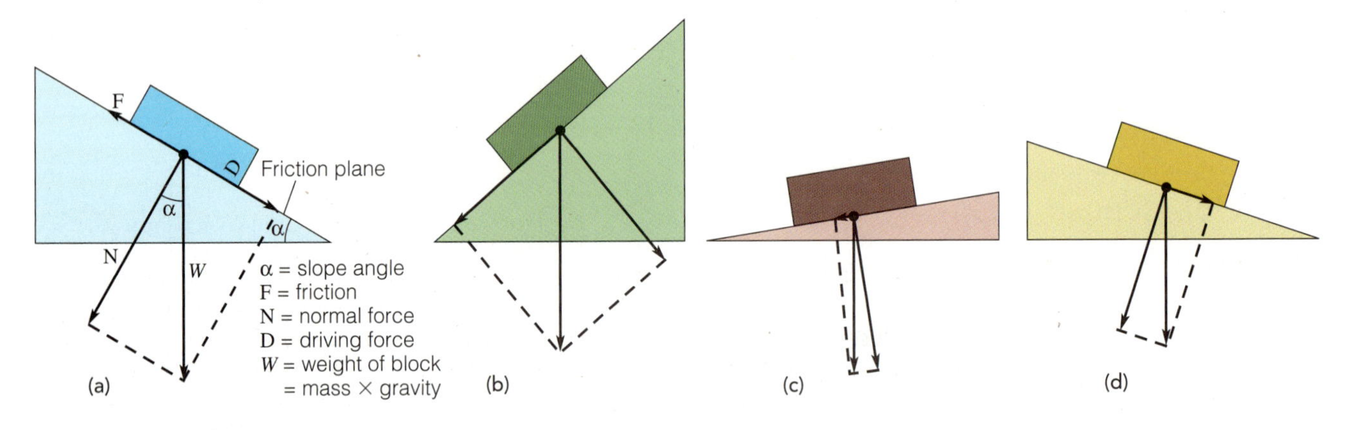

Figure 11.4

Vector Diagrams for Blocks on Slopes

(a) shows driving force and force of friction; (b–d) see Exercise 2 and Table 11.2.

Table 11.2

Forces on Slopes

Forces of friction resist downslope driving forces. See instructions in Exercise 2.

Diagram	Length of Driving Force D (mm)	Friction Coefficient f	Length of Normal Force N (mm)	Force of Friction F (mm)	Which Is Greater: the Force of Friction or the Driving Force?	Is This Configuration Stable? (Yes/No)
Figure 11.4a		0.45				
Figure 11.4b		0.96				
Figure 11.4c		0.33				
Figure 11.4d		0.20				

a. Label each force in Figures 11.4b through d as has been done in Figure 11.4a.

b. Measure the length of the driving force *D* in each diagram in millimeters and enter this in Table 11.2.

c. Measure the length of the normal force *N* in each diagram in millimeters and enter this in Table 11.2.

d. Calculate the length of the force of friction vector. To do this, multiply the normal force by the coefficient of friction given. Enter this in Table 11.2.

e. Draw and label an arrow for the force of friction on each diagram in Figures 11.4b, c, and d with the correct length in millimeters.

f. Is the force of friction arrow longer or shorter than the driving force in the same diagram? Use this to decide if the configuration is stable or not and enter your results in Table 11.2.

g. How does the slope angle influence stability?

____

____

h. If the coefficient of friction is higher, will the slope be more stable or less stable?

____

The Influence of Water

Water is another very important factor that affects the stability of a slope and helps determine whether mass wasting will occur. Water has two different effects, depending on the amount present. Sandcastles are a good demonstration of the influence of water on unconsolidated sediments (■ Figure 11.5). No one builds a sandcastle out of dry sand because the grains can only pile up until they reach the angle of repose, typically about 35°. That would make a poor castle. Instead, people who want to build an impressive, steep-sided sandcastle use damp sand, in which surface tension of the water binds the grains together. Notice the slopes steeper than 80° in places on the sandcastles in Figure 11.5—a much higher angle of repose than for dry sand. In fact, damp sand can hold a vertical slope under some conditions. The surface tension of the water between the grains acts like weak glue (■ Figure 11.6a). Damp conditions increase the stability of a slope compared to either dry or saturated conditions, so that mass wasting is less likely.

Matthew Sullivan

Figure 11.5

Sandcastle

Matthew could have built the flat sloping sides of this sandcastle much steeper if he had chosen to do so. Notice the round balls of sand showing greater than 90° slopes in places.

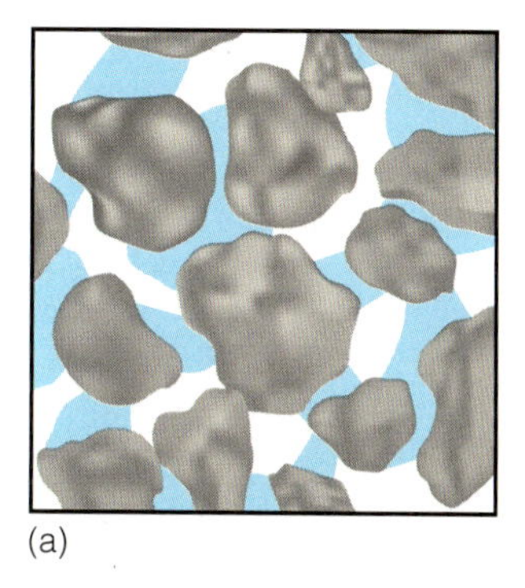

(a)

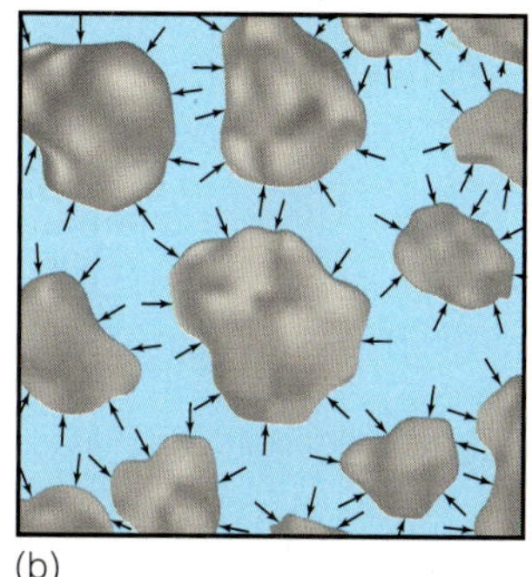

(b)

Figure 11.6

Water's Effects

(a) Damp sand has both air and water between the grains. The air–water boundary has greater strength than liquid-water due to *surface tension*. The surface tension helps hold the sand grains together. (b) Saturated sand has no surface-tension effect. Instead, pore pressure of the water may actually push grains apart, reducing friction between them, and causing them to lose cohesion.

Wet or saturated conditions reduce the stability of a slope. For dry or unsaturated masses, an increase in **load** (weight from natural materials, or buildings) will compress the air in the spaces between grains (called **pores**), bringing grains or rock fragments closer together. This increases both the friction and the driving force in equal amounts, so it does not change the stability of the material. However, when water saturates a rock mass it fills the pore space. Since water is relatively incompressible, an increase in load increases the pressure in the pores between the grains. This increase in **pore pressure** pushes against the grains, forcing them apart and reducing the friction between them. If waves or rain saturate the sand in the sandcastle, the sand castle collapses by flowing outward. Saturation leads to mudflows, debris flows, and earthflows. In rock, the water pressure may remove all of the support from grain contacts for the weight of the overlying rock mass; this would cause instability and initiate sliding.

Clay, in sediment and in shale, becomes plastic and very slick when wet. Some clays can absorb large amounts of water into their structure. Smectite clays, for example, can absorb so much water that they become a gel-like substance that is very weak. Such material fails easily on sloping surfaces. Clays commonly expand when wet and contract when dry, which increases the likelihood of mass wasting.

How Water Influences Rock Stability

3. Add a small amount of water to the top of a piece of sandstone and a piece of shale. Some shale that students look at in lab is unusually stable in water so it does not deteriorate when students put dilute acid (which is mostly water after all) on it to test for calcite. For this exercise, however, your instructor has selected shale that has a more normal interaction with water.

a. What happens to the water on the sandstone? ______________________
What happens to the sandstone with water on it? ____________________

b. What happens to the water on the shale? ______________________ What happens to the shale with water on it?

____

c. Place each rock in a container of water and let both sit for 10 minutes, while you do another activity. After 10 minutes, what has happened to each rock?

Sandstone: ______________________

____

Shale: ______________________

____

d. Which rock is less stable when wet?

____

e. Slide your fingers over the wet sandstone and over the wet shale. Describe how they feel.

Sandstone: ______________________

Shale: ______________________

f. How do you think these properties would influence mass wasting?

____

____

g. Imagine a situation with a sloping layer of sandstone resting on a parallel layer of shale, similar to the arrangement shown in ■ Figure 11.7. This is a fairly common geometry for rock layers. If rain fell on the sandstone, what would happen to the water?

____

As a result, what might happen to the mass of rocks?

____

Explain.

____

____

____

____

Sliding Can Experiments Use glass plates and warm and cold aluminum cans to demonstrate the effects of pore pressure.

Materials needed:

- Cooler with ice or a freezer
- Three empty aluminum cans—one at room temperature, two chilled—with the tabs removed or

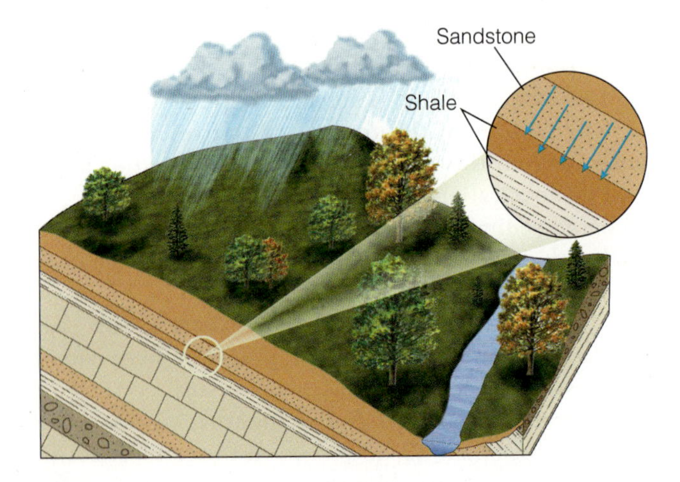

Figure 11.7

Water and Tilted Sandstone and Shale Layers

Permeable sandstone allows the shale to become saturated. See instructions in Exercise 3g.

pushed down so they do not protrude above the lip of the can
- Piece of plate glass or smooth flat plastic at least 6 in by 12 in
- Sponge, towels, or spray bottle to moisten surface
- Water for moistening
- Protractor or clinometer

4. Prop the plate glass at a low angle, moisten it, and place a room-temperature can upright on the glass. Tilt the glass at steeper angles until you establish at what angle the can starts to slide. ■ Figure 11.8 illustrates the setup.

a. Measure the angle with protractor or clinometer. ______________

b. Place a moistened glass plate on a flat surface. Put a chilled can with its top down on the glass. Tilt the glass slightly until the can moves. Measure the angle at which sliding starts. ______________ If the can warms to room temperature before you establish this angle accurately, put it away to chill again for other students and use another cold can to complete your experiment.

c. Once you have measured the angle, put a cold can at the top of the slope and let it slide all the way to the bottom. What happened when the can reached the bottom?

____

d. Why did the can have the behavior you observed in part **c**?

____

____

____

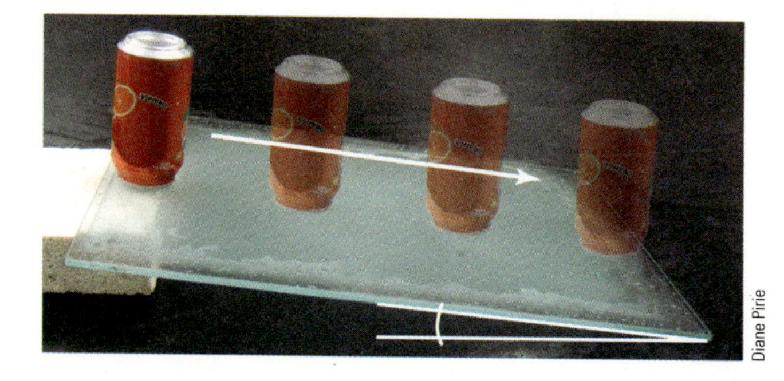

Figure 11.8

Sliding Cans Experiment

Can sliding on plate glass: experimental setup for Exercise 4. The angled lines and arc show the slope angle to measure.

e. On a separate piece of paper, explain the results for warm and cold cans. In your explanation, consider the fact that cold air expands when it warms up.

TYPES OF MASS WASTING

Factors such as the mode of movement and type of material determine the character of mass wasting and its effects. The term **landslide** sometimes refers to any rapid mass wasting, without regard to the material involved or the style of movement. Other terms for mass wasting, however, are more restricted in their meaning. During a mass wasting event, movement occurs in one or more of three possible styles: falls, slides, or flows. **Falls** move through the air when gravity pulls material from a steep cliff or overhang; **slides** move as coherent masses parallel to and in contact with the main surface on which the material slips; and **flows** move like fluids with different parts flowing at different speeds. Mass wasting incorporates different types of material, which can range from mud to solid rock masses. The material may consist of wet or dry loose particles of various sizes, or even masses of soil or sediment that move as a unit. A third variable in mass wasting is the speed of movement. ■ Figure 11.9 shows varieties of mass wasting grouped by style of movement, and ■ Table 11.3 summarizes some of the types of mass wasting.

Falls

Rockfalls result when cracks or fractures in steep cliffs fail suddenly. Such rock movements are rapid and common along highway roadcuts (Figures 11.9a and ■ 11.10a).

Talus (also known as **scree**) is an accumulation of fallen, angular rock fragments that pile up at the base of a steep slope or cliff (Figure 11.10b). Talus slopes accumulate gradually from thousands of rock falls, typically at an angle of about 45°, the angle of repose for this material. They consist of very angular, very coarse, loosely piled sediment with large pores (openings between the blocks).

Debris or Soil Falls occur when wave, stream erosion, or human activities undercut steep cliffs, allowing loose material to drop through the air as it collapses.

Slides

Rockslides are movement of one or more slabs of rock along a plane of weakness, such as a major joint (crack) in the rock, fault plane, foliation, or more commonly a

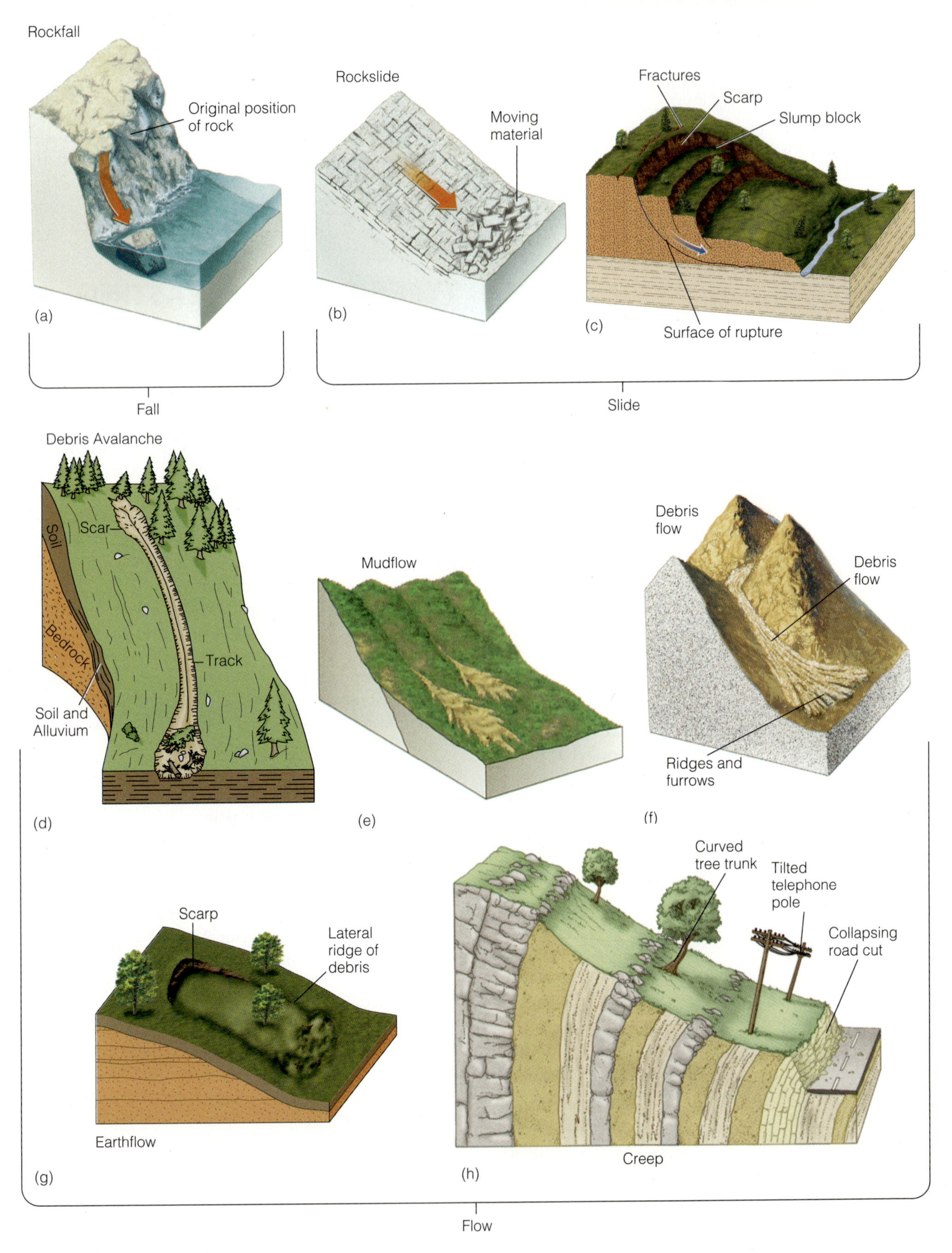

Figure 11.9

Types of Mass Wasting

Three categories: (a) is a fall, (b) and (c) are slides, and (d) through (h) are flows, in order of their speed, from the fastest, debris avalanches (d), to the slowest, creep (h). From MONROE/WICANDER, *The Changing Earth*, 5E. 2009 Brooks/Cole, a part of Cengage Learning, Inc. Reproduced by permission. www.cengage.com/permissions

Tonya Paul/Oroville Mercury Register

(a)

USGS

(b)

Figure 11.10

Rock Falls

(a) Rock fall onto Highway 70 near Rogers Flat, California, on July 25, 2003. The biggest pieces were dump-truck-sized and required blasting before the workers could move them. (b) Talus slopes form where many rocks fall off steep slopes and collect at the base. These talus slopes in Banff National Park, Canada, came about gradually as **frost wedging** (the action of water freezing in cracks and opening them) loosened rocks that then fell into aprons around the base of steep rocky cliffs.

Table 11.3

Types of Mass Wasting

Falls, slides and flows, their varieties, characteristics, and speeds.

Type of Movement and Its Characteristics		**Rock**	**Debris** (Coarse and Mixed Sediment/Soil)	**Fine-Grained Sediment/Soil**	**Speed** (with Typical Values Given in Parentheses)
Falls	Moves through the air		Debris fall	Earth fall	Very fast
Slides—movement as a unit	Translational slide—Movement on a slide plane	Rock slide or rock block glide	Debris slide or debris block glide	Earth slide	Fast to very fast
	Rotational slide—movement on a curved surface		Slump		Slow
Flows—different parts move at different speeds (chaotic movement). Usually the top and center moves faster than the bottom and sides.	Mixed with air	Rock avalanche	Debris avalanche	Mud avalanche	Extremely fast (about 70 km/h to over 400 km/h)
	High water content			Mudflow	Very fast (1 km/h to over 100 km/h)
	Moderate water content		Debris flow		Slow to fast (1 m/yr to 100 km/h)
	Low water content		Earthflow		Slow (from a cm to several hundred m per hour)
	Interstitial ice		Rock glacier		Slow (less than 5 m/yr)
	Dry			Grain flow	Moderate (from 0.1 to 35 m/s)
	Movement by expansion and contraction. Upper part moves faster than lower.		Creep	Soil creep	Very slow (from fractions of a mm to a few cm per year)

bedding plane (Figures 11.9b and ■ 11.11). Initially, the rock is one unit and breaks up into smaller units as it moves. The term **landslide** refers generally to any fast-moving mass wasting, but in another more restricted sense, it means rapid movement of surficial material on a sliding surface. An example is the landslide at Hong Kong's Po Shan Road in Figure 11.1.

Slumps involve movement of material in large blocks on a curved slip surface. The sliding surface curves both at the land surface and below ground in a spoon shape, or sloping bowl (Figure 11.9c). Typically, slumps occur where streams or ocean waves eroded the base of the slope of thick unconsolidated material. When movement occurs, the slump rotates downward from the higher toward the lower edge of the bowl.

Flows

Snow and Ice Avalanches involve air mixing into the flow of moving turbulent particles of frozen water. Commonly, an avalanche occurs where a weak layer of powder, sitting on top of hard, refrozen snow, begins to slide. You may not think of snow and ice as sediment and rock, but many geologists do, and so they consider avalanches as a type of mass wasting. If a large slab of ice and snow starts moving on a steep enough slope, it begins to break up, incorporates air, and turns into an avalanche. The deadly, fast-moving, ground-hugging flow also churns up relatively harmless, billowing snow that rises cloudlike from the avalanche base. Loose snow can also start moving to trigger an avalanche. Skiers moving across snow layers have triggered many avalanches, and even loud noises can act as a trigger. To reduce avalanche risk, patrols shoot large guns at the unstable snow-laden slopes to trigger avalanches under controlled conditions.

Debris Avalanches are the fastest-moving type of mass wasting (Figures 11.2, 11.9d, and ■ 11.12). They get their speed in a way similar to snow avalanches by the mixing of air in with the debris. This allows it to move very fast, generally faster than 100 miles per hour. Debris avalanches are mixtures of air, water, mud, and rock. Like snow avalanches, they may produce a billowing cloud above a more dangerous, ground-hugging flow. Both debris and snow avalanches commonly move along a narrow track and leave elongate scars (Figures 11.9d and 11.12).

Mudflows are fast-moving, fluid mixtures of water and predominantly fine to sand-sized sediment (■ Figure 11.13). If water exceeds about 60%, then the flow is a sediment-laden stream and not a mudflow. Although not as fast as debris avalanches, mudflows can still achieve speeds up to 80 miles per hour. In arid and semiarid areas, intense, sudden rainstorms easily trigger mudflows by mobilizing accumulated sediment into torrents of mud. News reports call these "mudslides," but geologists prefer not to use this term because the form of movement is a flow, not a slide. A special type of mudflow is a **lahar,** which occurs when water mobilizes the thick deposits of pyroclastic material and volcanic sediment that surround stratovolcanoes. A volcanic eruption sometimes, but not always, accompanies a lahar.

Debris Flows are slower than mudflows and contain less water and more large rock material (■ Figure 11.14). Generally, if more than half the debris in a flow is coarser than sand, then it is a debris flow. Debris flows may be similar to wet concrete in viscosity, texture, and flow characteristics.

Bernard Pipkin

Figure 11.11

Rockslide

Rock block glide at San Pedro, California. This slide has moved periodically since 1929.

USGS

Figure 11.12

Debris Avalanches

Heavy rain in October 1985 triggered these debris avalanches in Panuelos, Puerto Rico. The scars, tracks, and zones of deposition are all visible in this photograph.

Tonya Paul/Oroville Mercury Register

(a)

USGS

(b)

Figure 11.10

Rock Falls

(a) Rock fall onto Highway 70 near Rogers Flat, California, on July 25, 2003. The biggest pieces were dump-truck-sized and required blasting before the workers could move them. (b) Talus slopes form where many rocks fall off steep slopes and collect at the base. These talus slopes in Banff National Park, Canada, came about gradually as **frost wedging** (the action of water freezing in cracks and opening them) loosened rocks that then fell into aprons around the base of steep rocky cliffs.

Table 11.3

Types of Mass Wasting

Falls, slides and flows, their varieties, characteristics, and speeds.

<table>
<tr><th colspan="2">Type of Movement and Its Characteristics</th><th>Rock</th><th>Debris (Coarse and Mixed Sediment/Soil)</th><th>Fine-Grained Sediment/Soil</th><th>Speed (with Typical Values Given in Parentheses)</th></tr>
<tr><td>Falls</td><td>Moves through the air</td><td></td><td>Debris fall</td><td>Earth fall</td><td>Very fast</td></tr>
<tr><td rowspan="2">Slides—movement as a unit</td><td>Translational slide—Movement on a slide plane</td><td>Rock slide or rock block glide</td><td>Debris slide or debris block glide</td><td>Earth slide</td><td>Fast to very fast</td></tr>
<tr><td>Rotational slide—movement on a curved surface</td><td colspan="3">Slump</td><td>Slow</td></tr>
<tr><td rowspan="7">Flows—different parts move at different speeds (chaotic movement). Usually the top and center moves faster than the bottom and sides.</td><td>Mixed with air</td><td>Rock avalanche</td><td>Debris avalanche</td><td>Mud avalanche</td><td>Extremely fast (about 70 km/h to over 400 km/h)</td></tr>
<tr><td>High water content</td><td></td><td></td><td>Mudflow</td><td>Very fast (1 km/h to over 100 km/h)</td></tr>
<tr><td>Moderate water content</td><td></td><td>Debris flow</td><td></td><td>Slow to fast (1 m/yr to 100 km/h)</td></tr>
<tr><td>Low water content</td><td></td><td colspan="2">Earthflow</td><td>Slow (from a cm to several hundred m per hour)</td></tr>
<tr><td>Interstitial ice</td><td colspan="3">Rock glacier</td><td>Slow (less than 5 m/yr)</td></tr>
<tr><td>Dry</td><td></td><td></td><td>Grain flow</td><td>Moderate (from 0.1 to 35 m/s)</td></tr>
<tr><td>Movement by expansion and contraction. Upper part moves faster than lower.</td><td colspan="2">Creep</td><td>Soil creep</td><td>Very slow (from fractions of a mm to a few cm per year)</td></tr>
</table>

bedding plane (Figures 11.9b and ■ 11.11). Initially, the rock is one unit and breaks up into smaller units as it moves. The term **landslide** refers generally to any fast-moving mass wasting, but in another more restricted sense, it means rapid movement of surficial material on a sliding surface. An example is the landslide at Hong Kong's Po Shan Road in Figure 11.1.

Slumps involve movement of material in large blocks on a curved slip surface. The sliding surface curves both at the land surface and below ground in a spoon shape, or sloping bowl (Figure 11.9c). Typically, slumps occur where streams or ocean waves eroded the base of the slope of thick unconsolidated material. When movement occurs, the slump rotates downward from the higher toward the lower edge of the bowl.

Flows

Snow and Ice Avalanches involve air mixing into the flow of moving turbulent particles of frozen water. Commonly, an avalanche occurs where a weak layer of powder, sitting on top of hard, refrozen snow, begins to slide. You may not think of snow and ice as sediment and rock, but many geologists do, and so they consider avalanches as a type of mass wasting. If a large slab of ice and snow starts moving on a steep enough slope, it begins to break up, incorporates air, and turns into an avalanche. The deadly, fast-moving, ground-hugging flow also churns up relatively harmless, billowing snow that rises cloudlike from the avalanche base. Loose snow can also start moving to trigger an avalanche. Skiers moving across snow layers have triggered many avalanches, and even loud noises can act as a trigger. To reduce avalanche risk, patrols shoot large guns at the unstable snow-laden slopes to trigger avalanches under controlled conditions.

Debris Avalanches are the fastest-moving type of mass wasting (Figures 11.2, 11.9d, and ■ 11.12). They get their speed in a way similar to snow avalanches by the mixing of air in with the debris. This allows it to move very fast, generally faster than 100 miles per hour. Debris avalanches are mixtures of air, water, mud, and rock. Like snow avalanches, they may produce a billowing cloud above a more dangerous, ground-hugging flow. Both debris and snow avalanches commonly move along a narrow track and leave elongate scars (Figures 11.9d and 11.12).

Mudflows are fast-moving, fluid mixtures of water and predominantly fine to sand-sized sediment (■ Figure 11.13). If water exceeds about 60%, then the flow is a sediment-laden stream and not a mudflow. Although not as fast as debris avalanches, mudflows can still achieve speeds up to 80 miles per hour. In arid and semiarid areas, intense, sudden rainstorms easily trigger mudflows by mobilizing accumulated sediment into torrents of mud. News reports call these "mudslides," but geologists prefer not to use this term because the form of movement is a flow, not a slide. A special type of mudflow is a **lahar,** which occurs when water mobilizes the thick deposits of pyroclastic material and volcanic sediment that surround stratovolcanoes. A volcanic eruption sometimes, but not always, accompanies a lahar.

Debris Flows are slower than mudflows and contain less water and more large rock material (■ Figure 11.14). Generally, if more than half the debris in a flow is coarser than sand, then it is a debris flow. Debris flows may be similar to wet concrete in viscosity, texture, and flow characteristics.

Bernard Pipkin

Figure 11.11

Rockslide

Rock block glide at San Pedro, California. This slide has moved periodically since 1929.

USGS

Figure 11.12

Debris Avalanches

Heavy rain in October 1985 triggered these debris avalanches in Panuelos, Puerto Rico. The scars, tracks, and zones of deposition are all visible in this photograph.

Lyn Topinka/USGS

Figure 11.13

Mudflows

Mudflows are fairly fluid and tend to spread out if not confined in a channel. This is the May 18, 1980 mudflow from the east side of Mount St. Helens, looking from Upper Muddy River, Washington. The river cut down through the mudflow deposits exposing them in cross section in the foreground. Photograph date: October 24, 1980.

USGS

Figure 11.14

Debris Flow

A debris flow destroyed and buried this house at Slide Mountain, Nevada, in late May 1983. Rapid snowmelt picked up debris as the meltwater traveled through a canyon. When the debris flow emerged from the canyon, it destroyed five houses, covered a highway, and killed one person. Geologists had predicted this type of hazard for this location a decade earlier.

A Rock Glacier forms in cold, dry, mountainous regions where ice cements rock debris together below the surface (see also Lab 17). The rock and ice move downslope in a way similar to the way glaciers move, but at a slower speed, less than 5 meters per year. A rock glacier is typically tongue, or lobe-shaped (■ Figure 11.15), and may or may not have a glacier (or flowing ice) core.

F.H. Moffit/USGS

Figure 11.15

Rock Glacier

This rock glacier in the Valdez district, Alaska Gulf region, has pronounced ridges across its movement direction showing the direction of flow. Other rock glaciers may have smoother surfaces. The Kotsina River in the foreground is *braided.*

Earthflows are movements of soil and weathered rock confined by well-defined lateral boundaries (Figure 11.9g). Movement is parallel to the surface, not rotational as in a slump. Flow speeds vary between 1 meter per day and hundreds of meters per hour. The end of an earthflow is lobe-shaped. Rainfall, saturation and disruption of the surface can easily renew movement of an earthflow that has stopped advancing.

Grain Flows are movement of loose, dry grains down a steep slope, such as sand moving down the steep face of a sand dune (Figure 11.3b).

Creep is widespread slow movement of soil on sloping surfaces (Figures 11.9h). The upper parts of the soil move faster than the lower parts, causing telephone poles and fence posts to tilt over. After creep tilts trees, however, they grow up toward the light, producing bent trunks known as pistol butts. Freezing and thawing, and wetting and drying cause successive expansion and contraction of material on a slope which result in creep. The movement of material by burrowing organisms can also aid creep. Although creep is very slow, moving only a few feet a year, it is surprisingly damaging to structures built on the slope. The annual price tag to repair damage due to soil creep is comparable to that of a moderately large natural disaster.

Complex Movements Many mass-wasting events or processes involve multiple types of movement. For example, a slump will often develop earthflow at its toe. If several types of movement are involved and are about equally important, then the mass wasting is a **complex movement** (■ Figure 11.16).

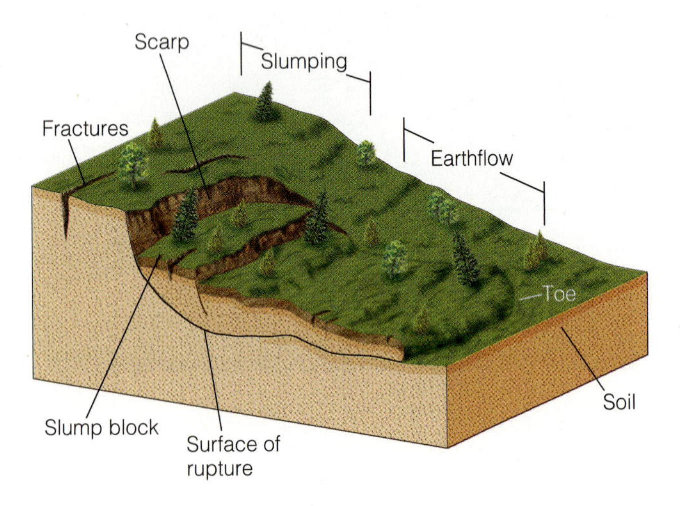

Figure 11.16

Complex Movements

This mass movement is a slump in the upper part, which turns into an earthflow in the lower part.

Mass Wasting in Sand Experiment

For this activity, read the whole experiment before you start. Work in groups of three to six students.

Materials needed:

- Large tray or stream table with damp but not soggy sand
- Model houses (optional)
- Water containers—beakers, squirt bottles, or watering cans
- Stopwatch
- Protractor or clinometer
- Centimeter ruler

5. Build a sandcastle "mountain" out of damp sand in the middle of the sand tray. ■ Figure 11.17 shows a reasonable configuration, but yours can vary somewhat.

a. How steep is the steepest slope of the "mountain"? ______ Would this pile of sand remain standing at the same slope once the sand dried out? ______ Why or why not?

____

____

____

____

____

(a)

(b)

Figure 11.17

Mass Wasting in Sand Experiment

(a) Setup for Exercise 5. (b) Experiment in progress, during the addition of a second beaker of water.

b. Flatten the top of the "mountain" and make a slight depression there. Add water to the top gradually and continuously so as not to create a stream flowing down the mountainside. Have your lab partners provide a constant water supply.

c. You should expect more than one type of mass wasting. For each mass-wasting event, one student should time it and another student should measure the distance moved. Record the data on a separate sheet of paper.

d. For rapid events, time the duration of the event. For events that are too quick to measure, record the time as less than the shortest time you can measure, <½ s, for example.

e. For slow events, as soon as you notice a slight change, write down the time. Note the time again when the event changes in character, and measure the amount of movement at that time. Continue watching as new styles of movement develop.

f. Select four different types of events for which you have the best data, or if you had less than four different styles of mass wasting, you may include two of the same style. Number these events 1–4. Enter your results in Table 11.4.

g. On a separate sheet, describe what happened and include sketches of each event listed in Table 11.4. Label each with the event number. In your description, include the shape and style of movement, its speed and distance traveled, and whether it started and stopped as you added water. Did cracks develop, or was it more fluid? Did it fall, topple, flow, or slide? Was it complex? If so, describe different parts: how they moved, and where each style of movement occurred.

h. Calculate the speed of each of the four mass-wasting events in centimeters per minute. You can calculate a minimum speed even if an event was too fast to time. In such a case, use your minimum measured time in the calculation, then put a greater than symbol (>) in front of the answer.

i. Refer to the diagrams of different types of mass wasting in Figure 11.9 and to Table 11.3. Which types of mass movement did you cause on the sand mountain? Enter this information in the last column of Table 11.4.

j. Write an essay on the separate sheet summarizing your results and noting potential hazards to life and property on slopes and surrounding valleys. What caused the mass wasting you observed? Explain this in terms of the factors that influence mass wasting discussed earlier in this lab. Also, discuss the speeds you observed and compare them to speeds of mass wasting covered in the section on types of mass wasting and Table 11.3.

Identifying Types of Mass Wasting

In the exercises that follow, you will have the opportunity to examine a number of events, to try to recognize different types of mass wasting, and to learn details about their causes, movement style, and history.

6. ■ Figure 11.18 shows an area of the San Juan Mountains, Colorado, in an aerial photo stereo pair. Use a stereoscope to study the area.

Table 11.4

Mass Wasting in Sand

See instructions in Exercise 5.

Mass Wasting Event Number	Duration of Event (s)	Distance Traveled (cm)	Speed of Event (cm/min)	Type of Mass Wasting

Figure 11.18

San Juan Mountains, Colorado

Aerial stereo pair of part of the region.

a. What different types of mass wasting do you see in the area? For each type you recognize, list the name of the type, the style of movement (fall, slide, or flow), and the evidence you see to support your hypothesis.

____

____

____

____

____

____

b. How many lobe- or tongue-shaped bodies of mass-wasting material do you see?

______

c. Outline areas on one of the photos where mass wasting is happening. Either color-code or label the outlined areas to distinguish different types of mass movement.

If coloring, make a color-coded key in the margin.

d. What natural geographic, geologic, slope, climate, or weather conditions in the area help to cause these types of mass wasting?

___

___

___

What evidence exists on the photographs to support this idea?

___

___

7. Examine the topographic map of Lake San Cristobal and Slumgullion Slide in ■ Figure 11.19a. This is another case of mass wasting in the San Juan Mountains in Colorado. A photograph from in front (Figure 11.19b) and a cross section along the length of the slide (Figure 11.19c) also show the slide. This is one of the largest active landslides in the United States. The geology of the area is mainly volcanic, consisting of andesite and pyroclastic material. The Slumgullion Slide started moving about 700 years ago and parts are still moving today, although very slowly.

a. On the map and photograph, outline the extent of mass movement and draw arrows indicating its direction.

b. Look for evidence of falling, sliding, or flowing. Which do you think is happening here?

___

c. What evidence do you see for this?

___

___

___

___

___

___

d. Is the movement confined or not? ___

e. What specific type of mass wasting is occurring in the bottom quarter or third of the mass? ___
Does this area qualify as one of complex mass wasting? ___

f. Examine the samples of andesite and volcanic ash provided and refer to earlier readings in this lab. Volcanic ash commonly weathers to clay, especially smectite. How does the local geology at Slumgullion influence the stability of the land?

___

___

___

g. Slumgullion Slide is still moving very slowly today. Outline the presently moving area on the map in Figure 11.19a. How did you recognize the moving area?

___

h. Figure 11.19c shows that some of the movement of Slumgullion Slide has produced faults. On this cross section, label the types of faults shown. Refer to fault definitions in Lab 8 as needed.

i. Along the sides of the landslide, between the area that is moving and the stationary area, what type of faults are present to accommodate the movement?

___

Draw the appropriate fault symbols for these faults on the map. Refer to Lab 8 again for fault symbols.

j. Lake San Cristobal is the largest natural lake in Colorado. How did the lake form?

___

___

k. Evidence of a delta (where the stream once entered the lake) that is now upstream, above the level of the lake, suggests that the lake was once 200 feet higher than it is now. What event could have caused the lake level to drop?

___

___

___

___

l. If people had lived downstream at the time, what problems could the event in part **k** have caused for them?

___

___

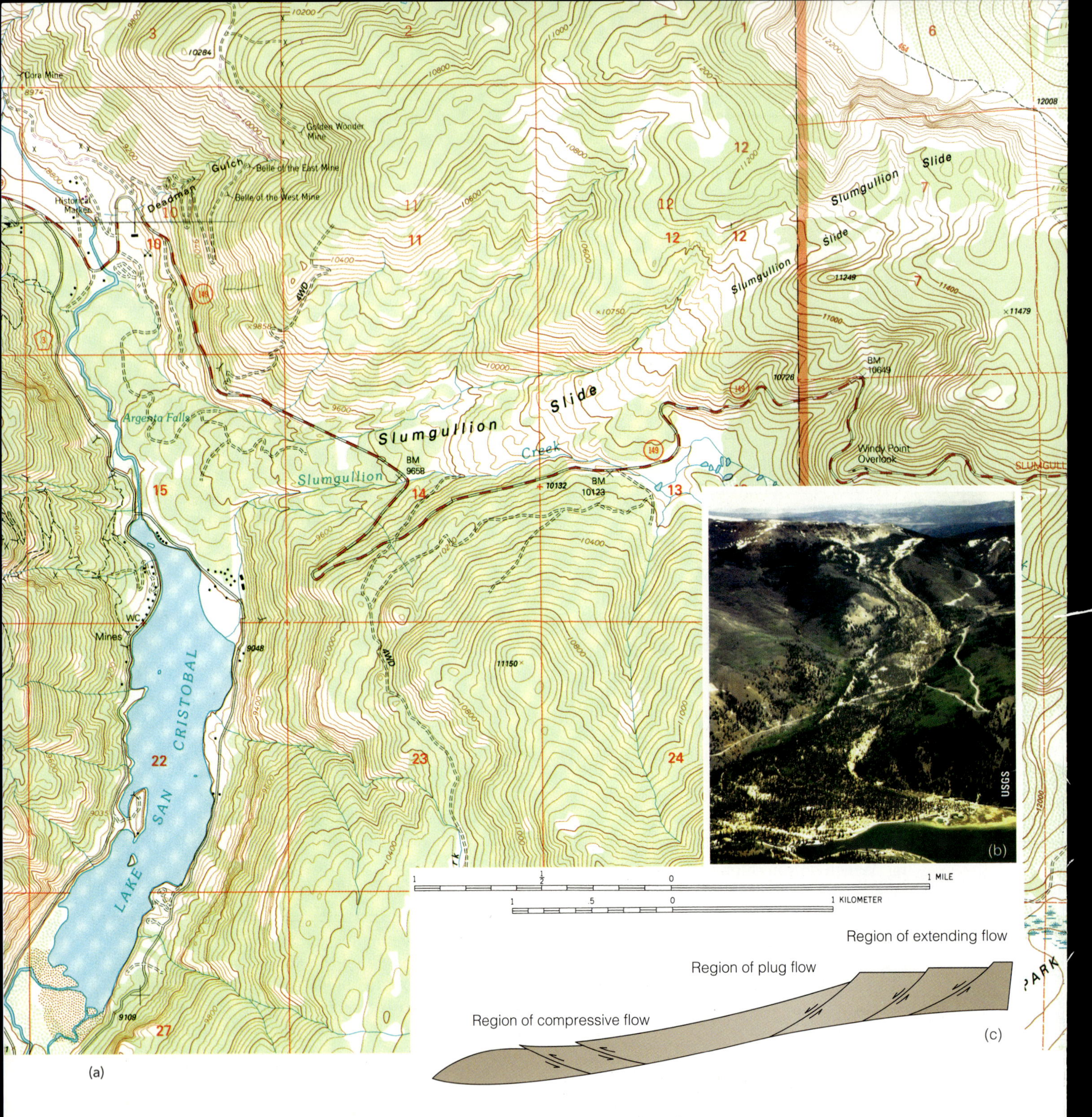

Figure 11.19

Slumgullion Slide, San Juan Mountains, Colorado

(a) Topographic map of Lake San Cristobal and Slumgullion Slide. Scale = 1:32,000, contour interval = 40′. (b) Photograph of Slumgullion Slide. (c) Longitudinal cross-section showing faulting where the slide is still moving.

m. Notice the houses on the part of the slide near the lake. What concerns would you have if you owned one of these houses?

____

____

____

n. What problems do you envision for Highway 149, which crosses the slide diagonally through Section 14, and the other road built on Slumgullion Slide by the lake?

____

____

____

8. Examine the first set of aerial photos in ■ Figure 11.20 a–b from 1968 and 1971 of apartments by the Hocking River near Ohio University. The rocks of the bluff are mostly horizontally layered Pennsylvanian siltstone and shale. As you work through these exercises, use stereoscopes to view the aerial photo sets that have two or three photos. Remember to line up the lenses with the same object on each of the two photos you are viewing. Stereoscopes will not work for single photos, as you need one photo for each eye to see the view in 3D.

a. In Figure 11.20b, using this photo triplet for 1971 alone, label the seven apartments with the highest level of risk from mass wasting. Give them ranks directly on the middle photo, starting with "1" for the apartment having the highest mass-wasting risk.

b. Now look at the whole photo set, including Figures 11.20c and d. How accurate was your assessment of risk in Exercise **a**? Did the apartment with the highest assigned risk from part **a** actually have the highest risk? How about the other apartments?

____

____

c. Study and observe the changes in the whole photo set, and in Figure 11.20f and describe them in ■ Table 11.5.

d. Were some of the apartments still in use in 1994? ________ How do you know?

____

____

e. Study the general area surrounding the apartments. What natural and what human factors led to mass wasting here?

____

____

____

____

____

____

f. If you inherited land above the highway nearby, what steps would you need to take before developing it?

____

____

____

How would the maintenance of the highway be affected?

____

____

____

9. In Figure 11.21, we see mass movement that has destroyed a number of homes in La Conchita, California, in 1995. Luckily, because of timely evacuation, no deaths or injuries occurred here the first time. However, On January 11, 2005, heavy rainfall for a long period in southern California reactivated the movement (Figure 11.22). This time people were unfortunate as there were ten deaths and the mass movement destroyed 13 houses and damaged 18 more.

a. Describe the shape of the surface on which the movement occurred in 1995 (■ Figure 11.21a).

____

b. For the main mass, was the movement in 1995 primarily in large blocks or chaotic and jumbled together with complete mixing of the parts?

____

Evidence?

____

____

____

(a)

(b)

Figure 11.20

Mass Wasting Near Hocking River, Ohio

A series of aerial photos and a map of apartments near Hocking River, south of Ohio University, Athens, Ohio. Stereo pairs and stereo triplets can be viewed in 3D, but single aerial photos cannot. (a) 1968 aerial photo; (b) 1971 stereo triplet.

(c)

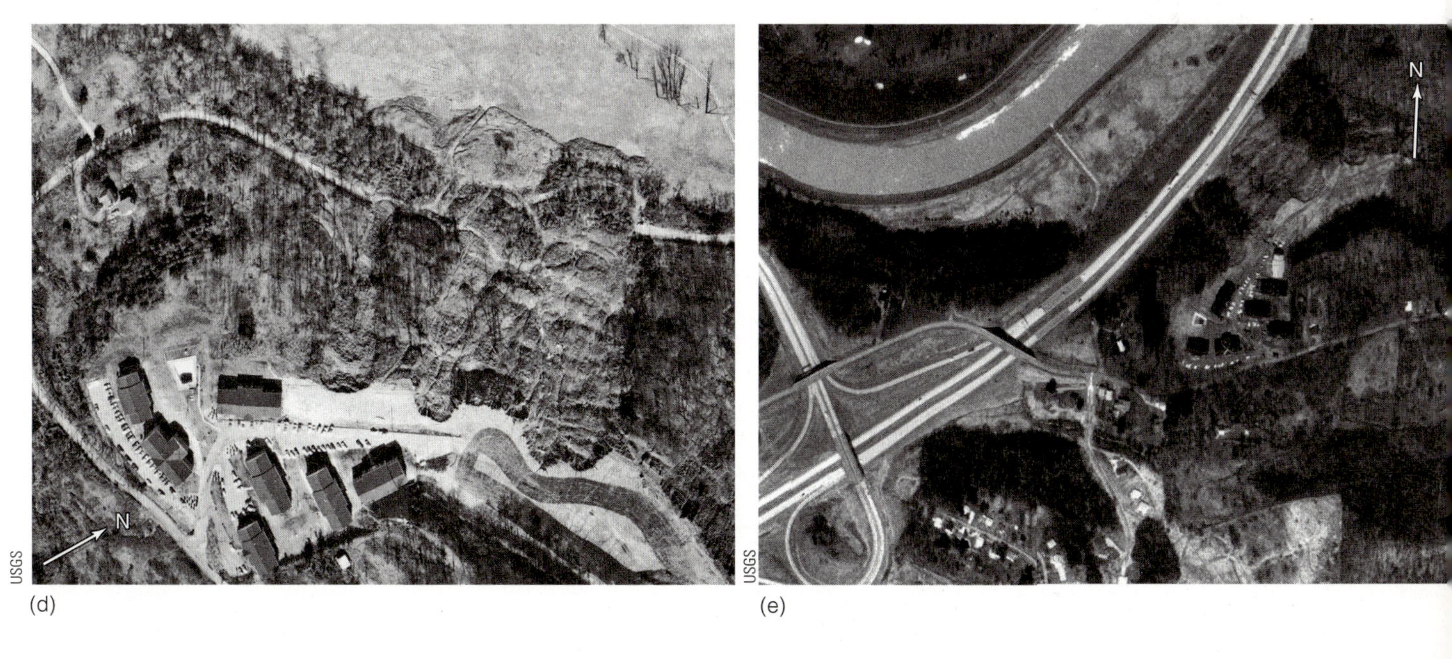

(d) (e)

Figure 11.20

Mass Wasting Near Hocking River, Ohio—*Continued*

A series of aerial photos and a map of apartments near Hocking River, south of Ohio University, Athens, Ohio. Stereo pairs and stereo triplets can be viewed in 3D, but single aerial photos cannot. (c) 1975 stereo pair; (d) 1976 aerial photo; (e) 1994 aerial photo.

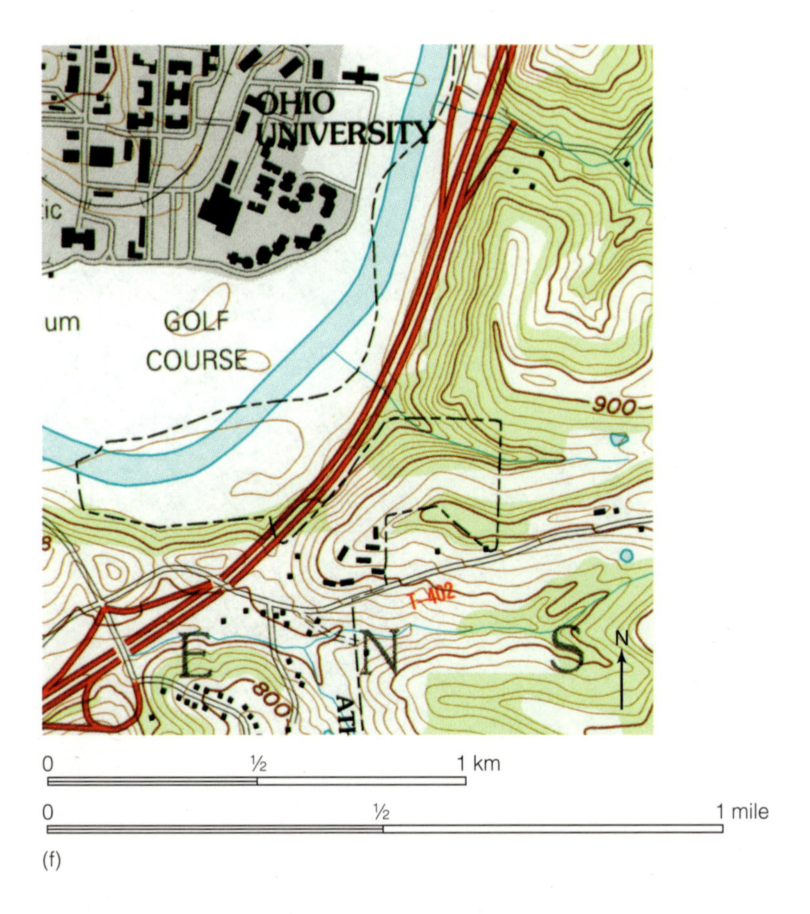

Figure 11.20

Mass Wasting Near Hocking River, Ohio—*Continued*

A series of aerial photos and a map of apartments near Hocking River, south of Ohio University, Athens, Ohio. (f) 1995 topographic map; contour interval = 20′.

c. Was the movement in 1995 of the main body a fall, slide, or flow?

____

What type of material seems to make up the 1995 mass?

____

d. Was the 1995 movement rotational or parallel to the original slope?

____

e. What would you call this type of mass wasting?

____

f. Compare the map in Figure 11.21d with the area in 11.21b and c, and outline where the landslide occurred on the map. The map was made before the landslide occurred. Using the photographs and the map, sketch a cross section of the landslide on a separate piece of paper. Label various aspects of the landslide and its deposits on your sketch, including the surface of fracture along which the material moved. Refer to Figure 11.9 to help you.

g. One reason this region has steep slopes is that it is experiencing an overall uplift rate of 0.1–0.7 mm/yr. Use the map to determine the angle of the slope before it failed. You can calculate the angle using trigonometry: the angle is equal to the arctangent (arctan, atan, or $\tan^{-1}$ on a calculator) of the rise in feet ________ over run (horizontally) in feet. ________ (1 mile = 5,280 feet). Or you can measure the angle by drawing the rise and run to scale in a triangle, and measuring

Table 11.5

Progression of Mass Wasting at Hocking River, Ohio

See instructions in Exercise 8.

Year	Number of Landslides Visible	Relative Area of Landslides Compared to 1968	# of Apartments Lost/Removed Due to Mass Wasting	Types of Mass Wasting How Have the Landslides Changed?	Describe Road Below the Apartments
1968		1 × same area			
1971					
1975					
1976					
1994					
map 1995		*			

*Use vegetation to judge landslide size roughly.

the slope angle with a protractor.

Angle = ___________________

Although it may not initially seem very steep, remembering what you have learned about angle of repose will indicate that this is a very steep angle for saturated unconsolidated material.

h. Use all three photographs for 1995 to examine the vegetation cover and lack of it in different parts of the mass-wasting feature in Figure 11.21, especially on the right side of the displaced mass. What suggests different type of movement here?

____

____

i. What type of mass wasting occurred on the right side of the slide in 1995?

____

j. Examine Figures 11.21c and d. How many houses did this mass movement in 1995 destroy? ______

k. What evidence exists that mass wasting has occurred in the general vicinity before?

____

____

____

Outline the older slump in Figure 11.21b and c.

l. Examine the new mass wasting that occurred in January 2005 in ■ Figure 11.22. On another sheet of paper, compare and contrast this movement with the 1995 movement.

m. Circle the area of La Conchita in Figure 11.21a that was not buried under mass-wasting deposits in 1995 but was

Figure 11.21

Mass Wasting at La Conchita, California

(a) Oblique aerial photograph: In the Spring of 1995, mass movement traveled into the seaside town of La Conchita, near Santa Barbara. Evacuation prevented any deaths or injuries. (b) La Conchita in October 2006 in a wider oblique aerial photograph. (c) Vertical aerial photograph showing the 1995 scar and older mass wasting scars to its left (NW). (d) Topographic map 1:16,000, contour interval = 20′.

Kevork Djansezian/AP Photo

Figure 11.22

More Mass Wasting at La Conchita

On January 10, 2005, the mass movement at La Conchita reactivated, burying several houses and killing 11 people. Photo taken January 11, 2005 while rescue workers attempt to find survivors among the debris of homes and sediment (yellow box outlines the area). A retaining wall, built between 1995 and 2005, is visible near the base of the slope.

buried in 2005 in Figure 11.22. Determine the number of houses that were destroyed or severely damaged in 2005.

______

n. Figure 11.22 shows the barrier that was built between 1995 and 2005 intended to protect homes from the landslide. Evaluate its effectiveness. Were any homes saved by the barrier? ______ Were any homes damaged specifically because of the barrier? ______ What is your evidence for this? Do you think the barrier was beneficial or detrimental overall? What factors in addition to loss of life and building destruction should you consider in your answer? Write the answers to these questions on the separate sheet of paper.

o. If you look closely at Figure 11.22, within the yellow box, you can see the debris from houses the day after the landslide. What does the condition and character of the building debris tell you about how this part of the mass-wasting event moved in 2005? Write the answer to this question on the separate sheet of paper.

p. What are the different alternatives for the people and the government of the town of La Conchita regarding the landslide? Discuss the pros and cons of political and business considerations with your fellow classmates. On the separate sheet of paper, write an essay detailing your discussion.

10. Search the Internet for a recent landslide event that is close to your area, or use one that your instructor provides. Describe the event and then answer the questions below.

____

____

____

____

____

____

____

____

____

a. What factors most influenced this event? Be as specific as possible.

____

____

____

____

b. Name the type of mass wasting if information is available.

____

c. Were there damages to life or property?

______ If yes, list the damages; if no, explain why not.

____

____

____

____

____

d. Look again at the bulleted section at the beginning of this lab. Having learned about mass wasting through experimentation and calculation, how has your perspective changed toward these hazards and why?

____

____

____

____

____

11. In the space below, enumerate and describe the types of triggering events and processes that initiated mass wasting, as discussed or explored in this Lab.

____

____

____

____

____

____

____

____

LAB 12

Streams and Rivers

OBJECTIVES

- **To understand the variations and origins of different types of stream valleys and channels**
- **To understand the characteristics of streams and how they are interconnected**
- **To understand the processes that form the main stages of stream (fluvial) erosion and to identify these stages from maps, air photos, and topographic profiles**
- **To understand stream flooding and recurrence interval**

Parvinder S. Sethi.

Figure 12.1

A Stream

Little Pigeon River, Smoky Mountains National Park, Tennessee, has a moderate gradient here and coarse sediment that makes up its *bedload.* Even the largest boulders move sometimes.

A **stream** (■ Figure 12.1) is any body of water that flows in a natural channel. For geoscientists, a **river** is simply a large stream (■ Figure 12.2). The word **fluvial** is often used in discussions about streams as an adjective meaning "having to do with streams or rivers"; for example, fluvial erosion means erosion by a stream. Streams may vary in size and appearance, but a few simple rules govern the resulting *landforms*:

- Gravity causes water in streams to flow down slope. The motion of the water induces the stream to erode rock, transport and deposit sediment, and modify the shape of its valley.
- The stream transports sediment by rolling, sliding, or hopping particles along the bottom (**bed load, Figure 12.1**), and by suspension in the water (**suspended load, Figure 12.2**). The chemical properties of water also permit it to carry sediment in solution (**dissolved load**).
- When a stream's velocity or its volume increases, its ability to carry sediment increases, and it will erode material from its bottom and sides. The most dramatic changes in a stream's channel shape occur during flooding.
- When a stream slows, or when infiltration or evaporation reduces the amount of water, its capacity to carry sediment diminishes, and fluvial deposition begins.

The area occupied by the water in a stream is the stream's **channel.** A stream with multiple channels is **braided.** If a stream has a very curving course it is **meandering.** The **floodplain** (Figures 12.2 and ■ 12.3a) is a low area of land adjacent to a stream channel, which holds the overflow of water during a flood. Natural meandering streams flood on average every 2 to 3 years. When they do, they carry sediment onto the floodplain. Where the river first overflows its banks, the water slows and deposits the largest quantity of sediment. This process builds up **natural levees** along the side of the channel. A stream's **valley** is the region that the stream directly and indirectly erodes, and it extends well beyond the stream channel. Mass wasting (see Lab 11) by landslides, earthflow, or soil creep, erodes the valley until the debris reaches the channel, where the stream carries it away. The area drained by a stream is the **drainage basin** (Figure 12.3b). The boundary separating one drainage basin from another is known as a **drainage divide** and occurs above the stream channel. Lesser divides separate different branches or **tributaries,** which are smaller streams that join the stream along its course.

Figure 12.2

Meandering River

An aerial photo of the Mississippi River in Missouri showing meanders and an extensive floodplain. The water is muddy due to the *suspended load.*

STREAM GRADIENT AND SINUOSITY

Because gravity is the driving force for stream flow, the steepness of the slope of a stream, known as the **stream gradient** (Figure 12.3a), is important in determining certain characteristics of the stream.

$$\text{stream gradient} = \frac{\text{change in elevation}}{\text{horizontal distance}}$$

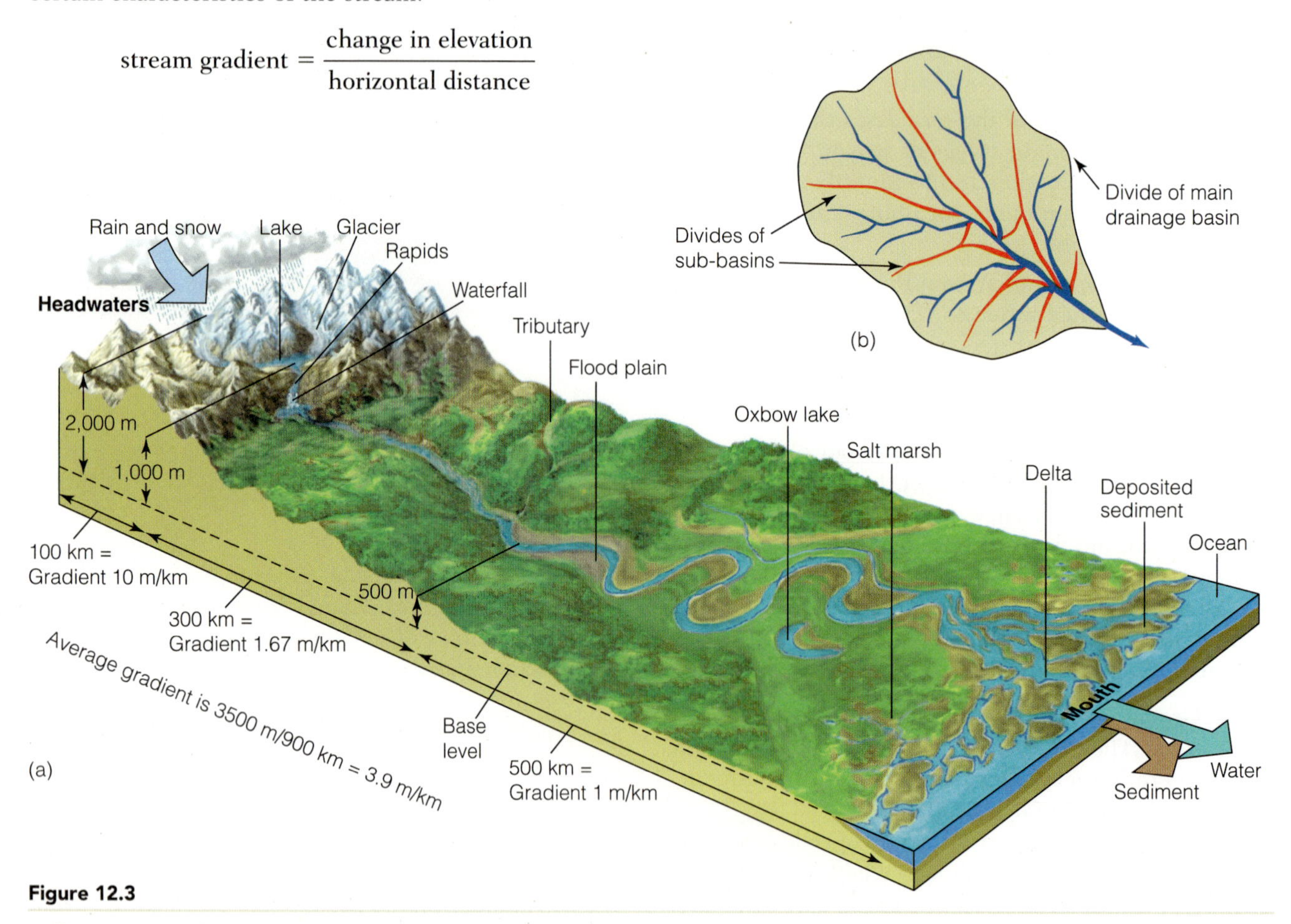

Figure 12.3

Changes along a Stream's Course

(a) Changes in a stream from its headwaters to its mouth. *Base level* is sea level for this stream. Stream gradients are the drop in elevation divided by the horizontal distance the stream travels. The gradient can be determined for the whole stream—the average gradient—or for a segment of a stream. (b) Schematic diagram showing a drainage basin and its **drainage divides,** the boundaries that separate the drainage basin from other drainage basins. Lesser divides in red separate branches or **tributaries.**

Figure 12.4

Mancos River and Several Canyons

From USGS, 15-minute topographic map of Soda Canyon, Colorado. Scale 1:62,500. 1 mi = 1.014 in. CI = 50 ft.

The gradient is typically in ft/mi or m/km. The changes in a stream along its course are primarily due to the change in gradient. The **head** of the stream is at its beginning or source, and this is also the highest point in the stream. Therefore the gradient near the headwaters of a stream is steep, and diminishes gradually to very low values in the floodplain and delta near the **mouth.** The mouth of a stream is its end, where it enters a large body of water or another river. Notice the changes in valley and channel shape and width with gradient in Figure 12.3a.

1. In the exercises that follow, we will first observe each stream, its valley, and its gradient to see whether there is a pattern before we start actual measurements. Notice that the maps involved have several different scales and contour intervals.

a. Examine the stream in Johnson Canyon near the center of the map in ■ Figure 12.4 and compare it to the

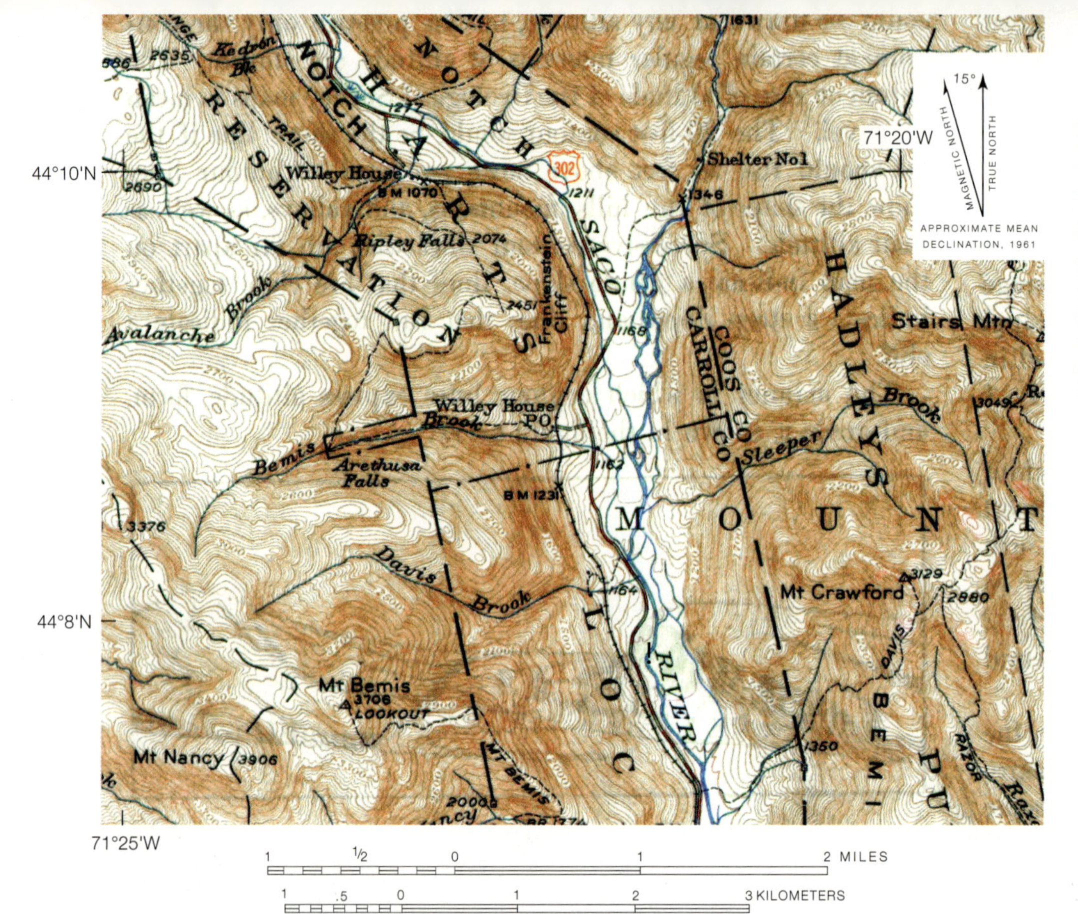

Figure 12.5

Map of Saco River

From USGS, topographic map of the Crawford Notch 15-Minute Quadrangle, New Hampshire. Scale 1:62,500. CI = 20 ft.

Saco River on the map in ■ Figure 12.5. In your own words, describe how these streams and their valleys are different.

____

____

____

b. Now, compare the Patoka River in ■ Figure 12.6 to the first two streams. What is different about this stream and its valley?

____

____

c. Of the three, which stream has
(a) the steepest gradient?

____

(b) the widest valley?

____

(c) the steepest valley walls?

____

(d) the most winding course?

____

(e) the most hilly divides?

____

Sinuosity is a measure of how much the course of a stream winds or curves. To determine sinuosity, measure the length of the channel along the stream's path and divide by the straight-line distance down the valley. ■ Figure 12.7 shows three streams and their sinuosities. A **meandering stream** has a sinuosity of greater than 1.5 and has the most winding pattern. In the next exercises, we look at some measurements of gradient and sinuosity of a few streams.

2. Before you examine the measurements, think about gravity and slope and how they would affect a stream. What *hypothesis* (see Appendix) can you formulate that would relate the sinuosity of a stream to its gradient? If a stream is flowing on a gently sloping surface, how would you expect its sinuosity to compare to a stream traveling down a steep slope?

____

____

____

____

____

____

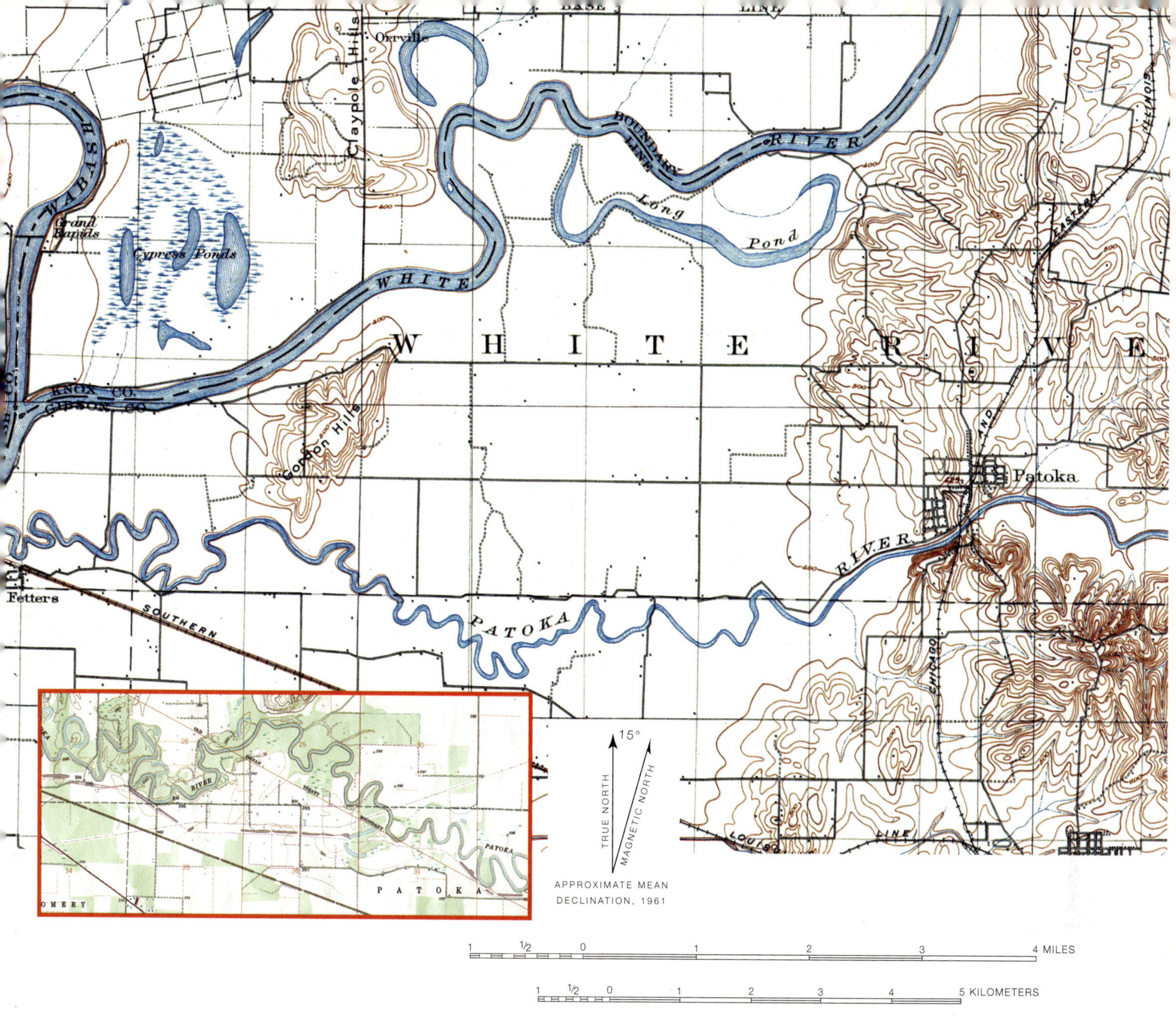

Figure 12.6

Maps of Patoka and White Rivers

USGS topographic map of the Princeton 15-Minute Quadrangle, Indiana-Illinois; surveyed in 1901. Scale 1:80,000. CI = 20 ft. *Inset:* part of the same area from the East Mount Carmel 7.5-Minute Quadrangle, Indiana-Illinois reduced to a matching scale. This map was revised in 1989. CI = 10 ft.

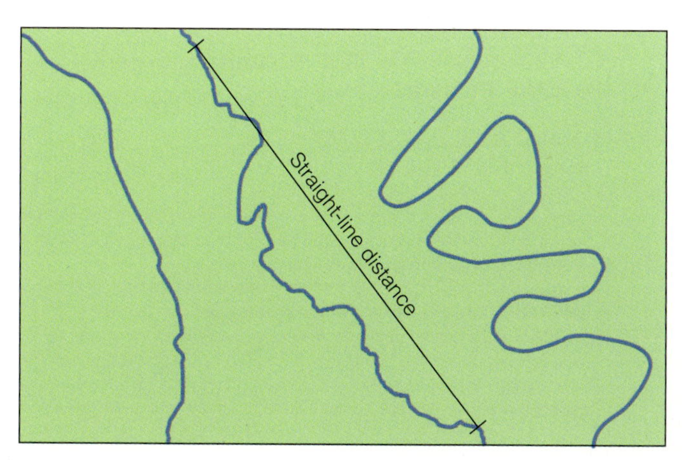

Figure 12.7

Three Streams of Different Sinuosity

The sinuosities from left to right are 1.11, 1.37, and 3.10, respectively.

Gradient versus Sinuosity

Materials needed:

- String
- Graph paper

3. Examine the maps for each of the streams mentioned in Table 12.1 and mapped in Figures 12.4 to 12.6, and 12.8 to 12.10,

a. What is the shape of the stream: Is it relatively straight, fairly sinuous, sinuous, or very sinuous? Enter this in Table 12.1.

b. For each stream fill the blanks in Table 12.1, using the calculations indicated in the table footnotes.

c. How do you measure the channel length of a winding stream on a map?

______

______

______

______

d. Using the data in Table 12.1 and graph paper, graph stream gradient (horizontal axis) versus sinuosity (vertical axis). Completely label your graph. Should zero be included on the vertical axis? ______

Table 12.1

Stream Characteristics

Stream	Elevation Change (feet)[1]	Channel Length (miles)[2]	Straight-line Distance (miles)[3]	Gradient (feet/mile)[4]	Sinuosity[5]	Channel Width (feet)[6]	Floodplain Width (feet)[7]
South Platte River, Colorado (Fig. 12.8)	5090 − 5060 = ______	3.4	2.57			90	~4300
		Shape of stream:					
Henrys Fork of the Snake River, Idaho (Fig. 12.9)	3.75	between +s at "Henry's" & "Mile 5" = ______	0.91			230	~12,000
			Shape of stream:				
North Fork of the Swannanoa R., North Carolina (Fig. 12.10)	3000 − 2300 = ______	6.7	5.9			20–30	~700
		Shape of stream:					
Patoka River, Indiana-Illinois (Fig. 12.6)	390 − 380 = ______	8.5	4.87			85	~14,000
		Shape of stream:					
Schoharie Creek, New York	1880 − 1580 = ______	6.5	6.1			60	~1300
		Shape of stream: fairly straight					
Saco River, New Hampshire (Fig. 12.5)	1260 − 1000 = ______	4.16	3.52			30−130	~575−2300
		Shape of stream:					
Mancos River, Colorado (Fig. 12.4)	5720 − 5560 = ______	6.02	3.91			30	~1000
		Shape of stream:					

[1]To determine elevation change, subtract the elevations of the two contours (listed in feet). Each stream crosses the two contours listed.
[2]Channel length is measured along the winding path of the stream. Lay a piece of string along the winding stream path, mark the start and stop locations, then straighten the string and measure it using the scale.
[3]The straight-line distances were measured between the two points of known elevation, indicated in the second column, where contours cross each stream. This is the same part of the stream measured for channel length.
[4]To determine the gradient, divide the elevation change by the channel length.
[5]Sinuosity is measured as channel length divided by the straight-line distance.
[6]Channel width was measured from the width of the water at an average spot along the stream segment.
[7]Floodplain width was measured at an elevation of 20′ above the stream.

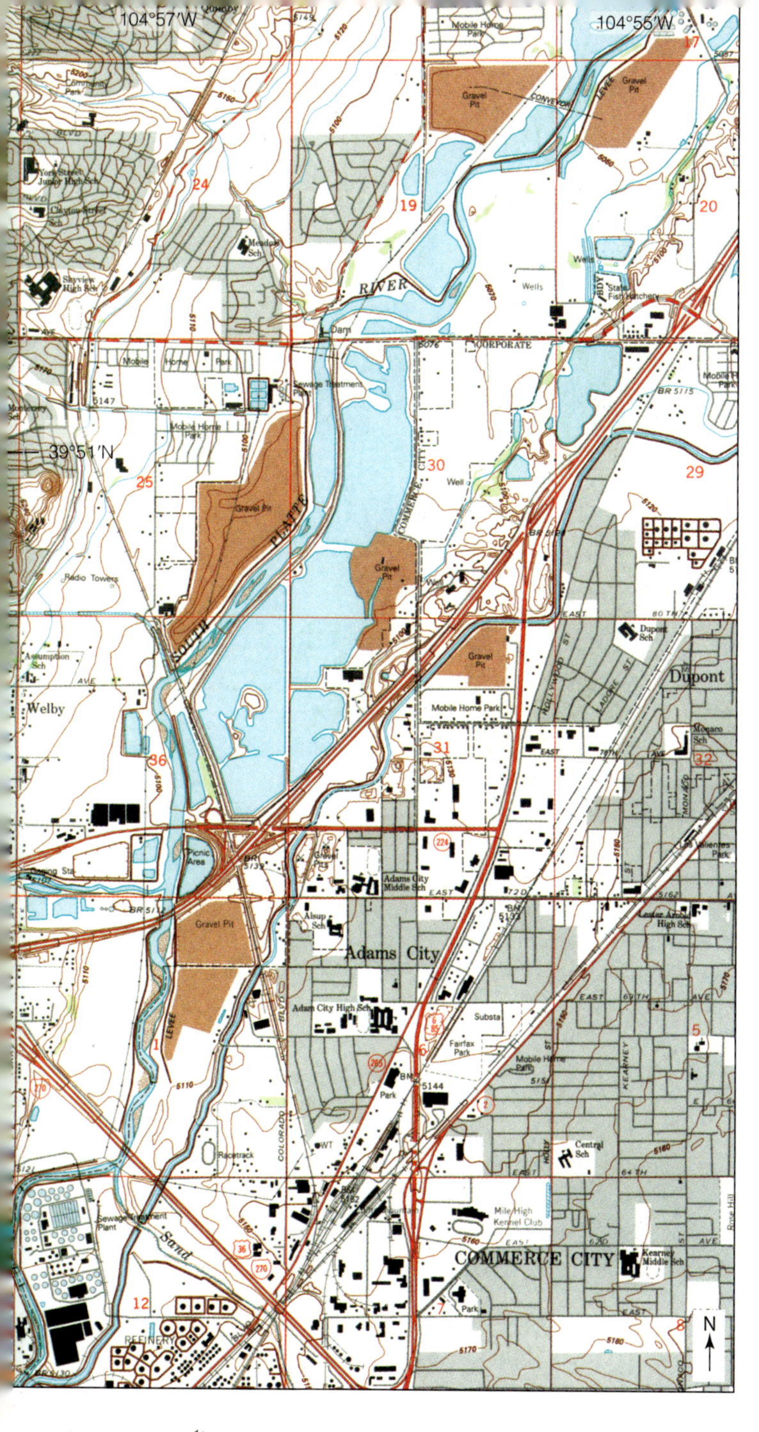

Figure 12.8

South Platte River

USGS 7.5-Minute topographic map of Commerce City, Colorado. Scale = 1:44,000. CI = 10 ft.

Why or why not? (*Hint:* Think about how sinuosity is calculated and what the possible values might be.)

____

____

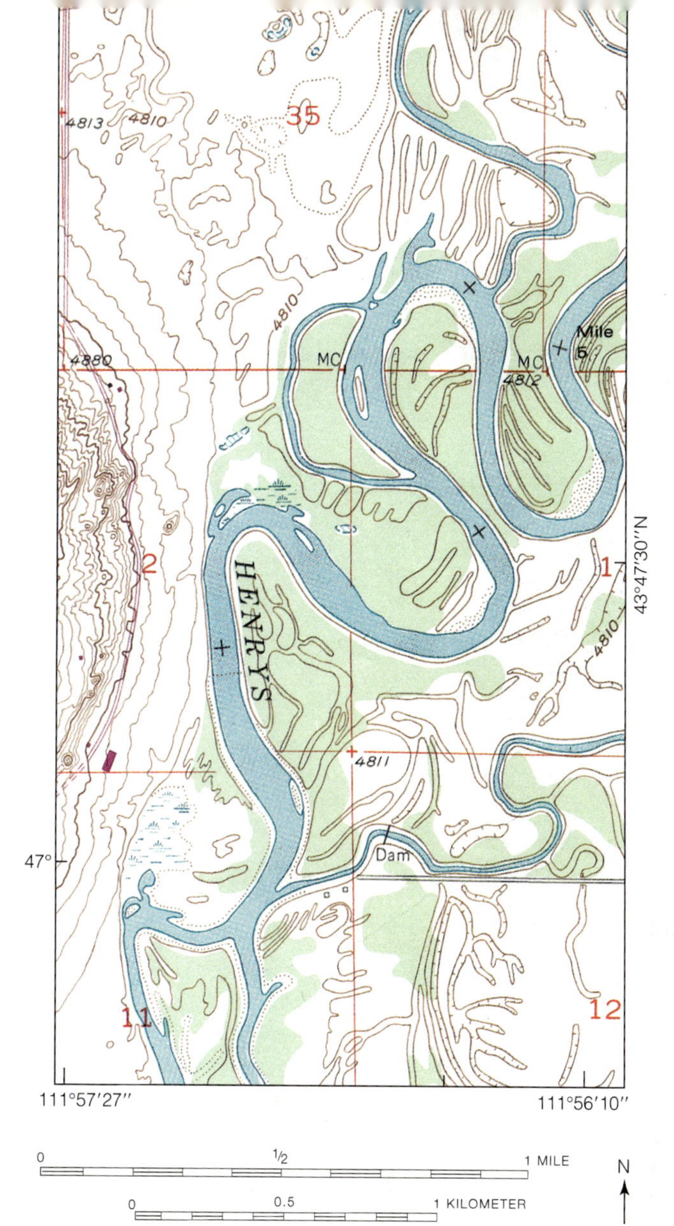

Figure 12.9

Henrys Fork of the Snake River

USGS 7.5-Minute topographic map of Menan Buttes, Idaho. Scale 1:24,000. CI = 10 ft.

e. Was your hypothesis in Exercise 2 correct?

______

f. Modify your hypothesis if necessary by answering the following question: What is the general relationship between stream gradient and sinuosity as seen in your graph?

____

____

____

____

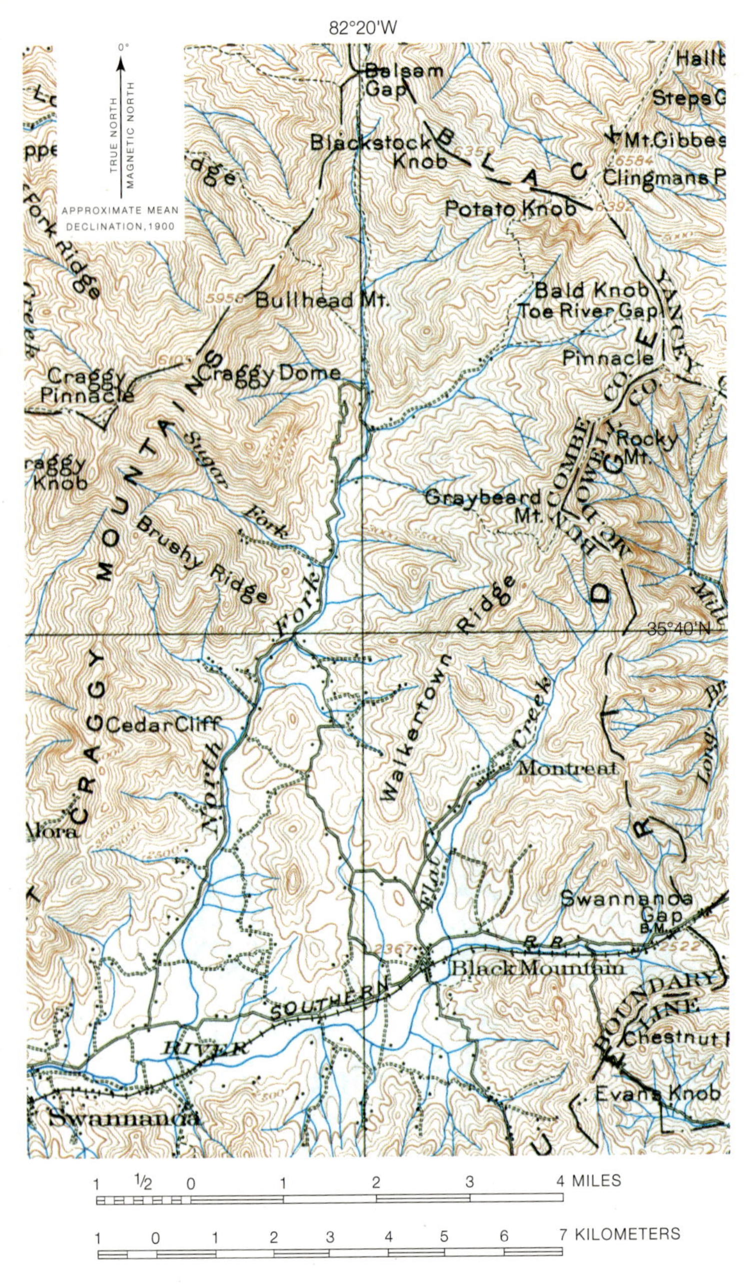

Figure 12.10

North Fork of the Swannanoa River

USGS topographic map of Mount Mitchell, North Carolina. Scale 1:125,000. CI = 100 ft.

4. Examine the data in Table 12.1 and make some additional observations:

a. What is the general relationship between stream gradient and channel width?

____

____

____

b. What is the general relationship between stream gradient and floodplain width?

____

____

____

STREAM EROSION AND ITS STAGES

Streams develop in three stages, commonly called *early, middle,* and *late stage,* as shown in ■ Figure 12.11a, b, and c, respectively. Streams are very effective at shaping the landscape, and cause more erosion than any other natural or human process. They can carve through an upland, forming canyons (early stage), widen those canyons into broad valleys in a hilly terrain (middle stage), and then reduce the hills to a broad plain (late stage). If the broad plain undergoes uplift, a fourth distinctive stage of stream erosion, *rejuvenation,* can develop, as shown in Figure 12.12.

Imagine an area uplifted to a plateau. After recent uplift, a stream has a steep gradient and plenty of power to cut downward through rock. This produces **early-stage erosion** (Figure 12.11a and d). Stream erosion in such an area undergoes changes with time. For example, downward erosion cannot continue indefinitely because a stream cannot erode downward below the level of its mouth. The level at the mouth is **base level,** which is sea level for those streams that flow into the ocean. Hard rock ledges may create a local base level and often produce cascades or waterfalls (Figure 12.11d). Lakes and reservoirs create temporary base levels for streams, but eventually fill with sediment. When enough sediment accumulates, the stream can begin to erode down through the lake deposits, and base level shifts downstream to the next lake or reservoir, or to the ocean.

If base level remains the same, then as the stream erodes downward, its gradient must decrease with time, and the stream progresses to **middle stage** (Figure 12.11b and e). With decreasing gradient, downward erosion slows, and sideways erosion (*lateral erosion*) becomes more important, producing a **late-stage** stream (Figure 12.11c and f). If uplift occurs or base level drops, a stream may again be able to cut downward; this produces a **rejuvenated** stream (■ Figure 12.12). These stages and their processes may overlap and are continuous. It may help you to refer to ■ Table 12.2 as you go through the next few pages, because it summarizes the differences in these four stages of stream erosion.

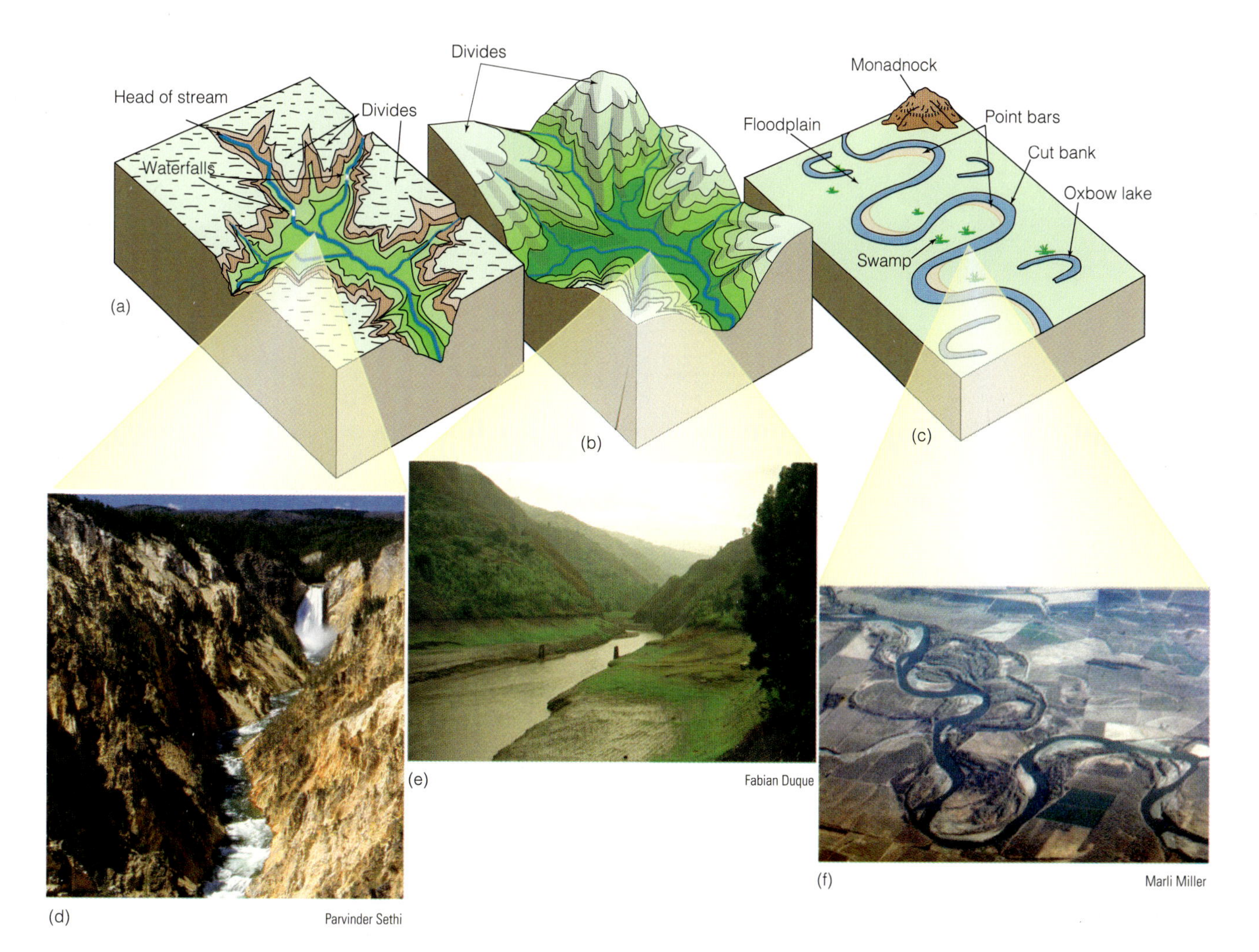

Figure 12.11

Stages of Fluvial Erosion

(a) Early stage, (b) middle stage, and (c) late stage. (d) Lower Falls and the Yellowstone River in the Grand Canyon of the Yellowstone. The Yellowstone River is in the early stage. (e) Valleys and hills around the Garagoa River in Columbia. This is middle-stage development. (f) This part of the Sacramento River in Northern California is a late-stage meandering stream.

5. Use the information from the preceding paragraphs and exercises, and also the relationship between sinuosity and stream gradient, for the next questions.

a. How would you expect a stream's valley to change with time in terms of the channel width, the valley width, and the sinuosity?

____

____

____

____

b. Consider a large block of land that is uplifted high above its surroundings to form a plateau. At first, the area is high above its base level and can have a steep stream gradient. While the stream gradient is steep, which type of erosion is more likely—downward or lateral? (Circle one.)

c. Later, the gradient decreases. Which type of erosion is more likely then?

____

Figure 12.12

Rejuvenation, Entrenched Meanders, and Natural Bridges

The process of rejuvenating a stream with development of entrenched meanders and natural bridges. (a) From a late-stage stream as in Figure 12.11c, downward erosion begins as base level drops or the area undergoes uplift. (b) Rejuvenated stream in entrenched meanders. (c) Somewhat like meander cutoffs in a late-stage stream, rejuvenated streams can cut off their entrenched meanders and develop a natural bridge. (d) 600-ft-deep entrenched meanders of the Colorado River at Dead Horse State Park, Utah. (e) A natural bridge over 200 ft above the bottom of the canyon at Natural Bridges National Monument, Utah. This is Sipapu Bridge (see Figure 12.24) and spans 267 ft.

Downcutting

A stream can only erode material where the water is in contact with the substance. If the stream were the only eroding force, it would develop a slot canyon with vertical cliffs, as shown in ■ Figure 12.13. This is an extreme case of downcutting. However, vertical cliffs such as this are not common; mass wasting usually causes the slopes to fall, slide, slump, flow, or creep, making the slope gentler and bringing sediment to the stream, which then carries it away. These processes contribute to the erosion of the surface, producing a valley that looks quite different from the canyon shown in Figure 12.13. Let's simulate this with a simple experiment.

Materials needed:

- Tray or pan, small enough to manipulate when half full of sand
- Dry sand

6. Tip and jiggle the tray of dry sand until the sand is smoothly sloping from one end to the other. Imagine a straight stream running down the middle of the sloping sand. Use your finger to simulate the stream's erosive force and move or scrape away sand along the stream's imaginary path. Do this repeatedly until a valley forms.

a. Does your finger touch every area where sand moves? _______ The same is true of the water in a stream. What other force moves the sand?

____

What do we call this general process?

_______________________________ This same process acts to make stream valleys.

Table 12.2

Features of Different-Stage Streams

Stream Stage	Early	Middle	Late	Rejuvenation
Valley shape	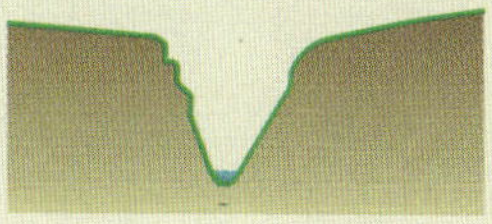	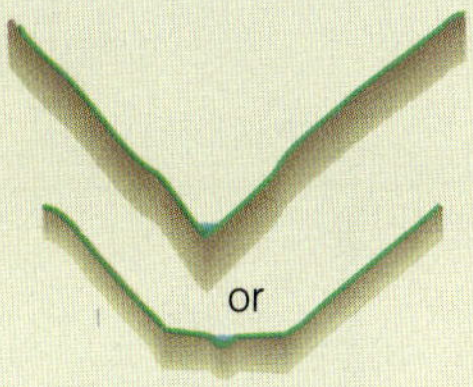 or		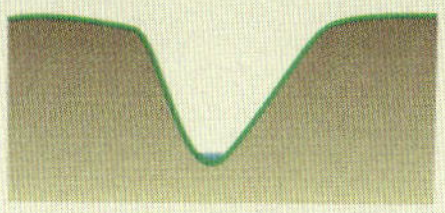
Stream shape (sinuosity)	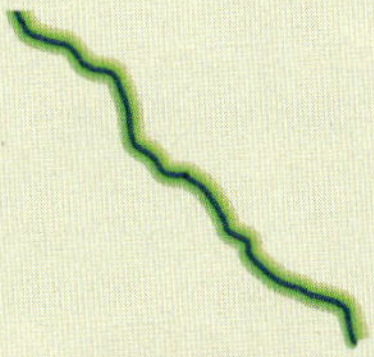		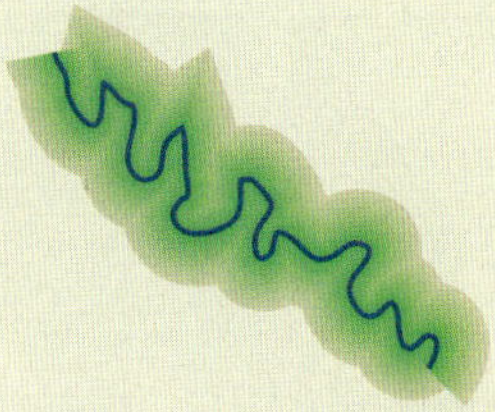	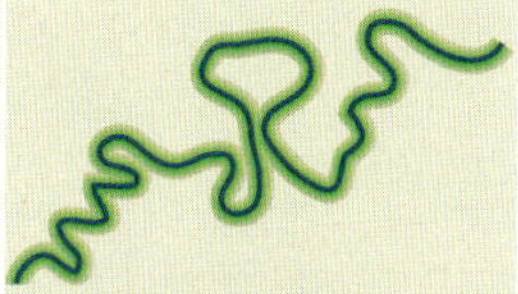
Shape of divides	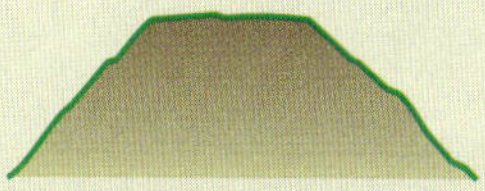	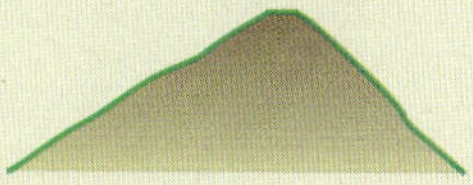		
Proportion of sloping area	• Small	• Largest	• Smallest	• Small to medium
Processes	• *Downcutting,* fluvial erosion and abrasion that deepens the channel, occurs when the stream has a steep gradient. • *Mass wasting* creates a steep V-shaped valley sides. • *Headward erosion* lengthens the stream valley at its head.	• Less *downcutting* • *Slope retreat* widens the stream valley. • *Mass wasting* by soil creep and landslides, is a dominant process. • *Reduction of divides* • *Widening of valleys.* • Headward erosion diminishes or ceases.	• *Sedimentation.* lower gradient toward base level leads to deposition of sediment in point bars and floodplains. • *Lateral erosion* and *meandering*—wandering or migration of the stream across its floodplain. • *Flooding* and deposition of fine grained floodplain sediments and *natural levees* (raised areas of courser sediment near the channel forming during floods)	• *Rejuvenation*—uplift of a mature (late stage) meandering stream • *Downcutting* and *entrenchment* of meanders results from a relative drop in base level. Lateral erosion is abandoned. • Initial stages of *mass wasting* create steep V-shaped valleys.
Resultant stream characteristics and landforms	• Low sinuosity • *Steep gradient* • *Rapids and waterfalls* • Coarse sediment • *Narrow V-shaped valleys* • *Plateaus* between stream branches make wide flat divides • Drainage of area incomplete—fewer streams per unit area	• Low to moderate sinuosity • *Floodplains* begin to develop • A *moderate gradient* • Valleys still retain a V-shape—the *V becomes wider* with a slight rounding of the channel. • Rounded to sharp divides—areas between tributaries are *mountainous or hilly*, not plateaus. • *Maximum relief* with fewer flat horizontal land areas • More complete drainage—an increase in streams per unit area results in an increase in drainage.	• High sinuosity • Low gradient allows meandering. • *Deltas* form when currents lose their velocity upon entering a large body of water (a gulf, lake, or ocean) and drop their sediment load. • *Meanders, point bars,* and *cut banks.* Stream may be *braided* with mid-channel sandbars. • Cutoff meanders and *oxbow lakes* (Figure 12.17) form • *Floodplain* becomes well developed, drainage decreases as land flattens and swamps form • *Monadnocks*—isolated hills or erosional remnants of more resistant rock • Divides become nearly flat, close to stream elevation with low relief.	• High sinuosity—inherited from the stream when it was in the late stage. • *Steep gradient*—high sinuosity and steep gradient is an unusual combination for a stream. • *Entrenched* or *incised meanders* (meanders in V-shaped canyons) form terraces or plateaus along the steep bank. The meandering shape of the path remains as a clue to the stream's past. • *Plateaus*—wide flat divides

Claudia Owen

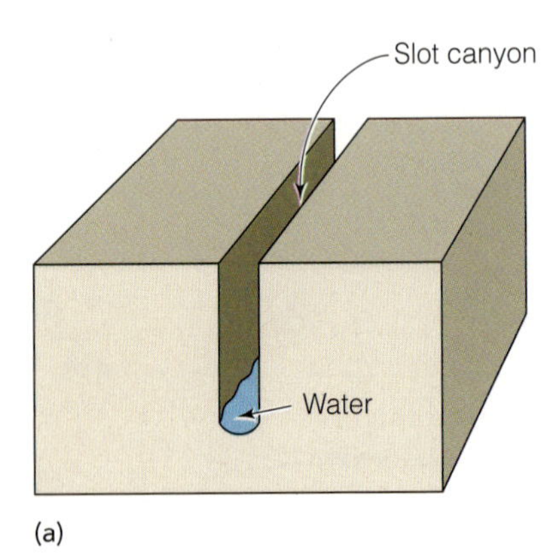

Jonathan Turk

Figure 12.13

Slot Canyons

(a) Block diagram of a downward cutting stream with no mass wasting of the valley walls. (b) Slot canyon in Utah. A flash flood killed 11 people hiking in a similar slot canyon in 1997. The rapid water rise and the steep sides of the canyon made it impossible for the hikers to avoid the flood.

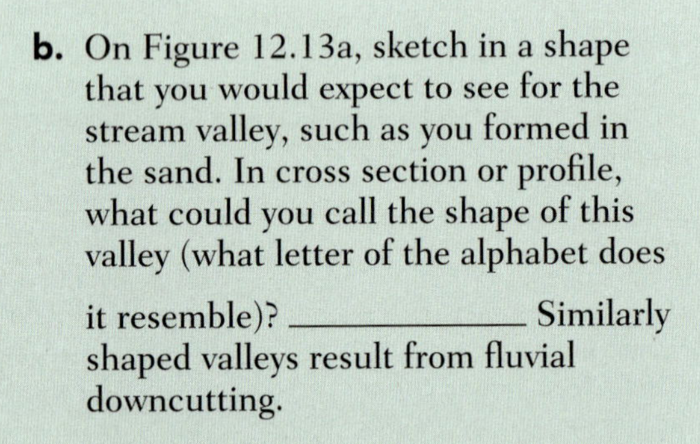

b. On Figure 12.13a, sketch in a shape that you would expect to see for the stream valley, such as you formed in the sand. In cross section or profile, what could you call the shape of this valley (what letter of the alphabet does it resemble)? ____________ Similarly shaped valleys result from fluvial downcutting.

7. Examine the topographic map of an early-stage stream in Figure 12.4. Describe a typical early-stage stream by circling the correct words in parentheses to complete the following sentences: In topographic maps of early-stage streams, contours are closer together near (*divides* / *streams*) and farther apart near (*streams* / *divides*). The stream is fairly (*sinuous* / *straight*).

8. Draw a topographic profile for this stream for the line from A to A′ in Figure 12.4 using ■ Figure 12.14. The line spacing is designed to give 3× vertical exaggeration where each line corresponds to an index contour line. Add your own vertical lines at the correct end positions for the profile (A and A′) on Figure 12.14.

9. Imagine the cliff dwellers who lived in the area in Figure 12.4 just above the C in Johnson Canyon. Using your knowledge of stream valleys and topographic maps, mark the easiest path (no cliffs) to travel from there to the dwellings on the north side of the canyon closest to the line A–A′.

Figure 12.14

Topographic Profile

For Exercise 8, Johnson Canyon, an early-stage stream in Figure 12.4 (3× vertical exaggeration).

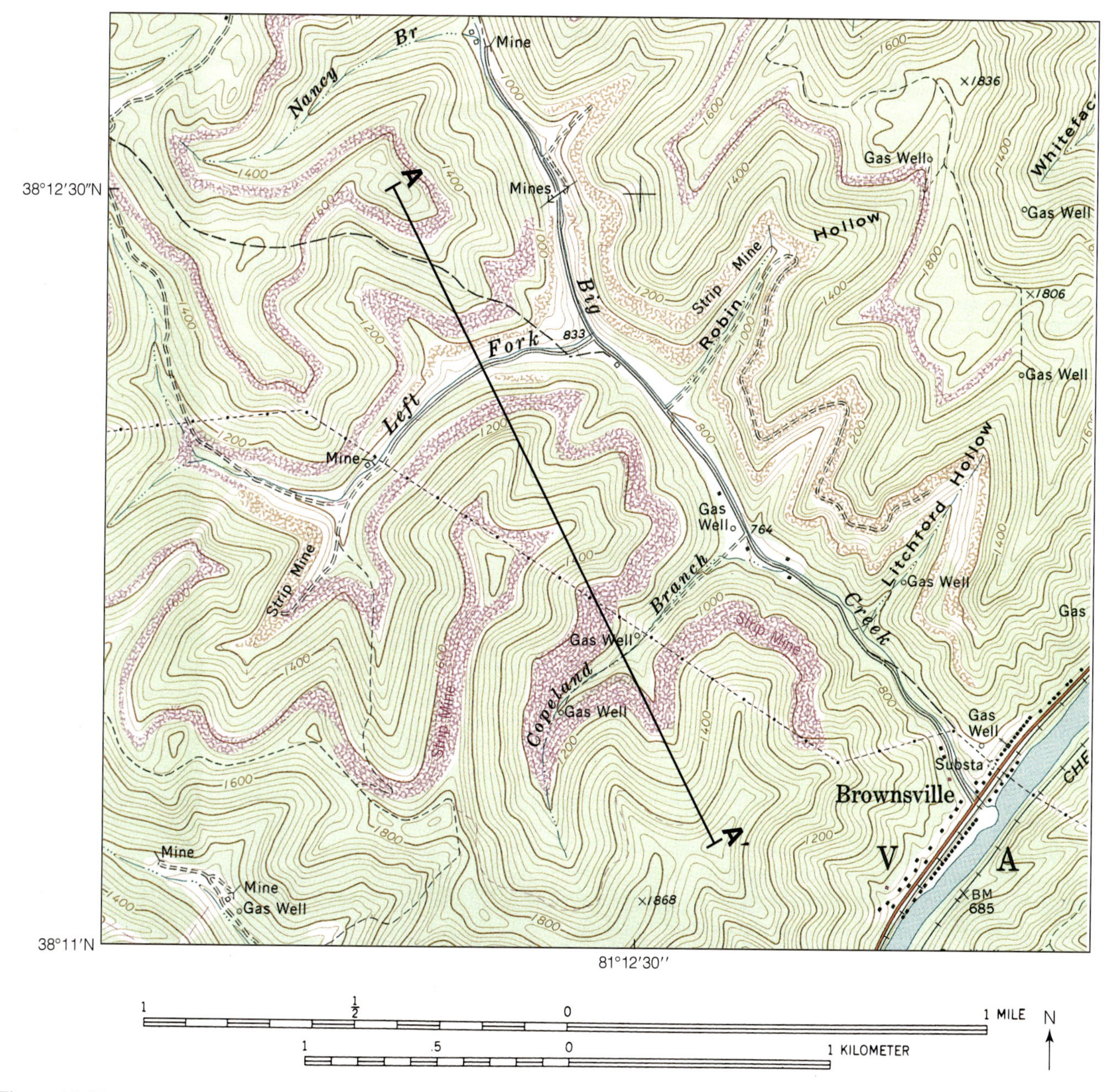

Figure 12.15

Map of Middle-Stage Streams

Topographic map near Fayetteville, West Virginia. Scale 1:24,000. CI = 40 ft.

10. Describe a typical map view of a middle-stage stream (■ Figure 12.15) by circling the correct words in parentheses in the following sentences: In maps of middle-stage streams, the spacing between contours varies (*greatly* / *little*) across the map compared to a map of an early-stage stream. Slopes near the stream are (*steeper than* / *the same as* / *gentler*) than slopes on the divides. Flat areas are (*rare* / *common*).

11. Compare early- and middle-stage streams:

a. Using primarily the index contours, draw a topographic profile for a middle-stage stream for A to A′ in Figure 12.15, using ■ Figure 12.16. Determine and plot the elevations at each low point (valley) and high point (ridge) that crosses the profile line before connecting the points on your profile.

Figure 12.16

Topographic Profile

For Exercise 11a, a middle-stage stream in Figure 12.15 (2× vertical exaggeration).

b. Compare differences between profiles of the early- and middle-stage streams that you completed in Figures 12.14 and 12.16.

Lateral Erosion and Meandering Streams

Streams that cause lateral erosion can also deposit sediment. This occurs when a stream with a low gradient near base level has a sinuous course known as **meandering** (Figure 12.2). Late-stage streams are meandering streams. **Meandering streams** migrate across their main flow direction as they meander. This causes lateral erosion where the flow is fastest, on the outside of meander bends forming **cut banks.** Deposition of sediment occurs in the areas of slowest flow, on the inside of the bends along **point bars.** As meanders migrate, they tend to get larger. At some point, the meander size becomes too great for the stream flow to maintain and a **meander cutoff** occurs where water takes a shorter route bypassing the meander loop. As sediment fills in the old channel next to the cutoff, the isolated meander becomes an **oxbow lake.** ■ Figure 12.17 shows this process.

12. Use the map of the Mississippi River in ■ Figure 12.18 to do the following exercises.

a. Describe the landscape before people modified it and during a season of low precipitation.

b. Describe the landscape during a season of very high precipitation.

c. Select five features commonly associated with late-stage streams, find them on the map in Figure 12.18, and sketch them on a piece of paper. You may be able to show multiple features on one sketch. Label each feature.

d. In the areas near McKinney Bayou in the central eastern part of the map, and near Park Place in the northwestern part of the map, are a series of contours that have an arcuate shape. What are the features here called? ______________

What do they tell you about the migration of the river?

e. On separate paper, roughly sketch the river's position at five times in the past using evidence on the map. Start with the river before it was situated at McKinney Bayou, and then progress in four more steps to the present river position.
Refer to late-stage stream processes in Table 12.2. What two main processes were occurring to move the river in this way?

f. How do the elevations right at the banks of the river compare with those of the general landscape farther from the river?

Why?

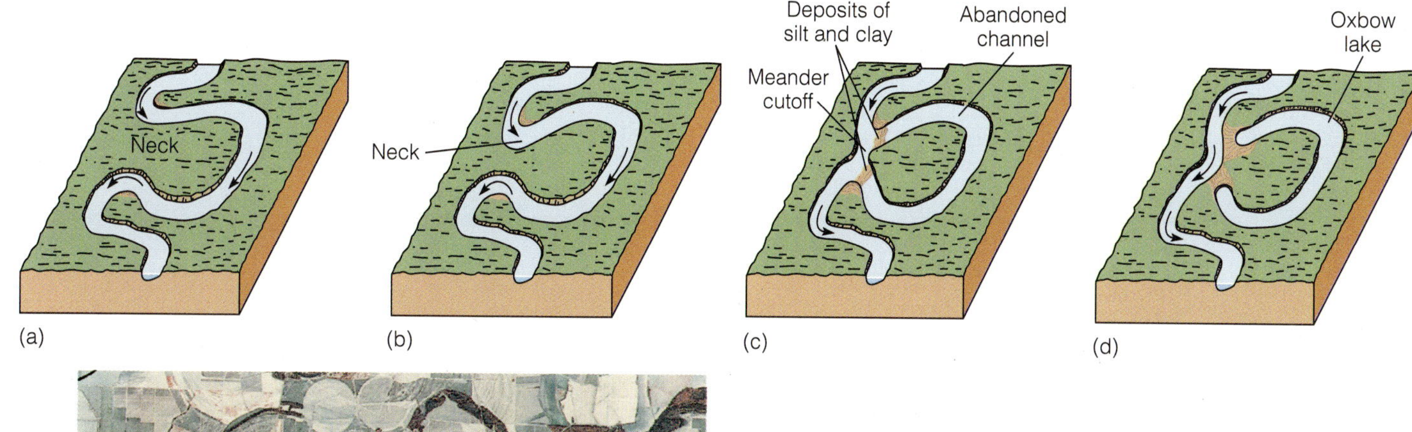

Figure 12.17

Meander Cutoff and Oxbow Lake Formation

(a) As a stream meanders, erosion on the outside of the bends, where the flow is the fastest, causes the meanders to grow. (b) Eventually, the meanders become so large that part of a bend upstream impinges on one downstream. During the subsequent flood events, the water overflows the channel and carves a cutoff between the two bends. (c) Deposition along the sides of the new cutoff channel blocks flow into the old channel, causing the stream to abandon it. (d) The abandoned channel develops into an oxbow lake. (e) Oxbow lake left by the Tallahatchie River, Mississippi. Inside the curve of the oxbow lake, old, still-visible point bars show where the river gradually migrated before the cutoff.

g. Roughly, sketch a generalized topographic profile across the stream in Figure 12.18 without trying to construct an accurate one for any one place.

Observing a Meandering Stream in the Stream Table

Materials needed:

- Stream table with sand and a flow-regulated water supply
- Sand bucket
- Scoop or spoon
- Clear acrylic sheet and washable marking pens, or toothpicks
- Small floating objects, such as parts of toothpicks
- Model houses or buildings (small boxes or other objects decorated to represent houses)
- Colored pencils

This experiment should help you understand how meandering steams change their course. Read the whole experiment and all of the questions before you start. Set up the stream table as follows and as shown in ■ Figure 12.19:

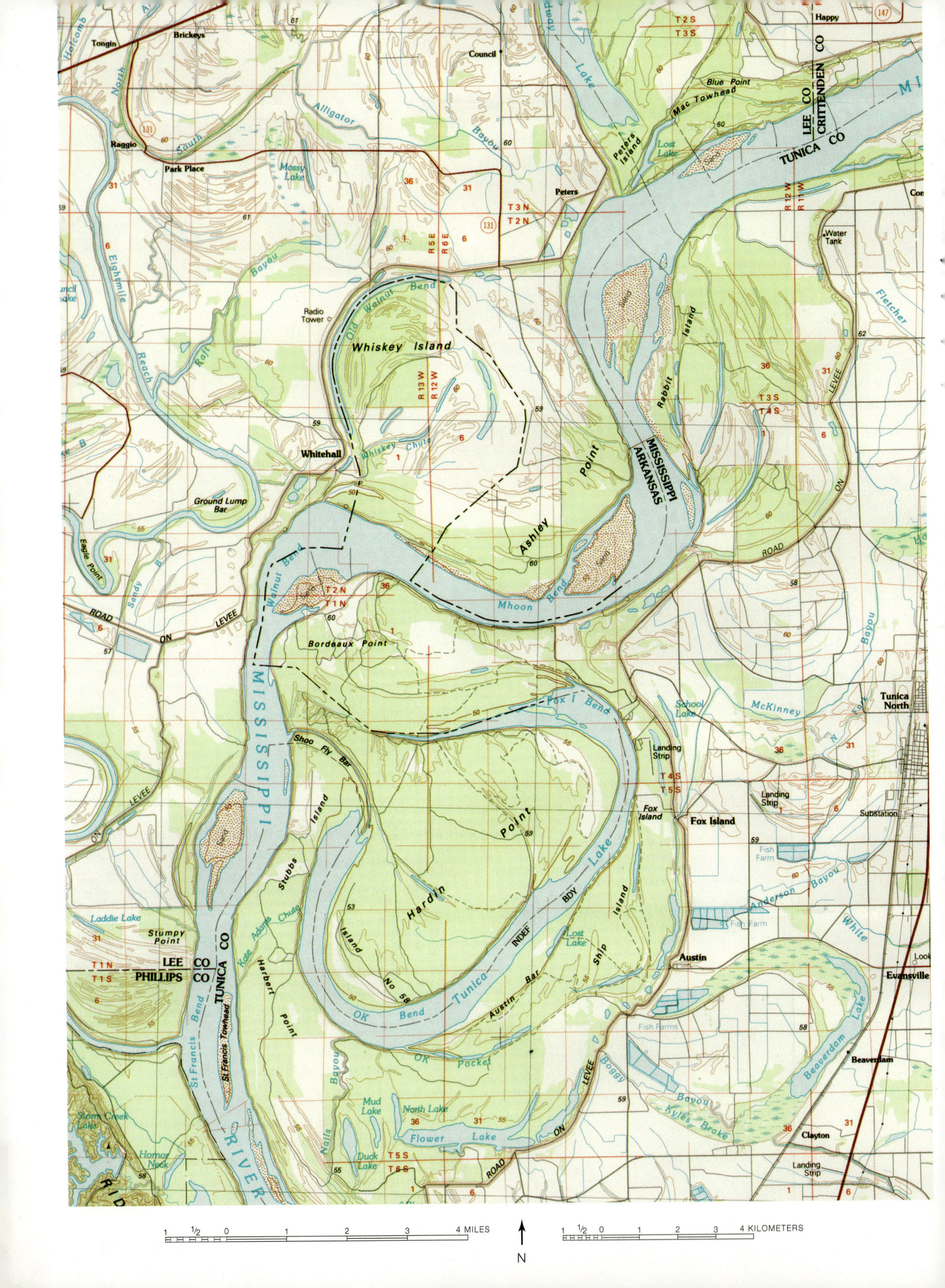

Tongin
Brickeys
Council
Happy
T2S
T3S
Raggio
Park Place
Alligator
Bayou
Mossy Lake
Blue Point
Mac Towhead
Lost Lake
Peters Island
Peters
LEE CO
CRITTENDEN CO
TUNICA CO
T3N
T2N
R5E
R6E
R12W
R11W
Water Tank
Eightmile
Reach
Radio Tower
Old Walnut Bend
Whiskey Island
R13W
R12W
Whiskey Chute
Whitehall
Rabbit Island
Fletcher
T3S
T4S
LEVEE
Ground Lump Bar
MISSISSIPPI
ARKANSAS
Ashley Point
Eagle Point
Sandy B
Walnut Bend
T2N
T1N
Mhoon Bend
ROAD
ON
LEVEE
Bordeaux Point
MISSISSIPPI
Fox I Bend
School Lake
McKinney
Bayou
Tunica North
Shoo Fly Bar
Landing Strip
T4S
T5S
Fox Island
Substation
Island
Point
Hardin
Lake
Fish Farm
Anderson Bayou
Stubbs
Kate Adams Chute
BDY
INDEF
Island
Laddie Lake
Stumpy Point
LEE CO
PHILLIPS CO
T1N
T1S
TUNICA CO
Harbert Point
No 58
Lost Lake
Ship Island
Austin
White
Evansville
Tunica
Austin Bar
OK Bend
St Francis Bend
St Francis Towhead
Fish Farms
Beaverdam Lake
Beaverdam
OK Pocket
Boggy Bayou
Kyles Brake
Bayou
Mud Lake
North Lake
Flower Lake
Clayton
Storm Creek Lake
Horner Neck
Nails
Duck Lake
T5S
T6S
ROAD
RIVER
Landing Strip
1 1/2 0 1 2 3 4 MILES
N
1 1/2 0 1 2 3 4 KILOMETERS

Figure 12.19

Setup for Stream-Table Experiment

For Exercises 13 to 15. The plastic sheet shows a tracing of the initial stream shape and size before the current is turned on.

- Shape the stream bed: With the water off, create a smooth, gentle slope from the inflow to the outflow lake. Dig out a small lake at the inflow end. Carefully scoop out a meandering path for the stream about 2 to 3 in wide (or as directed by your instructor) from the inflow to the outflow lake, placing the excess sand in the sand bucket and leaving the surface smooth. Be sure that the depth of channel along the length is constant at about 1 in.
- Mark the position of the stream channel for future reference. Choose one of the following methods:
  - You can lay the acrylic sheet over the stream table, and use the washable markers to mark placement of the sheet on the stream table and also the stream table position on the sheet. Then trace both edges of the stream channel on the acrylic.
  - Or you can place toothpicks on both sides of the bank right next to the edge, along its entire length, spaced close enough together to mark the location of the channel clearly. Push the toothpicks well into the sand so they will not wash away immediately. If your sand is not deep enough to lodge the toothpicks deeply, you will need to use the first method.

13. Draw the original position of the stream on a separate sheet of paper.

- Your instructor will tell you how to set the flow regulator.
- Have your small, floating objects ready to distribute into the flow once it starts and be ready to observe:
  - The line of fastest flow of the water along the length of the stream, paying particular attention to its behavior around meander bends
  - Erosion
  - Deposition
  - How the stream influences the houses or their front and back yards
- Start the flow and drop tiny, floating objects on the water to better mark the location of fastest flow.

14. Observe and record the behavior of the stream. You may need to set up the stream table a second time to make all of your observations.

a. Where is fluvial erosion occurring?

____

____

b. Where is fluvial deposition occurring?

____

____

Show both of these on your sketch with labels.

c. If you took a canoe on a stream similar in shape and flow to this one, what path would you follow so your canoe wouldn't get stuck on sandbars or rocks?

____

____

____

____

d. After about a minute (or when advised by your instructor), stop the water. Retrace the banks of the stream on the acrylic with a different color pen, or observe where the banks are now, compared to the position of the toothpicks.

Figure 12.18

Topographic Map of the Mississippi River (*facing page*)

Topographic map of the Mississippi River north of Helena, Arkansas. Scale 1:120,000. CI = 5 m.

e. Overlay the new stream position on the sketch of the original position of the stream you drew in Exercise 13. Use a different color to show clearly how the stream changed. Create a key indicating the significance of each color.

f. Based on your observations, formulate a hypothesis about how this stream will migrate in the future. Record your hypothesis on the same sheet that you used for your sketch in part **e**.

g. Place houses at various places right along the stream bank, especially where you think the stream might wash them away. Select one of the houses to draw onto your sketch and circle the letter from the choices below pertaining to its position. If the location of your house is between these points, circle the two it is between: (a) outside a meander, (b) inside a meander, (c) downstream side of a straight stretch, or (d) upstream side of a straight stretch.

15. Turn the water back on for another minute or two.

a. What happened to your house?

____

____

b. How good was your prediction from earlier? Was your hypothesis correct, or do you need to modify it? Write the modifications, if needed, alongside your prior hypothesis.

Long-Term Behavior of Meandering Streams There is insufficient time in a single lab to observe the long-term behavior of a meandering stream in the stream table, besides most stream tables do not have the correct configuration and sediment size to maintain a meandering pattern for long. Instead, to demonstrate some of the patterns of meander migration over many years, you can perform an experiment in which meanders migrate very rapidly.

Materials needed:

- A stiff, flat sheet of metal or plastic
- A container or sink to catch water
- A plastic squeeze bottle containing plain tap water

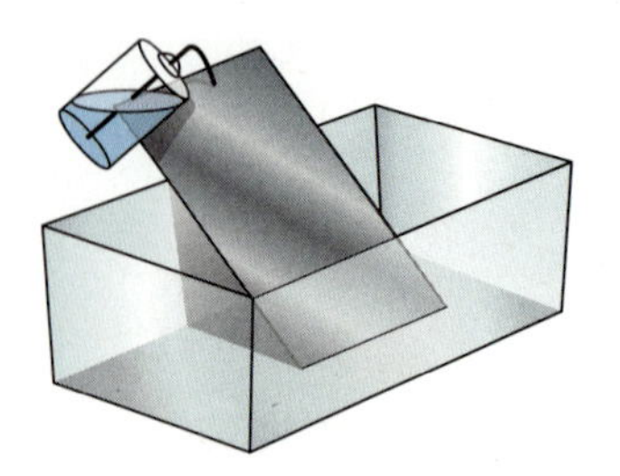

Figure 12.20

Rapid Meander Migration Experiment

Setup for the squeeze bottle experiment in Exercises 16 to 18.

16. Place the sheet tilted at an angle in the basin, as shown in ■ Figure 12.20. Place the tip of the squeeze bottle against the upper part of the sheet, and squeeze to produce a steady stream of water. Carefully observe what happens.

17. Observe the gradual migration of meanders (curves) as time passes. They migrate: (a) outward, (b) inward, (c) upstream, or (d) downstream. Circle two.

18. What happens to cause a meander to straighten out (known as a *meander cutoff*; Figure 12.17)? When does this happen? *Hint:* Carefully observe the volume of water at the time of the meander cutoff.

____

____

____

____

Natural meandering streams migrate and behave in a similar manner. One example is the Mississippi River. The Arkansas-Mississippi state boundary was established at the Mississippi River in 1836, when Arkansas was admitted to the Union (■ Figure 12.21), and its position has been governed by the **Boundary Rule** since that time. When a river changes course *gradually* over time, a state or county boundary on the river *moves with the river*; with a *sudden change* in the course of the river, the boundary *stays* where it was.

19. Examine the map in Figure 12.21 of the Arkansas-Mississippi state boundary and the Mississippi River.

a. Find how many places on this map the river is not at the state boundary. ________

b. What fluvial event or process would cause the river to depart from the state line? Use terminology from Table 12.2. Review the Boundary Rule above if necessary.

____

c. Where the river and state boundary are in the same place, you can see from the Boundary Rule that the river may not have necessarily been fixed in position since the boundary was established. They both may have changed gradually together. What kind of change in the river's course would allow it to move away from the state boundary, according to the Boundary Rule described above? *gradual / sudden* (Circle one.)

d. What fluvial processes have caused the course changes that *keep* the boundary with the river?

____

____

e. Since 1836, have any meander cutoffs occurred along these places where the river and state boundary correspond? ______

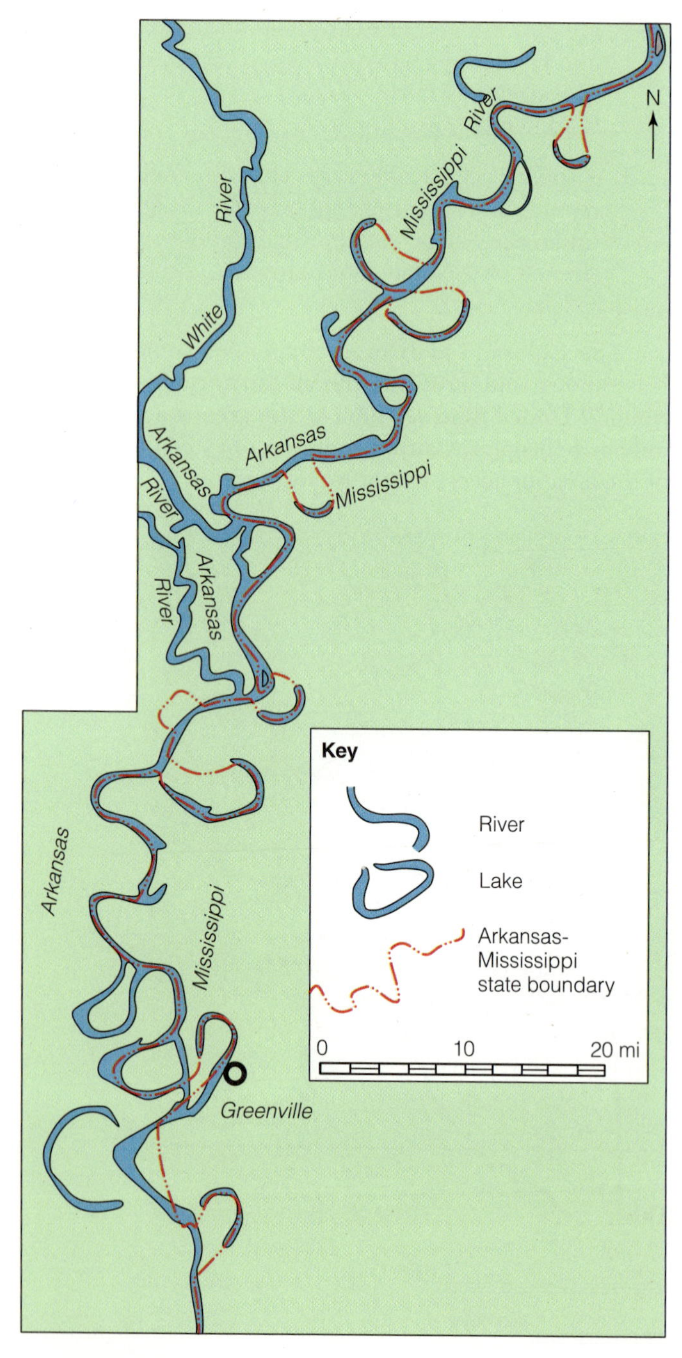

Figure 12.21

Map of the Mississippi River at the Arkansas-Mississippi Border

Mark Twain, in his book *Life on the Mississippi* (1883), wrote about the meandering Mississippi:

> The Mississippi between Cairo and New Orleans was 1215 miles long 176 years ago. It was 1180 after the cutoff of 1722. It was 1040 after the American Bend cutoff: It has lost 67 miles since. Consequently, its length is only 973 miles at present. . . .
>
> In the space of 176 years, the lower Mississippi has shortened itself 242 miles. That is an average of a trifle over one mile and a third per year. Therefore, any calm person, who is not blind or idiotic, can see that in the Old Oolitic Silurian Period, just over a million years ago[1] next November; the Lower Mississippi River was upwards of 1,300,000 miles long, and stuck out over the Gulf of Mexico like a fishing rod. And by the same token any person can see that 742 years from now, the lower Mississippi will be only a mile and three-quarters long, and Cairo and New Orleans will have joined their streets together, and be plodding comfortably along under a single mayor and a mutual board of aldermen. There is something fascinating about science. One gets such wholesale returns of conjecture out of such a trifling investment of fact.

[1] We now know the Silurian was more than about 416 million years ago.

> **20.** Explain Twain's reasoning on another sheet of paper. Any problems?
>
> **21.** Although Twain is pulling your leg, his descriptions of meander cutoffs are accurate. However, the Mississippi has not actually changed its length much, except where people have put in artificial cutoffs. How is this possible? What other natural processes are occurring over the same time period to counteract this shortening of the river?
>
> ______________________________
>
> ______________________________

Rejuvenated Downward Erosion

Uplift resulting from plate motions or a drop in sea or base level can revitalize downward erosion by increasing the gradient of a stream. A fourth stage of stream erosion only occurs under these special circumstances that revive the erosive downward cutting power of the stream. This stage, **rejuvenation** (Figure 12.12), has similarities to both the early and late stages. The early-stage-like characteristics include plateaus and deep canyons, but these canyons curve in the shape they inherited from the stream when it meandered in its late stage.

> **22.** Describe a typical map view of a rejuvenated stream (■ Figure 12.22) by circling the correct words in parentheses to complete the following sentences: In topographic maps of rejuvenated streams, contours are closer together near (*divides* / *streams*) and farther apart near (*streams* / *divides*). The stream is fairly (*sinuous* / *straight*) (Circle one in each pair.)
>
> **23.** A topographic profile across the channel of a rejuvenated stream would be most similar to (a) an early-stage stream, (b) a middle-stage stream, or (c) a late-stage stream. (Circle one.)

The Colorado plateau encompasses some of the least-known and most unexplored territory in the continental United States. Much of the area is accessible only by helicopter. Canyon walls are too steep in much of the area for access except by air or water.

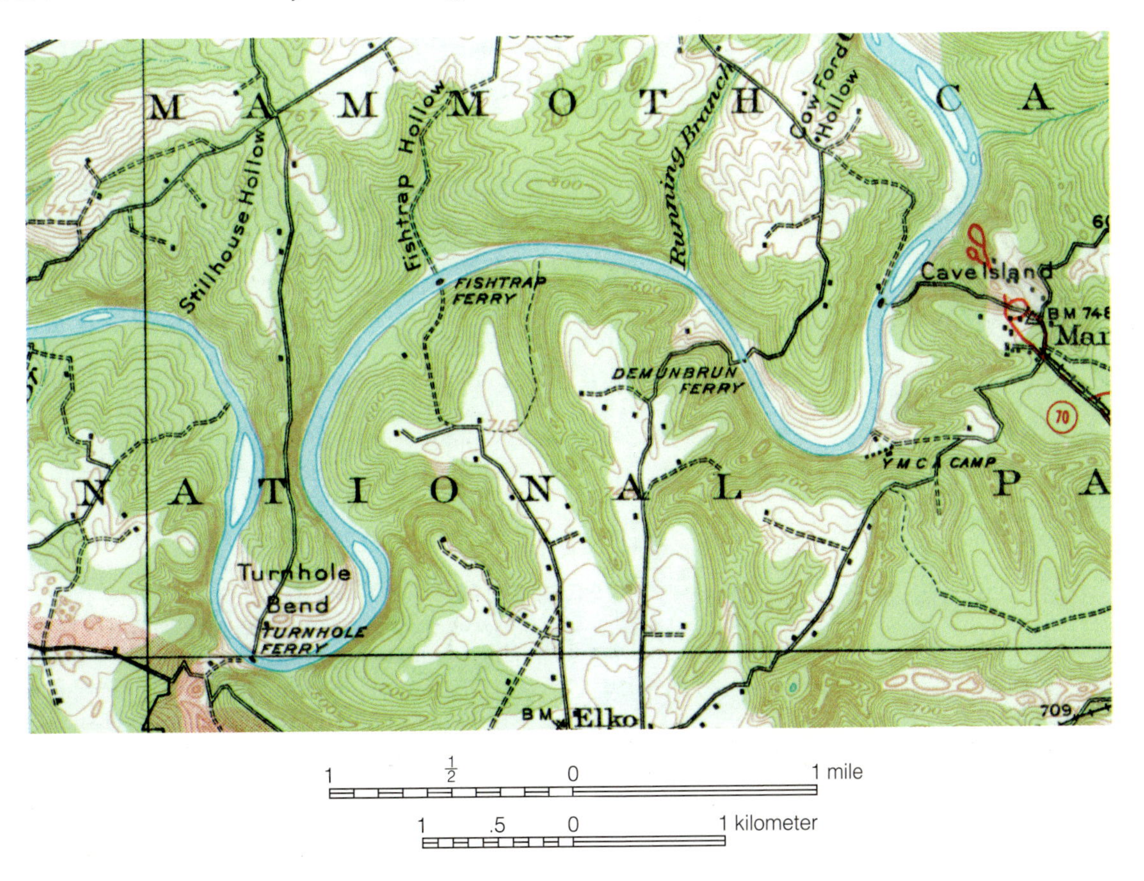

Figure 12.22

Rejuvenated Stream

From USGS, topographic map of the Green River, Kentucky. Scale 1:48,000. CI = 20 ft.

(a)

(b)

Figure 12.23

Colorado River

(a) Aerial photo pair of the Colorado River in Canyonlands National Park, Utah. Scale 1:45,000. (b) Aerial view of Glen Canyon cut by the Colorado River.

24. Examine the aerial photo pair of the Colorado River, Utah, in ■ Figure 12.23a and the single photo in Figure 12.23b. What is the stage of stream erosion? Stage:

____

List evidence:

____

____

____

In certain circumstances, rejuvenation can lead to the formation of natural bridges such as those at Natural Bridges National Monument (Figure 12.12c and e). The process is somewhat similar to the formation of a meander cutoff in a late-stage stream, only the stream cuts a tunnel through a canyon wall between two parts of an entrenched meander instead of cutting through a thin layer of sediment separating parts of a meander in a floodplain. Natural bridges of this type can only form where a powerful stream crosses a desert. The scarcity of erosive forces aside from the stream allows the cliffs and bridge to remain standing where they would have long since eroded in a moist climate.

25. In ■ Figure 12.24, locate each of the following features and label them on the map using the letters below:

a. a place where a natural bridge may be in the process of development

b. a natural bridge

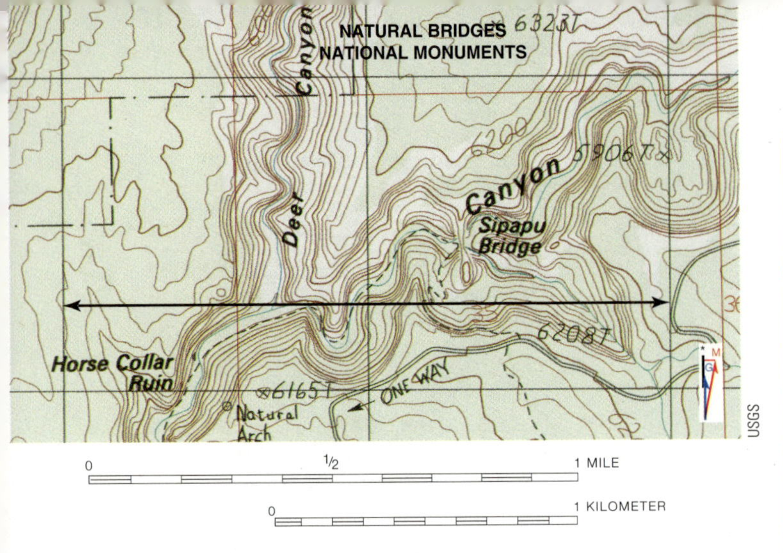

Figure 12.24

Map of a Natural Bridge

Supapu Bridge (see also Figure 12.12e) at Natural Bridges National Monument, Utah. Scale 1:24,000. CI = 40 ft.

c. a place where a natural bridge probably existed but has since collapsed

d. a place where the stream abandoned a meander-loop canyon when a natural bridge formed

e. On a separate sheet of paper, list the steps in the development of a natural bridge starting with an entrenched meander.

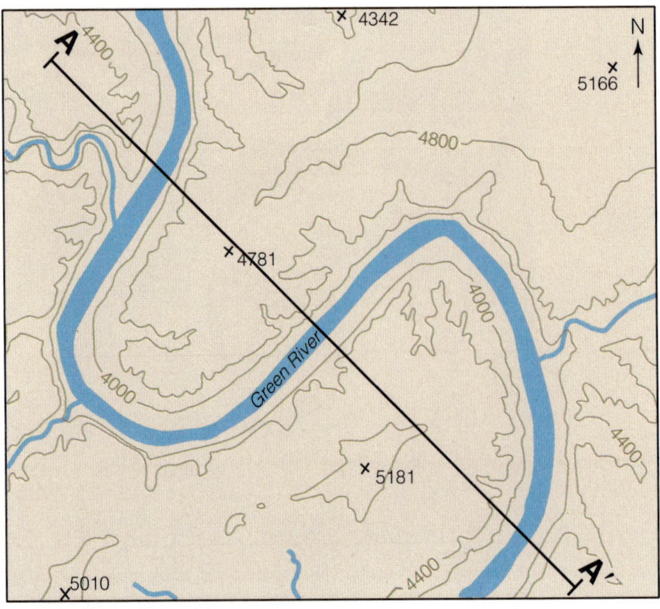

Figure 12.25

Green River, Utah

Simple topographic map of meanders in the Green River. Scale 1:45,000.

Practice Recognizing Stages in Stream Erosion and Rejuvenation

26. What is the stage of stream erosion for:

a. Saco River (Figure 12.5)?

____

b. Henrys Fork of the Snake River (Figure 12.9)?

____

c. North Fork of the Swannanoa River (Figure 12.10)?

____

27. Examine the map of Green River, Utah, in ■ Figure 12.25.

a. What is the contour interval for the map?

________ ft

b. Label the elevation of each contour on the map.

c. Use ■ Figure 12.26 to construct a topographic profile from A to A′ along the reference line on Figure 12.25.

d. Would a person (using binoculars, if necessary) standing at A be able to see another person at point A′? ________

e. Determine the stage of stream erosion by looking at both your profile and the map (Table 12.2 may help).

____

Figure 12.26

Topographic Profile

For Exercise 27c, Figure 12.25. Vertical exaggeration = 2×.

3600

f. Name at least two features that support your conclusion:

____

____

28. What is the stage of stream erosion for Lewis Creek (NE in Figure 12.4)?

____

Name at least two features that support this conclusion:

____

____

29. Look back at the map in Figure 12.6 again.

a. What is the stage of stream erosion for the White and Patoka Rivers in the western two-thirds of the map?

______________________ Label the features associated with this stage that you see on this map.

b. How has the Patoka River changed between 1901 and 1989? Circle a meander cutoff and oxbow lake in the inset map that was not cut off in 1901. Draw squares around two locations where the meanders have grown considerably larger since 1901.

Humans very commonly build next to rivers, sometimes in hazardous or unhealthy ways.

30. Examine the Platte River in Commerce City, Colorado, on the map in Figure 12.8.

a. List human activities next to the river that may either change the river's behavior or damage its natural environment.

____

____

____

b. Do the lakes next to the river look natural or artificial to you?

____

What shape would you expect to see in a lake so close to a natural late-stage stream?

____

FLOODING

It is natural for streams to flood—that's why the area next to a meandering stream is called a floodplain. The amount of water flowing in a stream is quite variable. **Base flow** is the *discharge* of water typically flowing in a stream when it is at its lowest level and only groundwater feeds it. A **flood** occurs when water spills out of its natural confines, over the stream banks, and covers areas not normally covered with water (■ Figure 12.27). Elevations near a stream vary, so that only the lowest regions commonly flood while higher floodwaters inundate other areas less frequently (■ Figure 12.28). The term **100-year flood** refers to an event that happens, on average, once every 100 years and is said to have a **100-year recurrence interval.** The area flooded is the **100-year floodplain.** When people build in the floodplain of a river, it is useful to know how often the water will submerge areas that have previously been dry. Flood frequency or recurrence graphs help establish that level.

A **natural levee** is a higher bank next to a stream where the stream has deposited sediment as it overflows or **floods.** Stream-deposited sediment is called **alluvium.** Floods and high water levels, with their greater water volume, occur less often than lower-level floods, but can be much more devastating. The water level in a stream is called its **stage,** and the volume of water passing a point in a particular period of time is its **discharge. Hydrologists**—people who study natural water systems—commonly estimate the probability of having a flood with a particular stage or discharge by using past history of stream levels. The nature of flood-

Parvinder S. Sethi.

Figure 12.27

Flooding

Failure of levees along the Mississippi River in south-central Iowa during the flood of 1993 damaged this family farm. The recurrence interval for this size flood is greater than 100 years.

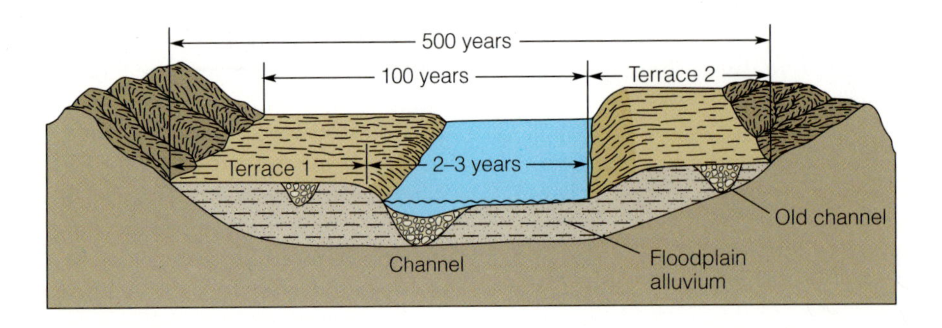

Figure 12.28

Stream Floodplains

Different levels next to a stream experience flooding at different *recurrence intervals*. Lower elevations flood more often than higher ones. The terrace levels resulted from different amounts of uplift and stream erosion.

ing depends on various stream characteristics. It can be confined but fast in narrow valleys upstream near the **headwaters.** Alternatively, it can be widespread but rising slowly in downstream areas (Figure 12.27), such as in the floodplain near base level where the slope of the stream is low and its path winding.

During a flood event, the water increases its velocity well above its average. The increase in velocity and volume allows the stream to transport more and larger sized pieces of debris. In other words, the size and amount *bedload* of the stream increases, but so does the *suspended load.* This allows the stream to cause a great deal of erosion during a flood. Large quantities of water and sediment flow over the banks and into the floodplain where fast-flowing water can pick up vehicles, undermine and destroy bridges, and wash away houses. Driving through flowing high water is a serious threat and commonly causes drowning, which is in part due to water velocity. Even if the flood does not destroy objects in the floodplain, the water and sediment can do a lot of damage that is often difficult to repair. When the floodwaters subside, people commonly come home to floors covered with thick mud.

Graphing Stream Discharge and Calculating Recurrence

A graph of flood recurrence (Figure 12.29) shows the highest water flow of each year plotted against its frequency of recurrence. To make such a graph, simply determine which year had the highest magnitude discharge or stream stage. Recall that *discharge* is the volume of water flowing past a point in a given amount of time and, in this context, stream *stage* means the height of water in a stream gauge, not the stage of stream erosion. Assign the highest magnitude flow rank 1 (= **magnitude rank** 1). Continue to sort through all of the discharges (or stages) for different years finding the next highest magnitude. Give each magnitude flood a rank down the list, so that the second highest is ranked 2 and third, 3, and so on. The next step is to determine the recurrence interval by using the following formula:

$$R = (n + 1) / m$$

where R = recurrence interval
n = number of years of record
m = magnitude rank

Using semilogarithmic graph paper (as in Figure 12.29), generally the higher flows will fall on nearly a straight line, allowing **extrapolation** (extending the line past the data points). By looking at this extrapolated line, you can find a recurrence estimate for the flood levels that no one has observed yet. This way, you can tell how often these larger floods would be likely to occur—information that is valuable in agriculture, the insurance industry, and for disaster planning in the area.

On the graph in ■ Figure 12.29 a 7-year flood has a discharge of just over 10,000 ft^3/s. This means that a flood of this size, on average, will occur once in 7 years. Another way to look at it is to say that this size flood has a probability of 1 in 7 of happening in any one year.

31. The data in ■ Table 12.3 give peak annual discharges for the years with the 10 highest flows for the Patoka River just upstream from the area shown in Figure 12.6. Figure 12.29 shows additional data for lower peak flows for this same stream for the years 1935 to 2003.

a. Rank the highest 10 discharges (streamflows) from Table 12.3 and enter their ranks in the table. The highest discharge gets a rank of 1 and the lowest of these 10 gets a rank of 10.

Table 12.3

Peak Flows for the Patoka River

The 10 highest peak flows for the Patoka River near Princeton, Indiana. The number of years of record (n) is 69. Drainage area is 822 square miles. Values of streamflow in the table are discharge. Flooding occurs at about 4000 ft³/s.

Water Year	Date	Stream-flow (ft³/s)	Magnitude Rank m	Recurrence Interval $R = (n+1)/m$
1937	Jan. 26	18,700		
1945	Mar. 08	15,400		
1950	Jan. 15	12,600		
1961	May 16	12,900		
1964	Mar. 16	15,200		
1972	Apr. 21	10,200		
1983	May 06	12,100		
1989	Apr. 08	10,200		
1996	May 04	13,900		
2002	May 17	10,500		

b. Calculate the recurrence intervals using the formula above for each of these discharges and enter them in Table 12.3.

c. Plot 10 more points by adding these data to the graph of *Streamflow* versus *Recurrence Interval* in Figure 12.29.

d. With a ruler, draw a best-fit straight line through your data points. A best-fit line should minimize the distance of all of the points from the line. The line should lie among the points with some points above it and some below it along its entire length. Use a clear ruler and position it as much in the center of the points as possible.

e. Extend the line to a recurrence interval of 100 years.

f. What is your estimated discharge for a 100-year flood?

____

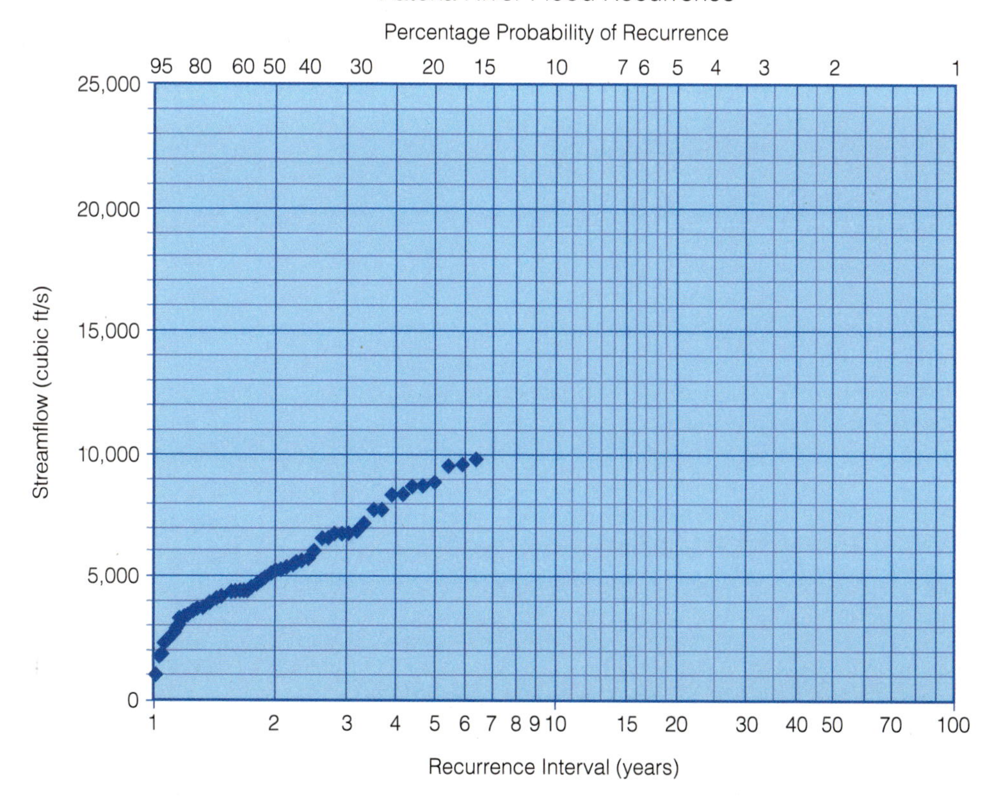

Figure 12.29

Flow Frequencies for the Patoka River Near Princeton, Indiana

Graph excluding the 10 highest flow events from a 69-year record from 1935–2003. Data points are for the peak flow for each year and are from USGS. *Streamflow* is another word for discharge.

g. What is the probability that a 100-year flood would occur in any one year?

____

32. For this stream, the discharge (called streamflow on the graph in Figure 12.29) in cubic feet per second can be determined from the stage, in feet above flood stage, by using the following equation:

$$streamflow = 2{,}000 \times (stage) + 4{,}000.$$

a. Imagine that your house is 7 ft above flood stage for this same stream. According to the equation above, what minimum discharge would flood the house? ____________________
Show how to do the calculations.

b. How frequently, *on average,* would your house be flooded? once every ______ years.

c. What is the probability that your house will be flooded in any one year?

____

d. Suppose your house was flooded one year: Does this mean that it is safe from flooding the next year? ______ Explain.

____

____

____

____

e. After a flood, what might you expect to see when you go back to your house?

____

____

____

____

____

____

LAB

Groundwater and Karst Topography

OBJECTIVES

- **To learn the difference between porosity and permeability**
- **To understand how groundwater flows, its importance as a resource, and risks of groundwater contamination**
- **To draw water-table contours and determine the direction of flow of groundwater and contaminants**
- **To understand the origin and geologic hazards of karst topography and their relationship to groundwater**
- **To be able to recognize karst features in photos and topographic maps**

George Billingsley, USGS

Figure 13.1

Springs at Marble Canyon, Arizona

Springs issuing from Redwall Limestone at Vasey's Paradise in Marble Canyon, Colorado River, 32 miles downstream of Lees Ferry, Arizona. From MONROE/WICANDER, *The Changing Earth*, 5E. 2009 Brooks/Cole, a part of Cengage Learning, Inc. Reproduced by permission. www.cengage.com/permissions

Groundwater is a substantial resource of water below the Earth's surface that started as rain, snowmelt, or streams. It rushes, trickles, or seeps from *springs* (■ Figure 13.1), or slowly and silently moves underground even as we extract it through wells for our use. Water moves downward through soil, sediment, and into porous rock until it reaches the zone where water fills all spaces—**pores** (Figure 13.2)—in rock and sediment. This zone is the **zone of saturation** and its top is the **water table** (Figure 13.4 on p. 299). The term **groundwater** refers to any water in the ground, but especially to water in the zone of saturation.

Groundwater is the largest reservoir of fresh liquid water on Earth and is a resource for drinking, irrigation, and industrial purposes. In many cases, groundwater takes a long time to accumulate underground; thus, once people deplete it, recovery may not come quickly or easily. Yet, we are dependent on this resource: In the United States more than 50% of the population relies on groundwater. The percentage increases to 75% in metropolitan areas. After it reaches the water table, groundwater continues to flow in response to gravity at a rate that is slow in comparison to streams—generally meters per day or less for groundwater compared to kilometers per hour for streams. Groundwater may flow through pores, cracks, and caves underground (■ Figure 13.2). As it flows through soluble rocks, groundwater dissolves the rocks, making the openings larger and eventually forming caves. In caves, the flow of water can be as fast as in surface streams, but cave flow is only common in karst regions where limestone or other soluble rocks exist.

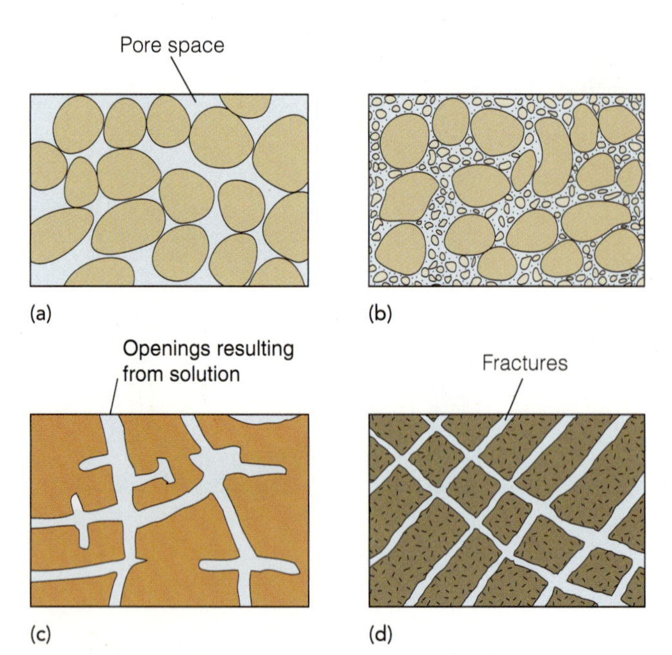

Figure 13.2

Different Types of Porosity

(a) Pores between grains of sand or other sizes of well-sorted clasts. (b) Porosity is considerably reduced in poorly sorted sediment. (c) Soluble rocks such as gypsum or limestone have porosity in solution cavities. (d) If present, porosity in crystalline igneous or metamorphic rocks generally takes the form of fractures.

Table 13.1

Porosity of Two Sediments

See instructions for Exercise 1.

Sediment Number	1	2
Brief description of sediment (size, composition, shape of grains)	coarse sand	fine sand
Volume of water (ml)	27	35
Volume of sediment (ml)	80	80
Porosity (%): $\frac{\text{vol. water}}{\text{vol. sediment}} \times 100$	33.75%	43.75%

$\frac{35}{80} \times 100$

POROSITY

Porosity is the percentage or proportion of pore spaces in a volume of rock or sediment. Let's determine the porosity of some sediment.

Materials needed:

- 2 100-ml beakers
- 2 containers of dry sediment, labeled "Dry Sediment 1" and "Dry Sediment 2"
- 2 containers for wet sediment, labeled "Wet Sediment 1" and "Wet Sediment 2"
- 50-ml graduated cylinder
- Water
- Hand lens or binocular microscope

1. Read the whole exercise before you start.

a. Examine both Sediment 1 and Sediment 2 and enter brief descriptions in Table 13.1. Based only on your visual examination of the sediments, circle the one that you would expect to provide the greater porosity and therefore hold the most water. Sediment 1 / Sediment 2.

b. Measure the porosity of Sediment 1 by placing some of it in a 100-ml beaker up to the 80-ml mark. Tap the beaker gently on the table to pack the sediment down. Add or remove sediment if necessary and repack, measuring carefully. Fill a graduated cylinder with exactly 50 ml of water. Slowly pour water from the cylinder down the side of the beaker into the sediment and measure how much water you need to reach the 80-ml level. As you pour, be sure the water is soaking in, not just running across the surface. Don't pour too quickly, or you will trap air bubbles in the sediment and not get an accurate measure of the porosity.

c. Read the volume remaining in the graduated cylinder and enter it in the following equation to determine how much water you added to the sediment:

50 ml − __23__ ml = __27__ ml.

Enter this number in Table 13.1.

d. Divide this volume of water by volume of sediment and multiply by 100 to get a percentage. Fill in the information in Table 13.1.

e. Measure the porosity of Sediment 2 in the same way as for Sediment 1. Fill in this information in Table 13.1.

f. How do the porosities of the two sediments compare? Make a numerical comparison.

1 has less porosity than 2

Do your results agree with your initial judgment of porosity from part a above? yes

When you are finished with the sediment, empty the beakers into the wet sediment container for the *correct* sediment, and clean the beaker for the next person. ***Do not mix*** wet and dry sediments or Sediments 1 and 2.

PERMEABILITY AND FLOW RATE

How fast water travels through rock or sediment depends on (1) the porosity, (2) the size of the pores, (3) how well they are connected, and (4) the pressure of the water. **Permeability,** related to the first three, is a measure of the ability of the rock or sediment to allow fluids to move through them. Generally, more porous materials are also more permeable, but the pores must be connected and large enough for the water to flow through rather than cling to the pore surface. Larger pores allow water to flow more readily. The higher the water pressure, the faster the water will flow through permeable material. It is common among sedimentary rocks for groundwater to flow faster parallel to bedding than across the layers.

Materials needed:

- Rock samples of coquina and pumice
- Squirt bottle
- Water
- Sink or plastic tray to collect water

2. Examine the coquina limestone, shale, and the pumice samples. The solid parts of these rocks have similar densities, yet there is a dramatic difference in the density of the whole rock.

a. Complete this sentence, choosing from the samples provided:

pumice has a lowest density because it has more porosity

b. Now compare the relative permeability of these rocks. With a plastic squeeze bottle, squirt water on the samples as you hold the rocks over a plastic tray or over the sink. Do not allow any water to flow around the edge of the rocks. Which is most permeable? coquina
Why? most connected pores

c. Next check whether the coquina has higher permeability parallel or perpendicular to the bedding. Gravity is pulling water downward, so use this fact when testing the sample. Hold the sample so the bedding is vertical to test how fast the water flows parallel to bedding. Then hold the rock horizontally to test for flow perpendicular the bedding. Which flows faster? horizontal.

d. Explain why you think both porosity and permeability are important for determining whether a rock layer will be a good reservoir for fluids—either water or oil. Consider not only how much fluid might be in the rock layer but also how easily someone could extract it.

Because you need to know how fast the fluid would travel through and how much would be retain.

Experimental Determination of Flow Rate in Sediment

For groundwater to be a useful resource, *recharge* must occur consistently and water must flow readily. **Recharge** occurs when water flows into the ground. High permeability allows a high flow rate, but if no pressure is forcing the water through the permeable material, no flow will occur. **Hydraulic head** is the height of a column of water above the discharge point or spring

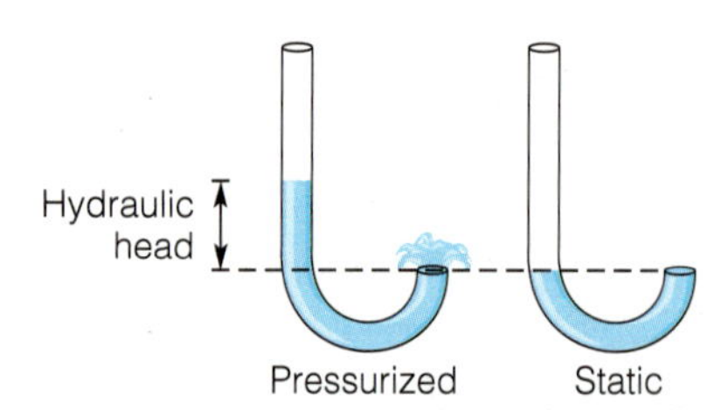

Figure 13.3

Hydraulic Head

Hydraulic head (difference in water level) creates water pressure because of gravity and the weight of water above a point of interest, in this case, the outlet.

(■ Figure 13.3). Gravity acting on this column of water creates the pressure that propels the water flow. The next experiments examine the effects of permeability and hydraulic head on flow rate. These experiments use two columns of either soft plastic tubing held up with ring stands or hard Plexiglas cylinders containing the same sediments for which you measured porosity for Exercise 1. In one, the sediment is finer grained than in the other. The columns have some method of retaining sediment without restricting water flow—this might be as simple as cheesecloth and cotton or fine mesh sieves attached to the bottom.

Materials needed:

- 2 columns
- Each column is marked with a centimeter scale or has a ruler attached alongside.
- One column has Sediment 1 in the bottom and the other has Sediment 2
- 3 beakers
- 2 funnels (if the columns are narrow)
- Small graduated cylinder
- Water
- Graph paper

3. Read all the instructions before you start this experiment.

a. In some cases, your lab setup may have soft plastic tubing. It is important **not** to squeeze the tube if you are impatient for the water level to drop so you can start your experiment. By squeezing the tube, you will be purposely altering the very thing you are setting out to measure. What characteristics of the sediment would this squeezing change?

______________________________ How?

____

____

____

b. Your lab instructor will assign you two or three measurements to make from the same column. Place a beaker under the appropriate column. If not already saturated, saturate the sediment by adding water until it is flowing (or dripping) out of the column and into the beaker. Once the sediment is saturated and dripping, fill the column with water slightly above the level assigned. When the level of the water reaches the assigned level, immediately place a graduated cylinder so it will catch every drop of water flowing out of the column during a 30- or 60-second period. At the end of the assigned time, stop measuring into the graduated cylinder by replacing it with the beaker. Enter the volume and time in the appropriate places in ■ Table 13.2.

Table 13.2

Flow Rate of Water through Sediment

See instructions for Exercises 3 and 4.

Sediment Number	Unit of Measure	Hydraulic Head				
		10 cm	20 cm	30 cm	40 cm	50 cm
1	Volume (ml)					
	Time (s)					
	Flow rate (ml/s)					
2	Volume (ml)					
	Time (s)					
	Flow rate (ml/s)					

c. Now allow one more drop to fall into the graduated cylinder. Read the new volume and subtract the first volume measurement to get an idea of how accurately you are able to measure the volume. If you see no difference, examine the cylinder graduations to determine the smallest increments you are able to measure with your graduated cylinder (commonly ½ to ⅕ of the size of marked graduations). Whichever is greater of these would be the minimum error in your experiment:

_____________ To establish the error in your experiments accurately, you would have to repeat the experiment a number of times to see the variation in the results. (However, this is a time-consuming process.)

d. Calculate the flow rate by dividing the volume by the time. Enter these figures in the table on the board, overhead, or computer spreadsheet to share with the rest of the class. (Think of this as the communication stage of the scientific method; see the Appendix.) If more than one group measures the same column and hydraulic head, average these results together.

e. Copy the data that your classmates collected into Table 13.2.

4. Use the graph paper to graph your results from Table 13.2. If your instructor prefers, use a spreadsheet's graphing feature.

a. It is conventional to place the independent variable (hydraulic head in this case) horizontally and the dependent variable (flow rate) vertically. After choosing your scale for each axis, double-check that the spacing and difference between 0 and the first number is equivalent to the spacing and difference for the next two numbers (this is a very common graphing error). Label 0 on each axis of your graph. Label axes and title your graph. Provide an appropriate key.

b. Draw a straight line through your data for each sediment. Should the line pass through the origin (0, 0) or not? _______

5. Determine the equation for each line as follows, or have the spreadsheet's graphing feature add a best-fit line to the graph and give you the equation:

a. What is the flow rate on the line at 0 hydraulic head? Sediment 1

____

Sediment 2

____

b. What is the flow rate on the line at 60-cm hydraulic head? Sediment 1

____

Sediment 2

____

c. Next, use the equation for a line given below. The flow rate and hydraulic head are your two variables; the other items are numbers you should substitute into the equation. The final equations should have the form $y = mx + b$, where y is flow rate and x is hydraulic head. Slope (m) and y intercept (b) are numbers.

$$y = [(\text{flow at } 60 - \text{flow at } 0) / 60] \times x + \text{flow at } 0$$

d. What is your equation for Sediment 1?

____

e. What is your equation for Sediment 2?

____

6. Think of your equations as hypotheses (see Appendix) based on each of the sediments. You are hypothesizing that a linear relationship exists between hydraulic head and flow rate as expressed by your equations.

a. Now use the equations to predict the flow rate at 70 cm, if your columns go that high, or 45 cm, if they don't. What flow rate do you predict?
Sediment 1

____

Sediment 2

____

b. Test your hypothesis for either Sediment 1 or Sediment 2. Pour water into the sediment column to set the hydraulic head to the appropriate value and see whether you get the flow rate you predicted.
Volume _______ ml; time _______ s; flow rate _______ ml/s.

c. How close was your measured flow rate? Subtract your answer from the predicted value. _______ Is it within the error of the experiment? _______ If not, the next step would be to reevaluate the error in the experiment or reevaluate the hypothesis.

7. Look back at the results in Table 13.1 and at your graph.

a. How do the flow rates at 50-cm hydraulic head compare for the two sediments?

____

____

____

____

____

b. Is the difference in the porosity of the two sediments sufficient to explain the difference in their flow rates? _______

c. What else could account for the difference in the flow rates?

____

____

8. Summarize your results:

a. How does porosity influence flow rate?

____

____

____

b. How and why does sediment size influence flow rate?

____

____

____

____

c. How and why does hydraulic head influence flow rate?

____

____

____

____

d. Is there a mathematical relationship between hydraulic head and flow rate? _______ If so, what is it?

____

____

WATER TABLE, GROUNDWATER FLOW, AND WELLS

We have been experimenting with saturated sediments in this lab, but remember that the **water table** is the top of the saturated zone underground (■ Figure 13.4). Where the water table intersects a natural hole or depression in the land surface, a lake exists. In stream valleys of humid regions, groundwater may flow out of the ground (**discharge**) into the stream. Because stream flow is faster than groundwater flow, the stream may draw down the level of the water table to the stream level. For this reason, it is common in humid regions for the water table to mimic the surface topography, but with lesser slopes.

Well water is an important source of water in many regions. A **well** is a hole drilled from the surface into the zone of saturation of an *aquifer* (■ Figure 13.5). An **aquifer** is a permeable body of rock or sediment below the water table that can sustain a productive water well. The lining of the well, if it has one, is permeable so water flows into the well from the surrounding rock or sediment. The level of water in the well will correspond to the water table (unless the water is under pressure). When water is pumped out of the well, the water level drops, and the water table around the well also drops, forming a **cone of depression** (Figure 13.5). The water level in lakes, swamps, streams, and wells in a region may all indicate the level of the water table. Knowing the level helps people judge how deep to drill for water when they put in a well. Since the price of drilling a well is usually by the foot, this could be very useful information.

Groundwater flows from high pressure to low pressure. The experiments with the sediment columns show that the pressure is higher when the water level in the columns is higher and also that the increased pressure produces faster flow of water in the sediment. However, underground, it is not just pressure but a difference in pressure from one place to another that causes flow. A sloping water table produces a difference in pressure and causes groundwater to flow from high to low areas of the water table. When a cone of depression forms,

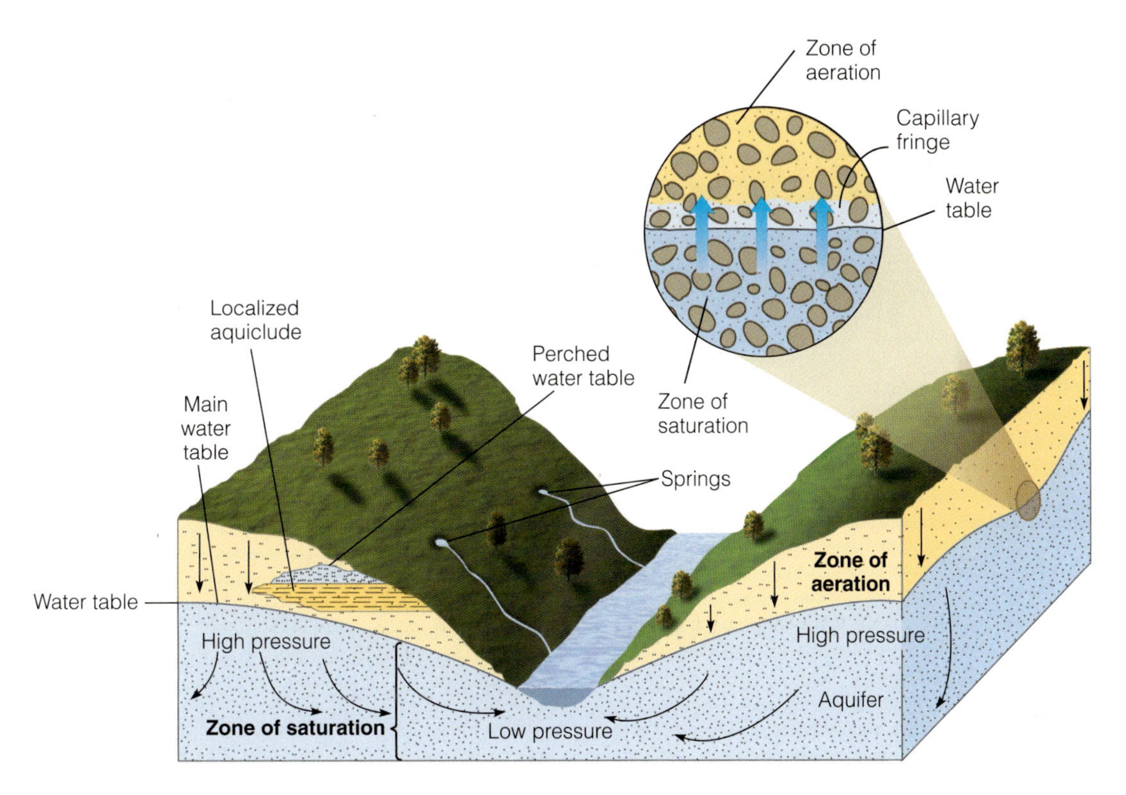

Figure 13.4

Water Table and Groundwater Flow

Block diagram showing the *water table* drawn down by the faster flow of the stream, and a **perched water table** above an impermeable layer, with **springs** issuing from it. Groundwater flows toward the stream because the pressure is lower there. Inset: *Pores* in the **zone of aeration** have air and water, the pores in the **zone of saturation** are full of water, and the *surface tension* of water draws water up into the **capillary fringe.**

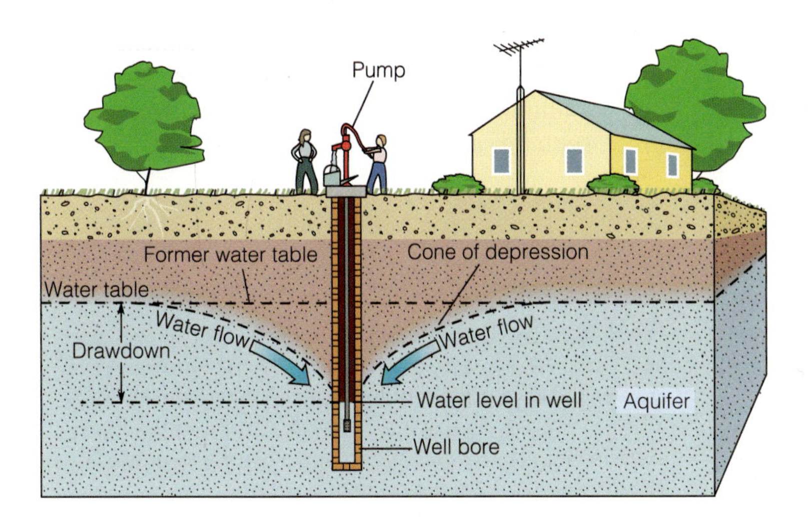

Figure 13.5

A Cone of Depression

Pumping a well causes a cone of depression in the water table. The drawdown creates a hydraulic head causing water to flow toward the well.

the difference in pressure between the well and the surroundings is what brings water to the well.

A Groundwater Model

The column experiment demonstrated that some sediment allows water to flow through it faster than other sediment. Highly permeable materials make good aquifers because they allow faster flow. If the material is less permeable or hardly permeable at all, then it is an **aquitard** or an **aquiclude,** respectively. An aquifer may be **confined** or sandwiched between aquicludes, which may allow the aquifer to become pressurized. Such an aquifer is an **artesian aquifer** (■ Figure 13.6). The pressure in an artesian aquifer forces the water to rise anywhere a natural break or a well intersects the

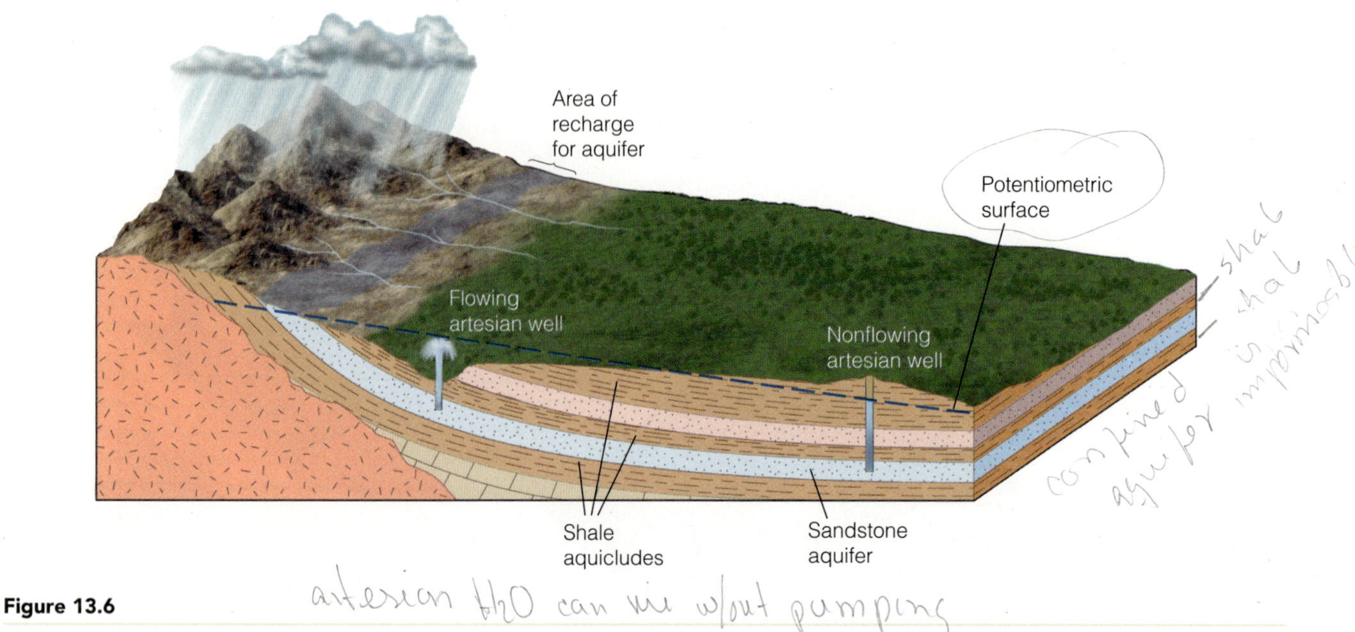

Figure 13.6

Artesian Aquifer

Where aquicludes confine (sandwich) an aquifer, and where the recharge area is at relatively high elevation, the water in the aquifer is under pressure creating an artesian aquifer. The potentiometric surface shows how high water will rise where the artesian aquifer is pierced by a well or fracture. From MONROE/WICANDER, *The Changing Earth*, 5E. 2009 Brooks/Cole, a part of Cengage Learning, Inc. Reproduced by permission. www.cengage.com/permissions

aquifer. The water could rise to a hypothetical surface height known as the **potentiometric surface.** It is analogous to the water table for an unconfined aquifer. Groundwater flows in response to a difference in pressure generated by a sloping water table. Similarly, water in an artesian aquifer flows from where the potentiometric surface is high—that is, from higher pressure to lower pressure.

For the next experiment, use the groundwater model provided, or answer these questions as a set of thought questions based on what you have learned thus far. If you use a model, it may be as elaborate as the groundwater model shown in ■ Figure 13.7, or as simple as a clear plastic box with layers of different-sized sediment and with clear straws or plastic tubing representing wells. Whatever style model you have, the lowest layer of sediment should have "wells" reaching to it with blue dye, such as in the model pictured. Shallower wells should have green dye. Your model should have drains leading from at least two levels, all of which should start closed. In addition, your model should have a way to recharge groundwater into the deepest sediment layer as well as the shallower layers.

Materials needed, or a thought experiment:

- Groundwater model as described above
- Colored dyes (food coloring works fine)
- Syringe or pipette
- Drain or bucket
- Water

9. Examine your model and the sediment used in it. Your instructor may have separate samples of the sediment to examine closely, otherwise look at the sediment within the model. If you are working this as a thought experiment, your instructor might give you additional information about the different sedimentary layers.

a. On a separate sheet of paper, carefully sketch your model. Make a key of the sediment layers, indicating which are aquifers, and which are aquicludes.

b. Examine the water level in all the wells and in any of the other features of your model that allow you to see its level, such as the tank (the circle) and any lakes or streams in our example model in Figure 13.7. Sketch the position of the water table on your drawing, using a color other than green or blue. In the key add a line this color, and label it "initial water table."

10. Now open the valve for the drain in the uppermost or upper middle sediment. In our example model, that would be the "stream." If necessary for your model, add recharge water into your sediment, using the method

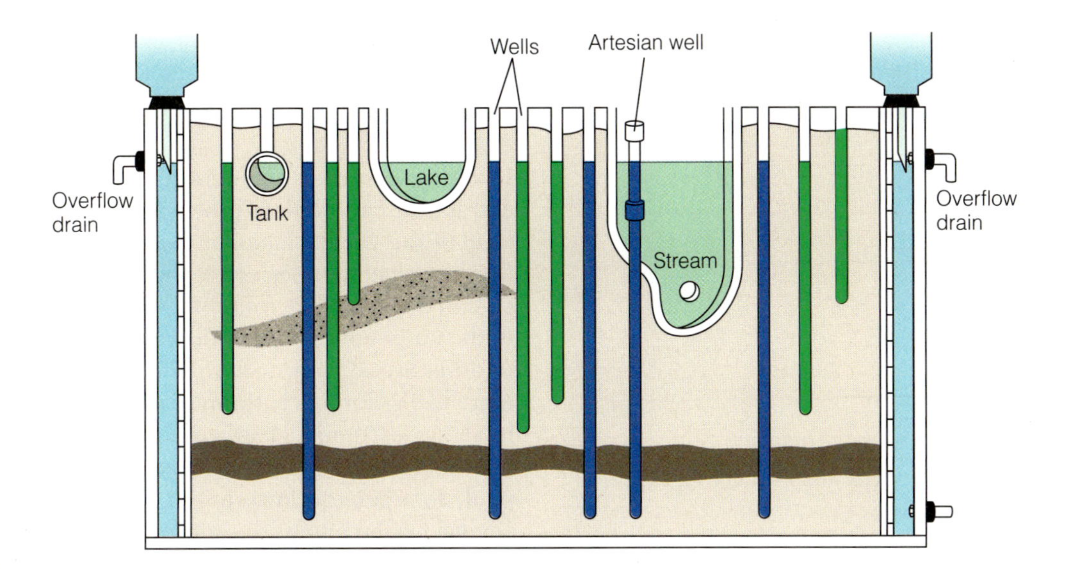

Figure 13.7

Example Groundwater Model

Your model may have quite a different setup, but should have similar components. It should have layers of sediment with different permeability, deep wells with blue dye, shallow wells with green dye, a drain from the upper sediment, and recharge for both upper and lower sediment. In this model, the two blue bottles at the upper left and right corners provide continuous recharge to the areas with sediment. The stream, tank (representing a leaking underground gas or chemical tank), and lake have perforations so that they interchange with the groundwater.

your instructor suggests. Continue to supply recharge to the same level throughout the experiment. In the model pictured in Figure 13.7, the two bottles at the top, left, and right recharge the groundwater automatically but may need to be refilled periodically. Watch the model for a minute or two. Periodically check that the stream is draining properly; sometimes air bubbles in tubing impede the flow. Jiggle or tap the tube to get the flow going again. If you are doing this exercise as a thought experiment, think through what you would expect for each of the following answers given the sediments and the conditions.

a. Sketch directly onto your drawing the level of water at each location where you can detect it, use different colors for wells with different color dyes, then sketch in the water table for each different color. So, for the example of our model in Figure 13.7, you would sketch two water levels: one green water table and the other the blue water level.

b. Label which water level is actually a potentiometric surface of an artesian system.

c. Label the highest and lowest hydraulic head on your sketch and explain, on the same piece of paper, why each is located as you've shown. How does the stream influence the water table and why?

e. Now that the water has run for a while, some of the dye should have spread from the wells. Color the areas in your sketch where the dyes have colored the sediment in your model.

f. In what direction is the groundwater flowing? Draw arrows on your sketch showing the direction of flow.

g. Summarize the relationship between the direction of groundwater flow and the slope of the water table.

____

____

____

h. Summarize the relationship between the direction of groundwater flow and the slope of the potentiometric surface.

(Look carefully to detect the slope in the potentiometric surface.)

____

____

____

i. What evidence shows that groundwater can flow upward?

____

____

11. Leave the stream outlet on the groundwater model open for the remainder of the experiments and replenish the supply of recharge water as needed. Use a syringe or pipette to pump water out of a centrally located green-dyed well.

a. What happened to the water level in the adjacent green-dyed wells and blue-dyed wells when you did this? *Hint*: If your model is set up similarly to Figure 13.7, then the answer will be different for the different-colored dyes.

____

____

b. What direction is the groundwater flowing when you extract water from the well?

____

c. What is the name of the cone-shaped drop in the water table around a well?

____

d. Try the same thing with a centrally located blue-dyed well. Why does the water in the different wells behave differently?

____

____

____

e. Based on what you have learned, what do you think will happen if two families have wells near each other that obtain water from the same aquifer, and one family starts extracting much more water from its well?

____

____

GROUNDWATER DEPLETION

In either confined or unconfined aquifers, excessive extraction can deplete groundwater. Declines in the level of the Ogallala aquifer have become quite serious in some places, as seen in ■ Figure 13.8a and c. If one person depletes the groundwater in their area, the drop in the water table, or potentiometric surface, may spread to other areas. How does water in an aquifer behave—more like water in a bathtub or more as if it is contained in an egg-carton-like configuration (Figure 13.8b)? The bathtub analogy applies if the aquifer acts as a common pool, so a general lowering of the water table occurs when a person pumps the aquifer from one location. For the egg carton model, the aquifer is compartmentalized, so a person pumping in one location does not influence neighboring areas.

12. Study the map in Figure 13.8a.

a. What do the blue and lavender colors indicate?

Rises

____

The orange and red?

Declines

b. Where was the aquifer most depleted?

canadian river

Where was the aquifer rising?

Plattet river

c. Does the whole aquifer act like a bathtub? No Roughly how large are single units or "compartments" of the aquifer?

100 miles

d. Would the groundwater use by farmers in one part of Kansas influence the water levels for people 25 miles away? yes

e. Does the aquifer appear to have uniformly sized compartments or do they vary in size in different regions?

Vary Where do they appear smaller?

Nebraska

Where do they appear larger?

Kansas & Texas

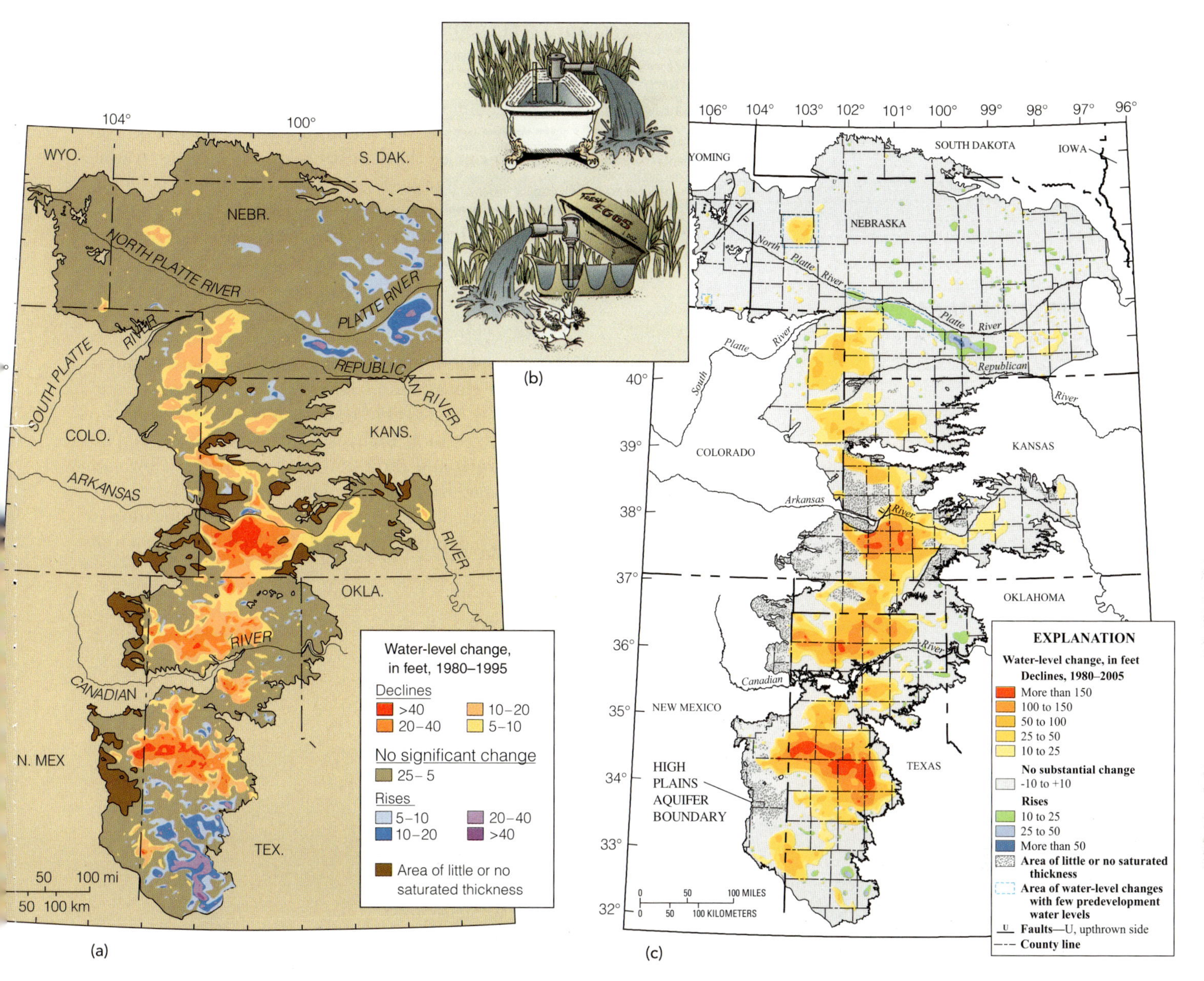

Figure 13.8

High Plains Aquifer

(a) Map of the High Plains Aquifer showing changes in the water level between 1980 and 1995. The aquifer is also called the Ogallala Aquifer; its extent is outlined in black on the map, and its water-level changes are indicated with different colors. (b) Two models for how aquifers work. (*Top*) Aquifers may act like bathtubs, where draining from one part may drain the whole thing; or (*bottom*) in the egg-carton model, one person pumping does not affect neighbors' water levels. Most aquifers are probably between these two extremes. (c) The same map area as in (a) but with water level change data up to 2005.

f. What problems can you envision for households, farms, and industries in areas colored orange on the map?

____

____

13. Study the map in Figure 13.8c, which is the same area but with more recently collected data. Notice that colors and water-level changes differ in the key. Compare this to the map in Figure 13.8a and answer these questions.

a. What is similar in these two maps considering both patterns and quantities?

____

____

b. What are the most marked changes between the two maps regarding declines and rises in water levels? In which state(s) do they occur?

____

____

GROUNDWATER CONTAMINATION

Although surface water is much more easily contaminated than groundwater because it is more accessible, a number of sources may contaminate groundwater, as illustrated in ■ Figure 13.9. In general, contaminants will tend to flow with the groundwater, although some float on top and some sink below it. Also, small pore spaces filter some contaminants, such as bacteria, and the rock or sediment they are flowing through bind others.

One common contaminant is *leachate* from landfills. If rainwater gets into a landfill, it can leach various toxic chemicals out of the landfill material into the water, forming a solution known as **leachate,** which you could think of as "garbage juice," an unappetizing thought. If the leachate leaks out of the landfill, it may enter the aquifer and flow along with the water.

Using Water-Table Contours

Where enough information is available about the water table from elevations of lakes, streams, swamps, and wells, it may be possible to draw a contour map of the water table as in ■ Figures 13.10 and ■ 13.11. The depth of wells and direction of groundwater flow can be determined from such a map.

The depth of the water table at any given location is less than the depth needed to drill a well, because a well

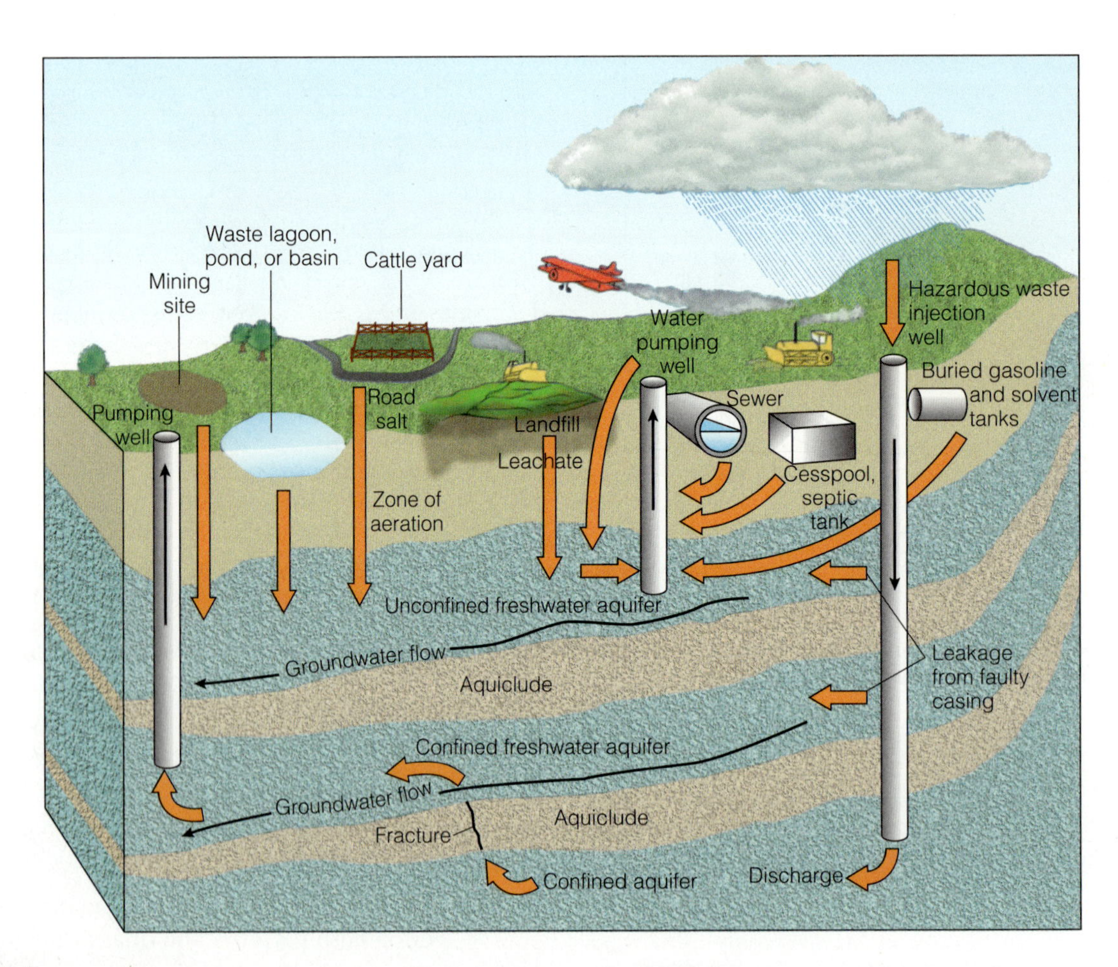

Figure 13.9

Groundwater Contamination

Contamination of groundwater has many possible sources. The two pumping wells extract groundwater contaminated by these sources.

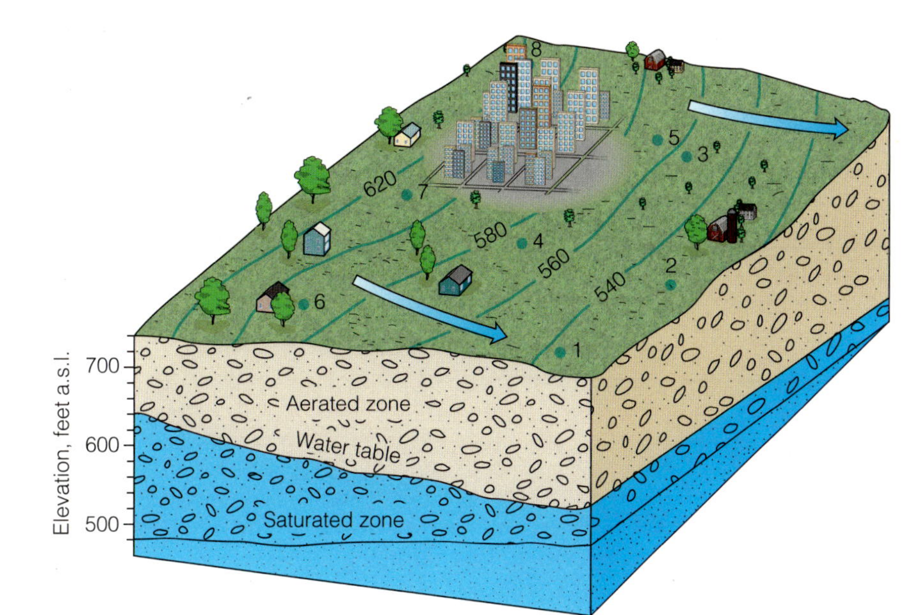

Figure 13.10

Water-Table Contours

Blue lines on the top of this block diagram are water-table contours showing groundwater elevations near a city. Numbered dots represent wells, which give information about the water table, allowing a hydrologist to draw the contours and determine the direction of groundwater flow (arrows), which is perpendicular to the contours. *a.s.l.* in the elevation scale = above sea level.

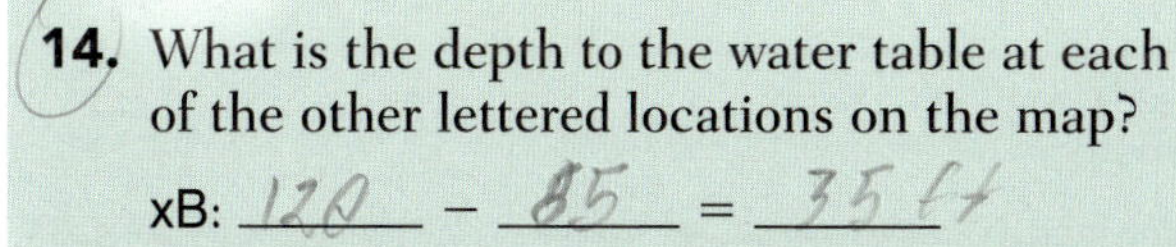

14. What is the depth to the water table at each of the other lettered locations on the map?

xB: 120 – 85 = 35 ft

xC: 130 – 95 = 35

xD: 110 – 107 = 3

xE: 100 – 95 = 5

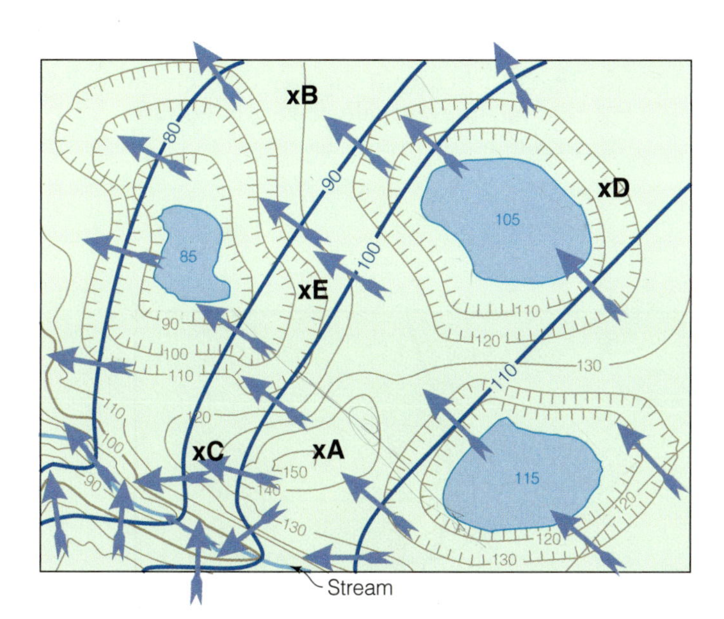

Figure 13.11

Groundwater Flow

Topographic map of an area with sinkhole lakes. Contours of the water table surface are heavy, dark-blue lines. Contour interval = 10 ft. Arrows show the direction of flow of the groundwater.

must penetrate the zone of saturation. To calculate the depth of the water table, first determine the elevation of the land surface at the location. Next, determine the elevation of the water table at the same point, and then subtract from the land elevation. For example, the location marked at xA in Figure 13.11 has a land elevation of 150 ft, and the water table is 104 ft. This makes the depth to the water table 150 ft – 104 ft = 46 ft.

The direction of flow of groundwater is important for a number of reasons. It helps to determine where water in a well comes from and what happens to contaminants and pollution that get into the groundwater (Figure 13.9). Because of gravity and the resulting pressure, water in the ground flows in the direction of the steepest downward slope of the water table, carrying contamination with it. Flow of groundwater, then, is *perpendicular* to the water-table contours. The arrows in Figure 13.11 show the groundwater flow direction.

15. Study Figure 13.11. If a company dumped toxic waste and contaminated the entire lake labeled 115, draw or shade in on the map the path that the toxic waste would take as it traveled through the ground in the groundwater.

a. Which, if any, of the Lakes 85 or 105 or locations labeled A, B, C, D, or E would become contaminated?

85, A, E

b. Would this toxic waste contaminate the part of the stream shown on the map?

NO

c. If toxic waste contaminated all of Lake 105 instead, which of the locations labeled A, B, C, D, or E would become contaminated? B

16. Groundwater flows much more slowly than streams. If the speed of groundwater flow in the area shown by Figure 13.11 is 2 m/d (a fairly high flow rate for groundwater) and the distance from Lake 105 to B is 0.5 km, how long would it take for contaminated water to reach a well at B? Assume that the contamination travels at the same speed as the water. Write out your calculations.

$2 \frac{m}{d} \left(\frac{1 km}{1000 m}\right) = 0.002 \frac{km}{day}$

$0.5 km \left(\frac{1 day}{0.002 km}\right) = 250 days$

17. Use the maps in ■ Figure 13.12, which show the elevation of water in wells.

a. Contour the water table in Figure 13.12a using the well data and a 5′ contour interval. Streams tend to draw down the water table so that the water table forms a V at the stream in a way that is similar to the V's that topographic contours form. The map already has one contour to give you the idea. The other contours will be similar.

b. Draw arrows on the map showing the direction of flow of groundwater. Remember that flow direction is perpendicular to water-table contours in the downslope direction.

c. Suggest possible sources of groundwater contamination based on the items on the map. List them:

houses
cattle yard
agricultural fields
industrial waste pound

d. Shade or highlight the area on the map that might eventually have groundwater contaminated from these sources.

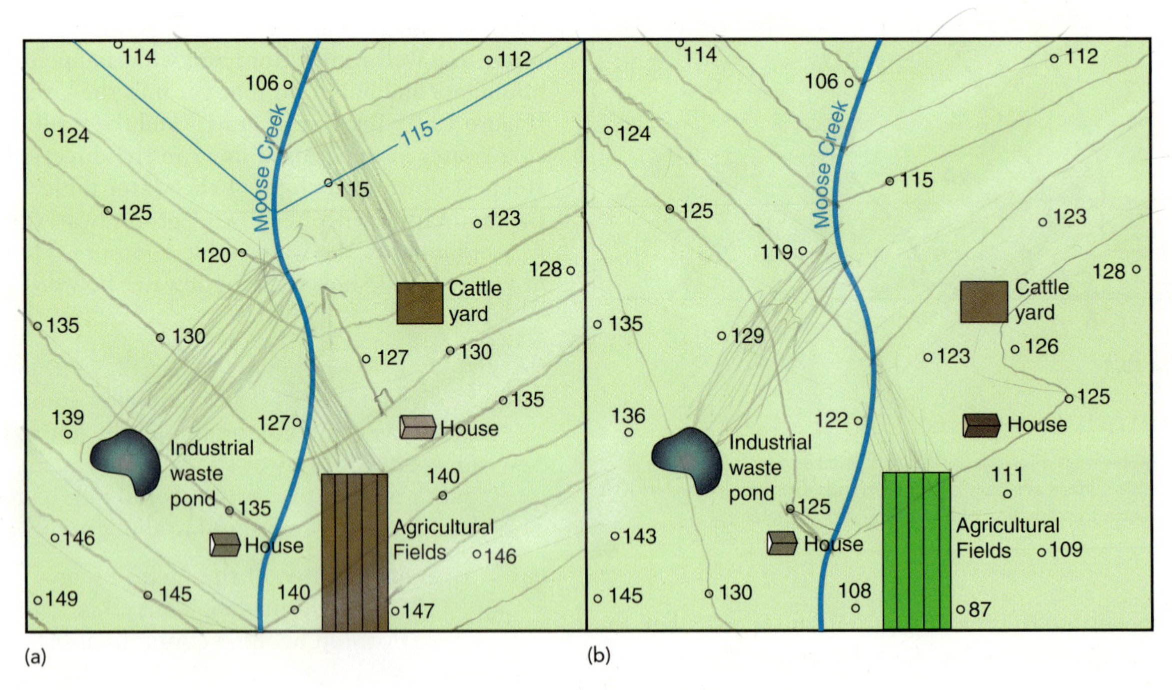

Figure 13.12

Map of Groundwater Elevations in Wells

Map of an area with numerous wells near Moose Creek, for an exercise in drawing groundwater contours. See instructions in Exercise 18. (a) Water elevations without pumping. One contour shows the water-table elevation of 115′. (b) One well is being pumped. Follow the directions in Exercise 18 to complete both maps with a contour interval of 5 feet.

e. Would the stream become contaminated? Yes If so, where? Shade that area of the stream.

f. In Figure 13.12b, the situation has changed because a farmer is irrigating the agricultural fields from a nearby well. This map shows the new well-water elevations. Use the new elevations to re-contour the water table and draw new arrows showing the new flow directions.

g. Which well is the farmer using? 87

h. Shade in areas of groundwater contamination with the new configuration.

i. What effect did pumping have on groundwater flow and spread of contamination?

contamination goes to the well and groundwater

GROUNDWATER CAUSES EROSION BY SOLUTION

As it moves, surface and groundwater may dissolve soluble rock, creating cavities and caves in limestone, dolomite, marble, rock salt, or gypsum, and resulting in topography called **karst.** Karst topography is different from stream-dominated topography because much of the high-volume drainage is underground and because solution produces certain unique landforms (■ Figure 13.13). Streams often disappear into closed depressions that lead to networks of caves, or caverns, commonly decorated with limestone precipitated from groundwater such as those shown in Figure 13.13b. A closed depression in karst regions is a **sinkhole** (■ Figure 13.14a), which may form by solution or when the roof of a cave collapses. Karst topography is most common in limestone regions. The sudden formation of sinkholes (**sinkhole collapse**) is one of the geologic hazards in some karst areas.

Because limestone is susceptible to dissolution, it may be highly permeable; if the porosity is on the centimeter scale rather than cavernous, the limestone may make an excellent aquifer. However, in many karst

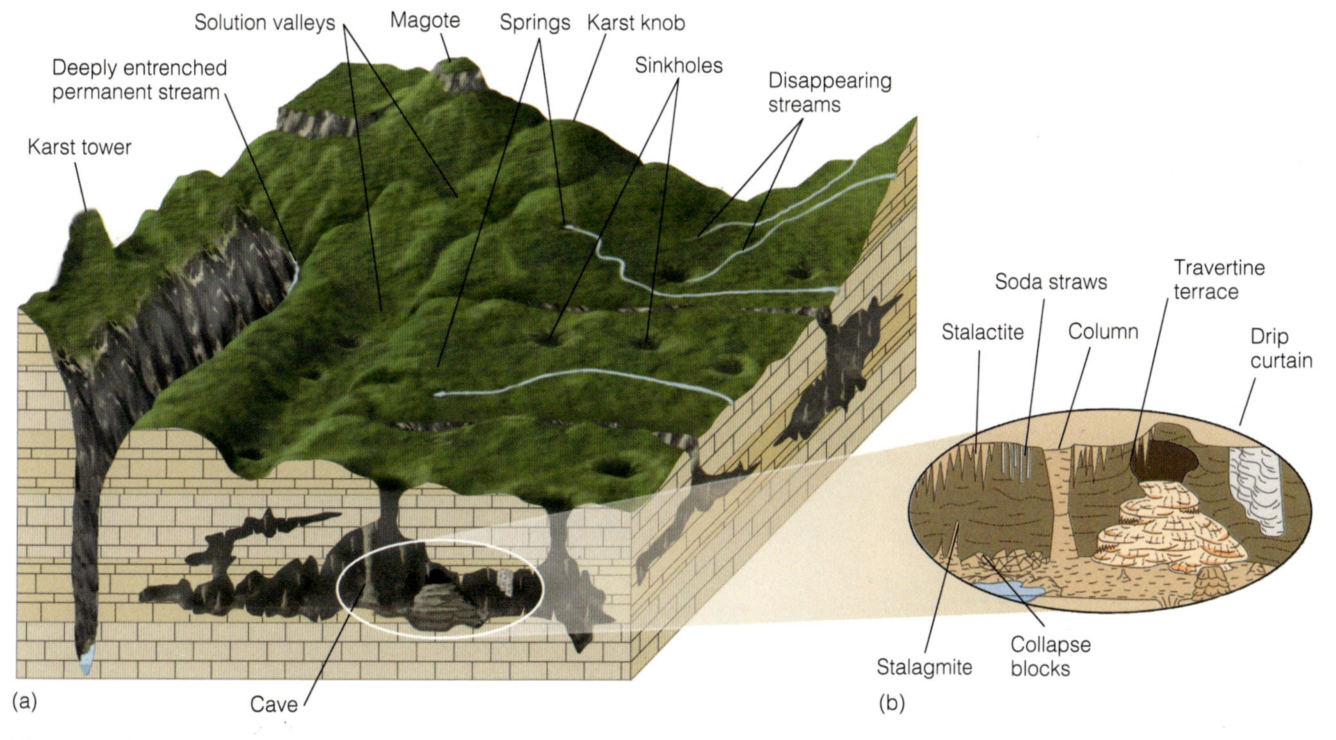

Figure 13.13

Karst Topography and the Formation of Caves

(a) Groundwater dissolves the limestone, forming a system of openings and passageways. The surface develops sinkholes and surface streams disappear through them. Sinkholes gradually enlarge and coalesce into solution valleys leaving behind remnant hills in the form of karst knobs, magotes, and karst towers. With uplift and downward erosion and entrenchment of surface streams (left), the water table drops, leaving interconnected caves above the water table. (b) Once air is present in the caves, water evaporating from dripping groundwater builds **speleothems,** such as shown here. From MONROE/WICANDER, *The Changing Earth*, 5E. 2009 Brooks/Cole, a part of Cengage Learning, Inc. Reproduced by permission. www.cengage.com/permissions

flow line ⊥ to H_2O Table contour

Bogdan Onac

USGS

Figure 13.14

Karst Features: Sinkhole and Mogote

(a) This sinkhole formed on May 8 and 9, 1981, in Winter Park, Florida, and had a diameter of 100 m and a depth of 35 m. The blue rectangle on the top right side of the sinkhole is a car. The vertical line across the sinkhole is a sewer pipe. The large white area on the bottom was a swimming pool. Water in the sinkhole shows the level of the water table. (b) Three mogotes, steep-sided karst hills, are intermediate between karst knobs and karst towers in steepness and height. Mogote means stack in Spanish. West-central Puerto Rico.

regions, the limestone has large caves where groundwater flow may be very irregular, making a search for groundwater a very hit-or-miss proposition.

Dissolution Experiment

As rainwater falls on the ground, it picks up a small amount of acid by dissolving carbon dioxide from the air. This carbonic acid is the same thing that gives soda pop its fizz. Over time, acidic water can dissolve holes in limestone and dolomite just as the dilute hydrochloric acid reacted with the carbonate minerals that you tested in Lab 2. In the following experiment, you will be able to observe the solubility of limestone.

Materials needed:

- 2 samples of the same limestone
- 4 steep-sided beakers large enough to hold one sample
- Labeling supplies
- 10% HCl
- Water
- Rubber gloves and tongs
- Paper towels
- Ruler
- Hand lens or microscope

18. Look at the limestone samples with a hand lens or microscope.

a. Place both of the samples of limestone in beakers. Label one *acid* and the other *water*. Very slowly, pour enough 10% HCl over the surface of one and water over the surface of the other to allow the sample to sit partially in the solution. The acid is much stronger than carbonic acid in a natural setting, but we want to speed up the process. Record what you see.

____

____

b. Return to the soaking samples after you have finished the rest of this lab, or after they have been soaking for 30 minutes. Using tongs or wearing rubber gloves, pick up the sample and place each in a beaker of water to rinse it. Remove the sample immediately and place it on a paper towel. Dry it well.

c. Examine the samples side-by-side with a hand lens or microscope and record any differences between the two samples. Pay close attention to void spaces and cement between grains.

____

____

____

d. Look at the bottom of the beakers that had the samples. How does the one with acid compare with the one with water?

____

____

Where did the particles that you see come from and why did they come loose?

____

____

____

Dispose of the liquid as instructed.

e. What does the change in the sample indicate about the change in the rock's porosity?

____

____

Karst Topography

Naturally acidified rainwater and groundwater act on an area of limestone causing karst topography to evolve. At first, a karst area may be very flat with a high water table. Near the surface of the water table, large openings may form in the rocks under water. These openings are underwater caves, which drain with a drop in the water table. At first a few **sinkholes** develop by dissolution and by collapse of caves, such as seen in Figure 13.14a and the right front area of Figure 13.13a. Where the water table is shallow or where the sinkholes penetrate to the water table, **sinkhole lakes** form. **Sinking streams** disappear into underground cave systems. As time passes and the water table drops, more dissolution occurs at deeper levels and individual sinkholes enlarge and merge to form **solution valleys.** Further erosion and lowering of the water table leads to the development of broad low areas with erosional remnants of low **conical hills,** moderate **karst knobs,** steeper **magotes** (Figure 13.14b), or high, very dramatic **karst towers.** In some regions, insoluble rocks, such as sandstone, may cap limestone, as in the area surrounding Mammoth Cave, Kentucky (■ Figure 13.15). The insoluble rocks make uplands or plateaus with solution valleys containing sinkholes where erosion has cut down to the limestone. Where erosion has produced lowland areas, isolated remnants make karst knobs or conical hills in a landscape called **cone karst.**

19. Examine the maps of the area near Mammoth Cave, Kentucky, in Figure 13.15 and refer to the diagrams in Figures 13.13 and 13.16 as you answer the following questions. The first map (Figure 13.15a) shows more northerly area of Mammoth Cave National Park and the second map (Figure 13.15b) shows the area of the gently sloping Pennyroyal Plateau directly to the south.

a. Use both maps to find and sketch the contours of each of the following features on a separate sheet of paper:

i. a sinkhole

ii. a solution valley

iii. a karst knob

b. Use Figure 13.15a to locate the main and new entrances to Mammoth Cave and two other caves. What is the minimum length of Mammoth Cave system? *Hint:* These four caves are all connected underground.

____

c. In the northwestern part of the map in Figure 13.15a, the Green River is a rejuvenated stream (Refer to Lab 12 as needed). Which of the following describes streams or valleys adjacent to the Green River?

i. The streams are sinking streams and the valleys are sinkholes.

ii. The area lacks streams but contains mostly solution valleys.

iii. Valleys here have V shape and the streams flow into Green River.

iv. The streams are also rejuvenated streams.

v. The streams are in the late stage and meander across a broad floodplain.

d. What types of valleys are in the southern area of the same map (Figure 13.15a)? *Hint:* use the diagram in Figure 13.13 to help you.

____

Figure 13.15

Maps of Mammoth Cave, Kentucky and Areas South (*following two pages*)

USGS topographic maps of the area around and to the south of Mammoth Cave, Kentucky. (a) Mammoth Cave, Green River, Woolsey Hollow, and surrounding area (left). (b) area just south and slightly overlapping with the area in (a), from The Knobs in the north, south into the Pennyroyal Plateau. Both maps have a scale 1:45,000 and contour interval 20 feet.

Buffalo
Mill Branch
Hollow
Nappers
Good Spring Church
Maple Spring School
Cade
GREEN
Cow Ford Hollow
Three Sisters Island
M A M M O T H C A V E
Mammoth Cave School
Running Branch
Fishtrap Hollow
Stillhouse Hollow
FISHTRAP FERRY
Cave Island
BM 748
Mammoth Cave
Colossal Cave
DEMUNBRUN FERRY
YMC CAMP
A T I O N A L P A R K
Houchins Valley
Double Cellars
Hunts Sink
Turnhole Bend
TURNHOLE FERRY
BM Elko
765
712
709
818
606
864
EDMONSON CO
CAVE CITY
BM Sloans Crossing
834
770
New Entrance Mammoth Cave
Cedar Sink
723
MAMMOTH
Owens Valley
Woolsey Hollow
Cedar Spring Valley
Whiteoak School
Union City
CAVE
Chaumont
Cedar Hill School
Cedar Spring
Cedar Spring School
THE KNOBS
4°
TRUE NORTH
MAGNETIC NORTH
APPROXIMATE MEAN DECLINATION, 1961
(a)

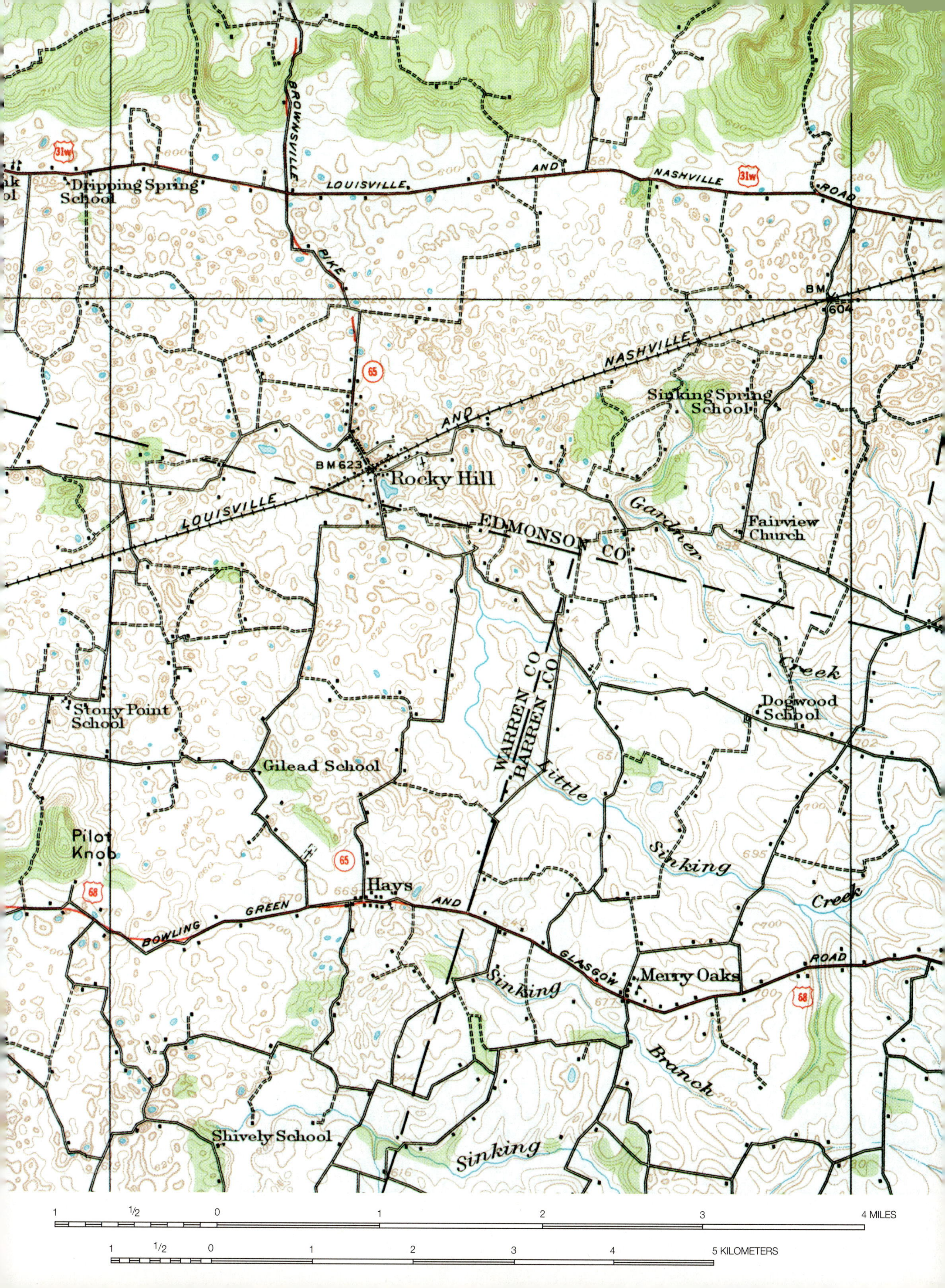
Dripping Spring School
LOUISVILLE AND NASHVILLE ROAD
31w
BROWNSVILLE PIKE
65
LOUISVILLE AND NASHVILLE
BM 604
Sinking Spring School
BM 623
Rocky Hill
EDMONSON CO
Gardner Creek
Fairview Church
Dogwood School
Stony Point School
WARREN CO
BARREN CO
Gilead School
Little Sinking Creek
Pilot Knob
68
Hays
BOWLING GREEN AND GLASGOW ROAD
Merry Oaks
Sinking Branch
Shively School
Sinking
1 1/2 0 1 2 3 4 MILES
1 1/2 0 1 2 3 4 5 KILOMETERS

What would happen to rain that fell into one of these valleys?

____

e. What type of feature is Woolsey Hollow?

____

f. Toward the southern edge of this map valleys and hollows give way to a different type of karst feature, exemplified by the green area about ¾ mi south of Cedar Hill School. What are these features?

____

g. The topography continues to change as we move south to the area shown in Figure 13.15b. Notice that depressions riddle the northern white part of the map. What are these features?

____

h. Use a highlighter, or colored pencil, to emphasize all the streams on map **b**. What is the most common word used in the name of these streams, besides the words *Branch* or *Creek*? ____________

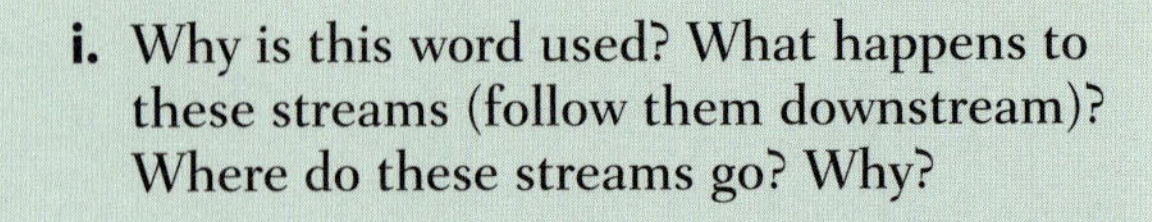

i. Why is this word used? What happens to these streams (follow them downstream)? Where do these streams go? Why?

____

____

____

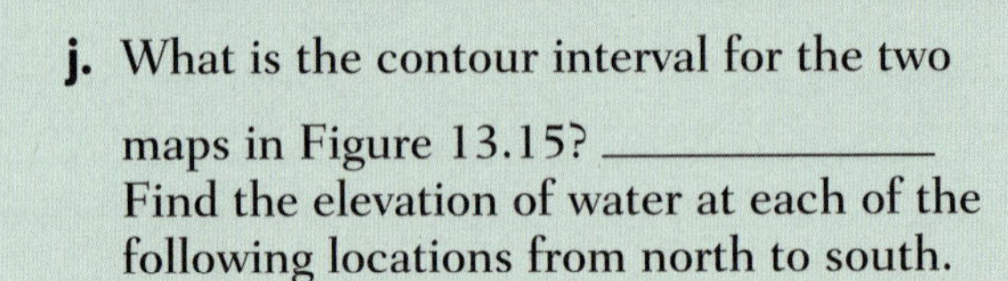

j. What is the contour interval for the two maps in Figure 13.15? ____________
Find the elevation of water at each of the following locations from north to south.

i. Green River ________

ii. Within Woolsey Hollow ________

iii. Near Highway 31w ________

iv. Green area of Little Sinking Creek

____

20. Study the cross section of the same general area around and south of Mammoth Cave in ■ Figure 13.16. This cross section is stylized and exaggerated and shows the area from southeast to northwest.

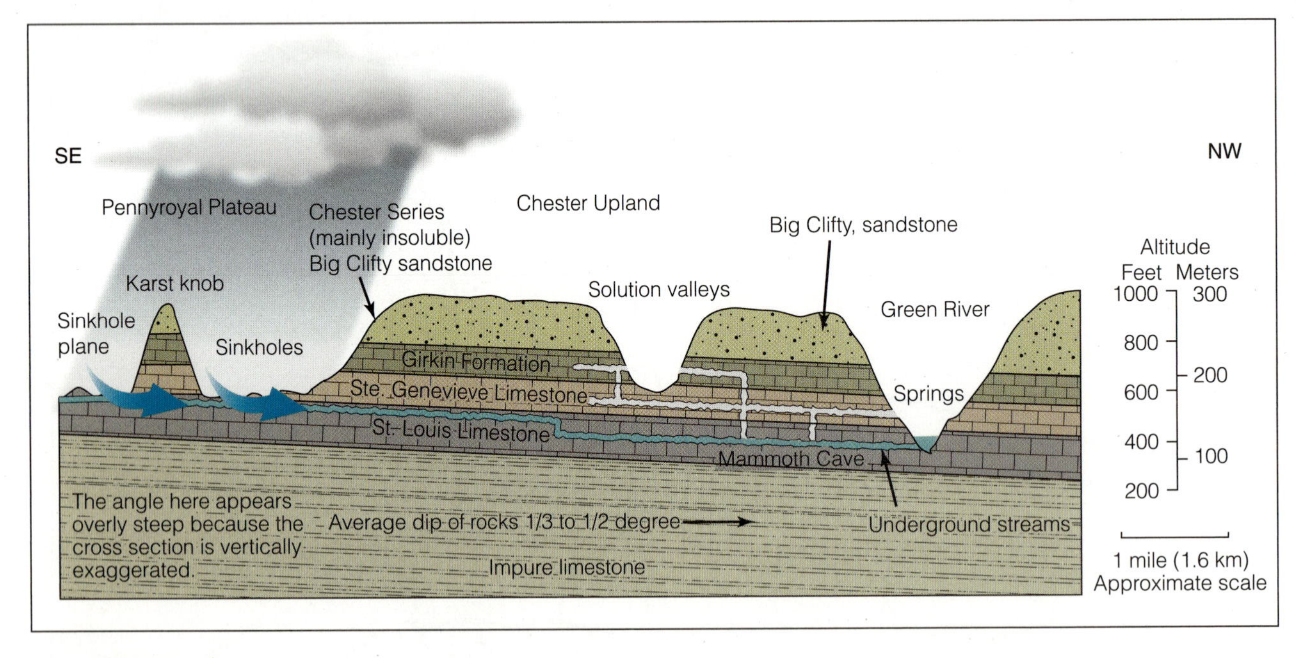

Figure 13.16

Mammoth Cave and Pennyroyal Plateau Areas

The cross section shows the area from southeast at the left to northwest at the right. Rainwater falling on the Pennyroyal Plateau sinks into the ground and travels on gently-sloping strata northwest, passing below and through parts of the Mammoth Cave system, and eventually making its way to come out as springs into Green River. The topography changes from sinking streams and sinkholes to karst knobs to plateaus of sandstone and other insoluble rocks with solution valleys between them, to the Green River with its entrenched meanders. The Mammoth Cave system is within the limestones of the St Louis, Ste. Genevieve, and Girkin Formations.

a. What specific rock type typical of karst topography is present in the area?

____

b. For each of the following features list the dominant specific rock type forming the feature:

Sinkholes: ______________________

Solution valleys: ______________________

The top of the karst knobs: ______________

Chester Upland: ______________________

Caves: ______________________

21. On a separate sheet of paper, answer the following questions:

a. Explain why each of the features in Exercise 20b form in these rocks.

b. The rocks here are dipping very gently to the northwest at about ⅓° to ½°, as indicated in Figure 13.16. In what way does the dip influence the changes seen in the topography from the maps in Figure 13.15a to b?

c. Explain the distribution of water levels that you determined in Exercise 19j.

d. Over time, what has happened to the water table in the area near Mammoth Cave? Describe at least three pieces of evidence to support your answer.

Another region of extensive karst topography occurs in Florida.

22. Examine the map of the area near Lake Wales, Florida (■ Figure 13.17).

a. What are depressions in this area that in some cases also form lakes?

karst Topography / sinkholes

b. How do these depressions form and what hazard is associated with their formation?

spring valleys

____

c. Why do some depressions have water in them while others don't?

it depends if they

are springs or valleys

d. What is the range of levels of the water table in this area?

112 to 116

e. If runoff from city streets pollutes the groundwater, what direction would this pollution flow?

Down Would the wetlands in the western part of the map become contaminated? Yes

f. What rock type do you think occurs in the Lake Wales area?

limestone

23. Examine the map of the area near Dade City, Florida, in ■ Figure 13.18a.

a. Draw a topographic profile from A to A′ using index contours and the lines in Figure 13.18b. With this geometry, your profile will have a 10× vertical exaggeration.

b. What are two karst features present here?

Bird lake

Bonet lake

c. Blue numbers on the map show the elevation of water at various locations on the map, in some cases giving a range of elevations. Contour the water table using a 5′ contour interval for contours below 100′, and a 25′ interval for elevations above 100′. The map already shows the 100′ contour.

d. Draw arrows showing the direction of groundwater flow.

e. Imagine you own land at the green rectangle just northwest of Kessler, near Orange Grove Villas. You want to drill a well on your land. If well drilling is $25/foot and you need to drill the well to 10′ below the water table, determine how much it will cost for the drilling footage you need?

$250 Show how to do all of the calculations.

(25)(10) = $250

f. Find the cheapest place to drill the well and place an x there on the map.

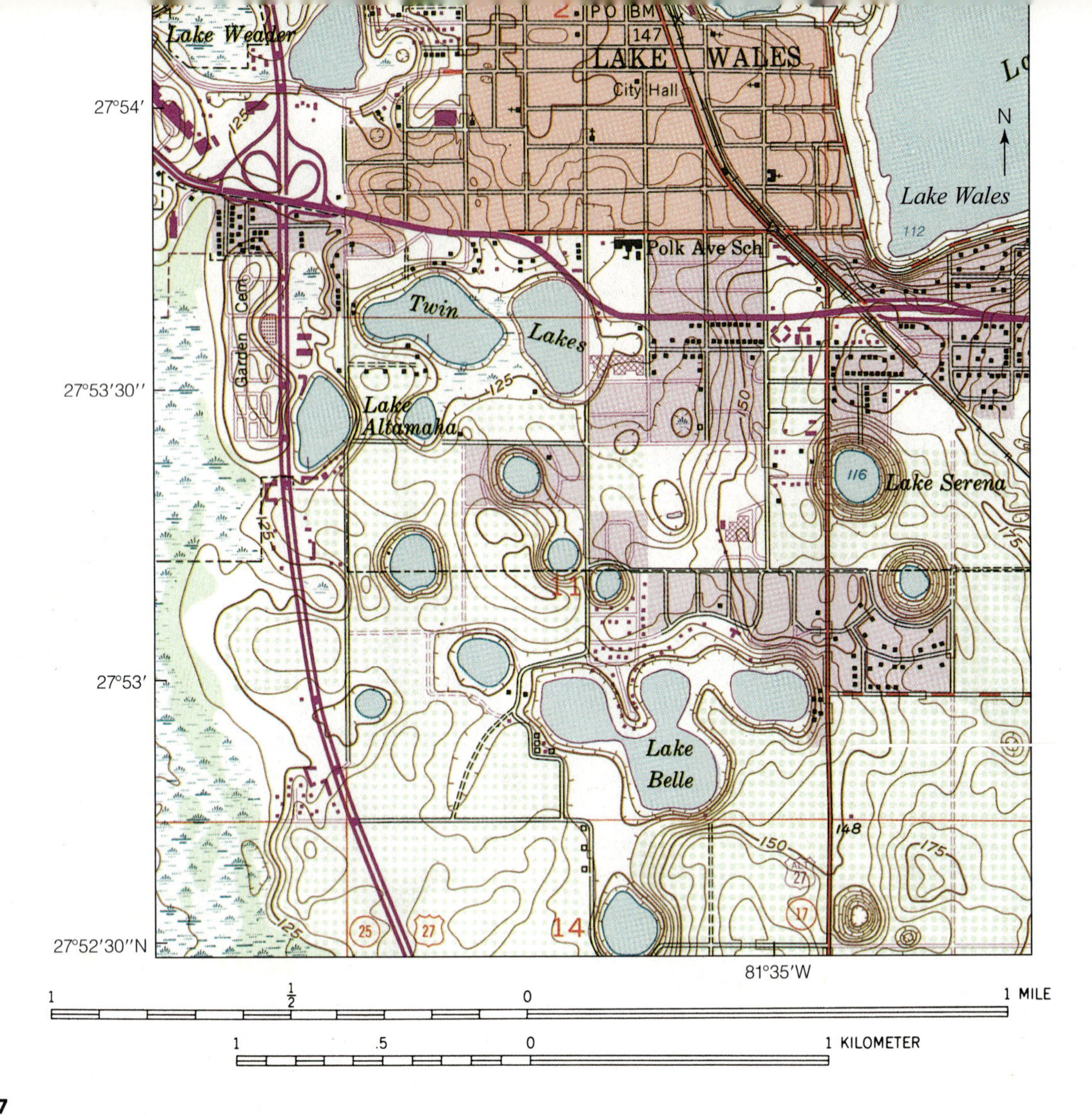

Figure 13.17

Lake Wales, Florida

USGS Topographic Map. Scale 1:24,000; contour interval = 5 ft.

g. What does the regular green-dot symbol on the map mean?

____

h. If people within a ½ mile all around your property use pesticides that get into the groundwater, color the areas on the map that would lead to contamination of your well. What color did you use to show this on the map? ________________

i. If you pump water out of your well fast enough to create a cone of depression with a radius of 2/10 of a mile, how does that change your answer in step **g**? To answer this question, use another color and shade in the additional area. What is the new color? ________________

j. Assume you like to go fishing in Bird Lake and you decide to convince the local orange growers to go organic to keep pesticides out of the lake. Use another color, which is ________________, to show on the map where you would start lobbying the growers.

Mogotes—karst cones, are present in the tropics and common in the Caribbean. *Cone* and *tower karst* is characterized by numerous steep-sided cone-shaped hills formed in later stages of karst development. They often rise out of lowlands and may be aligned in ridges, an alignment that can be due to deep fractures in hard limestones or a residual effect from underground rivers dissolving surrounding areas.

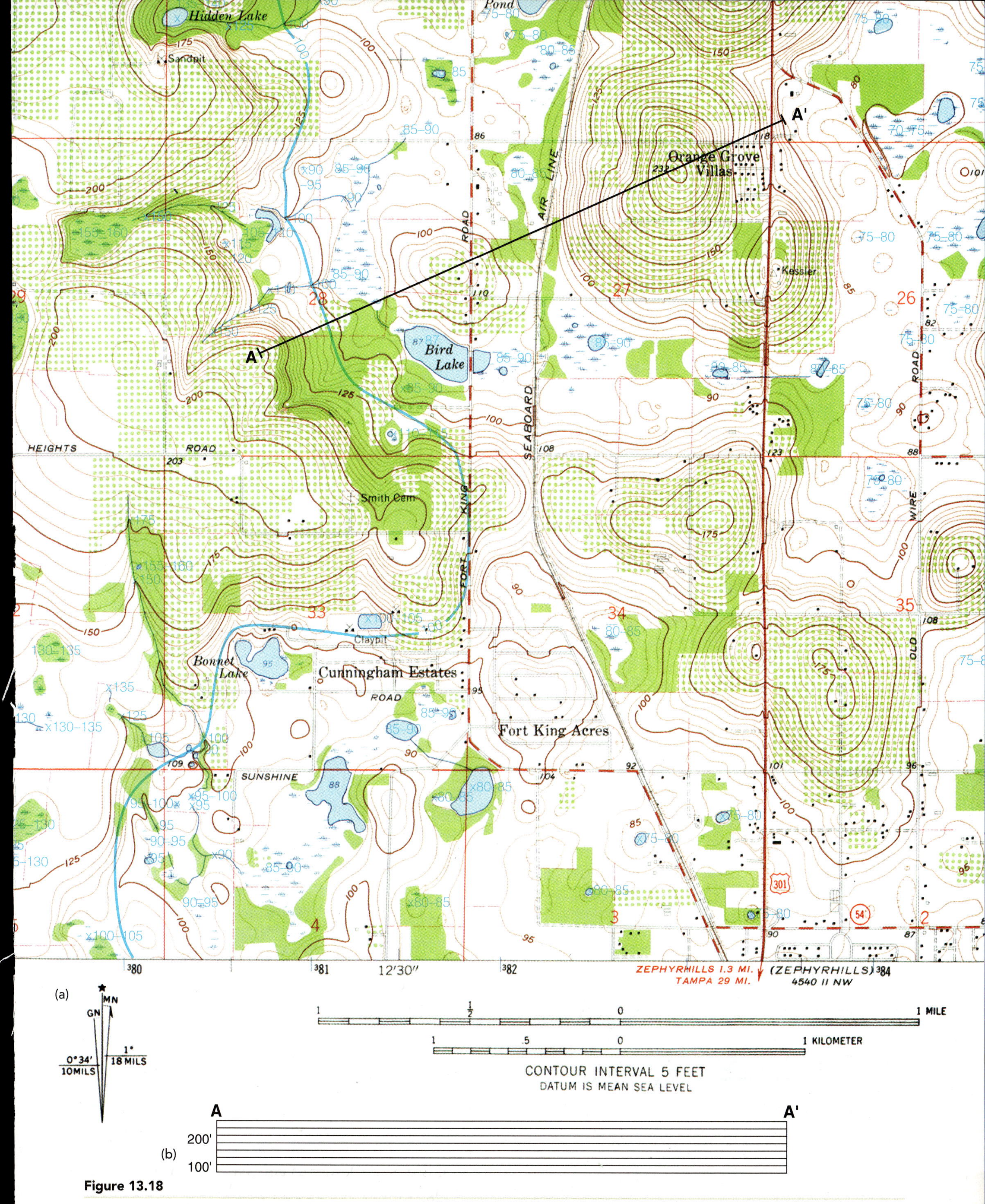

Figure 13.18

Orange Grove Villas, Florida

(a) From the USGS 7½-min Dade City Topographic Map with water-table elevations added in blue. Blue line is the 100-ft water-table contour Scale 1:25,000. (b) Lines for a topographic profile from A to A′ on the map in (a). This profile is set up to have a vertical exaggeration of 5×. See instructions in Exercise 24.

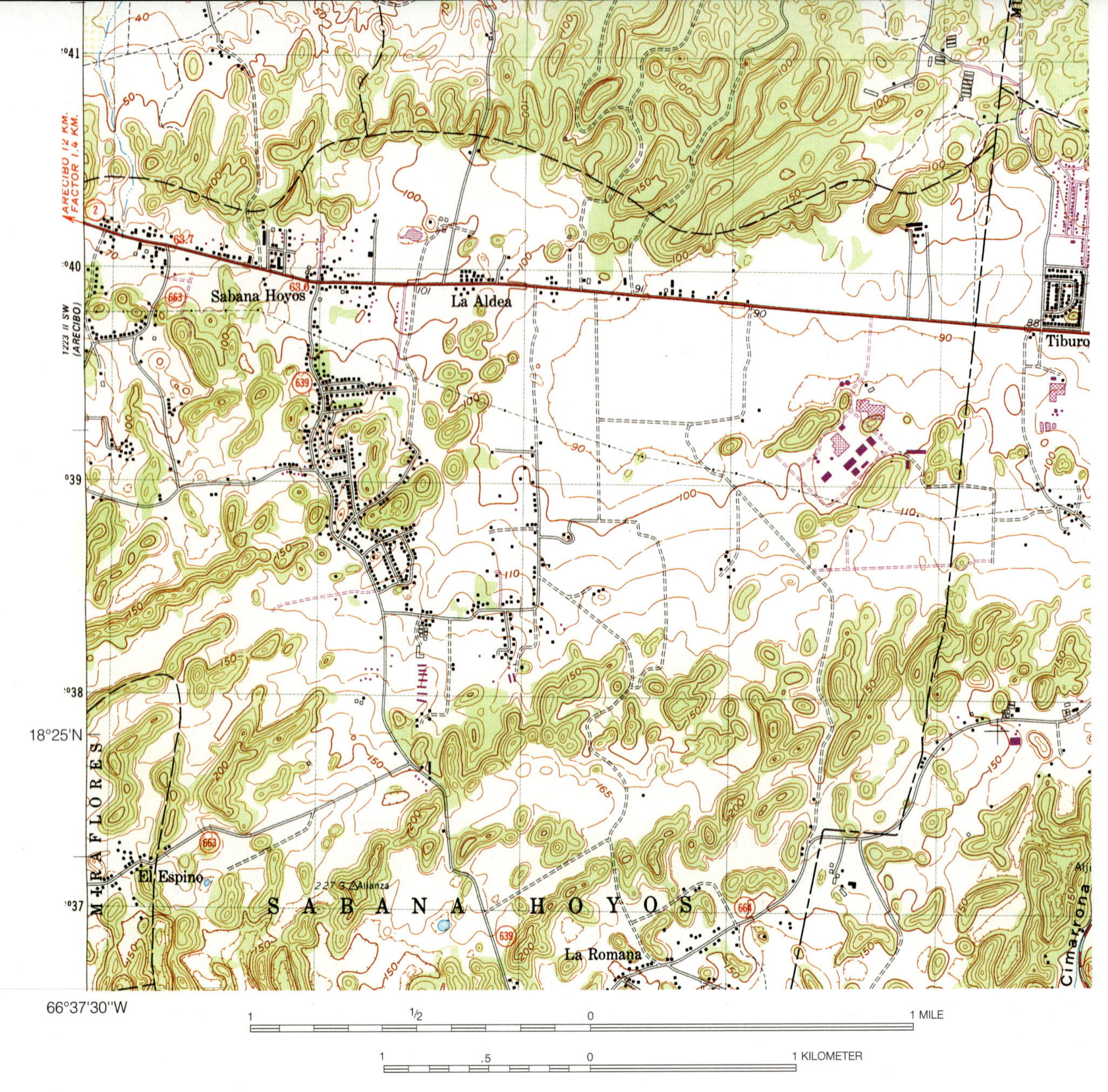

Figure 13.19

Map Near Barceloneta, Puerto Rico

From the USGS 7½-min Barceloneta, Puerto Rico Topographic Map. Scale = 1:30,000. Contour interval = 10 meters with intermediate dashed contour interval of 5 meters.

24. Examine the map of the area near Barceloneta, Puerto Rico, in ■ Figure 13.19.

a. What are the abundant karst features in this area?

____

b. Compare their height, size and shape with the karst knobs on the Mammoth Cave map (Figure 13.15) and the hills on the area near Dade City (Figure 13.18a)?

____

____

c. Describe the placement of roads and towns relative to karst features in the terrain.

____

____

How and why does this placement differ from that of artificial structures in other karst areas you have seen in this lab?

____

____

LAB 14

Shorelines and Oceans

OBJECTIVES

- **To learn the features of passive and active continental margins**
- **To understand coastal sediment transport processes and features**
- **To interpret coastal and ocean margin features on topographic and bathymetric maps**
- **To be able to locate, sketch a map of, and describe surface ocean currents**
- **To understand the formation of deep ocean currents**

Claudia Owen

Figure 14.1

Eroded Shoreline

This shoreline at Bandon, Oregon shows a number of erosional coastal features discussed later in this lab.

The oceans cover about 71% of Earth's surface and, yet, 95% of this vast area remains unexplored. The oceans interact in complex ways with other Earth systems. Plate tectonics shapes the depths and margins of the oceans. At the shoreline, where the oceans impinge on the land, the energy from waves powers erosion, sediment transport, sedimentation (■ Figure 14.1), and sometimes destruction of coastal properties. Waves form due to wind systems and wind comes from differential solar heating. Wind also creates predictable patterns of ocean currents and ocean currents strongly influence climate (see Lab 15) and weather (Lab 16).

THE EDGE OF THE OCEANS

The edges of the oceans, where ocean and continent meet, are known as **continental margins** and are where humans most commonly encounter the resources and hazards of the sea. It is also where vast quantities of sediment (16.5 billion tons per year by rivers alone) from the continents are transported to the sea and onto the continental margins. These sediments are the source of a substantial amount of oil and natural gas production. Eventually, over geologic time, the sediments accumulate, lithify, and become part of the continents, often by plate tectonic activity. A significant proportion of sedimentary rocks on continents originated at the edges of the oceans; movement of the Earth's plates repositioned many of them far inland. Both the Grand Canyon and the top of Mount Everest have lithified marine sediment from continental margins.

When studying plate tectonics (Lab 9), we learned how ocean crust formed. Because the oceanic crust is denser, and therefore lower in elevation than continental crust, the denser basins hold the oceans. In fact, the ocean water overfills the ocean basins somewhat and spills over onto the edge of the continental crust along the continental shelves. The edges of the oceans have a distinctive pattern of **bathymetry** (like topography, but under water) because of this interaction. Plate tectonics also influences the deep ocean with mid-ocean ridges occurring at divergent plate boundaries and oceanic trenches at convergent boundaries (Figure 1.1, p. 2). Continental margins may be of two basic types. A **passive continental margin** is one without a plate boundary, such as on either side of the Atlantic Ocean and the western part of the Indian Ocean. An **active continental margin** occurs where the continent's edge coincides with a convergent plate boundary. Active margins surround the Pacific Ocean. The eastern margin of the Indian Ocean is an active margin, as the December 2004 Sumatra earthquake and tsunami so dramatically demonstrated (refer to Lab 10). At a passive margin, the bathymetry progresses gradually from a *continental shelf* to a *continental slope* to a *continental rise* down to the *abyssal plain*; in contrast, the continental slope at an active margin drops off steeply to an *oceanic trench* and then up to the abyssal plain. ■ Figure 14.2 illustrates an ocean basin (a) and continent (b) with an active margin on the left and a passive margin on the right. The following features of the sea floor and continent edge are also shown in Figure 14.2:

- The **continental shelf** is a gently sloping area of shallow water adjacent to a continent, where the edge of the ocean water partially covers the edge of the continental crust. It holds the largest volume of sediment in the world. The shelf is also where most of human interaction with the ocean takes place as we experience hazards and use biological, mineral, and energy resources of the shelf. More environmental contamination and depletion occur here as well.
- Next is the **continental slope,** where the continental and oceanic crust meet (Figure 14.2b), causing the solid surface of the Earth to slope quite steeply downward away from the land, eventually to meet the deeper ocean basin.
- Along a passive continental margin, farther toward the deep sea from the continental slope, the seafloor slopes more gently again at the **continental rise** (Figure 14.2a, right side).
- The broad expanse of deep sea floor that is nearly flat is the **abyssal plain.** Over 70% of the ocean floor is abyssal plain (Figure 14.2a).
- At active margins with convergence, subduction occurs. Here the continental slope dives down into an **oceanic trench,** before rising again to reach the abyssal plain (Figure 14.2a and b, left sides). The deepest trench in the world is deeper than the highest mountain is tall.
- Farther out into the ocean, and not part of a continental margin, the seafloor may rise up to a **mid-ocean ridge** such as the Mid-Atlantic Ridge or the East Pacific Rise (Figure 14.2a and b). These *spreading centers* create new oceanic crust here.

Figure 14.2

Active and Passive Margins

(a) Bathymetric profile of a hypothetical ocean with an active margin on the left and a passive margin on the right. Vertical exaggeration is 60×. (b) Plate tectonic cross section of active and passive margins across the South American continent.

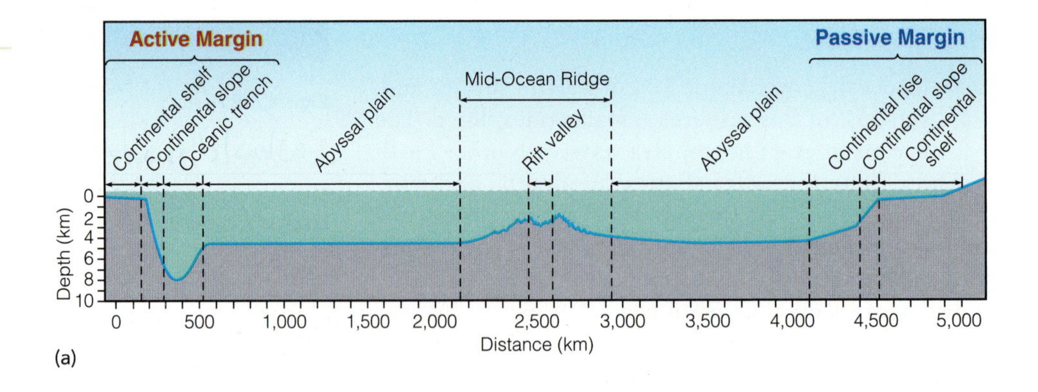

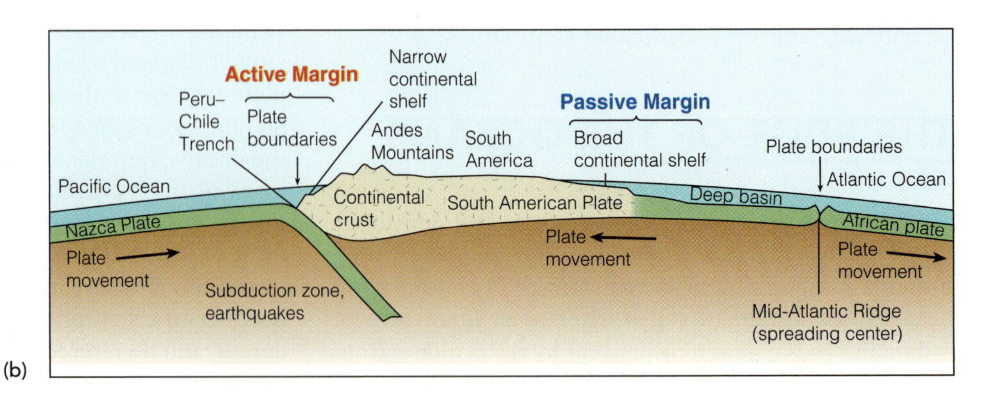

Ocean Margin Profiles

The profiles or cross sections of passive and active continental margins are quite different. In the following exercises, you will draw profiles of these margins, one near Long Island, New York, and the other near El Salvador, Central America.

1. Using ■ Figure 14.3, draw a bathymetric profile in ■ Figure 14.4 from A to A′ on the bathymetric map of the East Coast near Long Island, New York. The shoreline is the zero contour; place it at the top of the profile.

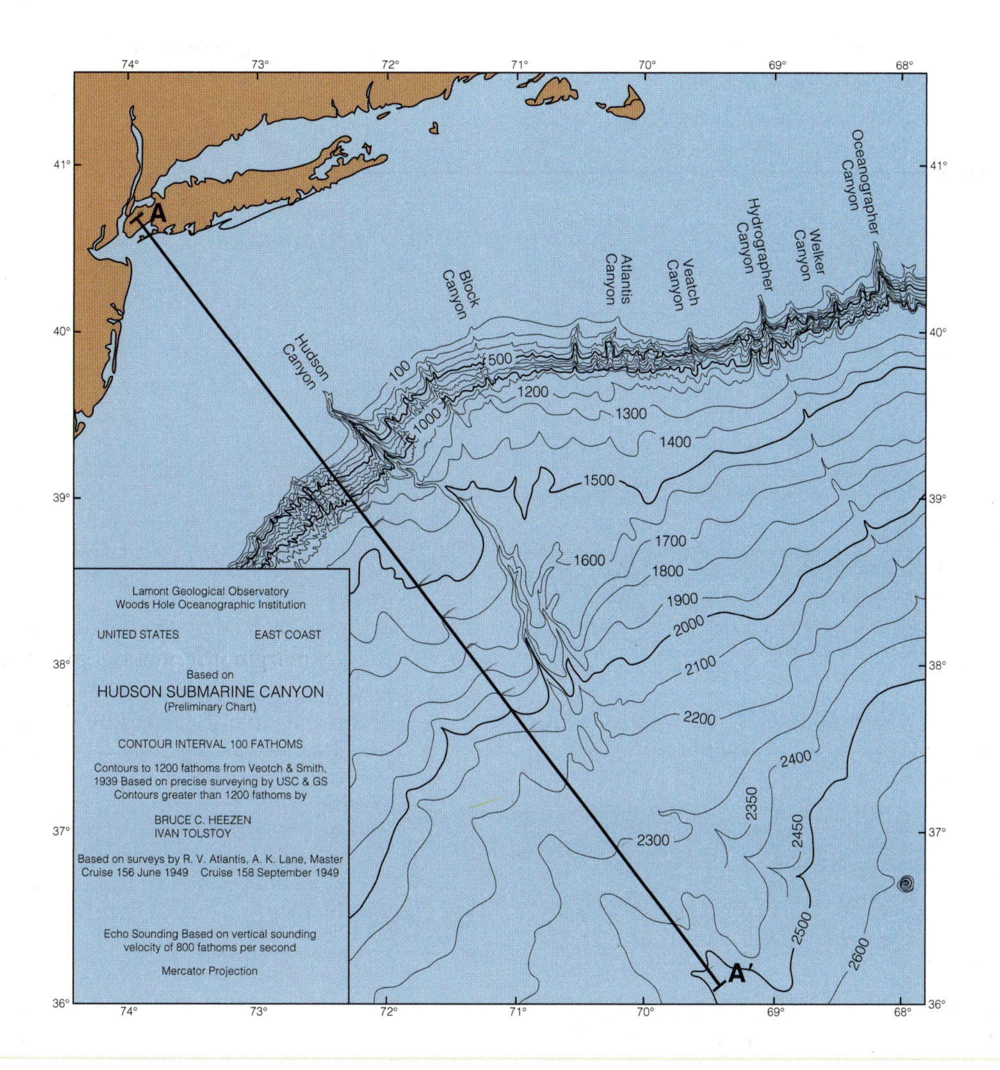

Figure 14.3

East Coast Bathymetry

Bathymetric map of part of the continental margin of eastern United States, near Long Island, New York. Scale = 1:5,000,000. Contour interval = 100 fathoms; 1 fathom = 6 ft.

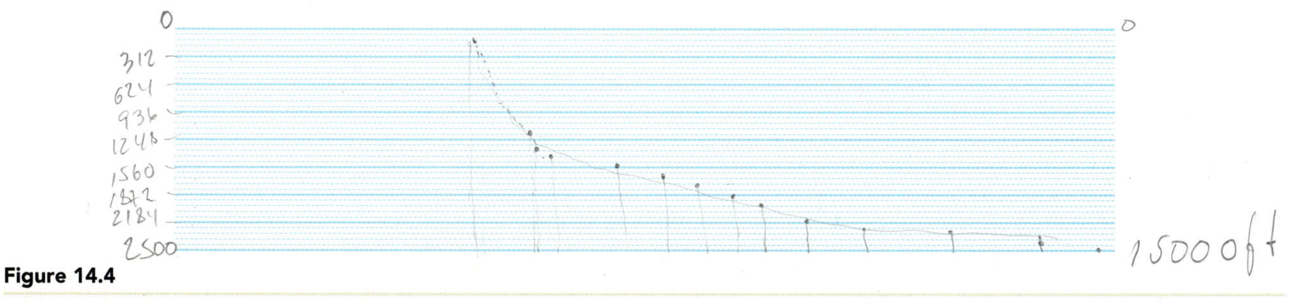

Figure 14.4

East Coast Profile

Bathymetric profile of the continental margin off the coast of Long Island, New York, corresponding to the line from A to A′ on Figure 14.3; see Exercise 1. Vertical exaggeration = 20×.

When you label the depth on the profile, label it in feet on one side and in fathoms on the other.

a. What is the vertical scale of your profile?

____

b. Label all the features listed in the bulleted section above that also occur on your profile.

c. Is this an active or passive margin?

active

2. Table 14.1 lists the thickness of sediments occurring along the continental margin off the East Coast of the United States.

Table 14.1

East Coast Sediments

Approximate thickness of sediments occurring off the east coast of the United States.

Location	Sediment Thickness (ft*)
Shoreline	300
In the middle of the continental shelf	2,400
At the shelf-slope break (the edge of the continental shelf)	15,000
At the slope-rise break	4,000
In the middle of the continental rise	>9,000
Abyssal plain	6,000

*The actual values vary from one location to another.

a. Use the information in the table to draw the location and thickness of sediments on your profile in Figure 14.4. Keep in mind that values in Table 14.1 are in feet.

b. Choose a color for sediments, and color them in on your profile.

c. Using different colors, color in the continental crust and oceanic crust where they should occur beneath the sediments. Include a key on Figure 14.4 explaining the colors.

d. What do you think would be the source of these sediments?

____

3. The Central American Pacific coast is shown rotated so the top is roughly northeast in Figure 14.5. Use this map for the following exercises:

a. Draw a bathymetric profile in Figure 14.6 across the Central American Pacific continental margin (Figure 14.5) from A to A′. You will have to insert the 100-fathom contour between the 0- and 500-fathom lines at the correct position in the profile.

b. Label all the features from the bulleted list above that occur on your profile.

c. Is this an active or a passive margin?

passive

d. How does the vertical scale of this profile compare to the profile you made in Exercise 1? ______________

4. Refer back to the plate map in Figure 1.1 on page 2 and locate the area shown in Figure 14.5. What type of plate boundary occurs here?

Continental margin

What are names of the plates involved?

Middle American Trench

____

SHORELINES

When you go to the seashore, you probably go to a beach that has sand. Where did all that sand come from? It may appear to be stable and immobile, but it moves continually as if it were a slow-moving river of sand. Above the high-tide level, wind moves some of the sand, but waves move the sand in the surf zone where they wash in and out.

Longshore Drift

Waves usually wash up the beach at an angle to the shoreline then run directly down the slope of the beach. Since the slope of the beach is perpendicular to the beach, the water does not wash back exactly where it washed up the beach. This continues repeatedly, causing the water to move along the shore, generating the **longshore current** (Figure 14.7). Each wave also carries some sediment along the shore. The movement of sediment parallel to the shore is the **longshore drift** or **beach drift.**

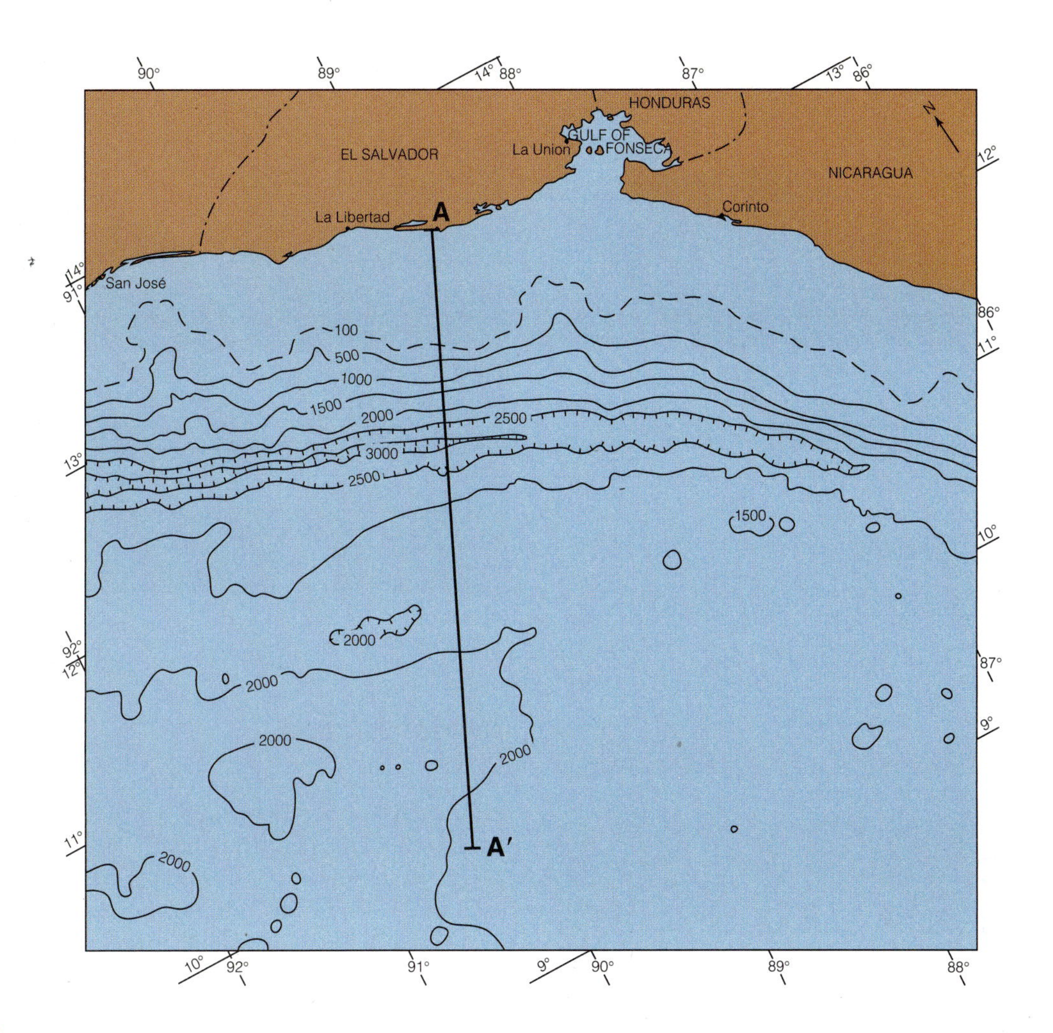

Figure 14.5

Central Americas Bathymetry

Bathymetric map of part of the Central Americas, Eastern Pacific Ocean. Scale = 1:5,000,000. Contour interval = 500 fathoms; 1 fathom = 6 ft.

Figure 14.6

Central Americas Profile

Bathymetric profile of the continental margin across the Central Americas, from A to A′ on Figure 14.5 for Exercise 3a. Vertical exaggeration = 20×.

At the mouths of bays or estuaries, longshore drift tends to deposit sediment across the opening, creating **spits,** which go part way across the bay mouth, or **bay-mouth bars,** which completely cross the mouth of the bay (■ Figure 14.8). Deposition occurs where an island close to shore blocks the wave action, creating a sandy connection between the island and the shore called a **tombolo** (Figure 14.8b). A breakwater parallel to the shore can act in a similar way and accumulate sand behind it.

5. Carefully study the shore angle and wave direction in Figure 14.7b. Then draw and label an arrow on the waves showing the direction the waves travel and another arrow on the beach showing the direction of the longshore drift.
6. Study Figure 14.8b and c. Look for features that might tell you the direction of the long-shore current and drift.
   a. Draw on Figure 14.8b the location and direction of the longshore drift and the probable travel direction of predominant waves.

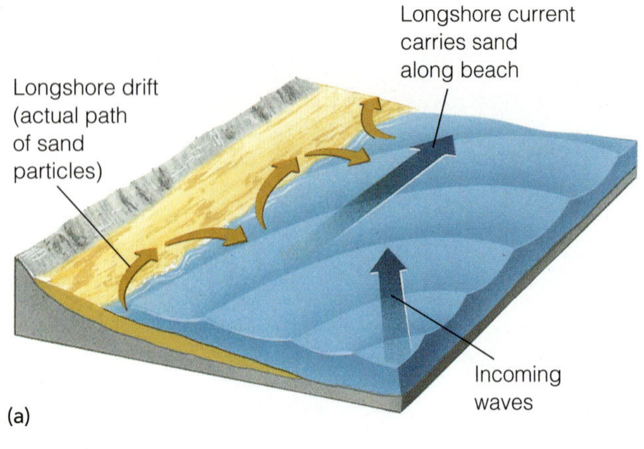

Figure 14.7

Longshore Current

(a) Development of longshore current. (b) Waves approaching a shoreline at an angle along the beach at Half Moon Bay, California.

b. Explain your reasoning and list the various pieces of evidence for the longshore drift and wave directions.

-the movement of the sediments longshore drift tends to deposit sediments across the opening creatin spit.

c. Explain why the tombolo has the orientation it has.

Deposition occurs when an island close to shore blocks the wave action creating a sandy connection

d. Draw an arrow on Figure 14.8c showing the direction of the longshore drift.

7. Look at the map of Morro Bay, California in Figure 6.14 on page ~~135~~. 131

a. What three coastal features, two natural and one constructed, indicate the direction of the longshore drift?

Breakwater

spit

____

b. What direction is the longshore drift there? North

Barrier Islands

Sandbars above sea level that are completely separate from land but are near and parallel to the mainland are **barrier islands.** Many of these long islands form when waves breach a long spit or when rising sea level turns wind-blown dunes, normally high up above the shoreline, into an island. Barrier islands typically migrate shoreward with time. One piece of evidence for this is that we see peat (deposited in the lagoon behind the island) at the ocean side of the island (■ Figure 14.9a–c). Barrier island migration is so rapid that it is historically detectable, as with Assateague Island in Maryland (Figure 14.9d). Despite the natural migration of barrier islands, people commonly build on them (Figures 14.9e, f and 14.10e, f). Not surprisingly, the beaches in front of resort hotels and other buildings on these islands are likely to erode, and then the buildings come under wave attack during **storm surges** (high water levels during storms), hurricanes, and cyclones.

8. Study Figure 14.9 to answer the following questions.

a. Explain in your own words how the peat bed (labeled lagoon sediments in Figure 14.9a–c) could be present under the entire barrier island in (c) when the peat only forms in the lagoon behind the island.

when low sea level the lagoon sediments move of the Barrier Island when high sea level Barrier Island goes back up

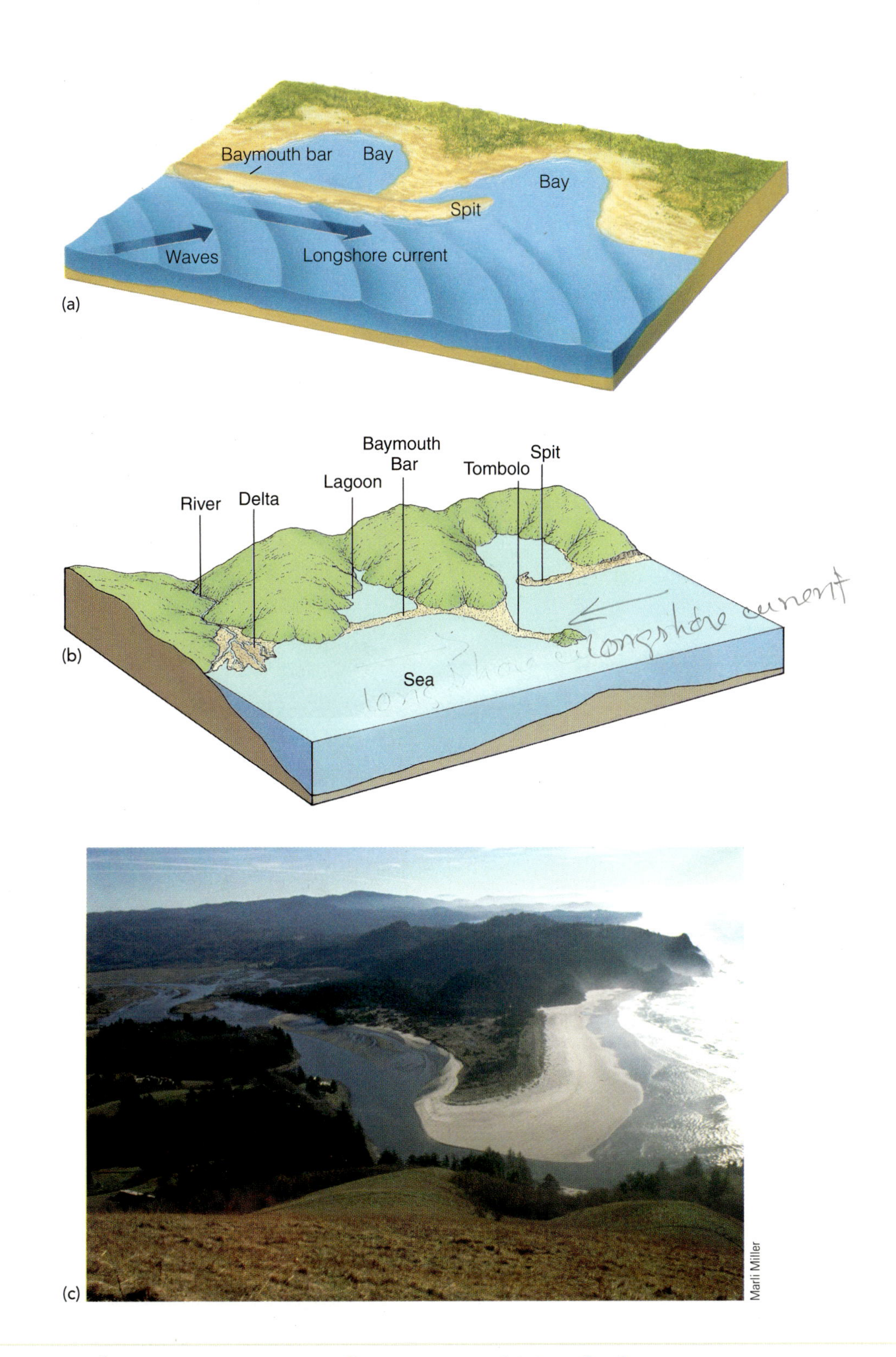

Figure 14.8

Development of Spits, Bay-Mouth Bars, and Tombolos

(a) Spits and bay-mouth bars form where the longshore current carries sand across the mouth of bays or estuaries. A spit crosses the mouth of the bay part way and a bay-mouth bar crosses all the way. (b) Depositional features along an imaginary coastline. Tombolos form behind islands where sand settles from the island blocking wave action. (c) A spit at the mouth of Salmon River, Oregon. (b) From GABLER/PETERSEN/TREPASSO. *Essentials of Physical Geography* (with CengageNOW Printed Access Card), 8E. 2007 Brooks/Cole, a part of Cengage Learning, Inc. Reproduced by permission. www.cengage.com/permissions

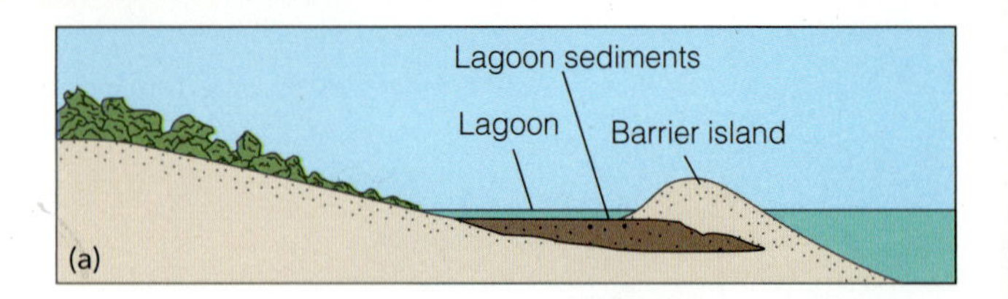

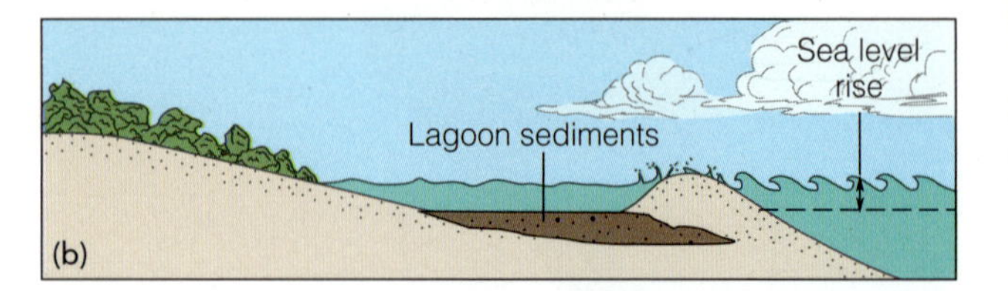

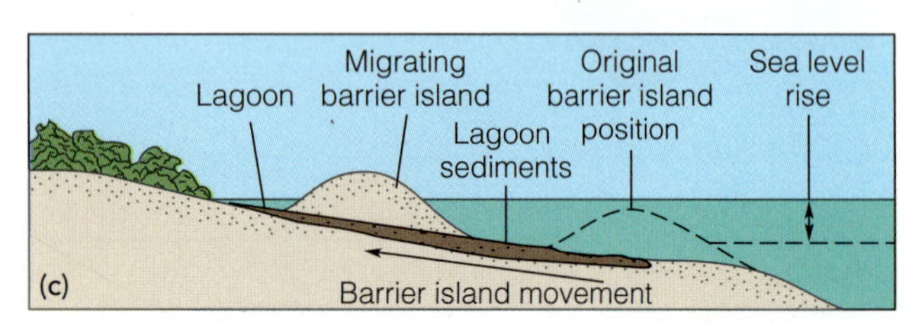

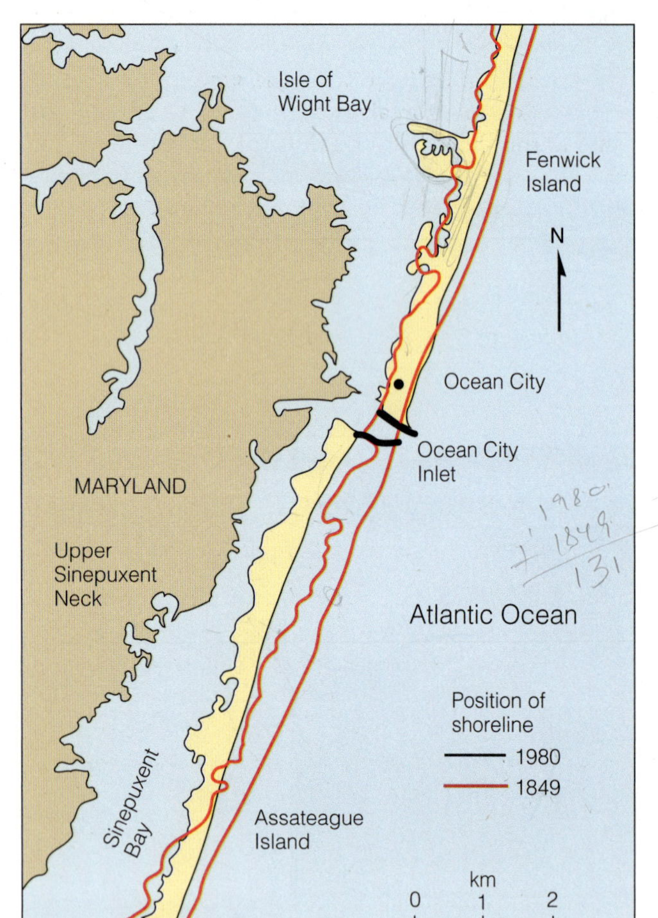

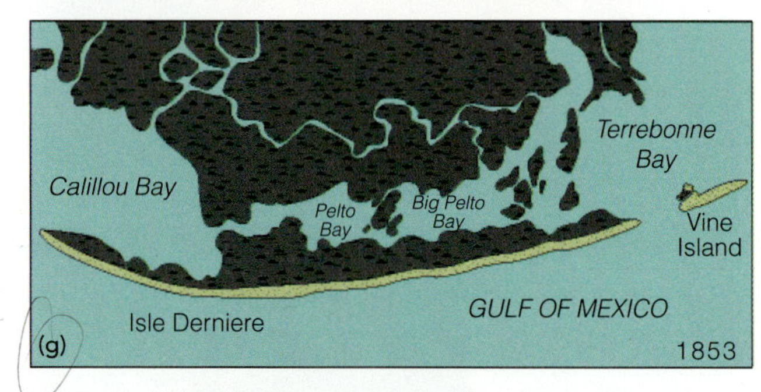

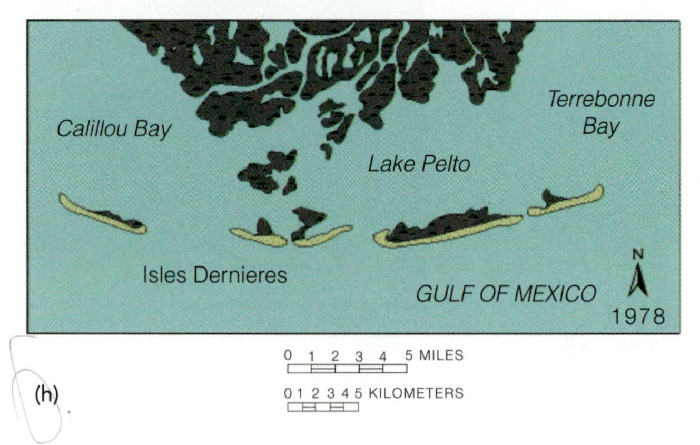

Figure 14.9

Barrier Islands

(a)–(c) Sequence showing sea-level rise and migration of a barrier island with a typical peat bed (lagoon sediments) beneath it. (a) Low sea level. (b) Sea level rises. (c) A barrier island showing how a peat bed can eventually underlie the whole island after it has migrated landward over the former location of the lagoon. (d) Migration of Assateague Island, Maryland. The island is retreating more rapidly than Fenwick Island (north) because of the jetties (black) at Ocean City Inlet. (e) and (f) Houses before and after showing the effects of a hurricane on a barrier island. (g) 1853 and (h) 1978 coastal subsidence and sea-level rise combine to change Isle Derniere and the various bays surrounding it in the Gulf of Mexico, Louisiana. (a-c) From MONROE/WICANDER/HAZLETT, *Physical Geology*, 6E. 2007 Brooks/Cole, a part of Cengage Learning, Inc. Reproduced by permission. www.cengage.com/permissions

b. Toward what direction has Assateague Island (Figure 14.9d) migrated?

c. Which part of Assateague Island (Figure 14.9d) has migrated the most?

d. If any buildings had been built on the north end of Assateague in 1849, could those buildings still be standing today? ______ If so, where; if not, why not?

e. What is the fastest rate of retreat of Assateague Island in meters per year?

______ Show how to do the calculations:

f. At this rate of retreat, how long would it take for the shoreline to erode back to a house built 200 ft from the shore (1m = 3.28 ft)? ______ Show how to do the calculations:

g. What happened between the time of photography of (e) and (f) in Figure 14.9? How could this have happened?

h. How much time passed between the two maps in (g) and (h)?

______ What happened to Isle Derniere in this time?

How have the bays changed?

Beach Erosion Control

Attempts to control beach erosion take advantage of longshore drift. A **groin** (Figure 14.10b) built perpendicular to the shore acts like a dam to the longshore drift so the sand builds up on the "up-current" side. However, this construction also tends to cause erosion on the "down-current" side because the groin has cut off the supply of sand. A breakwater (Figure 14.10c) built parallel to the shore helps sand deposition between it and the shore in a similar manner to a tombolo. For any structure added to a beach, an area of deposition will also cause an area of erosion where the structure interrupts the flow of sand. Another way to keep a shoreline from losing its beach is to import sand, called **beach nourishment** (Figure 14.10e and f).

9. Use ■ Figure 14.10 to answer the following questions:

a. Draw an arrow in Figure 14.10a showing the direction of the longshore current.

b. Outline, color and label on Figure 14.10b how the distribution of sand around the new groin on the left will change after a period of longshore drift, deposition, and erosion. Label all areas of deposition with 'D' and erosion with 'E.'

c. Label where deposition and erosion have occurred in Figure 14.10c. Explain the distribution of sand around the breakwater. Why has sand built up where it has and why has it eroded elsewhere?

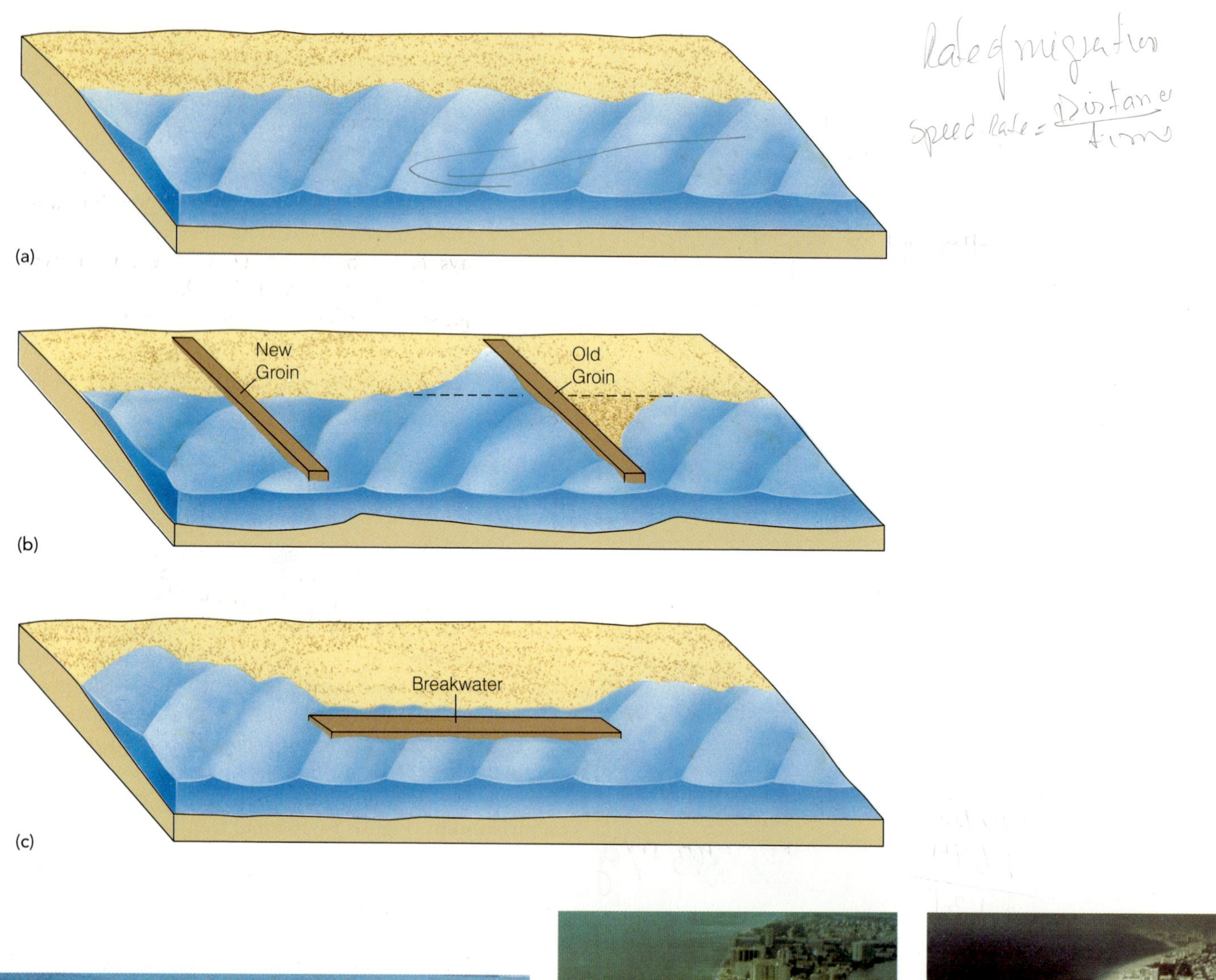

(d) USGS

(e) U.S. Army Corps of Engineers

(f) U.S. Army Corps of Engineers

Figure 14.10

Beach Protection Strategies

Groins, perpendicular to the shore, slow beach erosion. (a) A beach with oblique waves. (b) The same beach with an older groin at the right showing resulting deposition and erosion adjacent to the groin, and a newly built groin at the left. (c) A breakwater built parallel to the shore. Sand collects behind the breakwater somewhat like a tombolo. (d) Erosion control structures at shoreline at Norfolk, Virginia. (e, f) Built-up barrier islands of Miami Beach, Florida before beach replenishment. (f) The same location after beach replenishment.

d. What structures attempt to control beach erosion in Figure 14.10d?

groins, perpendicular to the shore

e. Based on the pattern of erosion and deposition, draw an arrow showing the direction of longshore drift in Figure 14.10d.

f. Use evidence from these figures and describe some of the problems with erosion control structures.

____

____

____

____

g. Clearly label the before and after pictures in Figures 14.10e and f (read the caption). Could the photo sequence be reversed in time? Explain.

____

____

10. Now look back at the map of Assateague Island in Figure 14.9d. The two jetties (two thick, black lines) at Ocean City Inlet were built for ship navigation. They act somewhat like groins at this location.

a. Draw an arrow on the map showing the longshore current direction here.

b. What evidence supports your drawing in part **a**?

____

____

____

c. What are the consequences to local beaches and barrier islands of building structures like this?

Instead of preventing erosion they bring more erosion on the opposite side of the structures.

Headland Erosion

Because of the persistence of longshore currents and wave **refraction**—the bending of waves toward the shoreline—wave energy and erosion are concentrated at **headlands.** Headlands are promontories that stick out into the water (■ Figure 14.11). The waves, impinging on both sides of the headland, cause the formation of **sea caves** and **sea arches** along zones of weakness in the rocks. After a number of years, the headland retreats and the sea arches collapse, leaving **sea stacks** (Figure 14.1, p. 317). The angle of the waves and their refraction cause the longshore drift to move sediment from the headlands toward bays. As the headlands erode, the bays fill with sediment, and the shoreline tends to become straighter and straighter. Other factors such as resistance to erosion, prevailing wind strength, seasonal variations, and the bathymetry also influence the ultimate shape of the coastline shape.

11. Assuming wave refraction and longshore drift are the dominant factors controlling changes in the shoreline in Figure 14.11, draw arrows on the figure showing the directions of the longshore currents in at least two places where they differ.

Stream Table or Sand Tub Experiment In a stream table (or a wide tub of sand), arrange the sand so it makes two headlands with a bay between them (■ Figure 14.12). Your lab instructor will explain how to generate waves in your "sea." Work as a group.

Materials needed:

- Basin or stream table with sand and water
- A flat, water-resistant plate or board that fits on end into the basin
- Toothpicks if your sand is deep enough or an acrylic sheet with washable markers
- Small houses or similar substitute object
- Streak plates for beach "protection" structures

12. Read the whole experiment before you start; then perform the following steps:

a. Place the acrylic sheet on the stream table, and mark the stream table and the acrylic with washable markers so you will be able to replace the acrylic in exactly the same position.

b. If you are using toothpicks, stick them deeply into the sand precisely at the shoreline, where the water meets the sand. Also, insert toothpicks along any cliffs. Alternatively, if you are using the

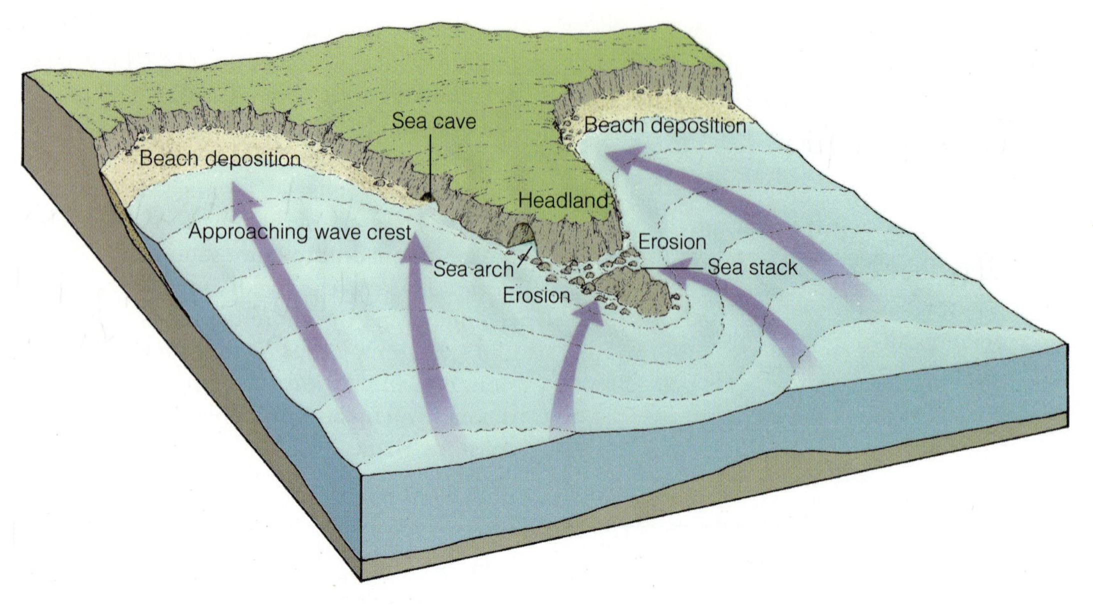

Figure 14.11

Headland Erosion and Sea Stack Formation

Wave refraction causes waves to bend toward a headland, concentrating wave energy there and eroding cliffs to sea caves, sea arches, and sea stacks. In the bays, sediment accumulates. Erosion of headlands and deposition in bays tends to make the whole shoreline straighter. From GABLER/PETERSEN/TREPASSO. *Essentials of Physical Geography* (with CengageNOW Printed Access Card), 8E. 2007 Brooks/Cole, a part of Cengage Learning, Inc. Reproduced by permission. www.cengage.com/permissions

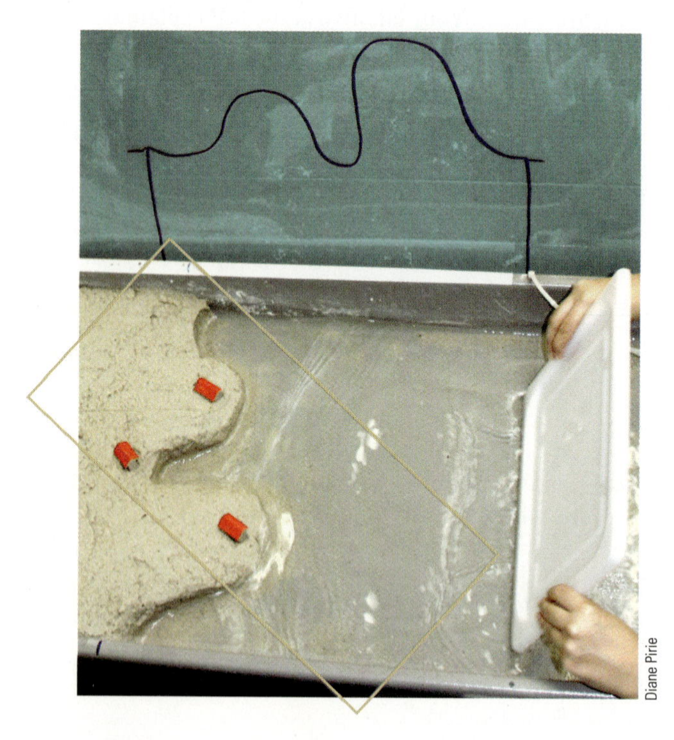

Figure 14.12

Shoreline Erosion Experiment

Sand coastline set up with two headlands and a bay. Yellow outline shows the area for a narrow stream table with a diagonal coastline arrangement.

acrylic sheet and markers, trace the shoreline onto the acrylic sheet, avoiding *parallax*[1] problems. Also, trace the edge of any cliffs. Remove the acrylic.

c. Have each person in the group select a waterfront home site. Place a few "houses" onto various sites. On a separate sheet of paper, sketch the configuration you drew on the acrylic or outlined with toothpicks and also draw the positions of the houses.

d. Start making waves as instructed. The waves should be as linear as possible, which is why we used a rigid plastic box lid in our experimental photo in Figure 14.12.

e. Observe wave refraction (bending of the waves) and draw in the wave crests (tops) on your sketch both far from, and near your shoreline.

[1] Parallax is the apparent visual displacement of a closer object against a background because of a shift in the point of observation (your eye). In other words, you will not trace an accurate position of the shoreline on the acrylic if your eye is not directly above the point you are tracing.

f. After about 1 min, what happened to your house or its position relative to the coast?

____

____

____

Remove the houses, and replace and align the acrylic to its former position on the stream table.

g. Draw arrows on the acrylic showing the direction the waves came in toward the shore and the movement of the sand.

h. Trace the new shoreline on the acrylic or directly onto your sketch. Remember, for best accuracy, mark where the water meets the sand and have your head directly over the place you are drawing. Also trace the cliff positions.

i. Mark on the acrylic or on your sketch where erosion (X's) and deposition (dots) occurred.

j. Copy the arrows and erosion and deposition patterns drawn on the acrylic onto your sketch if you have not done so already.

k. On the separate sheet of paper, write a summary of coastal sediment transport, erosion, and deposition.

l. Set up a diagonal straight coastline with a series of three groins or two breakwaters. This should only be tried with a fairly large stream table—small ones will have a confusion of waves reflected off the sides. Sketch the starting configuration, then make waves for 30 seconds or more, and sketch the resulting arrangement of the sand. Show the dominant wave direction and longshore drift on your new sketch. If you have time, try the other barriers, either breakwaters or groins, or compare your results with others who tried them.

Figure 14.13

Chesapeake Bay

This estuary is a *ria*; it formed when sea-level rise drowned a river valley. White along the coastline is beach sand.

Submergent versus Emergent Coastlines

Sea level is not constant, neither is the elevation of the land. Different coastal features result when the land emerges from the sea than when it becomes submerged. As sea level rises relative to the land, creating a **submergent coastline,** drowned river valleys, called **rias,** normally form **estuaries,** which contain **brackish** water—mixed salty and fresh water. Rivers commonly form a branching, tree-like pattern with their tributaries, which means that the estuaries typically have a branching form. Chesapeake Bay is a classic example (■ Figure 14.13). Coastal plains develop *barrier islands* along submergent coastlines.

Since the last ice age, sea level has risen over 100 m and is continuing to rise due to modern global warming, so you might think that all coastlines worldwide would be submergent. However, at active margins you may find both submergent and emergent features. **Emergent coastlines** occur where the land is rising faster than sea level. Features of an emergent coastline include headlands, sea cliffs, sea stacks, sea arches, and *marine terraces*. When the land emerges from the sea, wave action creates a flat area—a **wave-cut platform**—close to sea level due to wave erosion. As uplift continues, the platform becomes a **marine terrace** (■ Figure 14.14). The flat areas above the cliffs and hills in this figure are the marine terraces.

13. Examine ■ Figure 14.15.

a. Give the geologic terms for emergent and submergent coastal

Figure 14.14

Marine Terraces

Also called uplifted wave-cut platforms, visible north of Point Arena, Northern California. Uplift here is a consequence of transform movement along the San Andreas Fault system.

features on this map. Emergent features:

____

____

____

____

Label two of each emergent feature on the map. Name a submergent feature and give its geographic name:

____

b. Draw a topographic profile along the line from A to A′ using the lines in ■ Figure 14.16.

c. Label sea level and the wave-cut platform on your profile. What are the approximate elevations of three marine terraces on this profile? __________ __________ __________ Label them with 'MT' on the profile too.

d. Locate the marine terrace between 200′ and 300′ along the profile line on the map in Figure 14.15. Follow the same terrace to the south, past Cape Arago. What is its elevation adjacent to South Cove?

____

e. Explain what must have happened to produce a marine terrace with different elevations like this.

____

____

____

f. What other coastal features are visible on this map?

____

____

g. Based on the orientation of North Spit, what is the direction of the longshore current along the northern part of the map? __________
Draw an arrow showing its direction and label it on the map.

h. Explain how this coast can have both emergent and submergent features.

____

____

____

14. Examine Figure 14.1.

a. What coastal features discussed in this lab do you observe in the photograph?

____

____

____

____

Also, label these on the photograph.

b. Except for mountains in the distance, why are the tops of the higher rocks all about the same height?

____

____

If this question leads you to think of an additional coastal feature, write it in part a above.

OCEAN CURRENTS

Winds drive ocean currents, and like winds, ocean currents help to redistribute solar energy from the equator toward the poles; all of these, in turn affect climate and weather. Because global winds have a particular pattern (see Lab 15, Figure 15.2, p. 343), ocean currents tend to develop a specific pattern as well. The **trade winds** are easterly winds; that is, they blow from the east. They move masses of air between about

Figure 14.15

Map of Cape Arago

Composite of USGS topographic maps of Cape Arago and South Slough, Oregon, showing both emergent and submergent coastal features. Scale 1:45,000. Contour interval = 40'

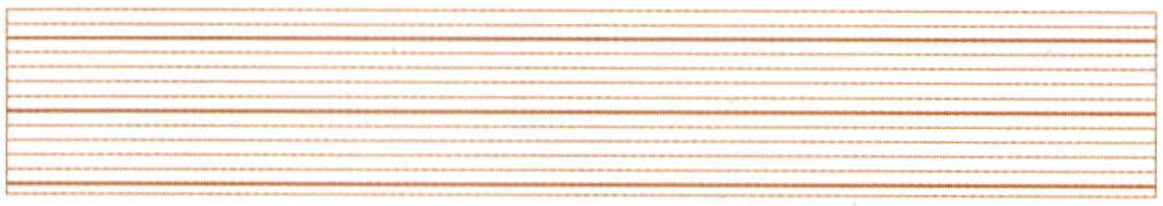

Figure 14.16

Profile of Cape Arago Area

Topographic profile from A to A′ in Figure 14.15 for Exercise 13b. Vertical exaggeration is 5×.

25° latitude and about 5° latitude, and generate ocean currents from east to west just north and south of the equator. The resulting currents, called north and south **equatorial currents** (■ Figure 14.17), are warm due to heating of the surface water from the more direct sunlight near the equator.

15. On the map of the Atlantic Ocean (■ Figure 14.18), use red and blue pens or pencils to make a key to indicate warm (red) and cold (blue) currents. Locate the equatorial currents on the map of the Pacific Ocean in Figure 14.17 and use them as a guide to draw in the warm north and south equatorial currents in their correct positions in the Atlantic Ocean on Figure 14.18. Label the currents.

As the equatorial currents move away from a continent on the east side of the ocean, **upwelling** will occur. Upwelling is when deep, cool ocean water moves toward the surface. This upwelling helps replace the water that the trade winds pushed away from the east side of the ocean.

When an equatorial current encounters a continent on the other (western) side of the ocean, two things happen: the warm water piles up, and the water is deflected north and south. At the coast, where the north equatorial current turns south and the south equatorial current turns north, waters from both equatorial currents collide, impinging on each other and an **equatorial countercurrent** develops. This current flows from west to east along the equator where winds are very light. So an equatorial countercurrent develops between the equatorial currents flowing in the opposite direction. The equatorial countercurrent also helps to replace the water that trade winds blow away from the east side of the ocean.

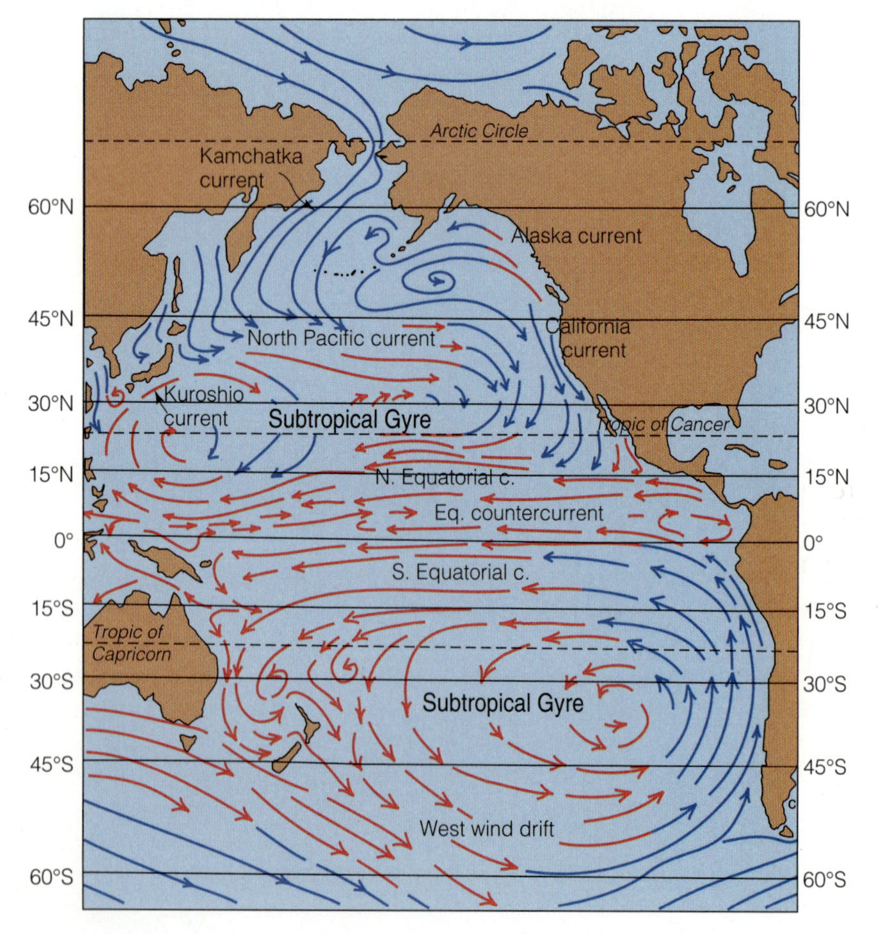

Figure 14.17

Pacific Ocean Currents

Red arrows indicate warm currents, and blue arrows cold currents.

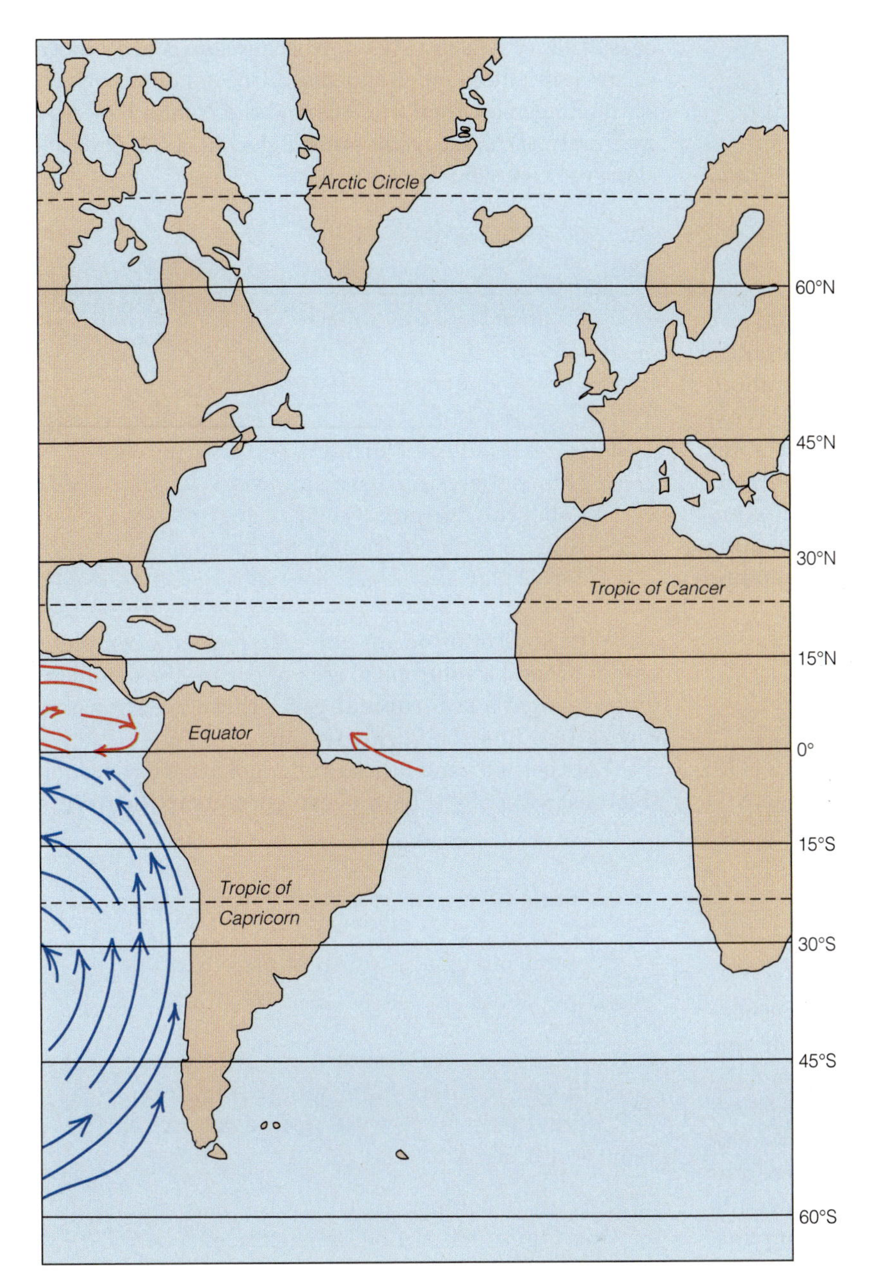

Figure 14.18

Map of the Atlantic Ocean

See instructions in Exercises 15–24.

Key:

16. Add colors to your key to indicate (1) where warm water from equatorial currents piles up because of running into a continent and (2) where cold water would occur due to upwelling. Color these warm and cold regions in on the Pacific and Atlantic maps in Figures 14.17 and 14.18.

17. Locate an equatorial countercurrent in the Pacific map in Figure 14.17 and use it as a guide for drawing the equatorial countercurrent on the map of the Atlantic Ocean (Figure 14.18). Label this current.

Where the two equatorial currents impinge on the continent and diverge away from each other, **western boundary currents** form. In the Northern Hemisphere, the western boundary current flows northward along the west side of the ocean, the east coast of the continent; in the Southern Hemisphere, it flows south along the west side of the ocean. The Kuroshio Current off the coast of Japan is the western boundary current for the North Pacific Ocean (Figure 14.17). The western boundary currents flow away from the equator and are, therefore, warm currents. They tend to warm the coastal areas and produce a warm, moist climate. The Gulf Stream in the North Atlantic is an excellent example of a western boundary current. It carries warmer water northward along the eastern coast of the United States, making the coastal region more temperate than areas farther inland.

18. On the map of the Atlantic Ocean (Figure 14.18), draw in the western boundary currents in the Northern and Southern Hemispheres. Label these and also add a color to your key and map to show the moist coastal areas.

The westerly winds that blow between 30° and 60° latitude (see Lab 15, Figure 15.2, p. 343) produce ocean currents that flow from west to east at about 40° to 50° north and south latitudes. The West Wind Drift and North Pacific Current are examples (Figure 14.17) of currents that westerly winds generate. Although these currents have cooled considerably compared to the equatorial currents, they are still coming from the equator and are warmer than the surrounding water. They are still considered warm currents.

19. Around what continent (not shown) does the West Wind Drift (Figure 14.17) flow?

________________________ On the map of the Atlantic Ocean (Figure 14.18), draw in the warm, east-flowing currents in the Northern and Southern Hemispheres.

Where the currents driven by the westerly winds encounter a continent on the east side of the ocean, they diverge to the north and south. Such currents that head toward the equator, in either the Northern or Southern Hemisphere, are called **eastern boundary currents,** because they move along the eastern portion of the ocean basin. As these currents head toward the equator, they carry colder water into warmer areas, so these are now considered cold currents. Where cold currents travel along a continental margin, a coastal desert is likely to develop because the cold air above them cannot hold much moisture. Coastal deserts, such as the Atacama Desert in Chile and the Namib Desert in southwestern Africa, result from such cold ocean currents along the coast (■ Figure 14.19).

20. Along what coast of a continent would such a desert result from an eastern boundary current: east / west / north / south (choose one)?

21. On the map of the Atlantic Ocean (Figure 14.18), draw in the eastern boundary currents in the Northern and Southern Hemispheres using the appropriate color. Label the currents. Add dry coastal areas to your key and color them on your map.

Where a set of ocean currents makes a complete circuit around a subtropical area of ocean, the complete loop is called a **subtropical gyre;** these are shown in Figure 14.17 for the Pacific Ocean.

Between 60° and 90° latitude, polar easterly winds (Figure 15.2, p. 343) blow ocean currents to the west.

22. Name two such westward-flowing currents in the Pacific Ocean in the Northern Hemisphere a bit south of 60°N latitude.

____

____

In the Southern Hemisphere, the East Wind Drift flows west around Antarctica, south of 60°S.

23. Draw in comparable cold west-flowing currents in Figure 14.18, one for each hemisphere.

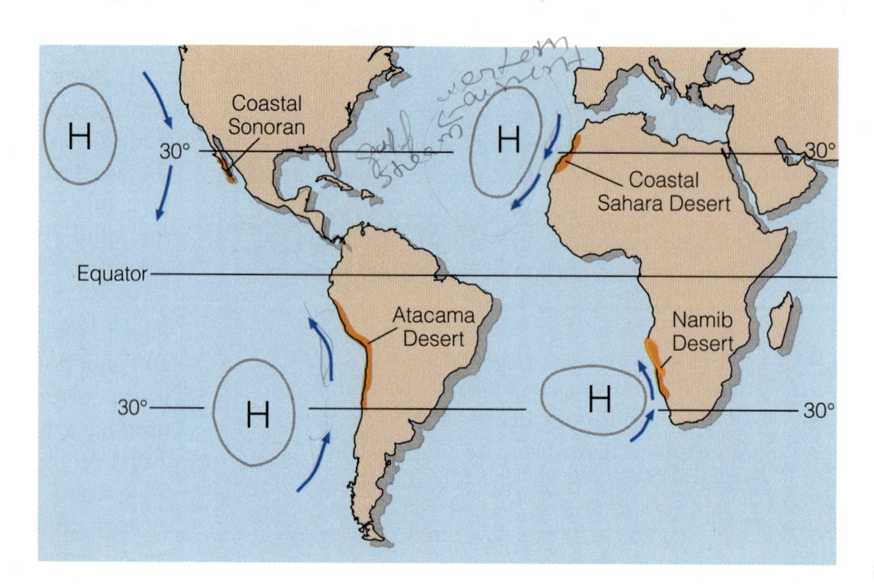

Figure 14.19

Coastal Deserts

Orange areas on this map show the location of coastal deserts, which form adjacent to eastern boundary currents (blue arrows). These cold currents chill and dry the air. **H** = high atmospheric pressure.

Any time a current encounters a continent, the coast can deflect it either north or south, or both. The geometry of the coastline may have some influence, as it does in eastern South America.

24. Notice the one current previously drawn in Figure 14.18 that flows from the south equatorial current in the Atlantic, northwest along the coast of Brazil. This is an unusual current because it crosses the equator. Explain why this current crosses the equator and has the direction that it does.

____

____

____

____

This equator-crossing component of the South Atlantic Equatorial current is an important component of the *thermohaline conveyor system*, which we will explore in the next section, because it helps complete the conveyor circuit.

25. Now that you are familiar with ocean currents, draw in the ocean currents on the map of the Cambrian world of about 500 Myr ago in ■ Figure 14.20. Use the following list to help you:

- Equatorial currents
- Equatorial countercurrents
- Western boundary currents
- Eastern boundary currents
- East-flowing currents between 30° and 60° latitude
- Subtropical gyres
- West-flowing currents between 60° and 90° latitude

OCEAN SALINITY AND DEEP OCEAN CURRENTS

Ocean currents do not only flow at the oceans' surface. Deep, slow-moving ocean currents flow where surface water cools, increases salinity and density, and then

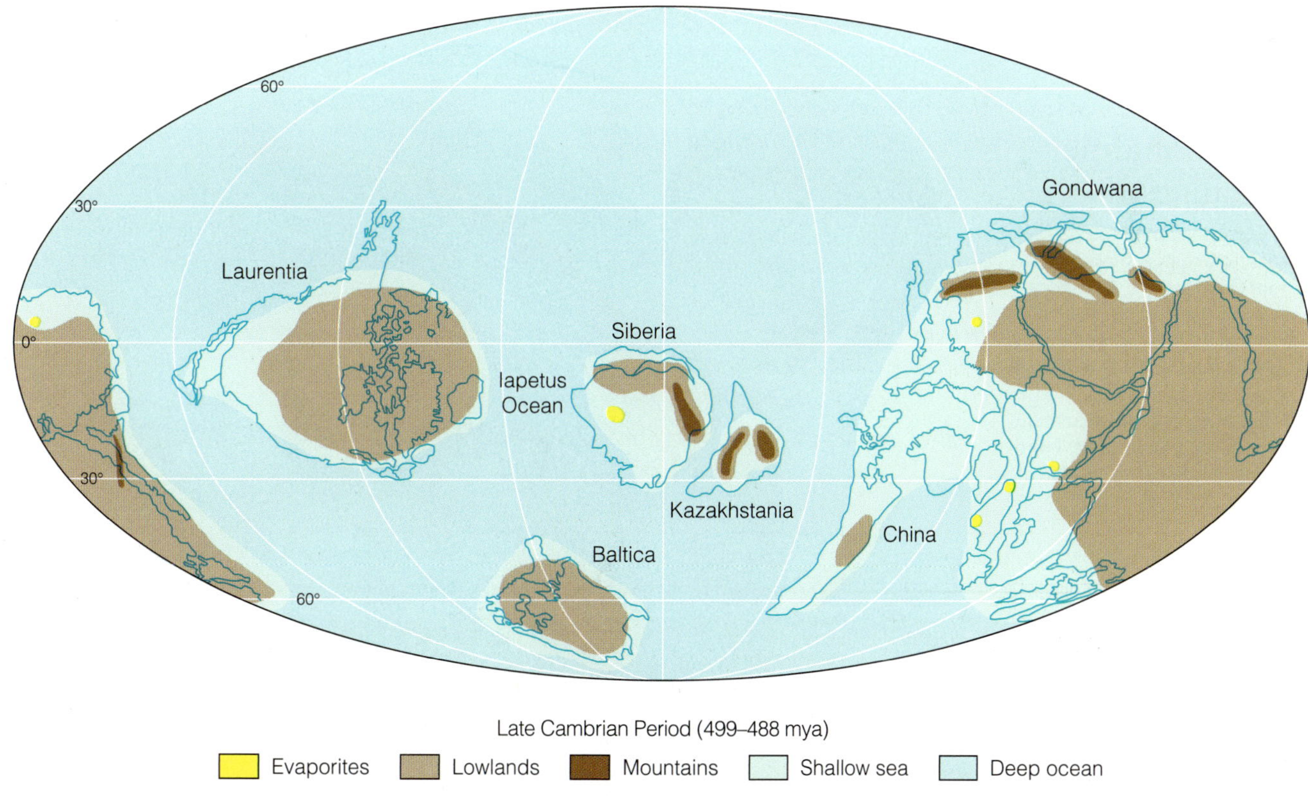

Figure 14.20

Late Cambrian Continents

Configuration of the continents during the Late Cambrian (499–488 Myr ago). See instructions in Exercise 25 to draw in the ocean currents as they would have existed at that time.

sinks. One place this happens is in the North Atlantic where the Gulf Stream becomes denser as it moves north. These currents are known as **thermohaline circulation,** or the **thermohaline conveyor system,** or the **global conveyor** for short. *Thermo-* is for the temperature difference and *-haline* is for the relative saltiness. ■ Figure 14.21 shows thermohaline circulation for the Atlantic Ocean.

Thermohaline circulation forms a complete circuit among the major ocean basins. Flow of deep and surface water in the Indian and Pacific Oceans completes the circuit not shown in Figure 14.21. Remember the current that was already drawn in on Figure 14.18 before you added anything to the map? That is another part that helps complete the circuit.

Thermohaline circulation helps to make the Gulf Stream as strong as it is. Evidence suggests that in the past, thermohaline circulation slowed. Without this circulation, the Gulf Stream slowed, cooling the North Atlantic and leading to cold periods in Europe. Such an event may have caused ice-age glaciers that were shrinking to grow larger for a time about 11,000 to 10,000 years ago during what is known as the **Younger Dryas.** A hypothesis of cooling caused by meteorite impact challenges this hypothesis, but the concept of possible cooling because thermohaline circulation slows is still valid.

Let's see if we can understand what drives thermohaline circulation.

Deep Ocean-Current Experiment

Materials needed:

- Two 250 ml Erlenmeyer flasks with flat rims
- Warm and cold water
- Thermometer
- Plastic bin or basin that can hold a liter or more
- 3×5 cards
- Salt
- Food coloring either red or blue, or both
- Teaspoon
- Stirring rod

26. Read all of the steps of the experiment before you start.

a. Fill one flask to the brim with warm water and the other with cold water. Add red food coloring to the warm water and/or blue food coloring to the cold water and stir.

Figure 14.21

Thermohaline Circulation

(a) Cross section of the Atlantic Ocean extending from Greenland to Antarctica showing temperature, salinity, and their effects on movement of deep waters. (b) The inset map shows the Atlantic loop of the thermohaline circulation with sinking of cold, salty North Atlantic water.

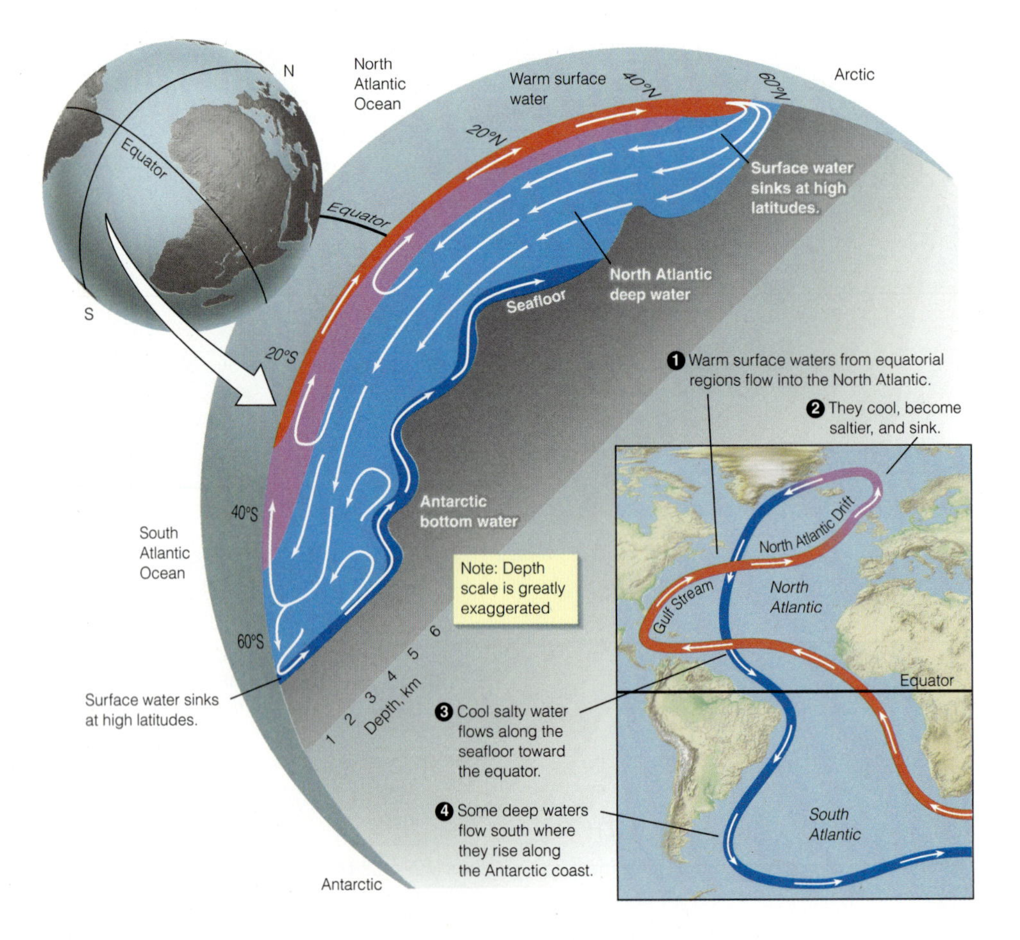

b. Measure the water temperature in the cold-water flask. ________

c. Measure the water temperature in the warm-water flask: ________ Place a 3×5 card on top of this flask.

d. Over the plastic bin, carefully invert the warm-water flask and card. Once inverted, the card will usually stay in place without holding it.

e. Place the flask and card on top of the cold-water flask (■ Figure 14.22a). Tilt the two flasks until they are horizontally side-by-side, in the position shown in Figure 14.22b.

f. Hold both in place just firmly enough to keep them from leaking while someone else pulls out the card.

g. Describe what happens to the warm water and to the cold water.

____

____

____

27. Repeat the experiment again, but this time use fresh water and salty water. Fill one flask to the brim with cold water, add 3 teaspoons of salt, and stir it. Add red dye to this salty water. Fill another flask to the brim also with cold water (of the same temperature). Instead of salt, add blue dye in this flask of fresh water and stir it. As before, join the two flasks horizontally as in Figure 14.22b and pull out the card. Describe what happens to the fresh water and to the salty water.

____

____

____

28. Summarize your results.

a. Which is denser—warm or cold water?

____

b. Which is denser—fresh or salty water?

____

c. *Circle* the following combination that would be the densest and draw a *box* around the one that would be the least dense: cold fresh water / cold salty water / warm salty water / warm fresh water

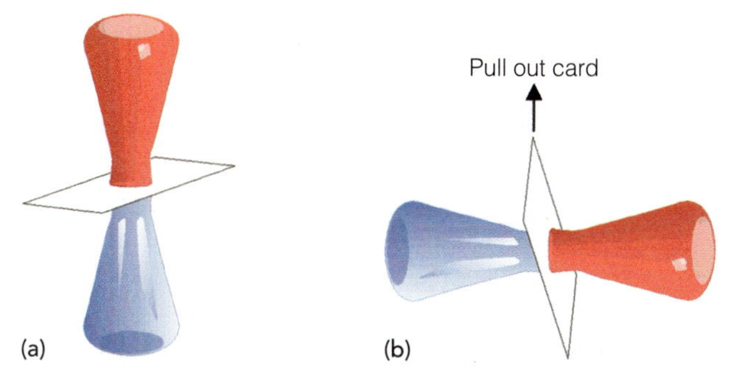

Figure 14.22

Experiment with Temperature, Salinity, and Density of Water

See instructions in Exercises 26–27.

Thermohaline Circulation

29. In the following problems, think about salinity, temperature, and density to answer the questions.

a. Today, as the warm water of the Gulf Stream travels northeast across the Atlantic Ocean, the water cools, and some evaporates, becoming colder and saltier. What happens to this water when it reaches the North Atlantic in the area of Greenland, Iceland, and Great Britain? ________________

b. If many glaciers melt in Greenland and Antarctica, how does the melt water's density compare with ocean water?

____

Will it float or sink? ____________

c. How will this influence what normally happens in the North Atlantic (your answer to part **a**)? ________________
How could this affect the climate of coastal areas near the Gulf Stream, especially in Europe? ________________

d. If rainfall increases due to global warming, what effect will this additional fresh rainwater have on the thermohaline circulation? ________________

30. Use Figure 14.21 and the results from your experiment to help you answer the following questions.

a. Ice that forms on the ocean surface is freshwater ice even though the ocean water is salty. The process of forming the ice makes the remaining water even saltier. Since it is cold enough to make ice, this water will be cold, extra salty, and therefore extra dense. What happens to this extra-dense water from the ocean surface in Arctic and Antarctic regions (Figure 14.21)?

____

b. What is the name used for the water mass at the equator that came from Antarctica?

____

c. What is the name used for the water mass at the equator that came from the North Atlantic?

____

d. Why is the water in part **b** below the water in part **c**? ______________

31. Use ■ Figure 14.23 to help you answer the following questions. Notice that salinity is in parts per thousand (‰) not percent.

a. What is the temperature of seawater at the surface near the equator? _________ What is its salinity? _____________ What is its density? _______________

b. What is the temperature of seawater at the seafloor? _________ What is its salinity? ______________ What is its density? ______________

c. Describe in detail how temperature changes with depth in the ocean near the equator.

____

____

____

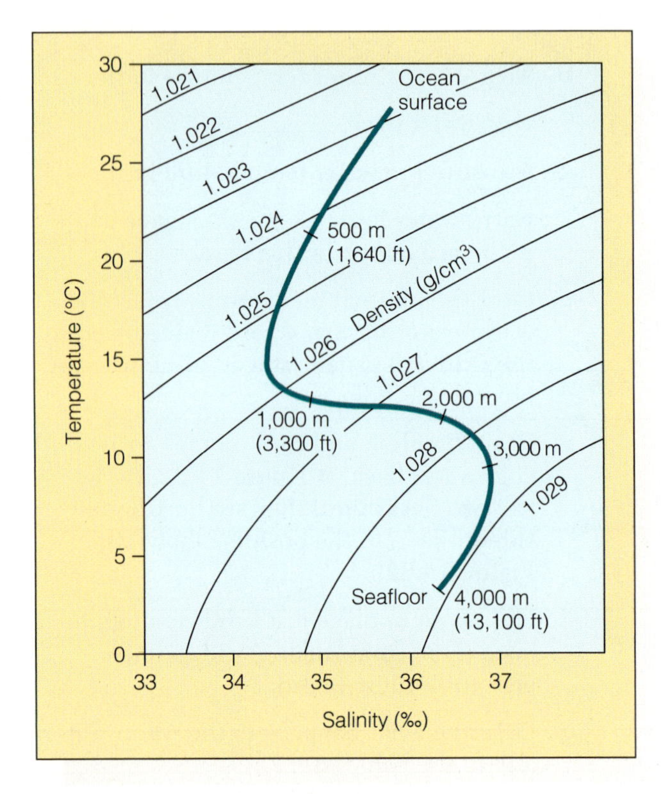

Figure 14.23

Temperature, Salinity, and Density of Water in Tropical Oceans

The heavy blue-green line (S curve) shows the temperature and density of seawater from the surface near the equator to the seafloor. The finer lines are density contours for water of the salinity and temperature given by the axes of the graph.

d. Describe in detail how salinity changes with depth.

____

____

____

e. Describe in detail how density changes with depth.

____

____

____

f. Given that the Earth has gravity and seawater can flow, could the density pattern you summarized be reversed in any parts of the ocean? _______. If so, how?

____

LAB 15

Atmosphere and Climate

OBJECTIVES

- **To learn about the origin of seasons and the differential heating of Earth's surface**
- **To understand the Coriolis effect and Earth's pattern of surface winds**
- **To understand the Milankovitch Cycles and their influence on Earth's climate**
- **To understand the greenhouse effect, radiative forcings, and to explore climate change**
- **To learn about the origin of the ozone layer and its thinning**

In this chapter and the next, we explore aspects of climate and weather. **Weather** describes the frequent changes in temperature, pressure, clouds, precipitation, wind, and humidity for an area, whereas **climate** is the average weather pattern in a particular place over many years. We start with the present general pattern of climate across the globe, and then study the pattern of past climate change. These patterns result from natural Earth–sun geometric relationships, Earth's energy budget, and the greenhouse effect. As we finish this chapter, we also explore issues connected with the atmosphere—global warming and damage to the ozone layer. Then, in the following chapter, Lab 16, we look at weather.

THE EARTH–SUN RELATIONSHIP

Because Earth is nearly a sphere, sunlight does not hit its surface with equal intensity everywhere. Where sunlight is more directly overhead, heating is greater (■ Figure 15.1). Where sunlight is at a low angle, solar heating diminishes as it spreads out over more area.

The angle of sunlight changes through the year causing the seasons to progress. The 23½° angle of the Earth's axis of rotation (relative to the plane of Earth's orbit around the sun) is responsible for seasonal changes because it changes the amount of heat delivered to the surface in different places. During the year, when the axis tilts most directly toward the sun, we have the longest day of the year: the **summer solstice** (the first day of summer). The hottest days occur somewhat later than the solstice, because more solar heating occurs with a high sun angle and it continues to warm the surface even as the sun angle starts getting lower. The same effect causes afternoons to be warmer than mornings even though noon (or 1 o'clock daylight savings time) has the highest sun angle. When Earth's axis tilts most directly away from the sun, the shortest day occurs: the **winter solstice** (the first day of winter). The coldest time occurs after the solstice as the still-short days allow continued cooling. Earth's axis of rotation is at right angles to the sun on the first day of spring (the **vernal equinox**), and the first day of fall (the **autumnal equinox**) and as a result, day and night are the same length. The Northern and Southern

Figure 15.1

Earth's Tilt and Solar Heating

(a) Earth at the June solstice. Where the sun is low in the sky, sunlight spreads out over a larger area (diffuse sunlight) than where the sun is directly overhead. (b) Map of Earth's heat distribution showing the net radiation received over the Earth's surface annually. Different amounts of net radiation correspond to different colors on the map. Yellow indicates no net gain or loss of energy.

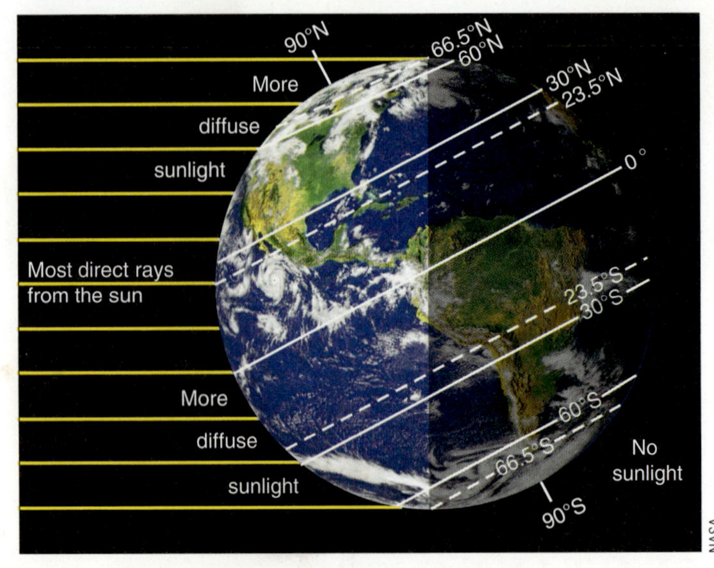

(a)

NASA

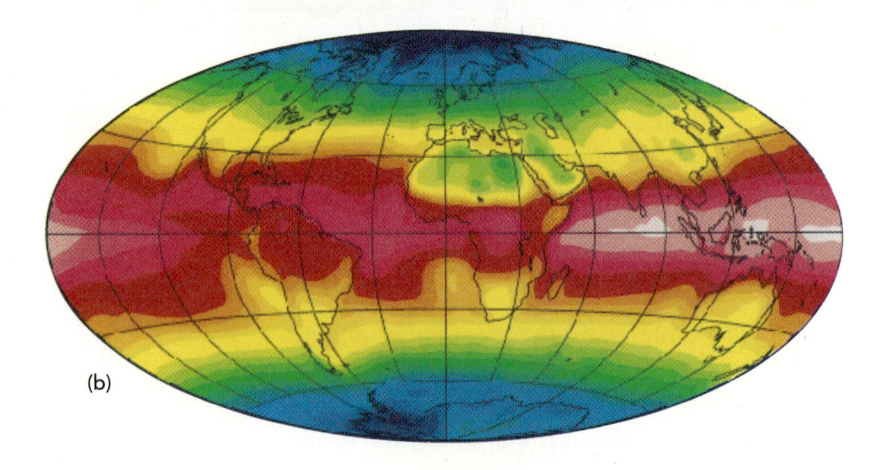

(b)

Hemispheres have opposite seasons because when the axis tilts toward the sun in one hemisphere, it tilts away in the other. A flashlight "sun" and a ball of clay with toothpicks for Earth's axis of rotation or a globe may assist working out these relationships in the following exercise.

1. Figure 15.1a shows the Earth at the June solstice. Use this figure to help you answer the following questions. Solar noon is occurring along the left edge of the Earth in this diagram, from the North Pole axis to the South Pole axis. Solar midnight occurs along the right edge.

a. At what latitude is the sun directly overhead at solar noon on the June solstice?

____

b. At what latitude is the sun on the horizon due north at solar noon? The sun rises briefly and immediately sets here.

________ How long is the night at this location? ____________

c. At what latitude is the sun on the north horizon at solar midnight? The sun sets briefly and then immediately rises again.

________ How long is the day at this location? ____________

d. At what latitude would solar heating be the greatest? ________

e. A solstice is the first day of two seasons. What season is just starting in the Northern Hemisphere in Figure 15.1a?

____

Southern Hemisphere? ____________

2. Figure 15.1b shows Earth's net radiation received. Use this figure to answer the following questions.

   a. What color signifies the highest amount of energy gained at Earth's surface from solar input over a year?

   ____

   b. What color signifies the most heat lost from Earth's surface in a year?

   ____

   c. What reasoning led you to these conclusions? ______________________

   ____

3. Table 15.1 gives the dates of the solstices and equinoxes.

   a. On a separate sheet of paper, draw the configuration of the sun and the Earth somewhat like Figure 15.1a, but for the December solstice and the March equinox. Label which solstice and which equinox you have shown for the Northern and Southern Hemispheres on each diagram.

   b. The March equinox diagram you drew could also be the September equinox. Which equinox would the September equinox be for the Southern Hemisphere?

   ____

   c. On the same sheet of paper, explain the seasonal variation and the relationship between the Earth and sun positions. Include in your explanation why the hottest part of the summer and the coldest part of winter do not correspond to the solstices.

Table 15.1

Equinoxes and Solstices

First Day of Season (Solstice / Equinox)	**Northern Hemisphere**	**Southern Hemisphere**
First Day of Spring (Vernal Equinox)	March 20 (or rarely 21)	September 22 or 23
First Day of Summer (Summer Solstice)	June 21 or 20	December 21 or 22
First Day of Fall (Autumnal Equinox)	September 22 or 23	March 20 (or rarely 21)
First Day of Winter (Winter Solstice)	December 21 or 22	June 21 or 20

WIND AND CORIOLIS EFFECT

Within the lower atmosphere (*troposphere* in Figure 15.8 on p. 351) differential heating of the air causes differences in air pressure. Air expands as it heats causing it to have lower density. Air contracts as it cools, which causes it to have higher density. Denser air creates more pressure due to the force of gravity. Air moves from regions of high atmospheric pressure with cool, sinking air, to regions of low pressure with warm, rising air. The horizontal component of this movement is what we call "wind." When air rises and sinks because of differences in density due to heating and cooling, we call the process **convection.**

Coriolis Effect Experiment

Wind cannot flow straight from high to low pressure because the Earth turns beneath it, causing an apparent deflection called the Coriolis effect. This effect influences objects moving across the Earth's surface. Together, these processes produce a global pattern of winds.

The Coriolis effect results from Earth's rotation. It causes moving objects to turn relative to the Earth's surface, just as a ball thrown on a turning merry-go-round appears to curve from the perspective of those on the merry-go-round. Actually, the Earth turns underneath the object as the object moves; thus, although it might start heading in one direction, before it gets very far the Earth has turned beneath it and the direction has changed. Therefore, the Coriolis effect also influences airplanes and ocean currents. This effect diminishes to zero at the equator and is strongest at the poles. Let's do some exercises that will help you visualize this effect and then go on to recognize how it influences winds and ocean currents.

Materials needed:

- Rotating platform such as a lazy Susan or two pie plates with a handful of marbles between them
- Ball-tracing medium: a powder such as talcum powder or flour, fine sand, or any surface that you can place on the rotating platform and that records the ball's path and is erasable
- Marble or large ball bearing
- Globe (optional)

4. This exercise demonstrates the Coriolis effect for the Northern Hemisphere.

a. Shake powder or sand so it is evenly and very lightly distributed across the platform. You should use very little sand. Keeping the platform stationary, gently flick a marble/ball so it crosses the center of the platform. Describe then erase the path of the ball.

____

b. What direction does the Earth turn if viewed from above the North Pole—clockwise or counterclockwise?

____

Hint: Use a globe to help if necessary. How must the Earth turn so the sun rises in the east?

c. Turn the rotating platform in the same direction as your answer in part **b.** Gently flick the ball so it crosses the center of the platform as you did in part **a.** Describe the path of the ball.

____

d. If you face the direction the ball was moving, what type of turn did the ball make—right or left? ____________ Draw the direction of rotation and the path of the ball and its direction of travel on the left circle below.

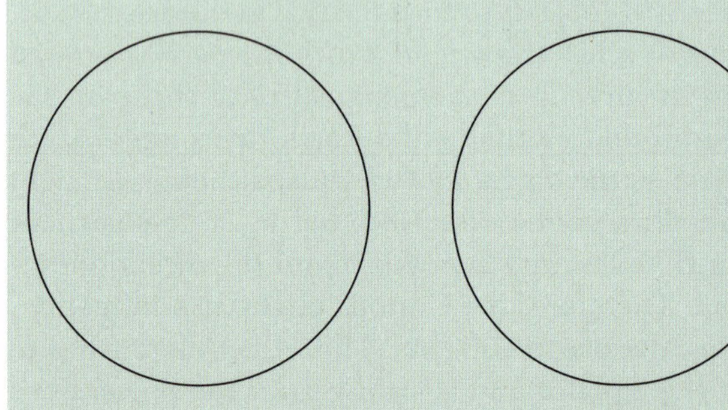

Northern Hemisphere **Southern Hemisphere**

e. Erase the ball's track. Try flicking the ball onto different locations on the rotating platform while continuing to turn the platform in the same direction. Is the direction the ball turns consistent if you always face in the direction the ball traveled? ______

5. This experiment determines the Coriolis effect for the Southern Hemisphere.

a. If you looked down at the Earth from a point in space above the South Pole, in what direction does the Earth turn?

____

b. Turn the rotating platform in this direction, and repeat your previous experiments. Now in what direction does the ball turn? ____________ Draw the direction of rotation of the platform and the path of the ball and its direction of travel on the circle at the right for the Southern Hemisphere in part **d.**

c. Explain why the ball travels along a curved path while the platform is turning.

____

____

____

____

Air Pressure and Global Winds

Although we don't notice it, air has weight. The weight of a column of air all the way to space is about 14.7 pounds for a square inch. This is air pressure—the force per unit area that the weight of air creates. The units, pounds per square inch (psi), illustrate the concept, but scientists more often use millibars. A **barometer** measures air pressure (or atmospheric pressure, also **barometric pressure**) in various units such as atmospheres, bars, or millibars (1 bar = 1,000 millibars). A bar and an atmosphere are very close in size and one bar is about equal to the pressure at sea level. Additional discussion of air pressure is in Lab 16.

The sun-warmed land or ocean heats the air directly above it, causing the air to expand, to become less dense, and to rise. The warm rising air causes an area of low air pressure because it has less weight. The more intense heating (Figure 15.1b) creates lower pressure at the equator. At the poles, where heating is less intense, the colder, denser air sinks, creating high pressure. Wind flows generally from the higher pressure toward the lower pressure, helping to redistribute heat on Earth.

Rising air leads to moist climates and clouds (■ Figure 15.2), and descending air leads to dry climates and clear skies. As the air rises and reaches higher altitudes, it cools off and loses moisture by condensation. This is because warm air has the capacity to keep more water vapor molecules moving, but when it cools, the water condenses. This cooling causes clouds and precipitation in the form of rain and/or snow. On the other

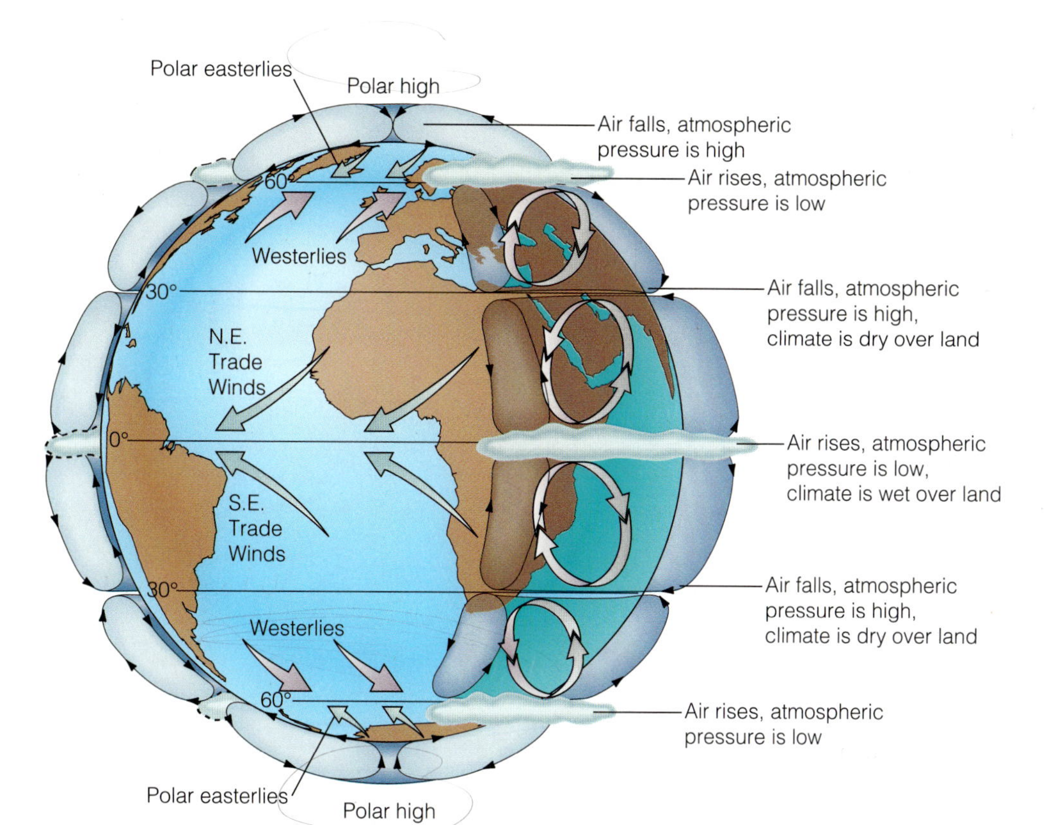

Figure 15.2

Global Air Circulation

Large convection cells of air circulate in the atmosphere within belts 30° latitude wide (shown on the left and right). The Coriolis effect adds an east and west component to the surface winds, as shown by the large arrows in the center of the diagram.

hand, when cold, denser air sinks in areas of high pressure, the air warms nearer the ground causing evaporation of water in the vicinity because the warmer air can hold more moisture. This warming results in a dry climate. The climate is dry even at the poles where the high-pressure, sinking air is frigid but warms slightly as it sinks, allowing it to take up more water vapor from the surroundings.

If the Earth did not rotate, winds would flow straight from the polar high to the equatorial low, but the winds turn because of the Coriolis effect, right in the Northern Hemisphere and left in the Southern Hemisphere. In fact, this turning is so pronounced that the winds, initially equator bound, do not reach the equator but are traveling from east to west (*polar easterlies*) by the time they arrive at about 60° latitude (Figure 15.2). At low latitudes, rising air at the equator produces clouds, and a rainy tropical climate. Sinking air at about 30° latitude creates high-pressure belts with dry climates. Near 60° latitude, winds from the high-pressure regions at 30° and 90° meet as *warm fronts* and *cold fronts* (discussed in the next chapter). The warmer air rises at 60° latitude, producing low pressure and precipitation.

Winds flow between the various belts and locations of high and low pressure. We name winds for the direction from which they come because we feel rather than see wind. The Coriolis effect will naturally influence winds coming from higher latitudes to be easterlies and winds coming from lower latitudes to be westerlies. Check this out with what you know about the Coriolis effect. In each hemisphere, north and south, 30° and 60° latitude separate three belts of winds: the **trade winds** (which are easterlies), the **westerlies** and the **polar easterlies.** The trade winds in both hemispheres come from higher latitudes (30°) heading for the equator, so they are easterlies. The westerlies in both hemispheres come from 30° and head toward 60°. The polar easterlies come from the poles (the highest latitudes).

6. Summarize the climate belts just discussed and shown in Figure 15.2.

a. Dry climate belts occur at 30°-90° latitudes.
Is the climate moist or dry at the polar highs? dry

b. Moist climate belts occur at 0°-60° latitudes.

c. The dry belts have high pressure and moist belts have low pressure.

The map in ■ Figure 15.3 has red arrows showing the direction and strength of Earth's average winds for January. The stronger winds have longer arrows, and weaker winds have shorter arrows.

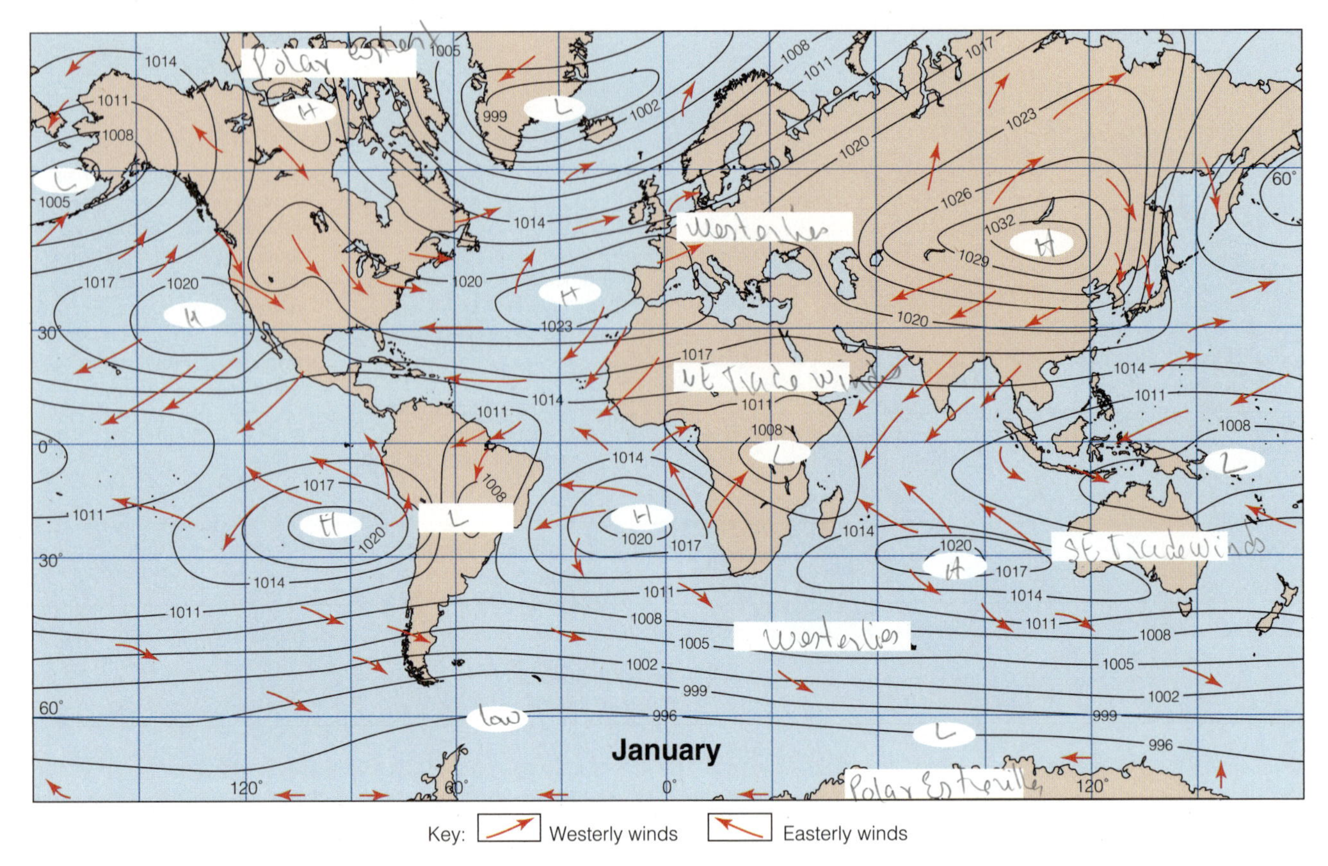

Figure 15.3

Global Average January Surface Winds

Red arrows show the direction and strength of the winds. Longer arrows are for stronger winds. Average surface barometric pressure of the **isobars** (lines of equal pressure) is in millibars.

7. Use the map in Figure 15.3 for this exercise set.

a. In the small white ovals on the map, write in the air pressure—either **H** for high or **L** for low based on the direction of the winds.

b. What are the fine, curving lines on the map, and what are the measurement units of the numbers on those lines? ___isobars, millibars___ Are your answers in part **a** consistent with these lines and numbers? ___yes___

c. Notice that the wind does not head directly toward the low-pressure areas or directly away from the high-pressure areas. What effect causes this?

___coriolis effect___

d. Does the wind rotate around high pressure clockwise (CW) or counterclockwise (CCW), in: the Northern Hemisphere? ___CW___, the Southern Hemisphere? ___CCW___ Around low pressure in: the Northern Hemisphere? ___CCW___, the Southern Hemisphere? ___CW___

8. On the same map:

a. Use different-colored pencils to highlight every easterly and westerly wind arrow on the map. Skip only ones headed directly north or south.

b. Fill in the key on the map showing what the colors represent. Remember that we name winds for the direction from which they come.

c. Describe the overall pattern of winds in terms of latitudes and wind directions: Approximately what four latitudes (counting both hemispheres) make the boundaries between easterly and westerly winds? What is the wind direction between each of these latitudes?

From 30°N to 60°N the winds are somewhat variable but generally easterly except
____

d. In the long white rectangles, label each belt according to the correct name of the winds there.

e. Look at the wind just north of the equator and just south of the equator. Generally, from what direction is the wind in the belt north of the equator?

0°-30°S the winds are generally

and from what direction is the wind northeasterly south of the equator generally?

southeast.

Proximity to oceans and seas also influences the climate because water has a high heat capacity. This means water requires a lot of energy to warm it up and it gives off a lot of energy when it cools. This property of water causes what is known as the **marine influence,** which is that coastal areas tend to have more moderate temperature changes than inland areas. This also influences the winds. During the warming of a day or a season such as spring and summer, the land warms faster than the sea and so winds tend to blow from the sea toward the land. This is a **sea breeze.** As the day or the seasons cool in fall and winter, the land cools faster than the ocean and winds tend to blow from the land toward the ocean, creating a **land breeze.**

9. On the map in Figure 15.3 again:

a. Notice the major deviations from the general pattern over Asia. During winter, is the land or the sea colder?

land

b. Would this create a land breeze or a sea breeze? land breeze

The land and sea breezes associated with the Himalayas here create the **Indian monsoon** (seasonal winds—dry in the winter and wet in the summer). These produce the deviation from the global wind patterns referred to in part **a.**

c. Difference in land and sea temperatures change the global wind patterns illustrated in Figure 15.2. From what direction would you expect the winds in India to come during the summer?

South west This is the summer monsoon, which brings heavy rain.

10. Draw five wind arrows around one of the highs in each hemisphere on the map in Figure 14.19 on page 334, keeping in mind how the Coriolis effect works around high pressures in each hemisphere. Refer back to Figure 15.3 for examples of wind directions around high-pressure systems.

Climate Types and Climographs

Back in Lab 1, we studied a map showing climate classifications for North America (Figure 1.7 on p. 9). The different climate classifications in that map differ in temperature, humidity, and precipitation. The diagrams in ■ Figure 15.4, called **climographs,** show the monthly average temperatures and precipitation for five of the climates designated on Figure 1.7.

11. Use what you see in Figure 1.7 on page 9, Figure 15.4, and what you now know about climate and climate belts to answer the following questions.

a. Find the location of the black dots in Figure 15.4 on the map in Figure 1.7, then name the Koeppen climate classification and give the symbol for each in the blanks in Figure 15.4.

b. As you travel north along the West Coast of North America, how does the monthly average temperature graph change in overall temperature?

____

In temperature variation throughout the year?

____

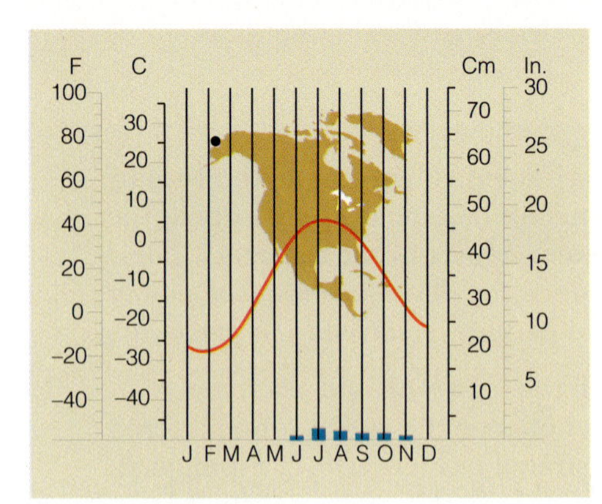

Figure 15.4

Climographs

Red curves show monthly average temperature variations with Fahrenheit and Centigrade scales on the left. Blue columns show precipitation in liquid water equivalents with scales in centimeters and inches on the right. Black dots on the maps of North America show the localities for these climographs. (a) Barrow, Alaska; (b) Portland, Oregon; (c) Duluth, Minnesota; (d) San Francisco, California; (e) New Orleans, Louisiana. From THOMPSON/TURK, *Earth Science and the Environment* (with ThomsonNOW(T) Printed Access Card), 4E. Brooks/Cole, a part of Cengage Learning, Inc. Reproduced by permission. www.cengage.com/permissions

(a) Koeppen classification symbol: ____________

and name: ______________________________

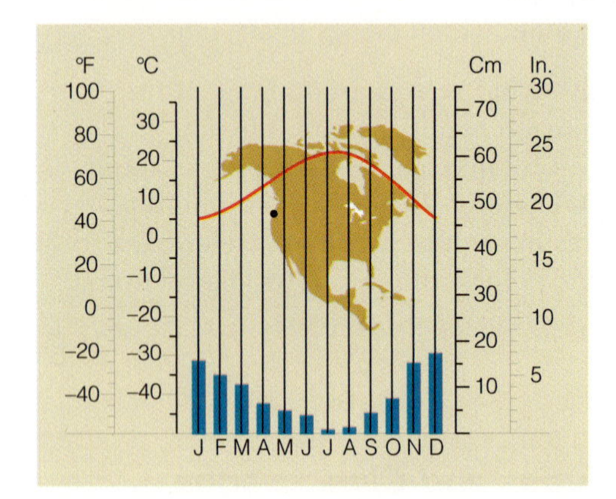

(b) Koeppen classification symbol: ____________

and name: ______________________________

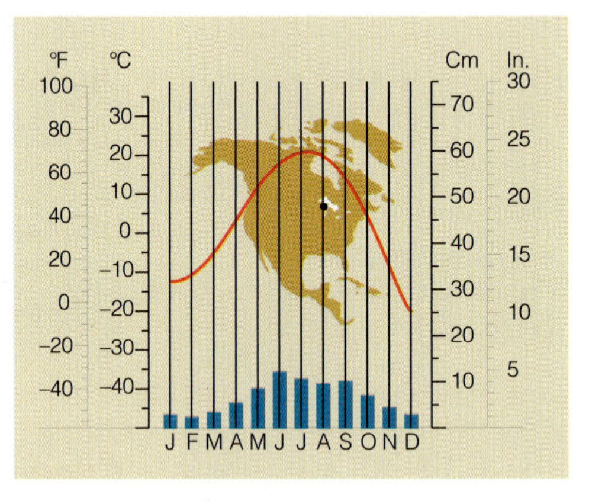

(c) Koeppen classification symbol: ____________

and name: ______________________________

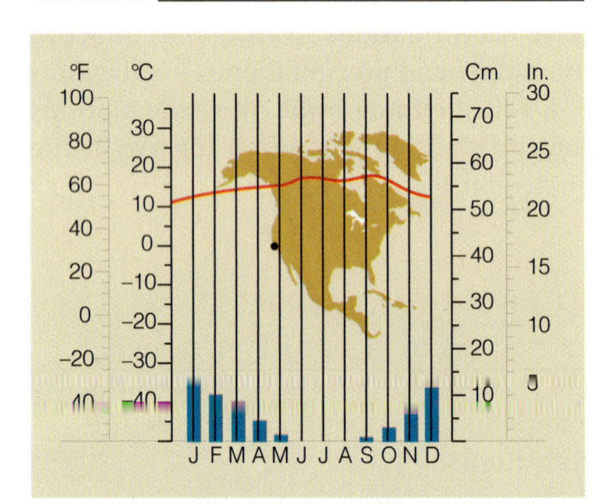

(d) Koeppen classification symbol: ____________

and name: ______________________________

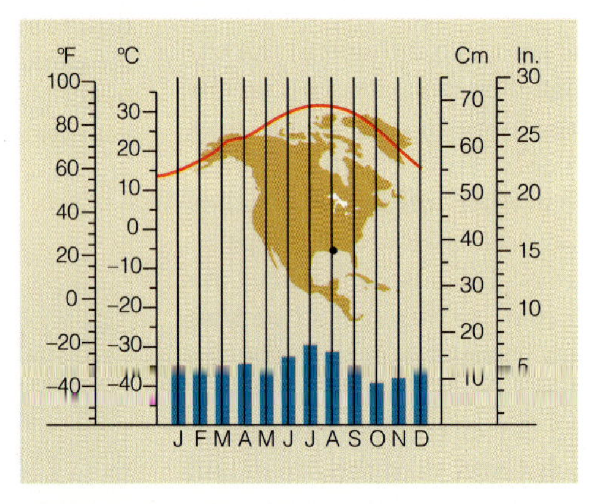

(e) Koeppen classification symbol: ____________

and name: ______________________________

c. As you travel inland from Portland, Oregon on the West Coast to Duluth, Minnesota, how does the temperature graph change?

____

____

How does the precipitation graph change?

____

____

d. Recalling cold and warm ocean currents and their effects on climate from Lab 14, which climate in Figure 1.7 would you expect to be drier: Csb or Cfa? _______
Is this consistent for San Francisco and New Orleans in Figure 15.4? _______

e. In Figure 15.4a, is it dry or moist at about 71° latitude in Barrow, Alaska?

____

f. In Figure 1.7, is it dry or moist at 55 to 60° latitude? _____________

g. In Figure 1.7, is it dry or moist at 25 to 30° latitude in Mexico? _____________

h. Match the places in parts e–g above with the diagram of climate belts in Figure 15.2. Label the appropriate belts with the corresponding letter e, f, or g on the left edge of Figure 15.2.

PAST CLIMATE FLUCTUATIONS

Presently, Earth is very likely (90% or greater probability) undergoing **anthropogenic**—or human-caused—climate change. However, in the past, climate has fluctuated naturally from hot to warm to cold and back again numerous times. It is important to understand past natural climate variation so we can recognize what parts of current changes are human-induced. Some of the past climate fluctuations caused a series of glacial and interglacial episodes during the Pleistocene. Evidence now indicates that cyclical changes in the geometry of Earth/sun system, called **Milankovitch cycles** (■ Figures 15.5 and ■ 15.6), resulting from gravitational effects of celestial bodies, drove these climate fluctuations of the past. However, it is very improbable that these variations are causing the rapid climate change we are seeing now.

Milankovitch Cycles

Why did Earth experience glaciations in the Pleistocene? Because the amount of energy coming in at the top of the troposphere (Figure 15.8) changed compared to the amount of energy leaving the troposphere. This kind of energy change, in general, is called a **radiative forcing.** If it is due to the input of solar energy, it is

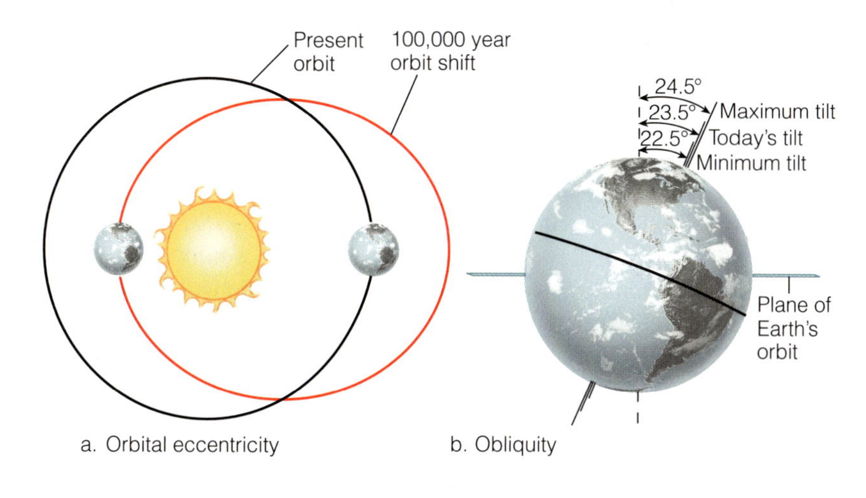

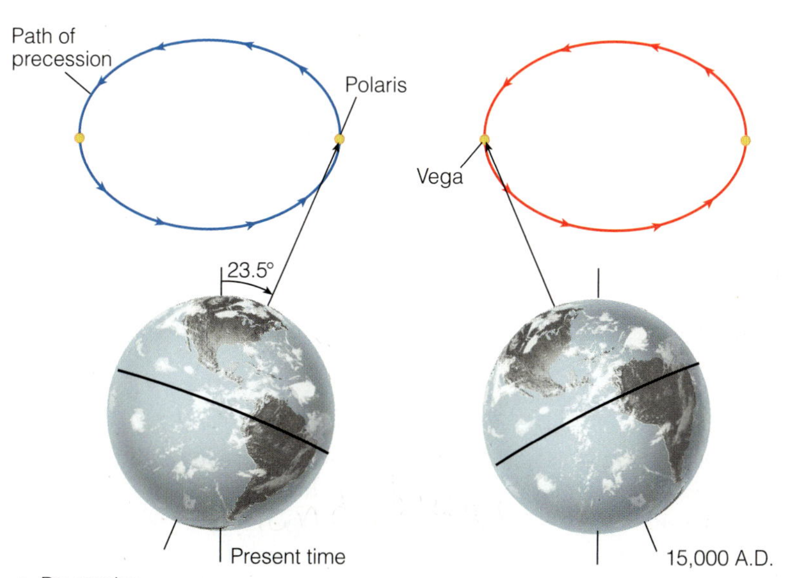

Figure 15.5

Milankovitch Cycles

Collectively, the following variations in Earth's attitude and movement are know as Milankovitch cycles: (a) Changes in Earth's orbit, **eccentricity,** from more circular to more elliptical (exaggerated). (b) Changes in the tilt of Earth's axis (**obliquity**) relative to the plane of Earth's orbit. (c) Changes in the direction Earth's axis points in space, **precession.** From THOMPSON/TURK, *Earth Science and the Environment* (with ThomsonNOW(T) Printed Access Card), 4E. Brooks/Cole, a part of Cengage Learning, Inc. Reproduced by permission. www.cengage.com/permissions

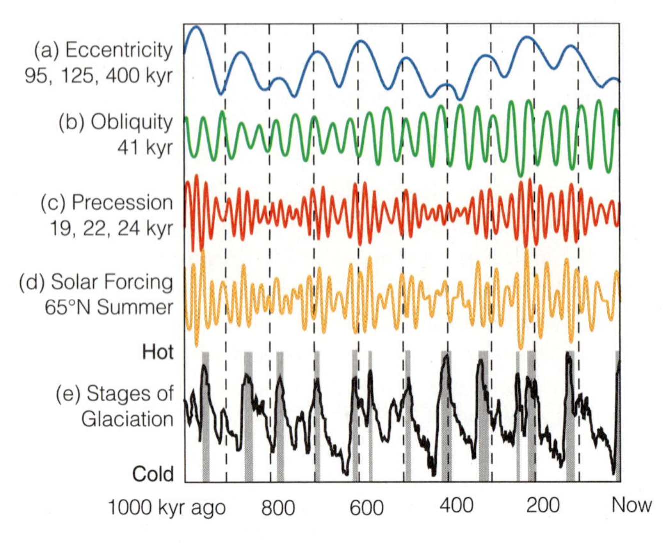

Figure 15.6

Periodicity of Milankovitch Cycles

Periodicity means oscillations or changes back and forth on a regular time interval. (a), (b), and (c) show the timing and size of oscillations of the various Milankovitch cycles. (d) Solar forcing shows how the Milankovitch cycles influence the amount of sunlight reaching Earth at the top of the troposphere. (e) Stages of glaciation shows similarity in the pattern of glacial stages with the Milankovitch cycles. Gray shading indicates interglacials (warm). k = 1,000.

a **solar forcing.** The word *forcing* is used because this kind of change forces the temperature of the troposphere, where we live, to change. The change in Pleistocene climate occurred because the shape of Earth's orbit changed, its tilt changed, and it wobbled on its axis. The three Milankovitch cycles describe these changes (Figure 15.5a, b, and c). **Eccentricity** is a measure of how much the orbit varies from a perfect circle. Earth's orbit has been more circular and more elliptical at various times in the past. An **ellipse** is an elongated circle. The orbit changed and then changed back on a cycle of about 100,000 years (95, 125 kyr, Figure 15.6a). When Earth's orbit is more elliptical, this eccentricity may enhance or subdue seasonal variation. When the orbit is more circular, these effects are less pronounced. Lower eccentricity (a more circular orbit) causes more yearly total influx of solar energy because Earth is closer to the sun for a longer time each year in this configuration. Earth's eccentricity is presently fairly low.

Earth's axis tilt, called **obliquity,** fluctuates on a cycle of 41 thousand years (41 kyr) from more tilted to less tilted and back again. As we saw at the beginning of this lab, tilt is what controls the seasons, so the obliquity in this Milankovitch cycle causes changes in seasonal contrast. The third cycle of change is known as precession. **Precession** refers to change in the direction of the Earth's axis without necessarily changing the tilt, and is similar to a slow-motion wobble of the axis of a spinning top. It changes the timing of the seasons relative to when Earth is closest to the sun. Earth's axis precession has cycles of 19, 22, and 24 kyr (Figure 15.6c). At present, Earth is closest to the sun in the Northern Hemisphere winter on January 3, but in about 11,000 years, it will be closest in summer. This makes Northern Hemisphere winters and summers milder now, but they will be more extreme in 11,000 years. The average annual heat from the sun will continue to increase for about 26,000 more years as Earth's orbital eccentricity diminishes to nearly zero (a nearly circular orbit).

Glaciations during the Pleistocene occurred when the combination of the Milankovitch cycles produced cool summers in the Northern Hemisphere for a sufficiently prolonged time. The Northern Hemisphere controls the timing of glaciation in both hemispheres because it has more land on which glaciers can grow and the glaciers influence Earth's energy balance by increasing how much energy the surface reflects back to space (higher *albedo*, discussed more later). Redistribution of heat between the hemispheres by the atmosphere and oceans keeps the planet's temperature in sync. Cool summers keep glaciers from melting. Of course, cold winters are also necessary for glaciers to grow, but they must be cold enough, but not too cold to snow.

The Solar Forcing graph in Figure 15.6d shows how the combined cycles effect the amount of sunlight coming in at 65°N latitude, the latitude at which glaciers are most likely to form. Notice that highly variable solar forcing coincides with gray (warm interglacials) in the Stages of Glaciation graph in Figure 15.6e. Variability does not allow the time glaciers need to grow large. Since cool summers allow glaciers to endure, what would make a cool summer? More tilt or less tilt? Precession so that Earth is closer to the sun or farther from the sun in summer? Orbit change (eccentricity) so the orbit is less or more circular? Let's investigate these geometries with an exercise.

Materials needed:

- Large sheet of paper at least 70 by 90 cm (or sheet of 8½ by 11 inch paper)
- Two pens and one pencil
- String
- Metric measuring tape or meter stick
- Scissors
- Modeling clay or Play-Doh® preferably in yellow and blue or green
- Toothpicks

12. The name of the geometric shape of Earth's orbit around the sun is an **ellipse.** The sun sits at one of the **foci** (plural of **focus**) of the ellipse (■ Figure 15.7). As you draw the ellipse in this exercise, you will understand better what a focus is. With your lab partners, use a large piece of paper to draw an exaggerated view of Earth's orbit as

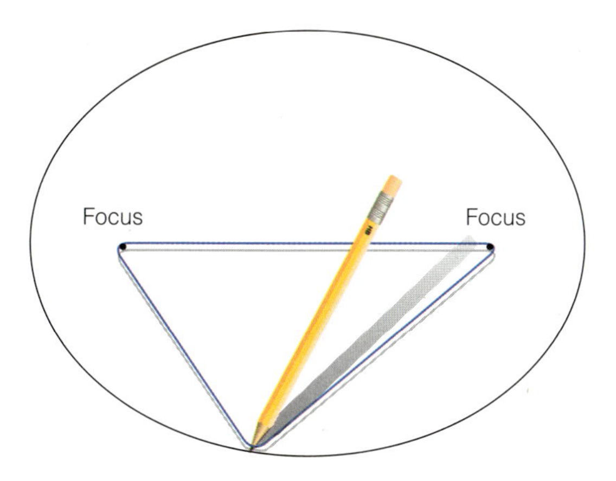

Figure 15.7

Drawing an Ellipse

Hold pens at the two foci. Stretch a string around the two foci and trace the ellipse with a pencil.

follows (numbers in parentheses are measurements for doing the exercise on 8½ by 11 inch paper):

a. Draw two dots near the center of the large paper 9 cm apart (3 cm for 8½ by 11 inch paper).

b. Measure and mark a piece of string 63 cm (21 cm) long with extra room to tie the ends.

c. Tie the ends of the string together to the 63 cm (21 cm) mark.

d. Place the string on the paper so the dots are within the loop.

e. With a partner, hold a pen at each dot.

f. Draw an ellipse as shown in Figure 15.7, by keeping the string tight and the pencil upright as you draw around the two pens. The two dots are **foci** (plural of **focus**).

13. Answer the following questions about your ellipse.

a. What does the ellipse represent?

____

b. Is the shape easy to distinguish from a circle with this spacing of the foci as they are? ______ If so, how do you distinguish it; if not, why not?

____

____

____

14. Now you will need some yellow and blue or green Play-Doh®.

a. Make a yellow ball to represent the sun. Place it at one of the dots inside your ellipse.

b. Make blue or green ball of Play-Doh® to represent Earth and put a toothpick through it to represent Earth's axis of rotation. Mark an N in the Play-Doh® next to one end of the toothpick to indicate the North Pole. The Earth will trace out the path of the ellipse you made.

c. The Earth's orbit around the sun is an ellipse that is more circular than the ellipse you made. What is the word that describes the measure of the orbit shape that changes with time?

____

d. What do we call the two ends of the toothpick? ____________

e. In one year, as the Earth revolves around the sun, the axis of rotation remains pointing in the same direction in space. What object in space is presently directly above Earth's North Pole?

____

The place where Earth is closest to the sun is the **perihelion** and the place where it is farthest from the sun is **aphelion.** The larger the difference between perihelion and aphelion, the greater the eccentricity of Earth's orbit. Earth is presently at perihelion on January 3 (it varies from January 2 to 4), and at aphelion on July 4 (varies from July 3 to 6).

15. Label perihelion, aphelion, and their dates on your ellipse from the prior exercise. Also, draw an arrow indicating the direction of revolution of the Earth around the sun, assuming you are looking down on the Northern Hemisphere.

a. Where in Earth's orbit and when during the year would Earth be coolest overall?

____

b. Use this information to label the approximate position of the June and December solstices on Earth's orbit. Sketch the Earth on the large sheet of paper at the correct position with its axis pointing in the right direction. Recall that now the tilt of the axis relative to a line perpendicular to the orbit is 23½°, but that this angle changes as obliquity changes. Keep the axis of your Play-Doh® Earth pointing in the same direction as you move it around the Play-Doh® sun since obliquity and precession changes will not happen in just one year.

c. Based on the present combination of axis tilt and timing of Earth's position in its orbit, are Northern Hemisphere summers cooler or warmer than they would be at other times when the orbit and tilt geometry was different? ____________

Remember that *precession* causes changes in the direction that Earth's axis points in space. In combination with the shape of Earth's orbit (*eccentricity*), and the location of *perihelion* and *aphelion*, precession can make warmer winters and cooler summers or colder winters and hotter summers.

16. On a separate sheet of paper, sketch the sun, Earth's elliptical orbit, perihelion, and aphelion.

a. Add Earth with its axis of rotation positioned for a hot summer. Label the North Pole *NP* and label it *hot summer*. Don't forget to use the direction of tilt of the axis relative to the sun to determine the correct position for summer.

b. Sketch in another Earth on the same diagram for 6 months later showing a cold winter for the Northern Hemisphere and label it.

c. On the same sheet of paper, sketch a cool-summer/warm-winter configuration for the Northern Hemisphere. Label the North Pole *NP* and label the two Earths that are 6 months apart with *cool summer* and *warm winter* as appropriate.

17. Measure and mark another piece of string for 84 cm (28 cm for 8½ by 11 inch paper) with room on both ends to tie them together. Use the same piece of paper you used to draw the ellipse in Exercise 12.

a. Draw a third dot in line with the other two but 30 cm (10 cm) from the sun past the other dot. The focus from the previous ellipse that did not have the sun should be in line between the new dot and the sun. The new dot will become a new focus.

b. Make a new ellipse for an orbit with a different eccentricity with the new loop of string and with the sun dot and the new dot as the two foci.

18. Answer the following questions about the second ellipse you just drew.

a. Does it have a higher or lower eccentricity than the first ellipse? ____________

b. On January 3, Earth is 147 Gm from the sun and on July 4 it is 152 Gm from the sun. G means billion so 1 Gm is a billion meters. How many kilometers are these numbers?

147 Gm: ____________________

152 Gm: ____________________

c. Earth's present eccentricity makes it look like the orbit is a circle if you were to draw it to scale. Is Earth's present eccentricity (i) greater, (ii) less than, or (iii) between the eccentricities of the two ellipses you drew?

d. Earth's obliquity changes from 22° to 24.5°. What is its present obliquity?

____

e. How would a greater obliquity influence Earth's climate?

____

____

As the climate cooled from an interglacial to a glacial episode in the past, snow accumulated and glaciers grew, turning dark areas of the planet white or light gray. This reflected more sunlight back into space and caused Earth to cool more. A phenomenon like this, which amplifies a small change in climate—perhaps due to Milankovitch cycles—is called **positive feedback.**

On the other hand, a change that tends to stabilize the climate, reducing the size of a change, is called **negative feedback.** A thermostat provides a good example of negative feedback. Although Milankovitch cycles appear to account for the timing of climate change, they provide insufficient amounts of change to account for the magnitude of past climate change. This much change needs positive feedback to amplify the variation in incoming solar energy that the Milankovitch cycles provide. To understand about important feedbacks, we will need to know more about Earth's energy budget and the greenhouse effect.

EARTH'S ENERGY BUDGET AND THE GREENHOUSE EFFECT

The amount of energy that comes in to the Earth from the sun must balance energy going out toward space (■ Figure 15.8). This is **Earth's energy budget.** When these two are out of balance, *radiative forcing* occurs, which forces temperature to change so that the balance is reestablished. A warmer Earth will emit more energy into space than a cooler Earth. The flow of radiant energy from the sun to the Earth and from the Earth out to space is part of the **electromagnetic spectrum,** which includes gamma rays, X-rays, ultraviolet rays, visible light, infrared radiation, microwaves, and radio (including broadcast television) waves (■ Figure 15.9a). All these different rays and waves are part of a continuum of radiant energy with different wavelengths. Gamma rays have the shortest wavelengths and are the most energetic and dangerous to living cells, and radio waves have the longest wavelengths and are the least energetic. In this section, we also discuss the influence of albedo and the greenhouse effect on Earth's energy budget.

Incoming and Outgoing Radiation

Incoming radiation from the sun that reaches Earth is mostly visible light and near infrared (Solar Spectrum in Figure 15.9a). This radiation passes readily through the atmosphere and adds energy when the Earth's surface absorbs it and converts it to heat. About 30% of the radiation that reaches Earth reflects back into space as the same wavelengths of visible light and near infrared. *Albedo* influences this part of Earth's energy budget. The Earth radiates the remaining ~70% toward space within the longer wavelengths of the far infrared (Earth's surface spectrum). The *greenhouse effect* influences this far-infrared part of Earth's energy budget (Figure 15.8).

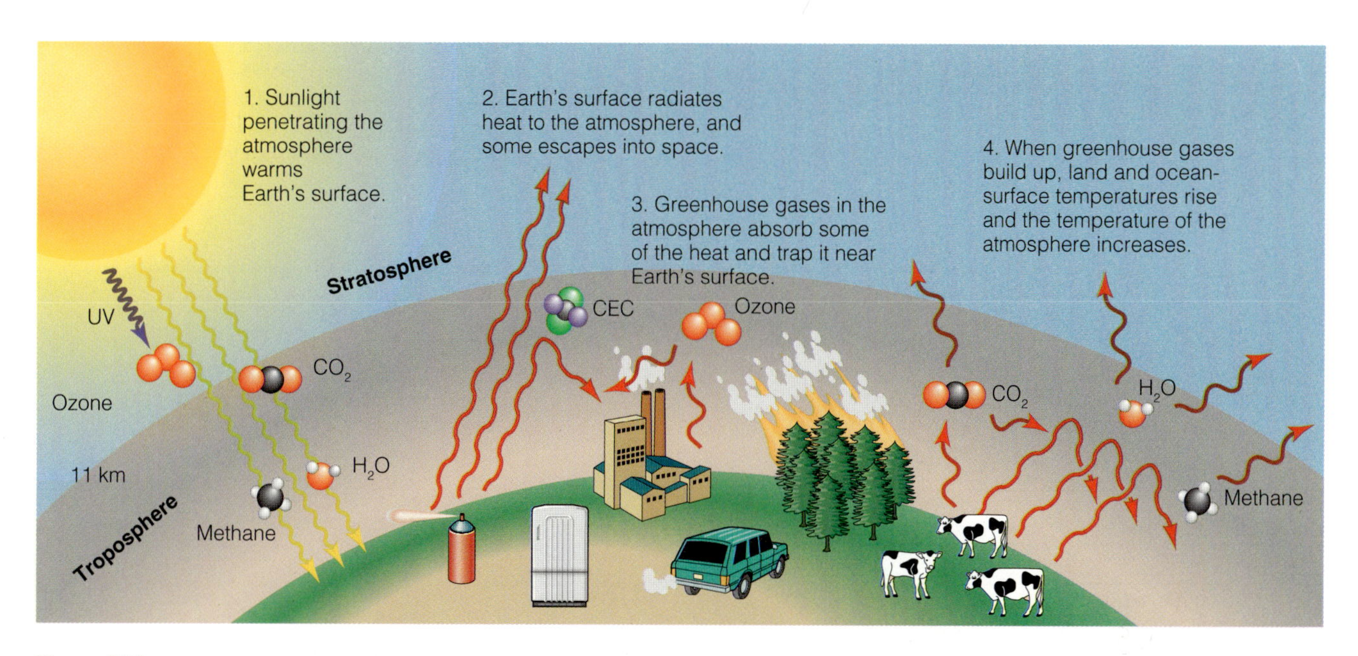

Figure 15.8

Greenhouse Effect

Greenhouse gases in the lower atmosphere (troposphere) absorb some of the outgoing far-infrared radiation and reradiate it in all directions. This produces the present average surface temperature of 15°C (59°F). Ozone absorbs ultraviolet (UV) rays in the stratosphere but far-infrared in the troposphere. Cars, fires, cattle, and industry release greenhouse gases into the atmosphere. Adapted from GARRISON. *Oceanography*, 6E. 2007 Brooks/Cole, a part of Cengage Learning, Inc. Reproduced by permission. www.cengage.com/permissions

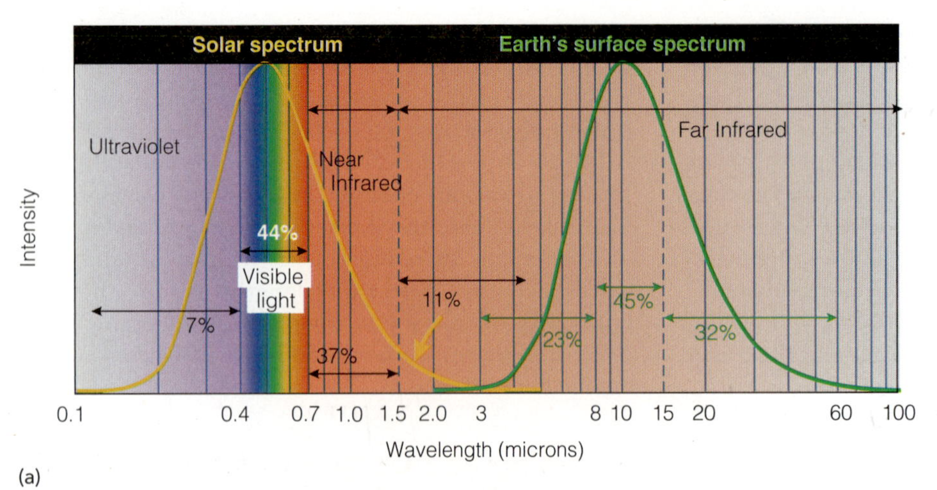

(a)

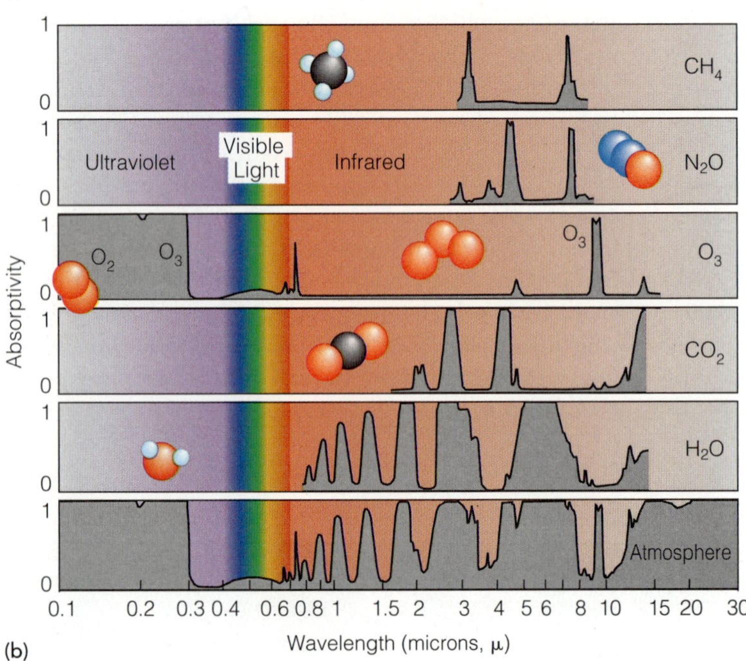

(b)

Figure 15.9

Spectra of the Sun, Earth, and Various Gases

(a) The sun's maximum radiation intensity is visible light at about 0.5 microns, equivalent to blue-green. Earth's surface maximum is at about 10 microns, in the far infrared. (b) Atmospheric absorption spectra of greenhouse gases in the atmosphere. For each gas, the vertical axis indicates the proportion of absorption from 0 to 1, where 1 would be 100% absorption. Gray areas show the wavelengths that the particular gas absorbs. Colored regions have little or no absorption. The lowest section of the diagram, labeled "atmosphere," shows the absorption of all the greenhouse gases and other atmosphere gases added together.

19. Use Figure 15.9a to answer the following questions:

a. More than 80% of radiation from the sun is what two types of radiation?

visible light

and

near infrared,

with wavelengths ranging from 0.4

to 1.5 microns.

b. What type of energy is reradiated (not reflected) from the Earth's surface (Earth's surface spectrum)?

far infrared from around 3 to around 70 microns

c. What is Earth's peak surface radiation wavelength?

10 microns

Albedo

Because **albedo**, the proportion or percentage of light reflected from a surface, has to do with how much radiation the Earth reflects back to space, it is a significant component of Earth's energy budget. We can do a simple experiment with albedo. A surface with high albedo will reflect most sunlight back into space. A surface with low albedo will absorb most of the light.

Materials needed:

- Black paper or container of black sand
- White paper or pale-colored sand
- Paperweights if needed
- Two thermometers
- Bright floodlight or sunlight
- Light meter (optional)

20. Under a floodlight or in sunlight, place a thermometer under the black piece of paper or insert it into the black sand just under the surface. Do the same with the second thermometer and a white sheet of paper or light-colored sand. If it is outside, use paperweights to keep the paper from blowing away. Check the temperature of each thermometer every few minutes until you observe a substantial difference between them.

a. Record the temperatures and time in Table 15.2 each time you check.

b. Describe your results.

____

____

____

21. If you have a light meter, measure the light level directly at the same distance from the floodlight or in sunlight.

a. Record the number and the measurement units:

____

b. Then measure and record the light reflected from the black paper or dark sand

____

and the white paper or light-colored sand.

____

c. Calculate the albedo of both surfaces by dividing the light off the surface by the direct light measurement and multiplying by 100. Dark surface: ________ Light surface: ________

Table 15.2

Albedo Experiment

Temperature change of surfaces with different albedos.

Time	Temperature (°C)	
	Dark Surface	Light Surface

22. Visible light passes through the atmosphere from the sun. If it reflects off a high albedo surface the light can travel back through the atmosphere and out into space without being retained as heat.

a. Which surface in your experiment had the higher albedo and why?

____

b. Which surface, if covering a large area outside, would cause more local warming?

____

23. The climate system has a number of different feedbacks, both positive and negative, and they are far too complex to go into all of them here. However, the ice albedo effect is very easy to understand.

a. Does clean ice have a high or a low albedo? high

b. If climate cools and glaciers grow, will Earth's albedo increase or decrease?

increase

c. How will that change in albedo modify Earth's energy budget?

decrease

d. Will this new energy budget cause glaciers to grow or shrink? grow

e. Explain why this is considered feedback.

Because it amplifies the glaciers.

____

f. Recall that positive feedback is an amplification of a change that has already happened, and a negative feedback dampens or diminishes the change. Is the ice albedo effect a positive or a negative feedback? positive

g. If the climate is warming and the ice is melting instead, does that change your answer about whether the ice albedo effect is positive or negative feedback? NO it is still + feedback

any, of these gases are greenhouse gases? all

d. Judging from these properties of greenhouse gases, if more of them collected in the atmosphere, how would you expect Earth's climate to change? ↑ inc. If the atmosphere had fewer of these gases, how would that influence the climate? ↓ dec.

Greenhouse Effect

The greenhouse effect influences the ~70% of the sun's energy that heats Earth's surface and that Earth gives off as far-infrared rays (Figure 15.9a). The **greenhouse effect** is a trap that keeps some of this energy from reradiating out to space. Without it, the oceans would have frozen billions of years ago, and life as we know it would not exist on Earth. Any trap needs a way to allow something in but not allow it out. The greenhouse gases allow sun*light* to pass through them, but they absorb far-infrared radiation as it leaves Earth's surface.

Greenhouse Gases and Temperatures of the Past

The **greenhouse gases** all have the property that they allow shorter-wavelength visible light through, but *also* they absorb the far-infrared rays of Earth's longer wave radiation.

24. Examine the graphs in Figure 15.9b. These graphs show the wavelengths that several different gases in the atmosphere absorb. Notice that the vertical scale is absorptivity. This is the reverse of intensity, and shows what the molecules soak up rather than what they give off. Each of these graphs is scaled from 0 to 1, or you can think of this as from 0% to 100%.

a. Which, if any, of these gases absorb more than 25% of any color of visible light? none

b. Which, if any, of these gases absorb more than 25% of any wavelength of far-infrared radiation? all

c. Review the properties of greenhouse gases. Based on these graphs, which, if

If we had a time machine, we could go back and collect past atmosphere samples. Time machines are science fiction, but we do have something that was in the past and did sample the atmosphere—ice. When ice forms in glaciers, it traps small bubbles of air—small air samples. The ice itself has hydrogen and oxygen molecules that can tell us about past temperatures (from analyzing the stable *isotopes*). The same kinds of data can also tell us how much water has evaporated from the sea to make glacial ice. All of this allows us to compare atmospheres of the past to that of the present and, as a result, to learn about past climates.

Two of the most important greenhouse gases are naturally occurring: carbon dioxide and methane. Later, we will look at their recent increases in the atmosphere; in the next exercises, however, we look at how the quantities of these compounds in the atmosphere have changed in the past in relation to temperature.

■ Figure 15.10 shows the relationships among age, over the last 400,000 years, carbon dioxide, methane, temperature, and sea level in the Vostok ice core extracted from Antarctica. Table 15.3 gives these values at depths every 200 m in the same ice core down to 2,000 m. Temperatures were determined from *deuterium*[1] values in different layers of the ice.

25. Examine the graphs in Figure 15.10c and d.

a. What three times had the warmest climate? 325,000, 238,000, and 130,000 years ago. These corresponded to interglacial stages in the Northern Hemisphere.

[1] Heavy hydrogen containing one proton and one neutron in the nucleus

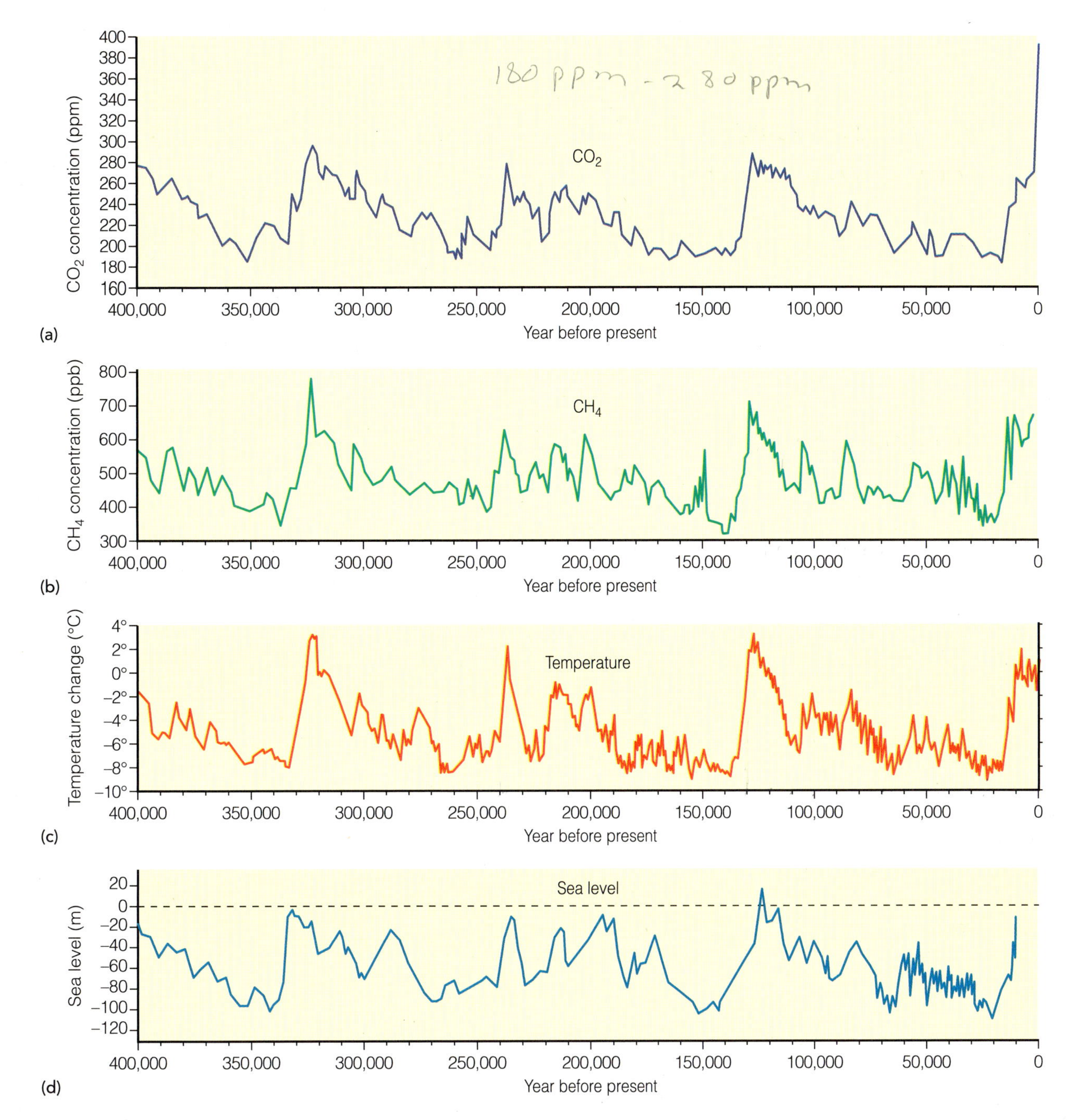

Figure 15.10

Vostok Ice Core Data

(a) Past changes in carbon dioxide levels in the atmosphere from ice core data. (b) Past methane values. (c) Temperature of the past in Antarctica measured as change from the present in Antarctica. Temperatures are determined from deuterium values in different layers of ice in the Vostok ice core extracted from Antarctica. (d) Past sea level. From MILLER. *Living in the Environment*, 16E. 2009 Brooks/Cole, a part of Cengage Learning, Inc. Reproduced by permission. www.cengage.com/permissions

b. What four times, more than 50,000 yrs apart, had the coldest climate?

22,000,

155,000,

265,000, and

335,000 years ago. These corresponded to glaciations in the Northern Hemisphere.

c. Why does sea level correlate with temperature in these graphs?

As Temp. rises sea level also rises.

26. Compare the graphs in Figure 15.10a, b, and c, looking at broad changes, not tiny peaks.

a. Describe the correlation or its lack between carbon dioxide and temperature.

CO_2↑ Temp↑

b. Describe the correlation or its lack between methane and temperature.

CH_4↑ Temp↑

c. Do these relationships match or contradict what you know about the greenhouse effect? match

27. Use the data in Table 15.3 and the graph paper in Figure 15.11.

a. Set up and label a temperature scale horizontally, and then graph carbon dioxide and methane against temperature. Make a key indicating what symbols/colors you used for carbon dioxide and methane. Add an appropriate title.

b. Why are the temperature numbers on this graph and in Table 15.3 so different from the numbers shown in Figure 15.10c, and why are they so low?

____

____

c. Draw an estimated best-fit straight line through the carbon dioxide points so that equal numbers of points along the entire length are above and below the line.

d. Also draw a best-fit line for methane. Does your graph confirm or contradict your observations in Exercise 26?

______________________________ Explain your results.

____

____

____

____

e. In a class discussion, speculate on why the correlation might be better, worse, or the same as you expected among these three variables. Summarize the discussion.

____

____

____

____

Table 15.3

Vostok Ice Core Data

2,000 m of Vostok, Antarctica, ice core sampled every 200 m.

Depth in Ice Core (m)	Average Temperature (°C)	CO_2 Concentration (ppmv[1])	Methane Concentration (ppbv[2])
200	−52.8	263.73	591.92
400	−64.4	220.02	424.48
600	−62.9	220.81	430.79
800	−61.3	207.70	484.84
1000	−62.7	192.84	425.41
1200	−59.1	220.98	524.93
1400	−59.7	227.18	406.50
1600	−60.2	242.53	457.25
1800	−53.8	267.22	594.07
2000	−63.9	191.69	369.66

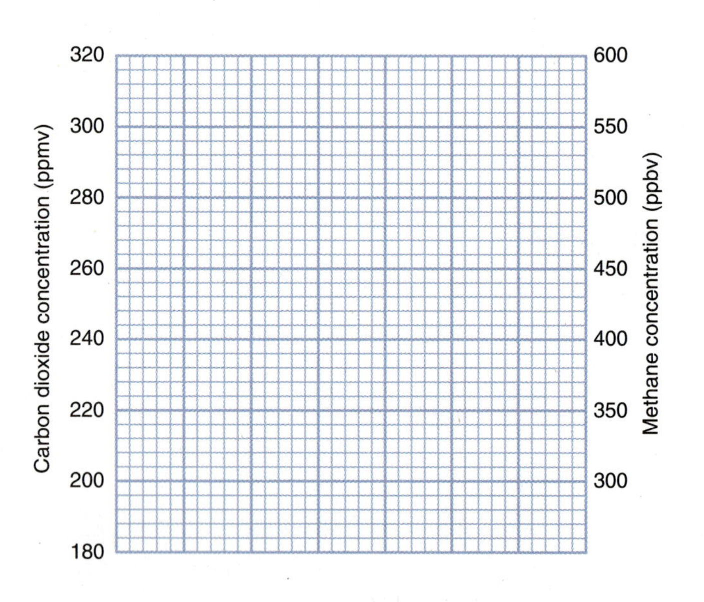

Figure 15.11

Graph Paper for Exercise 27

The relationships among past temperature and greenhouse gases help to reinforce the idea that these gases are indeed associated with climate change and that as we release more of them into the atmosphere, they are very likely to warm the planet.

RADIATIVE FORCING AND ANTHROPOGENIC WARMING

Recall that the changes in the amount of incoming and/or outgoing radiation at the boundary between the troposphere and the stratosphere are called **radiative forcings.** Radiative forcing is power per unit area, usually given as watts per square meter (W/m^2). If radiative forcing is positive, the climate will warm because more energy is coming in than going out. If it is negative, climate will cool because more energy is escaping from Earth than is arriving here. With its ability to absorb longwave radiation from Earth's surface, the greenhouse effect causes a radiative forcing when the amounts of greenhouse gases in the atmosphere change.

A number of other factors also cause radiative forcing (■ Table 15.4). Ozone losses in the stratosphere actually cool the climate because it cools the stratosphere; this gives stratospheric ozone a negative forcing. However, ozone air pollution in the troposphere acts as a greenhouse gas, so it warms the climate. Land-use changes reduce dark forest cover and tend to increase the surface albedo. This gives a cooling effect (negative forcing). Aerosols are solid particles or liquid droplets suspended in the atmosphere. They tend to block sunlight and help cloud formation. Both of these are cooling effects (negative forcings). Airplane contrails are clouds that one might think would cool the climate, but they also reflect light back to Earth and keep heat from escaping at night—the net effect is warming.

28. How would soot on snow change the albedo? ____________ Would it warm or cool the climate? ________
Would this be a positive or negative forcing?

____

We have seen that some climate change is natural and that human activities can also influence the climate. Do human-induced effects cause most of the warming or is it mostly natural? This next exercise explores the relative contributions of different factors to the overall radiative forcing. The data analysis and numbers for this exercise come from the Fourth Assessment Report on climate change by the IPCC (Intergovernmental Panel on Climate Change) made in 2007 and built by consensus among many expert climate scientists.

29. Use the data in Table 15.4 and the graph paper in ■ Figure 15.12 for the following exercises.

a. What does W/m^2 mean?

____

b. For each of the factors in Table 15.4, draw a bar starting from zero, and if it has a positive forcing, draw the bar to the right, or if negative, to the left.

c. For each of the factors, draw an error bar on the bar showing the range of possible values. An error bar looks like this: ├──┤

d. If you compare the natural forcings to the total net anthropogenic forcings, what conclusions do you draw about whether humans are causing climate change or not? Explain your answer.

____

____

e. How much leeway do the error bars give you for your answer in the previous question?

____

Increasing Greenhouse Gases in the Atmosphere

As you can see from your radiative forcing diagram, humans are substantially responsible for climate forcings, especially in the production of carbon dioxide. We use many resources that release greenhouse gases into the air. The concentration of the most important greenhouse gas, carbon dioxide, has been rising since the beginning of the Industrial Revolution (Figure 15.14b).

30. Examine the graphs in ■ Figure 15.13.

a. Rank the five gases in order of their concentration. Pay particular attention to the units.

CFC 12, CFC 11, Methane, NO, CO_2

b. Divide the concentration of the most concentrated one by the next most concentrated. You will need to convert these to equivalent units before dividing. How many times greater is the concentration of the most concentrated?

2 times

Table 15.4

Radiative Forcings

Potential human (anthropogenic) and natural sources of climate change.

Factors That Cause Radiative Forcing			Radiative Forcing (W/m^2)	Range	
Anthropogenic	Long-lived greenhouse gases	CO_2	1.66	1.49	1.83
		CH_4	0.48	0.43	0.53
		N_2O	0.16	0.14	0.18
		Halocarbons	0.34	0.31	0.37
	Ozone	Stratospheric	−0.05	−0.15	0.05
		Tropospheric	0.35	0.25	0.65
	Surface albedo	Land use	−0.2	−0.4	0.0
		Black carbon on snow	0.1	0.0	0.2
	Total aerosol	Direct effect	−0.5	−0.9	−0.1
		Cloud albedo effect	−0.7	−1.8	−0.3
	Linear contrails		0.01	0.003	0.03
	Total net anthropogenic		1.6	0.6	2.4
Natural:	Solar irradiance		0.12	0.06	0.3

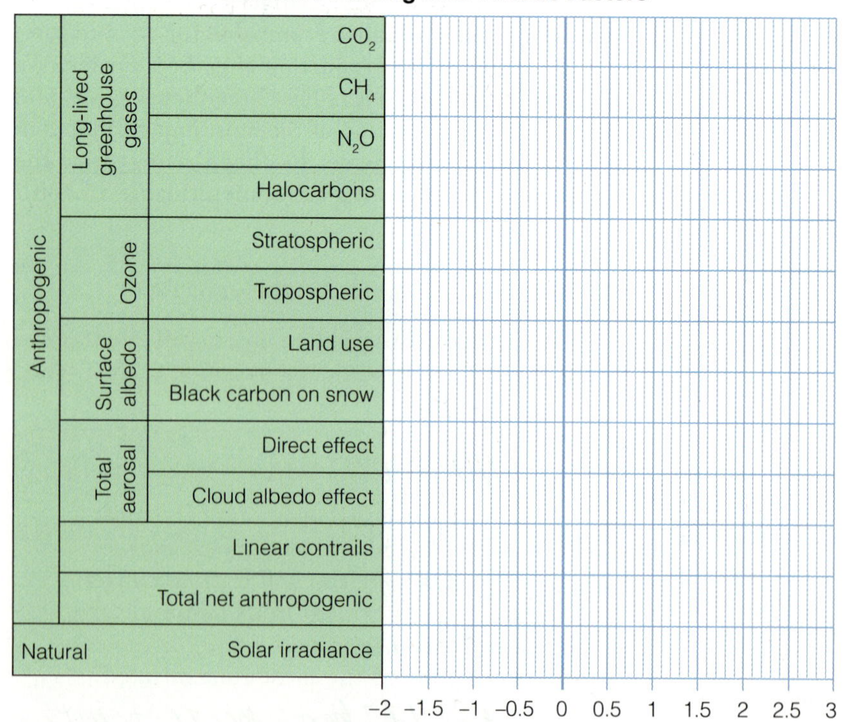

Figure 15.12

Graph Paper for Exercise 29

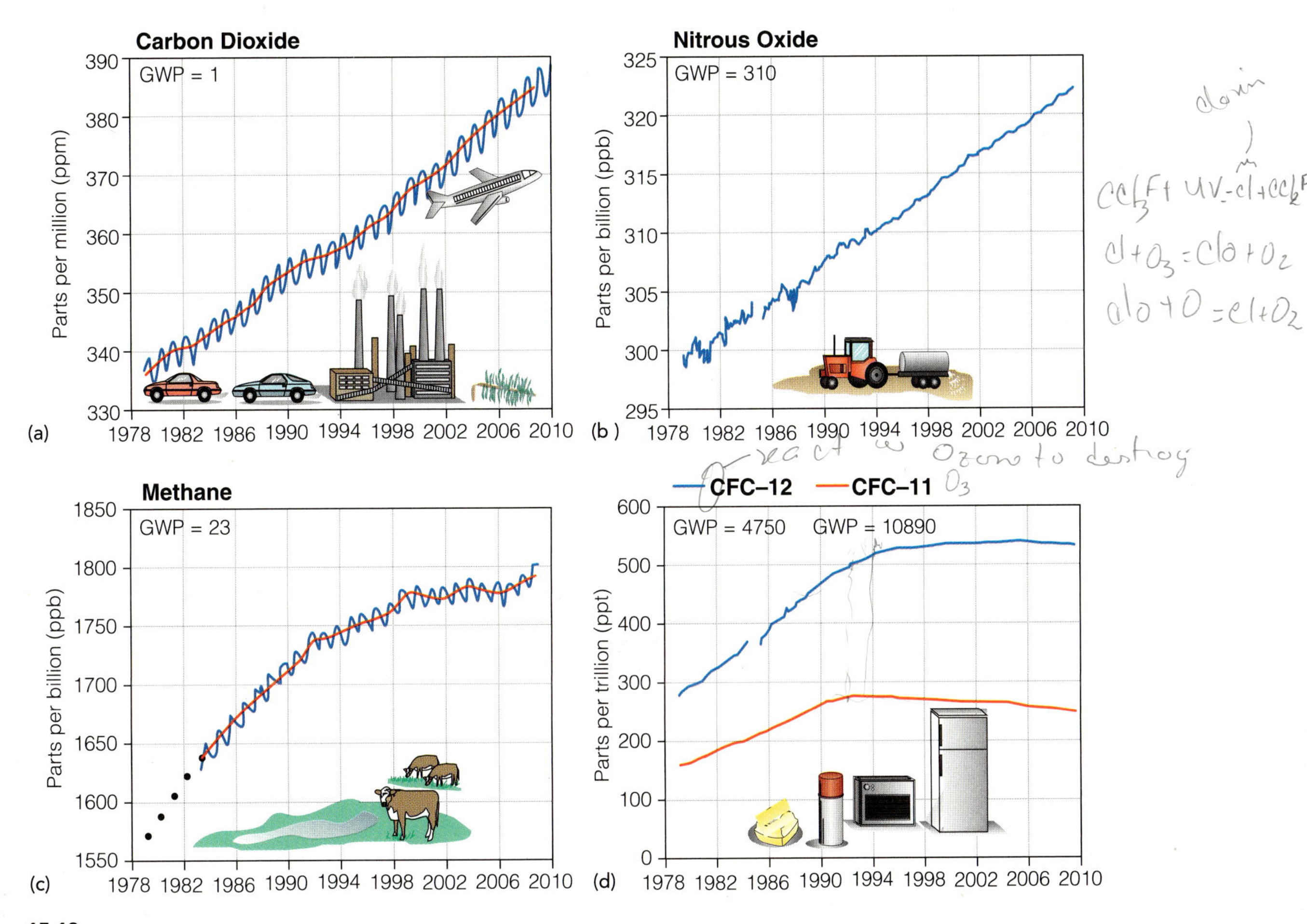

Figure 15.13

Changes in Greenhouse Gases

(a) Changes in carbon dioxide measured at Mauna Loa, Hawaii. This is known as the Keeling Curve, named for Charles D. Keeling, who took these measurements for many years. Burning of fossil fuels and of forests are major sources of carbon dioxide. 1 ppm = 1 molecule for a million molecules of air. (b) Levels of nitrous oxide (N_2O) in the atmosphere. One major source of nitrous oxide is fertilizer. (c) Levels of methane (CH_4) in the atmosphere. Livestock, are responsible for a substantial amount of methane released into the atmosphere. Wetlands and rice grown in water also release methane. (d) Levels of two CFCs in the atmosphere. Most uses of CFCs are banned worldwide so quantities are starting to decline although they may increase again as third world countries start to use more refrigerators and air conditioners.

c. Divide the concentration of the most concentrated one by the least concentrated. How many times greater is the concentration of the most concentrated than the least? 1.4 million times

d. List a source for each of the following gases. CO_2 factories, cars, methane cow feces, nitrous oxide agriculture, CFCs A/C, fridge

e. In light of the greenhouse effect, what is a likely future result of these human activities that release greenhouse gases into the atmosphere? It's going to keep increasing.

31. Examine the graphs in Figure 15.14.

a. Is global temperature rising? yes

b. Has the behavior of temperature been as consistent as the rise of CO_2 over the last 120 years? no

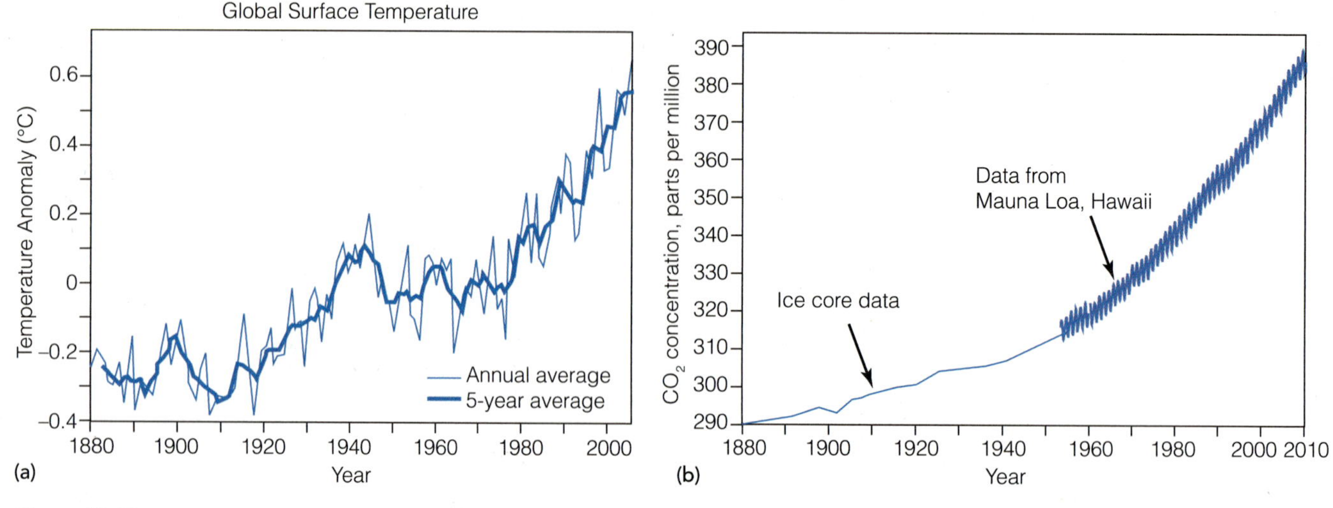

Figure 15.14

Average Global Warming and CO_2 Increase from 1880

(a) Average global surface temperature change showing annual and 5-year averages. (b) Carbon dioxide measurements from ice cores (smooth curve) and from direct atmosphere measurements from Mauna Loa, Hawaii from 1958–2010 (zigzag curve).

c. Check the difference in temperature you see in Figure 15.14a from 1880 to 2000 on the temperature graph in Figure 15.10c. Why do you think the temperature has not climbed nearly as much as the CO_2 amounts have risen? This is a challenging question because scientists are still trying to puzzle it out, so it might be a good topic for a class discussion.

Many things affect both rates in some way.

Changes in Climate

Climate scientists predict that some of the fastest changes in the climate will occur at high latitude. Let's see how climate has already been changing in Juneau, Alaska.

32. Examine the graphs in ■ Figure 15.15.

a. Summarize the climate changes you see in the two graphs including short-term variability and overall trend.

______

b. If you use the blue regression lines for each graph, how much has temperature and precipitation changed from the beginning of the data (Year: ______) to the end of the data (Year: ______) Change in temperature: ______ Change in precipitation: ______

c. How does the temperature change in Juneau compare with the worldwide average temperature change over the same time period? (A difference of 1° F = a difference of 5/9° C.)

______

d. Is this consistent with scientist's predictions mentioned above? ______

e. Would you expect rainfall to increase everywhere on Earth as a result of increasing greenhouse gases? ______ Explain your reasoning.

______

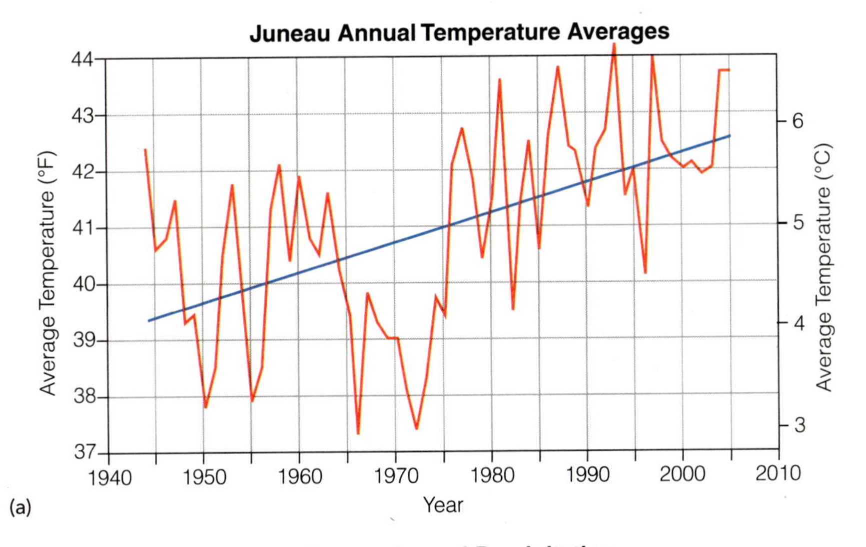

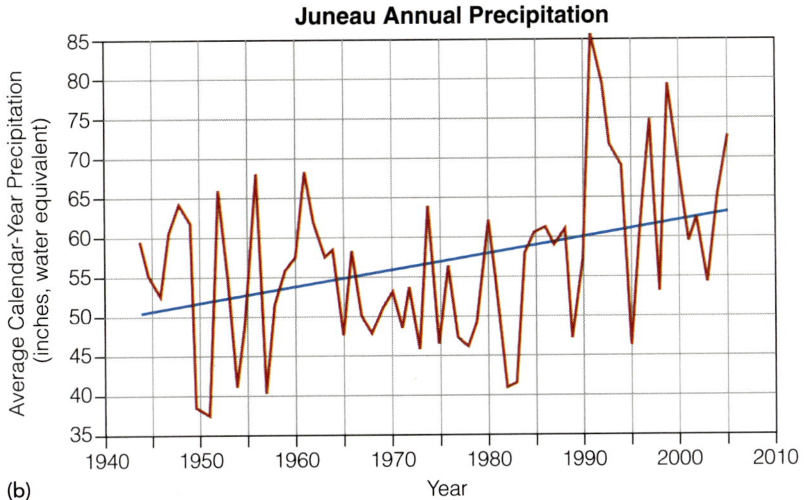

Figure 15.15

Climate Change in Juneau, Alaska

(a) Annual average temperatures at Juneau International Airport (red) and best-fit regression line (blue). (b) Annual precipitation at Juneau International Airport (red) and best-fit regression line (blue). (National Weather Service, Juneau).

OZONE LAYER

Above the troposphere is the **stratosphere** (Figure 15.16). Here ultraviolet radiation from the sun heats the air and creates the **ozone layer.** In the stratosphere, ozone and oxygen absorb much of the ultraviolet radiation and keep it from reaching Earth's surface, where it could do considerable damage. The sun gives off a range of radiation wavelengths that reach Earth's outermost atmosphere. The radiation is part of the electromagnetic spectrum, which is shown in Figure 15.9. Shorter wavelengths like UV radiation are more energetic and more harmful. Ultraviolet radiation damages chromosomes (RNA and DNA), causes sunburns and skin cancer, harms plants, and impairs the marine food web. The United Nations estimated that a drop of 16% of ozone from the ozone layer worldwide would cause a loss of about 7 million tons of fish per year.

The layer containing 97% of atmospheric ozone, the stratosphere, is about 30 km thick (Figure 15.16). Ozone is dilute in the ozone layer (has a low concentration). Even where ozone is most concentrated, at about 25 km up, it is less than 0.002% of the atmosphere. Nonetheless, along with oxygen, just this little bit of ozone prevents most ultraviolet rays from reaching Earth. However, a decrease of ozone concentration as little as 1% allows enough increase in UV rays to result in an increase in the incidence of skin cancer and of other health problems related to UV exposure.

33. Use ■ Figure 15.16.

a. Where is the ozone layer?

stratosphere ______________________

b. What is the maximum concentration of ozone in the ozone layer shown in Figure 15.16? ___12 ppm___ What is this number in percent? ___0.0012___ (*Hint*: Percent is parts per hundred and ppm is parts per million.)

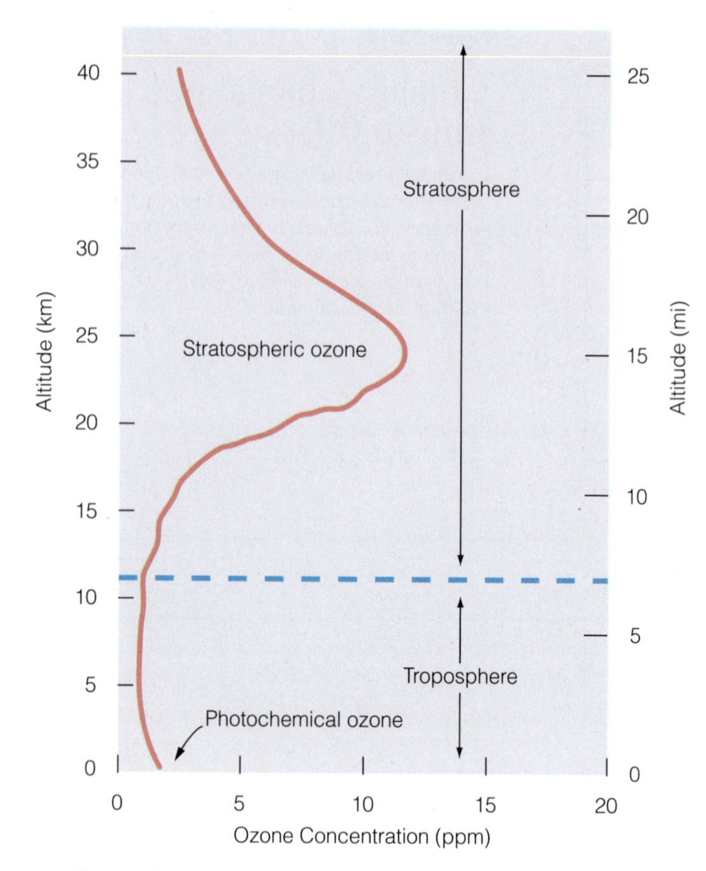

Figure 15.16

Ozone Layer

Average distribution of ozone in mid-latitudes above the Earth's surface. The ozone layer is the region of the stratosphere where ozone concentrations are higher. Even the highest concentrations of ozone are quite low. Ozone in the troposphere is a component of air pollution.

Formation of the Ozone Layer

How does the ozone layer form? It forms when a normal oxygen molecule, O_2, combines with an oxygen atom to make an **ozone** molecule, O_3. O_2 is also simply called oxygen. Both oxygen and ozone absorb ultraviolet rays (Figure 15.9b on p. 352). Oxygen in the stratosphere absorbs UV rays from the sun and forms ozone in a two-step process.

It takes high-energy UV rays to break apart the strong double chemical bonds in oxygen. A **chemical bond** is simply any force that tends to hold two atoms together. Each chemical bond in this case is a pair of electrons that oxygen atoms in the molecule share between them. Once broken apart, the separate oxygen atoms are ready to bond with other oxygen molecules to form the ozone.

The useful aspect of ozone is that it naturally absorbs ultraviolet radiation and in doing so breaks back into an oxygen molecule and an oxygen atom. Although this reaction destroys ozone, this is just part of the natural recycling of ozone that takes place continuously in the ozone layer and not the ozone-destroying reaction that artificially thins the ozone layer. Instead, this natural reaction frees an oxygen atom, making it ready to bond with an oxygen molecule again and form more ozone.

Ozone Recycling Activity

Materials Needed:

- Molecular model kits
- You may need special oxygen "atoms" to make ozone molecules. These need at least three holes.

34. Build molecules to simulate the natural reactions that form and recycle ozone in the stratosphere. As you go through this exercise, sketch and label the molecules and atoms that make up the chemical reactions in the appropriate places in ■ Figure 15.17. Include the chemical bonds in your sketches. Ozone has one double bond and one single bond and does not form a ring.

a. First build at least two oxygen molecules (O_2) with double bonds to be used in the first chemical reaction. Do not sketch this step.

b. Next, simulate ultraviolet radiation from the sun that breaks apart oxygen molecules. Sketch the molecules and atoms in Figure 15.17 in most suitable location. Write out this chemical reaction.

____

c. React an oxygen atom with an oxygen molecule. What does this produce?

______________________ Continue sketching the molecules and atoms in the correct places in Figure 15.17. Write out this chemical reaction.

____

d. Then simulate ultraviolet radiation from the sun that breaks apart ozone. Sketch the molecules and atoms with their correct bonds. Write out this chemical reaction.

____

e. Which one atom or molecule in these reactions would be readily available in the atmosphere because it is plentiful?

______________________ Which two would be rare?

____

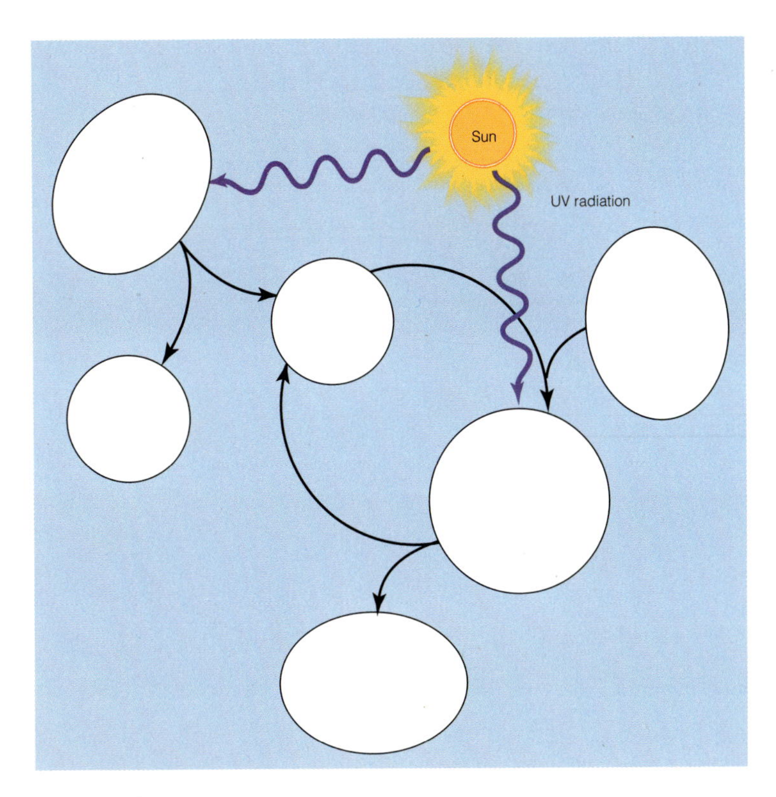

Figure 15.17

Ozone-Recycling Chemical Reactions

Sketch the ozone forming and recycling reactions. See instructions for Exercise 34.

f. Which two of these reactions are involved in recycling ozone? the reaction from part **b**, **c**, or **d**? ________ and from part **b**, **c**, or **d**? ________

g. In your own words, on a separate sheet of paper, completely recap the natural chemical cycle of ozone production and breakdown. Make sure to mention what step makes this a cycle.

Artificial Ozone Layer Depletion

The breakdown of chlorofluorocarbons (CFCs, also called *freon*) in the stratosphere ultimately causes ozone layer depletion. CFCs are manufactured molecules containing chlorine, fluorine, and carbon (HCFCs also contain hydrogen) and once came from coolants in refrigerators and air conditioners, solvents to clean electronic components, blowing agents in the production of plastic foams, and propellants in aerosol spray cans. CFCs have been banned in the United States as refrigerants since 1994 and as propellants since 1978. This is because CFCs cause damage to the ozone layer. HCFCs were substituted for CFCs because they caused less ozone layer thinning. The stability of CFCs allows them to thoroughly mix in the troposphere without breaking down. From there they diffuse upward into the stratosphere. At about 30 km altitude, ultraviolet radiation breaks off a chlorine atom (Cl). Figure 15.18 shows this reaction and the others that destroy ozone. The chlorine atom removes an oxygen atom from ozone and becomes chlorine monoxide (ClO), and the ozone turns into ordinary oxygen. The chlorine monoxide combines with a free-floating oxygen atom, releasing the chlorine atom. This step prevents an other ozone molecule from forming.

The chlorine is a **catalyst**—not used up in the reaction—and reacts repeatedly until it drifts into the troposphere where water quickly washes it out of the air. One Cl atom destroys on average 100,000 ozone molecules! As the ozone layer has diminished in the last 45 years, the incidence rate of skin cancer has increased 200% in the United States, and the increase in mortality from skin cancer is up 150%, although some of this increase may be due to increase in sun tanning and beach activities over this same time period.

Ozone Depletion Activity

35. Use the molecular model kits to enact the chemical reactions pictured in ■ Figure 15.18 so that you thoroughly understand the cycle of ozone destruction shown. Then answer the following questions.

a. Write each chemical reaction in Figure 15.18. List the reactions that take place by listing their letters in order. ______ After three reactions have occurred, what are the 4th ______ 5th ______ 6th ______ 7th ______ and 8th ______ reactions?

b. What molecule provides a chlorine atom?

____

c. What atom destroys ozone?

____

d. What molecule combines with a lone oxygen atom?

____

e. What might this oxygen atom have made if it hadn't been used up by the molecule in the previous question?

____

Figure 15.18

Ozone Destruction by Chlorine from CFCs

Chemical reactions that release chlorine atoms from CFCs and destroy ozone in the ozone layer. Sketch HFC 134 below, as directed in Exercise 35h. From MILLER. *Living in the Environment*, 16E. 2009 Brooks/Cole, a part of Cengage Learning, Inc. Reproduced by permission. www.cengage.com/permissions

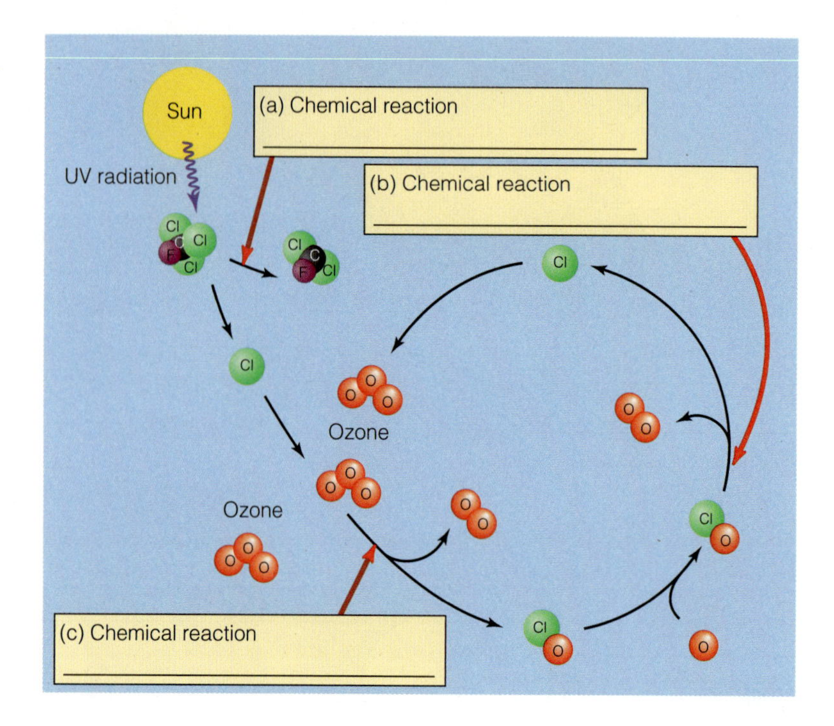

f. What atom does reaction b in Figure 15.18 release? ____________

g. How many chlorine atoms does it take to destroy, say, 100 ozone molecules and prevent another 100 ozone molecules from forming? ______, 1,000 ozone molecules? ______, 100,000 ozone molecules? ______

h. One of the most widely used refrigerants, as a substitute for CFCs, is HFC134. Build HFC134 with 2 hydrogens, 4 fluorines, and 2 carbons and sketch it next to Figure 15.18 in the space above. Explain why this molecule might make a suitable substitute for CFC to help protect the ozone layer.

____

However, HFC134 is still a strong greenhouse gas.

CFCs are not naturally occurring gases, so the destruction of the ozone layer is anthropogenic. The lifetime of CFCs in the atmosphere averages between about 50 to 100 years, so their ozone destruction is long lasting. In 1987, many nations signed an agreement called the Montreal Protocol to diminish the release of CFCs in the atmosphere. In 1992 this agreement was strengthened to reduce CFCs at a faster rate. Although CFCs are decreasing, the ozone hole is still occurring in the spring over Antarctica most years.

36. Examine Figure 15.13d on page 359.

a. Are CFCs still increasing in the atmosphere today? NO

b. When did the trend change visibly in both graphs? 1994

c. Do the treaties seem to be working?

Yes

Antarctic Ozone Hole As chlorine destroys more ozone, very low ozone concentrations over Antarctica, known as the **ozone hole,** occur as shown in Figure 15.19. Factors such as stratospheric winds, ice crystals, and the absence and return of sunlight over Antarctica, all play a role in concentrating chlorine and creating the Antarctic ozone hole. Ice crystals in polar stratospheric clouds that develop in the winter accumulate chlorine and release it when they dissipate with the return of the sun. The Arctic usually does not have an ozone hole, but values there have declined as well.

37. Examine ■ Figure 15.19a.

a. What do the colors mean on the diagram?

the amount Ozone

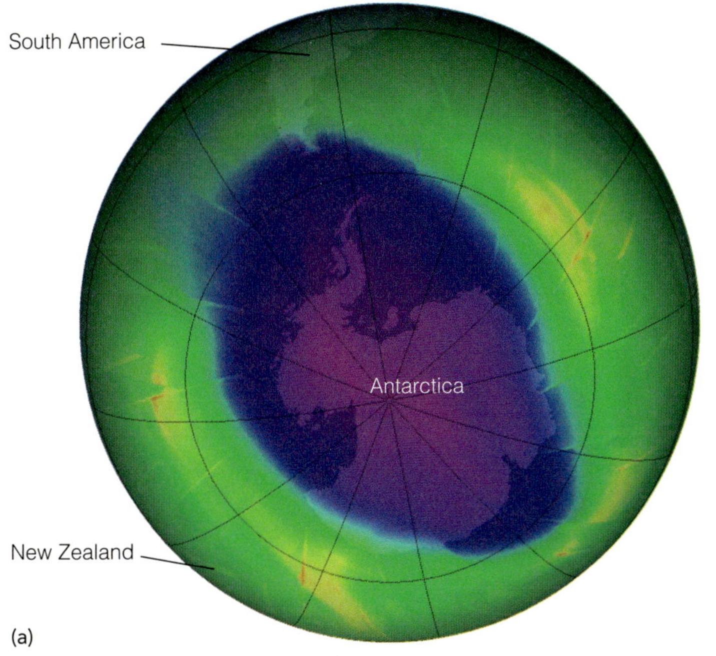

Total Ozone (Dobson Units)

110 220 330 440 550

September 25, 2009

Figure 15.19

Ozone Hole over Antarctica

(a) Ozone hole on September 25, 2009. Values are in Dobson units (DU). One Dobson unit is a 0.01-millimeter thick layer of ozone at standard temperature (0°C) and pressure (1 atm), which would be as if the ozone were all concentrated at Earth's surface. For example, 400 DU would make a 4 mm thick layer of ozone at the surface. (b) Ozone over Antarctica for a sequence of months in 2009. Each image is for the 27th of the month starting in July and ending in December 2009.

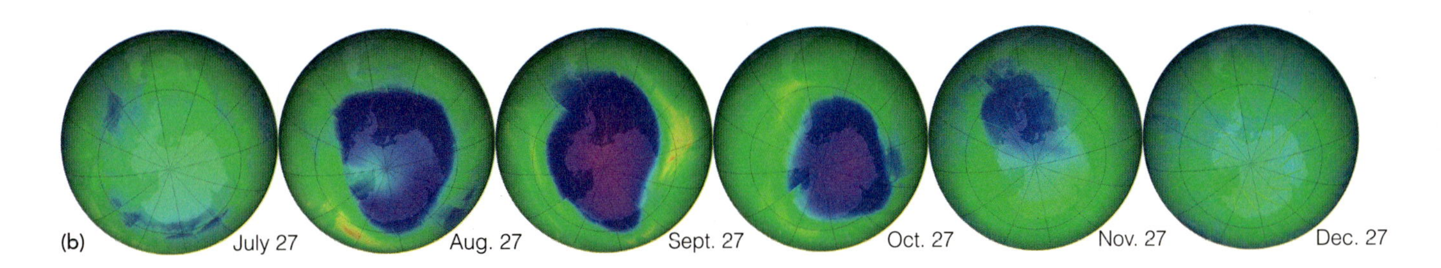

b. What is a Dobson unit?

____

____

____

c. What is the concentration of ozone at the southern tip of South America?

____

d. What is the highest concentration of ozone shown on the map (not just on the scale) on September 25, 2009?

__________________ Lowest concentration over Antarctica?

____

e. What is the percentage of the lowest ozone over Antarctica compared to the highest on the map?

____

f. In Figure 15.19b we see the ozone hole develop and then dissipate over Antarctica from July 27 to December 27, 2009. This pattern is typical for the ozone hole for now and is only expected to improve slowly due to the banning of use of CFCs over the next hundred years. In what season does the ozone hole begin to develop?

___Winter___ In what season does the ozone hole grow to the largest size and have the lowest concentrations of ozone? ___Spring___

g. Explain the timing and development of the ozone hole in connection with the sequence of pictures in Figure 15.19b.

____

____

____

Scientists have found that the effects of depletion in the ozone layer are more far-reaching than skin cancer and cataracts. The excess UV radiation resulting from a thinner ozone layer impairs the molecular chemistry of photosynthesis both on land and at sea, has direct consequences for human health, and can affect world food production. Even though the ozone layer may eventually recover from the destruction that the use of CFCs has brought on, new research by NASA scientists suggests that changes in the dynamics of the atmosphere due to greenhouse warming will alter the future ozone layer. Twenty-first-century climate change will alter atmospheric circulation, increasing the movement of ozone from the stratosphere to the troposphere and shifting the distribution of ozone within the stratosphere. Other research suggests that increasing greenhouse gases would delay or even postpone the recovery of parts of the ozone layer. The conclusions are in progress as we increase our knowledge of these complex interrelationships.

38. In the space below, outline the similarities, overlaps, and differences among the greenhouse effect, global warming, and ozone layer depletion.

LAB

16

Weather

OBJECTIVES

- **To understand relative versus absolute humidity and their effects on weather**
- **To recognize cloud types and understand how they may indicate approaching weather**
- **To understand weather fronts and their characteristics**
- **To be able to read a weather map and interpret it in relation to past and future conditions**
- **To learn the effects of El Niño and La Niña ocean surface temperatures on weather patterns in contrast to normal conditions**

Claudia Owen

Figure 16.1

Cumulonimbus Cloud

This cumulonimbus ominously moves in over the Painted Hills of John Day National Monument, Oregon, and threatens to dampen a geology field trip with heavy downpours, thunder, and lightning.

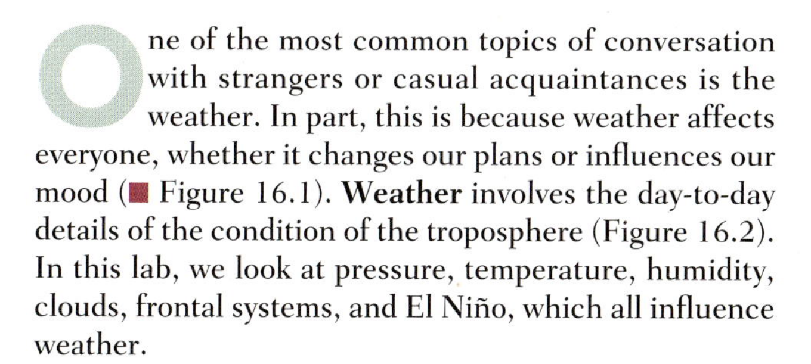

One of the most common topics of conversation with strangers or casual acquaintances is the weather. In part, this is because weather affects everyone, whether it changes our plans or influences our mood (■ Figure 16.1). **Weather** involves the day-to-day details of the condition of the troposphere (Figure 16.2). In this lab, we look at pressure, temperature, humidity, clouds, frontal systems, and El Niño, which all influence weather.

CONDITIONS OF THE AIR

The source of energy for our weather is solar heat radiation and sunlight. The more thermal energy the air has, the faster its molecules are moving and the higher its **temperature** is. Earth's rotation, differences in temperature, wind, and pressure belts all have strong influences on climate, as we saw in Lab 15, and therefore also on weather. The energy transferred when water evaporates from the oceans and then condenses provides heat to propel air upward and to help power our weather systems. Temperature, pressure, and humidity are three measurements that are important in determining conditions of the air in the lower atmosphere—we need to understand these three measurements to understand weather.

Temperature and Air Pressure

The troposphere is the lower layer of the atmosphere that contains water vapor, clouds, and weather. In the **troposphere,** temperature generally decreases with altitude (■ Figure 16.2).

Before weather satellites and television forecasting, people often tried to predict the weather using clouds, wind, and a **barometer,** an instrument that measures air pressure. Air pressure drops in connection with weather fronts, thunderstorms, tornadoes, and hurricanes, so it can be an indicator of imminent unstable weather. Pressure can be measured in millibars[1] (mb), inches of mercury (in. Hg), millimeters of mercury (mm Hg), or pounds per square inch (psi or lb/in.^{2}):

$$
\begin{aligned}
1.000 \text{ in. Hg} &= 25.40 \text{ mm Hg} \\
&= 33.86 \text{ mb} \\
&= 0.4912 \text{ psi}
\end{aligned}
$$

Pressure also changes with altitude (Figure 16.2), starting at about 1 bar at sea level and decreasing exponentially with height.

A thermometer measures temperature using the property that materials expand when they get warmer. Alcohol or mercury in the thermometer expands as the temperature rises, giving a higher reading on the thermometer. Properties of air respond to changes in temperature because air also expands when it warms and contracts when it cools. When it warms, the molecules move more rapidly and cause either the pressure to increase or the air to expand. The opposite is true when it cools. This relationship even holds when the volume of air changes; with decreased volume, when air compresses: it heats up; with increased volume, when air expands, it cools down. These properties play a role in the movement of air and in cloud formation. However, for clouds to form the air also needs humidity.

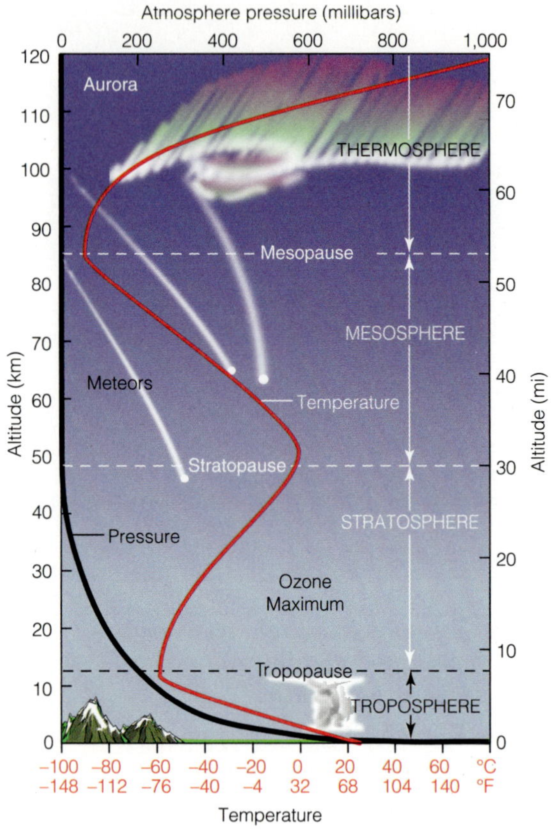

Figure 16.2

Layers of Earth's Atmosphere

Earth's atmosphere consists of layers defined based on temperature changes occurring within each layer. Temperature decreases with altitude in the troposphere and mesosphere, and increases with altitude in the stratosphere and thermosphere. The exact altitudes of the various layers depend on the latitude and the time of year. Most clouds and weather occur within the troposphere. Atmospheric pressure decreases with altitude. From AHRENS. *Meteorology Today,* 9E. 2009 Brooks/Cole, a part of Cengage Learning, Inc. Reproduced by permission. www.cengage.com/permissions

Humidity

Humidity is the water vapor content of the air. We measure it in two different ways. **Absolute humidity** is the total mass of water contained as vapor in a certain volume of air, measured for example, in grams per cubic meter (g/m^3). **Relative humidity** of a parcel of air is the percentage of water vapor relative to the total amount of water vapor that air can hold. When air starts to condense to form fog or clouds, it has 100% relative humidity. Relative humidity inversely relates to evaporation rate. That is, as the relative humidity goes up, the rate of evaporation goes down. When you feel the humidity or the dryness of the air on your skin, you are detecting relative humidity and its relationship to cooling from evaporation.

Warmer air is able to contain more moisture than colder air because it is more energetic. Faster-moving molecules prevent water vapor molecules from condensing. If air starts cold, at 100% humidity and warms up, the relative humidity drops even though the amount of water vapor in the air (the absolute humidity) stays the same. Conversely, warm air that cools down increases its relative humidity and condenses into clouds or fog if it reaches 100% relative humidity. Cold water molecules are less energetic than warm ones and more likely to condense than evaporate in cold air. ■ Figure 16.3 shows the absolute humidity for water-saturated air at different temperatures. When air is **saturated,** the air cannot hold any more moisture.

1. Read the graph in Figure 16.3. How much moisture can 1 m^3 of air at 30°C (86°F) contain? 30g How much water can air at

[1] One millibar = 0.02953 inch of mercury.

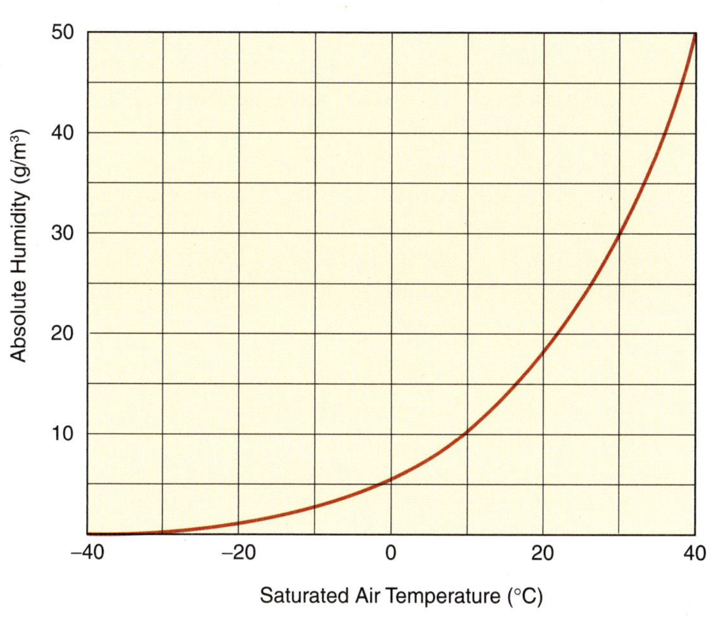

Figure 16.3

Absolute Humidity of Saturated Air

The relationship between absolute humidity and the temperature of water-saturated air. Air is saturated at the dew point, so the horizontal axis could be labeled "dew point." From GABLER/PETERSEN/TREPASSO. *Essentials of Physical Geography* (with CengageNOW Printed Access Card), 8E. 2007 Brooks/Cole, a part of Cengage Learning, Inc. Reproduced by permission. www.cengage.com/permissions

10°C (50°F) have in it? 10 g About how many times as much is the first number compared to the second? 3 What would happen if saturated 30°C air cooled to 10°C? humidity decrease How would the thermal energy in this mass of air have changed? decrease energy

2. From your own experience, in winter, when the air outside is cold, what is the humidity like inside where the air has warmed? Does it seem moist, or does it dry out your skin? Moist

Most people are aware that the air thins and cools at higher altitudes (Figure 16.2). Would the air pressure be higher or lower at the top of a mountain than at the base? lower How would upward movement and cooling affect the relative humidity of an air mass? upward movement low humidity

After a warm day, the temperature drops and, as it cools, the relative humidity goes up. At the temperature where the relative humidity reaches 100%, dew or fog forms. This temperature is called the **dew point.** At the dew point, the air is saturated. Every batch of air with its particular absolute humidity has a dew point. Simply heating or cooling air without adding or removing moisture does not change the dew point. The dew point alters when the absolute humidity changes—when water condenses by cooling or evaporates by warming, and changes the water content of the air. Humidifiers and evaporative coolers (also known as swamp coolers or desert coolers) artificially add moisture, and dehumidifiers and air-conditioners extract moisture. Breathing and using water indoors increases the absolute humidity.

Measuring the Dew Point

The following experiment teaches one way to measure the dew point.

Materials needed:

- Small glass beakers or metal cups
- Insulating cups
- Ice water
- Tap water
- Two thermometers

3. Place some ice water (no ice) in a glass beaker or metal cup without getting the outside of the container wet. Do the same with some tap water.

a. After a minute or two, check for condensation on the outside of the containers. If condensation occurs on one, but not the other,[2] place a thermometer in each container. When the thermometer has equilibrated with the water (reached a fairly constant temperature), record your observations in Table 16.1.

b. Mix ice water and tap water in two to six similar containers to make water of various temperatures. Keep the outsides of the containers dry while you pour and mix. After a minute, look for the pair of containers whose water temperatures are closest together and still have two different results—one with condensation and the other without. Measure the temperature in each of these two containers and record the results in Table 16.1.

[2] If moisture condenses on the tap water container, the dew point is higher than your tap water temperature and you will need to use warmer tap water, or heat the water. Do not use water above room temperature.

Table 16.1

Dew Point Determination

Measuring Dew Point Inside		Measuring Dew Point Outside	
Inside Air Temperature (°C) →		**Outside Air Temperature (°C) Is It Foggy? →**	
Water Temperature (°C)	**Condensation or Not**	**Water Temperature (°C)**	**Condensation or Not**
	= dew point		= dew point
Temperature – dew point		Temperature – dew point	
Relative humidity inside (from Figure 16.4)		Relative humidity outside (from Figure 16.4)	
Absolute humidity inside (from Figure 16.3)		Absolute humidity outside (from Figure 16.3)	

c. Continue mixing between the condensing and not-condensing containers to obtain intermediate temperatures until you have narrowed the dew point to about 2°C. Record your measurements in Table 16.1 for each condensing and noncondensing pair. When you find the pairs with the closest temperature but with opposite results, record their average temperature as the dew point in Table 16.1.

4. If the outside temperature is above freezing, measure the dew point outside. Mix up batches of water at about 5°C increments between 0°C and the outside air temperature and put them each in insulating cups. Take some additional, preferably metal, containers and a thermometer with you, taking extra care not to drop or chip the thermometer. Using the batches of water at 5°C intervals, pour them into the metal cups and test to see if condensation occurs on their outside surface. Measure the temperatures and record your results in Table 16.1. When you have the dew point narrowed down to between two temperatures, mix these two water batches to get intermediate temperatures and narrow the dew point measurements further.

5. Examine and summarize your results. Describe how the inside air is different from the outside air by answering the following questions:

a. Is the dew point different? ________ If so, where is it highest and why?

____

____

Dew point could also be called

____________________ air temperature.
Determine the absolute humidity for both locations using the graph in Figure 16.3 and enter the numbers in Table 16.1.

b. What is the relationship between dew point and absolute humidity? Answer this by completing this sentence: When the dew point is higher the absolute humidity is ______________.

c. Which location has the highest absolute humidity? ____________________

d. Subtract the air temperature minus the dew point for both locations and enter this information into Table 16.1.

e. Use the graph in ■ Figure 16.4 to determine the relative humidity. Do this by reading the difference you just determined from the x-axis and reading up and across to get the relative humidity off the y-axis. Enter this information into Table 16.1.

f. Which location has the highest relative humidity? ____________________
Without having an instrument or carrying out the experiment to measure it, how do you detect differences in relative humidity?

____

____

g. Assuming the outside air is the starting point for air inside, has moisture been

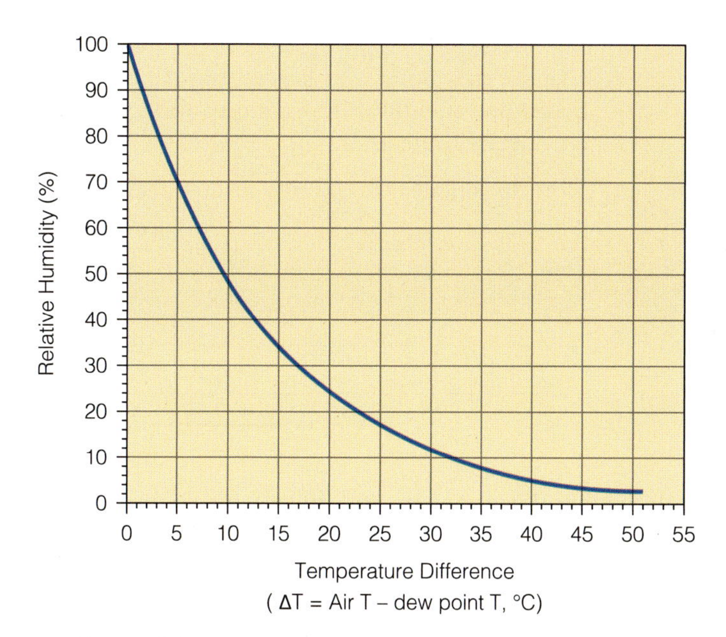

Figure 16.4

Relative Humidity of Unsaturated Air

As air temperature deviates from the dew point, the relative humidity drops. From this graph, you can determine the relative humidity of air based on its difference in temperature from the dew point, or you can determine the dew point based on the relative humidity and the temperature of the air. The horizontal axis is the difference between the air temperature and the dew point temperature.

added to or extracted from this to produce the inside air? How do you know?

____

____

Which air has the highest thermal energy?

____

Determining the Dew Point from Temperature and Relative Humidity

Another way to determine the dew point is by measuring relative humidity with a hygrometer or sling psychrometer. **Hygrometers** are instruments used to measure humidity. One way to measure relative humidity uses two thermometers, one with a wet bulb and the other dry. The thermometers measure cooling by evaporation on the wet bulb compared to the dry bulb temperature. An apparatus that does this is a **sling psychrometer,** which you swing in a circle to cool the wet bulb. You can make your own hygrometer (with no swinging required) using two thermometers. Put a small piece of wet gauze over the bulb of one thermometer, and use a fan to cool it. Because evaporation causes cooling, and the amount of cooling depends on the humidity of the air, the temperature differences in the wet and dry bulbs indicate the relative humidity. We discuss these concepts more below.

Materials needed:

- Sling psychrometer or other type of hygrometer

6. Measure the relative humidity in a number of locations inside and outside using a hygrometer/sling psychrometer, whichever instrument your instructor provides. If you use a sling psychrometer or any other hygrometer type that gives dry-bulb and wet-bulb readings, enter the data in those columns in ■ Table 16.2. Your dry-bulb reading is the temperature. Use the relative humidity chart that accompanies your sling psychrometer to determine the relative humidity, or find a Website to calculate it (search online

Table 16.2

Humidity Determination

Use a sling psychrometer or another type of hygrometer. If using a hygrometer that does not give wet bulb temperature, leave these spaces in the table blank.

Location	Dry-Bulb Temperature °C	Wet-Bulb Temperature °C	Relative Humidity (%)	Dew Point °C	Absolute Humidity (g/m³)

for "relative humidity calculator"). If your hygrometer gives you relative humidity readings directly, fill in the dry-bulb temperature for the air temperature, skip the wet-bulb column, and enter the humidity readings under relative humidity in the table.

7. An equation approximating the dew point from temperature and relative humidity is as follows:

$$D = T + 30.18\ (\log h - 2)$$

where D = dew point in °C
T = temperature in °C
h = relative humidity in percent

a. Determine the dew point for all locations by calculation using the equation or reading off the graph in Figure 16.4. If you are reading the graph, you need to use the relative humidity to determine a temperature difference (ΔT) from the graph then subtract it from the air temperature (dry bulb) to obtain the dew point. Enter the dew point in Table 16.2.

b. How do the dew point values you just obtained in part a compare to the values obtained in Exercises 3–5?

____

____

c. Use the dew point and Figure 16.3 to determine the absolute humidity and enter it in Table 16.2.

d. How do you account for any discrepancies?

____

____

____

____

Relative humidity goes up and down with changes in air temperature because the increase or decrease in the thermal energy of the air changes how much moisture the air can contain. With a lot of thermal energy, the air can keep more water molecules in a gaseous state. Recall that relative humidity is the amount of moisture the air has in it, divided by how much it can contain. If the temperature goes up the air can hold more moisture so the relative humidity goes down.

8. In ■ Figure 16.5, use a ruler to draw a line vertically through each temperature peak from the bottom to the top of the graph.

a. At what time of day does the temperature peak? 4 pm

Figure 16.5

Daily Air Temperature and Relative Humidity Changes

Temperature and relative humidity changes during five days in Hillsboro, North Dakota.

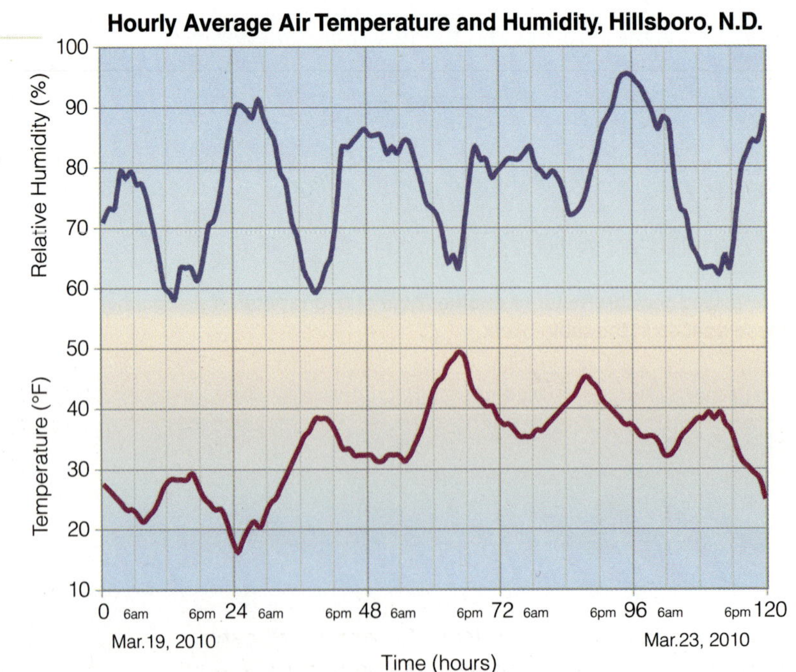

b. At this same time of day what has happened to the relative humidity?

humidity decreases

c. Draw a vertical line through each peak in relative humidity using a different color. What is the temperature usually like when relative humidity peaks?

Temperature is at the lowest

d. Explain the pattern of ups and downs in the two graphs. Why is the pattern like this?

The ↑ the temp. the ↓ the humidity.

Clouds versus Clear Skies

Two basic contrasting surface weather systems occur at atmospheric low and high pressure (Figure 16.6). Rising air makes a low-pressure system. Sinking air creates a high-pressure system. Let's consider these two types of systems more closely.

9. Examine ■ Figure 16.6.

a. In Figure 16.6a, winds blow toward the low where air is rising. As the air rises, does it warm or cool? cold

Do you think that the water vapor in this air will condense or will this air mass cause evaporation?

air mass decreases

b. In Figure 16.6b, winds blow away from a high where air is sinking. Why is it turning in this direction? (It may help to review p. 341 in Lab 15.)

Because when it hits the ground it diverges

As the air sinks, does it warm or cool?

warms Which do you think will happen to the water vapor in this air? (i) Will it condense or (ii) will this air mass cause evaporation, increasing its water vapor content? condense

The relationships explored in Exercise 9 are the key to understanding weather systems. Low-pressure systems are stormy and tend to have clouds and rain (or snow).

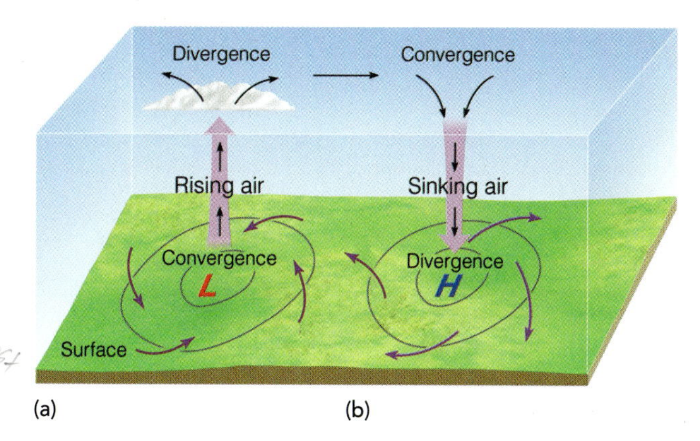

Figure 16.6

Cyclone and Anticyclone

(a) Cyclone (L = low pressure) for the Northern Hemisphere showing wind directions and the influence of Coriolis effect. Low pressure is associated with rising air that condenses to form clouds. (b) Anticyclone (H = high pressure). High pressure is associated with sinking air that produces clear skies. For the Southern Hemisphere the rotation directions for each system would be reversed (mirror image).

A low-pressure system[3] at latitudes higher than about 30° is called an **extratropical cyclone** and has rising and cooling air where condensation occurs. Conversely, air can hold more moisture as it warms up, so warming air tends to dry out its surroundings. Thus, high-pressure systems have dry, stable air and tend to have clear skies and no precipitation. These high-pressure systems are **anticyclones.**

Knowledge of air temperature and humidity relationships is also important in understanding cloud formation and different types of fronts.

Clouds

Clouds form when air cools, as the relative humidity increases to 100% (= dew point) and condensation occurs. Condensation and cloud formation may result from cooling due to air rising into higher altitudes where the temperature is lower, expansion of air as it rises, loss of energy at night, and contact with colder air. Cold land or cool water may each cause cooling and therefore may result in condensation and cloud formation.

The basis of cloud classification is their form and altitude. Clouds come in three basic forms (■ Figure 16.7):

Cirrus: thin, high clouds of ice crystals
Stratus: form horizontal layers (■ Figure 16.8)
Cumulus: puffy, cotton-ball-like clouds that may have tall vertical dimensions (■ Figure 16.9)

[3] Tropical cyclones—including tropical storms, hurricanes, and typhoons—are low-pressure systems originating at low latitudes, usually between 5° and 20° latitude. These extremely powerful storms obtain their energy directly from condensation and indirectly from warm seawater.

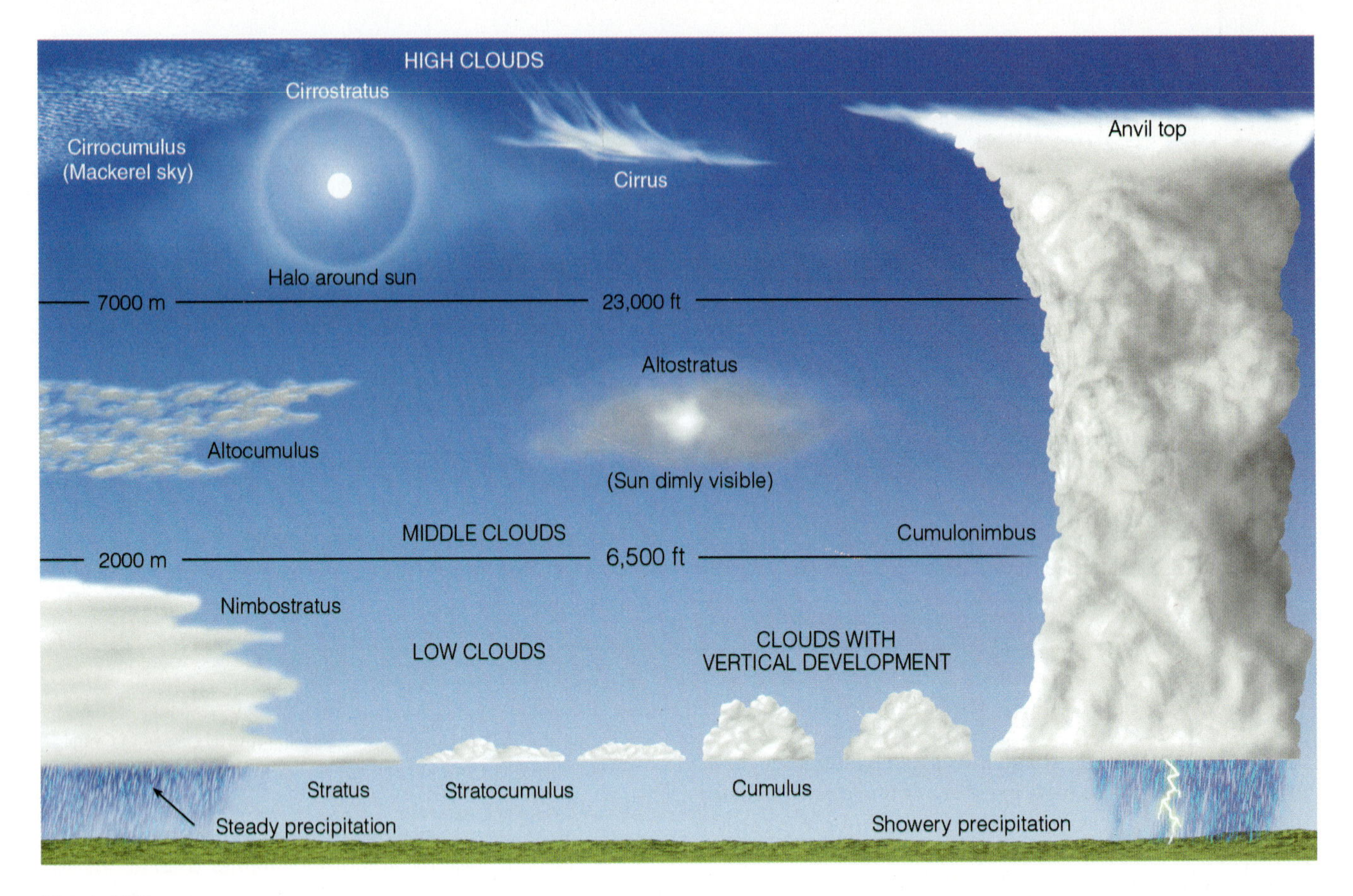

Figure 16.7

Classification of Clouds

Cloud types at various altitudes. From AHRENS. *Meteorology Today*, 9E. 2009 Brooks/Cole, a part of Cengage Learning, Inc. Reproduced by permission. www.cengage.com/permissions

Different cloud types form at different altitudes, as shown in Figure 16.7. **High clouds** all have *cirr-* as part of their name and generally form above 23,000 ft (7,000 m). These include cirrus, cirrostratus, and cirrocumulus clouds (Figures 16.8a, b and 16.9a). **Cirrus** (Ci) clouds, in the more restricted sense, are wispy with separate, hair-like strands that may curl. If they curl or have tufts at the ends, they are casually referred to as *mare's tails* (Figure 16.8a). **Cirrostratus** (Cs) clouds form a thin, smooth veil over the sky, producing a halo around the sun (Figure 16.8b) or a flat sheet of cirrus clouds or glaring white sky where it is difficult to separate one part from another. If mid-level clouds follow high clouds, rain or snow may be approaching. **Cirrocumulus** (Cc) clouds are small puffs sometimes appearing in long rows (Figure 16.9a). Individual puffs are all white (except at sunset) and appear much smaller than the width of your thumb held out at arm's length.

Middle clouds, which include altostratus and altocumulus clouds, form between about 6,500 and 23,000 ft (2,000 to 7,000 m); they consist of water droplets and, sometimes, some ice crystals. **Altostratus** (As) clouds form a light gray sheet through which you can vaguely

NOAA Central Library, Photo by Albert E. Theberge

Figure 16.8

Cirrus and Stratus Clouds (*above and facing page*)

(a) Cirrus clouds forming "mare's tails." (b) Cirrostratus clouds are smooth, flat, and allow the sun and moon to be clearly visible through them. When you see a halo around the sun or moon, you are looking through cirrostratus clouds. (c) Altostratus clouds allow you to see the sun and moon faintly through them. The sun may appear as a brighter spot or disk through the light gray sheet of these clouds. (d) Stratus clouds form a flat gray sheet of clouds usually too thick for the sun to be visible. (e) Nimbostratus clouds are raining, gray, sheet clouds.

(b) Cirrostratus, high

(c) Altostratus, middle

(d) Stratus, low

(e) Nimbostratus, low

Figure 16.8

Stratus Clouds—*Continued*

(a) Cirrocumulus, high

(b) Altocumulus, middle

(c) Stratocumulus from above, low

(d) Stratocumulus, low

(Continued)

Figure 16.9

Cumulus Clouds

(e) *Cumulus humilis*, low

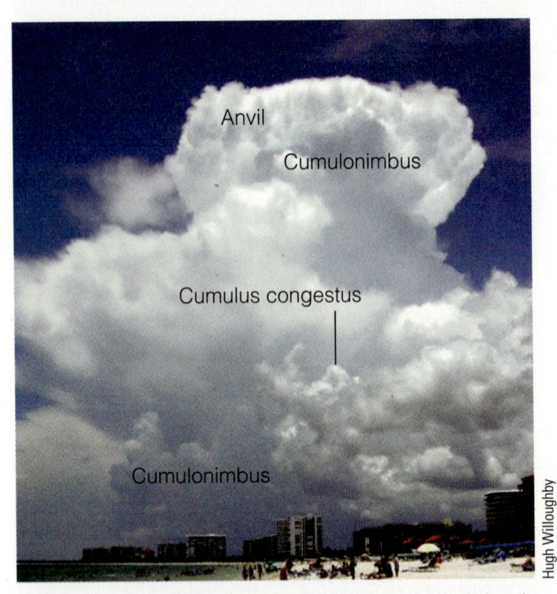

(f) Cumulonimbus and cumulus congestus, vertical development

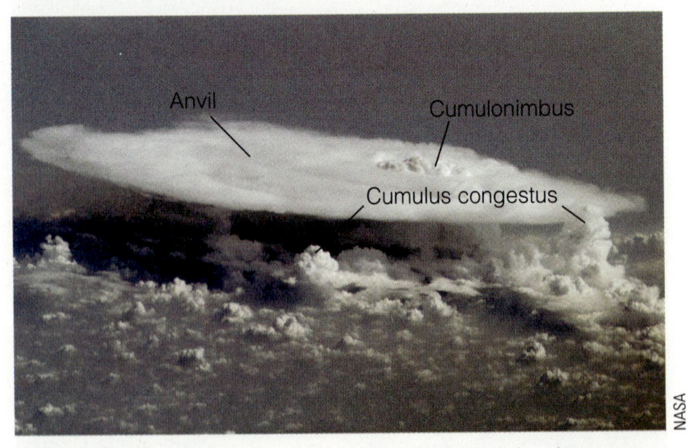

(g) Cumulonimbus and cumulus congestus, vertical development, with anvil

Figure 16.9

Cumulus Clouds—*Continued*

(a) Cirrocumulus clouds—see text for how to distinguish different cumulus types. (b) Altocumulus clouds appear to be about the width of your thumb held out at arm's length. (c) Ship Rock (a remnant volcanic neck in NW New Mexico) protrudes from a layer of stratocumulus clouds. As seen from a plane, the cloud sheet moves in and its advance is initially divided; later the layer filled the divide leaving only the topmost portion of the neck visible. (d) Stratocumulus clouds form sheets of cumulus puffy clouds that individually appear to be about the size of your fist or at least the width of three fingers held out at arm's length. (e) Fair weather cumulus, called *cumulus humilis*. (f) Towering cumulonimbus at the beginning of anvil formation and smaller cumulus congestus clouds building upward with tight, rounded, growing tops and flat bottoms. The cumulonimbus cloud has strong vertical development extending from near ground level all the way to the anvil at the top at the base of the stratosphere. (g) Cumulonimbus cloud seen from space with anvil and overshooting top. Surrounding it are smaller cumulus congestus and cumulus clouds.

still see the sun or the moon (Figure 16.8c). **Altocumulus** (Ac) clouds form multiple fleecy puffs, often uniform in size or in rows, bands, or rolls. Individual puffs are commonly thick enough to give the lower part a grayer appearance. An easy way to distinguish altocumulus clouds from cirrocumulus and stratocumulus clouds is to hold out your thumb at arm's length; the altocumulus clouds appear to be about the same width as your thumb (Figure 16.9b, inset); the other cumulus clouds appear smaller or larger.

Low clouds have bases below about 6,500 ft (2,000 m) and include stratocumulus, stratus, and nimbostratus clouds. They are generally made of water droplets and, occasionally, when it is cold, they may also have snow and ice crystals. **Stratus** (St) clouds are extensive sheets of low gray clouds (Figure 16.8d). **Nimbostratus** (Ns) are flat, dark gray rain clouds composed of many layers; *nimbus* means rain (Figure 16.8e). They generally have continuous, steady, but not intense rain. **Stratocumulus** (Sc) clouds form a layer of puffy clouds (Figure 16.9c and d) that appear to be about the size of your fist held out at arm's length. You can distinguish stratocumulus from altocumulus by the latter's apparent smaller size. Worldwide, stratocumulus clouds are the most commonly seen cloud type.

Some clouds develop large vertical structures, so they classify as clouds with **vertical development.** These are **cumulus** (Cu) clouds, and include small, fair-weather cumulus called ***cumulus humilis*** (Figure 16.9e), with little vertical growth but dome-shaped rather than the flatter tops of stratocumulus. **Cumulus congestus** and **cumulonimbus** (Cb) clouds have strong vertical growth and become tall cloud masses (Figure 16.9f, g). Figure 16.1 shows how dark the bottom of a towering cumulonimbus cloud can get. Cumulus clouds have distinct outlines and generally look like cotton or heads of cauliflower. The flat bottom of these clouds marks the altitude at which the rising air cooled and reached its dew point. As condensation occurs the air is sufficiently warm to continue rising. A tight cauliflower-shaped top of a tall growing cloud is a **cumulus congestus,** which produces showery precipitation. These clouds may develop into huge towering structures called **cumulonimbus** clouds, which may individually reach from a flat-bottomed base at 2,000 ft to 35,000 ft (600 m to 11 km) in a flat, wind-swept top called an **anvil.** *Nimbus* in the name refers to rain; these clouds are associated with thunderstorms and intense showers of rain, snow, snow pellets, and/or hail. Because the cloud cannot continue to rise into a warmer air mass, a flat anvil top of a cumulonimbus may form at the boundary between the troposphere and the stratosphere (Figure 16.9g). The air temperature in the stratosphere increases with altitude (Figure 16.2) so is warmer than the top of the troposphere.

Weather Fronts

Masses of air with varying characteristics such as warmer or colder temperatures, and more or less moisture, move across Earth's surface, helping to disperse heat from the equator and cold from the poles. Commonly, these air masses have fairly sharp or distinct boundaries.

- A **front** is a boundary or transition zone between two air masses of different densities. Usually, the density differences are due to a difference in temperature, with cold air being denser than warm air.
- At a **cold front,** colder denser air advances toward and underneath a mass of warmer air. This type of front tends to be a steep, nearly vertical boundary (■ Figure 16.10) and is associated with intense precipitation, sometimes including large hail, thunderstorms, and occasionally tornadoes.
- At **warm fronts,** warmer air moves into a region of colder air and tends to ride up slowly over the cooler air mass. Condensation and precipitation occur where these air masses are in contact and as the warmer air rises (■ Figure 16.11). The front is a gradually sloping boundary.
- If neither cold nor warm air masses displace the other, their boundary is a **stationary front.** Airflow generally parallels this type of frontal system.
- If the cold front catches up to the warm front, the warm air is wedged upward, and an **occluded front** forms (■ Figure 16.12).

On opposite sides of a front, the wind direction, air temperature, and absolute humidity reflected by the dew point are all likely to be different. If sufficient moisture is present in either air mass, clouds are likely at the front.

Where a cooler air mass displaces warmer air, a cold front is present. The steep boundary and typically faster speed of a cold front (Figure 16.10) cause it to have briefer but more dramatic precipitation, with *cumulonimbus* clouds, possible thunderstorm activity, heavy showers, and gusty winds. A cold front occurs in a trough of low pressure so that the pressure drops as the front approaches and rises behind it. Wind directions shift from southwest ahead of the front to northwest after it passes (in the Northern Hemisphere). Cold fronts and their associated weather tend to pass quickly, in a matter of a few hours, although many factors can influence their movement.

If a warmer air mass pushes into a region that a cooler air mass occupies, a warm front marks the boundary. Because of the broad wedge shape of a warm front (Figure 16.11) and its slower movement (on average about half as fast as a cold front), the associated precipitation tends to spread out and last longer as steady, sometimes drizzly, rain or light snow. Clouds are likely to be flattish *stratus* varieties (Figure 16.8c, d, and e) that get lower and lower as the front approaches. Winds change direction as the front passes, from southeast to southwest (in the Northern Hemisphere). The clouds from a typical warm front are likely to cover a location for about 2½ days as the front approaches and passes.

One helpful observation in weather prediction is to pay attention to clouds that are forming—their type

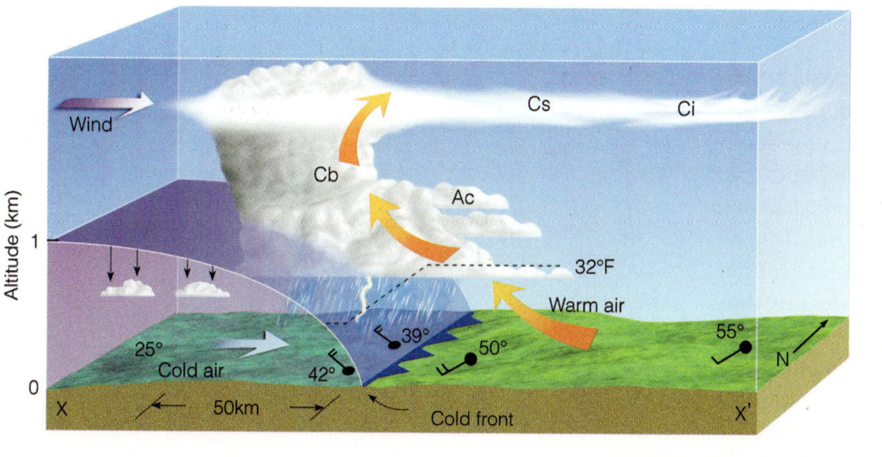

Figure 16.10

Cold Front Block Diagram

The cold air mass on the left moves in under the warm air mass on the right. The cold front is the boundary between them. Cumulonimbus clouds, showers, and heavy precipitation develop. Winds (special wind symbols in black) shift from SW to NW as the front passes. Notice that the wind directions do not correspond to air mass movements. Cloud abbreviations correspond to the cloud discussion in the text. From AHRENS. *Meteorology Today*, 9E. 2009 Brooks/Cole, a part of Cengage Learning, Inc. Reproduced by permission. www.cengage.com/permissions

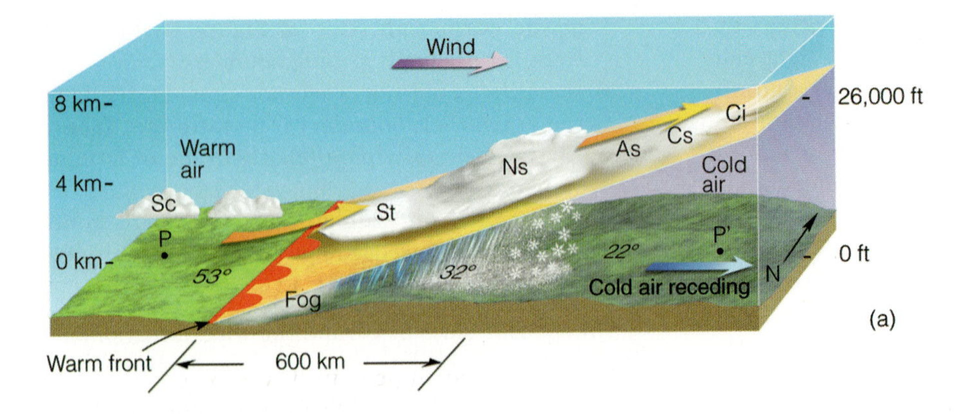

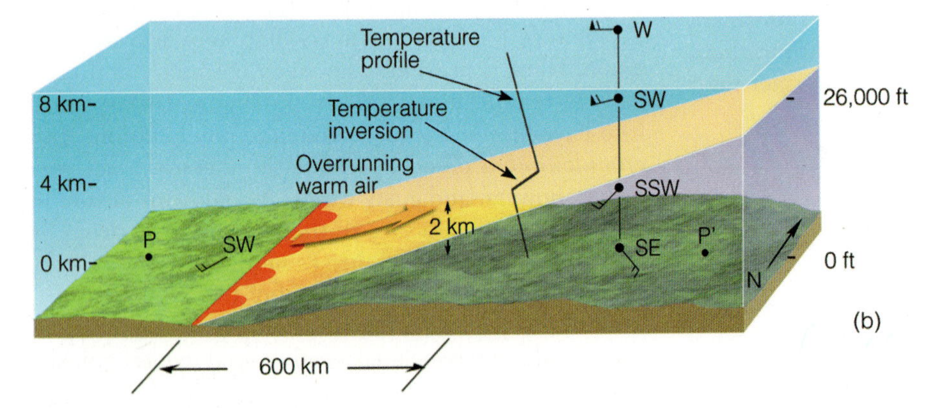

Figure 16.11

Warm Front Block Diagram

The warm air mass on the left is moving up over and replacing the cold air mass on the right. (a) Stratus clouds predominate, descending to lower and lower levels as the front approaches. Light, widespread, steady and warming precipitation is likely. Warm front precipitation covers a wider area (600 km) than that of cold fronts. (b) Same warm front with clouds removed to show other details. Temperature inversion (with warmer air above colder air) makes the air stable preventing vertical development of clouds. Winds change from SE to SW as the front passes. Notice that the wind directions do not correspond to air mass movements. From AHRENS. *Meteorology Today,* 9E. 2009 Brooks/Cole, a part of Cengage Learning, Inc. Reproduced by permission. www.cengage.com/permissions

often indicates the weather to come (Figures 16.10 and 16.11).

- Cold front clouds tend to have cumulus forms near the front. Ahead of the front, *cirrus* and then *cirrostratus clouds* give warning of its approach (Figure 16.10). Then altocumulus, *cumulus congestus,* and *cumulonimbus clouds,* with *cumulus humilis* following the front. Frequently, thunderstorms accompany cold fronts.
- Warm front clouds commonly progress from *cirrus* well ahead of the front (1,200 km or 750 mi), to *cirrostratus,* to *altostratus,* to *nimbostratus,* to *stratus,* and perhaps fog right at the front. A few *stratocumulus clouds* and some fog occur behind the front.

At a *stationary front,* the usually light (or absent) precipitation and weather disturbance may continue for some time in the same location. A stationary front may gradually develop into a combination warm front ahead and cold front behind in an area of low pressure (Figure 16.12) and eventually into an occluded front. At an *occluded front,* people on the ground feel two cooler air masses—either cool followed by cold or cold followed by cool—but they do not experience the warmer air mass because it has lifted off the ground (Figure 16.12c). The weather patterns at an occluded front tend to be similar to a warm front or to a warm front with a cold front at the tail end.

10. If it is a cloudy or partly cloudy day, go outside and observe and identify the clouds you see. Refer to Figures 16.7, 16.8 and 16.9 to classify the clouds. Write down your observations in Table 16.3. Observe and record as many of the other weather variables in the table as you have equipment to measure them. From the clouds and other information you observe, what type of front, weather system, or pattern might be passing through, and where/when are you within that system or pattern?

____

____

____

____

11. Observe the clouds and weather for the next few days to support or disprove this hypothesis. Write down your additional observations in Table 16.4 and evaluate them and summarize your conclusion on a separate sheet of paper.

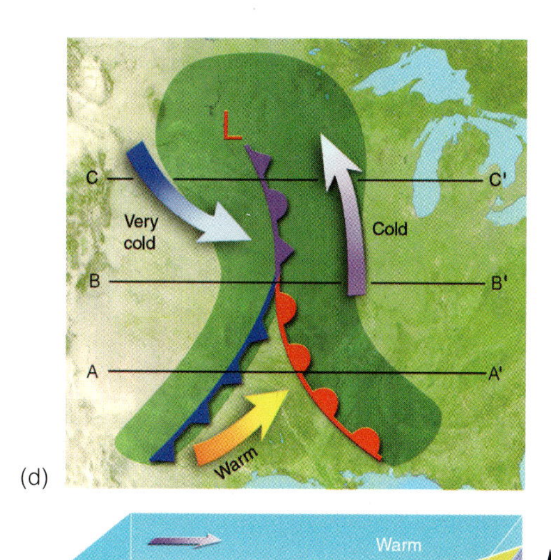

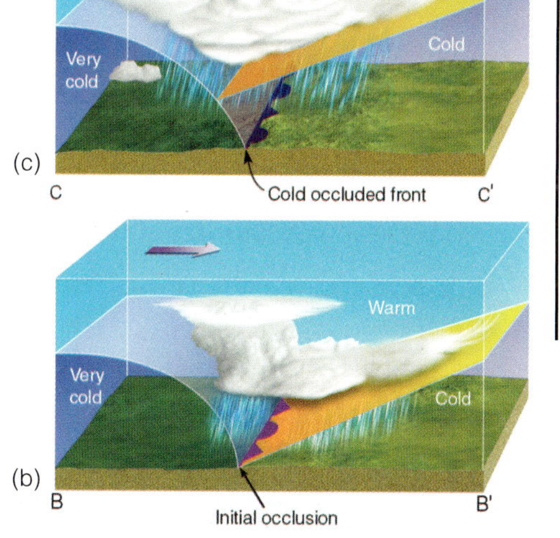

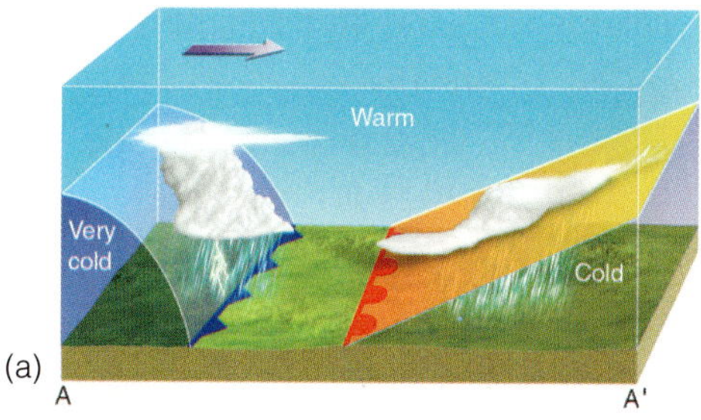

Figure 16.12

Occluded Front Block Diagram and Map

Three air masses and converging cold and warm fronts produce an occluded front. Cold air in the east (right) is first replaced with a warm air mass (middle, above), then a colder air mass in the west (left) moves in more rapidly. We can view (a), (b), and (c) as a time sequence from bottom to top—first (a), then (b), then (c)—or as cross sections through different places on the map in (d). (a) A warm air mass moves in behind a cold air mass; with another cold air mass behind (A–A′ in d). (b) Initial occlusion with the warm air mass above the two colder, just touching, air masses (B–B′ in d). (c) The warm air wedges off the ground, producing the occluded front (C–C′ in d). (d) A map showing an occluded front near the low pressure (L) of an extratropical cyclone, and separate cold and warm fronts to the south. From AHRENS. *Meteorology Today*, 9E. 2009 Brooks/Cole, a part of Cengage Learning, Inc. Reproduced by permission. www.cengage.com/permissions

Weather Maps

Weather maps are summations of the main weather features for a given area at a given time period (■ Figures 16.13 to 16.16). These features are listed in Figure 16.14 and may also include high- and low-pressure indicators and **isobars**—lines of equal pressure. Weather maps use various symbols for precipitation but will generally give a key showing the meaning of the symbols used. Small clusters of symbols will be placed around the location of weather recording stations. If the map is in color, different types of precipitation may appear in different colors, as in ■ Figure 16.15. Alternatively, many weather maps use pictorial representations of precipitation such as raindrops for rain and snowflakes for snow. These maps may use color to represent temperature.

A weather map generally indicates atmospheric pressure in some way, commonly with ***H*** for a region of high pressure and **L** for a region of low pressure. *Isobars*, lines of equal pressure, are commonly fine lines on the map just as contours are lines of equal elevation. Where isobars are close together, there is a rapid change in pressure that leads to high winds. Where they are far apart, the region is calm or has light winds. At the surface, the winds cross the isobars at a slight angle, heading away from highs and toward lows. Recall that we learned about the reason for this in the section on the Coriolis effect in Lab 15. The symbols for an individual weather station may show the **pressure tendency**, which is the change in pressure over the last 3 hours.

12. Study the map in Figure 16.13 and the map symbols in Figure 16.14.

a. Which types of precipitation occur somewhere on the weather map in Figure 16.13? What colors and symbols represent each of these types of precipitation?

[3] – Thunderstorm
- Rain shower
•• – light Rain

b. What type of front occurs on this weather map? Cold front

c. What is the range of air temperatures ahead of the front (in the southeast)?

59

Behind the front (in the northwest)?

34

Table 16.3

Weather Observations

Date and Time	Identified Clouds and Their Probable Altitudes	Temperature (°C)[1]	Humidity or Dew Point	Precipitation	Wind Speed and Direction	Pressure (mb) and Pressure Trend[2]

[1]To convert from Fahrenheit to Celsius, first subtract 32 then multiply by 5/9: Temperature in °C = (°F − 32) × 5/9.
[2]Rising, falling, or steady.

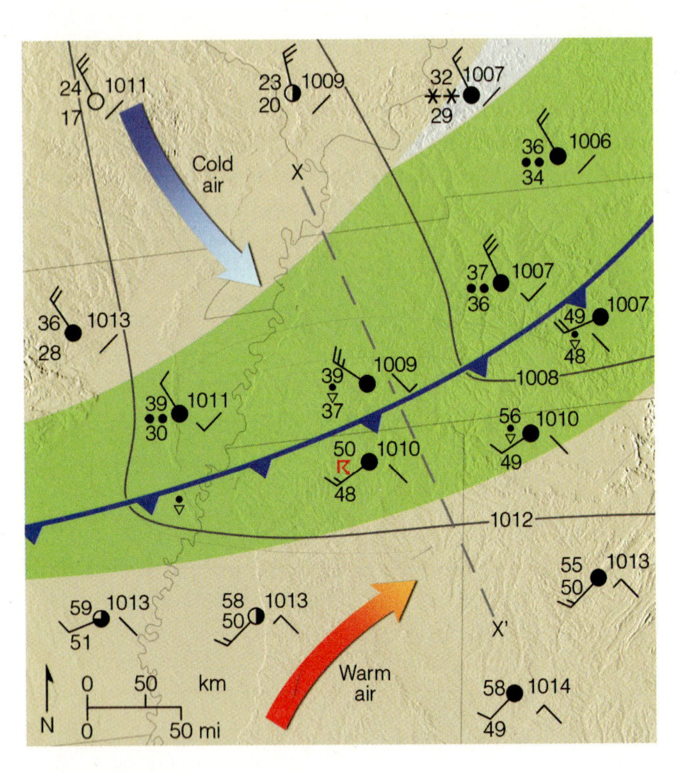

Figure 16.13

Weather Map

The cold air mass in the northwest is moving into and replacing the warm air mass in the southeast. Refer to Figure 16.14 for the meanings of the map symbols. From AHRENS. *Meteorology Today*, 9E. 2009 Brooks/Cole, a part of Cengage Learning, Inc. Reproduced by permission. www.cengage.com/permissions

d. What is the range of air pressures for the 4 stations closest to the front?

steady or rising What is the highest air pressure on the map?

Rising steady

e. What is the pressure tendency at the farthest west station where light rain is falling?

Rising unsteadly

f. In what general direction is the front moving? west

What is the wind direction ahead of the front? NW

What is the wind direction behind the front (in the northwest)?

NW What is the fastest wind speed on the map in knots? 29-37 and in miles per hour? 51-60

g. What are the air temperature 15°C and dew point 10.9°C at the station with a thunderstorm? Use the graph in Figure 16.4 (p. 371) to determine the relative humidity here. 44

h. What are the air temperature 58 and dew point 50 at the station with the lowest relative humidity? Use the graph in Figure 16.4 to determine the relative humidity here. 60

i. What is the relative humidity where the sky is clear? 60 How much

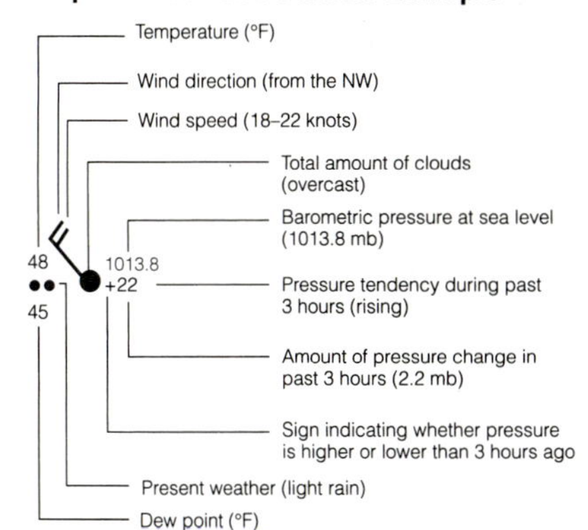

Wind Symbols

WINDS	MILES (STATUTE) PER HOUR	KNOTS	KILOMETERS PER HOUR
	Calm	Calm	Calm
	1–2	1–2	1–3
	3–8	3–7	4–13
	9–14	8–12	14–19
	15–20	13–17	20–32
	21–25	18–22	33–40
	26–31	23–27	41–50
	32–37	28–32	51–60
	38–43	33–37	61–69
	44–49	38–42	70–79
	50–54	43–47	80–87
	55–60	48–52	88–96
	61–66	53–57	97–106
	67–71	58–62	107–114
	72–77	63–67	115–124
	78–83	68–72	125–134
	84–89	73–77	135–143
	119–123	103–107	144–198

Figure 16.14

Weather Map Symbols

These symbols are standardized among the weather maps that meteorologists use. Often the barometric pressure is reported in a short form by leaving off the 10 or 9 in front and leaving out the decimal (138 would signify 1013.8). From AHRENS. *Meteorology Today,* 9E. 2009 Brooks/Cole, a part of Cengage Learning, Inc. Reproduced by permission. www.cengage.com/permissions

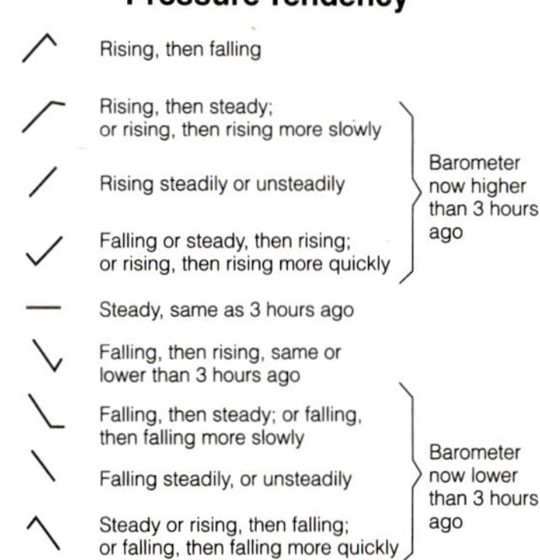

Cloud Coverage

- Clear
- 1/8
- Scattered
- 3/8
- 4/8
- 5/8
- Broken
- 7/8
- Overcast
- Obscured
- Missing

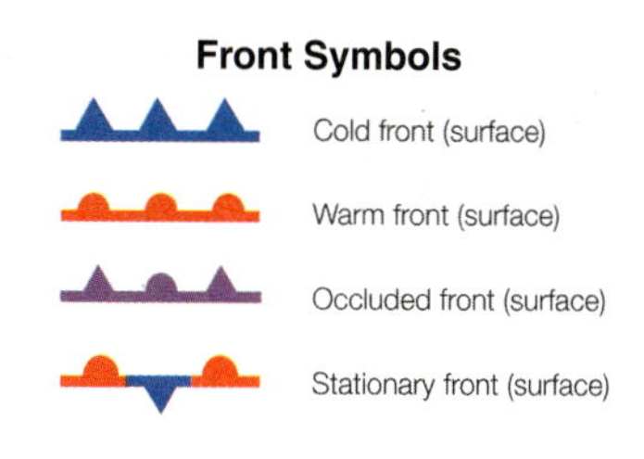

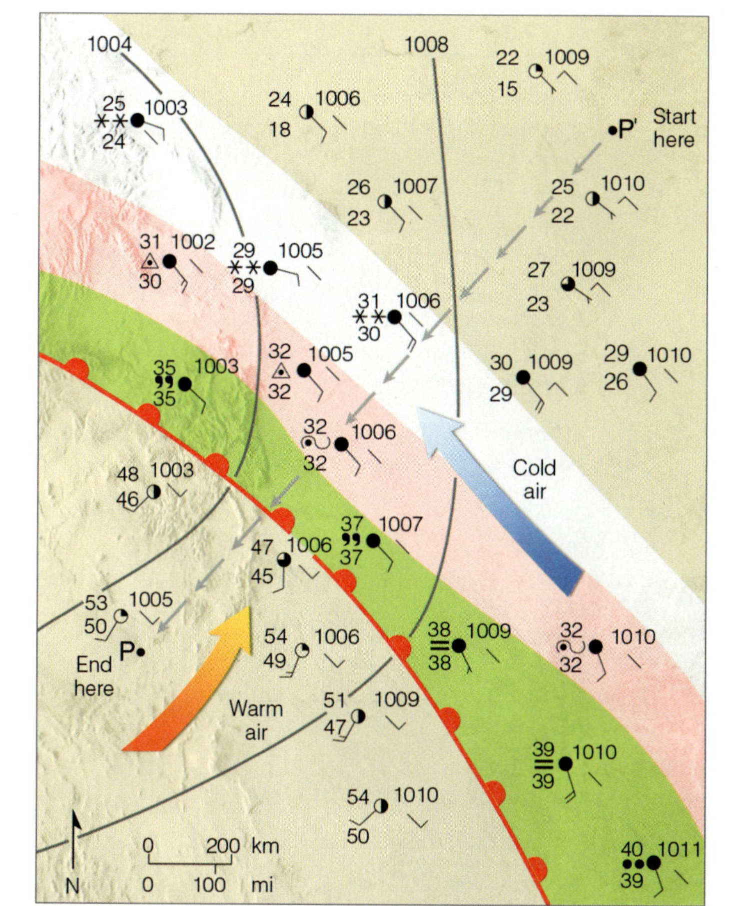

Figure 16.15

Second Weather Map

The warm air mass in the southwest is moving over and replacing the cold air mass in the northeast. Refer to Figure 16.14 for the meanings of the map symbols. From AHRENS. *Meteorology Today,* 9E. 2009 Brooks/Cole, a part of Cengage Learning, Inc. Reproduced by permission. www.cengage.com/permissions

cloud cover is present at the station directly to the east of this location?

over coast How far behind the front is the closest station with partly cloudy conditions? 60 mi

j. From what you know about clouds and fronts, what type of clouds are likely to be present at the thunderstorm location?

comulonimbus Ahead of the front, what middle clouds do you think are present? nimbus stratus

13. Study the map in Figure 16.15 and the map symbols in Figure 16.14.

a. Which types of precipitation occur on the weather map in Figure 16.15? What colors and symbols represent these types of precipitation?

____

____

____

____

____

b. What type of front occurs on this weather map? ________

c. What is the range of air temperatures ahead (northeast) of the front?

____

Behind (southwest of) the front?

____

d. In what general direction is the front moving? ________

e. What is the wind direction ahead of the front? ________
What is the wind direction behind the front? ________
What is the fastest wind speed on the map in knots? ________ In miles per hour? ________

f. What is the range of air pressures for the five stations closest to the front?

____

g. What is the highest air pressure on the map? ________

What is the pressure tendency where the sky is clearing behind the front?

____

h. What is the pressure tendency ________ and the relative humidity ________ at the station with freezing rain that is centrally located on the map?

i. What is the air temperature ________ and the dew point ________ at the station with the lowest relative humidity? What is the relative humidity here? ________

j. What is the relative humidity at the most northerly station on the map? ________

k. From what you know about clouds and fronts, what type of clouds are likely at the drizzly-rain locations?

____

Ahead of the front, what middle clouds do you think are present?

____

l. Use the scales on the maps to measure the width of the areas of precipitation for both fronts in Figures 16.13 ________ and 16.15 ________. How do they compare numerically?

____

14. Study the map in ■ Figure 16.16, its key, and its caption. Also use Figure 16.3 on page 369.

a. Use the temperature and dew point values and other symbols on the map to help you fill in the blanks below with one of the words: *the highest / intermediate / the lowest*:

Air east of the fronts (mP) has intermediate temperature and intermediate absolute humidity.

Air south of the fronts (mT) has highest

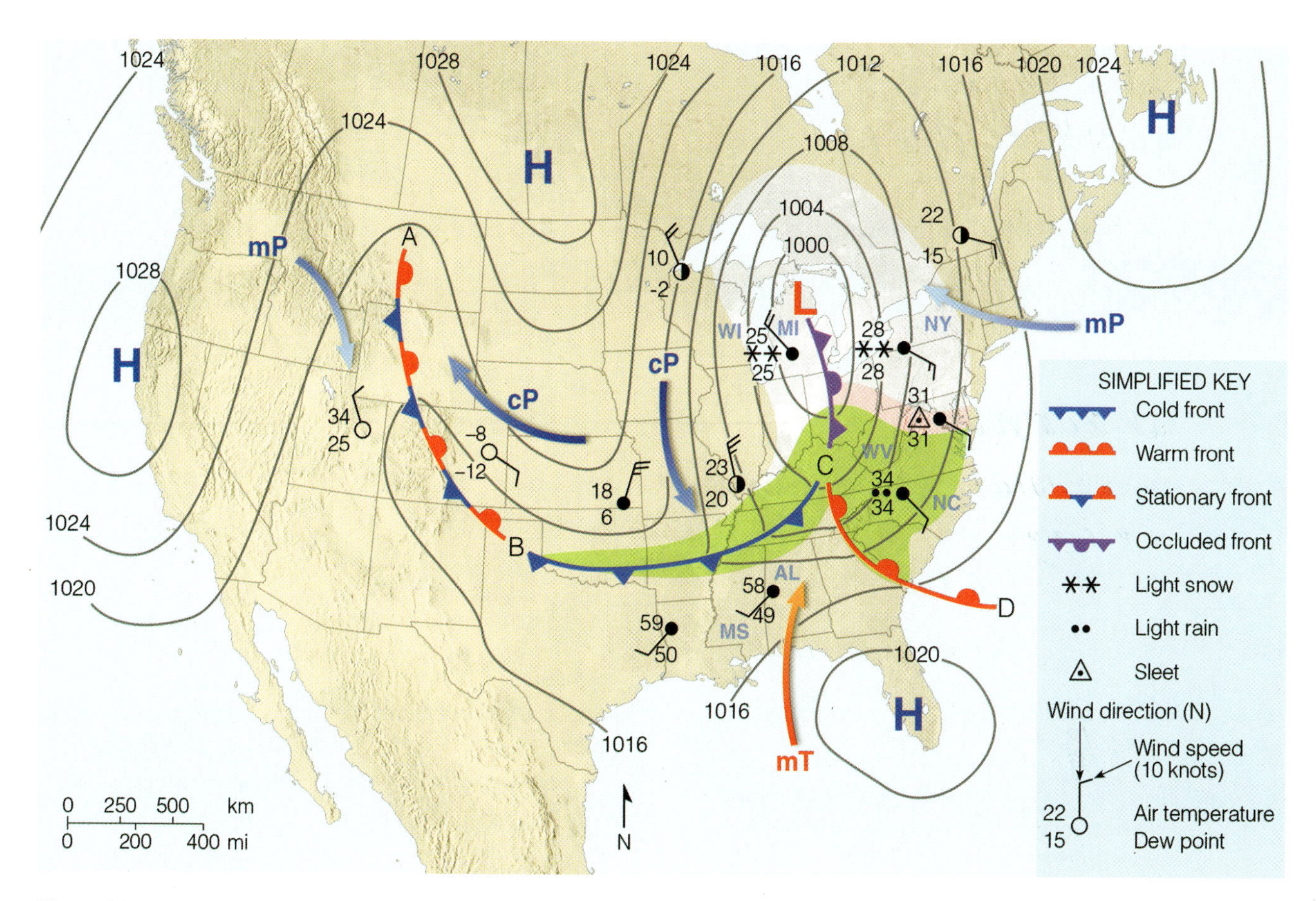

Figure 16.16

Winter Weather Map for the Continental United States

Showing surface pressure isobars in millibars and air masses. m = maritime (moist); c = continental (dry); T = tropical (warm); P = polar (cold); mP = maritime polar; cP = continental polar; mT = maritime tropical. Colored arrows show airflow directions. Green, pink, and white shading indicate areas of precipitation. From AHRENS. *Meteorology Today*, 9E. 2009 Brooks/Cole, a part of Cengage Learning, Inc. Reproduced by permission. www.cengage.com/permissions

temperature and

___highest___ absolute humidity.

Air in the north-central part of the map (cP) has ___lowest___ temperature and

___lowest___ absolute humidity.

b. What type of front occurs at the low pressure?

___cold front___

c. Describe the weather in western New York.

Temperature? ___28___

Relative humidity? ___28___

Absolute humidity (refer to Figure 16.3)?

___100___

Type and style of precipitation if any?

___N/A___

Pressure (use the isobars)? ___1008___

Assume the low in Figure 16.16 is moving eastward. Is the pressure rising or falling in western New York?

___rising___

What is the wind speed? ___13-17k___

Amount of cloud cover if any?

___over coast___

Likely cloud types if any? white clouds

d. Describe the weather in northern Wisconsin.

Temperature? -2°

Type and style of precipitation if any? none

Pressure? 1010 Is it rising or falling? rising

Wind speed? 10 knots

Amount of cloud cover if any? light

Likely cloud types if any? white

e. How do you expect the weather to change within the next day or so (assume the low in Figure 16.16 is moving eastward, the cold front is advancing toward the southeast, and the warm front is advancing toward the northeast) at each of the following locations? Simply answer *increase, decrease,* or *stay the same*:

Temperature change in North Carolina? increase

Temperature change in northern Alabama? decrease

Pressure change in West Virginia? increase

Pressure change in Michigan? stay the same

f. How do you expect the weather to change within the next day or so, using the same assumptions as in the previous question? This time describe the change from what it is presently to what it will be next:

Wind direction in North Carolina?

Present: SE

Next: NW

Wind direction in northern Alabama?

Present: SW

Next: NW

Clouds in central Mississippi?

Present: over cast

Next: clear

Clouds in North Carolina?

Present: over cast

Next: over cast

g. What fourth type of front is shown in Figure 16.16? cold front

What types of clouds and precipitation are associated with this front? over cast

15. Figure 16.17 is a map showing scattered pressure readings in millibars in and around the United States.

a. Draw isobars on the map using a 5-mb isobar interval. Locate and label the lows (L) and the highs (H).

b. Draw cold front symbols for any pressure troughs (which would look like valleys if this were a topographic contour map).

c. Where are the winds the fastest? Write *windy* there on the map. Where is it least windy? Write *calm* there on the map.

d. Use a green pencil to lightly shade in where you think precipitation is taking place.

e. Examine wind directions around the lows and highs and across contours in Figures 16.13, 16.15, and 16.16 to get a feel for the relationship between isobars and wind direction in the Northern Hemisphere. Remember that wind blows from high to low pressure, with a Coriolis turn to the right in the Northern Hemisphere. Then use your isobars in Figure 16.17 to help you draw wind arrows on the map. Before you start, you may want to review the discussion of isobars and wind in the section above.

16. Using a weather map for today from the newspaper or from an Internet site (your instructor will assist you with this):

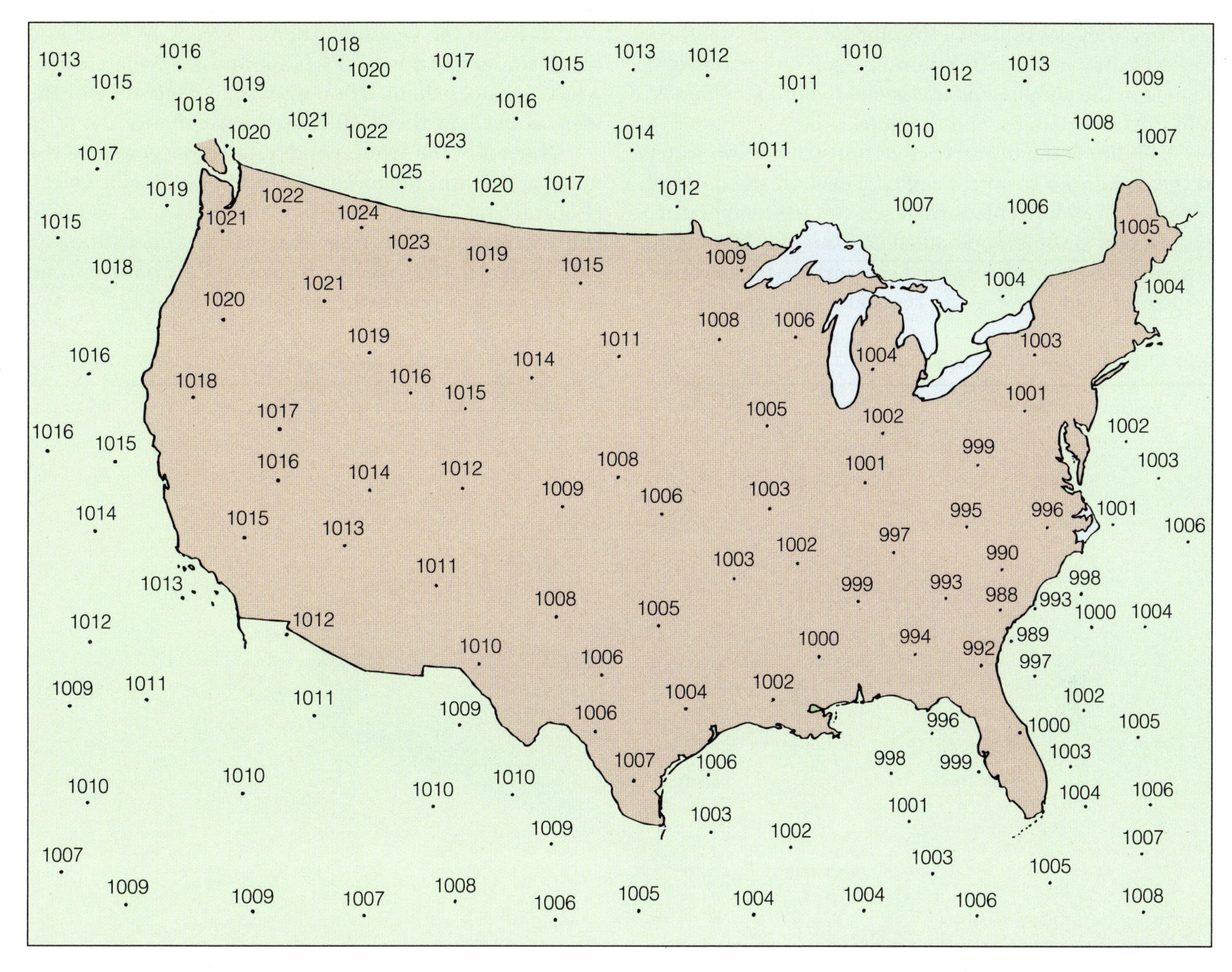

Figure 16.17

Barometric Pressure Readings across Continental United States

Values are in millibars adjusted to sea level.

a. Predict what the weather will be like tomorrow, without looking at the weather forecast.

____

____

____

____

b. Tomorrow, come back to this and evaluate your prediction.

____

____

____

____

The patterns of warm fronts and cold fronts that cross the United States and other parts of the world are strongly influenced by seasons, of course, but also by longer-term, temporary climate fluctuations such as El Niño and La Niña. Let's look at those fluctuations to see what they are, how they originate, and how they influence the weather.

EL NIÑO—LA NIÑA

The normal patterns of air pressure, wind, and ocean currents sometimes fluctuate, causing sea-surface-temperature variations such as El Niño, which influence worldwide weather patterns. Discussions in the news and science media on this subject have become so prevalent that the person on the street may blame unusual weather on El Niño or La Niña. While El Niño and La Niña are not weather phenomena or climate patterns, they do influence the climate and the weather. In fact, they cause a temporary 2 to 7 year cyclical change in the climate and are linked to droughts, floods, wild fires, hurricanes, and tornados.

El Niño is the occurrence of unusually warm water in the equatorial eastern Pacific Ocean, off the coast of Peru and Ecuador. The **normal** sea-surface-temperature pattern is also sometimes called *La Nada*. **La Niña** is an extreme version of the normal sea-surface-temperature pattern of cold water along the same coast, but is also disruptive to the weather when it occurs. We will first try to understand what El Niño and La Niña are and what produces them. Then we will look at some of the ideas about how they influence the weather.

Normally the trade winds blow and generate the north and south *equatorial currents* in the Pacific Ocean (Figure 14.17 on p. 332) that pile up warm water in the western Pacific near Australia, and Indonesia (■ Figure 16.18a). Water from the Peru Current, an *eastern boundary current* (Lab 14 and Figure 14.17), and

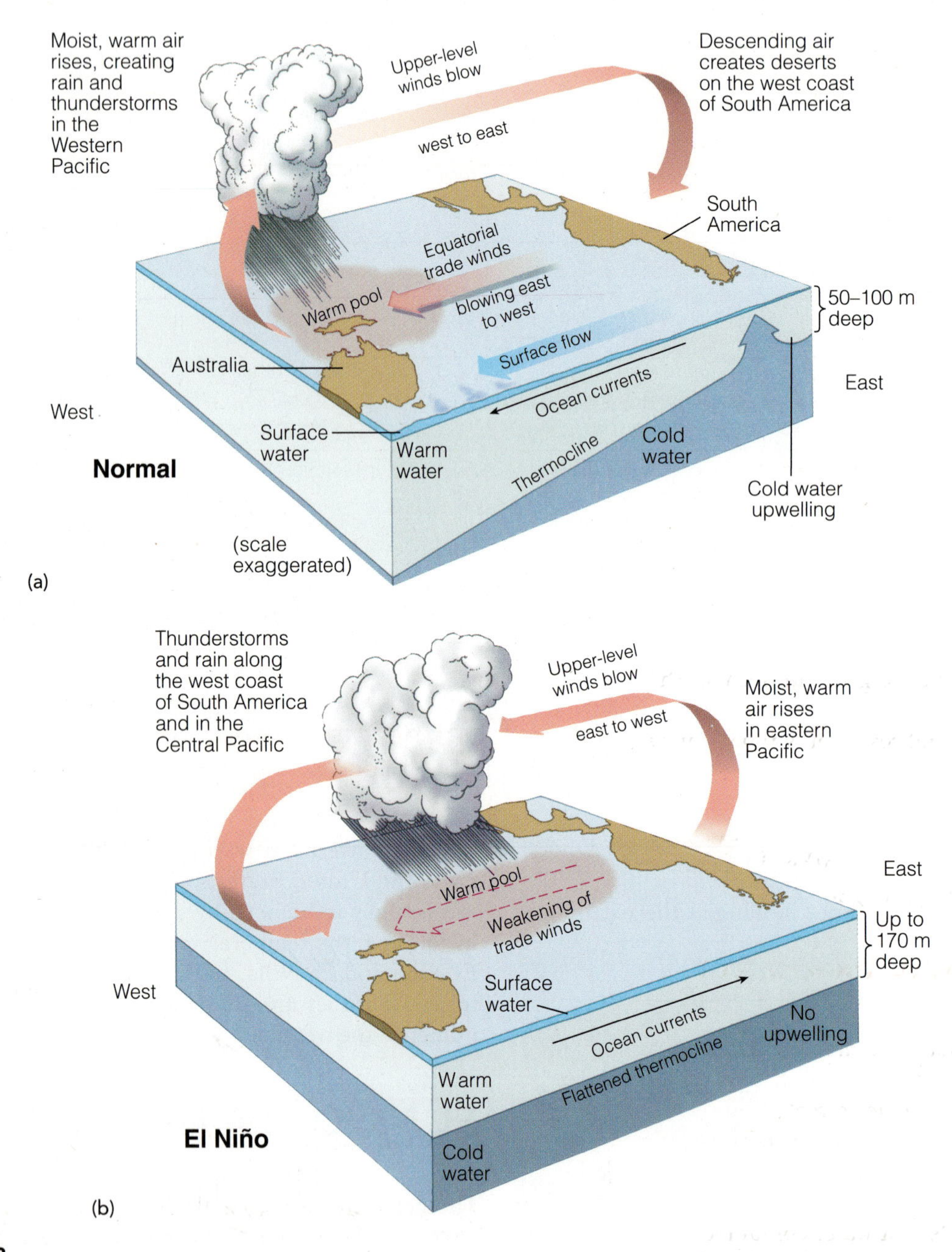

Figure 16.18

Normal and El Niño Conditions

(a) In normal conditions, the trade winds blow strongly, forcing warm water westward and causing a warm moist climate in the western Pacific. Cooler upwelling waters in the eastern Pacific normally produce a dry climate there. The **thermocline** in the ocean occurs where seawater temperatures change considerably from warm water above to cold water below. (b) During El Niño conditions, the trade winds are weak, the warm pool moves east along with clouds and unusually heavy rains. Upwelling ceases in the eastern Pacific.

some from *upwelling* replace the water that the wind moved away from South America. The **upwelling** water, which is colder, deep water rising from below, brings nutrients to the surface. The nutrient-rich waters are beneficial for fish, making fishing normally excellent in Peru. Trade winds, upwelling, and cold water are normal features of eastern Pacific circulation and **normal conditions.**

El Niño (Figure 16.18b) occurs when the trade winds slacken, the *equatorial currents* (Lab 14) slow, and upwelling diminishes. Some of the warm water that piled up in the western Pacific, near Indonesia now has a chance to slosh back across the ocean toward the South American coast. Other weather conditions associated with this change are a rise in air pressure with drought in Indonesia and a drop in air pressure with increased rainfall over the central equatorial Pacific, Ecuador, and northern Peru. Since the upwelling stops and the water warms up, the fish catch plummets. Because the warming water and declining fish were first noticed in December, Peruvian fishermen named it after the Christ Child—El Niño. The warmer water is what powers the rising air that condenses bringing increased rainfall to the eastern Pacific and South American west coast. It also creates a rising bubble of air at higher altitude, displacing the subtropical jet stream. A **jet stream** is a very fast westerly (eastward flowing) wind at the boundary between the troposphere and the stratosphere. Movement of jet streams is associated with changing weather.

In contrast to El Niño, La Niña is an extreme case of the normal sea-surface temperatures. When the trade winds blow more strongly than normal, the equatorial currents are stronger, piling more warm water up in the western Pacific, and results in stronger upwelling in the eastern Pacific during La Niña. La Niña makes the normally low-pressure and wet weather in Indonesia even wetter and brings flooding there. It makes the normally high-pressure and dry weather in Ecuador and Peru even drier, causing drought.

El Niño/La Niña Experiment

The El Niño/La Niña experiment that follows will help you to understand the conditions of El Niño and La Niña. As you work through this experiment, keep in mind how these changes affect air masses and weather above large bodies of warm or cool water.

Materials needed:

- Clear, wide, flat water container
- Food coloring: yellow and blue
- Salty water in ice bath with about 7% salt (about 7 grams of salt for each 100 ml of water) in a quantity enough to fill the flat water container to about half an inch deep.
- Stirring rod
- Hot plate or hot water
- Two beakers
- Thermometer
- Straw
- White-background cloth or paper

17. *Set up:* First read the whole experiment so you know what to do. To set up for the experiment you will need some information from your instructor.

a. Find out from your instructor how much

salty water you will need: ________ and how many drops of food coloring you will need for your two solutions so that

the yellow is fairly intense: ________ drops of yellow; and the blue is distinctly colored but not quite medium or dark:

________ drops of blue.

b. Your instructor may want you to measure and mix the salt for the saltwater solution.

c. Add the specified amount of blue food coloring to the saltwater solution in the ice bath and stir it.

d. Obtain the same quantity of 40°C tap water either directly from a hot water tap or heated with a hot plate. Stir in the specified number of drops of yellow food coloring.

18. *The experiment:* Place the water container on the white background.

a. Pour the warm yellow solution into the container. Prop up one end of the container so the solution drains away from that end. Smoothly pour the blue solution into the container on the propped up end. Smoothly lower the propped up end.

b. If you do this correctly, you should have two layers of liquid, one warmer and the other cooler; but the fact that you also have a blue layer and a yellow layer may not be immediately apparent, so don't give up.

c. Quickly measure the temperature of each layer. You will want to submerge the bulb of your thermometer within the layer, so tilt it sideways so it remains mainly in that layer:

top: ______ bottom: ______

d. Now simulate normal conditions (or La Niña). Without touching the water with the straw, blow across the top of the two liquid layers. This represents the trade winds in the Pacific Ocean. What do you observe in the water?

____

____

e. If you look closely, you may be able to see a color difference between the top and the bottom. Which color is on top?

__________ Why is one layer on top and the other on the bottom? Why didn't you just get completely mixed water?

____

____

f. While you were blowing, you should have observed a patch of differently colored water; where did it come from?

______________________ What is the term used for this movement in the ocean? ______________

g. From your measurements of water temperature, what is the water temperature probably like in this patch compared to the surrounding water? __________
How would this affect the air mass directly above it if it were a large body of water?

____

How would the air pressure change?

____

How would rainfall change?

____

h. Circle the term that correctly completes the sentence. This would be analogous to (normal/El Niño/La Niña) conditions in the ocean.

i. When you stopped blowing, what happened to the patch of water?

____

j. How would you expect the water temperature would change as the patch goes away and the other water comes back?

____

____

How would this affect the weather conditions such as rising versus sinking air, air pressure, and rainfall if this was a large body of water?

____

____

k. Circle the term that correctly completes the sentence. The flowing back of other water would be analogous to (normal/El Niño/La Niña) conditions in the ocean.

Fish like the cool upwelling water because it is high in nutrients and near the surface where phytoplankton (marine photosynthesizers) can grow to feed the fish. When upwelling stops, the fishing is bad, for both sea birds and Peruvian fishermen.

l. If your experiment were actually the ocean, where and when would you go fishing?

____

____

19. If you were a National Oceanographic and Atmospheric Administration (NOAA) scientist:

a. What and where would you measure to detect El Niño?

Eastern pacific ocean
unusually warm water

b. You would be aware that the trade winds blow strongly during which conditions?

Normal

c. What would you measure to predict that an El Niño was coming?

wind from east to west How would you measure it?

graph the pattern

____

How would the study of weather systems help you predict El Niño or La Niña and vice versa?

movement from
west to east

____

____

El Niño—La Niña Temperature Maps

NOAA scientists have installed buoys in the equatorial Pacific Ocean to measure atmospheric and oceanic conditions. Seventy buoys are arranged along lines of longitude spanning 8,000 mi of the equatorial Pacific Ocean in a grid or array called the Tropical Atmosphere-Ocean (TAO) Buoy Array. The map in ■ Figure 16.19a shows ocean surface temperatures for normal conditions such as existed in January 1993, measured by the buoy array of NOAA's TAO Project in the Pacific Ocean.

20. Enhance the readability of the water temperatures on the map in Figure 16.19a by coloring them with 2°C color intervals as described next.

a. Select six gradational colors from cool colors (purples, blues, and greens) to warm colors (yellows, oranges, reds) to represent water temperatures for every 2°C from 20° to 32°C. Use light-colored pencils when you fill in the boxes in the key below Figure 16.19b.

b. Use the same color scheme for each of the following maps. Do not color the round even numbers (e.g., 26.0) yet. Start by coloring all dots next to the numbers within one temperature range but not including the end members of the range; so, for example, color in yellow for numbers from 26.1 to 27.9.

c. Color the area both east-west and north-south between any two adjacent numbers within the same range. Switch to another color and do the same for the new range, until all but the round even numbers (e.g., 22.0, 24.0, 26.0) have been colored.

d. Interpolate (estimate the number) between any two locations that have different colors and mark a dot where the round even number should be. For example, if you find 25.6 and 26.5, the 26.0 dot should fall 4/9 from 25.6 and 5/9 from 26.5; you would place a dot there.

e. The even numbers are the edges where the temperature contour lines should go, so use them to draw in your temperature contours at 2° intervals. *Important hint*: Keep in mind that since temperatures are continuous, large jumps in temperature will still have a continuous temperature change; so, for instance, at a place where 23° is next to 27°, the temperatures 24°, 25°, and 26° must be in between so the colors should be too.

f. Color the two appropriate colors up to your temperature contours. If time permits or as indicated by your instructor, also draw in temperature contours using a 1°C contour interval.

21. Use Figure 16.18 and the map you colored in Figure 16.19a to answer the following questions:

a. What are the most striking features of normal sea-surface temperatures in the equatorial Pacific Ocean?

____

____

____

and the weather conditions in the west and east that accompany them?

____

____

____

b. Where is the water warm and where is it cooler? Explain why the contours from 160°W to 90°W are so much different in pattern than those from 150°E to 160°W.

____

____

____

____

c. What temperature is the surface water near the Galápagos Islands? ______ °C. To convert from degrees Celsius to degrees Fahrenheit, multiply the °C by 9/5 and add 32°F. What is the temperature in Fahrenheit? ______ °F.

22. The next map (Figure 16.19b) shows ocean surface temperatures for El Niño conditions during January 1998.

a. As before, color (and contour) the temperatures using the same colors as in the previous map.

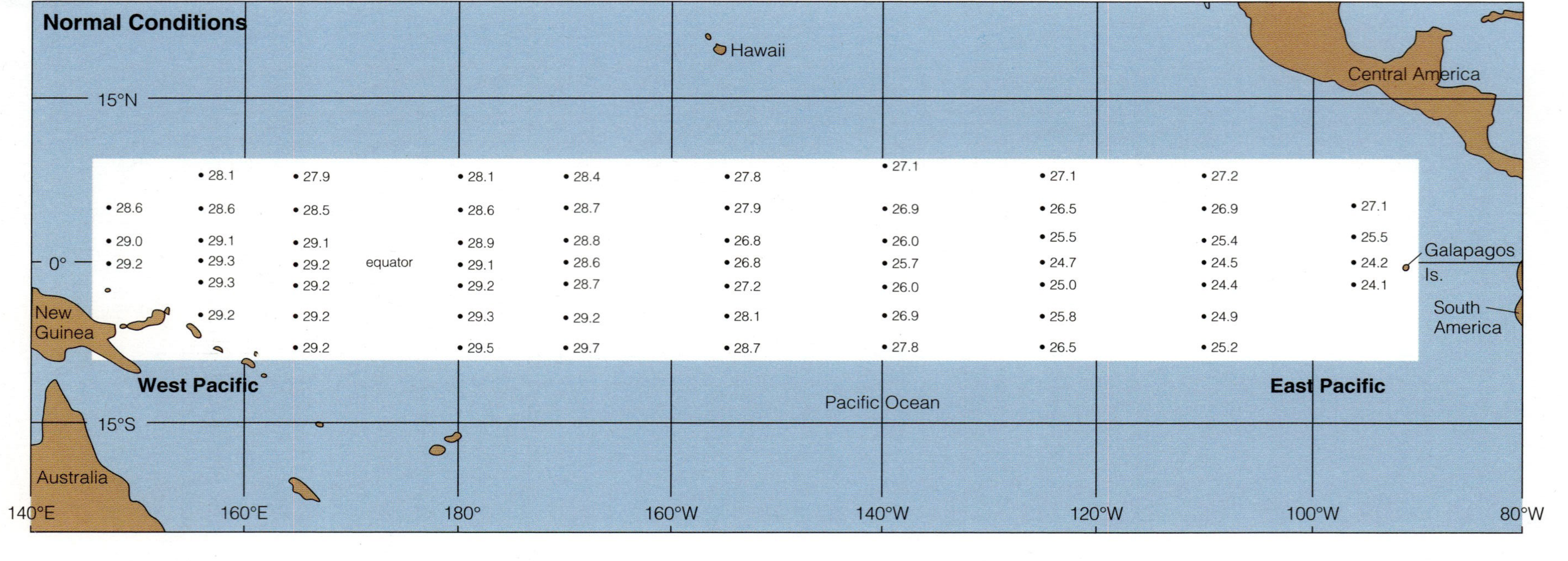

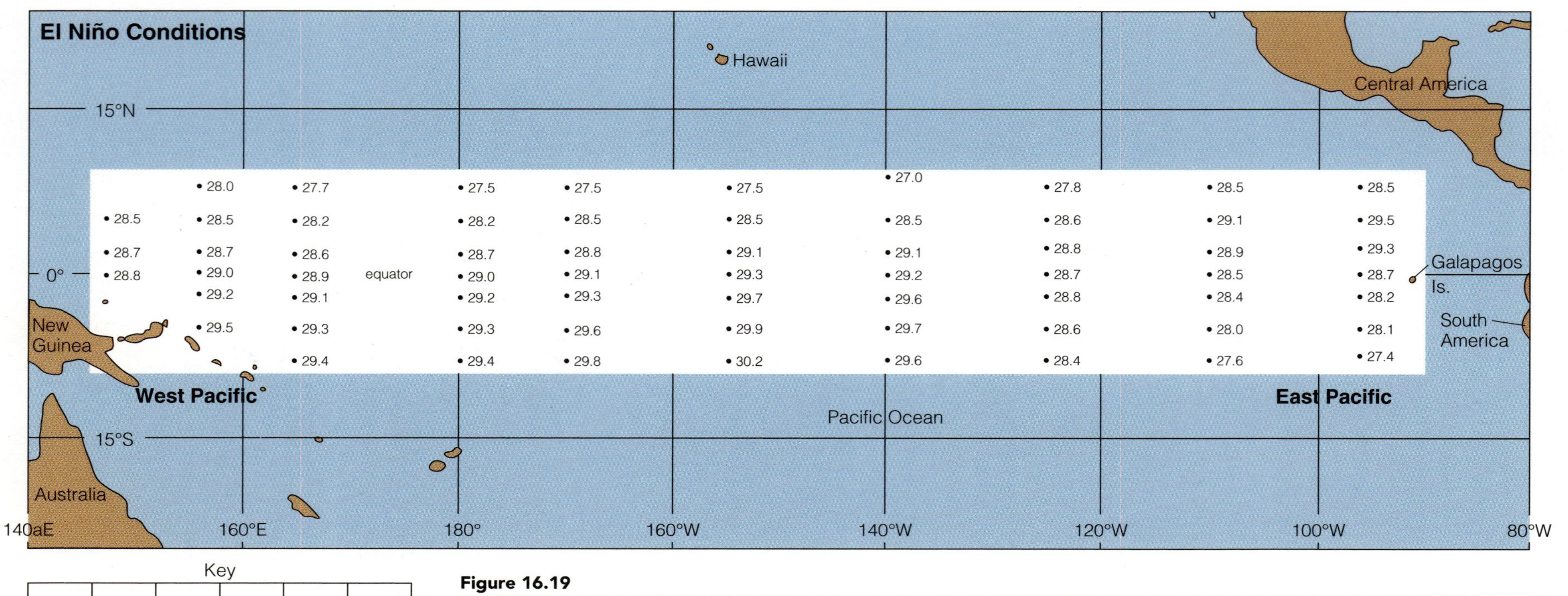

Figure 16.19

Sea Surface Temperatures for Normal and El Niño Conditions

(a) Data from NOAA TAO Array for January 1993 and (b) January 1998.

b. What are the most striking features of El Niño sea-surface temperatures?

____

____

____

and the weather conditions in the west and east (see Figure 16.18) that accompany them? How do they differ from normal conditions?

____

____

____

____

c. What temperature is the surface water near the Galápagos Islands? ______ °C. What is the temperature in Fahrenheit? ______ °F.

Recall that La Niña is similar to normal only more extreme. Now try your hand at determining which conditions exist using water temperature data.

23. The map in ■ Figure 16.20 shows ocean surface temperatures for September 1996.

a. Color (and contour) the temperatures using the same colors as in the previous maps.

b. What temperature is the surface water near the Galápagos Islands? ______ °C. What is the temperature in Fahrenheit? ______ °F

c. What conditions were prevalent at that time?

(a) Normal conditions (b) El Niño conditions (c) La Niña conditions

d. What are the most striking features of these water temperatures and the weather conditions in the west and east that accompany them?

____

____

____

____

How do they differ from normal conditions?

____

____

____

Weather, El Niño, and La Niña

Beside the changes in the seasons, El Niño and La Niña produce the largest short-term, worldwide climate change (Figure 16.21). The energy contained within the warm water of El Niño is equivalent to 500,000 20-megaton hydrogen bombs, and that is enough energy to change the climate. Where the warm water normally occurs, in the western Pacific near Indonesia and Australia, the weather is normally wet. During 1997, warm water that would result in the next El Niño traveled 150 miles each day toward the west coast of South America. Warm water like this creates a mass of warm, moist air above it, which rises and condenses into clouds. The rain usually associated with the warm water migrates eastward with the warm water. Australia and Indonesia experience a drop in seawater temperatures, drought, and an increase in forest fires. Rain falls in the central and eastern Pacific and western South America where normally the weather is dry, causing flooding and mudflows. The warm moist air rises 6–10 mi into the atmosphere, displacing the jet streams that have to flow around it. The deflected jet streams change worldwide weather (Figure 16.22).

La Niña is even more poorly understood than El Niño, but some of its effects are more severe than those of El Niño. For example, it tends to be associated with an increase in strength and frequency of hurricanes in the North Atlantic in the summer.

24. Summarize the weather patterns associated with the El Niño/La Niña phenomenon: fill in ■ Table 16.4 with the correct water temperatures, atmospheric pressures, wind, and rainfall patterns for normal conditions, El Niño, and La Niña. Note that drought signifies unusually dry not just normally dry conditions. Words in parentheses in each column indicate choices.

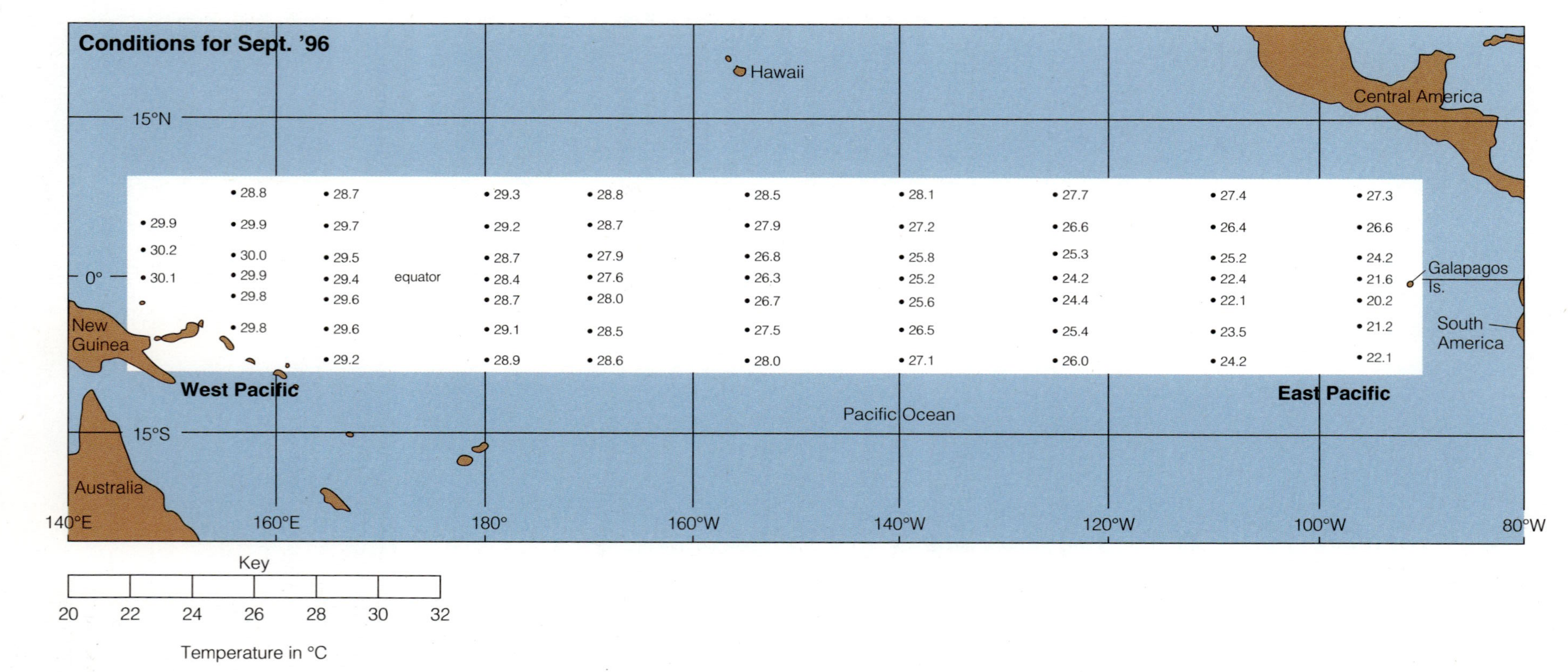

Figure 16.20

Sea Surface Temperatures for Unknown Conditions

Data from NOAA TAO Array for September 1996.

Table 16.4

Summary of Normal, El Niño, and La Niña Conditions

Conditions	Western Equatorial Pacific			Eastern Equatorial Pacific		
	Water Temperature (warm, cool, or cold)	Atmospheric Pressure (high or low)	Rainfall (heavy rain, dry, or drought)	Water Temperature (warm, cool, or cold)	Atmospheric Pressure (high or low)	Rainfall (heavy rain, dry, or drought)
Normal						
El Niño						
La Niña						

25. Examine the weather changes that El Niño causes (■ Figure 16.21) and indicate below the changes relative to normal weather patterns.

a. How does El Niño influence Indonesia on both maps?

Because water goes from east to west

b. How does El Niño influence Ecuador in December–February?

The air temp. drops

c. How does El Niño influence the U.S. South in December–February?

Changes temp & high winds

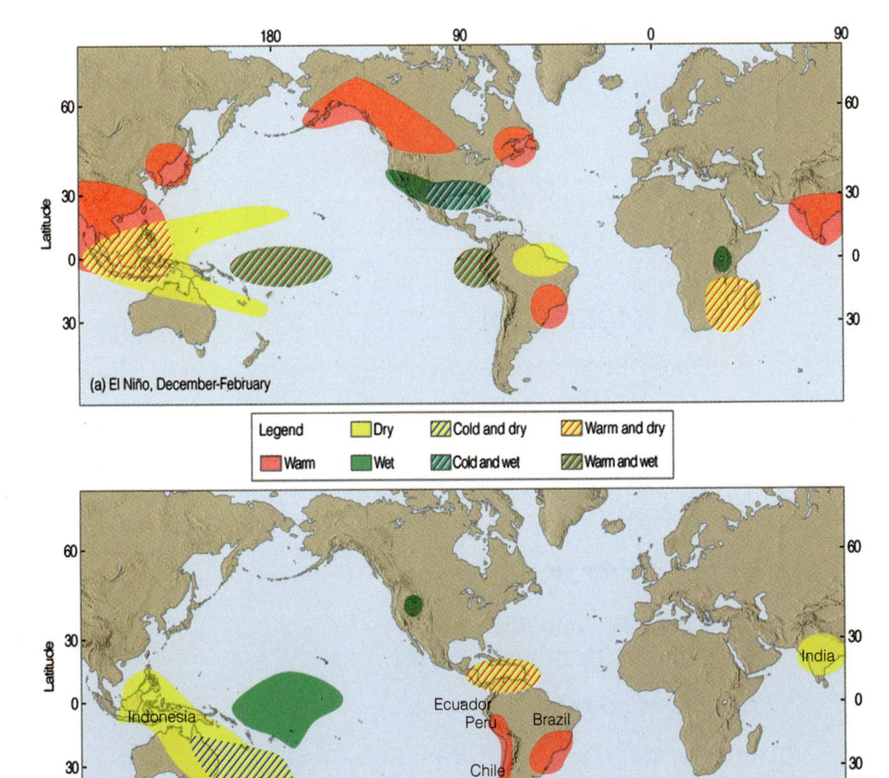

Figure 16.21

Worldwide Climate Abnormalities during El Niño

(a) Northern Hemisphere winter and Southern Hemisphere summer.
(b) Northern Hemisphere summer and Southern Hemisphere winter. From AHRENS. *Meteorology Today*, 9E. 2009 Brooks/Cole, a part of Cengage Learning, Inc. Reproduced by permission. www.cengage.com/permissions

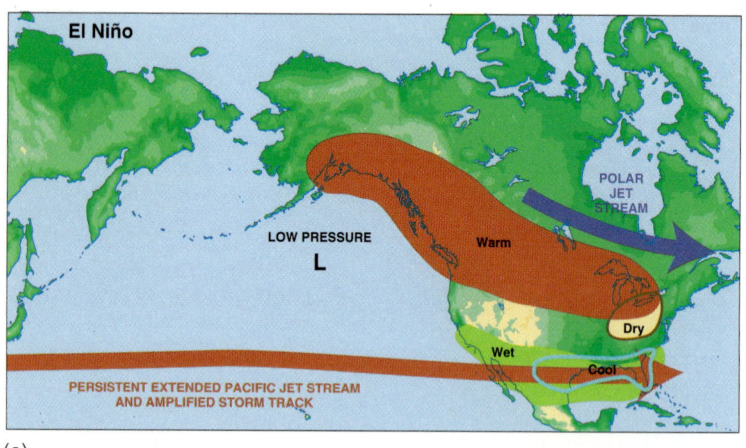

(a)

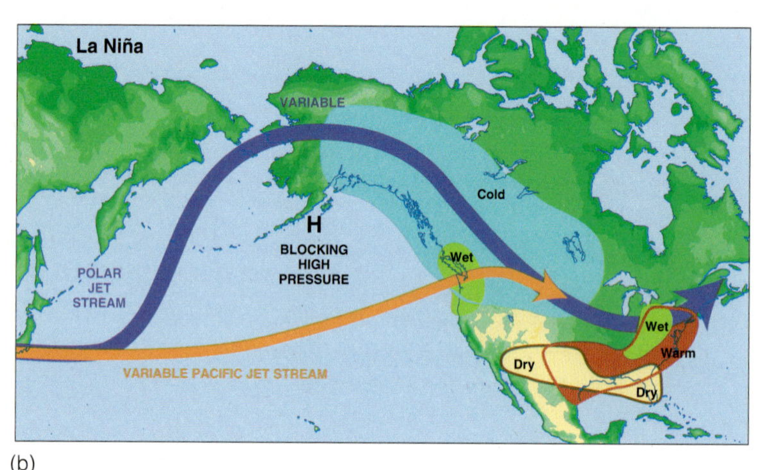

(b)

Figure 16.22

El Niño and La Niña Jet Stream Patterns for North Pacific and North America

(a) El Niño jet stream and resulting weather.
(b) La Niña jet streams and resulting weather. From GARRISON. *Essentials of Oceanography*, 5E. 2009 Brooks/Cole, a part of Cengage Learning, Inc. Reproduced by permission. www.cengage.com/permission

Southwest? warm & dry

North? warm

and the Northwest and New England?

Nothing

d. How does El Niño influence northern Brazil in December–February?

hot

e. How does El Niño influence India and SE Asia in December–February?

warm

f. How does El Niño influence Peru and Chile in June–August?

cold

g. How does El Niño influence India in June–August? dry

What are some potential serious repercussions of this influence?

leaves dry impact

26. Notice the changes in the Pacific and polar jet streams as a result of El Niño and La Niña in ■ Figure 16.22.

a. If the jet stream stays persistently straight across the Pacific as during El Niño, storms will also tend to travel straight across the Pacific and impact Southern California and Baja, Mexico. How does this influence those coastal areas?

warm & dry

b. If the jet stream takes great north and south swings as during La Niña, weather fronts will also tend to travel far south and far north. How does this influence the La Niña weather?

cold

LAB 17

Glaciers and Glaciation

OBJECTIVES

- **To understand the processes of glacier movement**
- **To recognize features of alpine and continental glaciation resulting from erosion and deposition**
- **To determine the direction of ice movement from maps showing oriented glacial features**
- **To understand glacial mass balance and its response to climate change**

In some climates, temperatures in winter are cold enough for abundant snowfall and in summer are sufficiently cool for snow to remain. Where this happens year after year, the snow accumulates and becomes ice that begins to flow, resulting in the formation of a glacier. A **glacier** is a mass of ice, persisting for multiple years, that flows due to its weight (■ Figure 17.1). The ice of a glacier forms when snow compacts and recrystallizes first into **firn** (a substance intermediate between snow and ice) and then into ice. Glacial ice flows when a sufficient thickness develops (about 40 m or more). The weight of the ice that piles up causes it to flow down slope or to spread out. Regions where glaciers form include high latitudes, both north and south, as well as high elevations. Even near the equator, very high elevations may be cold enough for ice and snow to form a glacier. In this lab, we will explore glaciers, the landforms that they produce, and the influence of climate change on glaciers.

GLACIER TYPES AND THEIR MOVEMENT

Typically, a glacier consists of two parts (■ Figure 17.2). The highest part of the glacier is the **zone of accumulation,** where the amount of snow, firn, and ice that accumulates is more than the amount that melts each year (Figure 17.1). As the snow turns to firn, the increasing residual of firn enlarges the glacier and eventually the firn becomes flowing ice. The lower part of the glacier is the **zone of ablation** or **wastage,** where the glacier is losing mass, mostly by melting. The end of the glacier is the **toe** or **terminus.** Firn in the zone of accumulation has a lighter color than the light gray or pale blue of glacial ice in the zone of ablation. The edge of the firn in late summer, or early fall before the first seasonal snowfall, clearly marks the boundary between the two zones. This boundary is the **firn limit**—the boundary where rough or cracked gray ice makes a transition to smoother lighter-colored firn (Figure 17.17b on p. 417).

If accumulation and ablation on a glacier are approximately the same, then the two processes balance and the terminus of the glacier temporarily remains fixed (Figure 17.2a). If ablation is more rapid than accumulation, then the terminus of the glacier **retreats** (Figure 17.2b). Even then, the ice continues to flow down toward the terminus; the position of the terminus retreats as the ice melts back. If, on the other hand, a glacier is accumulating faster than it is ablating, then the glacier grows—that is, it increases in thickness, length, and volume—and the terminus **advances** (Figure 17.2c). We refer to the

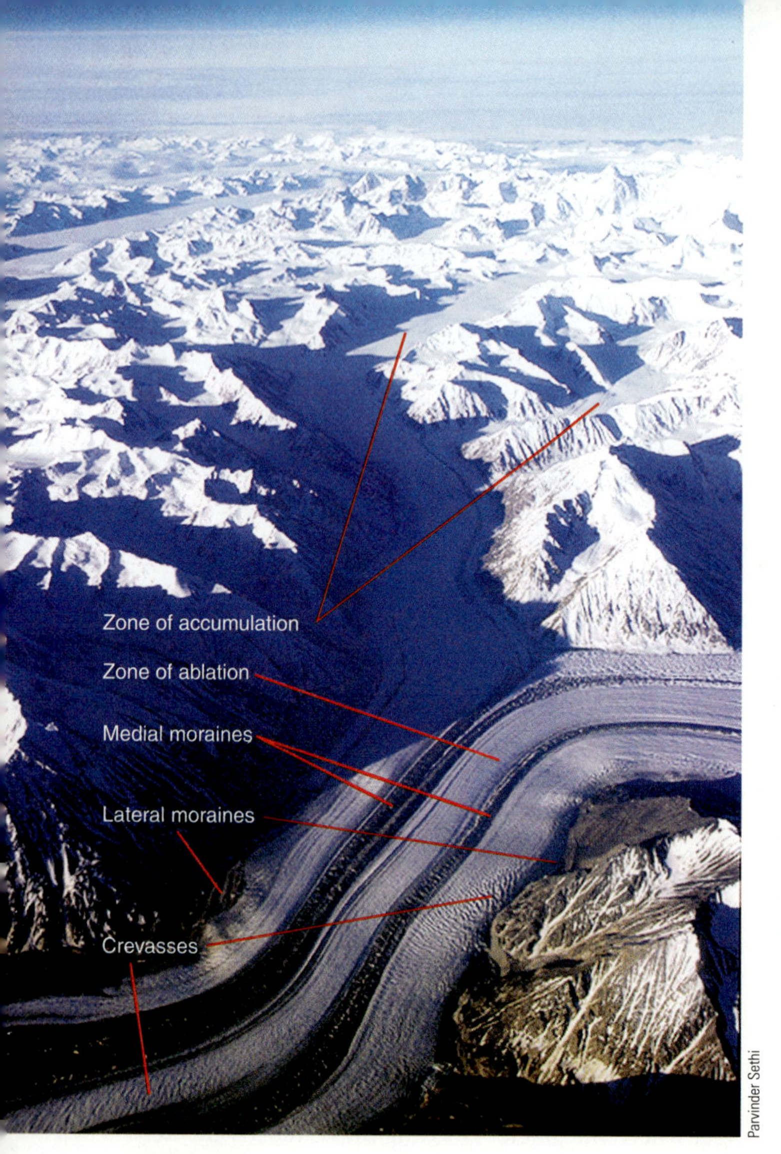

Figure 17.1

Alpine Glacier

Alpine glaciers on Queen Elizabeth Islands, Canada, near northwestern Greenland, showing multiple tributary glaciers and medial moraines. The upper part of the glacier is within the zone of accumulation where more snow accumulates than melts each year. The lower part of the glacier is within the zone of ablation. The previous winter's snow has all melted and the ice here shows crevasses. Large medial moraines stripe the middle of the glacier and lateral moraines accumulate along its sides.

resulting combination of accumulation and ablation in any of these scenarios as the **glacier budget** or the **mass balance** of the glacier.

Glaciers advance or retreat in response to changes in climate. In general, a warming climate, especially average summer warming over a number of years, may cause a glacier to retreat. A cooling climate or average summer cooling may produce glacier advance. Change in precipitation also influences a glacier's mass balance: if snowfall increases for some years, then the glacier may advance; if a drying trend occurs and precipitation decreases, then the glacier may retreat.

Alpine Glaciers

Alpine glaciers flow in the valleys of mountainous regions (Figure 17.1) and are also known as **valley glaciers** or **mountain glaciers.** The glacier flows down the valley from the zone of accumulation toward the zone of ablation at a lower elevation. Many alpine glaciers have tributary glaciers and, just like streams, can form a coalescing network.

An alpine glacier flows more rapidly in the center and top than at its sides and bottom because friction with the valley walls and floor slows the motion (■ Figure 17.3). Normal rate of flow in the upper center of a glacier ranges from a few centimeters to a few meters per day.

The flow pattern lower in the ice produces deformation that may produce **crevasses**—large fissures in the ice above (Figure 17.1). The flow of the glacier tends to curve the initially straight crevasses so that they commonly bow out in the direction of flow (becoming convex to the terminus). Crevasses commonly result from changing flow speeds and are especially prominent in places where a glacier moves over a changing slope. Although alpine glaciers may be large enough to bury large valleys, they are miniscule compared to continental glaciers.

Continental Glaciers

In addition to alpine glaciers, extensive, flowing ice fields and continent-scale ice sheets are present in some areas. A **continental glacier** or **ice sheet** is a thick glacier with an area of more than 50,000 km^2 that spreads outward in all directions. Currently, ice sheets only exist in Antarctica and the interior of Greenland where the ice may be over 4,000 m thick (Figure 17.4a). Yet glacial landforms are widespread, even in areas that are thousands of miles from any glaciers today and resulted from Pleistocene glaciations (Figure 7.24 on p. 158). With careful study of sea-floor sediments and ice cores, as we saw in Lab 15, glaciologists now know that some of the past glaciations in North America and Europe (Figure 17.4b and c) encompassed multiple, distinctly separate, glacial advances. These glaciers have since melted, but their distinctive landforms remain.

Like alpine glaciers, continental glaciers flow from the zone of accumulation to the zone of ablation, but—unlike their smaller cousins—they do not need to flow down a slope. The build-up of ice in a continental glacier creates its own gently sloping "hill," which the force of gravity causes to spread outward.

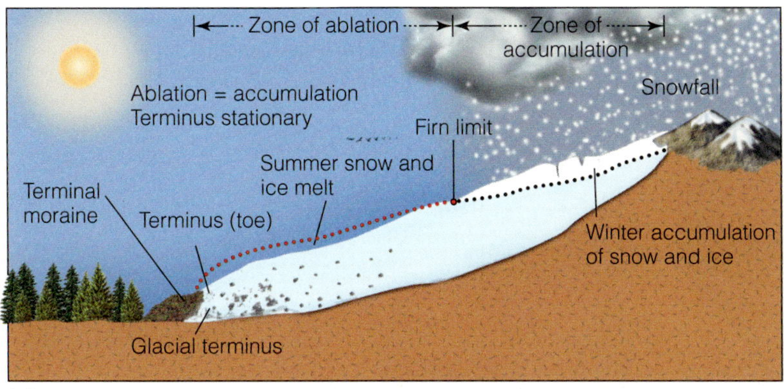

(a)

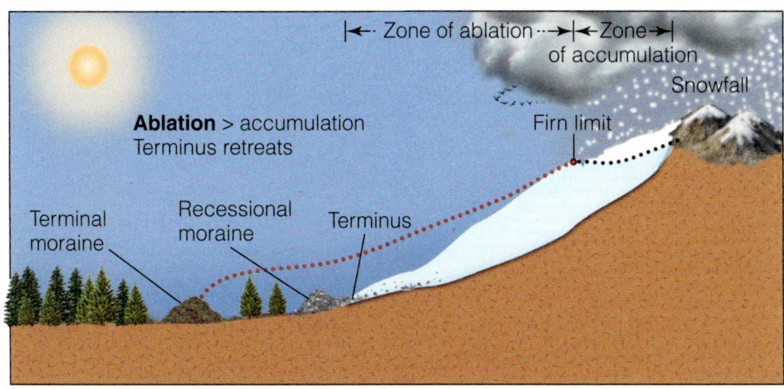

(b)

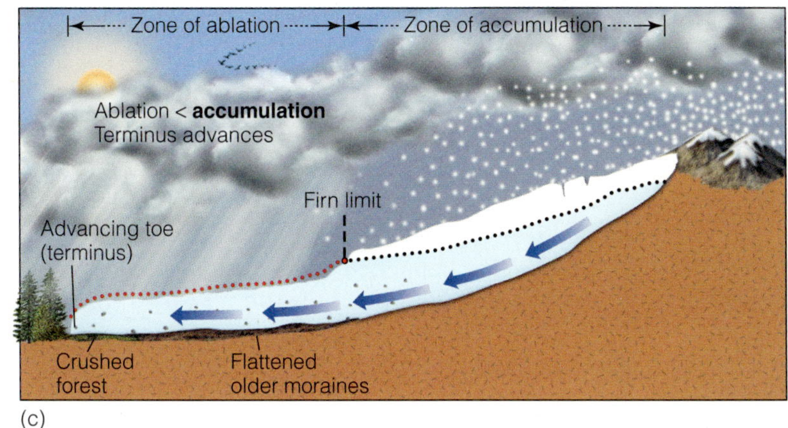

(c)

Figure 17.2

Glacier's Budget

(a) Additions in the zone of accumulation match losses in the zone of ablation causing a stationary glacier terminus. Since ice continuously moves down to the same location before melting, a large pile of till accumulates as a terminal moraine. (b) Additions in the zone of accumulation do not keep up with losses in the zone of ablation and the glacier retreats. A pause in the retreat of the glacier allows accumulation of a recessional moraine. (c) Additions in the zone of accumulation exceed losses in the zone of ablation and the glacier advances, flattening older moraines in its path. From MONROE/WICANDER/HAZLETT, *Physical Geology*, 6E. 2007 Brooks/Cole, a part of Cengage Learning, Inc. Reproduced by permission. www.cengage.com/permissions

Simulation of Glacier Flow

Materials needed:

- About 50 g of putty (this can be putty made for this purpose, or super-soft TheraPutty®, a compound used in physical therapy)
- Utility knife or razor blade
- Inclined flow channel representing a glacial valley
- Ruler
- Toothpicks
- Saucer, or other dish that is flat in the center

1. Simulate the flow of an alpine glacier by taking about 50 g of putty and placing it in the inclined channel provided. Make some simulated closely spaced (about 5 mm) crevasses in the putty perpendicular to the axis of the channel using a utility knife or razor blade. Insert a series of toothpicks vertically into the putty in a straight line across the flow channel.

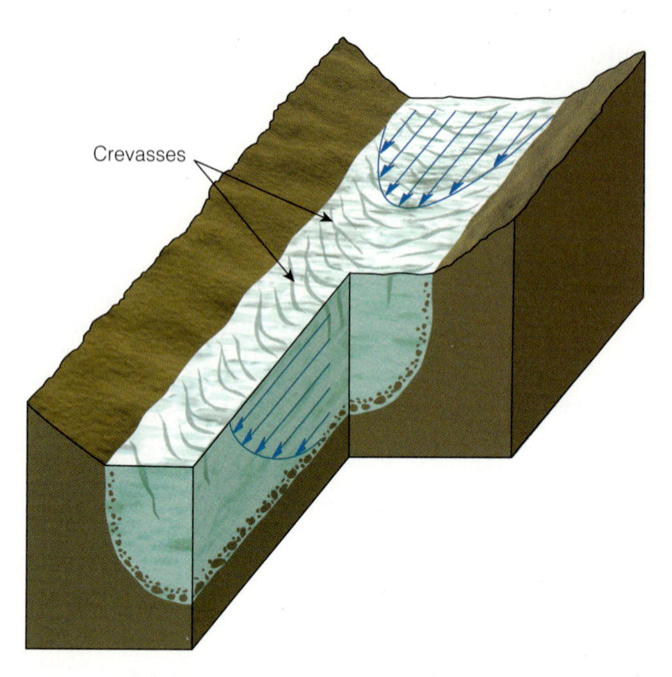

Figure 17.3

Flow in an Alpine Glacier

Speed of flow is proportional to the lengths of the arrows. Flow at the walls and floor of the glacier is slower than at the center and top because of friction between ice and bedrock. The curving lines in grey represent crevasses in the ice. Modified from *Physical Geology: Exploring the Earth*, 5th ed. by J.S. Monroe, and R. Wicander. Copyright © 2005 Thomson Higher Education.

a. Make a sketch of the putty as soon as you have done this.

b. After about 10 minutes, return to the experiment and make a new sketch.

c. How does the current shape of the putty compare to the arrangement at the beginning of the experiment?

____

____

____

d. What happened to the toothpick positions? Compare the center and the sides.

____

____

____

e. What happened to the tilt of the toothpicks? Describe how the tilt varies from the center to the sides depending on toothpick position.

____

____

____

f. How have the cuts in the putty changed?

____

____

g. Where was the flow the fastest? sides / center (circle one) and top / bottom (circle one)

h. How does the flow of the putty compare to the flow of a glacier (Figure 17.3)?

____

____

2. Simulate a continental glacier by placing a round mound of putty in the center of a saucer or shallow dish. Quickly insert a toothpick vertically in the center and six equally spaced in a circle halfway to the edge.

a. Quickly measure the approximate average distance between toothpicks.

____

b. Quickly make a sketch of the putty and toothpicks as soon as you have done this.

c. After about 10 minutes, return to the experiment and make a new sketch.

d. Measure the approximate average distance between toothpicks again.

____

e. What happened to the toothpick positions?

____

____

____

f. What happened to the tilt of the toothpick? Describe how the tilt changed with toothpick position.

____

____

____

g. Compare the flow of the putty with the flow of a continental glacier (Figure 17.4).

____

____

h. What four aspects of both alpine and continental glaciers' behavior are not represented by the putty experiment?

____

____

____

____

3. What force drives the flow of ice in all glaciers? ______________

4. The following exercises will help you to see how the flow of continental glaciers relates to elevation and accumulation.

a. Use the topographic contours (review contours in Lab 6 if necessary) on the Antarctic ice sheets shown in ■ Figure 17.4a to determine the direction of ice flow. Draw arrows on the figure showing the flow direction.

b. On Figure 17.4b, use the ice flow arrows to locate areas where the ice elevation would have been highest during

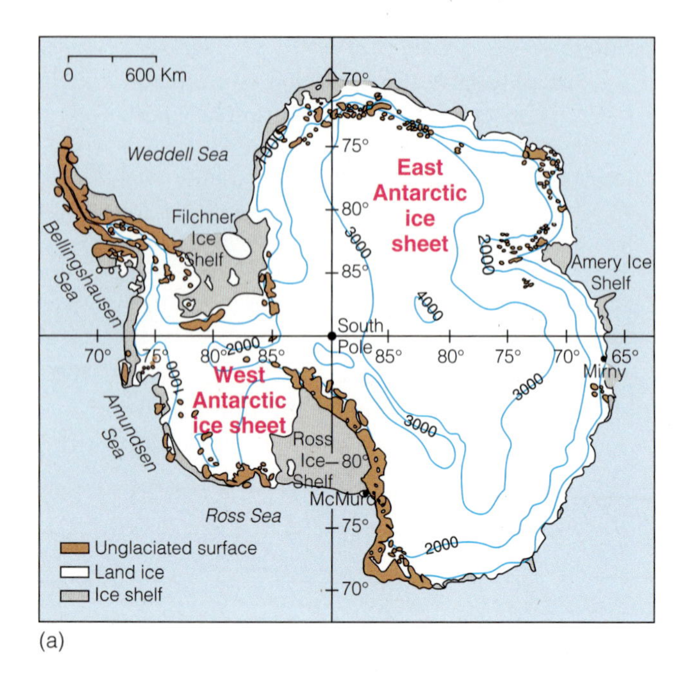

(a)

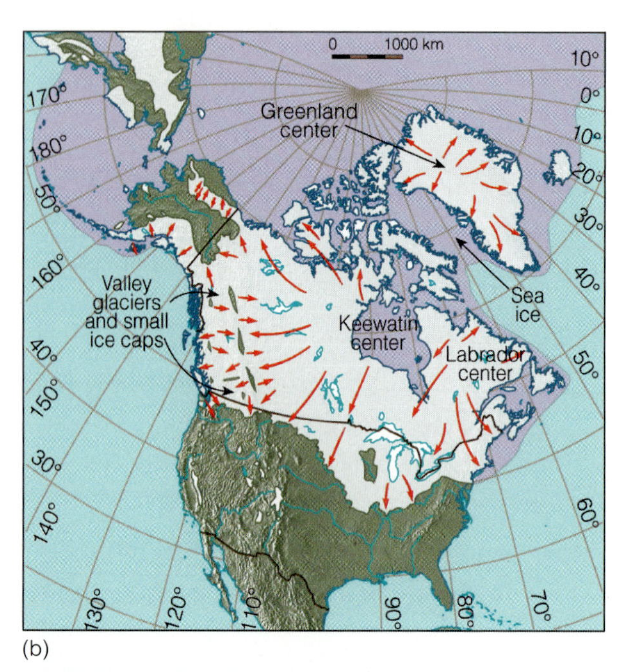

(b)

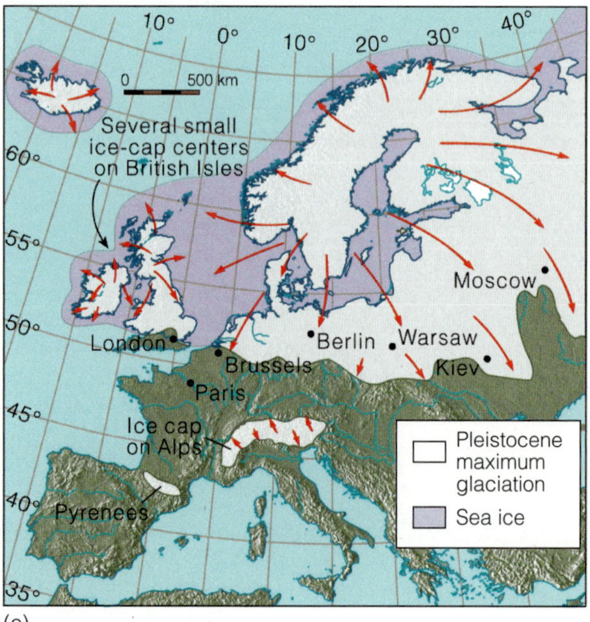

(c)

Figure 17.4

Continental Glaciers

(a) West Antarctic and larger East Antarctic ice sheets. (b) and (c) Pleistocene continental glaciers at their maximum 18,000 years ago. (b) North America and (c) Europe. From *Physical Geology: Exploring the Earth*, 5th ed. by J.S. Monroe, and R. Wicander. Copyright © 2005 Thomson Higher Education.

the Pleistocene. What three names does the map give for these locations?

____

____

____

c. Write the word "center" on the map of Pleistocene glaciers in Europe in Figure 17.4c at each glacial center where the ice elevation would have been the highest.

d. Similarly, label the glacial center on the Antarctic map (Figure 17.4a).

e. On all three of these maps (Figures 17.4a, b, and c), where is or was the zone of accumulation?

____

____

f. Where is/was ablation dominant?

____

LANDFORMS RESULTING FROM GLACIATION

Glaciers erode bedrock and deposit the resulting debris, producing two distinct classes of landforms. ■ Table 17.1 lists these landforms and also indicates which are common in alpine and/or continental glacial terrains.

Erosional landforms (Figures 17.5 and 17.6) result from the tremendous erosive power of a glacier. The rate of erosion depends on a number of factors. Erosion increases if the glacier is thicker and also if the glacier is not frozen to its bed. A glacier contains a great deal of broken rock material that it plucks from its base and sides as it moves. In alpine glaciers, debris also falls on the glacier from above and slowly sinks to the glacier's base. As the glaciers move, the fragments scrape over the bedrock and wear away the surface. The abrasive action produces polished surfaces and erosional grooves or striations parallel to the direction of ice movement (Table 17.1).

Depositional landforms (Table 17.1 and Figure 17.7) result when a glacier drops sediment directly as it melts, or when its melt waters transport and deposit the sediment. For example, as melting occurs at the terminus of a glacier, rock fragments in the ice pile up to form a ridge known as a **moraine**, which is one of the most common depositional features of both alpine and continental glaciation (Figures 17.1, 17.2, 17.7a, b, and f). The material in a moraine is sediment called **till**. Till is **very poorly sorted**, meaning it has a great assortment of sediment sizes, and it is **unstratified**, meaning it does not have layers (Figure 17.7e). Either continental or alpine glaciers may deposit **end moraines** (Figure 17.7a, b, f), which are any type of moraine deposited at the toe of the glacier. A glacier deposits a **terminal moraine** when it reaches its farthest extent. A glacier tends to obliterate previous depositional landforms as it advances, as is shown for an alpine glacier in Figure 17.2c. As meltwater flows away from the glacier's toe in *braided streams* (Figure 17.7g) it carries some of the finer fraction of the sediment away and deposits it in moderately sorted and stratified layers in an **outwash plain** (Figure 17.7b).

The type of glaciation may influence the nature of some glacial features. The next sections discuss how alpine and continental glaciations affect the landscape in very different ways.

Table 17.1

Landforms Produced by Glacial Erosion and Glacial Deposition

Deposition may be directly from the glacier itself or from glacial meltwater. Colors indicate what type of glacier typically produces which features: **alpine, continental, both.**

Erosional Landforms (Figure 17.5c and d)	Deposits and Depositional Landforms (Figure 17.7b)
Arête: a sharp ridge between cirques or glacial troughs. Arêtes commonly connect two horns, or multiple arêtes radiate out from a horn	**Drumlins**: smooth, streamlined hills that have oriented in the direction of glacial movement. The tapered end of the drumlin points in the direction of ice flow and the steeper, blunt end faces up glacier (or opposite the flow direction) (Figure 17.7d).
Cirque: a bowl- or amphitheater-shaped depression at the head, or origin, of a glacier	**Erratic:** an exotic boulder of a rock type different from the bedrock on which it is found. Usually deposited from continental glaciers or from an exceptionally long alpine glacier.
Col: a pass or low spot in an arête	
Finger lake: a long, thin lake occupying part of a U-shaped valley	**Esker**: a sinuous ridge that forms when a meltwater stream flowing through a tunnel in the ice deposits sediment (Figure 17.7c)
Fjord: a U-shaped valley occupied by the sea, creating a long, narrow steep-walled bay or inlet	**Kame:** any mound, knob, hummock, or short irregular ridge of stratified sediment that once rested against ice. Collapse of the sediment commonly occurred when the ice melted.
Glacial striations: grooves or scratches cut into bedrock by scraping action of rock fragments within the glacier	**Kettle:** a hole within outwash, kame, or till deposits, often containing a lake or a swamp resulting from melting of an ice block stranded as a continental glacier retreats
Hanging valley: a glacial tributary valley that is at a higher elevation than the valley into which it flows. You can recognize a hanging valley by the change in slope of the valley floor from gentle to steep as the tributary valley enters the main glacial valley (Figure 17.5c and d). Hanging valleys may have spectacular waterfalls at the change in slope.	**Morainal-dam lake**: a lake formed by water trapped behind an end moraine **Moraine** is a ridge of glacial debris or till deposited directly from the ice. **End moraine** forms at the end or terminus of the glacier. **Lateral moraine** forms at the side of alpine glaciers (Figure 17.7f).
Paternoster lakes: lakes arranged like beads on a string in a U-shaped valley	**Medial moraine** forms when two lateral moraines combine at the joining of two glaciers.
Tarn: a small lake in a cirque occupying an ice-carved rock basin	**Recessional moraine** forms as a retreating glacier temporarily halts its retreat.
Truncated spur: a ridge that stops abruptly at a glacial valley	**Terminal moraine** is a type of end moraine that forms at the farthest extent of the ice.
U-shaped valley or **glacial trough**: a valley with steep sides and gently sloping floor (U shaped) formed by glacial erosion of a pre-existing valley (usually by an alpine glacier, Figure 3.30, p. 66)	**Outwash plain** is a flat area of sand and gravel deposited from glacial meltwater, typically from braided streams. **Till** is poorly sorted material left behind where a glacier melts, composed of angular boulders in a sand-mud matrix.

Landforms Resulting from Alpine Glaciation

During the Pleistocene, alpine glaciers were more widespread in upland regions than they are at present. As the glaciers melted and retreated, they left a terrain dominated by erosional glacial features. Because the ice flows into valleys and is thickest in the lowest locations, alpine glaciation left a spectacular, rugged, and sharply sculpted terrain with wide, deep valleys and extreme relief. Recall that thicker ice erodes faster and, conversely, the highest points and steepest slopes, such as mountain sides and peaks, will not accumulate ice so they remain uneroded and even more prominent.

Alpine glaciers occupy upland valleys previously carved by streams. Within the map area in Figure 17.6, a former glacier, acting much like a gigantic rasp or file, produced the characteristic deep alpine glacial topography. Table 17.1 and ■ Figure 17.5 describe and show the different features of this topography. ■ Figure 17.6 shows how many of these features might appear on a topographic map.

Some depositional features develop from both alpine and continental glaciation. We will discuss these under continental glaciation. However, alpine glaciers have two special types of moraines that do not develop from continental glaciers. **Lateral moraines** form along the sides of an alpine glacier (Figure 17.7f). When a tributary glacier joins another alpine glacier, a **medial moraine** forms a dark stripe where the two glaciers flow together (Figure 17.1). Continental glaciers do not generally have sides, so they do not tend to produce lateral or medial moraines.

(a)

(b)

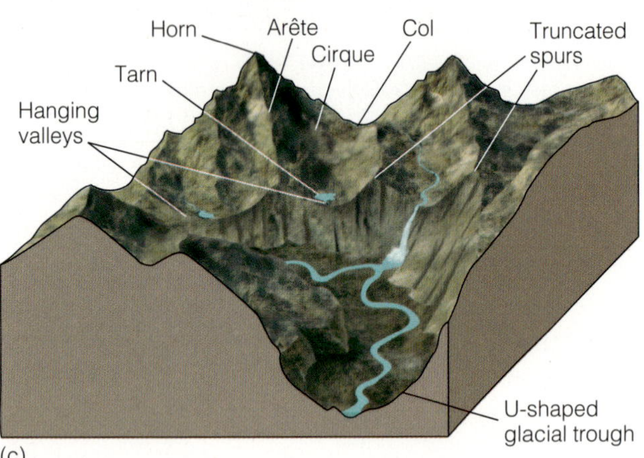

(c)

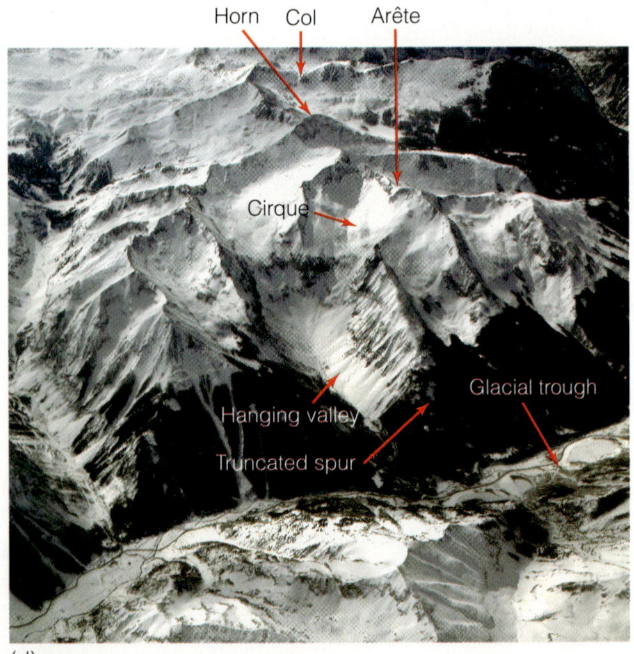

(d)

Figure 17.5

Erosional Alpine Glacial Landforms

Some erosional landforms resulting from alpine glaciers (see also Table 17.1). Drawings show a landscape (a) before glaciation, (b) during glaciation, and (c) after glaciation. (d) Horn, col, arêtes, cirques, hanging valleys, truncated spurs and glacial troughs in a glaciated area as seen from 30,000 feet over the Colorado Rockies.
(a, b, c) From *Physical Geology: Exploring the Earth*, 5th ed., by J. S. Monroe and R. Wicanden, Fig 17.14, p. 513. Copyright © 2005 Thomson Higher Education.

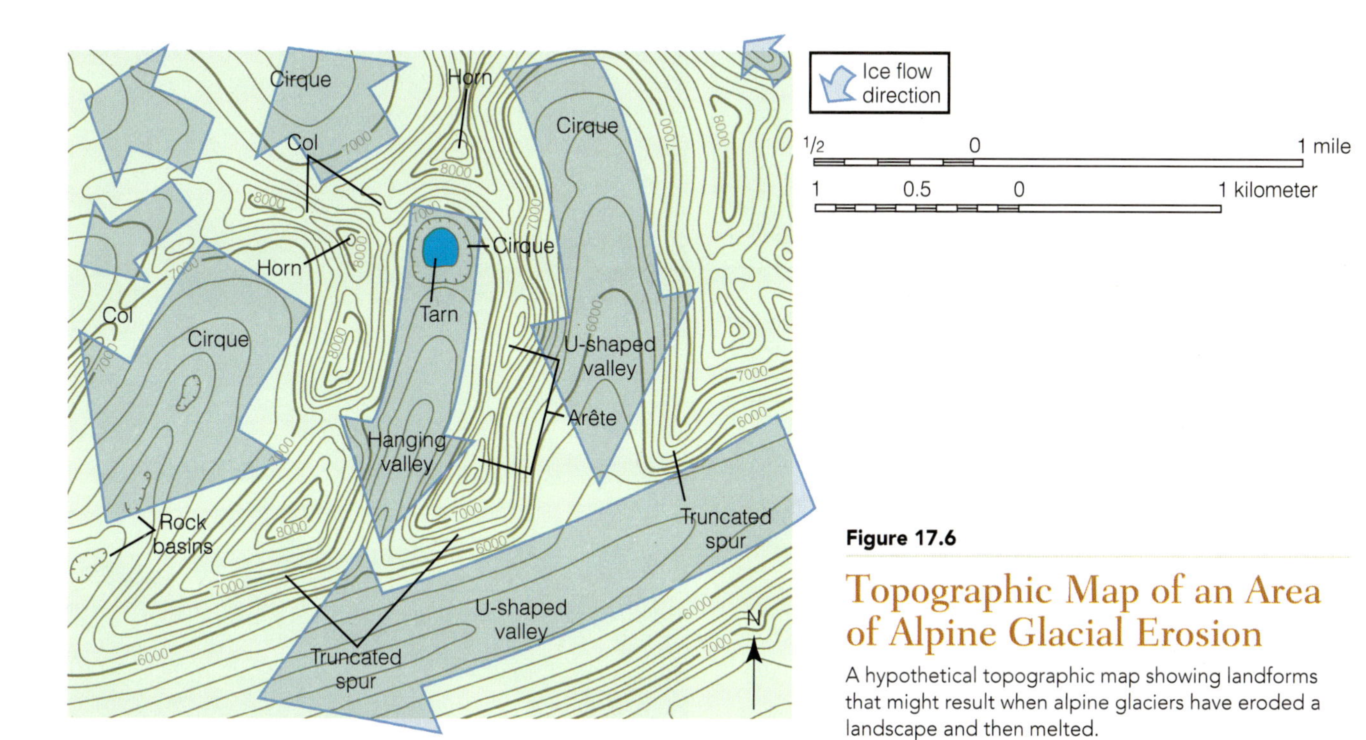

Figure 17.6

Topographic Map of an Area of Alpine Glacial Erosion

A hypothetical topographic map showing landforms that might result when alpine glaciers have eroded a landscape and then melted.

Landforms Resulting from Continental Glaciation

Because of the vast extent of continental glaciers and because they flow over both high and low parts of the topography, this type of glacier scours and rounds the land, and generally reduces the relief (■ Figure 17.7a–d). As ice sheets retreat, melting ice drops large quantities of the debris that it has accumulated. Thus, in contrast to the erosional landscape of Alpine glaciation, continental glacial terrains may be dominated by depositional features. ■ Figure 17.8 shows how several of these features would appear on a topographic map and Table 17.1 describes them.

Exercises Involving Glacial Erosion and Deposition

In the exercises in this section, refer to Table 17.1 and Figures 17.5 through 17.8 to help you answer the questions.

5. ■ Figures 17.9a and b show the Alaskan McCall Glacier in photos taken in 1958 and 2003.

a. What type of glacier is McCall Glacier?

____

b. In each photograph, draw and label the terminus of the glacier.

c. How has the mass of the glacier changed?

______________________ In addition to the change in the position of the terminus, what other evidence supports this answer?

____

____

d. Which process has been more active in the McCall Glacier during the period 1958–2003? ablation or accumulation (circle one).

e. What two different climate conditions could cause this type of change in a glacier?

____

____

f. What erosional landform has been exposed in Figure 17.9b?

____

g. What two depositional landforms exposed in Figure 17.9b are more evident than in Figure 17.9a?

____

____

Figure 17.7

Depositional Glacial Landforms

Features resulting from either continental or alpine glacial deposition. Some of these are more common for continental glaciers (see also Table 17.1). (a) During glaciation and (b) after glaciation. (c) Esker with a sharp crest in NW Manitoba, Canada. (d) Drumlins in Manitoba, Canada. (e) Glacial till near the Exit Glacier, in southern Alaska. (f) Lateral and end moraines formed during the Little Ice Age around the toe of this glacier on Bylot Island, Canada. (g) Glacial outwash seen in a braided stream, Brahmaputra River, Tibet. The light color of the water is a result of the presence of glacial flour, very fine glacially eroded sediment.

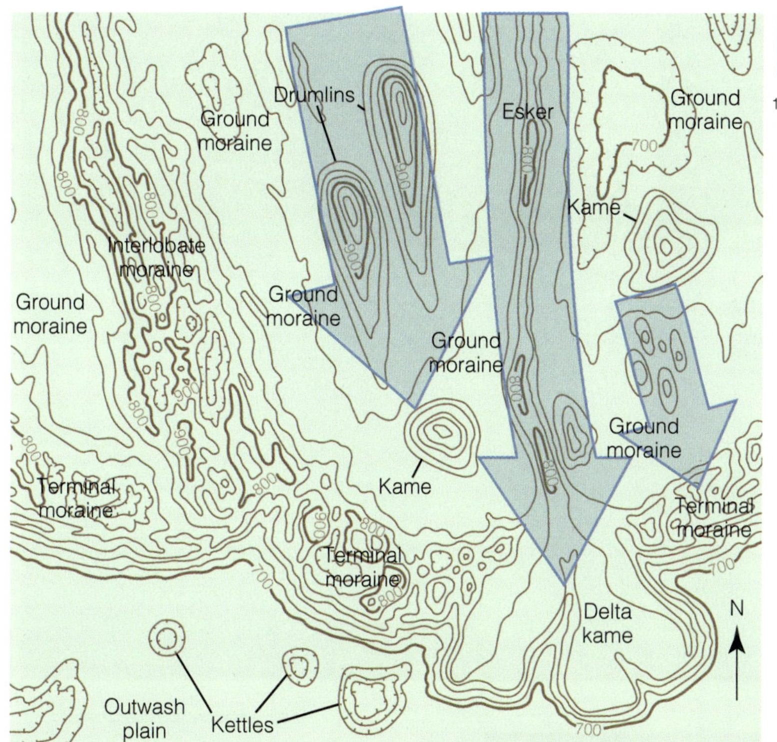

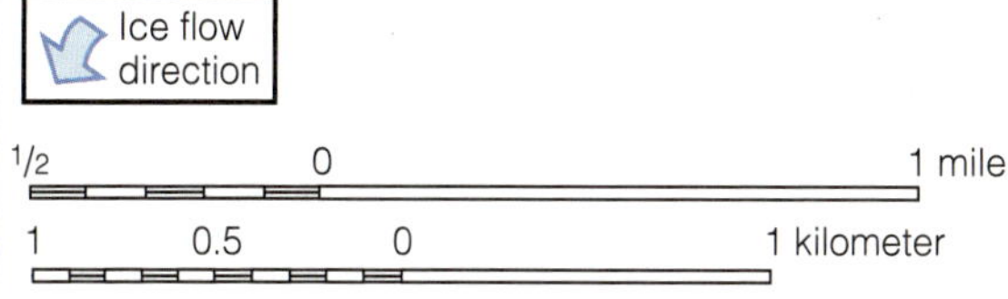

Figure 17.8

Depositional Landforms of Continental Glaciers

A hypothetical topographic map showing landforms that might have developed when a continental glacier deposited sediment across a landscape and then melted.

(a)

(b)

Figure 17.9

McCall Glacier

Photos taken in (a) 1958 and (b) 2003 of McCall Glacier, Alaska.

6. North Crillon Glacier in Glacier Bay National Park and Preserve, Alaska, is shown in the stereo pairs in ■ Figure 17.10. Use a stereoscope to view the air photos in three dimensions.

a. What formed the steep-sided valleys presently occupied by the two glaciers? Be specific:

____

____

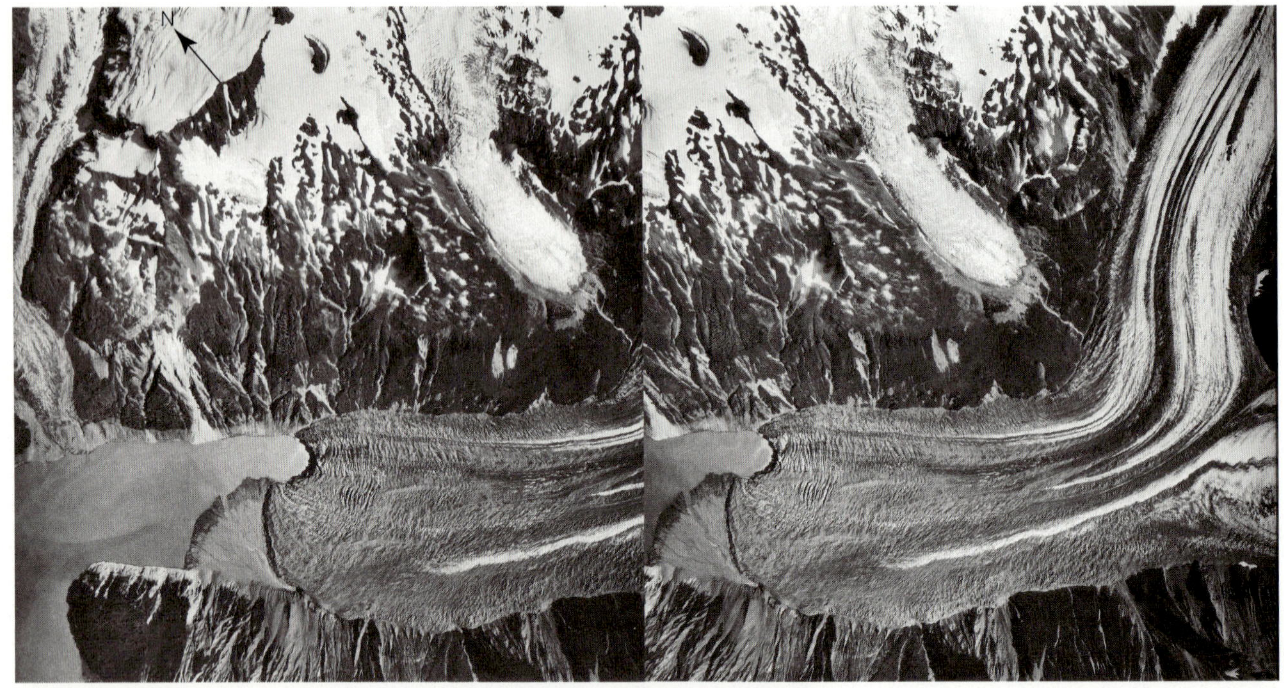

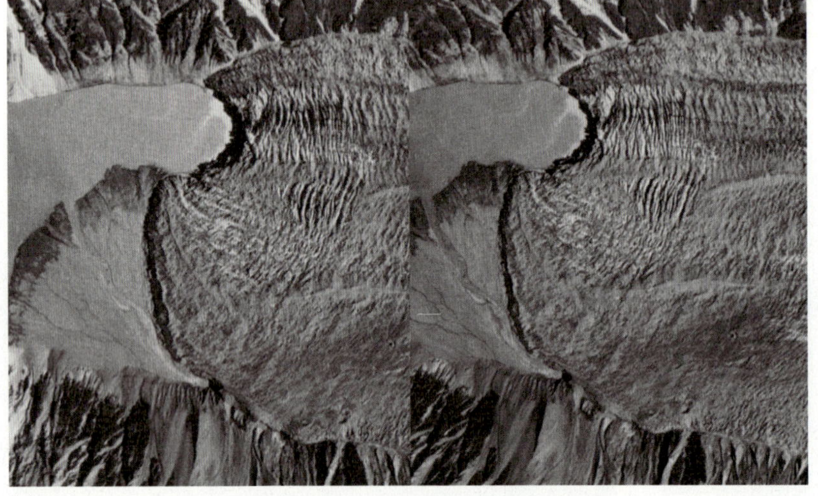

Figure 17.10

Crillon Glacier, Alaska

(a) Stereo pair of the terminus of Crillon Glacier, Glacier Bay National Park and Preserve, Alaska, with (b) close up version showing well-developed crevasses.

b. Which glacier is at a lower elevation? The big one or small one? (circle one) Why is this one so much lower than the other?

____

____

____

____

____

c. The white stripes on the Crillon Glacier are ice. What are the dark stripes?

____

What do they signify?

____

____

____

____

d. Examine the smaller glacier more closely. Why doesn't this glacier have any medial moraines?

____

____

e. What is distinctive about the location where the crevasses are most concentrated in the small glacier?

____

f. What evidence along its sides suggests that the smaller glacier was both longer and thicker not very long ago?

____

____

____

____

____

g. Is there any evidence that Crillon Glacier (the larger glacier) was longer in the fairly recent past? ________. If so, describe it.

____

____

____

h. The body of water in the air photo is a long narrow inlet of the sea called

______________________.

Examine the shoreline on the northeast side; is it steeply or gently sloping?

____

i. Examine Figure 17.10b more closely. A delta is forming at the end of Crillon Glacier on one side. Where are the crevasses near the terminus of the glacier most concentrated? (i) near the delta, or (ii) near the open water? _____ Why?

____

____

____

____

____

j. Refer to Figures 17.7b and 17.8; what type of glacial deposit is this delta?

____

7. Look at Tenaya Canyon on the topographic map of part of Yosemite National Park in the vicinity of Half Dome (■ Figure 17.11a).

a. Construct an unexaggerated topographic profile in Figure 17.11b for the line from A to A′ near North Dome to Half Dome. Use only the index (heavy) contours. The vertical scale is set up to match the map scale (1″ = 2,000′). Review topographic profiles in Lab 6 on pages 124–127 if necessary.

b. Describe the shape of the valley:

____

____

c. Do you think that Tenaya Creek carved the valley? _______

d. Give an explanation for the shape and size of the valley.

____

____

____

e. Is the face of Half Dome (the steepest part) the same slope as the rest of the valley? _______

f. The glacier did not reach as high as the shear face of Half Dome; instead, a major joint (crack in the rock) controls its location. Locate and draw in with bold lines on the map three other features with about the same orientation as Half Dome's face that might be controlled by nearly parallel joints.

8. Examine the map of Bird Ridge and Ship Creek in an area near Anchorage, Alaska in ■ Figure 17.12.

a. In each blank below, name the glacial feature at the corresponding numbered red dot on the map.

1. ______________________________
2. ______________________________
3. ______________________________
4. ______________________________
5. ______________________________
6. ______________________________
7. ______________________________
8. ______________________________
9. ______________________________
10. ______________________________

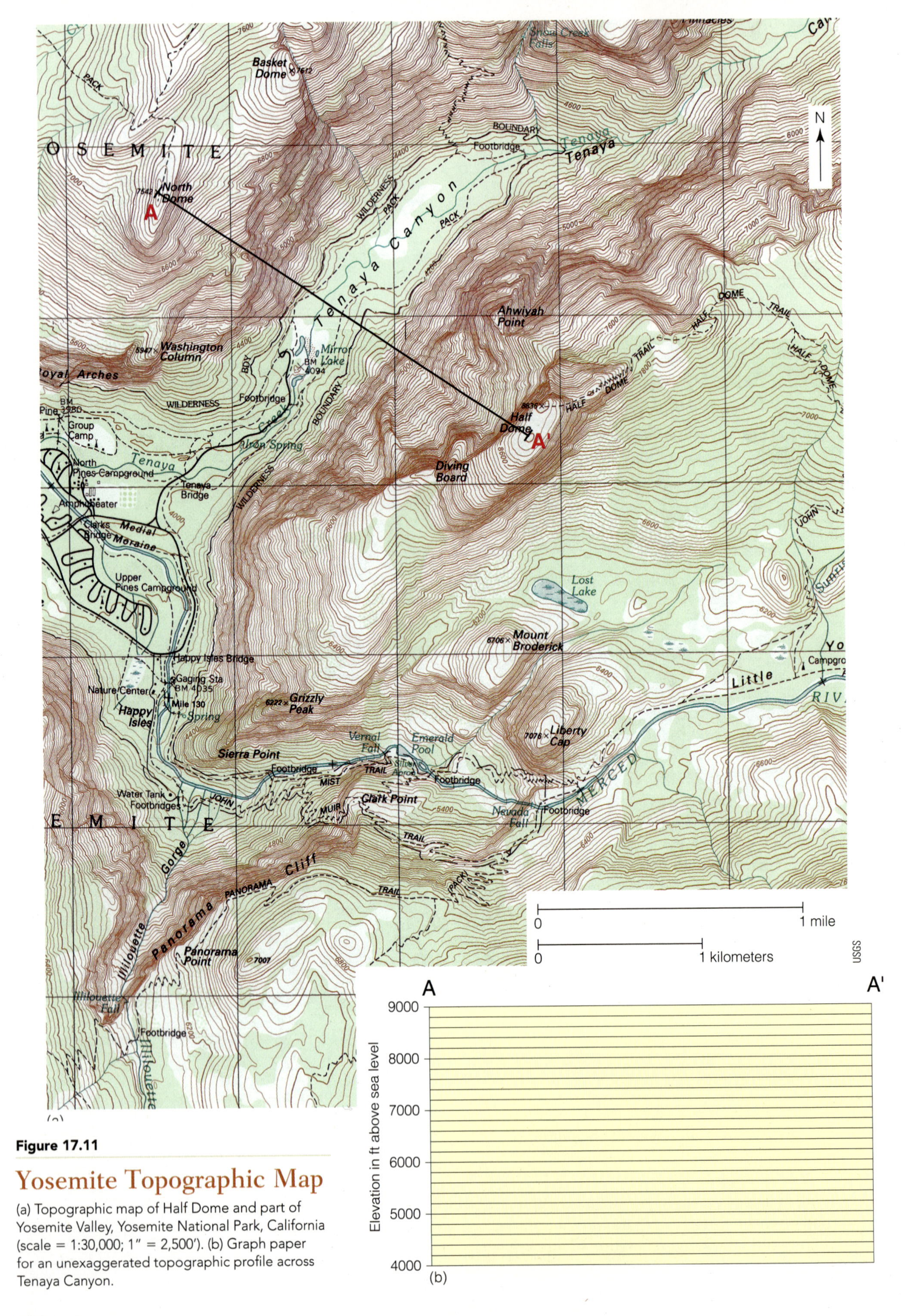

Figure 17.11

Yosemite Topographic Map

(a) Topographic map of Half Dome and part of Yosemite Valley, Yosemite National Park, California (scale = 1:30,000; 1″ = 2,500′). (b) Graph paper for an unexaggerated topographic profile across Tenaya Canyon.

Figure 17.12

Topographic Map Near Anchorage, Alaska

Bird Ridge and Ship Creek, Alaska (scale = 1:63,360; 1″ = 1 mile). USGS Anchorage (A-7), Alaska topographic sheet.

b. With a highlighter, draw four clear, bold arrows on the map indicating the direction of flow of the ice where the four largest glaciers once were.

c. Draw three additional arrows with a different color for ice flow where three tributary glaciers once were. Make a key on the map showing the two arrow types with appropriate labels.

9. Examine ■ Figure 17.13, showing part of Rocky Mountain National Park, Colorado, and answer the following questions.

a. What map units (write their letter symbols) indicate that glaciation has occurred in the park at each of the following times in the past?

- Upper Pleistocene glaciation, 35,000–13,000 years ago (Pinedale age) ______
- Upper Pleistocene glaciation, 150,000–130,000 years ago (Bull Lake age) ______
- Pleistocene glaciation, 550,000–400,000 years ago (pre-Bull Lake age) ______

b. What is the term for the type of material deposited during these three glacial episodes? ______

c. Draw the symbol from the map that indicates the crest (ridge top) of a glacial moraine. ______

d. What created or caused Shadow Mountain Lake and Grand Lake?

______

______

e. Consider the relative position of these lakes and the likelihood that glacial advance would mostly obliterate moraines in the area as the glacier moved over them (Figure 17.2c). Which of these two lakes is younger?

______

Explain.

______

______

______

______

______

f. Along the side of Shadow Mountain Lake, both on the west and east, are a number of symbols indicating the crests of moraines. What type of moraines are these? ______ Is the material making up the till in these moraines likely to be well or poorly sorted? ______

g. A few miles east of the area shown in Figure 17.13, 34 perennial snow fields and small glaciers were present in 1960 along the continental divide above 12,000′. What aspect of the continental divide allowed modern glaciers to be present there?

______

h. In contrast, the Pleistocene glaciers reached much lower elevations. How much lower did the Pleistocene glaciers reach within the map area in the figure? Use the elevation of Shadow Mountain Lake to make your calculation. (Note: the map gives the elevation in feet first, then meters.)

______

i. What caused the differences in the elevations and extent of glaciation between the Pleistocene and Holocene?

______

______

Sand, gravel, silt, and clay transported and deposited by glacial ice 2 million to 10,000 years ago comprise most of the surface cover in the map areas of Figures 17.14, 17.15, and 17.16. These surface deposits are resources for construction material. Deeper deposits are aquifers and affect water and land use due to their porosity. The rocky areas of glacial till make farming difficult.

Figure 17.13

Geologic Map of Part of Rocky Mountain National Park *(facing page)*

Glacial deposits around Shadow Mountain Lake and Grand Lake.
Source: From Geologic Map of Rocky Mountain National Park and Vicinity, Colorado; by William A. Braddock and James C. Cole, 1990; Miscellaneous Investigations Series Map I-197; USGS.

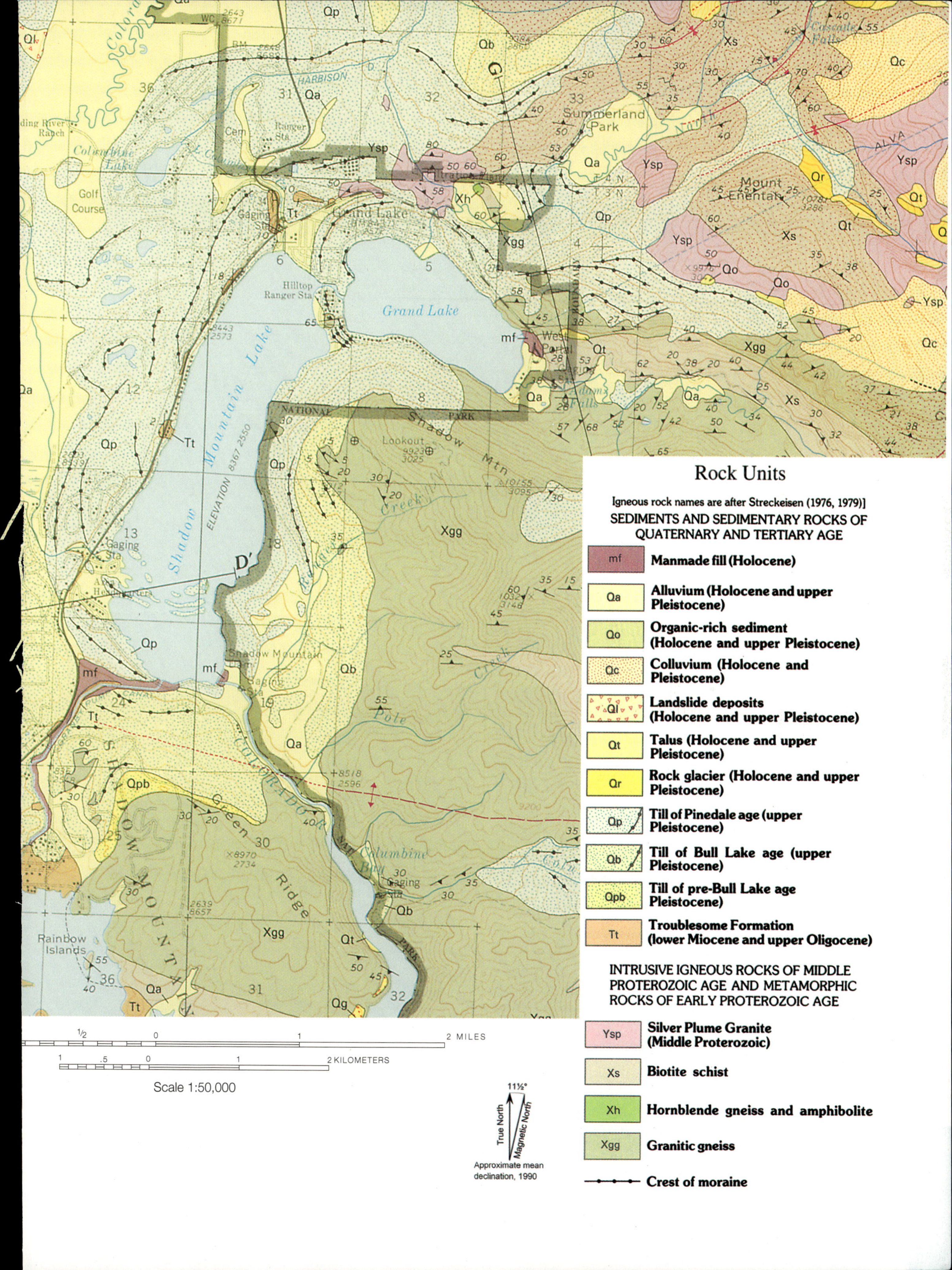

Rock Units
Igneous rock names are after Streckeisen (1976, 1979)]
SEDIMENTS AND SEDIMENTARY ROCKS OF QUATERNARY AND TERTIARY AGE
mf Manmade fill (Holocene)
Qa Alluvium (Holocene and upper Pleistocene)
Qo Organic-rich sediment (Holocene and upper Pleistocene)
Qc Colluvium (Holocene and Pleistocene)
Ql Landslide deposits (Holocene and upper Pleistocene)
Qt Talus (Holocene and upper Pleistocene)
Qr Rock glacier (Holocene and upper Pleistocene)
Qp Till of Pinedale age (upper Pleistocene)
Qb Till of Bull Lake age (upper Pleistocene)
Qpb Till of pre-Bull Lake age Pleistocene)
Tt Troublesome Formation (lower Miocene and upper Oligocene)
INTRUSIVE IGNEOUS ROCKS OF MIDDLE PROTEROZOIC AGE AND METAMORPHIC ROCKS OF EARLY PROTEROZOIC AGE
Ysp Silver Plume Granite (Middle Proterozoic)
Xs Biotite schist
Xh Hornblende gneiss and amphibolite
Xgg Granitic gneiss
Crest of moraine
2 MILES
2 KILOMETERS
Scale 1:50,000
11½°
True North
Magnetic North
Approximate mean declination, 1990
Grand Lake
Shadow Mountain Lake
Summerland Park
Mount Enentah
Shadow Mtn
Green Ridge
Rainbow Islands
Columbine Lake
Golf Course
Hilltop Ranger Sta
West Portal

Scale = 1:32,000

0.5 0 1 mile

0.5 0 1 kilometer

Figure 17.14

Glaciated Hills, Palmyra, New York

Topographic map of an area near Palmyra and Rochester, New York, showing the direction of the flow of the last glaciation. From USGS Palmyra, New York, 7½ Minute Quadrangle.

10. ■ Figure 17.14 is a topographic map of an area near Rochester and Palmyra, New York.

a. What are the most common glacial features shown on the map?

____

b. What type of glacier formed these features?

____

c. Examine the map and determine the direction of ice flow by checking which ends of the features are blunt and steep and which are more pointed and tapered. Draw an arrow on the map showing the average direction. What general direction is this? ______________________

d. Use a protractor to measure the trend of your arrow to get a more precise indication of the flow direction: ______________ (Review map trends and compass directions in Lab 1 if necessary.)

11. The topographic map in ■ Figure 17.15 shows an area in Michigan that experienced continental glaciation during the Pleistocene.

a. What type of glacial feature is Blue Ridge ______________________ and how did it form?

____

____

____

____

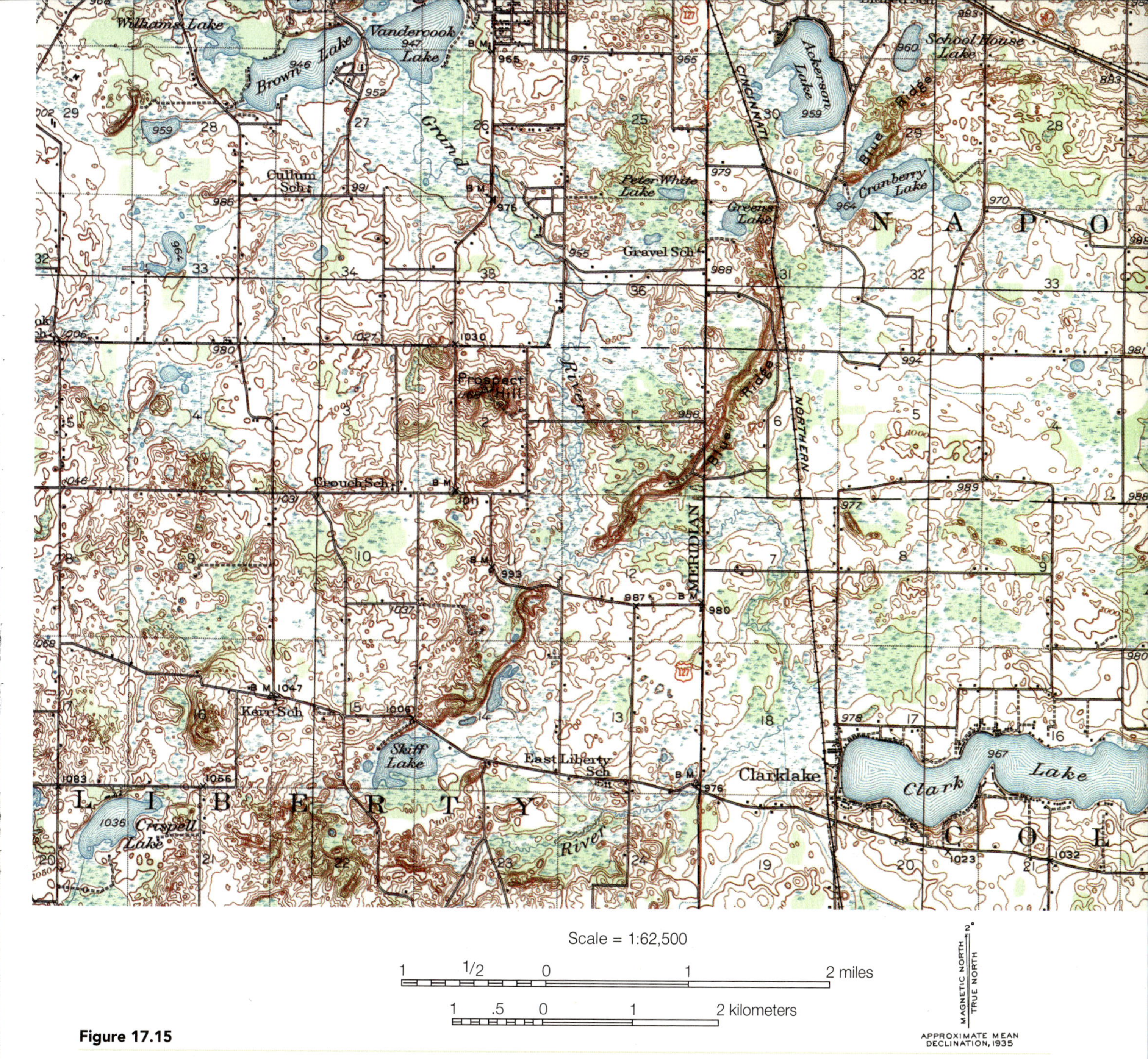

Figure 17.15

Blue Ridge Topographic Map

Topographic map of the area surrounding Blue Ridge, Jackson, Michigan. *Source:* From USGS Jackson, Michigan 15 Minute Quadrangle.

b. What type of glacial feature are the isolated hills such as Prospect Hill?

____

c. What is a likely glacial origin of these hills?

____

____

____

____

d. What type of glacial deposits occur in the area east of Blue Ridge?

____

e. Why is the land here so swampy?

____

____

____

____

f. What glacial feature is seen in the more hilly area in the southwest corner of the map?

____

g. Many small water-filled depressions occur in the map area, including Crispell Lake in the SW corner. What glacial feature are these? ______________

h. Explain how these depressions formed.

____

____

____

____

____

____

____

12. Examine the glacial geologic map of Maine in ■ Figure 17.16.

a. Look for evidence of continental glaciation in this section of Maine and list:

i) one erosional feature:

____

ii) three features deposited directly by the glacier:

____

____

____

iii) three features deposited by glacial meltwater:

____

____

____

b. Give the average approximate direction of the last ice flow

____

and list at least three pieces of evidence that show this clearly.

____

____

____

The last glacier was the Laurentide Ice Sheet, which was thick and heavy enough to depress the Earth's bedrock crust and submerge the coastline. As the massive glaciers melted starting about 18,000 years ago, sea level rose more than 300 feet worldwide. In spite of this, after the ice receded in Maine, isostatic rebound of the land caused more coastal areas to emerge from the ocean. **Isostatic rebound** occurs when the lithosphere rises after removal of a heavy weight from an area. In this case, the melting of ice removed the heavy weight of a continental glacier.

13. Continue working with Figure 17.16.

a. What does the thick blue line indicate that separates much of the orange and pale green **g** and **t** units from pink and blue **ms** and **m** units?

____

b. Were the pink and blue **ms** and **m** units deposited on land or in the sea?

____

c. Since sea level is now higher than it was during the Laurentide Ice Sheet, how is it possible for glacial marine deposits to be on land now? Shouldn't they be under water?

____

____

d. Compare the direction of the glacial flow features on land with the orientation of the moraine ridges in the glaciomarine deposits. Do both types of features indicate the same flow direction? ______
If not, what is the angle between the two flow directions? Measure this with a protractor: ______

In Maine, we saw the effects of isostatic rebound from retreating glaciers of thousands of years ago. Today we again see retreating glaciers, but most of the world will experience sea level rise not rebound. Again the cause

Figure 17.16

Surficial Geologic Map of Maine (*facing page*)

Map of surface glacial geology of southwestern Maine. *Source:* Maine Geological Survey, Department of Conservation, Edited by Thompson, MGS and Borns, University of Maine, 1985.

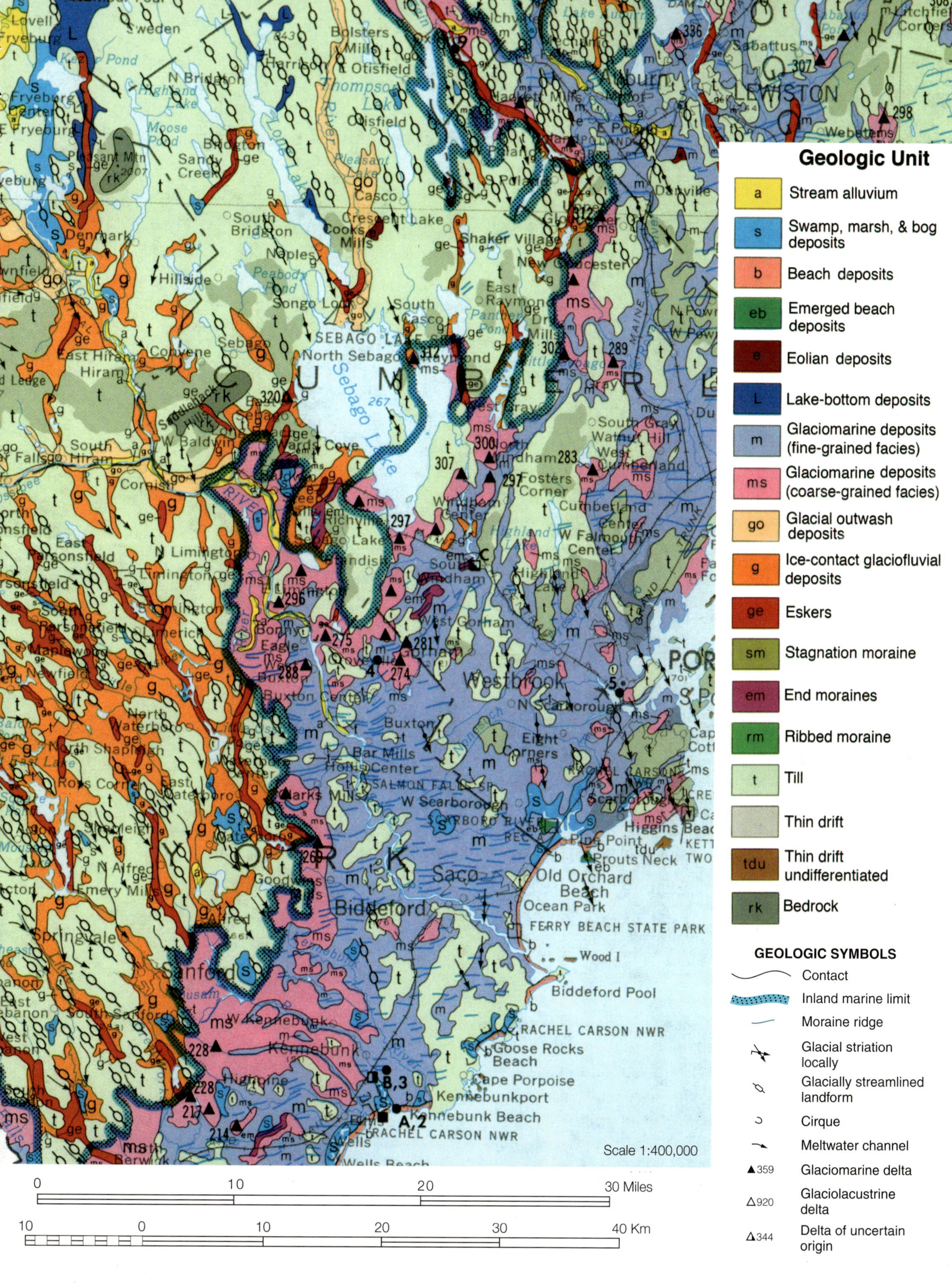
Geologic Unit
a Stream alluvium
s Swamp, marsh, & bog deposits
b Beach deposits
eb Emerged beach deposits
e Eolian deposits
L Lake-bottom deposits
m Glaciomarine deposits (fine-grained facies)
ms Glaciomarine deposits (coarse-grained facies)
go Glacial outwash deposits
g Ice-contact glaciofluvial deposits
ge Eskers
sm Stagnation moraine
em End moraines
rm Ribbed moraine
t Till
Thin drift
tdu Thin drift undifferentiated
rk Bedrock
GEOLOGIC SYMBOLS
Contact
Inland marine limit
Moraine ridge
Glacial striation locally
Glacially streamlined landform
Cirque
Meltwater channel
359 Glaciomarine delta
920 Glaciolacustrine delta
344 Delta of uncertain origin
Scale 1:400,000
0 10 20 30 Miles
10 0 10 20 30 40 Km
Sebago Lake
Westbrook
Saco
Biddeford
Old Orchard Beach
Ocean Park
FERRY BEACH STATE PARK
Biddeford Pool
RACHEL CARSON NWR
Goose Rocks Beach
Cape Porpoise
Kennebunkport
Kennebunk Beach
Kennebunk
Sanford
Springvale
Lewiston
Auburn
Windham
Gorham
Buxton
Bar Mills
Hollis Center
Limington
Cornish
Bridgton
Naples
Casco
Raymond
Gray
Prouts Neck
Higgins Beach

is climate warming, but this time the climate change is very likely anthropogenic. Although isostatic rebound is also occurring in areas such as Alaska, Canada, Scandinavia, etc., it is slow, and the more immediate and noticeable effect in glaciated areas as climate warms is the rapid retreat of glaciers.

CLIMATE CHANGE AND GLACIERS

Advancing and retreating glaciers are indicators of climate change. In Figures 17.9 and 17.17 we see glaciers that have retreated substantially.

Let's reexamine the concepts of advancing and retreating glaciers and their mass balance more quantitatively. Recall that **mass balance** is the difference between accumulation and ablation on a glacier. To establish the mass balance of a glacier we must determine the additions and losses of mass for the glacier each year. On the positive side of the balance are precipitation, wind drift, meltwater refreezing, and avalanches adding to the glacier. On the negative side of the balance are melting, **sublimation** (conversion of ice to water vapor), wind drift and avalanches off the glacier, and **calving** (breaking off) of ice from the glacier's margin. In practice, however, making these measurements is much more difficult and involves calculations for movement, volume, runoff, precipitation, and sublimation. Such data are available for some glaciers.

Table 17.2 shows mass balance estimates from the South Cascade glacier, Washington. Positive numbers reflect accumulation and negative numbers ablation (wasting). Figure 17.17 shows how the glacier has changed over time and displays a plot of the data. As you can see from the net balance plot (in green in Figure 17.17d), in some years there is a net loss from the glacier and in other years a net gain. It is difficult to see the overall change of the glacier over a given time period from the graph. To do this, let's analyze the data in Table 17.2 in a different way.

Table 17.2

Mass Balance for South Cascade Glacier, Washington, from 1980 to 2005

Year	Winter (m*)	Summer (m*)	Net (m*)	Cumulative Balance
1980	1.83	−2.85	−1.02	−1.02
1981	2.28	−3.12	−0.84	
1982	3.11	−3.03	0.08	
1983	1.91	−2.68	−0.77	
1984	2.38	−2.26	0.12	
1985	2.18	−3.38	−1.20	
1986	2.45	−3.06	−0.61	
1987	2.04	−4.10	−2.06	
1988	2.44	−3.78	−1.34	
1989	2.43	−3.34	−0.91	
1990	2.60	−2.71	−0.11	
1991	3.54	−3.47	0.07	
1992	1.91	−3.92	−2.01	
1993	1.98	−3.21	−1.23	
1994	2.39	−3.99	−1.60	
1995	2.86	−3.55	−0.69	
1996	2.94	−2.84	0.10	
1997	3.71	−3.08	0.63	
1998	2.76	−4.62	−1.86	
1999	3.59	−2.57	1.02	
2000	3.32	−2.94	0.38	
2001	1.90	−3.47	−1.57	
2002	4.02	−3.47	0.55	
2003	2.66	−4.76	−2.1	
2004	2.08	−3.73	−1.65	
2005	1.97	−4.42	−2.45	

*Units are meters of liquid water precipitation equivalent.

14. Use the data in Table 17.2 and the graph paper in Figure 17.18 to do the following activities and to answer the questions below.

a. What does a positive number in the Winter column in Table 17.2 mean?

____

____

____

b. What does a negative number in the Net column in Table 17.2 mean?

____

____

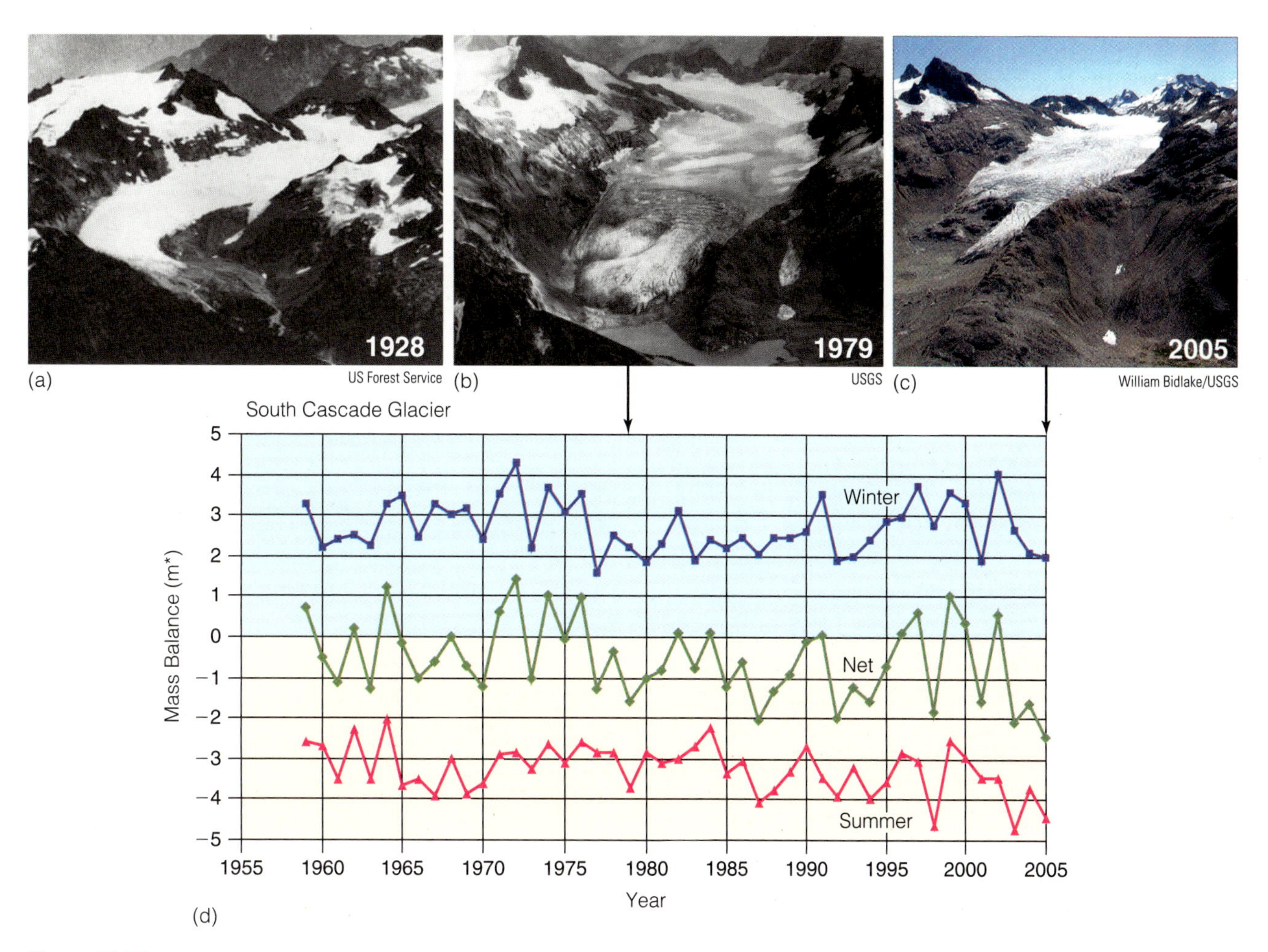

Figure 17.17

Photos and Annual Mass Balance of South Cascade Glacier, Washington

Oblique aerial photographs of South Cascade Glacier in Washington photographed in (a) 1928, (b) 1979, and (c) 2005. The angle is somewhat different among the three photos. (d) Graph of the annual mass balance of South Cascade Glacier, showing winter additions, summer losses and net ice accumulation or loss (combined total).

c. Determine the cumulative balance of ice in the glacier and enter the values in Table 17.2: The number for the first year is already given. For the next year, add the net for the present year to the balance for the previous year. For years with negative net mass, the cumulative balance will become more negative.

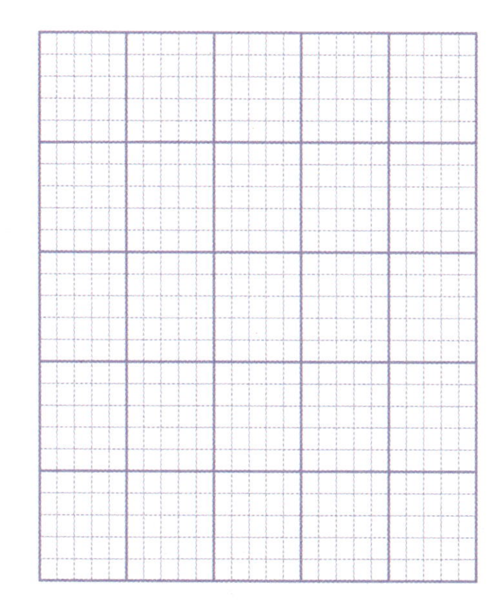

Figure 17.18

Graph Paper

Graph paper for plotting of cumulative mass balance of the South Cascade Glacier from 1980–2005.

d. Plot the values for cumulative balance for 1980–2005 on the graph paper in Figure 17.18. Connect the points with short straight lines. Don't forget to completely label your graph.

e. When you have finished graphing the data, look at the graph and compare this with what you observe on the photographs in Figure 17.17. Does the graph match what is happening to the glacier? ______

f. Now go back and compare what you see in your graph with the ***net*** data shown in the green graph in Figure 17.17d. Explain how the green line and the results you obtained in your graph are consistent with each other.

____

____

____

____

g. What is a likely explanation for what is happening to South Cascade glacier?

____

____

LAB 18

Deserts and Arid Landscapes

OBJECTIVES

- To learn the types of deserts, predict their location for different continent configurations, and understand what factors control their occurrence
- To understand erosional and depositional landforms of arid landscapes
- To be able to recognize desert landforms on topographic maps, and also on standard photos, satellite images, and aerial photos
- To understand the development of mountain fronts, alluvial fans, bajadas, pediments, and inselbergs in connection with normal faulting
- To understand the causes and consequences of desertification and analyze data to see correlations

The word **desert** originally meant deserted or uninhabited because people found these areas inhospitable. Today, the term refers to **arid** regions—areas with a dry climate. The climate either has low precipitation, less than 10 in/yr liquid equivalents[1] (25 cm) or has a higher rate of evaporation than precipitation. Vegetation covers less than 15% of the surface of deserts, leaving sediment and rock exposed. Two common misconceptions are that deserts are mostly covered with sand and sand dunes (■ Figure 18.1) and are in hot climates. However, examining the background of Figure 18.1 shows that large parts of deserts may include bare rock, or coarser sediment than sand. Streams in deserts are either few in number, as they travel through the desert from mountainous regions with higher precipitation, or are **ephemeral**—making brief appearances only to disappear again. In spite of their rarity, streams and running water have a surprising influence on desert landscapes. In this lab, we explore where deserts occur worldwide, how faulting, erosion, and deposition transform and sculpt desert landscapes, what surface features deserts commonly have, and how human activities may create deserts.

Parvinder Sethi

Figure 18.1

Desert at Death Valley

Desert dunes in the foreground near Stovepipe Wells, Death Valley National Park, California, with desert mountains in the distance.

[1] Inches per year of liquid equivalents means that any part of the precipitation that falls as snow is converted to the number of inches of liquid water as if the snow had fallen as rain or had melted.

TYPES OF DESERTS

Low *relative humidity* of regional air primarily controls the location of deserts. As we learned in Lab 15, as air warms it has the capacity to contain more moisture than when it is cooler, and it tends to dry out its surroundings. All deserts form in places where air warms, but the reasons for warming and the actual temperatures vary. Sinking air warms up as it reaches higher pressures and warmer temperatures, producing three major desert types dependent on location and topography: subtropical deserts, polar deserts, and rain-shadow deserts. When air masses move from over cold seawater to over a warmer continent, they create coastal deserts. A fifth type of desert occurs in continent interiors far from sources of moisture. Let's consider each of these types of deserts in more detail:

Subtropical Deserts occur between 15° and 30° N and S latitudes (■ Figures 18.2). These deserts are the most extensive and occur here because the air at about 30° latitude is falling and forming the subtropical high-pressure belts. As the air descends, it warms, decreasing its relative humidity and making the climate dry. The direct rays of sunlight further warm the air is as it flows toward the equator. This spreads the dry conditions to latitudes as low as about 15°. The term *tropical desert* refers to deserts at latitudes lower than 23½°—between the Tropics of Cancer and Capricorn.

Polar Deserts occur at the polar high near 90° latitude (Figure 18.2). We do not tend to think of polar regions as deserts because ice and snow lie on the ground. Precipitation, however, is extremely low and the surface ice contributes little to atmospheric moisture. The climate is also dry there because the air in polar regions sinks. Although it is very cold, it does warm up somewhat as it sinks, which reduces its relative humidity and causes dry conditions. Polar deserts occur in Antarctica, northern parts of Siberia, and some of Canada's northern islands (Figure 18.4).

Rain-Shadow Deserts result from air sinking and warming as it descends from mountains (■ Figure 18.3). Warm moist air moves up over mountains (**orographic uplift,** *oro* refers to mountains) on the windward side, cooling at higher altitudes and condensing, making clouds and precipitation. At this point the relative humidity is near 100%. The resulting rain or snow is **orographic precipitation.** As the air moves down the other side of the mountains, it warms. The warming reduces the relative humidity and allows the air to extract moisture from the surroundings. This produces a desert on the **leeward** side (downwind side) of the mountains, called a *rain-shadow desert.* Again, we see the necessary warming of the air and reduction of relative humidity to make desert conditions.

Continental Interior Deserts are far from oceans in the interior of continents. In many cases, these deserts result from repeated rain shadows as air moving inland

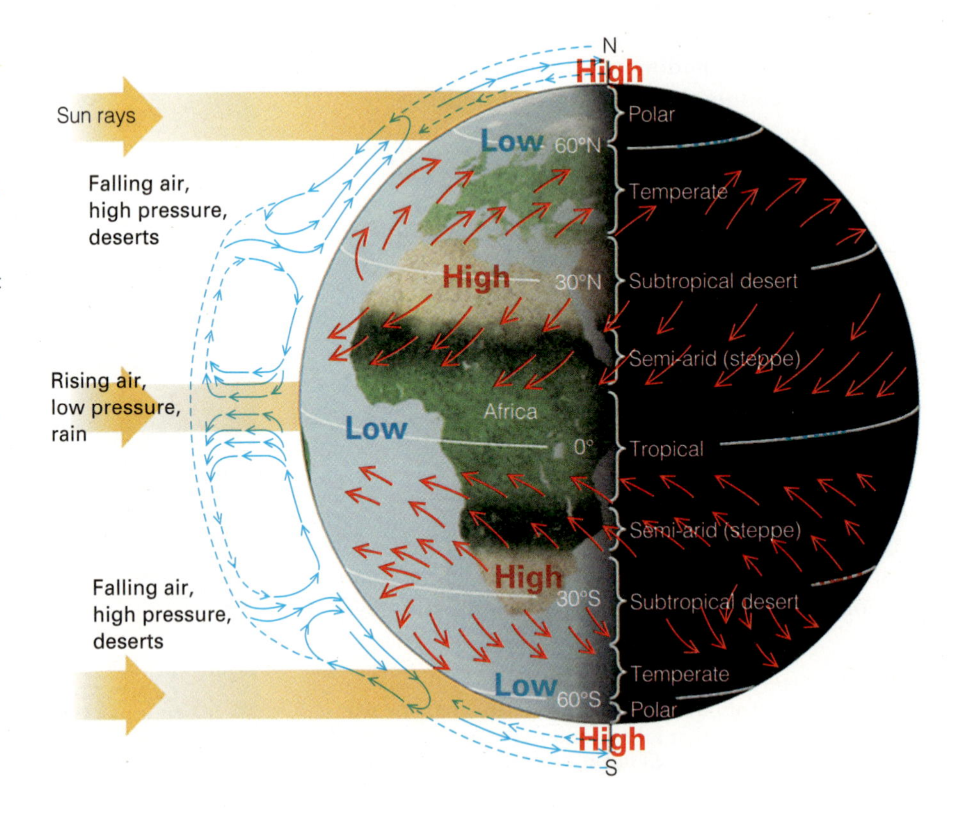

Figure 18.2

Solar Heating and Latitudes of High Pressure Desert Belts

Energy from direct rays of sunlight heats tropical air, causing the low-pressure, rainy belt near the equator. At 30° latitude, falling air in high-pressure belts warms as it reaches low altitudes, where it extracts moisture from the environment, making deserts. Polar regions also have high-pressure sinking air that dries out its surroundings as it warms somewhat from extremely frigid, creating cold polar deserts. Red arrows represent surface winds.

Figure 18.3

Rainshadow Deserts

As prevailing winds push up over mountains, they cool at the higher altitudes. The cooling causes condensation, reducing the absolute humidity (total water content of the air) as the relative humidity reaches 100%. On the other side, the sinking air warms, causing a drop in the relative humidity. The result is air that extracts moisture from the surroundings, creating a rainshadow desert. From MILLER. *Living in the Environment*, 16E. 2009 Brooks/Cole, a part of Cengage Learning, Inc. Reproduced by permission. www.cengage.com/permissions

flows over multiple mountain ranges. However, because there are no sources of moisture, deserts in remote continental interiors may not need high mountains to make them dry.

Coastal Deserts form along low-lying coasts that have cold, offshore ocean currents (Lab 14). Cool air from over the cold ocean moves over the warmer land. As the air warms and expands, its relative humidity drops, it absorbs moisture from the ground, and causes high evaporation rates. The Atacama Desert, the driest desert in the world, occurs along a coast with a bordering cold current.

1. Examine ■ Figure 18.4 to determine which deserts are which types.

a. Write the desert type(s) next to the name in ■ Table 18.1. If more than one type is present or more than one factor influences the climate for a desert, list all relevant desert types. Remember to consider the prevailing wind direction (Figure 18.2) when locating rain-shadow deserts.

b. Some deserts have a mixture of factors that cause them to be dry. Generally, the more factors involved, the drier the desert. For instance, the Atacama Desert fits into three desert categories, giving it three reasons for being dry. What three reasons make the Atacama so dry?

____

____

____

2. ■ Figure 18.5 shows the arrangement of continents some 260 million years ago. Use it for this exercise.

a. Sketch in the pattern of global winds. If necessary, refer to Figure 18.2 and Lab 15 for a discussion of the global wind patterns, page 342.

b. Use the sketched in winds patterns, the ocean currents from Lab 14 on the figure, and your knowledge of high-pressure belts in the atmosphere to determine where deserts were during the Late Permian. Indicate on the continents these probable desert areas using colors (or symbols) for different types of deserts and make a corresponding key.

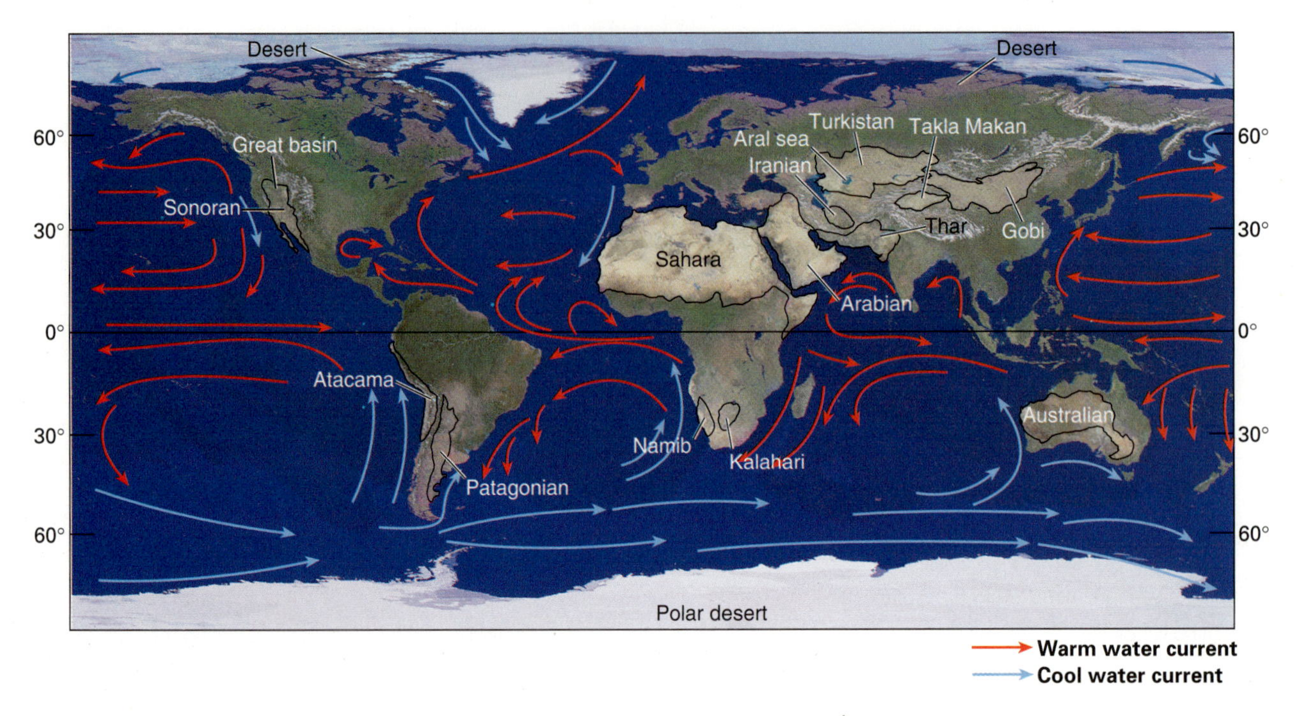

Figure 18.4

World Deserts

Locations of deserts worldwide. Recall the relationship between cold ocean currents and coastal deserts mentioned in Lab 14.

Table 18.1

Desert Types

See instructions in Exercise 1.

Desert	Desert Type(s)
Antarctica	
Arabian	
Atacama	
Australian	
Desert in Northern Canada	
Desert in Siberia	
Gobi	
Great Basin	
Kalahari	
Namib	
Patagonian	
Sahara	
Sonoran	
Takla Makan	
Thar	
Turkistan	

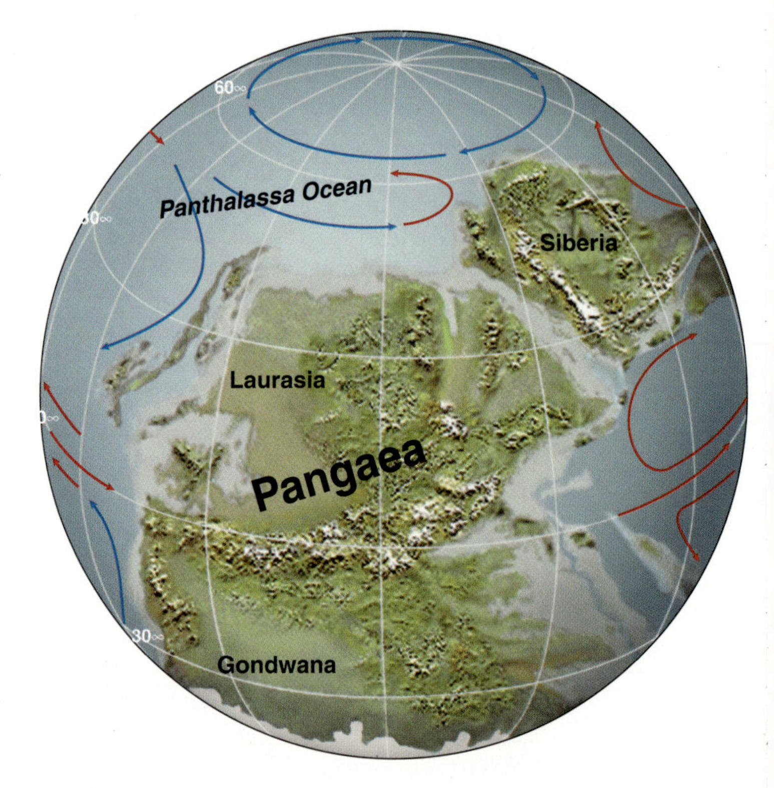

Figure 18.5

Pangaea in the Late Permian

Likely locations of warm (red arrows) and cold (blue arrows) ocean currents at the time are shown.

WIND TRANSPORT AND EROSION

Wind action is an important geological agent in deserts and at coastlines. In deserts, dry conditions allow more sediment transport because little moisture or vegetation holds sediment grains together. The lack of vegetation is also important because vegetation slows wind dramatically at the surface. In coastal areas, winds are especially strong or persistent because of the differential heating of water and land. This means that some of the processes observed in deserts, especially dune formation, may also occur in coastal regions as well. Wind has a much lower viscosity and density than water, so it picks up and carries much less sediment than streams or running water. Wind rarely picks up large pebbles (only hurricane-force winds and tornadoes are capable of doing so). Water causes major erosion in deserts and it is generally more effective in deserts than wind. Therefore, the flow of water is still the primary agent of erosion and deposition in deserts.

Nevertheless, deserts and coasts are where wind is more effective than in other regions, so it makes sense to study wind's influence in these locations. Wind is able to pick up small grains, which move in different ways depending on the strength of the wind and the size of the grains. ■ Figure 18.6 shows the relative height that wind lifts different-sized grains. Notice that only very fine sand reaches higher than about 5 cm (2 inches) above the ground. Some sand sizes move by **saltation,** which is a hopping or jumping motion. Coarser sediment moves by **creep** along the surface and from the impact of saltated grains. Turbulence can move finer sediment such as silt and clay in suspension. The clay size fraction can travel great distances, as seen during dust storms. Silt- and clay-sized dust can readily become suspended in light winds if the sediment surface is disturbed, such as with the passage of a vehicle on a dirt road, or a plow in a farmer's field. On the other hand, sand will suspend, and only temporarily, with winds stronger than about 20 km/hr.

Wind erosion takes two forms: **deflation,** which is the carrying away of sediment by wind; and *abrasion,* which occurs when wind-driven grains hit rock surfaces at high velocities. Deflation occurs where the wind lifts and blows away finer, loose material on the ground. **Deflation hollows** or basins form as a result of wind excavating sediment. Most deflation hollows are less than 2 m deep and 2 km long; but where sediment is especially susceptible to wind erosion, they can reach depths of 50 m or more.

Desert pavement is a surface found in many desert areas where larger sediments, such as pebbles and cobbles, densely cover the ground, as if someone intentionally paved the surface. Desert pavement occurs in areas where deflation removes the finer sediment from the surface and leaves the coarser sediment behind as a **lag** deposit, which then protects the surface, and the smaller-sized sediment beneath, from further erosion. Another mechanism that could form desert pavement is deposition of fine material on top of coarser sediment and then washing of the fine sediment down between the larger stones. Wetting, drying and shrinking of the surface material may allow the fine sediments room to move downward. Both mechanisms may occur.

The second mode of wind erosion is **abrasion.** This occurs when wind bombards a surface with saltating or suspended sediment grains. The effect is a kind of sandblasting that can remove a substantial amount of material over time. Since coarser sediment such as sand does not rise very high during wind transport, its effects are concentrated close to the ground. However, those effects can be very strong, so that during a single dust storm, certain items, such as fence posts, can be noticeably eroded. ■ Figure 18.7 shows rocks in Antarctica that polar winds have repeatedly sandblasted.

Ventifacts are wind-abraded rocks with a faceted (flattened) side (Figures 18.6 and 18.7). (*Ventifact* has the same root as vent and ventilation, from the Latin *ventus,* meaning wind.) The prevailing wind abrades, erodes, and sculpts the rock on one side. For some

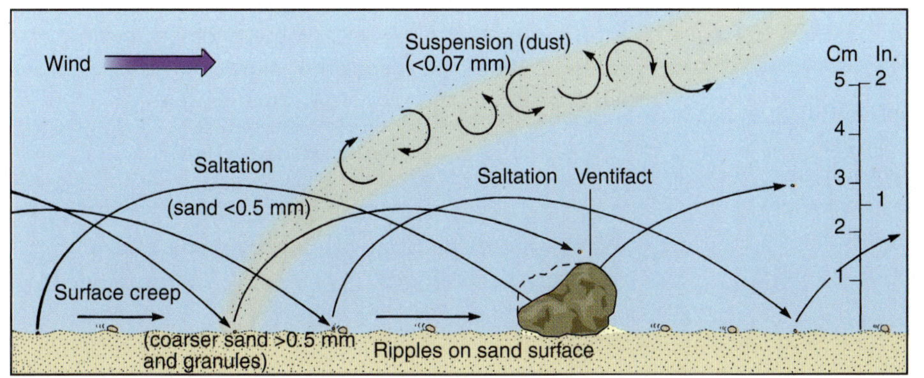

Figure 18.6

Movement of Sediment by Wind

Wind speed at the ground surface is quite low, but increases with height. At ground level, the larger grains (>0.5 mm, medium sand and coarser sediment) that can move at all, move by surface creep. Above these, smaller grains from 0.07 mm to 0.5 mm move by saltation. Only fine sediment becomes suspended in air, but when it does, its finest fraction can be blown thousands of miles. From GABLER/PETERSEN/TREPASSO/SACK. *Physical Geography*, 9E. 2009 Brooks/Cole, a part of Cengage Learning, Inc. Reproduced by permission. www.cengage.com/permissions

Figure 18.7

Ventifacts Near Lake Vida, Antarctica

The strong polar winds off the East Antarctic Ice sheet have eroded these rocks.

ventifacts, a seasonal shift in the wind or a movement of the rock allows the wind to sculpt another side giving these ventifacts a distinctive shape.

3. Consider desert surfaces discussed above.

a. What processes in a desert would lend themselves to forming desert pavement as lag deposits?

____

What processes would lead to downward sifting of fine sediment to produce desert pavement?

____

b. On Figure 18.7 sketch the wind directions that caused erosion for the largest five boulders. Highlight the most dominant wind for each boulder, either with a highlighting pen or by drawing that arrow bolder than you drew the others.

4. On a separate sheet of paper, draw sketches of the ventifacts that your instructor provides. Add arrows to your sketches showing the direction of the wind that produced the facets on the ventifacts. If one wind direction is more prominent, having caused more erosion than another, draw that arrow bolder than the others.

Erosional Desert Landforms

In humid regions, slopes tend to become more and more gently rounded because slope-reducing processes, such as rain splash and soil creep, are important erosional agents (■ Figure 18.8a). In arid lands, slopes tend to maintain a constant angle because slope-decreasing processes are not important. Instead, slope-undercutting processes such as abrasion and stream erosion dominate. With erosional forces stronger at the base of the slope, desert hills and mountains retreat parallel to the hillside, maintaining the slope steepness in a process called **cliff retreat** or **scarp retreat** (Figure 18.8b and c). As the cliff retreats, the base becomes a relatively flat bedrock plain or slope covered with a thin veneer of sediment called a **pediment.** Isolated, steep hills, often made of resistant rock types, may exist in the pediment and are called **inselbergs** (German for "island mountains," ■ Figure 18.9).

In arid lands with *flat-lying sedimentary beds,* a series of landforms develop that characteristically have steep cliffs (Figure 18.8c). Rocks that are resistant to erosion in desert climates, such as sandstone and limestone, form the steep cliffs whereas other rocks, such as shale, erode easily, forming gentle slopes. If these rocks are interbedded, they form **cliff and bench topography** (Figure 18.8c).

Wind and water commonly attack vertical cracks in the rocks called *joints,* eroding them faster than surrounding unjointed rocks. Steep cliffs of plateaus erode back, until lateral and headward stream erosion dissects the plateaus into smaller, nearly flat-topped units called **mesas** (Spanish for "tables"). A mesa is considerably wider than it is high. Stream processes may further erode mesas, isolating parts called *buttes* or the even narrower *monuments,* also called *spires* or *pinnacles* (■ Figure 18.10). **Buttes** are roughly one or two times wider than they are high and **monuments** are much taller than they are wide.

5. Examine the area of the Grand Canyon in Figure 8.13 on pages 176–179. Notice that some of the formation names in the Explanation include rock names. Name two with rock names where the formation stands out as steep cliffs in the map and cross section.

____

____

Name one formation with a rock name that makes gentler slopes.

____

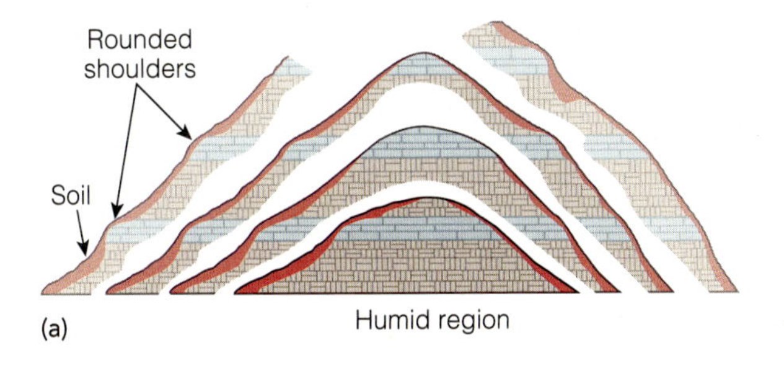

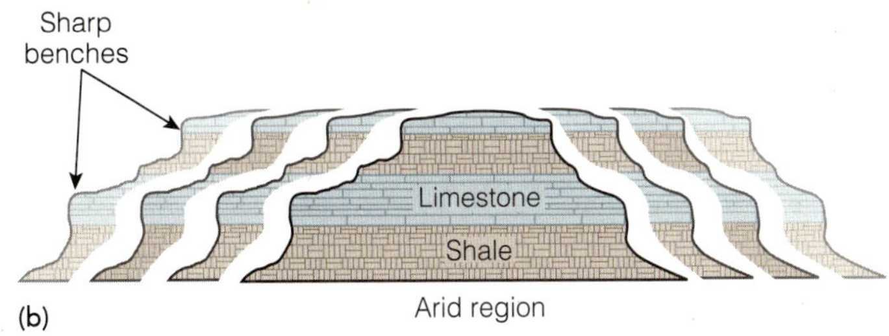

Figure 18.8

Slope and Cliff Retreat in Humid and Arid Climates

(a) In humid regions, soil development, slump, soil creep, rain splash, and possibly dissolution of limestones, wear down hills to gentler and gentler slopes. (b) In arid regions, cliffs retreat maintaining their slope. (c) Horizontal layers in the Grand Canyon erode by cliff retreat, leaving steep cliffs of harder rocks and gentler slopes of softer shale in *cliff and bench topography.*

(c)

What is the name for this type of topography?

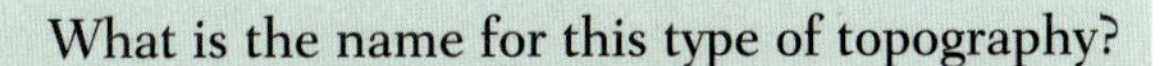

6. Label desert landforms as requested: On Figure 18.9b outline and label all the visible desert landforms. On Figure 18.10b, label all the monuments, buttes, and mesas. Some of these may appear in the distance.
7. Examine the photograph and topographic map of Monument Valley, Arizona in ■ Figures 18.11 and ■ 18.12.
   a. On the map, what type of feature is the totem pole?

   ____

   b. Match the photo with the map and label the monument names on the photo. Circle and label areas on the map that you can see in the photo foreground, middle ground, and background.
   c. Find named mesas on the map. What characteristics do they share that make them mesas?

   ____

   ____

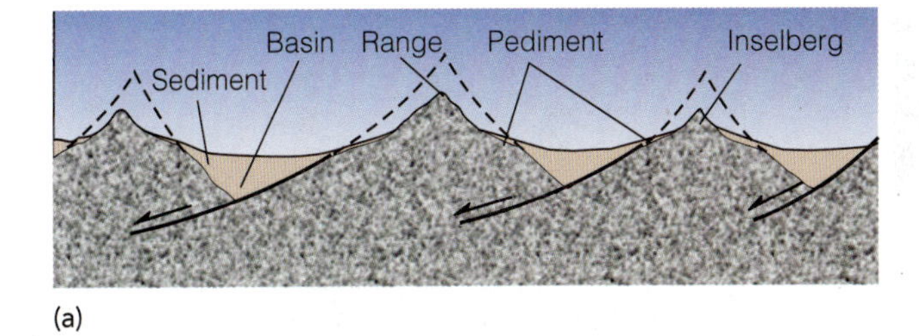

(a)

(b)

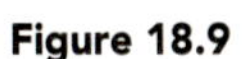

Figure 18.9

Inselbergs

(a) Inselbergs developing as erosional remnants of larger mountains isolated from each other and surrounded by bedrock pediment, covered with a thin veneer of sediment. (b) Photograph of an inselberg with surrounding pediment.

Figure 18.10

Mesas, Buttes, and Monuments

(a) Diagram showing how a mesa, a monument, and a butte were once connected as continuous, horizontal rock layers. Later erosion separated these dramatic desert features. The cliff-making rocks are sandstone and the sloping rock layers are shale. (b) West Mitten Butte in Monument Valley, Arizona. The spires, buttes, and mesas of Monument Valley were once a continuous layer of red sandstone (the Cutler Formation) deposited during the Permian Period (about 160 million years ago). It underwent later uplift and erosion as part of the Colorado Plateau.

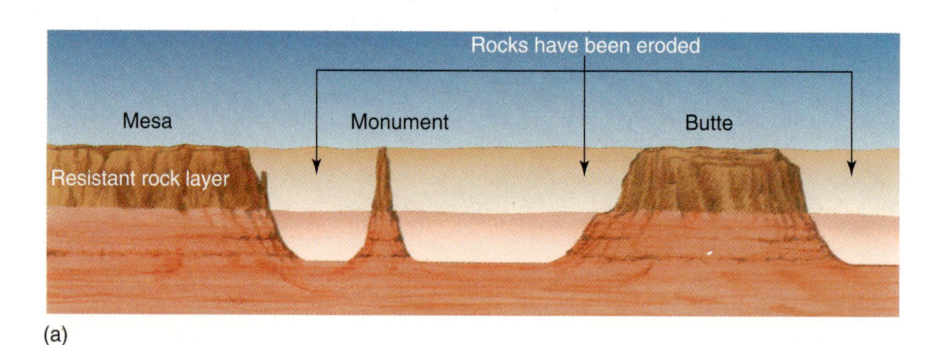

(a)

Parvinder Sethi

(b)

Bryan Brazil/shutterstock.com

Figure 18.11

Monument Valley

Navajo Tribal Park, Arizona, looking south to southwest not far from the area shown in Figure 18.10.

Figure 18.12

Topographic Map of Monument Valley Area (*facing page*)

Scale 1:24,000 USGS. Figure 18.11 is a photo of the central part of this map area and Figure 18.10 is a short distance to the north.

1 ½ 0 1 MILE

1 .5 0 1 2 3 KILOMETERS

Spearhea
5008AT
2-60
5902T
5820T
5921
Rain God Mesa
Gypsum
4879AT
5166T
5200
5800
5000
R E S E R V A T I O N
5185T
5092T
Sand Spring
Thunderbird Mesa
Totem Pole
5621T
Yel-Bichel
5631T
Big Chair
ncake Flats
5612T
Moqui Step Mesa
2-61
5703T
5242AT
5200
Tse Biyi
5758T
5780T
5926T
Big Hogan
5769T
5706T
5702T
5400
5808T
Hidden Arch
Yazzie
5735T
5786T
5753T
Natural Arch
6010T
5838T
2-62
6066T
5440AT
5600
6192T
4WD
HUNTS
MESA
Radio Tower
6346T
6376T

d. Label two more mesas that are not named on the map.

e. Locate and label two places where vertical or overhanging cliffs occur on the map. How do you recognize such places using contours?

____

f. Label three buttes. How do you distinguish a butte from a mesa?

____

____

____

g. Locate and label another monument on the map not visible in the photo.

h. About what percentage of the map area does sand (stippled pattern) cover? Here's how to compute it: For each kilometer square on the map that has sand, give a visual estimate of how much of that square kilometer is covered with sand and write the number in the square. Then add up all the fractions or percentages, and divide by the total number of square kilometers on the map.

____

i. Notice the elevations of the labeled summits of buttes and mesas in the area. What do you think caused many of them to have such similar heights?

____

____

j. Use Figure 18.10 to help you color in three geologic units on the four map squares in the area between Big Chair, Totem Pole, Hidden Arch, and the word "Yazzie" in Figure 18.12: 1) a resistant rock layer with very steep cliffs and tops of mesas, buttes, and monuments, 2) a moderately resistant rock layer with moderate slopes and 3) an easily eroded rock layer, gently sloping to almost flat. Assume that the rock layers are flat. Make a key or label the units.

k. Given the climate conditions here, name a rock type that might form the cliffs ______________________ and one that might form the gentlest slopes.

____

DEPOSITIONAL DESERT LANDFORMS

Deposition in deserts occurs in a number of ways. Wind-blown sediment includes sand and fine silt. Water deposits fans and aprons of coarser material at mountain fronts and finer sediment and salt in temporary lakes within enclosed basins.

Dunes

Many people think of deserts as made up entirely of sand dunes, but dunes actually occupy only a small proportion of deserts. Dunes make up only about 10% of the Sahara Desert, for example, and about 30% of the Arabian Desert. Other deserts have even smaller proportions of dune-covered areas. Typical desert scenes have no dunes and, instead, mostly consist of exposed bedrock or desert pavement. However, where sufficient sand is present, dunes can develop.

Wind forms sedimentary deposits called **eolian** deposits. If these are wind-blown deposits of sand, they create hills or ridges of sand know as **dunes** (■ Figure 18.13). Wind blows the sand up the gently sloping windward side of the dune by saltation. When the sand reaches the top, it blows off the top (Figure 18.13b) and falls onto the **lee side** (downwind) of the dune where the dune protects it from the wind. Sand here tends to build up until it reaches the angle of repose (Lab 11). Additional sand will make the sand "slip" downhill (actually this movement is *grain flow*), so the lee side of the dune is known as the **slip face** (Figure 18.13a and b). Dunes may be asymmetrical, with a much steeper slip face at the angle of repose, usually about 33° to 34°, and a more gently sloping windward side.

Dune shapes and sizes vary considerably, depending on wind strength, sand supply, and amount of vegetation (■ Figure 18.14). Table 18.2 summarizes similarities and differences among different types of dunes. Both **barchan dunes** and **parabolic dunes** have a crescent-like shape, but the wind is from opposite directions (Figures 18.15a and b), with tips downwind for barchan and upwind for parabolic. Barchan dunes form with low sand supply. Parabolic dunes commonly form where vegetation helps to anchor the dune tips and are especially common in coastal dunes. Two other dune types have relatively straight crests, but again the wind that produces them is from different relative directions. **Transverse dunes** form perpendicular to a moderate wind, but **longitudinal dunes** form parallel to strong winds (Figures 18.15c and d). **Star dunes** have multiple dune tips pointing in different directions due to varying wind directions and form the highest of all the dunes.

Some dunes migrate downwind because the sand erodes off the windward side, and settles on the downwind side (Figure 18.13b). If the wind continues

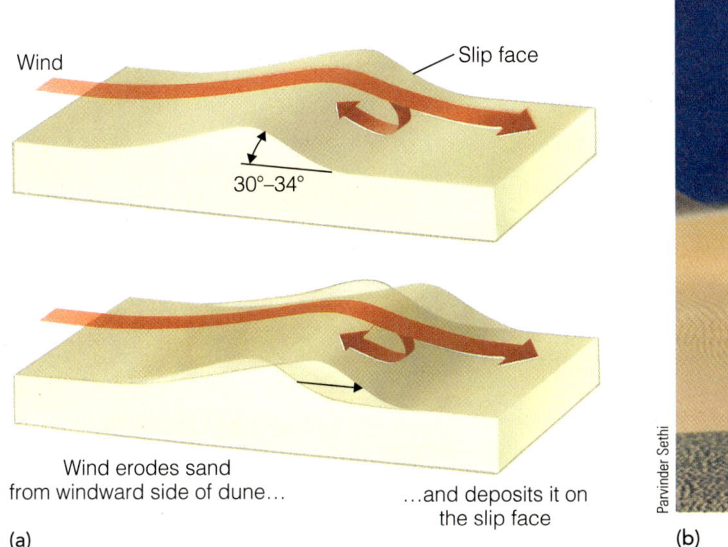

Figure 18.13

Sand Dune Migration

(a) A transverse dune. The slip face of the dune is downwind and steeper than the windward side. On the leeward side air flow produces a back eddy. The dune migrates downwind. (b) Dune in Death Valley, California. The rippled side is upwind and the steeper right side of the dune is the slip face. Wind is blowing sand off the top of the dune and onto the slip face.

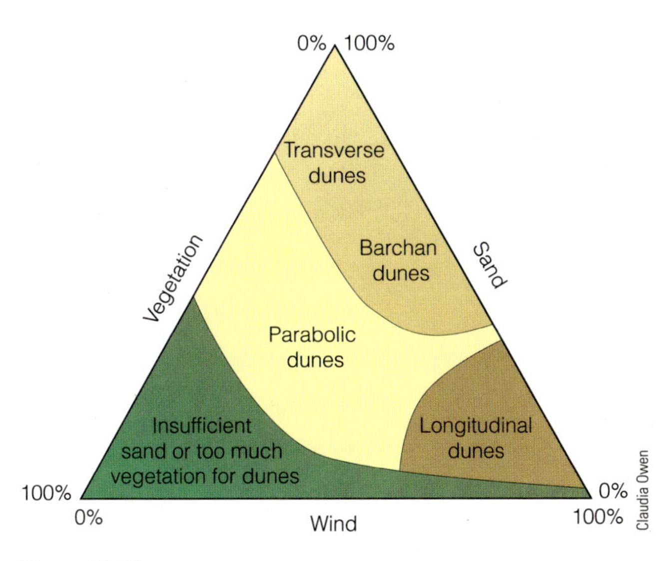

Figure 18.14

Triangle of Dune Types

Different dune types form under different conditions of wind strength, sand supply, and vegetation cover. The apex of the triangle represents the highest sand supply, and the base has no supply of sand. The strongest winds are on the lower right corner, and the left edge corresponds to areas with no wind. The lower left corner represents abundant vegetation cover, where no dunes would exist. The opposite side is for no vegetation. Dunes do not form if sand is unavailable or if too much vegetation stabilizes the sand. Adapted from J.T. Hack, 1941.

from the same direction, the whole dune will eventually move. Barchans move at speeds up to 25 m/yr, which is fast for a dune. Dune migration forms cross bedding as discussed in Lab 4. Some dunes migrate considerably when wind is consistently coming from one direction, but may migrate back as wind shifts, so that over time the dune's net migration is very small. The center of star and longitudinal dunes do not tend to migrate although their sand does shift position.

8. Examine Figures 18.1, ■ 18.15, and ■ Table 18.2.

a. What type of dune is visible in Figure 18.1?

____

b. Based on information in Table 18.2, plot and label the approximate location of this dune type on the triangle in Figure 18.14.

9. Examine the dunes at White Sands, New Mexico, in ■ Figure 18.16. Unlike most dunes, which are quartz, these dunes consist of grains of the mineral gypsum.

a. What type of dune is present in Figure 18.16a?

____

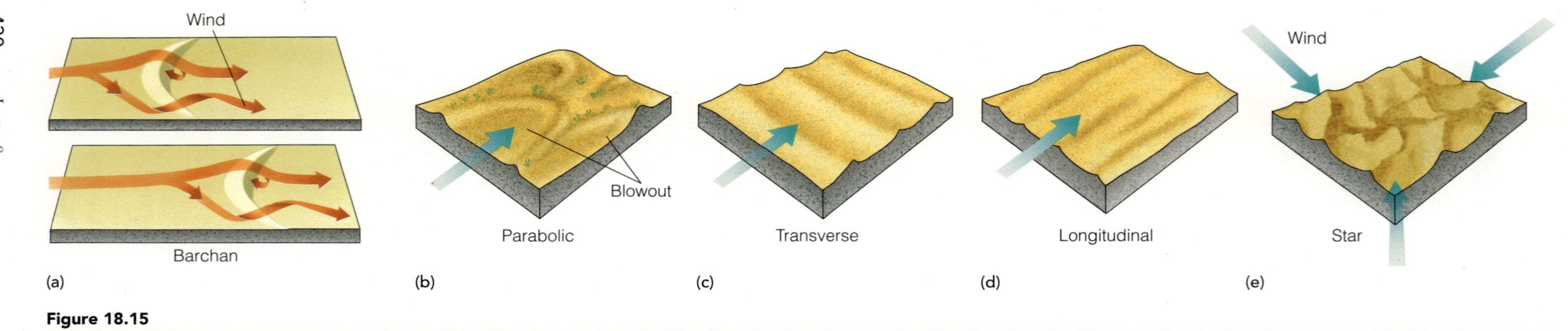

Figure 18.15

Some Types of Dunes

(a) Barchan dunes have their tips pointing downwind, with the slip face on the inside of the crescent. (b) Parabolic dunes have tips pointing upwind, commonly stabilized by vegetation. The slip face, if it is not embedded in vegetation, is on the outside of the crescent. (c) Transverse dunes are linear and perpendicular to the wind. (d) Longitudinal dunes are also linear but parallel to strong winds and in an area of low sand supply. (e) Variable wind directions make star dunes. From GABLER/PETERSEN/TREPASSO/SACK. *Physical Geography,* 9E. 2009 Brooks/Cole, a part of Cengage Learning, Inc. Reproduced by permission. www.cengage.com/permissions

Table 18.2

Dune Types

And their characteristics.

Dune	Geometry				Wind	Dune Migration	Sediment Supply	Vegetation/Surface
	Size	Shape	Cross Section	Orientation				
Barchan dunes	Small up to 30 m high	Crescent	Asymmetrical, steeper on inside curve than outside curve	Dune tips downwind	Nearly constant direction	Fastest Some > 10 m/yr, up to 25 m/yr	Limited	Little vegetation, flat dry surface
Parabolic dunes	Typically 10s of m high. Longest arms—12 km long	Parabolic (reverse crescent)	Steeper on outside curve than inside curve	Dune tips upwind	Strong—causes blowouts and deflation	Migrate	Abundant to moderate	Form in vegetation gaps/Some blowouts down to water table
Transverse dunes	200 m high and 3 km wide	Long, straight	Asymmetrical	Perpendicular to wind	Moderate	Migrate	Abundant	Little or no vegetation
Longitudinal (self) dunes	3 to 100 m high, up to nearly 200 km long	Long parallel ridges	Symmetrical	Parallel to prevailing wind	Strong from slightly different directions	Stationary	Low to moderate	Little or no vegetation
Star dunes	Some > 100 m high	Star-shaped dunes	Variable	Prongs point in multiple directions	Multiple directions	Stationary	Limited	Little vegetation

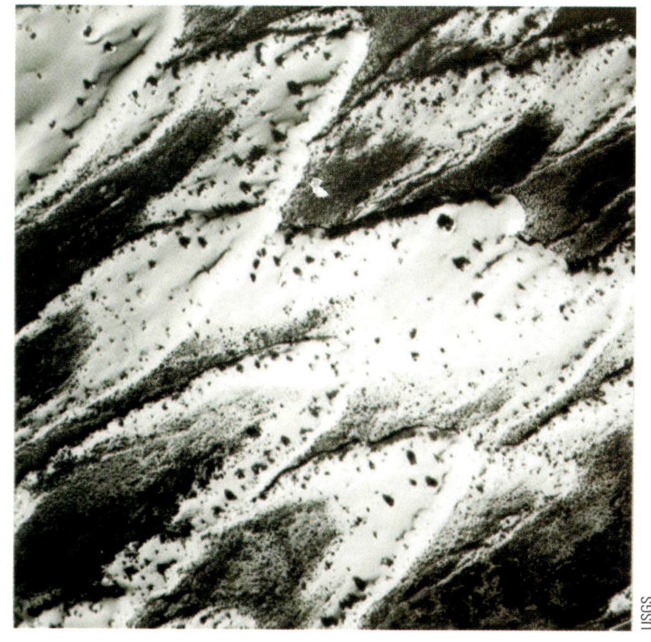
USGS
(a)

National Park Service
(b)

Figure 18.16

Dunes at White Sands, New Mexico

(a) Aerial photograph of White Sands dunes. The mineral gypsum makes up this unusually colored sand. Dark spots on the photograph are vegetation. (b) A White Sands dune of a different type than in (a) in another region of the dune field.

b. How can you tell these are not the other type of crescent-shaped dune?

____

____

____

c. The small dark specks in Figure 18.16a are vegetation. In what way does vegetation influence the formation of this type of dune?

____

____

d. Draw the prevailing wind direction on the aerial photo in Figure 18.16a.

e. Examine the curvature and steepness of the slope of the dune in Figure 18.16b, which is also in White Sands, New Mexico. What type of dune is this?

____

How can you tell?

____

____

10. Examine the dunes near the Salton Sea, California, in ■ Figure 18.17. Use a stereoscope with Figure 18.17a.

a. What type of dune is present here?

____

b. How would you recognize the slip face of a sand dune?

____

c. Study the dunes in Figure 18.17a with the stereoscope. What direction do the slip faces of these dunes face?

______________________ Are they all consistent? _______

d. What is the prevailing wind direction?

____

e. Figure 18.17b shows the same dunes about 44.4 years later. Compare them with those in Figure 18.17a. Some of these have numbers so you can match them up, but others are for you to match. On Figure 18.17a, label the dunes that match dunes 1, 3, 5, 7, and 8 in Figure 18.17b.

f. The red X's on the two photos occur at matching locations. Draw a line connecting the two red X's and extend it beyond to the full lengths of both photos.

g. Measure the shortest distance to this line in mm from the top of the slip face in the middle of the bend of the crescent

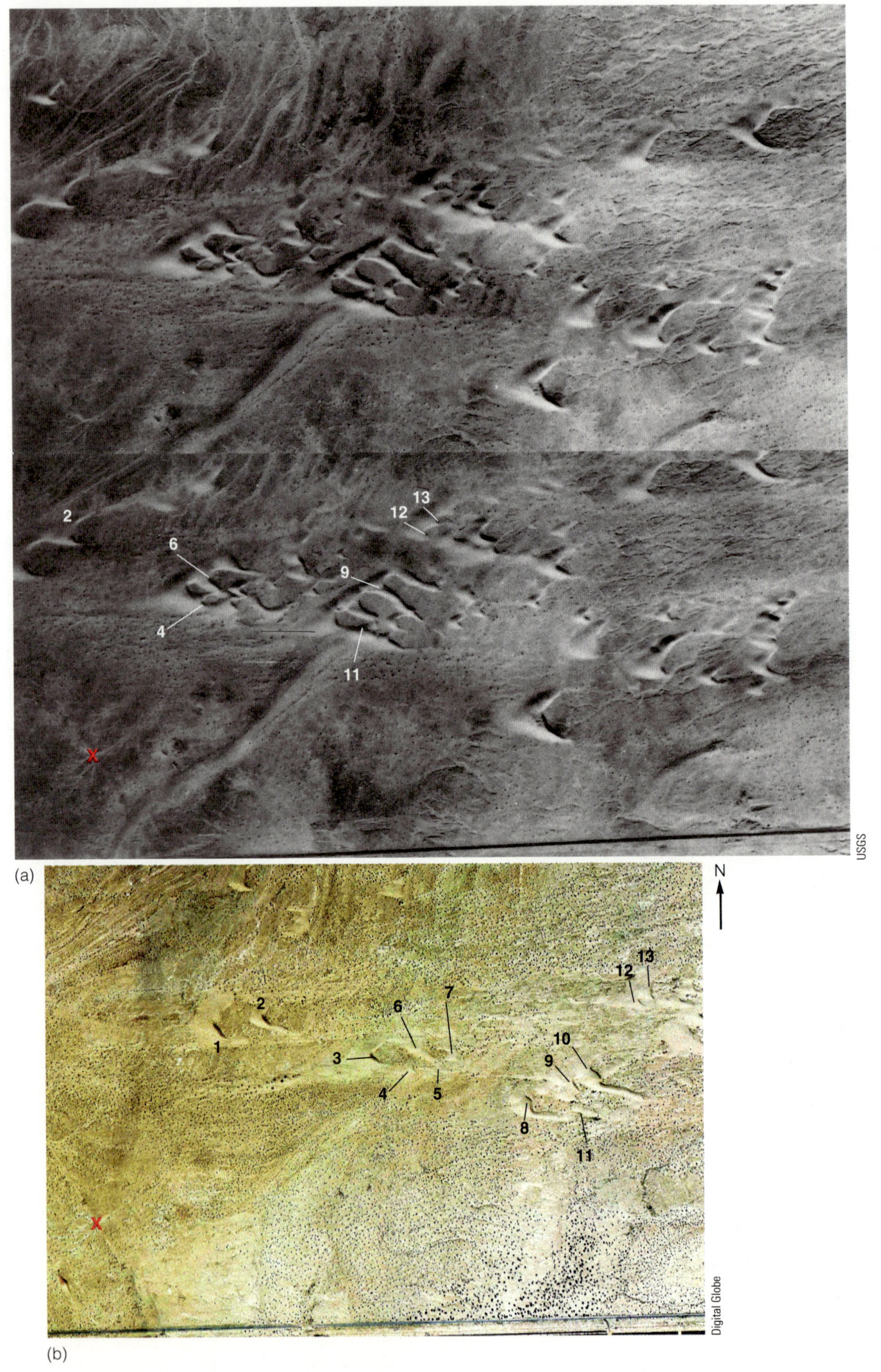

Figure 18.17

Dunes Southwest of Salton Sea

A dune field southwest of the Salton Sea in Imperial Valley, California. (a) A stereophoto pair taken in November 1959. Scale 1:16,000. (b) GlobeXplorer satellite image of part of the same area taken in April 2004. Red Xs are matching locations in the two sets of photographs.

Table 18.3

Dune Migration Measurements

1959 to 2004. See instructions in Exercise 10 to complete parts of this table.

Number of Matching Dunes from Salton Sea	Distance from the Line on 1959 Air Photo (mm)	Distance from the Line on 2004 Air Photo (mm)	Distance Migrated on Air Photo between 1959 and 2004 (mm)	Distance Migrated on the Ground between 1959 and 2004 (m)	Speed of Dune (m/yr)	N-S Dimension of Dune (m)
1						168
2						134
3						180
8				555	12.5	108
9				529	11.9	167
10				514	11.6	160

for each of the dunes 1, 2, and 3 in each photo. Enter the measurements in ■ Table 18.3, recording a negative number if the dune is left of the line and a positive number if it is right of the line.

h. At a scale of 1:16,000, 1 millimeter on the photo is equal to how many meters on the ground? ________

i. Calculate how many millimeters each dune migrated on the air photos (by subtraction) between 1959 and 2004. Then convert this to meters on the ground and enter both numbers in Table 18.3.

j. Calculate and record in Table 18.3 the speed of dunes movement in meters per year for the 44.4-year period. Since you calculated these speeds over the whole span of 44.4 years, they are averages for each individual dune.

11. Use your numbers and the other measurements in Table 18.3 with Figure 18.18. Use a stereoscope with Figure 18.17a.

a. Complete the graph in ■ Figure 18.18 for all six dunes listed in the table. The other points on the graph are measurements made for the other numbered dunes in Figure 18.17.

b. Examine and analyze the graph. What size dunes move the fastest?

____

c. What is the general relationship between how long a dune is and how fast it moves?

____

d. Hypothesize why this is true.

____

____

Stream Deposits in Deserts

We have just examined the action of wind in deserts, but what about water? Water action in deserts is surprisingly influential in modifying desert landscapes. Although rainfall is low, water is still a highly effective erosive and depositional force in deserts, partly because of the shortage of plants and partly because when water does arrive, it often does so in a deluge. In fact, water is a much more effective agent at sculpting the landscape than wind is. Many streams in deserts, called **ephemeral streams,** are dry most of the time, but when a rainstorm arrives, streams rise in moments. Flash flooding is common in deserts after heavy rainstorms. Debris flows and mudflows can occur in deserts as well.

Alluvial fans form where mountain streams abruptly issue from a narrow canyon and spread out into wide, gently sloping basins or valleys (■ Figure 18.19a). These are fan-shaped deposits of poorly sorted sediment that can contain from fine clays up to large boulders. Stream deposition, mudflows, and debris flows aid the growth of alluvial fans. **Bajadas** occur where multiple alluvial fans join together to form a wide apron of sediment at the base of a mountain range (Figure 18.19b).

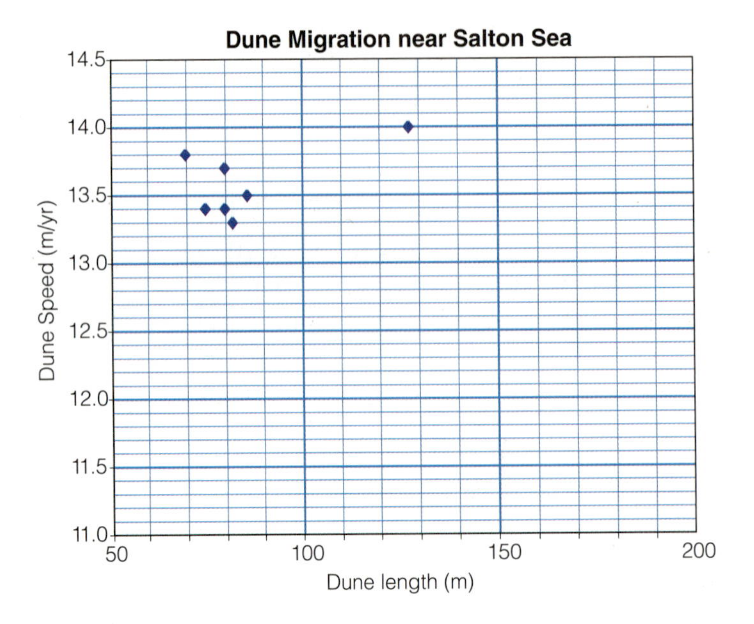

Figure 18.18

Graph of Dune Migration

A few small dunes have been plotted on the graph. See Exercise 11 for instructions for plotting additional dunes on this graph.

(a) (b)

Figure 18.19

More Desert Depositional Features

(a) Alluvial fan in Death Valley, California. This fan has developed as a result of recent uplift along a normal fault at the front of the range. (b) Alluvial fans coalesce (combine) to form a bajada along a fairly eroded mountain front. Pediments occur in areas near the mountains at the same slope as the bajada.

Desert Lakes

Playas are expanses of salt and fine-sediment deposits that form in ephemeral lakes in desert regions. As streams bring water to desert basins, they carry dissolved sediment. Where the water collects in temporary lakes, it evaporates, leaving behind deposits of salts including gypsum (hydrous calcium sulfate) and halite (sodium chloride). Figure 4.1b on page 72 shows an extensive playa in Death Valley, California. During the Ice Age, greater rainfall produced vast lakes, which have since mostly dried, leaving large amounts of salt. The gypsum in White Sands, New Mexico, also formed by evaporation but from an ancient shallow sea 250 million years ago. Later, erosion stripped the gypsum from the local hills and deposited it in sand dunes.

COMBINED LANDSCAPES

Major desert areas within North America occur in the Basin and Range Province in Nevada, Arizona, and parts of Utah, where *normal faults* are actively uplifting ranges and dropping down basins (valleys). As this happens, a distinctive landscape develops. ■ Figure 18.20 shows the stages of development of mountain fronts, alluvial fans, bajadas, pediments, and inselbergs. In the first stage, with normal faulting and rapid uplift (Figures 18.20a and 18.21a), landforms include steep, triangular-shaped fronts of the mountains, and deep, V-shaped canyons, alluvial fans, and playas (■ Figure 18.21a).

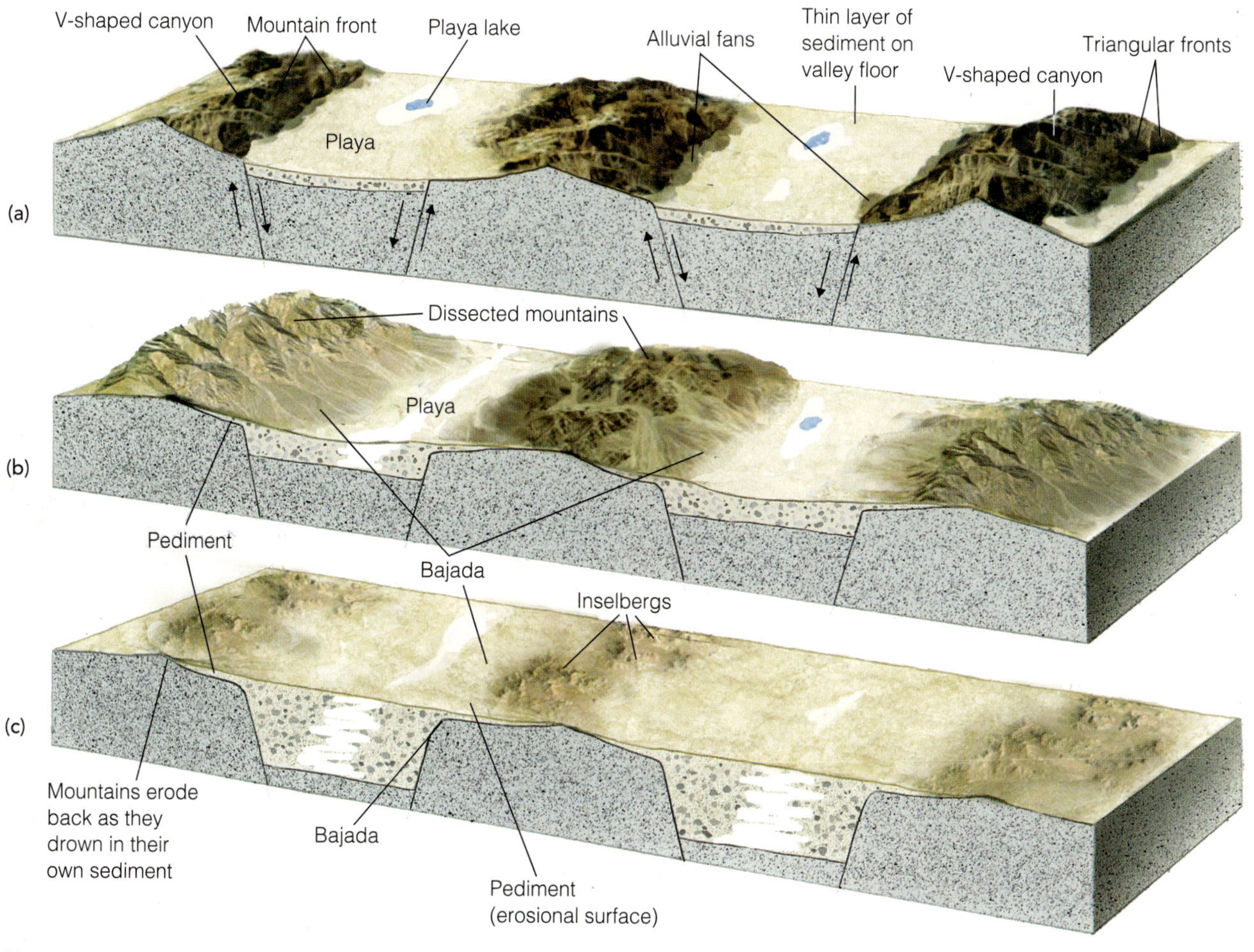

Figure 18.20

Stages of Basin and Range Erosion

Formation of alluvial fans, bajadas, pediments, and inselbergs in a series of stages of uplift and erosion. (a) Uplifted blocks create steep mountain fronts along normal faults (see Lab 8 for discussion of fault types). During the *early stage* of erosion of the uplifted mountains, desert streams produce individual and distinctly separate alluvial fans of coarse sediment. Erosion of V-shaped canyons into the mountain fronts form triangular faces. Basins are wide playas with salt and fine sediment deposits. (b) Continued stream erosion in the *middle stage* lowers the mountains. Scarp retreat begins to produce pediments. Alluvial fans coalesce into bajadas. Streams create both pediments and bajadas, the first by erosion and the second by deposition. (c) In the *late stage*, bajadas and pediments become indistinguishable at the surface but have vastly different thicknesses of sediment below the surface. All that remains of the mountains are small inselbergs surrounded by pediment. Renewed faulting can return the landscape to the stage shown in (a).

(a)

(b)

Figure 18.21

First and Third Stages of Basin and Range Erosion

(a) First stage: Two alluvial fans along a well developed and little eroded mountain front. A normal fault separates the mountains from the basin. V-shaped canyons dissect the steep mountain slope, forming a triangle-shaped mountain front. This landscape results shortly after uplift. White areas contain salt deposits from playa lakes that remain dry most of the time. (b) Inselbergs and pediment are all that remain of a mountain range in this location in the Mojave Desert, California. In comparison to Figure 18.9b, note that erosion is more advanced and the color of the eroded area clearly shows where the pediment occurs.

12. Examine the satellite image of Badwater Basin, Death Valley, California (■ Figure 18.22). This is the lowest place in the United States at 86 meters below sea level. Figure 18.21a is a closer view of the east side of the basin.

a. This is a composite of several photos, some of which were taken at different times. Carefully examine the areas of lowest elevation where diagonal brown and white areas were pieced together. What type of desert landform is this valley?

____________________________ In what ways, besides the color, was the area different at the time the brown photos were taken compared to the white photos?

____

____

What is the white material?

____________________________ (See also Figure 18.21a.)

b. Identify, outline, and label in Figure 18.22 alluvial fans, a bajada, playa.

c. Mountains rise up on both sides of the valley, but their geometry is quite different. Also, outline, and label sloping uplands deep V-shaped canyons, and dissected mountains.

d. The triangular mountain fronts are difficult to spot in Figure 18.22 so find, outline, and color them in Figure 18.21a first. You may be able to find them as small, steep triangles, mostly in shadow, between the alluvial fans on the satellite image in Figure 18.22.

e. Use Figure 18.20 to help you determine what the stage of development the west side of Badwater Basin is in Figure 18.22.

____

What about the east side?

____

f. Which side had more recent faulting?

______________ How can you tell?

____

____

13. Examine the topographic map of Devils Golf Course in ■ Figure 18.23, which is the area directly north of Figure 18.22.

a. Identify, outline, color, and label on the map each of the following features: alluvial fans, a young bajada, old bajada, playa, playa lake, mountains, and triangular mountain fronts. In an area of white space on the map, make a key indicating what each color means.

Digital Globe

Figure 18.22

Satellite Image of Badwater Basin, Death Valley, California

Scale = 1:180,000. DigitalGlobe composite posted on Google.

b. Identify the stage of development, using Figure 18.20, of the mountains on the west side of Figure 18.23 ____________

and on the east side ____________

c. Which side of the map had more recent faulting? ____________

14. Examine the shaded relief topographic map of Antelope Peak, Arizona, and vicinity in Figure 18.24.

a. Use colored pencils to color all inselbergs in the area and label them with an "IB."

What two other features are common in areas with inselbergs?

____

____

b. On a topographic map, bajadas and pediments have the same slope. What are the geologic differences between the two?

____

____

____

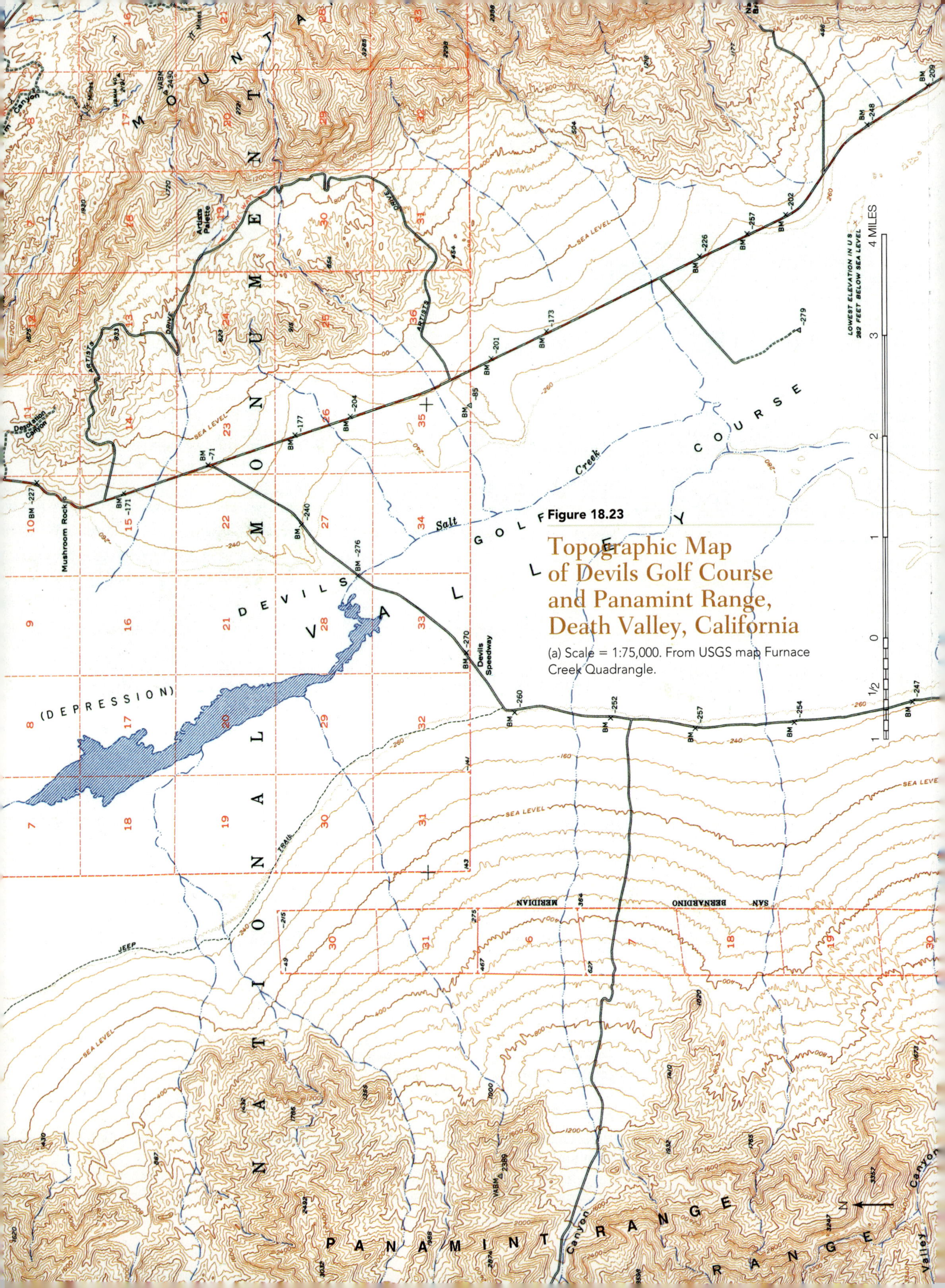

Figure 18.23

Topographic Map of Devils Golf Course and Panamint Range, Death Valley, California

(a) Scale = 1:75,000. From USGS map Furnace Creek Quadrangle.

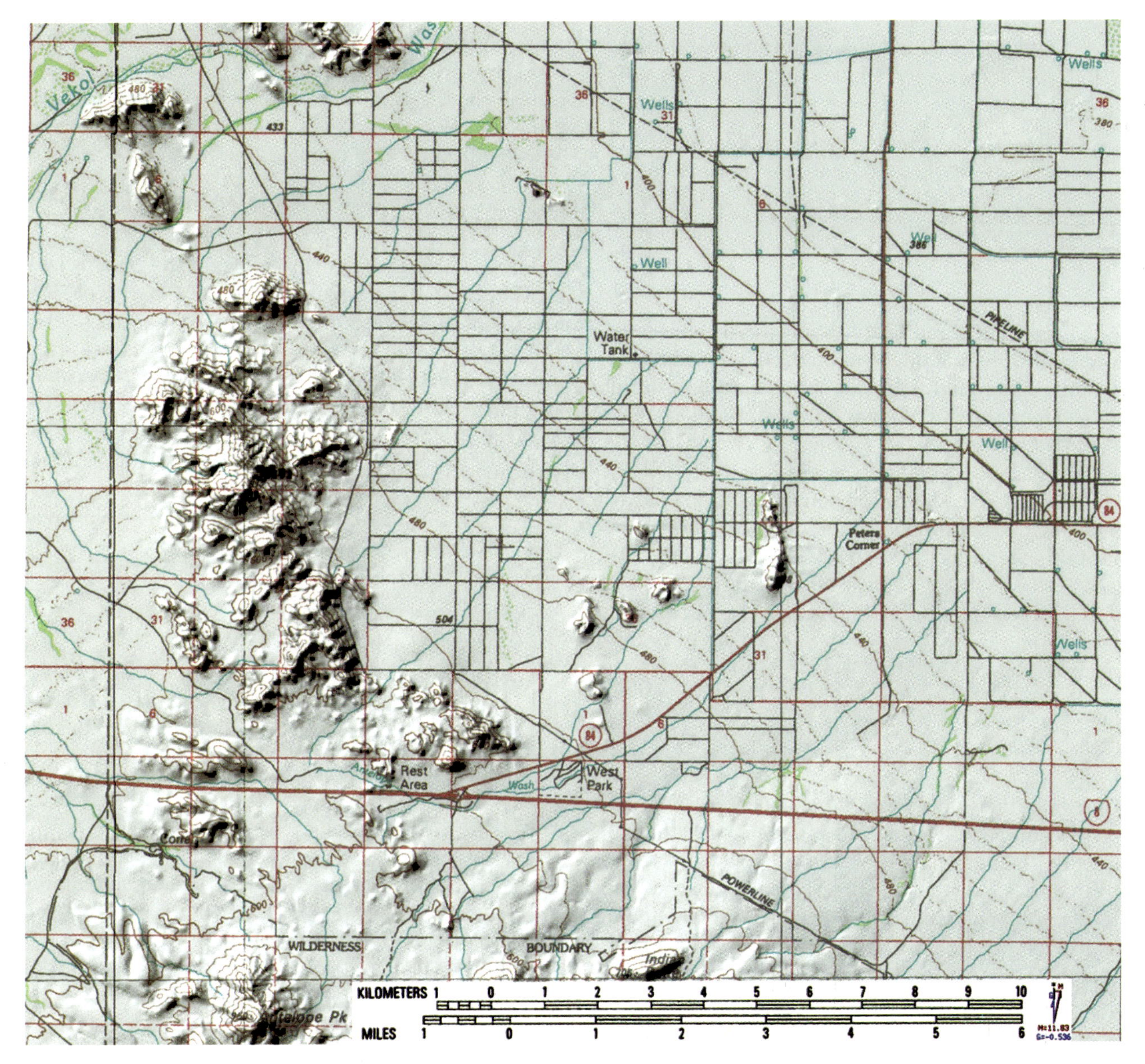

Figure 18.24

Topographic Map of Antelope Peak, Arizona, and Vicinity

Scale = 1:120,000, contour elevations are in meters.

c. Since the thickness of sediments is important in distinguishing bajadas and pediments, one way to tell them apart is by drilling. Which of the two would be a better place to drill a well?

_______________________ Why?

____

____

____

d. First use colored pencils to highlight all the wells in the area. Then color between the wells to estimate where the type of feature you mentioned in part **c** occurs. Label this area with the feature's name.

e. Since pediment results from erosion of mountains, and inselbergs are erosional remnants of the mountains, you should expect the two to be associated. Use the locations of the inselbergs to help you determine the location of the pediment.

Color this in and label it on the map. At this stage, you should have colored most of the map.

f. Use Figure 18.20 to help you determine the stage of basin and range development of this area. ______________________

Figures 18.20b and 18.19b show the second stage, where landforms include well-developed bajadas and *pediment,* sloping directly and imperceptibly into the bajada. The colors of bajadas and pediments seen in the field may be distinctive (see Figure 18.21b), or the transition may be invisible (Figure 18.19b).

In the last stage, bajadas completely fill the basins and pediments and inselbergs are all that are left of the mountains (Figures 18.9b, 18.20c, and 18.21b). All stages of development of this style of landscape occur within the Basin and Range Province in the Great Basin of North America.

DESERTIFICATION

Desertification is the conversion of nondesert areas into desert. A number of natural and **anthropogenic** (human generated) factors cause desertification; some of these are listed in ■ Table 18.4. Natural climate changes can cause gradual desertification in some regions. Human interference increases the pace of desertification or may cause it in places where it would not otherwise have occurred (■ Figures 18.25, 18.26b). In most places that are undergoing desertification, more than one factor is likely to be involved. Table 18.4 also lists some of the consequences of desertification and its impact on humans.

African Sahel

The area of northern Africa south of the Sahara Desert is known as the Sahel (■ Figure 18.26a). Levels of precipitation between 4 and 12 inches per year (10–30 cm) are normal for the region. In the early 1970s, severe drought after nearly a decade of low rainfall, devastated agriculture and natural vegetation in the Sahel. The Sahara Desert expanded into the Sahel during that time (Figure 18.26b). In some years, less rain fell than 30% of the average rainfall from 1960 to 1990, or a drop of 9 in of rain (24 cm) below the 1950–1979 mean (Figure 18.26c). Some of the suggested causes of the desertification include:

- From 1935–1970, the population doubled.
- Overgrazing destroyed grass cover.
- Drought in early 1970s and again in 1980s killed the local grasses and other vegetation.
- Greenhouse gases or air pollution from industrial countries changed global rainfall patterns.
- Changing sea surface temperatures changed rainfall patterns.

As a result of the drought, grass that supported many millions of cattle withered, the cattle died, and famine and starvation occurred in Ethiopia and Sudan. More

Table 18.4

Desertification

Some causes and consequences.

Natural Causes	Anthropogenic Causes	Symptoms of Desertification
• Natural long-term climate change caused by changes in Earth's geometric relationship with the sun • Prolonged drought • Short-term natural climate change related to oscillating atmospheric conditions which cause shifts in global wind patterns, such as the southern oscillation (related to El Niño, La Niña) or the northern oscillation • Natural changes in air and ocean currents and sea-surface temperatures in response to other climate-changing factors • Gases or aerosols in the atmosphere from volcanic eruptions	• Overgrazing, as in Figure 18.25 • Slash and burn practices • Poor soil management during agricultural activities, especially destruction of natural sod in grasslands • Water depletion by irrigation • Deforestation • Aerosols (tiny, airborne particles) from industrial nations that change the physics of clouds and precipitation • Anthropogenic climate change—global warming, mainly related to Increasing carbon dioxide In the atmosphere	• Sand dune encroachment • Lowering water table • Increase salinity of water and topsoil • Decrease in surface water (Figure 18.28) • Increase in number and severity of dust storms • Increase in soil erosion • Reduction in soil permeability and increased surface runoff • Destruction of native vegetation • Soil salinization • Crop failure • Famine • Loss of biodiversity • Extinction of some species

Figure 18.25

Overgrazing Can Cause Desertification

The area on the left of the fence was overgrazed and the area on the right was only lightly grazed.

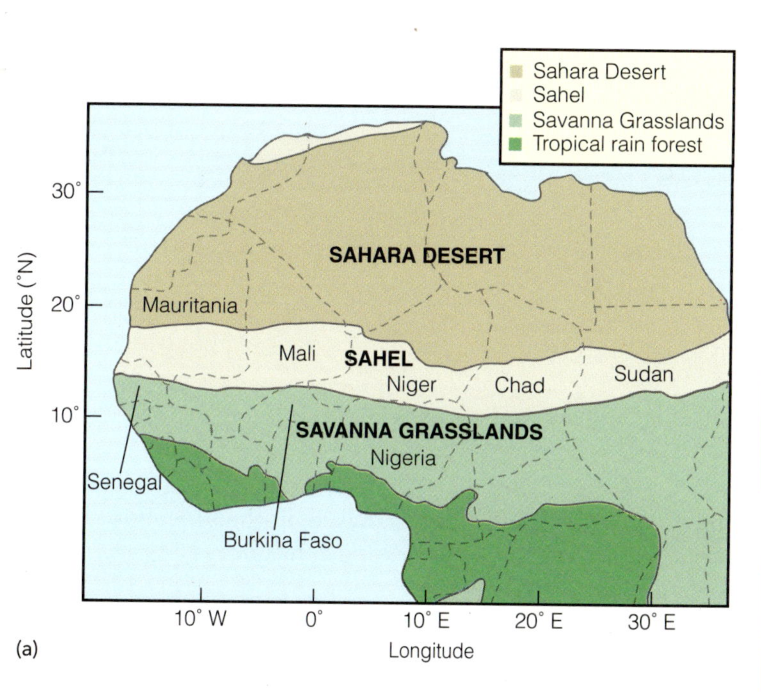

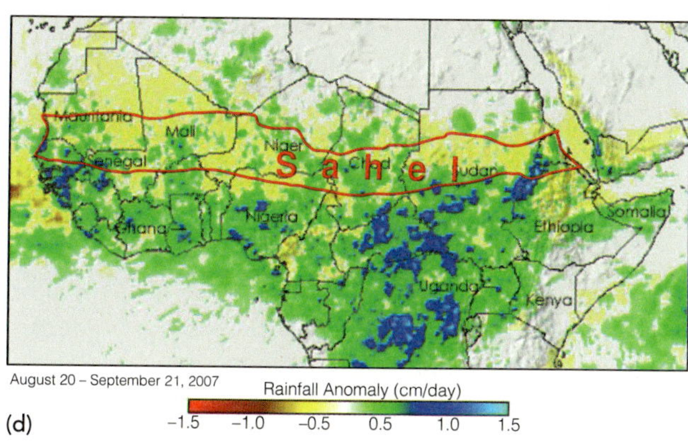

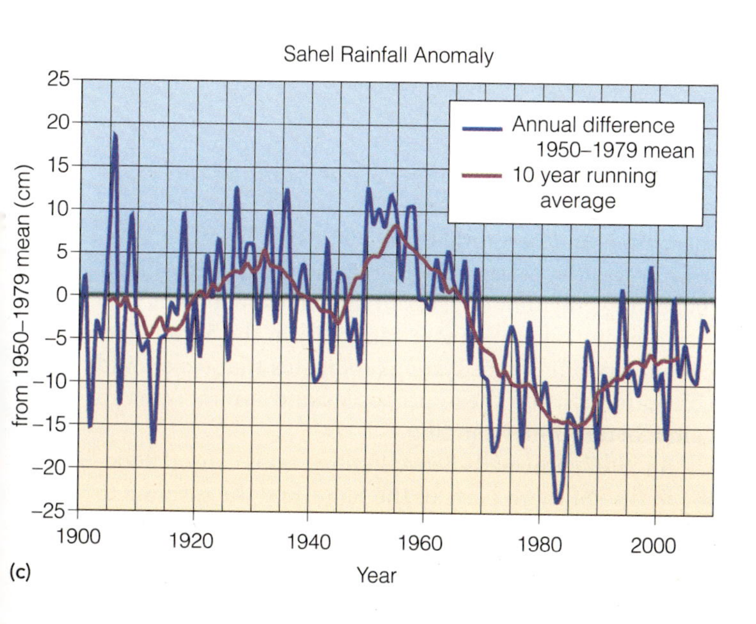

Figure 18.26

The Sahel

(a) Map of northern Africa showing the Sahara and the Sahel, a semi-arid band between the savanna (sparsely treed grassland) and the desert. (b) Sahara sand dune encroaching on natural grazing land in the Sahel, Mali, West Africa. (c) Rainfall in the Sahel compared to 1950–1979 average. Blue curve gives annual figures. Red curve is a running 10-year average for the 5 years previous and 5 years after the position on the curve. Notice the years of drought during the 1960s to 1980s. The Sahara Desert spread southward during that time. (d) Rainfall differences (anomaly) from normal for parts of Africa including the Sahel and Savanna grasslands.

recently, rainfall levels have risen and fallen. Overall, moderately dry conditions have continued, but a significant greening of the Sahel since 1994 is evident in carefully studied and quantified satellite images.

15. Let's look at the relationship between rainfall and desertification in the Sahel. Use the data in Table 18.5 to do the following:

a. On a piece of graph paper, draw a graph of how the size of the Sahara Desert changed over the 11-year period from 1980 to 1990. Leave enough room for a second graph. Time should be on the horizontal (x) axis and desert size should be vertical. Label the axes of your graph and do not forget the units. Include a title (as usual).

b. The two variables besides time in the table are the size of the Sahara Desert and the rainfall anomaly. An **anomaly** is just the difference from an average. Changes in the values of other variables do not influence an **independent variable.** The **dependent variable** will change in response to the values of the other variables. Which of these is the independent variable and which is dependent? Independent variable:

____

Dependent:

____

c. On the piece of graph paper, draw a second graph using the desert size and the annual rainfall anomaly. As is standard procedure when drawing graphs, plot the independent variable horizontally and the dependent variable vertically.

i. Completely label your graph.

ii. Draw an approximation of a best-fit line through the data.

iii. Discuss the significance of this graph and its comparison to the graph you drew in part **a.**

d. Examine the graph of Sahel rainfall in Figure 18.26c and compare it to the graph you made in part **a.** How has the size of the Sahara has changed since 1980?

____

Table 18.5

Sahel Rainfall and Sahara Size

1980 to 1990.

Year	Annual Rainfall Anomaly*	Size of the Sahara (km^2)
1980	4.04	8,633,000
1981	10.56	8,942,000
1982	−10.75	9,260,000
1983	−19.63	9,422,000
1984	−34.81	9,982,000
1985	−11.67	9,258,000
1986	−17.83	9,093,000
1987	−30.76	9,411,000
1988	22.71	8,882,000
1989	4.88	9,134,000
1990	-29.68	9,269,000

*Rainfall data are annual total rainfall percentage difference from the 1961–1990 mean for the rainy season months June–September.

Was desertification, based on the size of the Sahara, progressing continuously in one direction? ________ Based on your graph in part **c**, was Sahel rainfall a major factor in determining the size of the Sahara? ______

At first, people blamed poor management of resources and overpopulation for drought conditions in the Sahel, and certainly these were significant factors. More recently, studies have suggested that aerosols or increasing greenhouse gases in the atmosphere may play a significant role in climate variability in the Sahel. The complexity of climate systems indicates that we should continue to study these processes and interactions to gain a better understanding.

In 2007, unusually heavy summer rains caused flooding in the southern parts of the Sahel and the savanna to the south (Figures 18.26a and d). Drier than normal conditions that same year occurred in the northern Sahel.

16. Examine the rainfall map in Figure 18.26d for the Sahel and savanna grasslands of northern Africa.

a. Roughly what percentage of the Sahel had higher than normal rainfall in 2007 from August 20 to September 21?

____

b. How does this compare with the proportion of higher than normal rainfall in areas south of the Sahel?

____

Aral Sea

The Aral Sea in Kazakhstan, prior to the 1960s, was one of the largest inland seas in the world. It supported a diverse ecosystem and abundant fish populations. It was a haven to the shipping trade and supported the economy of hundreds of thousands of people. To grow new crops of cotton and rice, people in upstream desert areas started using water from rivers that flow into the Aral Sea, including the Amu Dar'ya and Syr Dar'ya Rivers (■ Figure 18.27). About 10 to 15% of the reduction in flow of these rivers is a result of climate change; the remaining 85 to 90% of the reduction is attributed to irrigation upstream. This irrigation dramatically reduced the water influx and caused shrinking of the Aral Sea (Figure 18.28). The Aral Sea loses water by evaporation, so when the amount of inflow decreases, the sea shrinks. As a result, every year the shallow sea loses more and more of its volume, reducing its size (Table 18.6). With less influx of water, boats were stranded (Figure 18.29, inset), the fishing industry collapsed, and shipping stopped. Today, only a small percentage of the water in the Aral Sea remains. Storm winds sweep the dried lakebed, now called the Aral-Karakum Desert, carrying at least 150,000 metric tons of salt and dust to the surrounding area, and producing severe, negative effects on agriculture and human health.

17. Examine Figures 18.4 and 18.27.

a. In what major desert (Figure 18.4) does the Aral Sea lie?

____

b. Roughly how far does the Amu Dar'ya River travel through desert before it reaches the Aral Sea? _______________

c. Use your knowledge of desert lakes to suggest what else, besides decline in the size of the lake, has been happening in the Aral Sea as a result of these changes.

____

____

____

____

18. Examine ■ Figure 18.28 and ■ Table 18.6.

a. On separate graph paper, make one graph of the area, volume, sea level, and salinity of the Aral Sea against time from 1960 to

Figure 18.27

Drainage Basin for the Aral Sea

Regional map showing the outline (red dots) of the Aral Sea drainage basin. The two largest rivers draining into the Aral Sea are the Amu Dar'ya and the Syr Dar'ya.

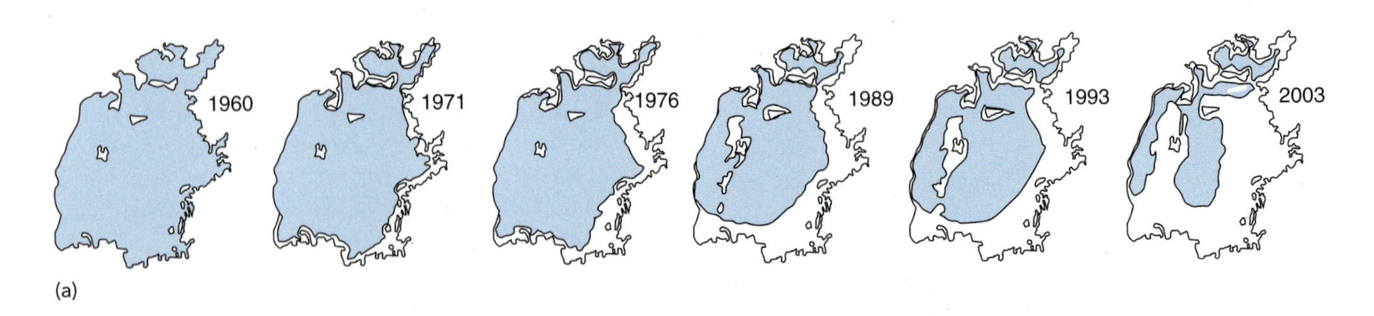

(a)

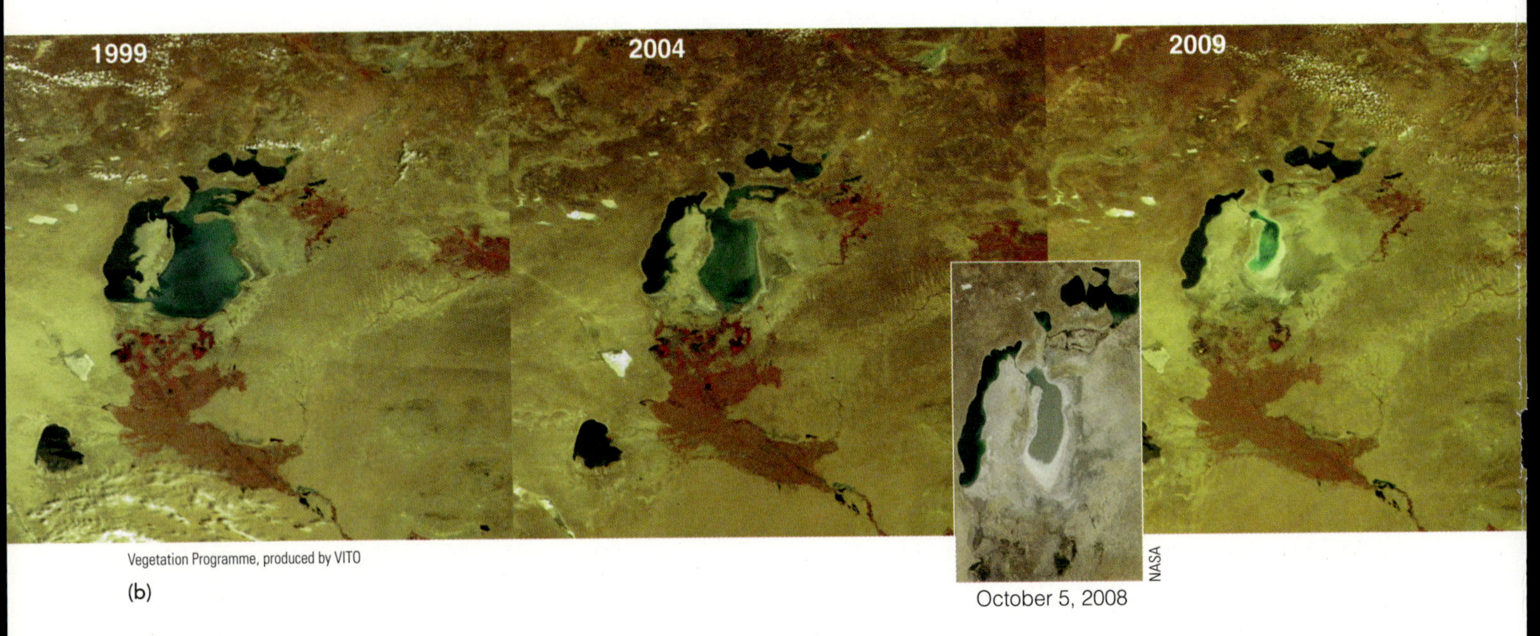

Vegetation Programme, produced by VITO
(b)
October 5, 2008

Figure 18.28

Shrinking of the Aral Sea

(a) Maps of the Aral Sea showing how much it has shrunk between 1960 and 2003. Blue indicates the extent of the sea, and white shows where the sea no longer occurs. (b) Infrared satellite images of the Aral Sea and the deltas of the Amu Dar'ya and the Syr Dar'ya Rivers for 5-year increments from 1999 to 2009. Red color indicates vegetation, especially around the river deltas. Inset from NASA shows October 2008.

2009. Plot the years on a linear scale; do not just plot the years listed in the table.

b. To what percentage of its 1960 size did the Aral Sea shrink in area by 1993? ________ In volume? ________ In sea level? ________

c. How did the salinity change between 1960 and 2009? Be specific.

____

Did this answer agree with your answer in Exercise 17c? ______

d. Read the caption for Figure 18.28b. What produces the red color in the

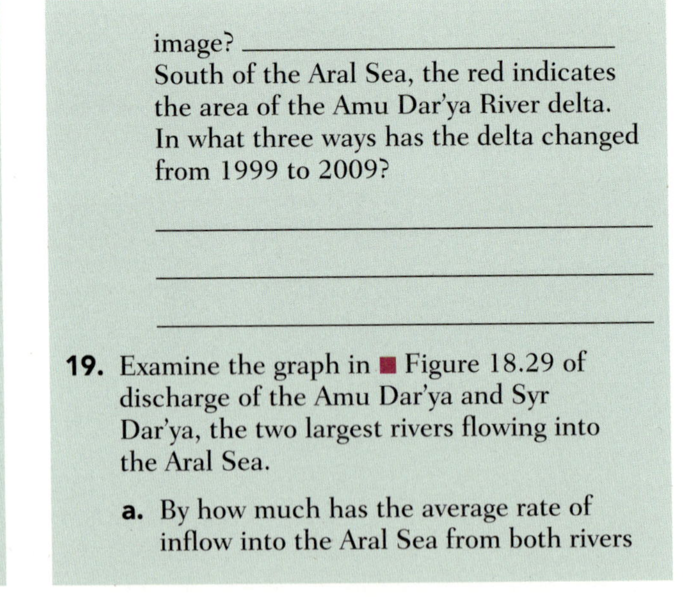
image? ______________

South of the Aral Sea, the red indicates the area of the Amu Dar'ya River delta. In what three ways has the delta changed from 1999 to 2009?

____

____

____

19. Examine the graph in ■ Figure 18.29 of discharge of the Amu Dar'ya and Syr Dar'ya, the two largest rivers flowing into the Aral Sea.

a. By how much has the average rate of inflow into the Aral Sea from both rivers

Table 18.6

Data for the Aral Sea

Area, volume, water level, and salinity data for the Aral Sea at uneven intervals from 1960 to 2009.

Year	Area (km²)	Volume (km³)	Sea Level (m)	Approximate Salinity (g/l)
1960	68,000	1,040	53	10
1971	60,200	925	51	—
1976	55,700	763	48.3	14
1985	45,713	468	41.5	23
1986	43,630	380	40.5	—
1987	42,650	354	40	27
1988	41,134	339	39.5	—
1989	40,680	320	39	30
1990	38,817	282	38.5	—
1991	37,159	248	38	—
1992	36,067	231	37.5	—
1993	35,654	248	37	—
1994	35,215	248	37	—
1995	35,374	248	37	—
1996	31,516	212	36	—
1997	29,632	190	35	—
1998	28,687	181	34.8	45
2000	23,400	162	34*	35
2001	—	—	33	—
2003	—	—	32	—
2006	—	—	31	—
2009	—	—	30	>100

*Data from 2000 on (from LEGOS) is about 1 meter higher than the data before 1999.

changed over time? Show the math:

_____________________ km³/yr.

Has it increased or decreased?

____

b. Estimate approximately when divergence of river water for irrigation first began at a scale large enough to cause a noticeable change in the river flows. Amu Dar'ya:

__________ Syr Dar'ya: __________

c. Draw two best-fit straight lines through parts of the data for the Amu Dar'ya River: one for the average flow before irrigation began to extract river water, and one for later changes in river flow due to irrigation. Where these two lines meet or cross should approximate when major irrigation started to extract water from the river. When did this occur? __________
How closely does this match your estimate in part **b**?

____

d. Now do the same thing for the Syr Dar'ya River. Where these two lines meet or cross should approximate when major irrigation started to extract water from the river. For what year do the two lines meet or cross (start of irrigation)? __________
How closely does this match your estimate in part **b**?

____

e. Predict the future for the Aral Sea.

____

____

f. What do you think will happen to the people who live there? List four or more consequences.

____

____

____

____

____

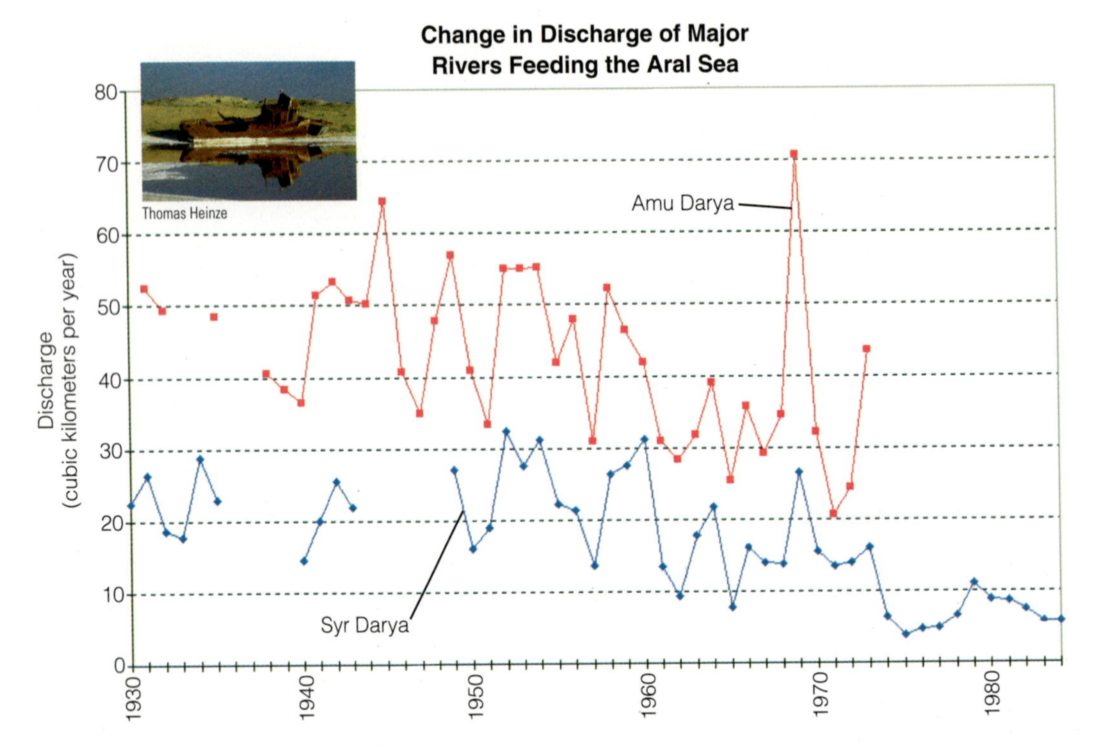

Figure 18.29

Amu Dar'ya and the Syr Dar'ya Discharge

Annual average discharge of the Amu Dar'ya and the Syr Dar'ya Rivers into the Aral Sea from 1930 to 1994. Data for some years are missing. Inset: Retreat of the sea stranded many boats. The fishing industry has collapsed as a result of the shrinking of the sea.

Appendix

SCIENTIFIC METHODS

The nonscientist may be tempted to think that science is merely the accumulation of "facts." Indeed, facts are the bricks in the structure of science, but science also consists of hypotheses, models, and theories that connect facts and make them meaningful. The processes scientists use to build this structure are *scientific methods.* The purpose of your laboratory course is not only to introduce you to the materials and the processes active on the Earth, but also to give you insight into the workings of science. People commonly refer to "the scientific method" as if only one method exists. Since science involves many methods, each constructed for specific purposes, we will refer to them as scientific methods. Here, as we outline the process of scientific methodology, you may find it is in some ways very much like natural human inquisitiveness in its origins.

Scientific methods have evolved into sophisticated systems to investigate the natural world. The systems use a rational and logical framework that does not make appeals to the supernatural to explain nature's phenomena. The structure of scientific methodology has several components: gathering data; formulating hypotheses, models, and theories; and testing in all of these phases. Communication continues throughout these processes.

Data Gathering

A preliminary stage in scientific investigation is to gather information through observations about the particular natural circumstances or processes of interest. Investigators strive to be as objective as possible. They must distinguish facts from fiction, rumor, guesses, and preconceived notions. The best observations are those that independent investigators can verify. Careful documentation of these observations is therefore necessary.

A second aspect of data gathering is to collate and classify the information to see whether any patterns are present in the data. At this stage, the researcher may be able to formulate various **laws,** which are generalizations derived from looking at many particular observations. For example, Newton's law of gravity does not explain gravity, but summarizes its action.

Hypotheses and Hypothesis Testing

Hypotheses are attempted explanations of observed natural occurrences. They try to connect diverse circumstances by stating a common, underlying principle. A hypothesis should be more than a wild guess; it should explain all the observed data and also make predictions about features and occurrences not yet seen. The predictions are not necessarily to predict future events, but to predict what to expect in an experiment or what one will find by gathering more data. These predictions provide a framework for testing the hypothesis by seeking further observations or by designing and conducting experiments. If new observations or experiments confirm predictions, then they *support* the hypothesis. It is important to understand that this support does not necessarily prove the hypothesis; the hypothesis is not *proven* because more extensive observations and improved experiments may eventually show the hypothesis to be wrong. Scientists strive to look for observations and design experiments specifically to try to test a hypothesis to its limits. In other words, they try to refute, or **falsify,** a hypothesis. One falsification indicates that a hypothesis is wrong and needs reformulating. If hypotheses are unsuccessful, then scientists discard them and invent new ones.

At the early stages of an investigation, several equally reasonable hypotheses may seem to fit the data. Scientists often use the method of **multiple working hypotheses** until they can eliminate the hypotheses that do not work. This is good scientific practice as it helps scientists to remain objective. After making further observations, scientists should throw out a hypothesis that proved incorrect or modify a hypothesis to fit new information, but they should *never* change the data or recorded observations to fit the hypothesis. Scientists also need to avoid only noticing phenomena or data that fit the hypothesis and ignoring information that refutes it.

Let's take a classroom example. In an experiment to explore cloud seeding, students observe whether large or small droplets freeze first. Suppose a technical difficulty causes you to start your experiment after the rest of the class. Quite a few students have already posted their data on the blackboard, and all of them reported that larger droplets froze first. You run your experiment but notice that smaller droplets froze first. What should you do? Should you do the experiment over until you get the same results? Should you throw out your results because they are obviously wrong? Should you report your data regardless of what others have reported? The correct answer is that you have done everything right but got an unexpected result. These are real data and you should report them.

Here's another example. Let's say you are measuring the amount of water draining from a tube during a certain period of time, but you accidentally let two additional drops fall after the time for measuring is up. What should you do? Should you throw out your results because they are obviously wrong? Should you report your data regardless of the problem? Should you do the experiment over until you do it correctly? In this second situation, you know you made an error in your measurements, so you need to do them over. The second case is not one of changing data, but correcting the error. Most scientists in this situation would write down the result in their notebook with a note about what went wrong, rather than completely ignoring it.

The concepts you learn in this appendix form an important foundation for Earth science and you will use them at various times in different labs. You will apply the scientific method to develop and test hypotheses.

Models

A **model** in science is a simplification, or representation, of some aspect of nature. The model is an analogy (or likeness) which aids in the visualization and understanding of natural phenomena. Occasionally, scientific models are actual physical replicas of some location or feature, analogous to a model airplane. Unlike physical models, scientists often "construct" models of ideas, or mathematical models, that imitate the aspect of nature being studied. This construction of models provides a test for theories or hypotheses.

Modelers often make certain simplified assumptions. If the model is unsuccessful at explaining a natural occurrence, the scientist reformulates it using more-complicated assumptions. If a model, in its simplified form, imitates nature fairly well, then scientists know that factors left out of the model are less important than those included. For particular investigations, models help researchers determine which factors of a system are significant and which are not.

Theories

A scientific **theory** is what a hypothesis or group of hypotheses becomes if their predictions are repeatedly successful. A theory also has a wide scope; it links together many diverse phenomena and has wide and profound explanatory power. The disparaging comment "It's only a theory" is a contradiction and results from misuse of the word "theory." In scientific practice, a theory is a comprehensive, encompassing concept that scientists have repeatedly tested and evidence has repeatedly supported. A theory, therefore, should not be casually disregarded.

Nonetheless, theories are not carved in stone and may be temporary, but less so than hypotheses. As scientists gather new data that the theory does not explain, they may have to modify the theory; they rarely simply abandon a theory. The new adjusted theory is likely to have broader scope than the old theory; in fact, most commonly the old theory was an approximation of the new theory, applicable in certain restricted circumstances.

When many related hypotheses have withstood the test of time and combine into a single theory that explains a vast range of data and observations within a field of science, a **unifying theory** is born. For the solid Earth sciences (geology), the *theory of plate tectonics* is such a unifying theory.

Communication

In addition to conducting research, scientists need to distribute and publish their results. The disclosure of their work in progress is important not only in developing their data, ideas, hypotheses, and models, but also for approval and acceptance within the scientific community. They submit the results for **review** as an article or **paper** to specialized magazines called **journals.** The journal editor sends copies of the paper to experts, called **referees,** who decide whether the work is new and valid science and therefore publishable. Editors reject bad data and bad hypotheses as well as those that have already been published elsewhere and are pretending to be something new. Students are not the only ones who are graded on their papers!

Once a journal publishes the article, other scientists can examine the research and the observations or new hypotheses and incorporate them into new research. This process spreads science and introduces it to the scientific community, as well as to society. Conferences are also important for transmitting results, and the Internet is increasingly being used for this process. The Internet has its dangers, however, as cranks can "publish" data and hypotheses that have not undergone the rigorous review process of professional science. In science, books are less important in transmitting knowledge than they are in the humanities, for example. This is due, in part, to the evolving, ever-expanding character of scientific ideas, hypotheses, and theories.

Cyclic Aspect of Scientific Methods

As implied earlier, science is a dynamic process. Observations lead to laws and hypotheses, hypotheses and models to theories, and theories to newer theories or unifying theories. Scientists constantly evaluate hypotheses, models, and theories and compare their predictions with new observations. As a result of the comparison of hypotheses with the "real world," they are adjusted or reformulated to provide a better explanation of the workings of nature, and then tested again. There is no room, therefore, for dogma in science. Scientists know that even their most treasured theories may eventually be replaced by other (and better) ones. Scientists, by training, are precise people. They continue testing hypotheses for weak points and making new observations. Scientists are also critical when examining each other's data because hypothesis testing depends on accurate data that truly reflect nature. Scientists do not view themselves as discovering the "truth," but as acquiring ever more accurate approximations of the truth. Their work is never complete.

SPECIAL FEATURES OF THE EARTH SCIENCES

Earth scientists, in general, conduct research in a similar way to how other scientists do research, but the Earth sciences have several features that distinguish them from the other sciences. In the experimental sciences (physics, chemistry, aspects of biology), scientists design and observe experiments specifically to test hypotheses. In the Earth sciences, astronomy, and certain fields of biology, they often rely on direct observation of the natural world. They must seek "case studies" as preserved in rocks in the field or as observed in the oceans or atmosphere as their primary source of information. This is why fieldwork and field trips are important in the Earth sciences. This means that Earth scientists conduct their research in the manner of detectives or forensic scientists, who must try to infer events from what remains at the "scene of the crime." They often have to deduce events logically from clues left within rocks, sediments, ice cores, or water and air samples. They must therefore be astute observers and have "Sherlock Holmes-type" minds. Although this is a demanding intellectual exercise, it also makes the practice of the Earth sciences great fun.

APPLY THE SCIENTIFIC METHOD TO YOUR OWN LIFE

You can use the scientific method in everyday life to find out whether your assumptions are correct. Let's take an example. Imagine you are looking for something that you have lost. Maybe it is a valuable ruby ring or your car key. You say to yourself, "I know it is somewhere in my room." You turn your room upside down and keep telling yourself, "It has to be here somewhere." After the fifth time you have gone through everything in your room and it just isn't there, you go into the living room and find it on the end table, just where you left it.

How could the scientific method help you here? By saying the object is in your room, you formed a hypothesis. You should then test that hypothesis by thoroughly searching your room. Once should be enough to make you reevaluate your hypothesis and start searching elsewhere, such as the living room. The stubbornness of holding onto your one hypothesis leads to an excessive waste of time. It would be better to compose multiple working hypotheses: "It is in my room or the living room." You could do a quick search of both rooms first, then a more thorough search if it doesn't turn up right away. This approach would be the most effective method for this situation.

Glossary

Italic terms below are also defined in this glossary.

A

ablate To become smaller. When a *glacier* ablates it loses *mass* by a combination of melting, *sublimation*, and *calving*.

ablation The processes by which a glacier loses *mass*.

ablation, zone of A part of a *glacier* where more solid water dissipates than falls. The solid water includes snow, sleet, hail, *firn*, and *glacial ice*. Dissipating, or *ablating*, processes are melting, *sublimation*, and/or *calving*.

abrasion The processes of wearing down rock by grinding, rubbing, scraping, scratching, scouring, chipping, and pitting.

absolute age The numerical age or *isotopic age*, usually of a *rock*, and with some degree of error.

absolute humidity The *mass* of water contained as vapor in a certain volume of air, usually given in units of grams per cubic meter (g/m^3).

abyssal plain The broad expanse of the ocean floor that is nearly flat and at a depth of about 5000 m.

accretionary wedge A wedge-shaped volume of thrusted and folded *rock* and *sediment* scraped and *thrust-faulted* off a subducting plate onto the edge of a continent. See *subduction*.

accumulation, zone of (a) (*glaciology*) The region of a *glacier* where more snow and other solid *precipitation* falls than melts and *sublimates* in a year (compare *zone of ablation*). (b) (*soil*) The part of a soil where *weathering* products and *minerals* washed down from above accumulate; B horizon.

acid mine drainage Acidic water draining from a mine.

active continental margin The edge of a continent with a current *plate boundary*.

adamantine luster A bright mineral *luster* that resembles the way a diamond shines.

advance Growth of a *glacier* by extending the *terminus* down the valley or by spreading out, generally resulting from positive net *mass* balance (see *glacier budget*).

aftershocks Smaller earthquakes that occur after and are associated with an *earthquake*.

aggregate, mineral A group or cluster of *crystals* with at least some of their edges touching one another.

alluvial fan A fan-shaped deposit of poorly-sorted, gravelly sediment formed where a stream canyon opens into a wide basin. *Braided streams*, *mudflows*, and *debris flows* deposit *sediment* in alluvial fans.

alpine glacier A type of glacier that forms and flows in valleys in mountainous regions, a *mountain glacier*, a *valley glacier*.

amorphous Said of a substance that has a random or disorderly arrangement of atoms; not *crystalline*.

amygdaloidal texture The *texture* that results when *vesicles* or gas cavities fill with secondary *minerals*. The secondary minerals were deposited after the *solidification* of the original *rock*.

amygdule A *vesicle* or gas cavity filled with secondary *minerals*. The secondary minerals were deposited after the *solidification* of the original *rock*.

angle of repose The steepest angle a slope can achieve without *mass wasting*.

angular unconformity An *unconformity* where the older *rocks* below the unconformity are tilted or *folded sedimentary/volcanic rocks* and are not parallel to the unconformity surface.

anhedral An adjective describing a *grain* lacking well-formed *crystal faces*.

anhydrous Said of a substance that has no water or OH in its chemical makeup.

anthropogenic Caused by humans.

anticline An upward-arching *fold*. See also *syncline*, *monocline*, *non-plunging fold*, and *plunging fold*.

anticyclone A high-pressure weather system with dry, sinking, stable air and clear skies with a clockwise wind rotation in the Northern Hemisphere and a counterclockwise wind rotation in the Southern Hemisphere.

aphanitic texture A *texture* of *igneous rocks* with *crystals* so *fine grained* that they can only be seen with magnification.

aquiclude A *rock* or *sediment* adjacent to an *aquifer* that is impermeable and thus prevents the flow of water through it. Compare with *aquitard*.

aquifer A *permeable* body of *rock* or *sediment* below the *water table* that can sustain a productive water *well*.

aquitard A *rock* or *sediment* that has lower *permeability* than an adjacent *aquifer* and retards the flow of water through it. Compare with *aquiclude.*

arête A sharp, steep, bedrock ridge separating two *cirques.*

arid A climate where rainfall is less than 10 in (25 cm) in one year or where *evaporation* rates exceed *precipitation.*

artesian aquifer A *confined aquifer* that is under pressure.

ash Sand-sized to powdery *volcanic* material produced by explosive *volcanic eruptions* when a spray of magma and particles of *rock* spew out of the *volcano.* Short for *volcanic ash.*

asthenosphere A layer of the Earth that is made of weak *plastic rock,* below the *lithosphere* and above the *mesosphere.* The asthenosphere is part of the upper *mantle.* See Figure 1.2.

aureole, contact A zone surrounding an *igneous intrusion* where *contact metamorphism* has occurred.

avalanche A fall or slide of snow that mixes with air and flows down a slope at very high speeds. Billowing snow rises up from the base as the flow moves. See also *debris avalanche.*

avalanche, debris A type of *mass wasting* in which air mixes with debris that moves downslope at very high speeds in a way similar to a snow *avalanche.* Also the deposit or *landform* formed from such movement.

axes Plural of *axis.*

axis of a fold The line around which folded layers are bent; a *fold hinge.*

axis of Earth's rotation The line around which the Earth rotates.

B

bajada A depositional *landform* resulting from the joining of multiple *alluvial fans* to form an apron of *sediment* at the base of a mountain slope.

barchan dune A crescent-shaped *dune* with tips pointing downwind.

barometer An instrument used to measure air pressure.

barrier island A long, narrow, low island consisting mainly of sand, parallel to shore, and separated from shore by a long, narrow lagoon or sometimes a bay.

Barrovian zone A *metamorphic zone* encompassing an area with a distinct pelitic *mineral assemblage* in a region of progressive metamorphism of *pelitic rocks.* The Barrovian zone is named for one of the *index minerals* chlorite, biotite, garnet, staurolite, kyanite, or sillimanite.

base level The level, or elevation, below which a *stream* cannot erode its *bed,* usually sea level, the level of a lake the stream enters, or the level of an especially *erosion*-resistant *rock* the stream crosses. The latter two are temporary base levels.

base line For the *Township and Range System,* an east-west line through the origin or reference point of the system.

basin, drainage The region drained by a *stream.*

batholith A roughly equidimensional *igneous intrusion* of large size, with an *outcrop* area greater than 40 mi^2 (or about 100 km^2).

bathymetric map A map that depicts the shape of the bottom of the ocean. See also *topographic map.*

bathymetry The shape of the bottom of the ocean. See also *topography.*

bay-mouth bar An extension of a beach built out across a bay or inlet by the *longshore current.*

beach drift *Sediment* moving parallel to the shoreline as a result of the *longshore current; longshore drift.*

bed (a) (stratigraphy) a single layer of *sediment* or *sedimentary rock* with distinct surfaces separating it from other layers; *stratum;* (b) (hydrology) the bottom or base of any body of water, as in *stream* bed.

bedding The arrangement of *sediment* or *sedimentary rock* in parallel (or subparallel) layers formed at the time of deposition.

bedding plane A planar or nearly planar surface separating two touching *beds* of *sedimentary rock.*

bedrock Solid rock at or below the surface that is part of the Earth as a whole, not broken off, not sediment or soil.

biochemical sediment *Sediment* that results from the actions of organisms.

body fossils A *fossil* made of the actual remains of an organism or parts of the organism.

body wave A *seismic wave* that moves through the Earth.

brackish Slightly salty with saline content between that of streams and sea water.

braided stream A *stream* with multiple *channels*, bars, and islands usually formed where the stream has too large a supply of *sediment* or a highly variable supply of water, or both.

breakwater An artificial structure built in the water parallel to the shore to aid sand *deposition* between it and the shore.

brittle A variety of *tenacity* of a *mineral* in which the mineral shatters, breaks, or fractures rather than bends, flows, or dents when struck with sufficient *force* to deform it.

brittle deformation *Deformation* that occurs when *rocks* break, and have little or no flow or *ductile deformation.*

budget See *glacier budget; glacier mass balance.*

butte A nearly flat-topped hill or mountain with steep sides and horizontal surface dimensions about one to two times its height.

C

calving Breaking off of large chunks of a *glacier* at its *terminus.* Calving is usually most active where a glacier enters water.

carbonate mineral class (carbonates) A group of *minerals* with members containing carbon and oxygen and one or more metals. Examples are calcite, dolomite, malachite, and azurite.

carbonization A process of *fossilization* in which the original *organic* matter has been reduced to carbon.

cartography Map making.

casings The material that is used to keep the upper portion of a *well* stable and maintain the opening. Casing must be either steel or plastic and must be at least 20 ft into the ground.

cast A *fossil* made by filling a *mold* with *mineral* or *sediment.*

catalyst A substance that initiates or hastens a chemical reaction, without being used up in the reaction.

cement Material that precipitates between *sediment grains* and so holds the sediment together.

cementation A process involving water moving through *sediment* and precipitating minerals that essentially "glue" the *sediment* together. See also *cement.*

channel A long, narrow trough occupied by the water in a *stream* or a connection between two bodies of water.

chemical bond The connecting force between atoms or molecules brought about by interactions among electrons. The physical and chemical properties of minerals are attributable for the most part to the types and strengths of these binding forces.

chemical sediment *Sediment* that forms by chemical precipitation of compounds out of a water solution.

chilled margin Generally *finer-grained igneous rock* along the edge of an *intrusion* or *extrusion* produced where the *magma* comes in contact with the *country rock*, air, or water. The edges are cooled more quickly than the centers, producing the finer *texture.*

chronology In geology and archaeology, the process of determining an object's or event's place within a chronological or time ordered scheme.

cinder A piece of *vesicular pyroclastic* material (pumice or scoria) thrown from a *volcano* that is solid when it lands.

cinder cone A small *volcano* made up of *pyroclastic* blocks, *volcanic ash,* and *cinders.*

cirque An ampitheater or side-ways, bowl-shaped, bedrock basin carved out at the head, or beginning, of a *glacier* or by a small glacier (cirque glacier).

clast A piece of broken *rock* or *mineral.*

clastic sediment *Sediment* made up of broken *rock* or *mineral* pieces; *detrital sediment.*

cleavage A physical property of a *mineral* that breaks along planes of weakness within the *crystal* structure.

cleavage, slaty The property of a *fine-grained metamorphic rock* that breaks along planes of weakness created by parallel *mineral grains.*

cliff retreat *Erosion* of a cliff in a way that maintains close to the original *slope* of the cliff.

coarse-grained texture (a) (loosely defined) A *texture* for which the *rock* has visible *grains* (larger than about 1/16 mm). Also *granular texture.* The term *phaneritic texture* is also used for *igneous rocks* and *granoblastic texture* for unfoliated *metamorphic rocks.* This usage of the term *coarse-grained* includes medium-grained and is more commonly applied to *igneous* and *metamorphic rocks.* Coarse-grained igneous rocks include granite and gabbro, and metamorphic rocks include gneiss and marble; (b) (strictly defined) A texture for which the *rock* has grains larger than sand-sized, or larger than about 2 mm. Coarse-grained *sedimentary rocks* include breccia and conglomerate.

coastal desert A *desert* that forms along a coast. Coastal deserts result where cold ocean currents flow along coastlines.

coefficient of friction A unitless number expressing how strong the *force of friction* will be for a given *normal force.*

col A low spot or pass in an *arête.*

cold front A *front* where cold, dry air advances beneath warm, moist air. Where the colder air comes in contact with warmer air, *condensation* and *precipitation* occur. See also *warm front, occluded front,* and *stationary front.*

compass rose A graduated circle showing various directions such as north, south, northwest, etc. and measurements in between.

compaction Bringing *grains,* especially *sediment,* closer together so they take up less space.

composite volcano A steep-sloped *volcano* made of interlayered *pyroclastic* deposits and lava flows; a *stratovolcano.*

compression *Deformation* that involves forces moving toward each other that cause *shortening.*

conchoidal fracture A *mineral* or *rock fracture* where the broken surfaces are smoothly curved, sometimes with concentric ribbing (Figure 2.6).

concordant intrusion A *magmatic intrusion* that intrudes parallel to existing *bedding* or *foliation.* Examples include *sills* and *laccoliths.*

condensation The process in which water vapor molecules join together to form water droplets or ice *crystals,* thus forming clouds, fog, or moisture clinging to surfaces.

cone of depression A cone-shaped drop in the *water table* around a *well* that results when water is pumped out of the well.

confined aquifer An *aquifer* that is sandwiched between *aquicludes* or *aquitards.*

confining pressure The pressure experienced by a *rock* at depth caused by the weight of the overlying *mass* of *rock.*

conformal projection A map *projection* in which the shape of small areas are preserved and where different directions at any point have the same scale.

conical projection A map *projection* in which the *graticule* is projected onto a cone with an axis coincident with the *geographic axis.*

contact The boundary between *rock* units or *formations,* where one formation gives way to the next, depicted as a black line on most maps.

contact aureole A zone surrounding an *igneous intrusion* where *contact metamorphism* has occurred.

contact metamorphism *Metamorphism* resulting from the heating of *rock* near a *magmatic intrusion; thermal metamorphism.*

continental collision A boundary between two *continental* plates where the two *plates* are *converging* (moving toward each other), typically forming a mountain range; a *continent-continent convergent plate boundary.*

continental crust The part of the Earth's *crust* that underlies the continents and is chemically distinct (*felsic*) from *oceanic crust.* See Figure 1.2.

continental divide A *divide* that separates *drainage basins* for *streams* that empty into different oceans.

continental glacier A *glacier* that is at least 50,000 km^2 in area.

continental interior desert A *desert* that results due to the low moisture content of the air in the far interior of a large continent.

continental rift The separation at a *divergent margin* where a continent diverges, beginning to separate the continent into smaller pieces. If divergence continues, a continental rift develops into an ocean basin with a *mid-ocean, ridge-spreading center.*

continental rise An area along a *passive continental margin,* between the *abyssal plain* and the *continental slope,* where the seafloor slopes more gently.

continental shelf A gently-sloping area of shallow water between a continent and the *continental slope.*

continental slope An area between the *continental shelf* and *continental rise* with a distinctly-steeper slope than either adjacent area.
contour interval The difference in elevation between two adjacent different *contour lines.*
contour line A line that connects points on a map representing places on the Earth's surface that have the same elevation.
contour map See *topographic map.*
convergent margin A *convergent plate boundary.*
convergent plate boundary A plate edge where two *plates* move toward each other; *convergent margin,* or destructive plate margin. Types include *continental collision,* and ocean-continent and ocean-ocean subducting plate boundaries (see *subduction*).
coral polyp A small colonial animal that, as a colony, builds coral reefs; an individual coral animal.
core The center part of the Earth, below the *mantle,* that is chemically distinct, primarily made up of iron. The core consists of the *inner* and *outer core.*
Coriolis effect An apparent deflection, to the right in the Northern Hemisphere and to the left in the Southern Hemisphere, of freely moving objects or substances, such as ocean water or the atmosphere, as a result of the rotation of the Earth. The deflection is relative to the Earth's surface.
correlation The process of matching the ages of *rocks* from different localities by matching stratigraphic sequences, *fossil* assemblages, and/or distinctive stratigraphic time markers such as volcanic ash layers.
correlative An adjective that applies to two or more *rocks* that have the same age.
country rock The *rock* intruded by *magma,* or the rock surrounding a magmatic intrusion.
creep Slow movement (a) (*faults*) Fault creep is gradual movement along a fault without perceptible earthquakes; (b) (*soil*) Soil creep is very gradual downslope movement of soil with upper levels of the soil moving faster than lower levels. Soil creep is generally at speeds measured in cm or inches per year; (c) (*eolian* transport) Wind-driven movement of grains that crawl along the surface without becoming airborne.
crenulated Irregularly wavy or wrinkled with small folds of a few millimeters, a texture seen in metamorphic rocks; also crenulation.
crevasse A deep, nearly vertical crack in *glacier* ice caused when ice moves at different speeds over an uneven surface.
cross bedding Inclined *bedding* where the inclined layers formed at a low angle to the major *sedimentary* bedding, because the sediment is deposited on a sloping *dune,* bar, or ripple, cross-stratification. Typically, the top of the inclined layer has been truncated.
cross section A side view of the Earth's interior, generally near the surface, exhibiting the arrangement and compositions of *rocks* and rock layers; a *structure section.*
cross-cutting relationships, the principle of States that a geologic feature that cuts across another feature is younger than the feature it cuts. The most common cross-cutting features are *discordant intrusions, faults,* and *unconformities.*
crust The surface solid layer of the Earth that is chemically distinct from other layers below. It is primarily made up of oxygen, silicon, and aluminum and is generally *mafic* to *felsic.* The crust makes up the uppermost part of the *lithosphere.* See also *oceanic crust* and *continental crust.* See Figure 1.2.
cryology *Glaciology.*
cryosphere All of the ice on Earth.
crystal A single *grain* of a *mineral* in which the structural planes of atoms extend in the same directions throughout the grain.
crystal faces Planar surfaces of a well-formed *crystal* that grew during *crystallization* or *recrystallization* and reflect the *mineral's* internal atomic order and arrangement.
crystalline Having an orderly internal arrangement of atoms in three dimensions, or having a *crystalline texture.*
crystalline texture A *texture* where *mineral grains* crystallized in place with grain boundaries touching—an interlocking texture. Most *igneous* and *metamorphic rocks* have this texture, as do *sedimentary evaporites.*
crystallization The formation of a *crystal* with an orderly three-dimensional arrangement of atoms.

D

debris avalanche A type of *mass wasting* in which air mixes with debris that moves downslope at very high speeds in a way similar to a snow *avalanche.* Also the deposit or *landform* formed from such movement.
debris fall A type of *mass wasting* in which loose, unconsolidated material collapses through the air. Also the deposit or *landform* formed from such movement.
debris flow A type of *mass wasting* in which more than half of the material in the *flow* is greater than sand-sized (compare *mudflow*). Also the deposit or *landform* formed from such movement.
deflation Wind *erosion* that lowers the land surface by blowing mostly fine *sediment* away.
deflation hollow A low spot on the land surface caused by *deflation.*
deformation The process of change in shape or form of *rocks* after they formed—for example *folding, faulting,* stretching, *shortening,* and flattening.
density The *mass* per unit *volume* of a substance.
deposition The laying down or accumulation of *sediment* or other material.
depositional landforms Features of a landscape that formed by *deposition* of *sediment.*
depression A *topographic* feature that is lower in the middle and higher around its sides.
desert An *arid* region where, in its natural state, vegetation covers less than 15% of the surface.
desert pavement A surface in *desert* environments that is so completely covered with stones that wind cannot pick up any sediment. The desert pavement protects finer sediment below from the wind.
desertification The process of turning *semiarid* land into *desert.*
detrital sediment *Sediment* made up of broken *rock* or *mineral* pieces; *clastic sediment.*
deuterium A form of hydrogen atom with one proton and one neutron in its nucleus, thus having an atomic *mass* of 2.
dew point The temperature where the *relative humidity* reaches 100% and dew, fog, clouds, or other *condensation* form.

differential erosion The process of *erosion* in which harder *rocks* tend to erode less and stand out as hills or ridges and softer rocks erode more to become valleys or troughs.

differential stress The pressure or stress experienced by a *rock* undergoing *deformation* in which the *forces* on the rock are not equal in every direction; *directed pressure.*

dike A planar, *igneous intrusive* body that cuts across layers or cuts through unlayered *rocks;* a planar *discordant intrusion.*

dip The direction and angle of downward tilt of a geologic plane. Dip is measured from *horizontal,* perpendicular to the *strike,* along the steepest slope of the plane. Planes commonly measured include *bedding, foliation,* and *fault* planes. See *dip direction* and *dip angle.*

dip angle The angle between a *bedding, foliation, fault,* or other geologic plane and a *horizontal* plane. Dip angle is measured from horizontal to the steepest slope of the plane.

dip direction The approximate compass direction of the steepest downward slope of a geologic plane, especially a *bedding, foliation,* or *fault* plane, perpendicular to the *strike.*

dip-slip fault A *fault* with displacement parallel to the *dip* of the fault plane. Displacement may be up or down the dip of the fault plane.

directed pressure The pressure, or stress, experienced by a *rock* undergoing *deformation* in which the *forces* on the rock are not equal in every direction; *differential stress.*

disconformity An *unconformity* where the *rocks* above and below the unconformity are *sedimentary* and/or *volcanic* rocks and parallel to the unconformity surface.

discordant intrusion A *magmatic intrusion* that cuts across *bedding, foliation,* or existing *rock* masses. Examples include *dikes, stocks,* and *batholiths.*

divergent margin A *divergent plate boundary.*

divergent plate boundary A *plate* edge where two plates move away from each other; a *divergent margin, spreading center,* or constructive plate margin.

divide, drainage The boundary separating one *drainage basin* from another.

double refraction A *special property* of a *mineral* that breaks light passing through a clear *crystal* into two different rays, causing an image viewed through the crystal to appear double.

downwelling The movement of warmer surface seawater downward. See also *upwelling.*

drainage basin The region drained by a *stream.*

drainage divide The boundary separating one *drainage basin* from another.

driving force (*mass wasting*) The part of gravity that is directed along the surface and therefore can cause movement.

drumlin A glacial hill consisting of sediment smoothly streamlined as a *glacier* moved over it. The long axis is parallel to flow and the tapered end points in the direction of flow.

ductile A type of *tenacity* of a *mineral* in which the specimen can be pulled out into an elongated shape.

ductile deformation *Deformation* that occurs when *rocks* bend, flex, and/or flow and generally when the rocks are deep and possibly warm.

dull luster A description of the surface of a *mineral* that is not shiny; *earthy luster.*

dune A hill or ridge made of sand deposited by wind.

dynamothermal metamorphism *Metamorphism* involving heat and *differential stress.*

E

earthflow A type of *mass wasting* in which soil and weathered rock confined on both sides between specific boundaries shift in position downslope without rotation. Also the deposit or *landform* formed from such movement.

earthquake The vibration of the ground caused by natural geologic *forces.*

Earth's axis of rotation The line around which the Earth turns.

earthy luster A description of the surface of a *mineral* that is dull, not shiny; *dull luster.* It may also be rough and/or dusty.

eastern boundary current A cold ocean current that flows toward the equator near the eastern edge of an ocean.

eastings In a geographic grid, the north-south trending lines, with numbers increasing eastward.

effervescence A *special property* of a *mineral,* such as calcite, in which the mineral reacts with an acid solution by bubbling due to the release of carbon dioxide gas.

El Niño The occurrence of unusually warm water in the equatorial eastern Pacific Ocean, off the coast of Peru and Ecuador.

elastic (*mineral*) A type of *tenacity* of a *mineral* where the sample bends when *force* is applied but resumes its previous shape when the pressure is released; (*deformation*) to behavior of materials that snap back to their original shape after stress or pressure has been removed.

elastic rebound A snapping-back action as bent *rocks* return to their original shape during an *earthquake.*

electromagnetic spectrum A range of radiant energy of different wavelengths and energy levels that includes gamma rays, X-rays, *ultraviolet rays,* visible light, *infrared rays,* microwaves, and radio waves.

end moraine A ridge of *sediment* deposited directly from a *glacier* at its *terminus* (compare *terminal moraine*).

eolian Having to do with wind.

ephemeral Temporary.

ephemeral lake A temporary or short-term lake.

ephemeral stream A *stream* or part of a stream that flows only in response to precipitation or input from snow melt and is dry the rest of the year.

epicenter The point on the land surface directly above the *focus* of an *earthquake.*

equal area projection A map *projection* where equivalent areas on the ground are preserved as equal areas on the map.

equatorial countercurrent A warm ocean current that flows to the east at or near the equator between the *equatorial currents.*

equatorial current A warm ocean current that flows to the west just north or south of the equator.

erosion The wearing away and removal of *rock* or *sediment.*

erosional landforms Landscape features produced by *erosion.*

erupt To *extrude,* spew forth, or emanate *lava* as flows, fountains, or *pyroclastic* material from the Earth.

eruption, volcanic The extrusion or emanation of *lava* as flows, fountains, or *pyroclastic* material from the Earth.

esker A long, sinuous ridge of sediment deposited by a melt-water stream that flowed under a *glacier.*

estimated resource An estimate of the total quantity of a particular natural substance (*resource*) on the Earth.

euhedral An adjective describing a *mineral grain* with well-formed *crystal faces.*

evaporation The process by which liquid water turns into water vapor.

evaporite A *rock* formed by *precipitation* of *minerals* from water due to *evaporation.*

exaggerated profile A *topographic profile* where the *vertical* scale is larger than the *horizontal* scale. Exaggeration accentuates the topographic features by making the highs and lows more extreme and the slopes steeper than reality.

explanation, map See *map key.*

extratropical cyclone A low-*pressure* weather system with wet, unstable, rising air, cloudy skies, and *precipitation,* with a counterclockwise wind rotation in the Northern Hemisphere and a clockwise wind rotation in the Southern Hemisphere.

extrude To spew forth or emanate *lava* as flows, fountains, or *pyroclastic* material from the Earth; to *erupt.*

extrusive rock A type of *igneous rock* that forms at the surface of the Earth; *volcanic rock.*

extrusion The spewing forth, emanation, or *eruption* of *volcanic* material.

F

face Planar surface of a well-formed *crystal* that grew during *crystallization* or *recrystallization; crystal faces.*

facet A crystal *face* or other natural or artificial plane on a *mineral* or *rock.*

fall A style of *mass wasting* in which material drops through the air.

far infrared radiation Part of the *electromagnetic spectrum* with wavelength and energy level between *near infrared* and microwaves.

fault A *fracture* or zone of fractures across which displacement has taken place so the two sides remain in contact; also to form a fault.

fault scarp A cliff line or *offset* of the land surface where a recently-moved *fault* intersects the surface.

feel A *special property* of a *mineral* that is distinctive to the touch, such as the greasy feel (not *luster*) of graphite.

feldspar mineral group (feldspars) A group of minerals, the most abundant in the crust, that have the general formula $MAl(Al,Si)Si_2O_8$ in which M is most commonly some combination of Ca, Na, and/or K. Example feldspars with different substitutions for M are alkali feldspar (Na and K), potassium feldspar (K, an alkali feldspar), orthoclase (K), plagioclase feldspar (Na, Ca).

felsic A chemical composition term for *igneous rocks* that indicates the presence of low magnesium and iron content and high silica content (~70%). The term *felsic* is derived from feldspar (*fel*) and *silica* (*sic*). Felsic may also be used for *metamorphic rocks.*

fibrous fracture A *mineral fracture* where the broken surface has the appearance of many threads or fibers.

fine-grained texture A *texture* of a *rock* with individual *crystals* so small that they can only be seen with magnification. Fine-grained rocks include basalt, shale, and slate. See also *aphanitic texture.*

finger lake A long, narrow lake in a long *rock basin* formed by glacial *erosion* or in a valley dammed by a *moraine.*

firn A type of frozen water that is transitional between snow and *glacial ice* and has survived through a summer but is still *permeable* to liquid water.

firn limit The highest level on a *glacier* where all the previous year's snow has melted or ablated and the lowest level where *firn* is preserved in late summer or early fall before the first major storm deposits more snow.

fissile A property of a *rock* such as shale that splits or breaks into thin platy slabs, sheets, or flakes.

flexible A type of *tenacity* of a *mineral* where the specimen bends without breaking but will not spring back to its original shape when pressure is released. Compare with *elastic.*

flood basalts Extruded *mafic magma* that has spread out and *solidified* in wide, flat *lava* sheets over extensive areas.

flow A style of *mass wasting* in which material moves chaotically at different speeds like a fluid. Also the deposit or *landform* formed from such movement.

flow banding A banded or layered pattern produced by motion of *lava* in a *lava flow.*

fluorescence A *special property* of a *mineral,* such as some varieties of fluorite, by which the mineral emits light during exposure to *ultraviolet rays.*

fluvial Having to do with *streams* or *rivers.*

focus The place beneath the Earth's surface where an *earthquake* starts.

fold A bend in *rock* layers or other planar features, generally formed by *compression* parallel to the layering; also, to form such a bend. See also *anticline, syncline,* and *monocline.*

fold axis The line around which folded layers are bent; a *fold hinge.*

fold hinge A *fold axis.*

foliation A planar *texture* or *structure* of a *metamorphic rock* that was produced by differential stress. See also *slaty cleavage, schistosity,* and *gneissosity.*

footwall The block beneath an inclined *fault* plane. Compare with *hanging wall.*

force A push or pull on an object caused by its interaction with another object.

force of friction A *force* that tends to resist sliding or rolling motion between two surfaces in contact with each other. If motion is already occurring, friction tends to slow the motion. If no motion is occurring between the two surfaces, friction acts to tend to keep them from moving.

formation (a) (*cartography*) A continuous or once-continuous body of *rock—igneous, sedimentary,* or *metamorphic*—that can be easily recognized in geologic fieldwork and is the basic unit shown on geologic maps; (b) (*stratigraphy*) A rock body identifiable by its geologic characteristics and its position relative to the stratigraphic sequence in an area; a fundamental rock unit with distinct features.

fossil The naturally preserved remains or trace of life preserved in a *rock,* usually *sedimentary,* at the time the rock formed.

fossil fuel An energy *resource* that comes from the *organic* remains of organisms preserved in *rocks.*

fossil succession, principle of States that organisms evolve in a definite order, that species evolve and go extinct, never to re-evolve, so the evolution of a species and its extinction become time markers separating time into three units: one before the organism existed, one during the existence of the organism, and one after the organism went extinct.

fossiliferous Containing abundant *fossils.*

fossilization The process of formation of a *fossil.*

fractional scale A method of expressing a map *scale* by giving the fraction or ratio of a given distance between two points on the map to the distance between the corresponding points on the Earth's surface. The two distances are expressed in the same units, for example, inches on the map to inches on the ground; *representative fraction.*

fracture (a) (*mineral*) A property whereby the *mineral* exhibits irregular and nonplanar surfaces when broken; (b) (*structure*) A crack or break in *rock.*

front A boundary or transition zone between air masses of different density. See also *cold front, warm front, occluded front,* and *stationary front.*

frost wedging Frost *weathering*; mechanical weathering caused freezing and thawing.

G

geographic axis The *axis of Earth's rotation;* the line around which the Earth rotates.

geographic pole The end, either north or south, of the *axis of Earth's rotation.*

geologic map A map used to show the distribution of various *rock* masses, *formations, structures,* and their age relationships.

geologic resource A naturally-occurring substance that can be used and comes from the Earth but not directly from living things.

geologist A scientist who has been trained and works in the geological sciences—those sciences having to do with the study of the Earth.

geology The science that is the study of the Earth.

geophysics The branch of *geology* that studies the physics of the Earth, including *seismology,* geomagnetism, and aspects of volcanology and *oceanography.*

glacial An adjective referring to a *glacier.*

glacial flour *Sediment* produced by the grinding of rocks over bedrock at the base of a glacier.

glacial ice *Recrystallized* snow that forms when *firn* turns into ice. Glacial ice is denser than snow or firn and is impermeable to liquid water.

glacial lobe A tongue-shaped extension from the margin of a *continental glacier* that projects farther than adjacent parts of the glacier; *ice lobe.*

glacial trough A *U-shaped valley* carved by a *glacier.*

glacier A large mass of ice at least partly on land that survives from year to year and flows due to the *force* of its own weight (see also *alpine glacier, ice sheet, continental glacier, piedmont glacier*).

glacier balance *Glacier budget.*

glacier budget The balance (difference) between *precipitation* of new snow and *ablation* of snow, *firn,* and ice in a *glacier.* When the glacier budget is positive, the glacier *advances*; when it is negative the glacier *retreats;* and when it is in balance the *terminus* of the glacier is stationary.

glacier mass balance *Glacier budget.*

glaciology The study of *glaciers,* snow, and ice; *cryology.*

glassy texture The *texture* of glass, which has *vitreous luster* and no definite *minerals; hyaline texture.*

global conveyor The flow of cold, deep, and warm surface seawater that makes a complete loop from sinking in the North Atlantic, flowing south deep in the Atlantic, through the Indian Ocean, and into the North Pacific Ocean, where it rises and forms a returning warm surface current that flows back through these oceans to reach the North Atlantic again. Also called *thermohaline circulation.*

Global Positioning System (GPS) A satellite-based system of radio location.

global warming *Anthropogenic* world-wide average increase in temperature of Earth's climate.

gneissic banding See *gneissosity.*

gneissosity A type of *foliation* where the *rock* is *medium-* to *coarse-grained* and light and dark *minerals* are arranged in parallel bands, streaks, or layers (gneissic banding). Mineral segregations are at least about 2 mm wide.

Gondwana A *supercontinent* that existed hundreds of millions of years ago made of the southern continents assembled together; also called Gondwanaland.

Gondwanaland See *Gondwana.*

graded bedding A single layer or *bed* of *sediment* or *sedimentary rock* in which the largest *grains* are concentrated at the bottom, gradually decreasing in size upward to the smallest at the top of the bed.

gradient, stream The steepness of the *slope* of a *stream* usually given in feet per mile or meters per kilometer.

grain A *mineral* or *rock* particle or *crystal* that has a distinct boundary separating it from surrounding grains, *matrix,* or *groundmass.*

grain flow A type of *mass wasting* in which dry *sedimentary* particles move much like a liquid.

granoblastic texture A *coarse-grained metamorphic texture* in which the *mineral grains* are randomly oriented in the *rock.*

granular texture A *texture* where the *rock* has visible *grains; coarse-grained texture.*

graphic scale A way of representing the scale of a map (or diagram) by drawing a line on the map (or diagram) representing the actual distance; a *scale bar.*

graticule The crisscrossing *lines of latitude* and *longitude* that form a reference grid on the spheroidal surface of the Earth.

greasy luster The surface shine of a *mineral* that resembles the way petroleum jelly or a greasy surface reflects light.

great circle The circle produced by a plane passing through the center of the Earth intersecting with the surface of the Earth.

greenhouse effect A process that heats the Earth's surface and lower atmosphere as follows: Visible light from the sun passes through the atmosphere and heats the Earth's surface. The warmed surface radiates *far infrared rays,* which are partially absorbed by *greenhouse gases* in the lower atmosphere and radiated back to the Earth's surface.

greenhouse gas A gas that allows light through it but absorbs *far infrared radiation.* This traps heat in the atmosphere.

Greenwich meridian The *meridian* that has a value of zero degrees and passes through Greenwich, United Kingdom.

grid A system of lines intersecting at right angles to form rectangles.

grid north The orientation of the north-south set of *grid* lines of a regional coordinate system used on a map.

groin An artificial structure built at a beach perpendicular to the shore to slow beach *erosion.*

ground moraine *Sediment* dropped directly from a *glacier* over a wide area as it retreats, leaving a fairly thin covering of sediment rather than large piles or concentrations.

groundmass Smaller *grains* surrounding larger grains in an *igneous rock* with *porphyritic texture.* See also *phenocrysts,* the larger grains.

groundwater Water in the ground, mostly below the *water table.*

guide fossil A fossil of an organism that existed for a short period of time so that its occurrence in a *rock* suggests a narrow range of age; an *index fossil.*

H

habit The external shape a *mineral* typically displays if it was free to grow when it formed.

hackly fracture The fracture of a *mineral* that breaks to produce a sharp, jagged surface.

half-life The length of time it takes for one-half of a *radiogenic isotope* to decay.

halide mineral class (halides) A group of *minerals* with members containing one of the halogen elements, fluorine, chlorine, or bromine and one or more metals, but no other nonmetals. Examples are halite, fluorite, and sylvite.

hanging valley A valley that enters another larger valley high up on the valley walls so that a steep drop occurs at the point of connection between the valleys. Either *streams* or *glaciers* can produce hanging valleys, generally when the erosion of the *tributary* stream or glacier does not keep pace with the *erosion* by the trunk (main) stream or glacier.

hanging wall The block above an inclined *fault* plane. Compare with *footwall.*

hardness The resistance of a *mineral* to abrasion or scratching.

headland A promontory at a shoreline that sticks out into the water.

headward erosion *Erosion* that extends or lengthens a valley at its upper end.

high-grade metamorphism *Metamorphism* at high temperatures and pressures. Compare with *low-* and *medium-grade metamorphism.*

hinge, fold The line around which the *beds* of the *fold* are bent; *fold axis.*

horizontal An orientation, such as the horizon, that is exactly level or side-to-side, not up and down.

horn A sharp mountain peak formed by *headward glacial erosion,* with planar or concave sides and ridges and surrounded by three or more *cirques* or glacial valleys and three or more *arêtes.*

humidity The water vapor content of the air. See also *absolute humidity* and *relative humidity.*

hummocky A landscape with numerous small hills, called hummocks.

hyaline texture The *texture* of glass, which has *vitreous luster* and no definite *minerals; glassy texture.*

hydraulic head The difference between the height of the surface of water open to the air, or at atmospheric *pressure,* and water at a point in the subsurface.

hydrocarbons Chemical substances made up of hydrogen and carbon.

hydroxide mineral class (hydroxides) A group of *minerals* with members containing hydrogen and oxygen and one or more metals, but no other nonmetals (e.g., limonite).

I

ice lobe A tongue-shaped extension from the margin of a *continental glacier* that projects farther than adjacent parts of the glacier, *glacial lobe.*

ice sheet A *continental glacier.*

igneous intrusion A body of still molten or solidified *magma* beneath the Earth's surface; a *magmatic intrusion.*

igneous rock A *rock* that has formed by the *solidification* of *magma.* See Lab 3 for descriptions of individual igneous rocks.

immature sediment *Clastic sediment* that has a high proportion of easily weathered material (e.g., *rock* fragments, *mafic minerals,* and feldspar) and contains angular, *poorly-sorted grains.*

impermeable Having no *permeability.*

index contour A *contour* that is drawn with a heavier line and is labeled with the elevation; usually every fifth line.

index fossil A *fossil* of an organism that existed for a short period of time so that its occurrence in a *rock* suggests a narrow range of age; a *guide fossil.*

index mineral A *mineral* that is indicative of a *metamorphic zone.* The first appearance of the index mineral as the zonation progresses from lower to higher temperature marks the beginning of a metamorphic zone and coincides with an *isograd.* For example, the area that starts at the first appearance of biotite and ends at the first appearance of almandine garnet corresponds to the biotite zone. Biotite is the index mineral as long as garnet and other higher temperature index minerals are not present. If garnet is also present, then the *rock* belongs to the garnet zone or higher temperature zone.

indirect dating A dating method whereby the numerical age of a *sedimentary rock* can be approximately determined by using a combination of *relative* and *isotopic dating* techniques.

infrared rays Part of the *electromagnetic spectrum* that is longer in wavelength and lower in energy than visible light but shorter in wavelength and higher in energy than microwaves.

inner core The part of the Earth's *core* that is solid metal, mostly iron.

inorganic Not *organic.*

inselbergs Isolated rocky hills or knobs surrounded by *pediment,* erosional remnants of mountains in *desert* regions.

interlobate moraine *Moraine* formed between adjacent *ice lobes* of a *continental glacier.*

intermediate A chemical composition term for *igneous rocks* that indicates the presence of middle quantities of magnesium, iron and silica content (silica, ~60%).

International Date Line The line on the Earth's surface across which the date changes, located near 180° longitude.

intrusion A body of still molten or solidified *magma* beneath the Earth's surface; short for *magmatic* or *igneous intrusion*; see also *dike, sill, pluton, stock, batholith.*

intrusive rock An *igneous rock* that solidified beneath the Earth's surface; *plutonic rock.*

irregular fracture A type of *mineral fracture* that is uneven and not *conchoidal, fibrous,* or *hackly fracture.*

isobars Lines of equal pressure on a map or graph.

isograd A line of equal metamorphism, making a boundary between *metamorphic zones.*

isostatic rebound Uplift of a region due to unloading of the *crust* by *erosion*, or melting of an *ice sheet*.

isostasy The condition of gravitational equilibrium, where the underlying *asthenosphere* buoys up (supports) the *crust* and *lithosphere*.

isotope A type of atom of an element with a specific number of neutrons and protons in the nucleus.

isotopic age The age of a substance that was determined using *radiometric dating* (*isotopic dating*).

isotopic dating The process of determining the numerical or *absolute age* using *radiogenic isotopes*; *radiometric dating.*

J

joint A crack in a *rock* along which no relative movement has occurred parallel to the crack.

K

karst topography The *topography* resulting from the flow of *groundwater* through areas of soluble *rocks.* Features of karst include caves, solution valleys, *sinkholes,* karst towers, and underground drainage, which are caused by the dissolution of limestone or other soluble rocks, such as rock gypsum.

key bed A distinctive rock layer that can be used in correlation. Some key beds, such as a volcanic ash layer, can indicate equivalence in age between rock layers.

key, map See *map key.*

L

La Niña A more extreme version of the normal current pattern of cold water along the coast of Peru and Ecuador; the opposite of *El Niño.*

laccolith An *igneous intrusive* body that intrudes parallel to layers (*concordant*) and bulges upward to make a three-dimensional body, doming the layers above it.

lahar a *mudflow* produced by water mixing with *volcanic ash.*

landform The external form of a distinctively shaped rock mass; erroneously referred to a *"formation"* in common speech.

Land Office Grid System A *grid* system used in much of the United States based on 36-mi² grid, called *townships,* with square-mile subdivisions called *sections.*

landslide The event or process in which *rock* or debris moves rapidly down the surface slope in response to gravity; a type of *mass wasting;* also, the *landform* that results from this process; *Township and Range system.*

large scale The scale of a map where the ratio between an object on a map and the same object on the Earth's surface is large. For large-scale maps, a large area on the map covers a small amount of the Earth's surface, and represented objects appear large. The term can also be used for diagrams other than maps where scale is involved. Compare with *small scale.*

latitude, line of A locational reference line that is measured by the angular distance north or south of the equator; a *parallel.* Compare with *longitude.*

Laurasia A *supercontinent* that existed hundreds of millions of years ago made of North America, Europe, and Asia assembled together.

lava Molten *rock* at the Earth's surface.

lava dome *Extruded magma* that is so *viscous* that it does not spread out but instead forms a dome or mound; also the resultant body of volcanic rock.

lava flow *Extruded magma* that has flowed laterally and solidified in a tongue shape or as a surficial sheet; also the resultant body of volcanic rock.

lava fountain A dramatic extrusion of glowing magma that squirts or jets high into the air, similar to a fountain of water, only hotter, more viscous, and commonly much higher, sometimes over 500 ft.

lee *Leeward.*

leeward The downwind side.

left-lateral fault A *strike-slip fault* for which one side moves left along the *fault* from a viewpoint on the other side, looking across the fault plane.

legend, map See *map key.*

line of latitude See *latitude, line of.*

line of longitude See *longitude, line of.*

lineation A linear (arranged parallel to a line) *texture* or *structure* of a *metamorphic rock* that was produced by *differential stress.*

liquefaction Quicksand-like behavior of water-saturated *sediment* when vigorously shaken by an *earthquake.*

lithification The process by which loose *sediment* turns into a consolidated *sedimentary rock;* consolidation. The processes of *cementation* and *compaction,* and in some cases *crystallization,* aid the lithification process.

lithologic Pertaining to the physical character of a *rock* covering such aspects as composition and *textures;* usually pertains to properties that are visible rather than microscopic.

lithosphere An approximately 100-km-thick layer of the Earth at the surface that is made of strong brittle *rock,* underlain by the *asthenosphere.* The upper part of the lithosphere is made up of the *crust,* and the lower part coincides with the uppermost *mantle.* See Figure 1.2.

load (*mass wasting*) The weight of material piled on top of a slope. (*streams*) The sediment the stream caries, including bedload transported along the bed, suspended load transported as particles in the water and dissolved load dissolved in the water. (wind transport) The sediment transported by wind, including bedload (*creep* and *saltation*), suspended load (lifted and carried long distances in the wind).

loess Fine, windblown *sediment* usually derived from *deserts* but deposited outside desert areas.

longitude, line of Any of the half circles between the North and South Poles that are used as a means of specifying the east-west position of a location by the angular distance from the *prime meridian;* a *meridian.* Compare with *latitude.*

longitudinal dune A long, nearly straight *dune* oriented parallel to the wind direction.

longshore current A current of water moving parallel to the shoreline as a result of waves encroaching at an angle to the coastline.

longshore drift *Sediment* moving parallel to the shore as a result of the *longshore current.*

low-grade metamorphism *Metamorphism* at low temperatures and pressures.

luster How light is reflected from a fresh surface of a mineral. See also *metallic, submetallic, nonmetallic, vitreous, resinous, waxy, pearly, fibrous, splendent, adamantine, and earthy lusters.*

M

mafic A chemical composition term for *igneous rocks* that indicates the presence of high magnesium and iron content and low *silica* content (~50%).

magma Molten *rock.*

magmatic intrusion A body of still molten or solidified *magma* beneath the Earth's surface; an *igneous intrusion.*

magnetic declination The angle between *magnetic north* and geographic or *true north.*

magnetic north The direction toward which a magnetic compass (or magnetized needle) points.

magnetism A *special property* of a *mineral,* such as magnetite, that is attracted to a magnet.

mantle A chemically distinct part of the Earth that is below the *crust* and above the *core.* It is primarily made up of oxygen, silicon, and magnesium and is *ultramafic* in composition. The mantle includes the lower part of the *lithosphere,* the *asthenosphere,* and the *mesosphere.* See Figure 1.2.

map key A guide to the various colors and symbols on a map; also called map legend or explanation.

mass A measure of a body's resistance to acceleration. On Earth, the mass of an object is proportional to its weight.

mass balance, glacier *Glacier budget.*

massive Lacking *bedding,* parallel-layered *structure,* or *foliation.*

mass movement See *mass wasting.*

mass wasting The downslope movement of loose surface material due to gravity, without the aid of a transporting substance such as flowing water, glaciers, or wind.

matrix (sedimentary) Finer-grained material between and/or surrounding larger *grains* in *sediment* or *sedimentary rock;* (igneous) *groundmass.*

mature *Clastic sediment* that has no easily-weathered *grains* and contains well-rounded and *well-sorted* material.

meandering stream A winding *stream;* a *stream* with a *sinuosity* of greater than 1.5.

medial moraine A concentration of *sediment* in a stripe on/in *glacial* ice formed where two *alpine glaciers* come together.

medium-grade metamorphism *Metamorphism* at moderate temperatures and *pressures.*

medium-grained Having visible *grains* about the size of sand. Medium-grained rocks include sandstone and schist. Sometimes, especially for *igneous* and *metamorphic rocks,* medium-grain size is loosely included as part of the term *coarse-grained.*

Mercator projection A map *projection* where the global *graticule* is projected onto a cylinder aligned with the *geographic axis* with the cylinder touching the equator.

meridian The half a circle between the North and South Poles that is used as a means of specifying the east-west position of a location; a *line of longitude.*

mesa A nearly flat hill or high *landform* with steep sides and with horizontal dimensions several times greater than its height, a small plateau.

mesosphere (a) (solid Earth) A layer of the Earth that is made of rigid *rock,* below the *asthenosphere* and above the *outer core.* The mesosphere corresponds to the lower *mantle.* (b) (atmosphere) A layer of the atmosphere (between about 50 and 80 km altitude) above the *stratosphere* and below the thermosphere, where temperature decreases with altitude. See Figure 16.2.

metallic luster The surface shine of a *mineral* that resembles the way metals reflect light.

metamorphic facies The set of all *mineral assemblages* that may be found together in a region where the *rocks* have different chemical composition but were all *metamorphosed* at the same conditions of temperature and *pressure.*

metamorphic grade An approximate measure of the amount or degree of *metamorphism,* most closely tied to metamorphic temperature, but also related to metamorphic *pressure.* See also *low-, medium-,* and *high-grade metamorphism.*

metamorphic rock A *rock* that has undergone *metamorphism.* See Lab 5 for descriptions of individual metamorphic rocks.

metamorphic zone An area or region in which a distinctive *mineral assemblage* in *metamorphosed* shale indicates a specific range of metamorphic conditions, especially temperature.

metamorphism The combination of all of the processes that change a *rock* above 200°C in the solid state as a result of changes in temperature and/or *pressure.* Changes primarily involve *texture* and mineral content.

metamorphose (verb) To undergo *metamorphism.*

meteorology The science that is the study of the atmosphere, including weather and climate.

mid-ocean ridge A long, symmetrical mountain range or broad rise in the ocean, commonly but not always at the middle, with many small, slightly *offset* segments. The axes of mid-ocean ridges correspond to *spreading centers,* or *divergent plate boundaries.*

mineral A naturally-occurring, usually *inorganic,* chemically homogeneous *crystalline* solid with a strictly-defined chemical composition and characteristic physical properties. See Table 2.3 for descriptions of individual minerals.

mineral assemblage A group of *minerals* that grow or coexist together at the same temperature and *pressure.*

mineral class A set of *minerals* that have some formally defined common characteristics. The classes *silicates, carbonates, sulfates, sulfides, phosphates, oxides,* and *hydroxides* are defined by the nonmetallic or metalloid part of their chemical composition. Families of the silicates, such as chain or sheet silicates, are classified by the internal atomic arrangement of *silica tetrahedra* (Table 2.9). Another mineral class is the *native elements.*

mineral group A set of closely related minerals, having similar structure but a limited variation in chemical composition.

mold A *fossil* made of the imprint of an organism or part of an organism, such as when a clam shell dissolves, leaving a cavity with its clam-shell shape.

moment magnitude See *seismic moment.*

monocline A step-like or s-shaped *fold* in otherwise horizontal or gently dipping beds. Compare *anticline, syncline.*

monument A *rock* tower several times taller than it is wide; a pinnacle or rock spire.

moraine A pile, ridge, or accumulation of *sediment* deposited directly from *glacier* ice.

mountain glacier A type of *glacier* that forms and flows in *valleys* in mountainous regions; an *alpine glacier*; a *valley glacier*.

mudflow A type of wet *mass wasting* that moves as a fluid in which more than half of the material is sand size or smaller; the amount of water is not enough to support the grains (as in sediment transport by a stream). Also a deposit formed by this process. Mudflows generally contain more water and are faster-flowing than *debris flows*.

Mw See *seismic moment magnitude*.

N

native element mineral class (native elements) A group of *minerals* with members containing a pure naturally occurring element. Examples are diamond, graphite, copper, gold, silver, sulfur.

natural gas Gaseous hydrocarbons, primarily methane CH_4.

near infrared radiation Part of the *electromagnetic spectrum* with wavelength and energy level between visible light and *far infrared*.

no streak Said of a *mineral* that is too hard (H > 6 to 7) to leave a powder on a porcelain *streak* plate.

nonconformity An *unconformity* where the older *rocks* below the unconformity are *plutonic igneous* or *metamorphic rocks*.

nonmetallic luster The surface shine of a *mineral* that does not resemble the way metal reflects light.

non-plunging fold An *anticline* or *syncline* that has a horizontal *fold axis*.

nonrenewable resource A *resource* that has a limited quantity that is diminished by use.

normal fault A *dip-slip fault* with downward movement of the *hanging wall* relative to the *footwall*.

normal force A physical *force* directed perpendicular to a surface.

northings In a *grid*, the east-west-trending lines, with numbers increasing northward.

O

oblique-slip fault A *fault* with displacement (movement) betwEEn the *dip* and the *strike* of the fault plane.

occluded front A *front* where two cold air masses have come in contact and wedged out a warmer air mass above. See also *cold front*, *warm front*, and *stationary front*.

oceanic crust The part of the Earth's *crust* that underlies the oceans and is chemically distinct (*mafic*) from *continental crust*.

oceanic trench A deep trough in the *bathymetry* of the ocean floor. Trenches are associated with *subduction* zones.

oceanography The science that is the study of the oceans.

offset The amount of movement of one side of a *fault* relative to the other.

ore A *rock* or *sediment* containing one or more economic metal *resources*.

organic (a) (chemistry) Pertaining to a compound containing carbon or carbon and hydrogen, but not simply carbon and oxygen; (b) pertaining to or derived from living organisms.

original horizontality, principle of States that *sedimentary* layers are deposited *horizontally* or nearly so.

orographic precipitation Rain or snow that results from the clouds that form due to *orographic uplift*.

orographic uplift The lifting of air that results when wind blows toward mountains.

outcrop An area of *rocks* exposed at the surface without foliage, soil, *sediment*, or artificial structures covering them.

outer core The part of the Earth's *core* that is liquid metal above the *inner core*.

outwash plain A plain formed beyond the limits of a glacier where glacially-derived sediments are deposited by glacial meltwater.

overburden Unwanted *rock* above a valuable *mineral* or coal deposit.

overturned Said of *sedimentary* layers that have been tilted more than 90°.

oxide mineral class (oxides) A group of *minerals* with members containing oxygen and one or more metals, but no other nonmetals or metalloids. Examples are magnetite, hematite, and corundum.

ozone Oxygen gas with an extra oxygen atom: O_3.

ozone layer The part of the atmosphere, in the *stratosphere*, with a relatively high concentration of *ozone*.

P

P wave The fastest type of *earthquake* wave, with the vibration direction parallel to the direction the wave travels; a compressional *seismic body wave*.

Pangaea A *supercontinent* that existed hundreds of millions of years ago made of all of the continents assembled together. See also *Gondwana* and *Laurasia*.

parabolic dunes A sand *dune* that has a crescent shape (like a parabola) with the tips of the crescent pointing upwind and with a *slip face* on the outside of the curve.

parallax An apparent change of position of a closer object against a background of farther objects with respect to a reference point (typically ones eye).

parallel A locational reference line that is measured by the angular distance north or south of the equator; a *line of latitude*.

parent rock (a) (*metamorphic*) The *rock* from which a *metamorphic rock* forms; the *protolith*; (b) (*sedimentary*) the primary source rock from which a sediment has weathered.

parting The property of a *mineral* that breaks along nearly planar surfaces but that is not consistent or planar enough to be considered *cleavage*.

passive continental margin The edge of a continent without a *plate boundary*.

pearly luster The surface reflection of a *mineral* that resembles the way pearls shine.

pediment A gently-sloping *bedrock* plain that develops in a *desert* from *cliff retreat* of the mountain front.

pelitic rock A *rock* with overall chemical composition similar to clay-rich shales; a *metamorphic rock* with a shale *protolith*.

permeability A measure of the ability of *rock* or *sediment* to allow liquids or gases to flow through it.

permeable Having *permeability.*
petrifaction or petrification A process of *fossilization* in which original *organic* material is replaced by other *inorganic* substances.
petrology The science that is the study of *rocks* and how they form.
phaneritic texture A *texture* of an *igneous rock* where the *rock* has visible *grains; medium-* or *coarse-grained texture.*
phenocryst A large *grain* surrounded by smaller grains, or *groundmass,* in an *igneous rock.* A *rock* containing phenocrysts has *porphyritic texture.*
phosphate mineral class (phosphates) A group of *minerals* with members containing phosphorus and oxygen and one or more metals, but no other nonmetals or metalloids. A major phosphate mineral is apatite.
photosynthesis A chemical *synthesis* performed by plants using light from the sun to form *organic* molecules and by-product oxygen from water and carbon dioxide.
piedmont glacier A *glacier* formed where two or more *alpine glaciers* flow out and join together on a plain at the foot of a mountain range.
placer A concentration of a *mineral* that forms where turbulent flowing water segregates heavier minerals from lighter ones.
planimetric map A map that depicts the location of major cultural and geographic features such as towns, rivers, roads, and railroads.
plastic An adjective describing material that can flow, stretch, bend, and/or flex without breaking. In the Earth sciences this term is usually applied to *ductile rock,* which is not rigid.
plate A relatively rigid section of *lithosphere* that moves as a unit and relative to other plates.
plate boundary The edge or margin of a *plate.*
plate tectonics A field of earth science that involves the study of *plates,* their boundaries and movements, and their influences on other aspects of the Earth such as *rocks, structures, topography,* mountains, mountain building, and the Earth's interior; see *tectonics.*
plate tectonics, theory of The theory that the Earth's *lithosphere* is divided into a few pieces called *plates* that move relative to each other and relative to the Earth as a whole. *Deformation,* crustal destruction, and regeneration are concentrated around the margins of plates, and the interior of plates remain relatively rigid and do not tend to deform. See *tectonics.*
playa A low, flat plain lacking vegetation at the lowest part of a landlocked *desert* basin, commonly consisting of sand, silt, clay, and salt.
playa lake An *ephemeral* (temporary) lake in a *playa.*
plunge The angle of tilt down into the ground, measured from *horizontal,* of the *axis* of a *plunging fold.*
plunging fold A *fold* with an *axis* that is not *horizontal.*
pluton A body of rock resulting from crystallization of magma injected at depth in the Earth's crust; see also *dike, sill, pluton, stock, batholith.*
plutonic rock An *igneous rock* that *solidified* beneath the Earth's surface.
polar easterlies Winds between the pole and 60° latitude that blow from the east.
poorly sorted See *sorted, poorly.*
pore A void or space within *rock* or *sediment.*
pore pressure The *force* per unit area (*pressure*) from air or water on the inside of *pores* pushing outward.
porosity The percentage or proportion of the *volume* of *rock* or *sediment* made up of *pore* spaces.
porphyritic An adjective used to describe *igneous rocks* with *porphyritic texture* with less than 25% *phenocrysts.*
porphyritic texture The *texture* of an *igneous rock* with large *crystals* or *phenocrysts* embedded in a more finely *crystalline* or *glassy groundmass.*
porphyroblast A large *metamorphic mineral grain* or *crystal* surrounded by smaller grains in a *rock* with *porphyroblastic texture.*
porphyroblastic texture A *metamorphic texture* in which distinctly larger *grains* are surrounded by smaller grains.
porphyry The name of an *igneous rock* with *porphyritic texture* in which *phenocrysts* comprise 25% or more of its *volume.*
potentially renewable resource A *resource* that is replenished but may or may not be used faster than the rate of replenishment.
potentiometric surface The level that water in a *confined aquifer* would rise if the aquifer were punctured.
precipitation (a) (*meteorology*) The process in which liquid or solid water falls from clouds, forming rain, snow, sleet, hail, and freezing rain; (b) (chemistry) a chemical process in which solids form out of a solution.
prime meridian The *meridian* that has a value of 0° and passes through Greenwich, United Kingdom.
pressure *Force* per unit area.
principal meridian For the *Township and Range System,* a north–south line through the origin or reference point of the system.
projected lifetimes A crude estimate of the length of time a particular *mineral resource* will last, calculated by assuming that a *resource* will continue to be used at the current rate and dividing that into the *reserves* of that resource.
projection A geometric technique used to convert information on a three-dimensional object, such as a sphere, to two dimensions, such as a map.
protolith The *rock* from which a *metamorphic rock* forms; the *parent rock.*
psychrometer An apparatus that measures *relative humidity* by comparison of the cooling effect of evaporation on a wet thermometer bulb compared to the dry bulb temperature.
Public Land Survey A *grid* system used in much of the United States based on 36-mi^2 grid, called *townships,* with subdivisions, called *sections,* for every square mile; the *Township and Range System.*
pyroclastic Pertaining to *pyroclastics* or having *pyroclastic texture.*
pyroclastic texture The *texture* of *rock* or loose material made of *volcanic ash* and/or larger *rock* fragments exploded from a *volcano.*
pyroclastics Fragmental *volcanic* products such as *ash, cinders,* pumice pieces, *volcanic bombs,* and blocks, formed by aerial ejection or explosion from a *volcanic vent; tephra.*

Q

quarry A place where *rock* is extracted from the surface, commonly for road rock or other construction rock.

R

radioactive Possessing *radioactivity*.

radioactivity A process whereby the nucleus of an unstable atom spontaneously decays by losing or gaining a particle, thereby changing into the nucleus of another element.

radiogenic isotope An (unstable) *isotope* that is naturally *radioactive*.

radiometric dating See *isotopic dating*.

rain-shadow deserts A *desert* formed where cool, sinking air on the *lee* side of a mountain range causes drying conditions as it warms.

raised relief map *Topographic map* that is drafted onto a plastic sheet or other medium that is molded or shaped to give the viewer an idea of the *relief* in the map area in three dimensions.

recrystallization The formation of new *grains* of *mineral* material already present in a *rock*. The original material of an organism may recrystallize as part of the process of *fossilization*. During *metamorphism*, some minerals recrystallize from the *parent rock;* for example, the recrystallization of quartz as a quartz sandstone *metamorphoses* into a quartzite.

reduction A chemical reaction that removes oxygen from a substance or reduces the electrical charge on atoms to a lesser number. For example, reduction may change Fe^{3+} to Fe^{2+}.

refraction The bending of waves of all kinds, caused by changes in wave speed. For example, light waves refract when passing through a lens; ocean waves refract as they approach the shoreline at an oblique angle.

regional metamorphism *Metamorphism* over a wide area or region where *differential stress* combines with a wide range of temperatures and *confining pressure* at moderate to great depths. Regional metamorphism commonly results at *convergent plate boundaries*.

relative age The age, usually of a *rock* or *fossil,* established in comparison to other ages, using words such as older or younger rather than numerical ages.

relative humidity The percentage of water vapor relative to the total amount of water vapor that a particular *volume* of air at a particular temperature can potentially hold.

relict bedding *Sedimentary bedding* that has been preserved after *metamorphism* and is visible in a *metamorphic rock*.

relief The difference in elevation between the highest and lowest point in an area. For example, steep slopes and cliffs occur in areas of high relief.

renewable resource A *resource* for which its use does not deplete its quantity.

replacement *(fossil)* When the original minerals of a body fossil are replaced by a later mineral.

representative fraction A method of expressing a map *scale* by giving the ratio of a given distance between two points on the Earth's surface to the distance between the same two points as represented on the map; *fractional scale*.

reserves The quantity of a *resource* that has been found and is economically recoverable with existing technology.

reservoir (a) (water) A body of water stored in a valley behind a dam; (b) (petroleum geology) A body of *permeable rock* containing oil and natural gas.

resinous luster The surface shine of a *mineral* that resembles the way plastic reflects light.

resource (a) Any naturally-occurring substance that can be used by humans; see also *geologic resource;* (b) (numerical definition) The total quantity of a particular natural substance on Earth.

retreat *(glacier)* A decrease in the size of a glacier in which area covered by the glacier decreases and the terminus is closer to the ice source than previously. *Ablation*, mainly by melting and *calving,* is faster than forward flow of ice.

retreat, cliff *Erosion* of a cliff in a way that maintains close to the original *slope* of the cliff.

retreat, scarp *Erosion* of a *scarp* in a way that maintains close to the original *slope* of the cliff.

reverse fault A *dip-slip fault* with upward movement of the *hanging wall* relative to the *footwall*.

Richter magnitude A measure of the relative size of an *earthquake* indicating the amount of earthquake energy released, using a logarithmic scale of the amplitude of *seismic waves* measured and adjusting for distance from the earthquake. Local magnitude. Compare *seismic moment magnitude*.

right-lateral fault A *strike-slip fault* for which one side appears to move right along the fault from a viewpoint on the other side, looking across the *fault* plane.

river A part of a *stream* system where the stream is large.

rock An aggregate of *mineral grains* and/or mineral-like matter or a homogeneous mass of mineral-like matter. By mineral-like matter, we mean material such as opal or volcanic glass that is a natural *inorganic* solid and material such as found in coal that is the solid *organic* remains of organisms that have changed in some way since the organism's death.

rock basin A *bedrock depression*.

rock glacier A moving *mass* of *rock* with ice in the *pore* spaces or with a core of ice. Movement is similar to a small *alpine glacier*.

rockfall A movement of large blocks of *rock* downward at least partly through the air; falling rock. Also the deposit formed from such movement.

rockslide *Mass wasting* in which a large slab of *rock* moves along a planar slip surface. Also the deposit or *landform* formed from such movement.

rule of V's The principle that the surface trace of an inclined planar feature in a valley forms a V that points in the direction of *dip* of the feature.

S

S wave An *earthquake* wave with the vibration direction perpendicular to the direction the wave travels; a shearing *seismic body wave*.

salinity The proportion of the mass of salts in a solution to the total mass of the solution. For seawater the salinity is usually expressed in parts per thousand.

saltation Movement of *grains,* most commonly sand, propelled by wind or water where the grains are ejected by impact of other grains and then hop upward in an arc to collide with the surface where they are likely to propel other grains to move in the same way (Figure 18.6).

saturation, zone of The region under ground where water fills all of the *pores;* below the *water table*.

scale The size reduction needed to convert the actual feature into its representative on a map or diagram.

scale bar A way of representing the *scale* of a map (or any diagram) by drawing a line on the map (or diagram) representing the actual distance; *graphic scale.*

scarp (*mass wasting*) A cliff line or *offset* of the land surface at the top of a *landslide*, *earthflow*, or *slump* where the *mass movement* pulled material away. (*fault*) See *fault scarp.*

scarp retreat See *cliff retreat.*

scree The accumulation of broken rock that lies on a steep mountainside or at the base of a cliff; also, *talus.*

schistosity A type of *foliation* where the *rock* is *medium-* to *coarse-grained* and mica or other *minerals* are oriented parallel to each other but are fairly evenly distributed throughout the rock. The rock tends to split parallel to this foliation.

scoriaceous Term used to describe a basaltic flow with about 50% gas bubbles (vesicles).

sea arch A hole cut all the way through the sides of a headland along a cliff face so as to form a natural arch or bridge of rock.

sea cave A hole or cavity at the base of a sea cliff formed primarily by wave action.

sea stack A small steep-sided rocky projection formed of erosion resistant rock, above sea level and isolated from a cliff or coastline by erosion.

sectile A type of *tenacity* of a *mineral* where the sample can be shaved with a sharp blade as if it were wax or soap.

section One square mile in a *township* in the *Township and Range System.*

sediment Loose material at the Earth's surface from *rock* and *mineral* particles, from organisms and their remains, and/or from chemical precipitation.

sedimentary rock A *rock* made by the *lithification* or consolidation of *sediment.*

seismic moment magnitude A number that indicates the relative size of an *earthquake* in a way that makes it more accurate for large earthquakes although it is more difficult to calculate than *Richter magnitude;* indicated with the symbol M_w.

seismic wave An *elastic* wave or vibration generated by an *earthquake* or explosion.

seismogram The record of an earthquake recorded by a *seismograph.*

seismograph An instrument that inscribes the Earth's motion during an *earthquake* on a record called a *seismogram.*

seismologist A scientist who studies *earthquakes.*

seismology The study of *earthquakes* and their waves.

semiarid A climate where rainfall is between 10 inches and 20 inches (25 cm and 50 cm) per year.

serpentine minerals A group of soft, greasy or silky luster, soapy-feeling, green to black phyllosilicates minerals with the formula $(Mg,Fe)_3Si_2O_5(OH)_4$ formed by *low-grade metamorphism* of peridotite or dunite.

shear A *force* of *deformation* that has a scissor-like motion causing one *rock mass* to pass by another.

shield volcano A large and gently-sloping *volcano* with a shield shape, made up of basaltic *lava flows* and very little *ash.*

shortening A type of *deformation* in which parts of the deformed object move closer together from two directions.

silica Silicon dioxide (SiO_2), an essential constituent in *silicate minerals.* The mineral quartz is pure silica.

silica tetrahedron The shape of the arrangement of a silicon atom surrounded by four oxygen atoms, making the basic building block of *silicate minerals,* in which the oxygen atoms are centered at the apexes of a *tetrahedron* with the silicon atom in the center of the tetrahedron; singular of silica tetrahedra.

silicate mineral class (silicates or silicate minerals) A group of *minerals* with members containing silicon and oxygen as basic building blocks for their internal structures. Examples are listed in Table 2.9.

silky luster The surface shine of a *mineral* that resembles the way silk reflects light.

sill A planar *igneous intrusive* body that intrudes parallel to layers (*concordant*).

sinkhole A hole or closed *depression* in *karst* regions.

sinuosity A measure of how winding or *meandering* the course of a *stream* is.

skeletal texture A *sedimentary texture* in which the entire *rock* is essentially made up of visible *fossils.*

slaty cleavage The property of a *fine-grained metamorphic rock* that breaks along planes of weakness created by parallel *mineral grains.* A type of *foliation*; rock cleavage.

slickensides Smooth, slick striations on *rocks* along *faults* where the walls of the fault have slipped past each other. Slickensides show the direction of slip.

slide Movement of material down one or several planar or curved slip surfaces in such a way that it moves together in one or a few coherent masses with little deformation. Movement may be planar or rotational. Also the deposit or *landform* formed from such movement.

slip face Steeply sloping downwind surface of a *dune.*

slope The measure of the steepness of a feature such as a hill, a mountain, a line. The steepness of a surface calculated as the rise divided by the horizontal distance.

slump A rotational *slide* that moves on a curved slip surface. The slip surface is concave up, shaped like the bowl of a spoon. Also the deposit or *landform* formed from such movement.

small scale The *scale* of a map where the ratio between the map and the Earth's surface is a minute fraction. For small-scale maps, a small area on the map covers a large amount of the Earth's surface, and represented objects appear small. The term can also be used for diagrams other than maps where scale is involved.

smell A *special property* of a *mineral* that has a distinctive odor.

smectite A family of clays primarily composed of hydrated sodium calcium aluminum silicate that can absorb water between the sheets of its *crystal* structure. Specific varieties of smectite include montmorillonite, beidellite, and saponite.

snow avalanche See *avalanche.*

soil Loose material at Earth's surface that is the product of *weathering* (mainly chemical weathering) in place and can support rooted plants.

soil fall *Mass wasting* involving the collapse of soil from a steep cliff, usually caused by *undercutting* the cliff. Also the pile of material resulting from such a fall.

solidification The process where *magma* becomes solid *rock.* Solidification includes *crystallization* but also includes the hardening of magma that results in the formation of *volcanic* glass.

sorted, poorly A property of *sediment* that has a wide range of *grain* sizes. Compare with *well-sorted.*

sorted, well A property of *sediment* that has a narrow range of *grain* sizes. Compare with *poorly sorted.*

source rock A *rock* in which oil and/or *natural gas* originate and then migrate out. Compare *reservoir rock.*

special property A *mineral* property that is only possessed by a few minerals. *Effervescence* in acid and *magnetism* are two special properties. Since every mineral has *luster* of one sort or another or breaks in one way or another, *luster, cleavage,* and *fracture* are not special properties.

specific gravity The ratio of the *mass* of a substance to the mass of an equal *volume* of water; closely related to *density.*

spit An extension of a beach naturally built part way out across a bay or inlet by the *longshore current.*

splendent luster A brightly shiny *luster* resembling the shine of patent leather, for example, the luster of biotite mica.

spoil Broken and crushed waste *rock* from a mining operation.

spreading center A plate margin where two *plates* move away from each other; a *divergent plate boundary* or margin.

statement scale The scale on a map expressed verbally; *verbal scale.*

stationary front A *front* where air masses move in such a way that neither the warmer nor the cooler air displaces the other. See also *cold front, warm front,* and *occluded front.*

stereoscope A device with two lenses designed to aid in viewing aerial photographs so *topography* can be seen in three dimensions.

stock Roughly equidimensional *igneous intrusion* of small size, with an *outcrop* area less than 100 km^2 (or 40 mi^2).

strata Layers of *sedimentary rock* that are visually separable from other layers; *beds;* plural of *stratum.*

stratigraphic superposition, the principle of States that *sedimentary rock* layers are deposited in sequence one on top of the other, so that at the time of *deposition,* the oldest rocks are at the bottom of a sequence and the youngest rocks are on top.

stratigraphy The study of *strata,* or *sedimentary* layers.

stratosphere The part of the upper atmosphere (between about 20 km and 50 km altitude), above the *troposphere* and below the atmospheric *mesosphere,* where temperature increases with altitude. See Figure 16.2.

stratovolcano A steep-sloped *volcano* made of interlayered *pyroclastic* deposits and *lava flows; composite volcano.*

stratum Singular of *strata.*

streak The color of a mineral when powdered. Streak is tested on a small piece of porcelain, known as a streak plate. See also *no streak.*

stream Any body of water that flows under the *force* of gravity in a relatively narrow *channel.*

stream gradient See *gradient stream.*

strike The orientation of a *horizontal* line on a plane, especially a *bedding, foliation,* or *fault* plane; perpendicular to the *dip.*

strike-slip fault A *fault* with *horizontal* displacement (parallel to the *strike* of the fault plane).

strip mining A method of mining usually used for extracting shallow, *horizontal* or nearly horizontal *mineral* or coal deposits, where the *overburden* is removed from a strip and piled up next to it, and then the coal or mineral is removed. Overburden from the next parallel strip is piled in the previously mined strip, and the process continues.

structure The physical arrangement of a *rock* mass. Examples include *intrusive* bodies, *unconformities,* orientation of *rock* layers, and *deformational* features such as *faults* and *folds.*

structure section A side view of Earth's interior, generally near the surface, exhibiting the arrangement and compositions of *rocks* and rock layers; *cross section.*

subduction The process of movement of a slab of *lithosphere* downward into the *asthenosphere* at an ocean-ocean or ocean-continent *convergent margin.*

subhedral An adjective describing a *mineral grain* with some but not all well-formed *crystal faces.*

sublimate To convert from ice directly to water vapor.

sublimation The process of conversion of ice directly to water vapor.

subtropical desert A *desert* that occurs between about 15° and 30° latitude either north or south of the equator. Deserts between 15°. and 23½° are sometimes called tropical deserts.

submetallic A type of *metallic luster* with a surface shine similar to a tarnished or dull metal.

subtropical gyre A roughly circular flow of ocean currents with a center in subtropical latitudes at about 30°, made up of an *equatorial current,* a *western boundary current,* an eastward-flowing current near 45° latitude, and an *eastern boundary current.*

sulfate mineral class (sulfates) A group of *minerals* with members containing sulfur and oxygen and one or more metals. Examples are gypsum and anhydrite.

sulfide mineral class (sulfides) A group of *minerals* with members containing sulfur and one or more metals but no oxygen. Examples are pyrite, chalcopyrite, galena, and sphalerite.

supercontinent A continent such as Eurasia consisting of a large proportion of Earth's landmass assembled together.

surface tension Property of a liquid surface displayed by its acting as if it were a stretched elastic membrane; typically existing only where the liquid surface is in contact with gas (such as the air).

surface wave A *seismic wave* that only travels along the Earth's exterior.

syncline A *fold* that bows downward in the center.

synthesis Putting atoms together to form molecules or recombining atoms in one molecular combination to make another combination.

T

talus Sediment that gathers at the base of a cliff or very steep, rocky *slope* from *rockfalls.* Talus can be of all sizes but usually includes mostly coarse and angular *clasts.* Also called *scree.*

tarn A lake in a *rock basin* in a *cirque.*

taste A special property of a mineral that is bitter, sour, or salty to the tongue.

tectonic An adjective for *tectonics.*

tectonics A branch of geosciences having to do with the broad *structure* or architecture of the outer Earth, especially *plates* and the *lithosphere.*

tenacity The cohesiveness of a specimen, a description of a *mineral's* resistance to mechanical *deformation* (breaking, bending, crushing, and so on).

tephra See *pyroclastics.*

tension A *force* of *deformation* that pulls apart or has forces moving away from each other.

tephra *See pyroclastics.*

terminal moraine An *end moraine* that forms at the farthest *advance* or extent of the *glacier*. The term is sometimes used as a synonym for an end moraine.

terminus The end or farthest extremity of a glacier, the lower or outermost edge.

terrane A piece of crust bordered by *faults* that has a distinct history from neighboring pieces.

tetrahedron A regular geometric solid with four sides that are equilateral triangles of equal size; a triangular pyramid; singular of tetrahedra.

texture The arrangement and size of *mineral grains, rock* fragments, or glass in a rock.

thermal metamorphism *Metamorphism* resulting from the heating of *rock* near a *magmatic intrusion; contact metamorphism.*

thermocline A layer of ocean water, just below the thermally-mixed surface layer, that is generally less than 1 km deep and in which the water temperature decreases rapidly with depth.

thermohaline circulation Ocean currents that flow because of differences in *density* caused by differences in temperature and *salinity* of seawater. Also *global conveyor*.

thrust fault A low-angle *dip-slip fault* for which the *hanging wall* moved up.

till *Sediment* deposited directly from a *glacier*, which is *poorly sorted* and *unstratified.*

tombolo A sandy connection between an island and the mainland or between two islands.

topographic map A map with color, shading, or *contours* that indicate the shape of the land surface; if drawn with contours, also known as a contour map. See also *raised relief map.*

topographic profile A side view of the *topography.*

topography The shape of the physical features of the land surface. See also *relief.*

township A 6-by-6-mile square in the *Township and Range System.*

Township and Range System A *grid* system used in much of the United States based on 36-mi^2 grid, called *townships,* with square-mile subdivisions, called *sections.*

trace fossil A *fossil* that shows signs of an organism's existence or activities, but not involving the actual remains of the organism. Examples are footprints or animal burrows.

trace of a contact The line made on the Earth's surface by the intersection of the surface with the contact between two *rock* bodies.

trade winds Easterly winds (winds that blow from the east) that blow between 30°N and 30°S latitude, including the northeast trade winds in the Northern Hemisphere and the southeast trade winds in the Southern Hemisphere.

transform fault The type of *fault* found at a *transform-fault plate boundary.* Transform faults are also *strike-slip faults.*

transform-fault margin A *transform-fault plate boundary.*

transform-fault plate boundary A plate margin where two *plates* move *horizontally* past each other; *transform-fault margin.*

transverse dune A long, nearly straight *dune* that forms perpendicular to moderate winds.

tributary A smaller *stream* or *glacier* joining with a larger stream or glacier.

troposphere The part of the lower atmosphere (up to an altitude of about 20 km), below the *stratosphere,* where temperature generally decreases with rise of altitude. See Figure 16.2.

true north The direction of geographic north, toward the northern *axis of Earth's rotation.*

truncated spur The end of a ridge between glacial *hanging valleys* that was cut by the main *glacier*.

tsunami A *seismic* sea wave; a series of ocean (or lake) waves generated by a sudden disturbance of the seafloor (or lake bottom) such as an *earthquake,* submarine *landslide,* or *volcanic eruption.*

twinning When part of the atomic structure of a mineral changes to a mirror image of itself in response to volume reduction during cooling or imposed stress.

U

ultramafic A chemical composition term for *igneous rocks* that indicates the presence of very high magnesium and iron content and very low silica content (~40%).

ultraviolet rays Part of the *electromagnetic spectrum* with a shorter wavelength and higher energy than visible light but longer wavelengths and less energy than X rays; ultraviolet light.

unconformity A substantial time gap in the *rock* record where rocks either were *deposited* then eroded or were simply not deposited.

undercutting Removal of material from the base of a *slope* in such a way that it will tend to undergo *mass wasting,* especially when it creates an overhanging cliff.

uneven fracture A type of *mineral fracture* that is irregular and does not fit any of the other standard fracture terms: *conchoidal, fibrous,* or *hackly.*

unifying theory A guiding principle or integrative concept for an entire field of study.

Universal Time The time at the *prime meridian.*

Universal Transverse Mercator Coordinates A coordinate system commonly included on *topographic* maps. See Lab 1 for more details.

unstratified Lacking *bedding.*

up indicator A sedimentary feature or characteristic that shows which way up a *rock* or sedimentary layer was deposited. Up indicators are used to help recognize *overturned beds.*

upwelling The movement of deep, cool ocean water upward to the sea surface. Compare with *downwelling.*

U-shaped valley A valley with steep walls and gently-sloping floor that has a U-shaped cross section and was formerly occupied by a glacier.

V

valley glacier A type of *glacier* that forms and flows in valleys in mountainous regions, an *alpine glacier*, a *mountain glacier*.

vector A quantity having magnitude and direction.

ventifact A rock with facets eroded by wind abrasion.

verbal scale The scale on a map expressed as a statement; *statement scale*.

vertical A direction, such as a plumb line, that is exactly up and down, perpendicular (at right angles) to *horizontal* or to the horizon.

vesicle A small cavity in a *rock* that was originally a gas bubble in *magma*.

vesicular texture A *volcanic texture* that refers to the presence of small cavities in a *rock*, called *vesicles*, which were originally gas bubbles in the *magma*.

viscosity A fluid property referring to the fluid's resistance to flow.

viscous Having high *viscosity*.

vitreous luster The surface shine of a *mineral* that resembles the way glass reflects light.

volcanic ash Sand-sized to powdery *volcanic* material produced by explosive *volcanic eruptions* when a spray of magma and particles of *rock* spew out of a *volcano*.

volcanic bomb A large piece of *lava* thrown from a *volcano* that cooled as it flew through the air becoming streamlined.

volcanic eruption The extrusion or emanation of *lava* as flows, fountains, or *pyroclastic* material from the Earth.

volcanic rock A type of *igneous rock* that formed at the surface of the Earth; *extrusive rock*.

volcanic vent The opening in the Earth's surface where *volcanic eruptions* occur.

volcano A hill or mountain formed where material erupts frequently or repeatedly from a *volcanic vent*.

volume Size or extent in three dimensions; a measure of the combined width, depth and height.

V's, rule of The principle that a V formed by the surface trace of a planar feature in a valley points in the direction of the feature's *dip*.

W

warm front A *front* where warm, moist air wedges out cold air, typically accompanied by *condensation* and *precipitation*. See also *cold front, occluded front,* and *stationary front*.

wastage zone See *ablation, zone of*.

water table The top of the *zone of saturation*.

wave-cut notch A sharp angle, cut, or indention at the base of a sea cliff.

waxy luster The surface shine of a *mineral* that resembles the way wax reflects light.

weather front See *front*.

weathering The decomposition and breakdown of *rock* or loose material in place at Earth's surface by chemical or mechanical means.

weather map A map showing the distribution of aspects of the weather, such as storm systems, winds, *precipitation*, temperature, *pressure*, and *warm* and *cold fronts*.

well A hole drilled into the *zone of saturation*, below the *water table*.

well sorted See *sorted, well*.

westerlies Winds in a belt from 30° to 60° latitude blowing from the west.

western boundary current A warm ocean current that flows away from the equator near the western edge of an ocean.

X

xenolith A piece of foreign *rock* embedded in *igneous rock*.

Z

zone See *metamorphic zone*—an area or region in which a distinctive *mineral assemblage* coexists that indicates a specific range of *metamorphic* conditions, especially temperature. See also *Barrovian zone*.

zone of ablation See *ablation, zone of*.

zone of accumulation See *accumulation, zone of*.

zone of saturation See *saturation, zone of*.

zone of wastage See *ablation, zone of*.